THE ELEMENTS

7A 8A

3A 4A 5A 6A

| | | | | 1 **H** 1.00797 ± 0.00001 | 2 **He** 4.0026 ± 0.00005 |

f⁻¹
Ca⁺²
CaF₂

1B 2B

| | | | 5 **B** 10.811 ± 0.003 | 6 **C** 12.01115 ± 0.00005 | 7 **N** 14.0067 ± 0.00005 | 8 **O** 15.9994 ± 0.0001 | 9 **F** 18.9984 ± 0.00005 | 10 **Ne** 20.183 ± 0.0005 |

| | | | 13 **Al** 26.9815 ± 0.00005 | 14 **Si** 28.086 ± 0.001 | 15 **P** 30.9738 ± 0.00005 | 16 **S** 32.064 ± 0.003 | 17 **Cl** 35.453 ± 0.001 | 18 **Ar** 39.948 ± 0.0005 |

28 **Ni** 58.71 ± 0.005	29 **Cu** 63.54 ± 0.005	30 **Zn** 65.37 ± 0.005	31 **Ga** 69.72 ± 0.005	32 **Ge** 72.59 ± 0.005	33 **As** 74.9216 ± 0.00005	34 **Se** 78.96 ± 0.005	35 **Br** 79.909 ± 0.002	36 **Kr** 83.80 ± 0.005
46 **Pd** 106.4 ± 0.05	47 **Ag** 107.870 ± 0.003	48 **Cd** 112.40 ± 0.005	49 **In** 114.82 ± 0.005	50 **Sn** 118.69 ± 0.005	51 **Sb** 121.75 ± 0.005	52 **Te** 127.60 ± 0.005	53 **I** 126.9044 ± 0.00005	54 **Xe** 131.30 ± 0.005
78 **Pt** 195.09 ± 0.005	79 **Au** 196.967 ± 0.0005	80 **Hg** 200.59 ± 0.005	81 **Tl** 204.37 ± 0.005	82 **Pb** 207.19 ± 0.005	83 **Bi** 208.980 ± 0.0005	84 **Po** (210)	85 **At** (210)	86 **Rn** (222)

| 63 **Eu** 151.96 ± 0.005 | 64 **Gd** 157.25 ± 0.005 | 65 **Tb** 158.924 ± 0.0005 | 66 **Dy** 162.50 ± 0.005 | 67 **Ho** 164.930 ± 0.0005 | 68 **Er** 167.26 ± 0.005 | 69 **Tm** 168.934 ± 0.0005 | 70 **Yb** 173.04 + 0.005 | 71 **Lu** 174.97 ± 0.005 |

| 95 **Am** (243) | 96 **Cm** (247) | 97 **Bk** (249) | 98 **Cf** (249) | 99 **Es** (254) | 100 **Fm** (253) | 101 **Md** (256) | 102 **No** (253) | 103 **Lr** (257) |

Atomic Weights are based on C¹² —12.0000 and Conform to the 1961 Values   Printed in U.S.A.

WILLIAM L. MASTERTON

PROFESSOR OF CHEMISTRY
UNIVERSITY OF CONNECTICUT, STORRS, CONNECTICUT

EMIL J. SLOWINSKI

PROFESSOR OF CHEMISTRY
MACALESTER COLLEGE, ST. PAUL, MINNESOTA

ILLUSTRATED BY ALEXIS KELNER

W. B. SAUNDERS COMPANY · PHILADELPHIA · LONDON · TORONTO

SAUNDERS COMPLETE PACKAGE FOR THE TEACHING OF GENERAL CHEMISTRY

Masterton & Slowinski: CHEMICAL PRINCIPLES, *3rd Ed.*

Slowinski, Masterton & Wolsey: CHEMICAL PRINCIPLES IN THE LABORATORY, *2nd Ed. with teacher's guide*

Boyington: STUDENT'S GUIDE TO MASTERTON & SLOWINSKI'S CHEMICAL PRINCIPLES, *3rd Ed.*

Masterton and Slowinski: MATHEMATICAL PREPARATION FOR GENERAL CHEMISTRY

Peters: PROBLEM SOLVING FOR GENERAL CHEMISTRY

Slowinski and Masterton: QUALITATIVE ANALYSIS AND THE PROPERTIES OF IONS IN AQUEOUS SOLUTION

Masterton and Slowinski: COLOR SLIDES TO ACCOMPANY CHEMICAL PRINCIPLES, *3rd Ed.*

THIRD EDITION ☀ SAUNDERS GOLDEN SERIES

CHEMICAL PRINCIPLES

W. B. Saunders Company: West Washington Square
Philadelphia, Pa.   19105

12 Dyott Street
London, WC1A   1DB

833 Oxford Street
Toronto 18, Ontario

The front cover illustration is
VARIATION WITHIN A SPHERE, NO. 10, THE SUN
by Richard Lippold
courtesy of the Metropolitan Museum of Art

Chemical Principles

ISBN 0-7216-6172-6

Print No: 9   8   7   6   5   4   3   2

PREFACE

During the past few years, those of us who teach general chemistry have been made aware of some hard facts concerning the background, interests and objectives of our students. Despite the optimistic forecasts of a decade ago, upgrading of high school courses has not removed the need to stress fundamental principles in the beginning college course. Our students are not yet prepared to work problems in quantum mechanics. Equally important, they show little interest in abstract theory; instead, they want to know how the principles of chemistry apply to the world around them. Finally, we must realize that very few students in the general course are potential chemists. The overwhelming majority plan to use chemistry as a tool in medicine, the biological sciences, engineering or allied professions.

We have tried to keep these facts in mind in preparing this edition of *Chemical Principles*. For example, the chapters on thermochemistry and chemical thermodynamics (4 and 12) have been revised to emphasize the application of this powerful discipline in such areas as metallurgy, thermal pollution and the energy crisis. The last chapter (Chapter 23) has become an introduction to biochemistry in which we attempt to show how chemical techniques discussed in previous chapters can be used to deduce the structures of such vital natural products as DNA.

The worked-out examples in the body of the text and the problems at the end of each chapter have been recast to illustrate not only the principles of chemistry but also their practical application. To cite two examples, the student is asked to apply the rate laws to determine the rate of formation of ozone in an urban atmosphere (#14.31, p. 397) or to use the principles of chemical equilibrium to calculate the percentage of hemoglobin in the blood tied up as the carbon monoxide complex (#15.41, p. 427).

We have retained the general framework and sequence of topics of the second edition. As before, the first several chapters emphasize the quantitative, experimental aspects of chemistry. Descriptive inorganic chemistry is again organized about types of reactions (precipitation reactions in Chapter 16, acid-base reactions in Chapters 17 and 18, complex-ion formation in Chapter 19, and redox reactions in Chapters 20 and 21). The only major organizational change involved moving a discussion of organic chemistry up to Chapter 8, where it serves to illustrate the principles of chemical bonding and the relationship between physical properties and molecular structure.

A new chapter (Chapter 11) dealing with the unique properties of water and aqueous solutions has been added. Included here are discussions of the chemistry of water pollution, purification and desalination. A descriptive chapter on the properties of the atmosphere and its components (Chapter 15) has been substituted for one in the second edition that dealt with the more amorphous topic of the reactions of elements.

We are curious to learn the reaction of students and instructors to two innovations in this edition. Interspersed throughout the text are a few "historical perspectives" designed to illustrate the tortuous path by which the principles of chemistry have developed and the human characteristics of some of the gifted individuals who set forth these principles. In a more prosaic vein, we have arranged the problems at the end of each chapter in matched pairs, side by side. One member of each pair is supplied with an answer. In our classes, we prefer to assign only unanswered problems. The student checks his answer and, more important, his reasoning, when these problems are discussed in class. He can then go back and work the corresponding answered problem illustrating the same principle. Hopefully, this technique will encourage self-study; at a minimum, it should reveal correlations that may not have been entirely obvious.

One of the more enjoyable tasks of an author is to respond to the suggestions (and parry the criticisms) of students and instructors who use his text. Many of the ideas that have prompted this revision have grown from discussions with faculty and students at other colleges and universities. On a day-to-day basis, our colleagues at the University of Connecticut and Macalester College, notably Ray Boyington, Charles Waring, and Wayne Wolsey, have helped us to evaluate student reactions to the text. We are grateful to B. K. Shakhashiri of the University of Wisconsin, Wilmer Stratton of Earlham College, and Ronald Ragsdale of the University of Utah for detailed critiques of the second edition. Gordon Atkinson of the University of Oklahoma and Joe Deck of the University of Louisville provided us with a thoughtful and mercifully brief review of the manuscript. It was a pleasure to work with Alexis Kelner, who provided the illustrations for the text.

We acknowledge, at long last, the patience of our wives and children, who have suffered through three editions with a minimum of complaints. We are indebted to the staff of the W. B. Saunders Company for cajoling, coaxing and frightening us into submitting manuscript more or less on schedule. Finally, we can neither confirm nor deny the rumor that the production staff at Saunders took up a collection to send one of us to Poland when the page proof appeared.

WILLIAM L. MASTERTON

EMIL J. SLOWINSKI

contents

√1 BASIC CONCEPTS 1

   1.1 The Role of Chemistry Today, 1; 1.2 Measurements,
   2; 1.3 Separation of Matter into Pure Substances, 11;
   1.4 Identification of Pure Substances, 17; 1.5 Kinds of
   Substances: Elements and Compounds, 20; *Problems,
   22*

2 ATOMS, MOLECULES, AND IONS 25

   2.1 Atomic Theory, 25; 2.2 Components of the Atom,
   27; 2.3 Molecules and Ions, 31; 2.4 Relative Masses
   of Atoms: Atomic Weights and Gram Atomic Weights,
   33; 2.5 Masses of Atoms. Avogadro's Number, 38; 2.6
   Masses of Molecules. Molecular Weight. Gram Molecular Weight, 39; 2.7 An Historical Perspective; Atomic
   Weights in the 19th Century, 40; *Problems, 44*

√3 CHEMICAL FORMULAS AND EQUATIONS . . 48

   3.1 Simplest (Empirical) Formulas from Analysis, 49;
   3.2 Molecular Formulas, 52; 3.3 The Mole, 53; 3.4
   Translation of Reactions into Equations, 54; 3.5 Interpretation of Balanced Equations, 55; 3.6 Theoretical
   Yields. Percentage Yields, 57; *Problems, 59*

4 THERMOCHEMISTRY 63

   4.1 Thermochemical Equations, 63; 4.2 Heats of Formation of Compounds, 68; 4.3 Heats of Bond Formation, 71; 4.4 Calorimetry, 74; 4.5 Sources of Energy,
   77; *Problems, 84*

vii

√5 THE PHYSICAL BEHAVIOR OF GASES 88

    5.1 Some General Properties of Gases, 88; 5.2 Atmospheric Pressure and the Barometer, 89; 5.3 The Ideal Gas Law, 92; 5.4 Using the Ideal Gas Law, 99; 5.5 Mixtures of Gases: Dalton's Law of Partial Pressures, 105; 5.6 Real Gases, 107; 5.7 The Kinetic Theory of Gases, 109; *Problems, 117*

√6 THE ELECTRONIC STRUCTURE OF ATOMS . 121

    6.1 Some Properties of Electrons in Atoms and Molecules, 121; 6.2 Experimental Basis of the Quantum Theory, 123; 6.3 Theories of Electronic Structure, 127; 6.4 Modern Quantum Theory, 130; 6.5 Electron Arrangements in Atoms, 137; 6.6 Electron Configurations and Ionization Potentials, 149; 6.7 The Electronic Structures of Atoms and the Periodic Table, 150; 6.8 The Correlation of Atomic Radii with Electron Configurations, 156; *Problems, 158*

√7 CHEMICAL BONDING 162

    7.1 Ionic Bonding, 162; 7.2 Formation of the Covalent Bond, 167; 7.3 Properties of the Covalent Bond, 170; 7.4 Lewis Structures, the Octet Rule, 175; 7.5 Molecular Geometry, 181; 7.6 Hybrid Atomic Orbitals, 186; 7.7 Molecular Orbitals, 190; 7.8 The Metallic Bond, 196; *Problems, 199*

8 PHYSICAL PROPERTIES OF MOLECULAR SUBSTANCES; NATURE OF ORGANIC COMPOUNDS 203

    8.1 General Characteristics of Molecular Substances, 203; 8.2 The Nature of Organic Molecules, 210; 8.3 Macromolecular Substances, 225; *Problems, 230*

√9 LIQUIDS AND SOLIDS; CHANGES IN STATE 233

    9.1 Properties of Liquids, 233; 9.2 Liquid-Vapor Equilibrium, 235; 9.3 Nature of the Solid State, 244; 9.4 Phase Equilibria, 250; 9.5 Nonequilibrium Phase Changes, 254; 9.6 Defect Crystals, 255; *Problems, 259*

Vapor, 409; 15.7 The Upper Atmosphere, 411; 15.8 Air Pollution, 416; *Problems, 425*

16 PRECIPITATION REACTIONS 429

16.1 Net Ionic Equations, 429; 16.2 Solubilities of Ionic Compounds, 431; 16.3 Solubility Equilibria, 434; 16.4 Precipitation Reactions in Analytical Chemistry, 440; 16.5 Precipitation Reactions in Inorganic Preparations, 444; *Problems, 445*

17 ACIDS AND BASES 448

17.1 The Dissociation of Water; Nature of Acids and Bases, 448; 17.2 pH, 449; 17.3 Strong Acids and Bases, 451; 17.4 Weak Acids, 453; 17.5 Weak Bases, 460; 17.6 Factors Influencing Strengths of Acids and Bases, 465; 17.7 General Theories of Acids and Bases, 469; *Problems, 473*

18 ACID-BASE REACTIONS 477

18.1 Types of Acid-Base Reactions, 477; 18.2 Acid-Base Titrations, 483; 18.3 Buffers, 490; 18.4 Application of Acid-Base Reactions in Inorganic Syntheses, 493; 18.5 Applications of Acid-Base Reactions in Qualitative Analysis, 495; 18.6 An Industrial Application of Acid-Base Reactions: The Solvay Process, 498; *Problems, 502*

19 COMPLEX IONS; COORDINATION
COMPOUNDS 506

19.1 Charges of Complex Ions. Neutral Complexes, 507; 19.2 Composition of Complex Ions. General Principles, 508; 19.3 Geometry of Complex Ions, 512; 19.4 Electronic Structure of Complex Ions, 517; 19.5 Rate of Complex Ion Formation, 523; 19.6 Complex Ion Equilibria, 524; 19.7 Complex Ions in Analytical Chemistry, 525; *Problems, 531*

20 OXIDATION AND REDUCTION:
ELECTROCHEMICAL CELLS 535

20.1 Oxidation Number, 536; 20.2 Balancing Oxidation-Reduction Equations, 540; 20.3 Electrolytic Cells, 544; 20.4 Voltaic Cells, 554; *Problems, 560*

10 SOLUTIONS 263

10.1 Introduction, 263;   10.2 Concentration Units, 265;
10.3 Principles of Solubility, 269;   10.4 Effect of Temperature and Pressure on Solubility, 272;   10.5 Colligative Properties of Nonelectrolyte Solutions, 276;   *Problems, 285*

11 WATER, PURE AND OTHERWISE 289

11.1 Physical Properties of Water, 290;   11.2 The Structure of Water, 292;   11.3 Electrolyte Solutions, 295;
11.4 Water Pollution, 300;   11.5 Water Purification, 309;
Problems, 318

12 SPONTANEITY OF REACTION;
ΔG AND ΔS 321

12.1 Criterion of Spontaneity. Maximum Useful Work,
321;   12.2 Free Energy Change, ΔG, 326;   12.3 Entropy Change, ΔS, 329;   12.4 The Gibbs-Helmholtz
Equation, 334;   *Problems, 343*

13 CHEMICAL EQUILIBRIUM IN
GASEOUS SYSTEMS 347

13.1 The N_2O_4-NO_2 Equilibrium System. Concept of K_c,
348;   13.2 The General Form of K_c, 351;   13.3 Applications of K_c, 353;   13.4 Effect of Changes in Conditions
upon the Position of an Equilibrium, 357;   13.5 Relation Between the Free Energy Change and the Equilibrium Constant, 364;   *Problems, 367*

14 RATES OF REACTION 371

14.1 Meaning of Reaction Rate, 372;   14.2 Dependence
of Reaction Rate upon Concentration, 374;   14.3 Dependence of Reaction Rate upon Temperature, 380;
14.4 Catalysis, 384;   14.5 Reaction Mechanisms, 386;
Problems, 394

15 THE ATMOSPHERE 398

15.1 The Composition of the Atmosphere, 398;   15.2
Nitrogen, 400;   15.3 Oxygen, 403;   15.4 The Noble
Gases, 407;   15.5 Carbon Dioxide, 409;   15.6 Water

21 OXIDATION-REDUCTION REACTIONS: SPONTANEITY AND EXTENT 565

21.1 Standard Potentials, 565; 21.2 Spontaneity and Extent of Redox Reactions, 570; 21.3 Effect of Concentration on Voltage, 575; 21.4 Strong Oxidizing Agents, 580; 21.5 Oxygen; the Corrosion of Iron, 585; 21.6 Redox Reactions in Analytical Chemistry, 589; *Problems, 591*

22 NUCLEAR REACTIONS 595

22.1 Natural Radioactivity, 596; 22.2 Rate of Radioactive Decay, 600; 22.3 Bombardment Reactions, Artificial Radioactivity, 603 22.4 Mass-Energy Relations, 609; 22.5 Nuclear Fission, 614; 22.6 Nuclear Fusion, 618; *Problems, 620*

23 AN INTRODUCTION TO BIOCHEMISTRY 624

23.1 Carbohydrates, 625; 23.2 Proteins, 628; 23.3 Nucleic Acids, 643; *Problems, 656*

APPENDIX 1
CONSTANTS, CONVERSION FACTORS AND PROPERTIES OF WATER 659

APPENDIX 2
ATOMIC AND IONIC RADII 660

APPENDIX 3
NOMENCLATURE OF INORGANIC COMPOUNDS 661

APPENDIX 4
EXPONENTS AND LOGARITHMS 665

APPENDIX 5
SI UNITS 674

APPENDIX 6
ANSWERS TO PROBLEMS 681

INDEX 701

1 Basic Concepts

1.1 THE ROLE OF CHEMISTRY TODAY

For centuries man has struggled to control his immediate environment so as to satisfy his basic needs. The degree of success achieved in this struggle is due in no small part to contributions from the natural sciences, chemistry among them. Food production has been increased through the use of chemical fertilizers and insecticides. Many of the clothes we wear are made from synthetic fibers such as nylon and Dacron which are formed by chemical processes. Perhaps the most significant achievement of chemistry in recent years has been the development of antibiotics and other drugs to combat disease and pain.

Now, in the 20th century, we are faced with a new problem. For the first time man has the ability to change the global environment, hopefully for better, but quite possibly for worse. The development of nuclear weapons 25 years ago raised the frightening possibility that man might, by design or accident, destroy life on this planet. More recently we have found that many of the materials and processes developed to meet the needs of society have profoundly disturbed ecological systems. Chemical fertilizers promote the growth of algae in lakes and streams. Pesticides used to control agricultural insects and prevent disease have adverse, sometimes fatal, effects upon wildlife populations. Perhaps most serious of all is the problem of thermal pollution. Evidence is accumulating that the combustion of fuels to produce useful energy is raising the temperature, not only of the immediate surroundings, but of the earth itself.

Frequently the price that we must pay to clean up our air and water supplies cannot be measured in dollars alone. To cite but one example, consider the problem of atmospheric pollution by sulfur dioxide, much of which comes from the combustion of high-sulfur coal in power plants. Pollution could be reduced by using low-sulfur coal, which is in abundant supply in the western part of the United States. Shipping this coal across the country to cities along the eastern seaboard would increase the cost of electric power. More serious from an environmental standpoint, the deposits of low-sulfur coal lie in veins near or at the surface of the earth, so that strip mining would be required. Anyone who has witnessed the devastation that has resulted from strip mining in Appalachia can hardly be en-

1

thusiastic about applying it to the spectacularly beautiful wildlands of Idaho and Wyoming, even for so worthy a purpose as reducing atmospheric pollution.

There are, to be sure, many other ways to reduce sulfur dioxide emissions from power plants. Perhaps we should concentrate upon processes to remove sulfur compounds from the fuel before it is burned. Alternatively, we might develop methods of extracting sulfur dioxide from the combustion gases. At present, many chemists, along with engineers and other scientists, are actively engaged in research and development in these areas. A major task of chemistry for the remainder of this century must be to develop new approaches to the complex problems caused by pollution. Equally important, chemists and other scientists have a responsibility to educate the public as to the approaches now available, explaining as best they can the advantages and disadvantages of various alternatives.

In this, a general course, our major objective will be to introduce the principles that underlie all of chemistry. Wherever appropriate we shall point out how these principles can be applied to environmental problems.

Chemistry deals with the properties and reactions of the materials which make up the earth and the universe

Most often we shall work with simple systems, involving a single pure substance or a mixture of a few such substances. Systems of this type lend themselves to controlled laboratory experiments, several of which will be described in this chapter. Through such experiments, chemistry has developed, extrapolating from the simple to the complex, from the laboratory to the everyday world.

1.2 MEASUREMENTS

Progress in the development of chemical principles is based upon shrewdly designed experiments carried out under carefully controlled conditions. Virtually all these experiments involve quantitative measurements of such parameters as length, volume, mass, and temperature. In this section we shall consider the instruments used to measure these quantities and the units in which they are expressed.

MEASURING DEVICES

LENGTH. Most of us are familiar with a simple measuring device found in the general chemistry laboratory, the meter stick, which reproduces, as accurately as possible, the fundamental unit of length in the metric system, the **meter.** The meter was originally intended to be

The meter is defined with a precision of about 6 parts in one billion

1/40,000,000 of the earth's meridian that passes through Paris. This distance was fixed by two marks on a platinum-iridium rod stored at the International Bureau of Weights and Measures at Sèvres, near Paris. In 1960 the standard international meter was defined as being 1650763.73 times the wavelength of a certain line in the visible spectrum of krypton.

When we examine a meter stick, we see that it is divided into one hundred equal parts, each one *centimeter* in length (1 cm = 10^{-2} m). A centimeter is, in turn, subdivided into ten equal parts, each one *millimeter* long (1 mm = 10^{-3} m). A much larger unit, familiar to track and field runners, is the *kilometer* (1 km = 10^3 m). The prefixes "kilo," "centi," and "milli" are used in the metric system to designate units obtained by multi-plying by 1000, 0.01, and 0.001, respectively. Another unit widely used in chemistry to express the dimensions of tiny particles such as atoms and molecules is the **angstrom** (1 Å = 10^{-8} cm).

The diameter of an iron atom is about 2.52 Å

VOLUME. Units of volume in the metric system are simply related to those of length. The **cubic centimeter** (cc or cm³) represents the volume of a cube one centimeter on an edge. A larger unit is the **liter**, which was rede-fined in 1960 to be exactly 1000 cc; a **milliliter** has a volume exactly equal to that of one cubic centimeter.

The device most frequently used to measure volumes in the general chemistry laboratory is the graduated cylinder. When greater accuracy is required, we use a pipet or buret (Fig. 1.1). The volumetric flask shown at the right in Figure 1.1 is designed to contain a specified volume of liquid (e.g., 50, 100, . . . 1000 ml) when filled to a level marked on the narrow neck.

MASS. The amount of matter in a substance, its **mass**, is ordinarily

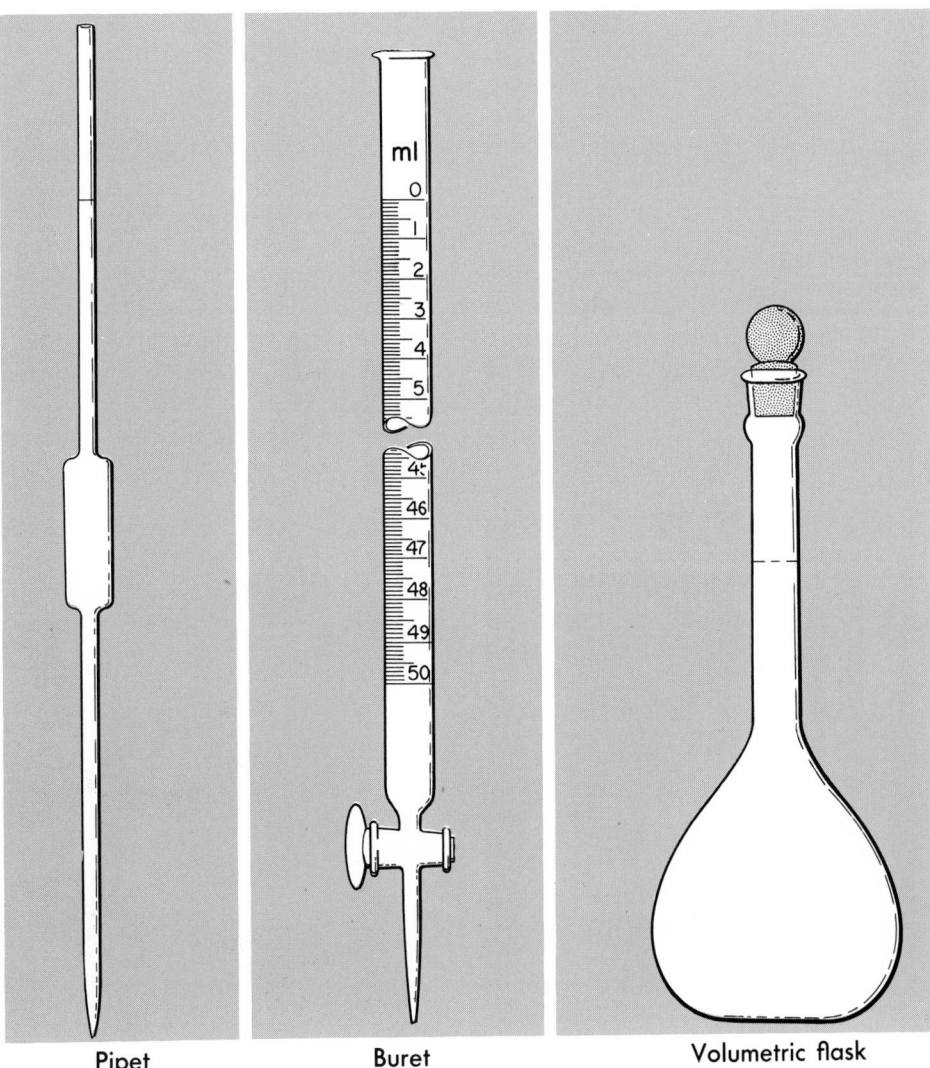

Pipet   **Buret**   **Volumetric flask**

Figure 1.1   The pipet delivers a fixed volume of liquid with an accuracy of about 0.1%. Variable volumes can be delivered with about the same accuracy from a buret.

measured by means of a balance (Fig. 1.2). To understand what is involved, consider the simple two-pan balance shown at the left of the figure. We will assume that it has been adjusted so that, with nothing on either pan, the balance comes to rest with the two pans at the same height. To weigh an object, we place it on the left-hand pan. Metal weights of known mass are then added to the right-hand pan to restore balance, i.e., to bring the pans to the same height. Under these conditions, the gravitational force acting on the sample, its **weight,** is equal to that acting on the pieces of metal.

$$\text{weight sample} = \text{weight metal}$$

But Newton's first law of motion tells us that gravitational force is directly proportional to mass

$$\text{weight} = k \, (\text{mass}) \tag{1.1}$$

where the proportionality constant k has a fixed value at a given location. It follows that, at balance,

$$k \, (\text{mass sample}) = k \, (\text{mass metal})$$

or $$\text{mass sample} = \text{mass metal}$$

We see, then, that the balance detects not only *equality of weight* but also *equality of mass.*

How much would a 12 g coin weigh on an analytical balance on the moon, where the gravitational pull is about ⅙ that on earth? Ans. 12 g. Why?

The mass of an object is a more fundamental property than its weight. The gravitational "constant" k in Equation 1.1 decreases as one moves away from the center of the attracting body. An astronaut moving out from the earth experiences an appreciable loss in weight as the gravitational pull of the earth decreases (Problem 1.29). His mass, on the other hand, does not change unless he eats a sandwich or in some other way exchanges matter with the surroundings.

The unit of mass used most frequently in chemistry is the **gram,** which

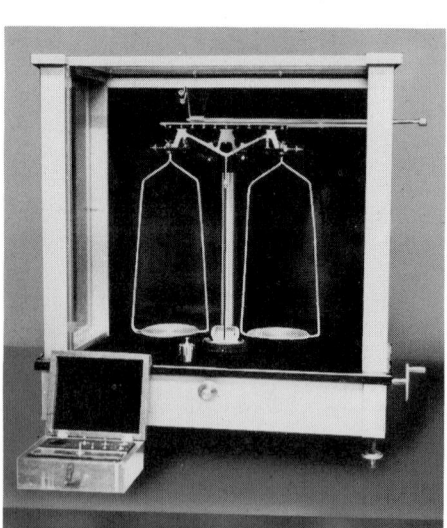

Two pan balance   Single pan balance

Figure 1.2 Two pan and single pan analytical balances. In the single pan balance, weights within the instrument are added or removed by turning dials.

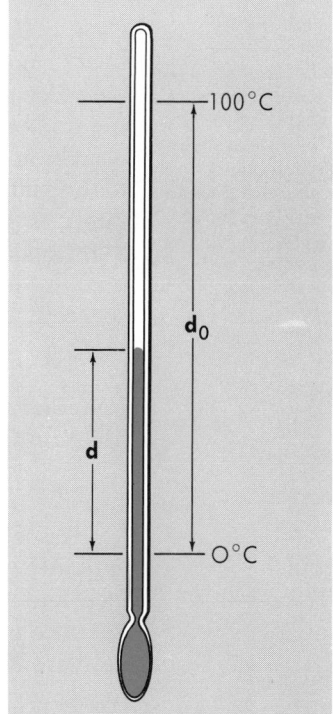

Figure 1.3 At 45°C, as read on a mercury in glass thermometer, d = 0.45 d$_0$, where d$_0$ is the distance from the 0° to the 100° mark.

represents one thousandth of the mass of a platinum-iridium block kept at the International Bureau of Weights and Measures (1 kg = 10^3 g). Many of the smaller weights we use with analytical balances have their masses specified in *milligrams* (1 mg = 10^{-3} g); for very tiny objects, we may quote mass in micrograms (1 μg = 10^{-6} g).

TEMPERATURE. The concept of temperature is familiar to all of us, largely because our bodies are so sensitive to temperature differences. When we pick up a piece of ice, we feel cold because its temperature is lower than that of our hand. After drinking a cup of coffee, we may refer to it as "hot," "lukewarm," or "atrocious"; in the first two cases at least, we are describing the extent to which its temperature exceeds ours. From a slightly different point of view, temperature may be regarded as the factor that determines the direction of heat flow. Anyone brave enough to swim in a Minnesota lake in January feels cold because heat is absorbed from his body. If he takes a hot shower afterward, which he certainly will, heat flows in the reverse direction. In general, whenever two objects at different temperatures come in contact with each other, heat flows spontaneously from the higher to the lower temperature.

This is really the definition of heat

To measure temperature with a mercury-in-glass thermometer, we take advantage of the fact that mercury, like other substances, expands as the temperature rises. The thermometer is designed so as to make readily visible a rather small fractional change in volume (Fig. 1.3). The total volume of the thin capillary column is only about 2 to 3 per cent of that of the mercury reservoir at the base.

A disadvantage of defining temperature this way is that it depends on the properties of mercury. An alcohol-in-glass thermometer set up the same way would agree with the described thermometer only at 0° and 100°

Thermometers used in the chemistry laboratory are marked in *degrees Celsius* (centigrade). On this scale, the freezing point of water is taken to be 0°C and the boiling point at one atmosphere pressure to be 100°C. When we place a good quality mercury-in-glass thermometer in a beaker containing crushed ice in equilibrium with water, the mercury will come to rest exactly at the 0° mark. In a beaker of boiling water, the mercury will rise to the 100° mark. The distance between these two marks is divided as accurately as possible into one hundred equal parts, each of which corresponds to a Celsius degree. Thus, a temperature reading of 45°C corresponds to a mercury level 45 per cent of the way from the 0° mark to the 100° mark.

UNCERTAINTIES IN MEASUREMENTS: SIGNIFICANT FIGURES

Associated with every experimental measurement is a degree of uncertainty whose magnitude depends upon the nature of the measuring device and the skill with which it is used. If, for example, we attempt to measure out 8 ml of liquid using a 100-ml graduated cylinder, the volume delivered is likely to be in error by at least 1 ml. With such a crude measuring device, we will be fortunate indeed to obtain a volume closer to 8 than to 7 or 9 ml. To improve on the accuracy of the measurement, we might use a narrow 10-ml cylinder on which the divisions are considerably farther apart. The volume we measure now may be within 0.1 ml of the desired value of 8 ml; i.e., in the range 7.9 to 8.1 ml. Using a buret, we might be able to do even better; if we are very careful, the uncertainty in our measurement may be reduced to 0.01 ml.

The person who carries out an experiment has a responsibility to indicate the uncertainty associated with his measurements. Such information is vital to anyone who wants to repeat the experiment or to judge its validity. There are many ways to do this; we might, for example, report the three volume measurements referred to above as

$$8 \pm 1 \text{ ml} \quad \text{(large graduated cylinder)}$$
$$8.0 \pm 0.1 \text{ ml} \quad \text{(small graduated cylinder)}$$
$$8.00 \pm 0.01 \text{ ml} \quad \text{(buret)}$$

Throughout this text, we will drop the ± notation and simply write

$$8 \text{ ml}; 8.0 \text{ ml}; 8.00 \text{ ml}$$

with the understanding that there is an *uncertainty of at least one unit in the last digit* (1 ml, 0.1 ml, 0.01 ml).

This method of specifying the degree of confidence in a measurement is often described in terms of **significant figures.** We say that in "8.00 ml" there are three significant figures; each of the three digits is experimentally meaningful. Similarly, there are two significant figures in "8.0 ml" and one significant figure in "8 ml."

Frequently we need to know the number of significant figures in a measurement reported by someone else. The manner in which this is done is illustrated in Example 1.1.

Example 1.1 An instructor asks his class in general chemistry to weigh a gold nugget. Among the masses reported are the following

$$20.03 \text{ g, } 20.0 \text{ g, } 0.02003 \text{ kg, and } 20 \text{ g}$$

How many significant figures should be assumed in each case?

Solution. The student who reports 20.03 g clearly believes that each of the four digits is meaningful; he is specifying 4 significant figures. Similarly, the student reporting 20.0 g indicates 3 significant figures. She has placed the zero after the decimal point to make it clear that the nugget was weighed to the nearest 0.1 g.

A moment's reflection should convince you that the third student, like the first, has shown 4 significant figures. The zero immediately to the right of the decimal point is not significant; it is there only because the mass is expressed in kilograms rather than grams. By the same token, there are 2 significant figures in the quantity "0.064 g" and only 1 in "0.007 g."

We cannot be sure of the number of significant figures in "20 g." Perhaps the student weighed the nugget to the nearest gram and wants to indicate 2 significant figures. Then again he may be trying to tell us that his balance weighs only to the nearest 10 g, in which case only the first figure in "20 g" is significant. This ambiguity could be avoided by giving the mass in exponential notation (Appendix 4) as either

$$2.0 \times 10^1 \text{ g; 2 significant figures}$$

or $$2 \times 10^1 \text{ g; 1 significant figure}$$

Most of the quantities that we measure in the laboratory are not end results in themselves; they are used to calculate other quantities. We might, for example, measure the mass and volume of a sample in order to determine its density (Example 1.2). Clearly, the precision of any such derived result is limited by that of the measurements upon which it is based. In particular, **when experimental quantities are multiplied or divided, the number of significant figures in the result cannot exceed that in the least precise measurement.**

Example 1.2 A student checks the purity of a water sample by determining its density (mass per unit volume) at 25°C. (The value given for pure water in a handbook is 0.9970 g/ml.) He measures out 25 ml of the sample from a cylinder and determines its mass to be 25.624 g. What should he report for the density?

Solution. Unfortunately the precision of the density determination is limited by that of the volume measurement, which has only 2 significant figures.

$$\text{density} = \frac{\text{mass}}{\text{volume}} = \frac{25.624 \text{ g}}{25 \text{ ml}} = 1.0 \text{ g/ml}$$

The five significant figures in the mass measurement are wasted; the student might as well have used a crude balance weighing to the nearest gram. More important, the uncertainty in the volume is so great that the experiment tells us nothing about the purity of the sample (*sea water* has a density, to 2 significant figures, of 1.0 g/ml). It would be advisable to repeat the experiment, substituting a pipet for the graduated cylinder.

$$\frac{25.624 \pm 0.001}{25 \pm 1} = 1 \pm 0.04$$

When experimental quantities are added or subtracted, the precision of the sum or difference is limited by the **absolute uncertainty** (rather than the number of significant figures) in the *least precise measurement*. Suppose, for example, we wish to calculate the mass of a solution prepared by adding 10.21 g of instant coffee and a "pinch" of sugar, 0.2 g, to 256 g of water. The implied uncertainties in these masses are

instant coffee:	10.21	± 0.01 g
sugar:	0.2	± 0.1 g
water:	256	± 1 g
total mass:	266	± 1 g

The sum of the masses cannot be more precise than that of the water (± 1 g). The total mass should be reported as 266 g rather than 266.4 g or 266.41 g.

CONVERSION OF UNITS

LENGTH, VOLUME, AND MASS. Frequently it is necessary to convert measurements expressed in one unit (e.g., mg) to another unit (g or kg). Conversions within the metric system, in which the units are related to each other by powers of ten, are readily carried out. The arithmetic is somewhat more complicated when we make conversions within the English system where, for example:

$$1 \text{ ft} = 12 \text{ in}; \quad 1 \text{ gallon} = 4 \text{ qt}; \quad 1 \text{ lb} = 16 \text{ oz}$$

Because of its simple, decimal interrelations, the metric system is universally employed by scientists and is in common use in most countries. Conservatism, coupled with a certain amount of stubbornness, is responsible for the retention of the English system in the United States and the British Commonwealth. Great Britain is committed to conversion to the metric system by 1975. In this country it now appears likely that a change will be made before 1980.

Conversions between metric and English units can be carried out using conversion factors such as those given in Table 1.1 (A more extensive list of conversion factors is given in Appendix 1, p. 659).

TABLE 1.1   Conversion Factors Relating Length, Volume, and Mass Units

METRIC	ENGLISH	METRIC-ENGLISH
Length		
1 km $= 10^3$ m	1 ft $= 12$ in	1 in $= 2.540$ cm *
1 cm $= 10^{-2}$ m	1 yd $= 3$ ft	1 m $= 39.37$ in *
1 mm $= 10^{-1}$ cm	1 mile $= 5280$ ft	1 mile $= 1.609$ km *
1 Å $= 10^{-8}$ cm		
Volume		
1 ml $= 1$ cc $= 10^{-3}$ liter	1 gal $= 4$ qt $= 8$ pt	1 liter $= 1.057$ qt *
1 μl $= 10^{-6}$ liter	1 qt $= 57.75$ in^3 *	1 ft^3 $= 28.32$ liter *
Mass		
1 kg $= 10^3$ g	1 lb $= 16$ oz	1 lb $= 453.6$ g *
1 mg $= 10^{-3}$ g	1 ton $= 2000$ lb	1 g $= 0.03527$ oz *
1 μg $= 10^{-6}$ g	(short)	

In this table the first quantity is taken to be exact

* These conversion factors, unlike the others listed, are inexact. They are quoted to four significant figures, which will ordinarily be more than sufficient.

With the aid of conversion factors, we can readily convert experimental quantities from one set of units to another. To illustrate the general approach, suppose we need to convert a length of 22.4 inches to feet. To do this we use the conversion factor 1 ft = 12 in. Dividing both sides of this equation by 12 in gives a quotient equal to unity.

$$\frac{1 \text{ ft}}{12 \text{ in}} = 1$$

Multiplying 22.4 in by the quotient 1 ft/12 in does not change the value of the length but does accomplish the desired conversion of units

$$22.4 \text{ in} \times \frac{1 \text{ ft}}{12 \text{ in}} = 1.87 \text{ ft}$$

The conversion factor, 1 ft = 12 in, can be used equally well to convert a length in feet, let us say 4.00 ft, to inches. In this case, we divide both sides of the equation by 1 ft to obtain

$$\frac{12 \text{ in}}{1 \text{ ft}} = 1$$

Multiplying 4.00 ft by the quotient 12 in/1 ft converts the length from feet to inches:

$$4.00 \text{ ft} \times \frac{12 \text{ in}}{1 \text{ ft}} = 48.0 \text{ in}$$

Multiplying a quantity by 1 or its equivalent will only change the quantity into its equivalent

Notice that a single conversion factor (e.g., 1 ft = 12 in) gives us two quotients (1 ft/12 in or 12 in/1 ft), both of which are equal to unity. In making a conversion, we choose the quotient which will enable us to cancel out the unit we wish to eliminate. Example 1.3 illustrates the application of the conversion factor approach to a slightly more complicated problem.

Example 1.3 On a certain day, the concentration of sulfur dioxide in the air over Philadelphia is found to be 2.74×10^{-7} g/liter. Express this concentration in

    a. g/ml   b. lb/ft^3

Solution
    a. Here all that is required is to convert volume in liters to milliliters, using the conversion factor 1 ml = 10^{-3} liter

$$2.74 \times 10^{-7} \frac{\text{g}}{\text{liter}} \times \frac{10^{-3} \text{ liter}}{1 \text{ ml}} = 2.74 \times 10^{-10} \frac{\text{g}}{\text{ml}}$$

    b. In this case, two steps are required. Logically we might
        (1) convert grams to pounds: 1 lb = 453.6 g
        (2) convert liters to ft^3: 1 ft^3 = 28.32 liters

We can set up the arithmetic in a single expression:

$$2.74 \times 10^{-7} \frac{\text{g}}{\text{liter}} \times \underset{(1)}{\frac{1 \text{ lb}}{453.6 \text{ g}}} \times \underset{(2)}{\frac{28.32 \text{ liter}}{1 \text{ ft}^3}} = 1.71 \times 10^{-8} \frac{\text{lb}}{\text{ft}^3}$$

The result is quoted to 3 significant figures, the number in the experimentally measured concentration. We could have rounded off the conversion factors to 3 figures before carrying out the calculations.

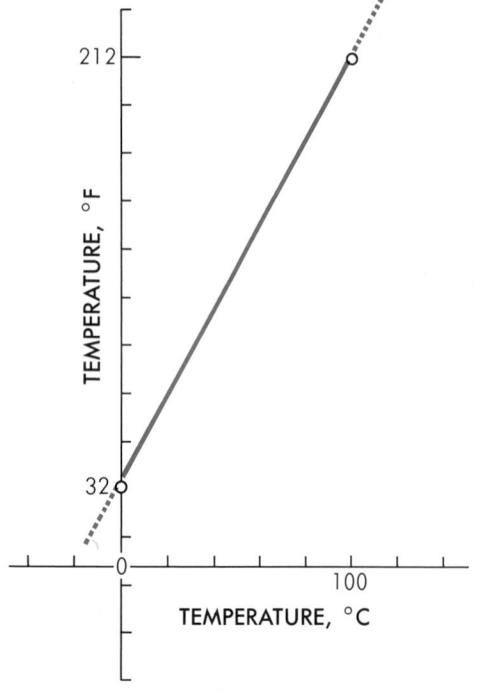

TEMPERATURE, °F

212

32

0   100

TEMPERATURE, °C

Figure 1.4   °F = 1.8 °C + 32
                      slope   intercept

The conversion factor approach will be used repeatedly throughout this text. If this is your first contact with it, it may seem awkward or artificial. You will find, however, that it is the most straightforward way to solve a wide variety of problems in chemistry. It is particularly useful in dealing with unfamiliar units or carrying out a multistep conversion as in Example 1.3b.

TEMPERATURE SCALES.   The common temperature scale used in the United States today was established by a German instrument maker, Fahrenheit, who took the normal freezing and boiling points of water to be 32° and 212°, respectively. He believed that with these two fixed points normal body temperature would be 100°F, while 0°F would represent the lowest temperature that could be achieved artificially in the laboratory (he used a mixture of ice and ammonium chloride).

As it happened, Fahrenheit was wrong on both counts, but his choice leads to a rather simple relationship between degrees Fahrenheit and degrees centigrade. From Figure 1.4 we see that °F is a linear function of °C. Since the equation of a straight line is y = ax + b, it follows that

All common temperature scales are related through linear functions. Can you see why?

$$°F = a°C + b$$

The constants a and b are readily evaluated; b is the intercept on the vertical axis, 32, while a is the slope of the line

$$a = \frac{\Delta y}{\Delta x} = \frac{212 - 32}{100 - 0} = \frac{180}{100} = 1.8$$

Making these substitutions we obtain

$$°F = 1.8°C + 32 \qquad (1.2)$$

Another temperature scale that we will find particularly useful in describing the behavior of gases is the absolute or Kelvin scale. The relation between °K and °C is

$$°K = °C + 273 \qquad (1.3)$$

The scale is named after Lord Kelvin, an English physicist who demonstrated by a mathematical development based upon both theory and experiment that it is impossible to achieve a temperature lower than 0°K.

1.3 SEPARATION OF MATTER INTO PURE SUBSTANCES

When a chemist carries out a reaction in the laboratory, he seldom if ever is fortunate enough to obtain a single pure substance as a product. Instead, he usually obtains a mixture of two or more substances, which may be liquids, solids, or gases. He then is faced with the problem of separating from this mixture the particular substance in which he is interested. In order to carry out meaningful experiments with this substance, it should be as pure as possible.

There are a great many different techniques which have been worked out to separate pure substances from mixtures. We will consider three of the more common methods:

The isolation and characterization of pure substances is a continuing problem for many chemists

1. *Distillation*, used to separate a solid from a liquid, and *fractional distillation*, by which two liquids can be separated from each other.

2. *Fractional crystallization*, a technique used routinely to purify solids.

3. *Chromatography*, one of the most versatile separation methods, which can be applied to solid, liquid, or gaseous mixtures.

DISTILLATION AND FRACTIONAL DISTILLATION

A mixture of two substances, only one of which is volatile, is readily separated by *distillation*. A simple distillation apparatus which can be used to separate sodium chloride from water is shown in Figure 1.5. When the solution is heated, the water boils off, leaving a residue of sodium chloride in the distilling flask. The water may be collected by passing the vapor down a condenser through which cold water is circulated. In many arid regions, particularly in the Mideast, distillation is used to convert sea water to fresh water.

If both components of a mixture are volatile, simple distillation will not achieve a complete separation. Suppose, for example, that we heat a solution of ether (bpt = 35°C) in benzene (bpt = 80°C). The mixture will begin to boil at a temperature somewhat above 35°C. As heating is continued, we find that the vapor entering the condenser is somewhat richer in the more volatile component, ether, than was the original mixture. It will still, however, contain an appreciable amount of the less volatile component, benzene. Similarly, the residue in the distilling flask will become enriched in benzene but will still contain some ether.

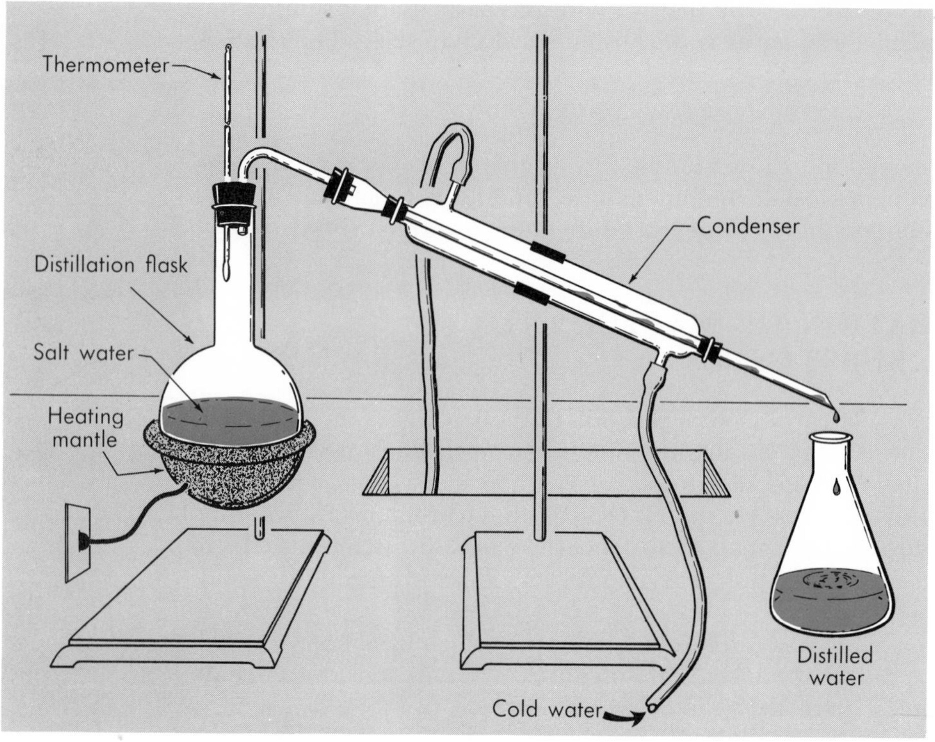

Figure 1.5 Separation of a liquid from a solid by simple distillation.

To make this argument quantitative, consider the distillation diagram at the left in Figure 1.6. Here the weight percentage of benzene in the vapor (upper curve) and in the liquid (lower curve) is plotted as a function of temperature. If we start with a 50-50 mixture of benzene and ether it will start to boil at about 50°C (point A). The temperature rises as the distillation proceeds until at 60°C the vapor has the composition given by point B (about 30 per cent benzene) and the remaining liquid a composition corresponding to point C (about 70 per cent benzene). Distillation of a 50-50 mixture until the boiling point was 60°C would leave a solution that was 70 per cent benzene and produce a distillate that was somewhat less than 30 per cent benzene, since the first portions of distillate would be very rich in ether. Clearly the distillate is enriched in ether, the residue in benzene; both fractions, however, still contain significant amounts of the other component.

We could improve upon this partial separation by subjecting both distillate and residue to further distillation. Each time we repeat the process, the distillate will become richer in ether, the residue richer in benzene. If we are patient enough, we will eventually end up with fractions which are pure or nearly pure. A more convenient way to carry out this process of *fractional distillation* is to insert between the distilling flask and the condenser a column of the type shown at the right in Figure 1.6. The glass beads in the column offer a surface upon which the less volatile component of the vapor (e.g., benzene) can condense and fall back into the distilling flask. The more volatile component (e.g., ether) more easily moves up the column and passes over into the distillate. If the rate of distillation is carefully controlled, a good separation can be achieved. Huge fractionating

Why doesn't some of the salt distill?

What would be the composition of the first distillate obtained by boiling the 50-50 mixture?

On a fractionating column the upper portions are kept cooler than the lower

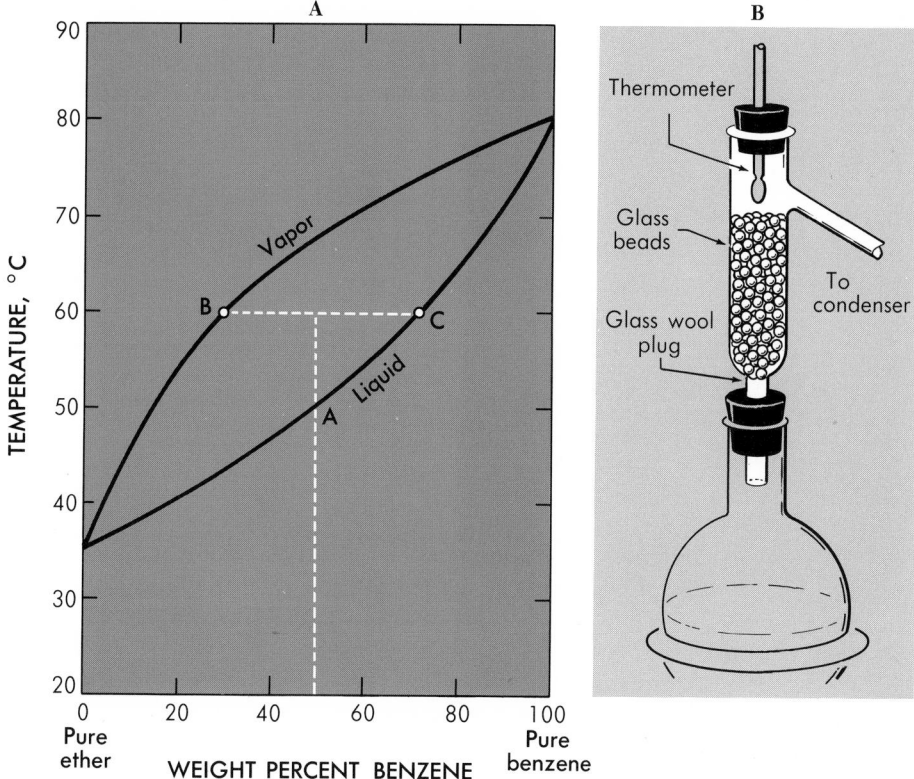

Figure 1.6 Fractional distillation. *A*, Distillation diagram for ether-benzene. Note that at all temperatures the vapor is richer in the more volatile component, ether. *B*, A simple fractionating column.

columns of very sophisticated design are used routinely in the petroleum industry to separate on the basis of varying volatility crude oil fractions such as gasoline (bpt 40–200°C), kerosene (bpt 175–325°C), and diesel fuel (bpt > 275°C).

FRACTIONAL CRYSTALLIZATION

Most of the reagent grade solid chemicals that we work with in the general chemistry laboratory have been purified by *fractional crystallization*. In this process a solid, A, containing a relatively small amount of an impurity, B, is dissolved in the minimum amount of hot solvent. The solution is then cooled, often to room temperature. If all goes well, the solid that separates out on cooling will be pure A, which can be filtered off and dried. The filtrate, which should contain all of the impurity B and relatively little A, is ordinarily discarded.

To illustrate the process of fractional crystallization, suppose we wish to purify a sample of tartaric acid, an organic compound isolated from grape juice, contaminated by a small amount of another organic substance, succinic acid. Specifically, suppose the sample contains 5 g of succinic acid mixed with 95 g of tartaric acid. Referring to Table 1.2, we see that this sample should dissolve completely in 100 g of water at 80°C (100 g of water at 80°C dissolves 98 g of tartaric acid and 71 g of succinic acid). Con-

Two similar organic molecules

tartaric acid   succinic acid

TABLE 1.2 Solubilities of Tartaric and Succinic Acids (g/100 g water) °

°C	TARTARIC ACID	SUCCINIC ACID
20	18	7
30	25	11
40	37	16
50	50	24
60	65	36
70	81	51
80	98	71

° Throughout the text discussion, it is assumed that the presence of one solid in solution has no effect on the solubility of the other. For example, at 80°C, the solubility of succinic acid is taken to be 71 g per 100 g of water, regardless of how much tartaric acid is present.

sider now what happens when this solution is cooled to 20°C. Since the solubility of tartaric acid at 20°C is only 18 g/100 g of water, we see that

$$95 \text{ g} - 18 \text{ g} = 77 \text{ g}$$

of tartaric acid crystallizes out of solution. All the succinic acid (5 g) stays in solution, since 100 g of water at 20°C can hold as much as 7 g of succinic acid. Hence, by dissolving this sample (95 g tartaric acid, 5 g succinic acid) in 100 g of water at 80°C and cooling to 20°C, we recover 77 g of *pure* tartaric acid, 81 per cent of the amount we started with.

This discussion implies two of the limitations of fractional crystallization.

1. The compound to be purified must be much more soluble at high than at low temperatures. Otherwise, much of it will be "lost"; i.e., it will remain in solution upon cooling. It would be futile to try to purify a sample of sodium chloride by dissolving in boiling water and cooling to room temperature, since its solubility at 20°C (36 g/100 g water) is nearly as great as at the boiling point (40 g/100 g water).

2. To obtain a pure sample by a single fractional crystallization, the amount of impurity must be relatively small. In the example discussed, consider what would have happened if the sample had contained 10 g of succinic acid instead of 5 g. Since the solubility of succinic acid at 20°C is 7 g/100 g water, it is evident that the tartaric acid isolated from the solution would have been contaminated by

$$10 \text{ g} - 7 \text{ g} = 3 \text{ g}$$

Alternately we could carry out the crystallization at 30°C

of succinic acid. A second recrystallization would have been required to give a pure product.

CHROMATOGRAPHY

The separation technique known as chromatography ° can be applied to extremely complex mixtures; as many as 20 amino acids in a hydrolyzed

---

° The word chromatography is derived from the Greek *chroma*, meaning color; in some of the early experiments the separated components were identified by their colors.

protein sample have been identified by this method. Moreover, it can be used for very small samples or for components present at very low concentrations. Chromatographic techniques have been developed to analyze for air pollutants at concentrations of one part per million or less.

To understand the principle behind chromatographic separations, let us consider the process of **column chromatography,** shown schematically in Figure 1.7. The column, which may be a simple buret, is packed with a finely divided solid such as silica gel or other chemically inert material which tends to adsorb other solids on its surface. A solution containing the sample is poured into the column, where it is retained by the packing at the top (stage 1). A solvent in which the components have different solubilities is then poured slowly through the column. The sample is gradually flushed down the column, continuously dissolving in the solvent and readsorbing on the packing. The component A, which is most soluble and/or least strongly adsorbed, moves most rapidly; a component such as C, which is more strongly attracted to the packing than to the solvent, moves very slowly. Gradually, the mixture is separated into zones or bands (stage 2). Finally, as more solvent is added, components A, B, and C pass successively from the bottom of the column (stage 3).

A variation on this technique, known as **paper chromatography,** is often used in the general chemistry laboratory. The solid adsorbent is a strip of

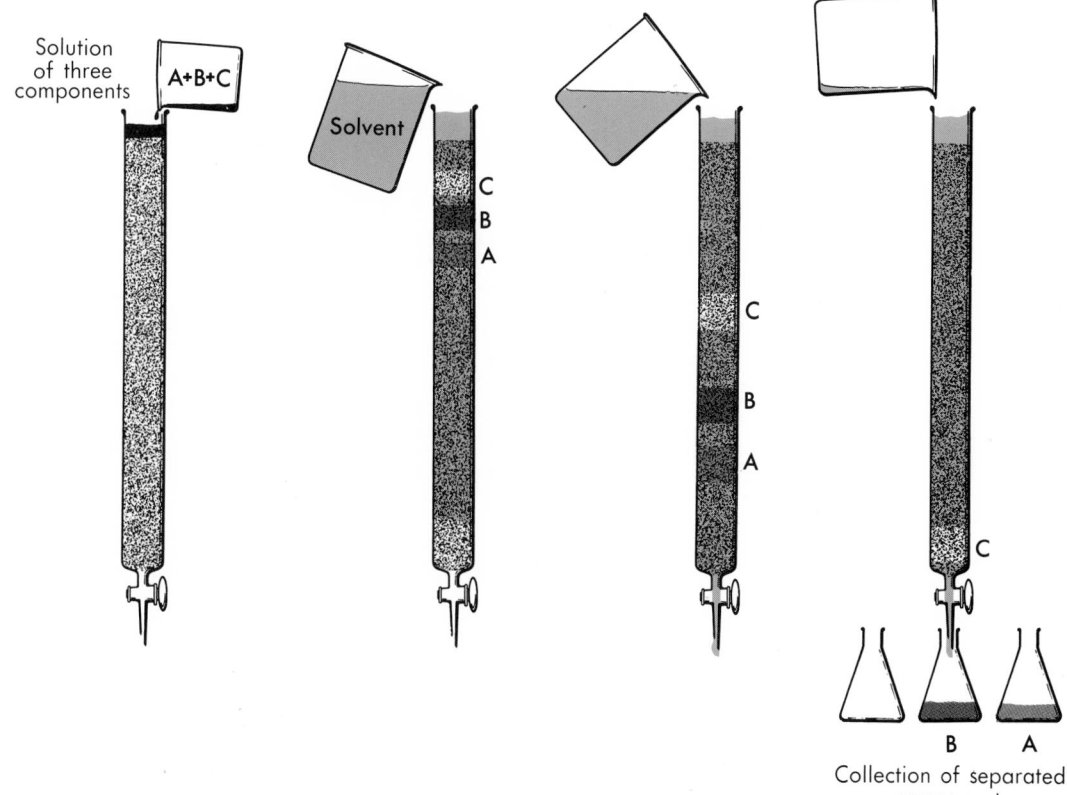

Figure 1.7 Column chromatography for a mixture of three components.

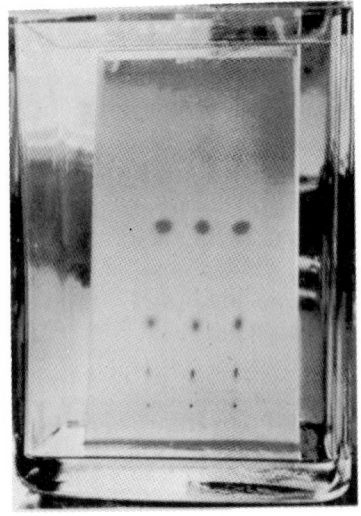

Figure 1.8 A thin-layer chromatogram showing the separation of three dyes (butter yellow, Sudan Red G, and indophenol) on activated Silica Gel G with benzene. Plate is shown in developing tank in solvent layer. (Photograph by L. A. Webb, courtesy of Dr. J. M. Bobbitt, University of Connecticut.)

The spots on the plate contain less than 1 mg of dye

Figure 1.9 Separation of chlorinated pesticides dissolved in hexane, using VPC.

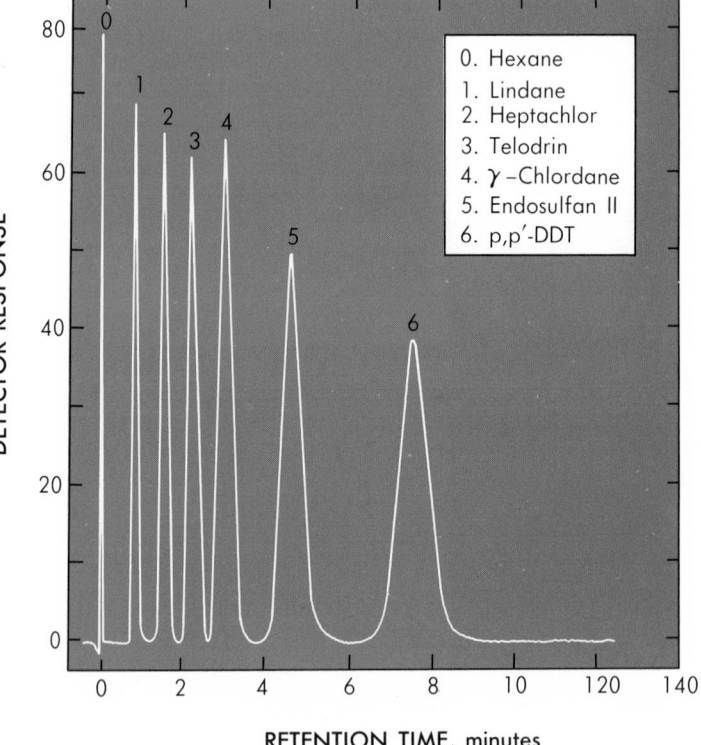

filter paper suspended vertically in a stoppered flask or cylinder. The solution containing the sample is introduced as a spot near the bottom of the paper. Enough solvent is added to almost but not quite reach the sample spot. As the solvent rises by capillary action, the components of the sample gradually separate. They can be made visible, as a series of spots, by adding a developing reagent which reacts with them to give a colored product. **Thin layer chromatography** (TLC) is used by organic chemists and biochemists to separate complex mixtures of natural products. The process resembles paper chromatography except that a glass plate covered with a thin layer of an adsorbent such as silica gel is substituted for the filter paper. After separation has been achieved, the components can be recovered by scraping them off the plate.

Perhaps the most widely used chromatographic technique is **vapor phase chromatography**, in which the components are separated as vapors. The column is similar to that previously described except that the solid packing is usually coated with a thin layer of a high-boiling, nonvolatile liquid such as silicone oil. A few microliters of sample ($1 \mu l = 10^{-6}$ liter) are injected into the top of the column. An unreactive "carrier" gas such as helium is passed through the column. The components of the sample gradually separate as they continuously vaporize into the helium and either adsorb onto or dissolve in the packing. Ordinarily the more volatile components move faster and come out of the column first. As successive fractions leave the column, they activate a detector and recorder to give a plot such as that shown in Figure 1.9.

Vapor phase chromatography offers the highest potential resolution of any method now known

1.4 IDENTIFICATION OF PURE SUBSTANCES

The chemist who isolates a substance from a mixture, perhaps by one of the techniques discussed in Section 1.3, has the further problem of identifying it and checking its purity. If the substance has never been prepared before, the task is a formidable one indeed. More frequently the substance is one whose properties have been established and recorded in the literature. In this case the problem is simpler but by no means trivial.

Substances are identified and tested for purity by measuring one or more of their **intensive properties** (i.e., properties which are independent of the amount of sample). One such property is density (mass per unit volume). If we find that a sample of a colorless liquid weighing 10.321 g occupies a volume of 11.76 ml at 20°C, its density is easily calculated:

$$\text{density} = \frac{\text{mass}}{\text{volume}} = \frac{10.321 \text{ g}}{11.76 \text{ ml}} = 0.8776 \text{ g/ml}$$

This value may give us reason to believe that we are dealing with nearly pure benzene (recorded density at 20°C = 0.8773 g/ml). It would, of course, be useless to measure a single **extensive property** of the sample such as mass or volume, which depends upon amount of sample.

The density of a pure substance is one of its characteristic, and easily measured, properties

There are a great many intensive properties which can serve to characterize a particular substance. Life is too short to measure more than a few of them. Common sense suggests that we concentrate upon properties which can be measured most precisely and are highly sensitive to small amounts of impurities. Two properties that meet these criteria are melting point and boiling point.

BEHAVIOR OF PURE SUBSTANCES DURING MELTING AND BOILING

An effective way to check the purity of a solid is to measure its melting point. A pure crystalline solid, if heated slowly, melts sharply at a characteristic temperature (mpt benzene = 5.50°C). The temperature stays constant as long as any solid remains (Figure 1.10A); only when it is completely melted does the temperature rise.

The melting point behavior of an impure solid differs from that of a pure substance in one important respect. It is ordinarily found, for reasons discussed in Chapter 10, that the solid starts to melt at a temperature below that of the pure substance and that the temperature rises steadily during the melting process (Figure 1.10B). Any evidence of a deviation from the horizontal in a temperature-time plot leads us to suspect the presence of impurities.

Sample A and sample B melt at the same temperature and look alike. What simple experiment might establish that they are identical?

The purity of a liquid can be tested in a manner quite similar to that described for a solid. In this case a constant temperature during the boiling process is the criterion of purity. The heating curve for a pure liquid looks very much like that shown at the left in Figure 1.10. If the liquid is impure, we ordinarily find that the temperature rises steadily during the boiling process (recall Figure 1.6).

COLOR: ABSORPTION SPECTRUM

Some of the substances that we work with in general chemistry can be identified at least tentatively on the basis of their color. Gaseous nitrogen

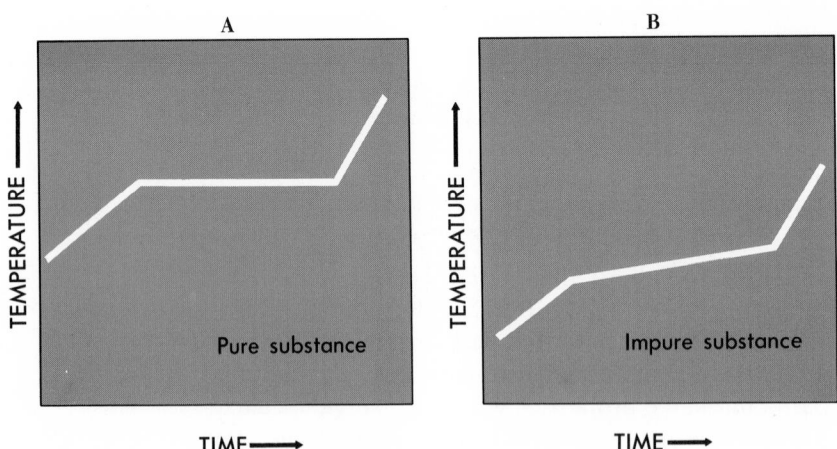

Figure 1.10 Time-temperature curves for the melting of a pure substance (left) and a mixture (right).

TABLE 1.3

WAVELENGTH REGION	COLOR ABSORBED	COLOR TRANSMITTED
<4000 A	Ultraviolet	Colorless
4000–4500	Violet	} Red, Orange, Yellow
4500–4900	Blue	
4900–5500	Green	} Purple
5500–5800	Yellow	
5800–6500	Orange	} Blue, Green
6500–7000	Red	
>7000	Infrared	Colorless

dioxide has a brown color; vapors of bromine and iodine are red and violet, respectively. An aqueous solution of copper sulfate is blue, a solution of potassium permanganate, purple, and so on.

The colors of gases and liquids are due to the selective absorption of certain components of visible light. Bromine, for example, absorbs in the violet and blue regions of the spectrum (Table 1.3). The subtraction of these components from visible light accounts for the red color that we see when we look through a sample of bromine vapor or liquid. The purple (blue-red) color of a potassium permanganate solution results from absorption in the green region.

White solids and colorless liquids do not absorb visible light

Many colorless substances absorb light in the ultraviolet or infrared regions of the spectrum. Ozone in the atmosphere absorbs harmful ultraviolet radiation from the sun. Carbon dioxide absorbs infrared radiation given off from the earth's surface, preventing excessive loss of heat to outer space. Most organic substances absorb at certain well characterized wavelengths in the infrared. By exposing a substance to infrared radiation covering a range of wavelengths and measuring absorption as a function of wavelength, a spectrum of the type shown in Figure 1.11 is obtained. Such an infrared spectrum serves as a "fingerprint" to identify the substance. Organic chemists use this technique routinely; one of its major advantages is that it can be applied to samples weighing a milligram or less.

The infrared spectrum is the best single property one can use to prove the identity of a pure substance

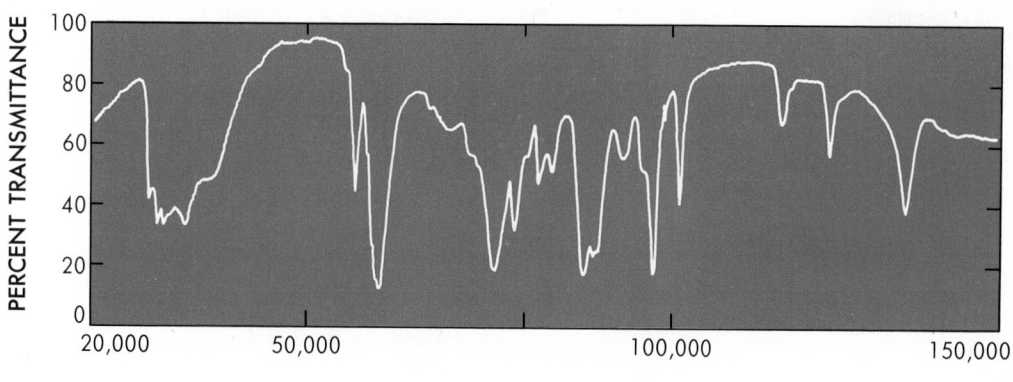

Figure 1.11   Infrared absorption spectrum of ascorbic acid (vitamin C).

1.5 KINDS OF SUBSTANCES: ELEMENTS AND COMPOUNDS

Of the many thousands of pure substances which have been isolated over the years, approximately 100 are unique in one important respect. No one has ever succeeded in resolving these substances, called **elements,** into two or more substances that differ in their properties from the original. All other pure substances are classified as **compounds.** A compound, by definition, is a pure substance that can be resolved into two or more elements.

All compounds, at sufficiently high temperatures, will break down into elements

unique in 1 respect

Many different methods have been used to resolve compound substances into elements. The English chemist Joseph Priestley decomposed mercuric oxide into the elements mercury and oxygen by exposing it to an intense beam of sunlight focused by a powerful lens. Sir Humphry Davy showed that ordinary quicklime, long believed to be an elementary substance, was a compound by passing an electric current through melted lime and demonstrating that two different substances (calcium and oxygen) were produced at the electrodes.

ISOLATION OF THE ELEMENTS

In Figure 1.12 is plotted the number of known elements as a function of time from 1750 to the present. The steep portions of the curve correspond to periods when new concepts or experimental techniques were introduced into chemistry. The first such period, centered around 1800, reflects two

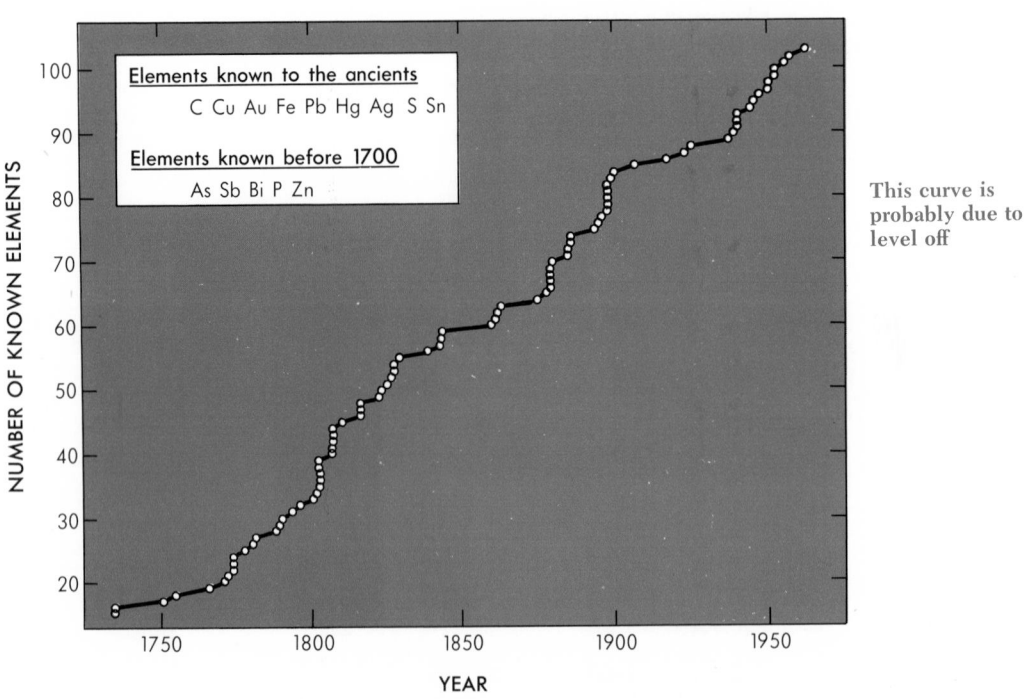

This curve is probably due to level off

Figure 1.12 Discovery of the elements from 1750 to the present.

factors: the evolution of quantitative analytical techniques and the discovery of the voltaic cell, used by Davy to isolate six elements (sodium, potassium, magnesium, calcium, strontium, and barium) in only two years (1807–1808).

Since 1940, 16 new elements have been discovered by chemists and physicists. These elements were not isolated from natural sources; instead they were synthesized from known elements by bombardment with high energy particles (Chapter 22). They are all unstable and decompose to form other elements, in most cases very quickly. Since the synthesis and decomposition of these elements involve far more energy than is usually associated with laboratory processes, we do not consider that such reactions invalidate our criterion for elementary substances.

ABUNDANCE OF THE ELEMENTS

At first glance, it seems surprising that many of our most familiar elements are relatively scarce in nature. The list of the 20 most abundant elements given in Table 1.4 does not include such "common" metals as nickel, copper, lead, and tin.

This feature of Table 1.4 emphasizes the distinction between *abundance* and *availability*. Gold, silver, and mercury, which together make up less than 0.00001 per cent of the earth's crust, have been known since antiquity because they are found in concentrated ore deposits from which they are readily extracted. Mercury is obtained from cinnabar (mercuric sulfide) by heating in air. Gold, which occurs in the elementary state, can be separated mechanically or by dissolving in mercury. In contrast, aluminum-bearing rocks and minerals are widely scattered in nature and highly resistant to decomposition. Even though aluminum is the third most abundant element, it did not come into commercial use until early in this century after an economical process was worked out for its extraction (Chapter 20).

The **symbol** of an element (Table 1.4) consists of one or two letters, taken most frequently from the name (O, Si, Al, Ca). Some symbols are derived from the Latin name of the element (iron, *ferrum*) or one of its compounds (sodium carbonate, *natrium*).

TABLE 1.4 Abundance of Elements (Earth's Crust, Oceans, and the Atmosphere)

Oxygen	O	49.5%		Chlorine	Cl	0.19%		If you don't know
Silicon	Si	25.7		Phosphorus	P	0.12		the symbols for the
Aluminum	Al	7.5		Manganese	Mn	0.09		elements, you
Iron	Fe	4.7		Carbon	C	0.08		should learn them,
Calcium	Ca	3.4	99.2%	Sulfur	S	0.06	0.7%	now
Sodium	Na	2.6		Barium	Ba	0.04		
Potassium	K	2.4		Chromium	Cr	0.033		
Magnesium	Mg	1.9		Nitrogen	N	0.030		
Hydrogen	H	0.87		Fluorine	F	0.027		
Titanium	Ti	0.58		Zirconium	Zr	0.023		
		All others	<0.1%					

PROBLEMS

1.1 Suggest experiments that might be carried out by chemists to

    a. reduce the sulfur content of natural gas (the principal sulfur-containing compound is hydrogen sulfide).
    b. determine the amount of heat discharged into a river by a nuclear power plant.
    c. check the sulfur dioxide content of the air over St. Louis.

1.2 Explain, in terms that would be understood by a student who has had no exposure to chemistry,

    a. the distinction between mass and weight.
    b. how a thermometer is constructed.
    c. why most of the familiar elements were isolated between 1770 and 1820.

1.3 Describe, in some detail, experiments by which you might determine

    a. the density of mercury to four significant figures.
    b. the volume of an Erlenmeyer flask to three significant figures.
    c. the boiling point of methanol, given a methanol solution containing about 10% of water.
    d. the melting point of DDT, given an impure sample.

1.4 Suggest ways of separating

    a. pure water from Great Salt Lake.
    b. the components of the residue left after the distillation of sea water.
    c. pure benzene from a solution with ether.
    d. carbon monoxide from the atmosphere.

1.5 Differentiate clearly between

    a. intensive and extensive properties.
    b. paper and thin layer chromatography.
    c. visible and infrared light.

1.6 An iron atom has a radius of about 1.26 Å.

    a. How many iron atoms are lined up along the edge of a bar 14.245 cm long?
    b. Estimate the volume of an iron atom in cc $\left(V = \dfrac{4\pi r^3}{3}\right)$.

1.14 Suggest several alternative methods to prevent emission of sulfur dioxide from power plants in addition to those discussed in Section 1.1. Comment on the merits and/or feasibility of these methods.

1.15 Explain the principle behind

    a. fractional crystallization.
    b. fractional distillation.
    c. identification by IR spectra.

1.16 Describe what you would observe in an experiment in which

    a. a solution of 20% benzene in ether is distilled.
    b. the melting point of tartaric acid contaminated by 10% succinic acid is measured.
    c. the solution in (a) is subjected to vapor phase chromatography.

1.17 Suggest how you would isolate

    a. pure aspirin from aspirin crystals wet with ether.
    b. elementary calcium from calcium oxide.
    c. nitrogen gas from water saturated with air.

1.18 Explain the difference between

    a. °F and °C.
    b. distillation and fractional distillation.
    c. abundance and accessibility of an element.

1.19 The radius of a copper atom is 1.28 Å; its mass is 63.5 amu (1 amu = 1.67×10^{-24} g).

    a. Calculate the volume of a copper atom in cc.
    b. Calculate the density of a copper atom in g/cc.
    c. Explain why the density of copper atoms is greater than that of a bulk sample of copper (8.96 g/cc).

1.7 A student calibrates a 20-ml pipet by filling it with pure water (d = 0.9970 g/ml) and allowing the water to run into a beaker which weighs 64.232 g. In four trials he obtains masses (water + beaker) of 84.136 g, 84.151 g, 84.141 g, and 85.279 g. What value would you advise him to report for the volume of the pipet?

1.8 What is the mass in grams of

a. an object balanced by weights labeled 10 g, 1 g, 500 mg, and 20 mg, assuming each weight is accurate to 1 mg?
b. a solution containing 20.00 kg of water, 12 g of sodium chloride, and 209.421 g of methanol?

1.9 Using the appropriate conversion factors and taking the density of water to be 1.00 g/ml, calculate

a. the density of water in g/liter.
b. the density of water in lb/ft³.
c. the mass in ounces of 1.00 pint of water.

1.10 Fish containing more than 0.50 part per million of mercury (e.g., 0.50 g of Hg per million grams) cannot be sold commercially. How many grams of mercury can be present in a half-pound fish without exceeding this limit?

1.11 Using the data in Table 1.4 and assuming a total mass of 3×10^{24} g, estimate the number of tons of titanium available to man.

1.12 Suppose Fahrenheit's scale had been set up taking body temperature to be exactly 100° (it is 98.6°) and the freezing point of water to be 32°. On the revised scale, what would be

a. the equation relating °F to °C?
b. the normal boiling point of water?

1.13 You are given a sample containing 65 g of tartaric acid and 12 g of succinic acid. Using the data in Table 1.2,

a. how much water should be used to dissolve the sample at 80°C?
b. how much tartaric acid will come out of solution on cooling to 20°C? how much succinic acid?
c. suggest two ways in which this procedure might be modified to avoid contamination of the final product with succinic acid.

1.20 A student determines the density of a piece of metal by weighing it (mass = 12.54 g) and adding it to a flask which has a volume of 25.00 cc. He finds that 21.62 g of water (d = 0.9970 g/cc) is required to fill the flask with the metal in it. What value should he report for the density of the metal?

1.21 What is the volume of the liquid remaining when

a. 4.0 cc are withdrawn from a bottle containing 12.20 liters of liquid?
b. 5.2 cc are withdrawn from a flask containing 52.643 g of benzene (d = 0.877 g/ml)?

1.22 The density of mercury is 13.60 g/cc. Calculate the pressure exerted by a column of mercury one cm² in cross section and 7.60×10^2 mm high, expressing your answer in

a. g/cm².
b. dynes/cm² (1 g/cm² = 980.7 dynes/cm²).
c. atmospheres (1 atm = 1.013×10^6 dynes/cm²).

1.23 On a certain day the concentration of carbon monoxide in the air over Denver reached 1.1×10^{-5} g/liter. How many tons of carbon monoxide were present, assuming a total volume of air of 1.5×10^{12} cubic feet?

1.24 The concentration of bromine in sea water is about 0.006%. Each year, about 1×10^5 tons of bromine are extracted from sea water. Taking the total mass of the oceans to be 1.3×10^{21} kg, how long will it be before all of this bromine has been extracted from sea water?

1.25 A temperature scale is set up on the basis of the properties of benzene, a common organic liquid. On the benzene scale, 0°B is set at the freezing point of benzene, which is 5°C. 100°B is taken to be the temperature at which benzene normally boils, which is 80°C. Find the equation relating °C to °B.

1.26 Suppose you have a mixture of 65 g of succinic acid and 20 g of tartaric acid. If this mixture is dissolved in 100 g of water at 80°C and cooled to 20°C, how much succinic acid will come out of solution? How much tartaric acid?

*1.27 A manufacturer of thermometers uses a capillary column with a diameter of 0.050 mm. He wants the degree markings on the thermometer to be exactly 2 mm apart. The density of mercury decreases by 0.018 per cent per °C. What must be the volume of the bulb at the bottom of the thermometer?

*1.28 Oil spreads on water to form a film about 1000 A thick. How many square miles of ocean will be covered by the slick formed when 100 barrels of oil are spilled (1 barrel = 31.5 gal)?

*1.29 The gravitational force exerted by a body such as the earth or the moon is directly proportional to its mass and inversely proportional to the square of the distance from the center of the body. An astronaut at the surface of the earth ($r = 4.0 \times 10^3$ miles) weighs 150 lb.

    a. What will be the gravitational force, to 0.1 lb, exerted by the earth on the astronaut when he is 10% of the distance to the surface of the moon, 2.43×10^5 miles from the center of the earth? 20%? 30% . . . 80%? 90% 100%?

    b. Repeat the calculations in (a), this time for the gravitational force exerted by the moon, which has a radius 0.27 that of the earth and a mass 0.012 times that of the earth.

    c. Combine your answers to (a) and (b) and plot the weight of the astronaut vs distance from the surface of the earth as he travels to the moon.

Answers to problems listed in the right column will be found in Appendix 5. Problems marked with an asterisk ordinarily require extra effort or ability.

2   aTOMS, MOLECULES, anD IONS

Every pure substance has associated with it a unique set of properties. We have seen how chemists use these characteristics of matter to isolate and identify elements and compounds. Chemists are equally interested in interpreting the properties of substances in terms of their particle structures. Many of the principles of chemistry can be understood most simply from this point of view. In this chapter we begin our study of the microstructure of matter by examining three of the building blocks of elements and compounds: the atom, the molecule, and the ion.

2.1 ATOMIC THEORY

The notion that matter ultimately consists of discrete particles is an old one. About 400 B.C. this idea was put forward in the writings of Democritus, a Greek philosopher, who apparently had been introduced to it by his teacher, a man named Leucippus. The idea was rejected by Plato and Aristotle, and it was not until about 1650 A.D. that it again was suggested, this time by the Italian physicist Gassendi. Sir Isaac Newton (1642–1727) supported Gassendi's arguments with these words:

. . . it seems probable to me that God, in the Beginning, formed Matter in solid, massy, hard, impenetrable, movable Particles, of such Sizes and Figures, and with such other Properties, and in such Proportions to Space, as most conduced to the End for which he formed them. . . .

Prior to 1800 the idea that matter is particulate in nature was based largely on the intuition of its adherents. Perhaps the most remarkable of these was the Russian chemist Mikhail Lomonosov (1711–1765), who speculated on the nature and motions of the tiny particles of which matter is composed. His ideas anticipated by at least a century the kinetic theory of gases and the concept of thermal energy.

In 1808 an English schoolteacher, John Dalton, using scientific insight of a remarkable quality, developed an explanation of many of the

Philosophical speculation seems to be easier than doing convincing experiments

25

then known laws of chemistry which became known as the **atomic theory.** He assumed that elements consist of tiny particles called **atoms.** He further postulated that all the atoms in a given elementary substance are alike and that compound substances are formed when one or more atoms of one element combine in a definite proportion with one or more atoms of another element. This theory, simple though it was, was very convincing in its ability to explain experimental facts and generalizations deduced from them.

While certain of Dalton's ideas proved untenable as chemists learned more about the structure of matter, the essentials of his theory have withstood the test of time. Three of Dalton's main postulates, which now comprise modern atomic theory, are given here with examples to illustrate the meaning of each.

1. *An element is composed of extremely small particles called atoms. All the atoms of a given element are chemically identical.* The element oxygen is made up of oxygen atoms, all of which behave chemically in the same way.

2. *Atoms of different elements have different properties. In the course of an ordinary chemical reaction, no atom of one element disappears or is changed into an atom of another element.* The chemical behavior of oxygen atoms is different from that of hydrogen atoms or any other kind of atom. When the elementary substances hydrogen and oxygen combine with each other, all the hydrogen atoms and all the oxygen atoms that react are present in the water formed, and no atoms of any other element are formed in the process.

3. *Compound substances are formed when atoms of more than one element combine. In a given pure compound the relative numbers of atoms of the elements present will be definite and constant. In general, these relative numbers can be expressed as integers or simple fractions.* In the compound substance water, hydrogen atoms and oxygen atoms are combined with each other. For every oxygen atom present, there are always two hydrogen atoms. Ammonia, a gaseous compound of nitrogen and hydrogen, always contains three hydrogen atoms for every nitrogen atom.

The second postulate offers an obvious explanation for the **Law of Conservation of Mass,** which, in modern form, states that **there is no detectable change in mass in an ordinary chemical reaction.** The third postulate explains the **Law of Constant Composition: a compound,** regardless of its origin or method of preparation, **always contains the same elements in the same proportions by weight.** Clearly, if the atom ratio of the elements in a compound is fixed, their proportions by weight must also be fixed. The validity of this law became generally recognized at about the same time that Dalton's theory appeared. Prior to 1808 many people agreed with the French chemist, Berthollet, who believed that the composition of a compound could vary over wide limits, depending on how it was prepared. Joseph Proust, a French expatriate working in Madrid, refuted Berthollet's ideas by showing that the "compounds" Berthollet had cited were actually mixtures.

The third postulate led Dalton to formulate another of the basic quantitative laws of chemistry, the **Law of Multiple Proportions,** which states that **when two elements combine to form more than one compound, the masses of one element which combine with a fixed mass of the other element are in a ratio of small whole numbers** such as 2:1. Dalton reasoned that elements A and B might form two compounds, in one of which two atoms of A were combined with one of B,

If atoms are conserved in a reaction, mass is also conserved

It was not a simple matter to prove unequivocally that pure substances exist

while in the other, one atom of A was combined with one atom of B. If this happened, the weight of A combined with a fixed weight of B in the first compound would be twice that in the second. The validity of this law was rapidly established, partly on the basis of experiments that Dalton himself carried out.

Despite the successes of Dalton's atomic theory, it was not immediately accepted by all scientists. Many chemists felt that it was a waste of time to speculate on the particulate nature of matter, and that natural laws should be based exclusively upon experimentally measured quantities. As late as 1900 the well-known German chemist, Ostwald, in writing a textbook of general chemistry, deliberately avoided all mention of atomic or other elementary particles.

"Be not the first by whom the new is tried, nor yet the last to cast the old aside"

2.2 COMPONENTS OF THE ATOM

Like any useful scientific theory, the atomic theory raised more questions than it answered. Even before Dalton's ideas had been generally accepted, philosophers and scientists were speculating as to whether atoms, tiny as they are, could in turn be broken down into still smaller particles. Nearly one hundred years were to pass before this question could be answered in the affirmative on the basis of experimental evidence. Three physicists did pioneer work in this area: J. J. Thomson, an Englishman working at the Cavendish laboratory at Cambridge; Ernest Rutherford, a native of New Zealand who carried out his research at McGill University in Montreal and at Manchester and Cambridge in England; and Robert A. Millikan at the University of Chicago.

ELECTRONS

The first convincing evidence for subatomic particles came from experiments involving the conduction of electricity through gases at low pressures. When an apparatus of the type shown in Figure 2.1 is partially evacuated and connected to a high voltage source such as a spark coil, an electric current flows through the tube. Associated with the flow are colored rays of light, which originate at the negative electrode (cathode). The properties of *cathode rays* were studied extensively during the last three decades of the nineteenth century. In particular, it was found that the rays were bent by both electric and magnetic fields. From a careful study of the nature of this deflection, J. J. Thomson demonstrated in 1897 that the rays consist of a stream of negatively charged particles, which he called *electrons*. Thomson went on to measure the mass-to-charge ratio of the electron, finding that

$$m/e = 5.69 \times 10^{-9} \text{ g/coulomb}$$

The fact that this ratio is smaller, by several orders of magnitude, than that of any other charged species, implies that we are dealing with a tiny subatomic particle.

In 1909 Millikan determined the charge on the electron, using the apparatus shown schematically in Figure 2.2. He measured the effect of an electrical field on the rate at which charged oil drops fall under the

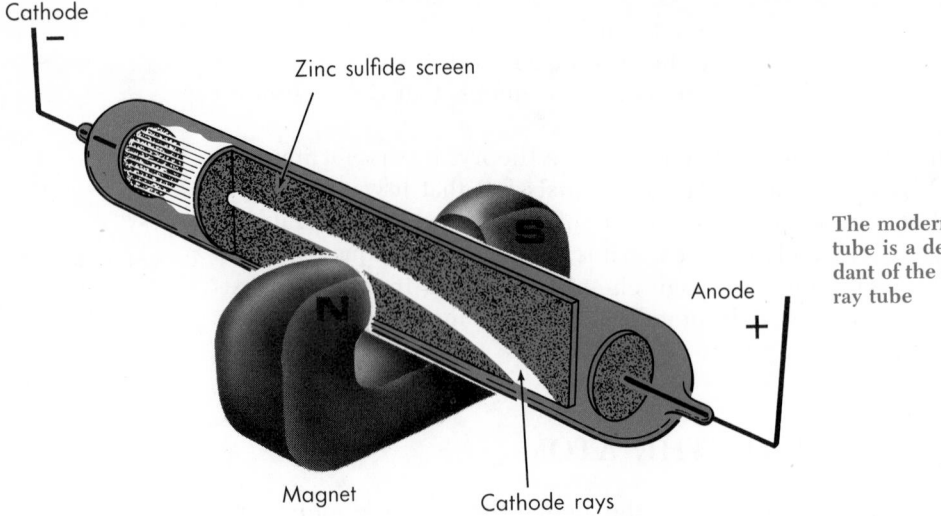

Cathode

Zinc sulfide screen

S

N

Anode
+

The modern TV
tube is a descen-
dant of the cathode
ray tube

Magnet

Cathode rays

Figure 2.1 Cathode ray rube. The beam is deflected by magnetic and electric fields in such a way as to indicate that it is negatively charged.

influence of gravity. From his data Millikan calculated the charge on the
drops, which he found always to be an integral multiple of a smallest charge.
Assuming the smallest charge to be that of the electron, he arrived at a
value of 1.60×10^{-19} coulomb. Combining this number with the mass-to-
charge ratio quoted previously, we obtain for the mass of the electron

$$m = (1.60 \times 10^{-19} \text{ coulomb}) \times (5.69 \times 10^{-9} \text{ g/coulomb}) = 9.11 \times 10^{-28} \text{ g}$$

This is only about 1/1837 of the mass of the lightest atom, that of the ele-

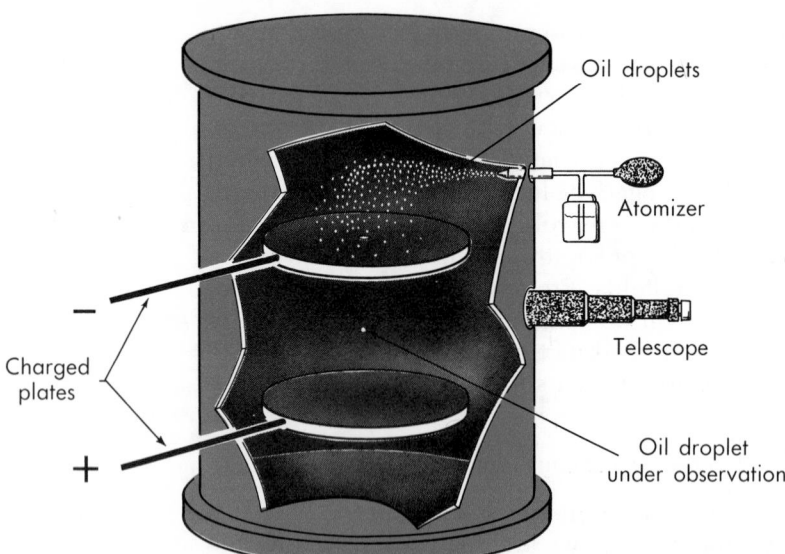

Oil droplets

Atomizer

Telescope

Charged
plates

Oil droplet
under observation

Figure 2.2 Oil drop experiment. By measuring the rate at which a charged oil
drop falls between the plates, it is possible to determine the charge on the drop.
Millikan found that this charge was always an integral multiple of a smallest
charge (about 1.60×10^{-19} coulomb), which he took to be that of the electron.

ment hydrogen. The electron is also much smaller than the hydrogen atom, with a diameter only 1/10,000 as great, or about 10^{-12} cm.

The existence of electrons has been confirmed in many experiments. Electrons are among the particles (β particles) given off by atoms undergoing radioactive decay. The ordinary radio vacuum tube depends for its operation upon the ability of metal atoms to give off electrons when they are heated to high temperatures (*thermionic effect*). Certain metals, notably cesium, emit electrons upon exposure to ultraviolet light; we make use of this phenomenon, known as the *photoelectric effect*, to operate automatic door openers and certain types of burglar alarm systems.

For various reasons it is difficult to assign a precise size to an electron

All atoms contain an integral number of electrons. This number, which may vary from 1 to over 100, is characteristic of an atom of a particular element. All hydrogen atoms contain one electron; all atoms of the element lawrencium contain 103 electrons. We shall have more to say in Chapter 6 about the way in which these electrons are distributed relative to one another. For the time being, it is sufficient to point out that electrons are found in the outer regions of atoms, where they comprise what amounts to a cloud of negative charge about the positively charged atomic nucleus.

THE ATOMIC NUCLEUS

In 1911 Ernest Rutherford and his students performed a series of experiments that profoundly influenced our ideas regarding the nature of atoms. Using a radioactive source, they bombarded a piece of thin gold foil with α-particles (helium atoms stripped of their electrons). With a fluorescent screen, they observed the extent to which the α-particles were scattered by the foil. Most of them went through the foil almost undeflected. A few, however, were reflected back from the foil at acute angles. The relative numbers of α-particles reflected at different angles were determined by counting on the screen the scintillations caused by individual particles. By a beautiful mathematical analysis of the electrostatic forces involved, Rutherford was able to show that the scattering was caused by a positively charged center within the gold atom which had a mass nearly equal to that of the atom but a diameter (about 10^{-12} cm) only 1/10,000 of that of the atom. The experiment was repeated, with similar results, with foils of many different elements. In this way Rutherford established that an atom contains a tiny, positively charged, massive center, called the **atomic nucleus.**

The results of this experiment were at first very surprising. Can you suggest why?

Since the time of Rutherford we have learned a great deal about the properties of atomic nuclei, although we still do not have a clear physical picture of the forces that hold the nucleus together. For our purposes, we may consider the nucleus of an atom to be made up of two different types of particles:

1. The **proton,** which has a mass nearly equal to that of the hydrogen atom and carries a unit positive charge, equal in magnitude but opposite in sign to that of the electron.

2. The **neutron,** an uncharged particle with a mass about equal to that of the proton.

All nuclei contain an integral number of protons, exactly equal to the number of electrons in the neutral atom. In the nucleus of every hydrogen atom there is one proton; the nucleus of every lawrencium atom contains 103 protons. The number of protons in the nucleus of an atom is a fundamental property of the corresponding element, known as its **atomic number.**

$$\text{atomic number} = \text{number of protons}$$

Thus, we would say that the atomic number of the element hydrogen is 1, while that of the element lawrencium is 103.

Atoms of the same element may vary in the number of neutrons found in the nucleus. In the element hydrogen, for example, we find three different kinds of atomic nuclei containing 0, 1, and 2 neutrons, respectively. These three species are often referred to as **isotopes** of the element hydrogen. They differ in mass; the heaviest isotope of hydrogen (tritium) has a mass about three times as great as that of the lightest isotope. A deuterium atom (1 proton, 1 neutron) is about twice as heavy as a "light" hydrogen atom (1 proton, 0 neutrons). As another example, two well known isotopes of uranium, "uranium-235" and "uranium-238," both contain the same number of protons, 92, but differ in the number of neutrons, 143 vs 146.

All atoms of hydrogen, irrespective of mass, contain one electron. All uranium atoms contain 92 electrons

The **mass number** of a nucleus is found by adding the numbers of protons and neutrons.

$$\text{mass number} = \text{number of protons} + \text{number of neutrons}$$

The three isotopes of hydrogen have mass numbers of 1, 2, and 3, respectively, while the mass numbers of the two isotopes of uranium are

"light isotope": mass number $= 92 + 143 = 235$ (uranium-235)

"heavy isotope": mass number $= 92 + 146 = 238$ (uranium-238)

We often indicate the composition of a nucleus by writing the atomic number as a subscript at the lower left of the symbol of the element and the mass number as a superscript at the upper left. For the species discussed above, we would write

$$_1^1\text{H}, \ _1^2\text{H}, \ _1^3\text{H}; \ _{92}^{235}\text{U}, \ _{92}^{238}\text{U}$$

Example 2.1

a. Write nuclear symbols for three isotopes of oxygen (atomic no $= 8$) in which there are 8, 9, and 10 neutrons, respectively.

b. One of the most harmful species in nuclear fallout is a radioactive isotope of strontium, $_{38}^{90}\text{Sr}$. How many protons are there in this nucleus? how many neutrons?

Solution

a. The mass numbers must be: $8 + 8 = 16$; $8 + 9 = 17$; $8 + 10 = 18$. Thus we have:

$$_8^{16}\text{O}, \ _8^{17}\text{O}, \ _8^{18}\text{O}$$

b. The number of protons is given by the atomic number, 38. To obtain the number of neutrons we subtract the number of protons from the mass number.

$$\text{number neutrons} = 90 - 38 = 52$$

2.3 MOLECULES AND IONS

Isolated atoms are rarely encountered in matter. Only in a very few elementary substances, the so-called noble gases (He, Ne, Ar, Kr, Xe, Rn), is the individual atom the structural unit of which the substance is composed. Most elementary and compound substances are made up of other types of structural units, two of the most important of which are molecules and ions.

Most isolated atoms are extremely reactive chemically

MOLECULES

The fundamental structural unit in most volatile substances, both elementary and compound, is the molecule, which is an aggregate of atoms held together by relatively strong forces called chemical bonds. In contrast, the forces between molecules are relatively weak. As a result, molecules do not interact strongly with one another but behave more or less as independent particles.

Hydrogen chloride is an example of a molecular substance. The hydrogen chloride molecule is a simple one consisting of a hydrogen atom joined by a chemical bond to a chloride atom. The structure of the hydrogen chloride molecule is often indicated as

$$H—Cl$$

where the dash represents the bond holding the two atoms together. Examples of elementary molecular substances are hydrogen and chlorine. At ordinary temperatures and pressures, the fundamental structural unit in both of these substances is a diatomic molecule and may be represented as

At ordinary temperatures H—Cl, H—H and Cl—Cl exist as gases

$$H—H \qquad Cl—Cl$$

Most of the molecular substances with which you are familiar are built up of molecules more complex than those just cited. The water molecule, for example, consists of a central oxygen atom bonded to two hydrogen atoms. In the ammonia molecule, a central nitrogen atom is bonded to three individual hydrogen atoms. Methane, the major constituent of natural gas, has as its basic structural unit a molecule in which a carbon atom is bonded to four hydrogen atoms. The structures of these molecules can be represented in two dimensions as

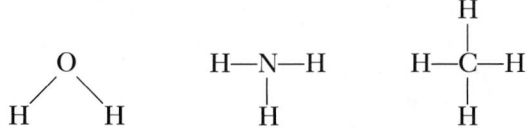

Still more complex molecules are found in substances such as ethane (2 carbon and 6 hydrogen atoms per molecule) and the insecticide DDT (14 carbon atoms, 9 hydrogen atoms, 5 chlorine atoms per molecule). The proteins that are built into the nerve and muscle tissues of our bodies consist of huge molecules containing thousands of atoms of several different elements, including carbon, hydrogen, nitrogen, oxygen, and sulfur.

Two-dimensional "structural formulas" do not, in general, represent the true geometry of any but the simplest of molecules. Space-filling

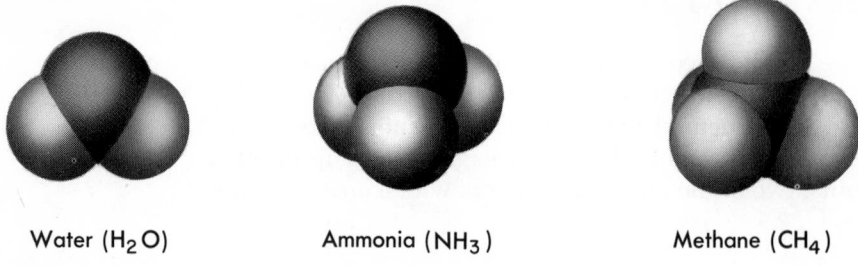

Water (H_2O)   Ammonia (NH_3)   Methane (CH_4)

Figure 2.3   Molecular geometry of water, ammonia and methane.

models of the type shown in Figure 2.3 give a more adequate representation of the orientations of atoms within a molecule. We will have more to say about molecular geometry and other aspects of molecular structure when we discuss chemical bonding in more detail in Chapter 7.

IONS

If sufficient energy is available, it is possible to remove one or more electrons from a neutral atom, leaving a positively charged particle that is somewhat smaller than the original atom. Alternatively electrons may be added to certain atoms to form negatively charged species that are somewhat larger than the atom from which they are derived. These charged particles are referred to as **ions.** Examples of positive ions (**cations**) include the sodium ion, Na^+, formed from a sodium atom by the loss of a single electron, and the Ca^{2+} ion, derived from a calcium atom by extraction of two electrons

$$Na \text{ atom } (11 \text{ p}^+, 11 \text{ e}^-) \rightarrow Na^+ \text{ ion } (11 \text{ p}^+, 10 \text{ e}^-) + e^-$$

$$Ca \text{ atom } (20 \text{ p}^+, 20 \text{ e}^-) \rightarrow Ca^{2+} \text{ ion } (20 \text{ p}^+, 18 \text{ e}^-) + 2 \text{ e}^-$$

Two common negative ions (**anions**) are the chloride ion, Cl^-, and the oxide ion, O^{2-}, formed when atoms of chlorine or oxygen acquire extra electrons.

$$Cl \text{ atom } (17 \text{ p}^+, 17 \text{ e}^-) + e^- \rightarrow Cl^- \text{ ion } (17 \text{ p}^+, 18 \text{ e}^-)$$

$$O \text{ atom } (8 \text{ p}^+, 8 \text{ e}^-) + 2 \text{ e}^- \rightarrow O^{2-} \text{ ion } (8 \text{ p}^+, 10 \text{ e}^-)$$

Since a macroscopic sample of matter must be electrically neutral, ionic compounds always contain both cations and anions as structural units. Ordinary table salt, sodium chloride, is made up of an equal number of Na^+ and Cl^- ions. The structure of a small portion of a sodium chloride crystal is shown in Figure 2.4; that of calcium oxide (quicklime), in which there are equal numbers of Ca^{2+} and O^{2-} ions, is quite similar. In the ionic compounds sodium oxide and calcium chloride there are unequal numbers of cations and anions. In order to maintain electroneutrality there must be two Na^+ ions for every O^{2-} ion in a macroscopic sample of sodium oxide; similarly two Cl^- ions are required to balance one Ca^{2+} ion in calcium chloride. It is important to recognize that in all these species and, indeed, in all ionic compounds, there is a continuous three-dimensional network of chemical bonds holding oppositely charged ions together. These bonds, which originate in the electrostatic attraction between cations and anions,

Ions may be present in solids and in solution in water or other solvents

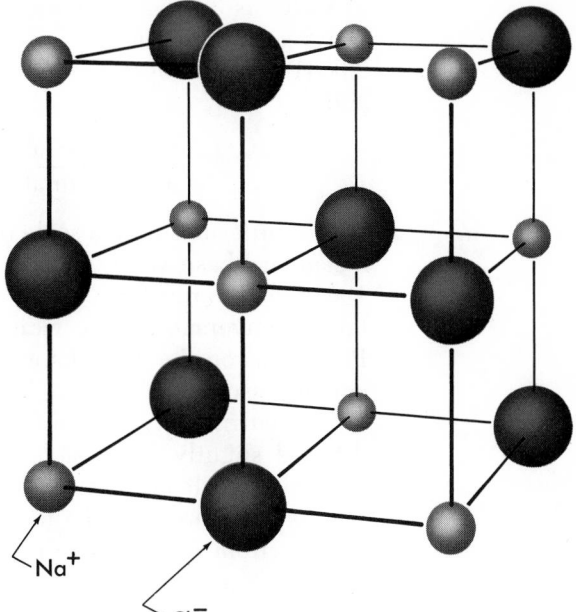

Figure 2.4 Crystal structure of NaCl, expanded for clarity. Each Na⁺ ion is in contact with 6 Cl⁻ ions, each Cl⁻ with 6 Na⁺ ions.

A salt crystal will contain many many ions in the pattern shown in the figure

are extremely strong. For this reason we ordinarily find that ionic compounds are nonvolatile and have relatively high melting points.

2.4 RELATIVE MASSES OF ATOMS: ATOMIC WEIGHTS AND GRAM ATOMIC WEIGHTS

The problem of determining the relative masses of different atoms is one which has occupied the attention of a great many chemists since the time of Dalton (Section 2.7). We will shortly consider one of the experimental approaches that has been used to determine the relative masses of atoms. Before doing so, it will be helpful to define two quantities, atomic weight (AW) and gram atomic weight (GAW).

ATOMIC WEIGHT

Relative weights of atoms of different elements are most often expressed in terms of their atomic weights. The atomic weight of an element is a number that indicates how heavy, on the average,* an atom of that element is compared to an atom of another element. The fact that hydrogen has an atomic weight slightly greater than 1 and oxygen an atomic weight of almost exactly 16 tells us that a hydrogen atom is about 1/16 as heavy as an oxygen atom. In another case we conclude that since the atomic weight

---

* The word "average" is used here because most elements consist of a mixture of isotopes. (See Example 2.2.)

of sulfur is 32.064, an atom of sulfur must be a little more than twice as heavy as an oxygen atom. To be exact, a sulfur atom is 32.064/15.9994 = 2.0041 times as heavy as an oxygen atom. In general, for two elements X and Y,

$$\frac{\text{mass of an atom of X}}{\text{mass of an atom of Y}} = \frac{\text{atomic weight of X}}{\text{atomic weight of Y}}$$

In order to set up a scale of atomic weights, it is necessary to establish a standard base value for one particular species. For many years prior to 1961 two different scales of atomic weights were in use. Chemists chose to take the average atomic weight of the element oxygen to be exactly 16. Physicists, on the other hand, found it more convenient to assign the most common isotope of oxygen, $^{16}_{8}O$, an atomic weight of exactly 16. Since naturally occurring oxygen contains traces of heavier isotopes, the two scales differed slightly from each other. To convert from the chemists' to the physicists' scale of atomic weights, it was necessary to multiply by 1.000275.

As the years passed, the existence of two atomic weight scales differing slightly from each other proved sufficiently annoying to cause scientists to work for the adoption of a single scale that would be acceptable to both physicists and chemists. This was accomplished in 1961 when it was agreed to set up a uniform scale in which the atomic weight 12 was assigned to the most common isotope of carbon. Such a compromise satisfied the physicists' requirement that a single isotope be used as a standard. For chemists, it resulted in a change of only about 0.004 per cent from the old chemical scale. The atomic weight of oxygen, for example, changed from 16.0000 to 15.9994 with the adoption of the new scale. The atomic weights used throughout this text are all based on this so-called "carbon-12" scale.

GRAM ATOMIC WEIGHT

The **gram atomic weight** of an element is the **weight in grams that contains the same number of atoms as twelve grams of carbon-12** (or 15.9994 g of oxygen, 1.00797 g of hydrogen, and so forth). If the atomic weight of an element is known, its gram atomic weight can be written down immediately. For example, consider sulfur: since the atomic weight of this element is 32, a sulfur atom must be 32/16 as heavy as an oxygen atom. If the number of atoms in 16 g of oxygen is represented by N, then the weight of an equal number, N, of sulfur atoms must be

$$\frac{32}{16} \times \text{wt of N oxygen atoms} = \frac{32}{16} \times 16 \text{ g} = 32 \text{ g}$$

In other words, the gram atomic weight of an element is numerically equal to its atomic weight. It must be kept in mind, however, that while the atomic weight of an element is dimensionless, its gram atomic weight has the unit of grams.

At wt S = 32
1 GAW of sulfur
weighs 32 g

DETERMINATION OF ATOMIC WEIGHT

In Section 2.7 we shall comment upon a few of the methods that have been used over the years to arrive at atomic weights. Today these

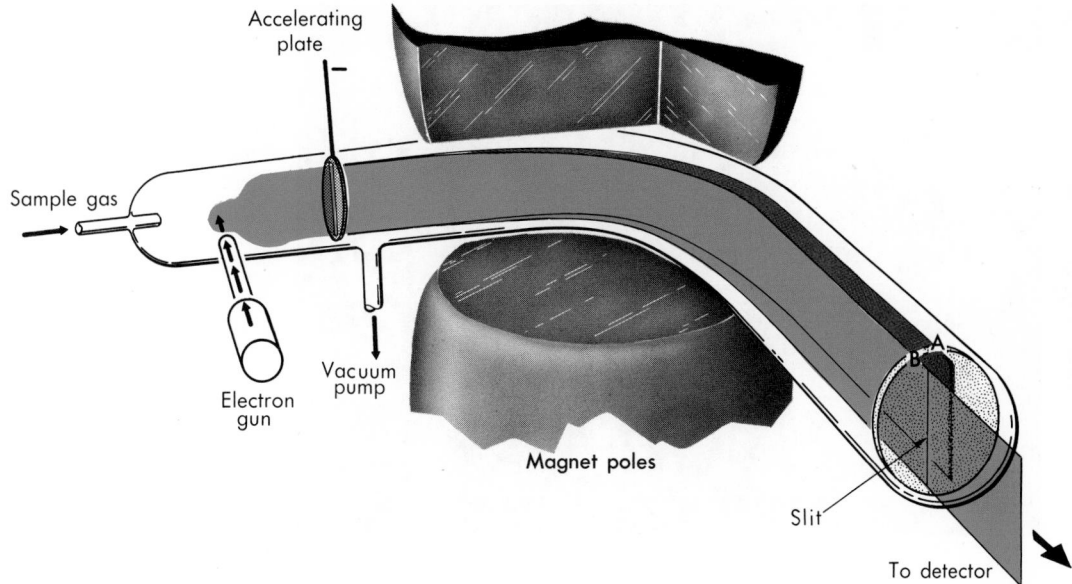

Figure 2.5 Mass spectrometer. A narrow beam of positively charged ions is split into multiple beams in the magnetic field. Ions with a high charge to mass ratio are deflected most strongly, striking the slit at B.

methods are largely of historical interest; they have been superseded by the development of the modern mass spectrometer, shown in Figure 2.5. This instrument measures the mass-to-charge ratio of charged particles. It can be used to investigate a wide variety of chemical problems; indeed, an early prototype of the mass spectrometer was used by J. J. Thomson to determine the mass-to-charge ratio of the electron. Before discussing how the mass spectrometer can be applied to atomic weight measurements, let us consider the principle upon which it is based.

In a mass spectrometer (Fig. 2.5) a gaseous sample in the ionization chamber is bombarded by a stream of electrons emitted from a heated filament. A few of the gas particles are converted to positive ions, which pass through a narrow slit and are accelerated by a voltage of 500 to 2000 volts toward a magnetic field. The field deflects the ions from their straight-line path toward the collector.

For a given accelerating voltage and magnetic field strength, the extent to which the ion beam is deflected depends upon the charge and mass of the ions. The greater the charge of the ion, the greater its deflection will be. If a +1 ion is bent to point A (Fig. 2.5), a +2 ion of the same mass might appear at B. The deflection of an ion is inversely related to its mass; the lighter the ion, the more readily it is pulled off its course by the magnetic field. Of two ions which have the same charge but different masses, the heavier one might arrive at A, while the lighter ion is deflected to B. By moving the slit from A to B, we can detect these two ion beams separately. In practice it is simpler to use a stationary slit and adjust the accelerating voltage or the magnetic field strength so that first one ion beam and then the other goes through the slit to the detector.

Depending upon the nature of the gas being analyzed, we may obtain a relatively simple or an exceedingly complex mass spectrum. With helium,

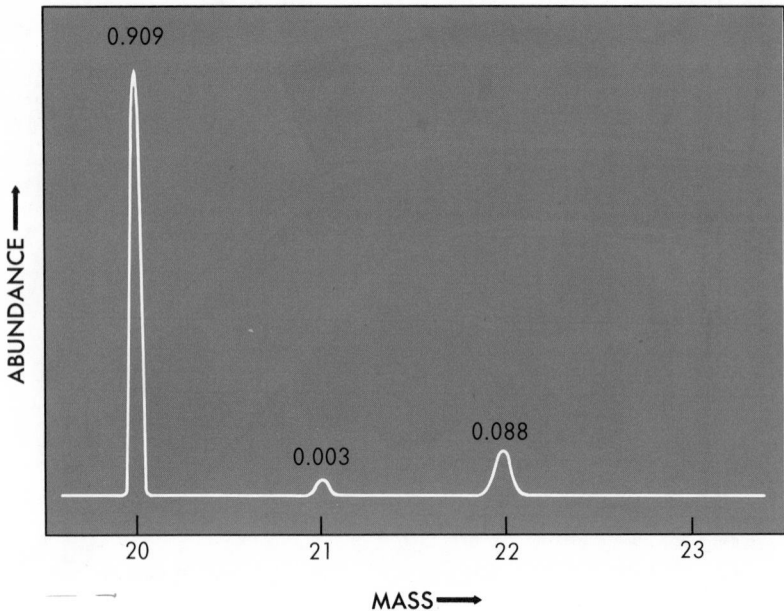

On a high resolution instrument relative atomic masses can be measured to at least four decimal places

Figure 2.6 Mass spectrum of neon (+1 ions only). The principal peak corresponds to the most abundant isotope, neon-20.

which consists of individual atoms almost all of which have a mass of 4 ($^4_2$He), the principal species formed is a +1 ion with a mass-to-charge ratio of 4. If the gas used is neon, the spectrum is more complex. Since ordinary neon consists of three different isotopes, a beam of +1 neon ions is split into three parts when it passes through the magnetic field. The most dense beam is one in which the ions have a mass-to-charge ratio of 20 ($^{20}_{10}$Ne$^+$). Two weaker beams, consisting of +1 ions derived from the less abundant isotopes $^{21}_{10}$Ne and $^{22}_{10}$Ne can also be detected (Fig. 2.6).

To illustrate how the mass spectrometer can be used to determine atomic weights, consider the element helium. In principle, it is possible to calculate exactly the mass of the $^4_2$He$^+$ ion by measuring the extent to which it is deflected by the magnetic field. The appropriate equation is

$$m = \frac{qH^2r^2}{2\ E} \tag{2.1}$$

where m is the mass of the ion, q is its charge, H is the magnetic field strength, r is the radius of curvature of the ion beam as it passes from the magnetic field, and E is the accelerating voltage. In practice, since both H and r are difficult to determine accurately, we measure the voltage necessary to bring a +1 ion of known mass such as that derived from carbon-12 to the same point, using the same magnetic field strength. Since q, H, and r are then the same for the two ions, Equation 2.1 tells us that

$$m\ He \times E\ He = m\ C\text{-}12 \times E\ C\text{-}12 \tag{2.2}$$

or

$$\frac{m\ He}{m\ C\text{-}12} = \frac{E\ C\text{-}12}{E\ He}$$

Experimentally we find that the voltage applied to the carbon-12 beam is 0.3336 times that required to bring helium-4 ions to the same point (i.e., E C-12/E He = 0.3336).

Hence

$$\frac{m\ He}{m\ C\text{-}12} = 0.3336;$$

or,

$$AW\ {}_{2}^{4}He\ isotope = 0.3336 \times 12 = 4.003$$

But, since naturally occurring helium contains essentially 100 per cent of this isotope, the number 4.003 must also represent the atomic weight of the element helium.

The determination of the atomic weight of an element such as neon, in which more than one isotope is present, is a somewhat more difficult experimental problem. Here we need to determine not only the mass but also the relative abundance of each isotope. The latter can be estimated by measuring the relative heights of the peaks shown in Figure 2.6. Knowing the masses of each isotope and their abundances, we can readily calculate the average atomic weight (Example 2.2).

Example 2.2 From mass spectra it can be determined that the element neon consists of three isotopes whose masses on the carbon-12 scale are 19.99, 20.99, and 21.99. The abundances of these isotopes are, respectively, 90.92 per cent, 0.25 per cent, and 8.83 per cent. Calculate an accurate value for the atomic weight of neon.

Solution. We can find the atomic weight of any element by adding the contributions of each of the isotopes. In the case of neon, we have:

$$mass \times fraction$$

neon-20	19.99×0.9092	$= 18.17$
neon-21	20.99×0.0025	$= 0.05$
neon-22	21.99×0.0883	$= 1.94$
	atomic weight of neon	$= 20.16$

Remember that the atomic weight of an element reflects the average relative mass of its atoms

---

When the substance entering the mass spectrometer is made up of polyatomic molecules rather than individual atoms, the pattern obtained is more complex. In Figure 2.7, we show the mass spectrum of the simplest gaseous hydrocarbon, methane

$$\begin{array}{c} H \\ | \\ H-C-H \\ | \\ H \end{array}$$

The parent peak, at mass 16, is formed by the loss of an electron from the methane molecule. The other peaks are attributed to fragments of the molecule formed by breaking C—H bonds. Ethane and other hydrocarbons, in which there are C—C as well as C—H bonds, show considerably more complicated mass spectra.

At a given voltage, the mass spectrum of a substance, like its infrared spectrum, is characteristic of that species and can be used to identify it.

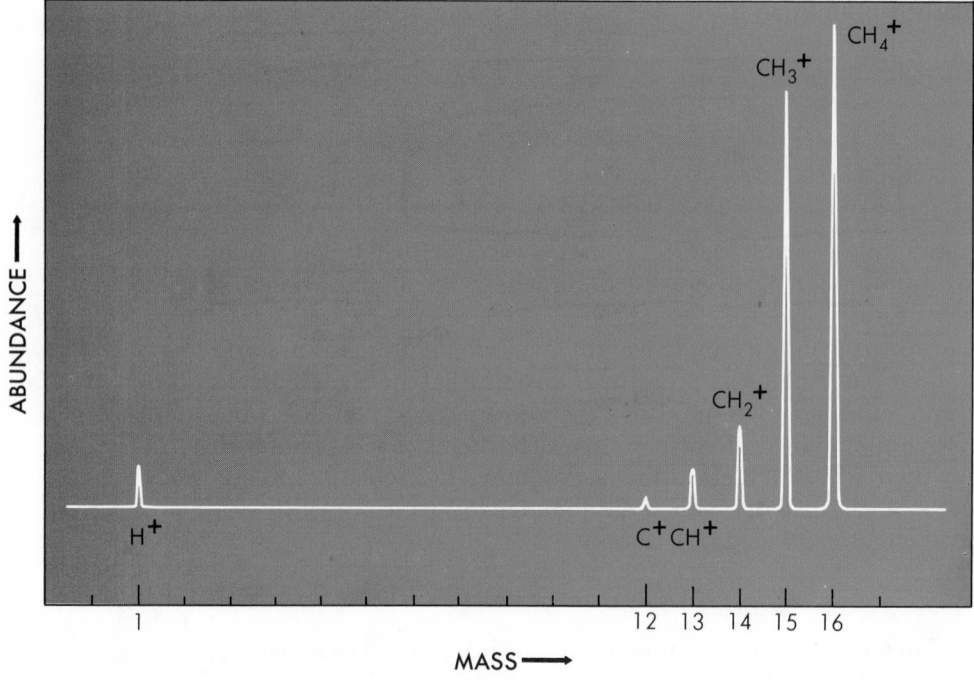

Figure 2.7 Mass spectrum of methane.

Mass spectroscopy is widely used to identify molecular substances. It is particularly useful for detecting trace constituents in mixtures, which show up as detectable peaks even when present at concentrations as low as a few parts per million. The nature and amount of air pollutants such as the oxides of sulfur and nitrogen are often determined by using the mass spectrometer.

2.5 MASSES OF ATOMS. AVOGADRO'S NUMBER

In order to calculate the mass of an individual atom, we must know the number of atoms in one gram atomic weight of any element. This number, known as Avogadro's number and given the symbol N, may be calculated in various ways; two different methods are illustrated by Problems 2.36 and 2.37 at the end of this chapter.

Avogadro's number, to four significant figures, has been found to be 6.023×10^{23} (i.e., 602,300,000,000,000,000,000,000). This is a number so large as to defy comprehension. Some idea of its magnitude may be gained when one realizes that if the entire population of the world were to be assigned to counting the number of atoms in one gram atomic weight of an element, each person counting one atom per second and working a 48-hour week, the task would require more than three billion years. From another point of view, the fact that Avogadro's number is so huge means that the atom itself is almost inconceivably small. Individual atoms are far too small

Why does Avogadro's number have the same value for all elements?

to be seen with the most powerful microscope or to be weighed on the most sensitive analytical balance (Example 2.3).

Example 2.3

    a. Calculate the mass of a hydrogen atom.

    b. Calculate the number of magnesium atoms in a sample weighing 1.0×10^{-6} g. (This is the smallest sample of magnesium that can be weighed on the most sensitive analytical balance.)

Solution

    a. Knowing that one gram atomic weight of hydrogen weighs 1.008 g and contains 6.023×10^{23} atoms, we have

$$1.008 \text{ g} \simeq 6.023 \times 10^{23} \text{ atoms H}°$$

$$\text{mass of 1 atom H} = 1 \text{ atom H} \times \frac{1.008 \text{ g}}{6.023 \times 10^{23} \text{ atoms H}}$$

$$= 1.674 \times 10^{-24} \text{ g}$$

    b. 1 GAW Mg = 24.31 g Mg = 6.023×10^{23} atoms Mg

$$\text{number of atoms Mg} = 1.0 \times 10^{-6} \text{ g} \times \frac{6.023 \times 10^{23} \text{ atoms}}{24.31 \text{ g}}$$

$$= 2.5 \times 10^{16} \text{ atoms } (25,000,000,000,000,000)$$

In some experiments with high energy particles it is possible to detect *single* atoms or ions!

2.6 MASSES OF MOLECULES. MOLECULAR WEIGHT. GRAM MOLECULAR WEIGHT

In dealing with substances such as water, elementary chlorine, or ethane, in which the fundamental building block is the molecule, it is convenient to define a number known as the **molecular weight** (MW), which tells us how heavy a molecule is compared to an atom of $^{12}_{6}$C. When we say, for example, that the molecular weight of water is "18," we mean that a water molecule is 18/12 as heavy as a carbon-12 atom (or 18/16 as heavy as an oxygen atom, or . . .).

If we know the composition of a molecule, we can readily calculate its molecular weight. For example, once it is established that the methane molecule consists of one carbon and four hydrogen atoms, it follows that

$$\text{MW methane} = \text{AW carbon} + 4 \text{ (AW hydrogen)}$$
$$= 12.01 + 4(1.008) = 16.04$$

The molecular weight is the sum of the atomic weights of all the atoms in the molecule

It is possible to determine molecular weights experimentally by using the mass spectrometer. We could, for example, deduce from the mass spectrum of methane shown in Figure 4.7 that its molecular weight must be 16.04. Two classic experimental approaches to the determination of molecular weights will be discussed later in this text (Chapters 5 and 10).

The **gram molecular weight** (GMW) of an elementary or compound substance is defined as the weight in grams which contains Avogadro's num-

---

° The symbol $\simeq$ is used to indicate equivalence rather than equality.

ber (6.023×10^{23}) of molecules. (Note the analogy to gram atomic weight defined previously.) The molecular weight of water is 18.02; its gram molecular weight is 18.02 g. Similarly, the molecular and gram molecular weights of methane are 16.04 and 16.04 g, respectively.

Knowing the gram molecular weight of a substance, we can readily calculate the weight of an individual molecule (Example 2.4).

Example 2.4 Calculate the mass in grams of a water molecule.

Solution. We know that the gram molecular weight of water is 18.02 g; one gram molecular weight of any substance contains 6.023×10^{23} molecules. Therefore

$$18.02 \text{ g water} \simeq 6.023 \times 10^{23} \text{ molecules water}$$

$$\text{mass} = 1 \text{ molecule} \times \frac{18.02 \text{ g}}{6.023 \times 10^{23} \text{ molecules}} = 2.992 \times 10^{-23} \text{ g}$$

2.7 A HISTORICAL PERSPECTIVE; ATOMIC WEIGHTS IN THE 19TH CENTURY

The problem of the determination of correct atomic weights puzzled chemists for a half century

Too often, the beginning student is left with the impression that scientific principles are developed in a beautifully logical sequence by a succession of gifted individuals, all of whom understand and appreciate the works of their predecessors. Occasionally this happens; more frequently it does not. Scientists, even great scientists, are capable of mistakes in reasoning. Sometimes they continue to maintain a point of view in spite of an overwhelming weight of evidence to the contrary. Perhaps we can illustrate the human qualities of scientists and the tortuous path that leads to scientific principles if we follow the atomic weight concept as it evolved in the 19th century.

DALTON. In his paper introducing the atomic theory in 1808, John Dalton wrote: "It is one great object of this work to show the importance and advantage of ascertaining the relative weights of the ultimate particles. . . ." Dalton realized that the problem of determining the relative masses of atoms of different elements was intimately related to that of finding the atom ratio in which they combine. Consider, for example, the elements hydrogen and oxygen, which react to form the compound water. Analysis shows that water contains about 1 part by weight of hydrogen to 8 parts by weight of oxygen. If we assume that the atom ratio in water is 1:1, it follows that a hydrogen atom must be 1/8 as heavy as an oxygen atom. If, on the other hand, two atoms of hydrogen are combined with one atom of oxygen, then a hydrogen atom need be only 1/16 as heavy as an oxygen atom to explain the observed combining weight ratio of 1:8.

Dalton saw no simple way of determining the atom ratio in which two elements combine. Consequently, he made what seemed to him to be a reasonable assumption:

"When only one combination of two . . . (elements) . . . can be obtained, it must be presumed to be a binary one (1 atom A + 1 atom B) . . . ."

Applying this reasoning to the water molecule, Dalton assumed it to be made up of one hydrogen and one oxygen atom. This would require that a hydrogen atom be about 1/8 as heavy as an oxygen atom; if hydrogen were assigned an atomic weight of unity, that of oxygen would have to be eight. From our vantage point it is easy to demonstrate the logical inconsistencies that Dalton's assumption leads to. Extended to the compounds ammonia and nitric oxide, for example, it leads to the absurd conclusion that a nitrogen atom must have one mass when it combines with a hydrogen atom and another quite different mass when it combines with an oxygen atom (Problem 2.33). Dalton soon became aware of paradoxes of this sort and modified his assumption accordingly. However, scientific ideas die slowly, especially when put forward by a man of Dalton's eminence. As late as 1890 we find journal articles in which the formula of water is given as HO.

DULONG AND PETIT, BERZELIUS. The first direct approach to the determination of atomic weights was proposed in 1819 by two French scientists, Pierre Dulong and Alexis Petit. They suggested that the amount of heat required to raise the temperature of an atom of a solid element by a given amount, let us say 1°C, should be independent of the type of atom; in their words, "the atoms of all simple bodies have the same capacity for heat." If this is true, then since 1 gram atomic weight of every element contains the same number of atoms, a fixed amount of heat should be required to raise the temperature of 1 gram atomic weight of a solid element by 1°C.

The Law of Dulong and Petit is expressed most simply in terms of the property known as specific heat, which is the amount of heat required to raise the temperature of one gram of a substance by 1°C. In modern form, the Law states that the product of the specific heat of a solid element multiplied by its gram atomic weight is approximately 6 cal/°C.

$$\text{GAW} \times \text{specific heat} \approx 6 \text{ cal/°C} \qquad (2.3)$$

Equation 2.3 is far from exact; for many elements the product of the specific heat and the gram atomic weight differs from 6 cal/°C by 10 per cent or more. It does, however, allow us to obtain approximate gram atomic weights which can then be refined by using precise analytical data giving the percentage by weight of the element in one or more of its compounds (Problem 2.16).

The argument put forth by Dulong and Petit was not immediately accepted by the scientific community. A major obstacle was the specific heat of carbon, which at 25°C is only about one third as large as the value predicted by Equation 2.3. Many of the chemists of the early 19th century, like some of their 20th century counterparts, were convinced that a law which permitted exceptions was worse than no law at all. On the other hand, the Swedish chemist Berzelius, generally recognized as the preeminent experimentalist of his day, was sufficiently impressed with the Law of Dulong and Petit to use it in revising his table of atomic weights,[*] excerpts of which are presented in Table 2.1.

This Law offers beginning students a way to find approximate atomic weights for many metals

---

[*] Throughout this discussion, we express atomic weights in terms of the modern scale. This should not be taken to imply that 19th century chemists were agreed on a common base for atomic weights. As a matter of fact, Berzelius at various times used three different scales, one based on H = 1, another on O = 100, and still another where H = 1/2. You can imagine how much this helped to clear up the confusion concerning atomic weights!

TABLE 2.1   Atomic Weights of Berzelius (Corrected to the C-12 Scale)

	1818	1826°	MODERN VALUE		1814	1826	MODERN VALUE
Ag	432.51	216.26	107.88	Fe	108.55	54.27	55.85
Al	54.72	27.39	26.98	Mg	50.68	25.34	24.32
Bi	283.86	212.86	209	Pb	414.24	207.12	207.21
C	12.05	12.25	12.01	S	32.19	32.19	32.07
Ca	81.93	40.96	40.08	W	193.23	189.31	183.86

° The values quoted under 1826, except for that of silver, took into account the Law of Dulong and Petit.

GAY-LUSSAC, AVOGADRO, AND CANNIZZARO.   The Law of Dulong and Petit, as useful as it was, did nothing to resolve the controversy about the atomic weights of gaseous elements. The key that would eventually unlock that door was provided by the Frenchman Gay-Lussac, who, in 1806, discovered the law of combining volumes of gases (Chapter 5). Gay-Lussac seems not to have grasped the theoretical implications of his work, or perhaps he was too polite to point them out (scientists are like that sometimes).

Dalton saw clearly where Gay-Lussac's Law led and he didn't like what he saw. Consider, for example, the reaction between hydrogen and oxygen, where Gay-Lussac showed that

2 volumes hydrogen + 1 volume oxygen → 2 volumes water vapor

Dalton realized that this simple, integral relationship between combining volumes implied an equally simple relationship between reacting particles

2 particles hydrogen + 1 particle oxygen → 2 particles water

At this point, Dalton, equating particles with atoms, was in trouble. One atom of oxygen could hardly yield two particles of water, both of which must contain at least one atom of oxygen (in Dalton's words, "Thou canst not split an atom"). Faced with what he took to be a direct challenge to the atomic theory, Dalton attempted to dispute Gay-Lussac's results by citing contradictory experiments of his own. His argument accomplished little except to prove that he was a better theoretician than experimenter.

In 1811 an Italian physicist with the improbable name of Lorenzo Romano Amadeo Carlo Avogadro di Quaregua e di Cerreto called attention to the error in Dalton's reasoning. He pointed out that Dalton had confused the concepts of atoms and molecules. If one assumed that the oxygen molecule were *diatomic* then two molecules of water, each containing one atom of oxygen, could be formed from one oxygen molecule. Avogadro interpreted Gay-Lussac's Law to mean that

2 molecules hydrogen (H_2) + 1 molecule oxygen (O_2) →
$$2 \text{ molecules water } (H_2O)$$

After thinking about this for a while, Avogadro concluded that Gay-Lussac's Law really implied that *equal volumes of all gases at the same temperature and pressure contain the same number of molecules.*

Avogadro's hypothesis suggested a simple method of determining molecular weight. If it were correct, then the densities of gases at the same temperature and pressure must be in the same ratio as their molecular weights. Unfortunately Avogadro's ideas made very little impact on the scientific community; examination of the literature of that period suggests that many eminent chemists did not bother to read Avogadro's paper (this still happens from time to time). Those who were familiar with Avogadro's ideas tended to dismiss them because they refused to believe that an element could form diatomic molecules.

Nearly 50 years later Avogadro's hypothesis was revived by a fellow countryman, Stanislao Cannizzaro, professor of chemistry at Genoa. He pointed out that it could be used to determine not only molecular weights but atomic weights as well. To see how this is possible, let us consider the problem of determining the atomic weight of hydrogen, which forms a large number of gaseous compounds with other elements. One liter of each of these compounds at, let us say, 25°C and 1 atmosphere pressure must, by Avogadro's hypothesis, contain a fixed number of molecules, which we shall call x. The number of atoms of hydrogen in one liter of a gaseous compound will be x if there is one atom of hydrogen per molecule, 2 x if there are two, and so on. With this idea in mind, let us consider the experimental data in Table 2.2.

Examination of Table 2.2 suggests that the compounds hydrogen chloride, hydrogen cyanide, and hydrogen fluoride each contain one atom of hydrogen per molecule and, hence, that x atoms of hydrogen weigh 0.0412 g. From a similar table for gaseous compounds of oxygen, we find that the smallest weight of oxygen is 0.654 g. Accordingly, 0.654 g must then be the weight of x atoms of oxygen. It follows that a hydrogen atom is

$$\frac{0.0412}{0.654} = \frac{1}{16}$$

as heavy as an oxygen atom, or that, on the modern scale, hydrogen has an atomic weight of about 1.

Cannizzaro presented his reasoning at a conference at Karlsruhe, called in 1858, to try to resolve the confusion concerning atomic weights of gaseous elements. His talk appears to have made few converts, but Cannizzaro was shrewd enough to hand out copies of his paper to all the delegates

To understand Cannizzaro's method requires some careful thought. Would you say that Cannizzaro's approach depends at all on luck?

TABLE 2.2   Weights of Hydrogen in One Liter
of Gaseous Compounds at 25°C, 1 atm

Compound	wt 1 liter	×	%H	=	wt H
hydrogen chloride	1.491 g		2.76		**0.0412**
ethylene	1.147 g		14.37		0.1648
hydrogen cyanide	1.105 g		3.73		**0.0412**
hydrogen sulfide	1.393 g		5.92		0.0824
methane	0.655 g		25.15		0.1648
ammonia	0.696 g		17.76		0.1236
hydrogen fluoride	0.818 g		5.04		**0.0412**
phosphine	1.390 g		8.90		0.1236

as they left the conference. At least one of them read it. Lothar Meyer, a prestigious German chemist, said later:

"I also received a copy which I put in my pocket to read on the way home. Once arrived there I read it again repeatedly and was astonished at the clearness with which the little book illuminated the most important points of the controversy. The scales seemed to fall from my eyes. Doubts disappeared and a feeling of quiet certainty took their place."

Due in no small part to the efforts of Meyer, Cannizzaro's arguments gradually became accepted. Within 20 years all but a few diehards had been convinced, and the atomic weight scale had been established.

PROBLEMS

2.1 Describe an experiment which would

a. demonstrate the Law of Multiple Proportions.
b. show that elementary hydrogen is a mixture of isotopes.
c. check the mass-to-charge ratio of the electron.
d. demonstrate that barium chloride is an ionic rather than a molecular substance.

2.2 Consider the isotopes $^{23}_{11}Na$ and $^{30}_{14}Si$. For each isotope, give

a. the number of protons in the nucleus.
b. the number of neutrons in the nucleus.
c. the number of electrons in the neutral atom.
d. the number of protons, neutrons, and electrons in +1 ions of these isotopes.

2.3 Criticize each of the following statements.

a. The molecular weight of sodium chloride is 58.44.
b. All the atoms of a given element are identical in all respects.
c. The nuclei of all atoms are composed of protons and neutrons.
d. The mass of a carbon-12 atom is exactly 12 g.

2.4 How many electrons are there in

a. a magnesium atom?
b. a gram atomic weight of carbon?
c. a cubic centimeter of liquid water?
d. a coulomb?

2.18 Describe an experiment designed to

a. demonstrate the Law of Conservation of Mass.
b. separate two isotopes of the element boron.
c. show that a sulfur atom is about twice as heavy as an oxygen atom.
d. show that a proton has a charge equal in magnitude to that of an electron.

2.19 Complete the following table.

Species	no. protons	no. neutrons	no. electrons	charge
$^{11}_{5}B$	5	6	5	0
$^{15}_{7}N^+$	___	___	___	+1
___	3	4	___	0
___	___	10	10	−1

2.20 Which of the following statements is(are) always true? sometimes true? never true?

a. A −1 ion is heavier than the atom from which it is derived.
b. The ratio of the molecular weights of two elementary substances is the same as the ratio of their atomic weights.
c. There are more atoms in 1 g of any substance than there are people in the world.

2.21 Give the total number of nuclear particles in

a. an atom of chlorine with a mass number of 37.
b. a water molecule (H_2O) (consider only the most common isotopes).
c. a gram atomic weight of fluorine, $^{19}_{9}F$.

2.5   The element iron forms two different chlorides in which the percentages by weight of iron are 44.0 and 34.4, respectively. Show how these data illustrate the Law of Multiple Proportions.

2.6   Polyethylene is a polymeric material in which there are two hydrogen atoms for every carbon atom. What are the weight percentages of hydrogen and carbon in polyethylene?

2.7   Suppose the atomic weight of oxygen were taken to be 100 (Berzelius took this as the basis of one of his scales). On this basis, what would be

    a. the atomic weight of nitrogen?
    b. the relative mass of oxygen and nitrogen atoms?
    c. the product of the gram atomic weight and the specific heat of a metal (on the modern scale, the product is 6.0 cal/°C)?

2.8   In a certain mass spectrometer, the voltages required to bring +1 ions of carbon-12 and a certain isotope of phosphorus to the same point are 704 and 273 volts, respectively. Calculate the mass on the C-12 scale of this isotope of phosphorus.

2.9   The element boron consists of two isotopes of masses 10.02 and 11.01 whose abundances are 18.83 and 81.17%, respectively. Calculate the average atomic weight of boron.

2.10   The element copper consists of two isotopes with masses of 62.96 and 64.96. Using the atomic weight table, estimate the abundances of these isotopes.

2.11   Calculate

    a. the number of atoms in 3.02 gram atomic weights of carbon.
    b. the mass in grams of a carbon atom.
    c. the number of gram atomic weights in 18.0 g of carbon.

2.12   Determine the number of

    a. atoms in one gram molecular weight of water.
    b. gram molecular weights of phosphorus (4 atoms per molecule) in 12.0 g of the element.
    c. molecules in a sample of methane (1 carbon atom, 4 hydrogen atoms per molecule) weighing one milligram.

2.22   A certain element, X, forms an oxide in which there is one atom of X for every two atoms of oxygen. The percentage by weight of X in this oxide is 78.8. What is the percentage by weight of X in another oxide in which there are equal numbers of atoms of X and O?

2.23   On a weight per cent basis, hydrogen is the ninth most abundant element (Table 1.4, Chapter 1). What position would hydrogen occupy in this table if the elements were ranked according to *atom* per cent?

2.24   At one time it was suggested that the atomic weight scale be set by taking the mass of the $^{19}_{9}F$ atom to be exactly 19. If this choice had been made, by what % would the atomic weights of elements differ from those on the carbon-12 scale where the mass of fluorine-19 is 18.9984?

2.25   Calculate the ratio of the voltages required to bring a proton (atomic weight = 1.007) and a +2 ion of nitrogen-14 (atomic weight = 14.01) to the same point in a mass spectrometer.

2.26   The element oxygen consists of three isotopes with masses of 15.9950, 17.00, and 18.00 whose abundances are 99.76, 0.04, and 0.20, respectively. Calculate the conversion factor from the (old) "physicists' scale" ($^{16}O = 16.00$) to the "chemists' scale" (average atomic weight of oxygen = 16).

2.27   Magnesium consists of three isotopes of masses 23.99, 24.99, and 25.99.

    a. Which do you think is most abundant?
    b. Given that the % abundance of the isotope referred to in (a) is 78.6, estimate the abundances of the other isotopes.

2.28   Work Problem 2.11, substituting titanium for carbon.

2.29   Arrange, in order of increasing mass,

    a. a molecule of oxygen.
    b. an atom of nitrogen.
    c. 1×10^{-10} gram molecular weights of oxygen.
    d. an O= ion.
    e. 1×10^{-10} gram atomic weights of copper.

2.13 When light strikes a photographic film covered with silver bromide, some of the Br^- ions are converted to bromine atoms which escape from the film. How many Br^- ions would have to be "lost" before one could detect a change in mass, using the most sensitive balance, which weighs to 1.0×10^{-6} g?

2.14 When a gallon of gasoline is burned, as much as 100 g of carbon monoxide may be produced. If this amount of carbon monoxide were spread out uniformly through the air of the Los Angeles Basin (total volume ≈ 6000 km³), how many CO molecules would there be per cubic centimeter?

2.15 For use in semiconductor devices, silicon crystals are prepared in which only one atom out of every billion is an impurity. If such a crystal weighs one mg, how many foreign atoms does it contain?

2.16 Berzelius found that bismuth oxide contained 10.1% by weight of oxygen.

    a. Use this data to obtain four possible values for the atomic weight of Bi, assuming Bi:O atom ratios of 2:1, 1:1, 2:3, and 1:2, respectively.

    b. From the specific heat of bismuth, 0.0293 cal/g°C, estimate its atomic weight.

    c. Combining the results of (a) and (b), obtain the atomic weight of Bi to 3 significant figures. What must be the atom ratio in bismuth oxide? What atom ratio did Berzelius assume in 1818 (Table 2.1)?

2.17 For many years prior to 1858, it was generally assumed that the atomic weights of elementary substances were directly proportional to their densities in the gas state. If this were correct, what would one obtain for the atomic weight of mercury, whose vapor has a density of 3.63 g/liter at 400°C and 1 atm as compared to 0.579 g/liter for oxygen? How do you explain the deviation from the true atomic weight of mercury? Interestingly enough, Berzelius quoted an atomic weight of mercury of about 200. How do you suppose he arrived at that number?

2.30 At an altitude of 200 km, there are about 5.0×10^{14} molecules/cc. Estimate the density of air at this height, taking an average molecular weight of 29. (The density of air at the earth's surface is about 1.1×10^{-3} g/cc.)

2.31 In a certain nuclear explosion, about 1.0 kg of the dangerous radioactive isotope strontium-90 was produced. If this amount of strontium-90 were spread out uniformly in the lower one mile of the atmosphere over the continental United States (area $\approx 3.0 \times 10^6$ square miles), how many atoms of strontium-90 would there be per cubic foot?

2.32 In Rutherford's experiments, a sample of a radioactive element, radium, was used as a source of α-particles ($^4_2$He nuclei). If α-particles were being produced at a rate of 2.2×10^9/min from a radium sample weighing 1.0 mg, and if each α-particle was produced by the disintegration of a radium atom, what fraction of the radium was decomposing per minute?

2.33 Consider the three compounds water (11%H, 89%O), ammonia (18%H, 82%N), and nitric oxide (47%N, 53%O). Taking the atomic weight of hydrogen to be one and following Dalton's assumption that the atom ratio in each of these compounds is 1:1, show that these data lead to two different values for the atomic weight of nitrogen.

2.34 A student looking at the data in Table 2.2 might object that if it were expanded to include other compounds, one of them might contain 0.0206 g of hydrogen. If this were to happen, what effect would it have on the atomic weight of hydrogen (based on O = 16)? If it were to happen, certain other masses of hydrogen which do not appear in Table 2.2 should also show up. What is one such number?

°**2.35** Taking the smallest weight of oxygen in one liter of any of its gaseous compounds at 25°C and 1 atm to be 0.654 and applying Cannizzaro's reasoning to the data given below, determine the atomic weight of nitrogen.

Compound	wt 1 liter (25°C, 1 atm)	%N	Compound	wt 1 liter (25°C, 1 atm)	%N
1	1.799	63.6	4	1.758	97.7
2	0.696	82.2	5	1.308	87.5
3	4.415	25.9	6	1.104	51.8

*2.36 One way to find Avogadro's number is to determine the number of electrons required to plate out a given mass of a particular metal. It is found that 3.04×10^3 coulombs are required to form one gram of copper metal from Cu^{2+} ions. Using the known atomic weight of copper and the charge of the electron in coulombs given in this chapter, calculate the number of atoms in one gram atomic weight of copper (Avogadro's number).

*2.37 By X-ray diffraction (Chapter 9) it is possible to determine the geometric pattern in which atoms are arranged in a crystal and the distances between atoms. It is found that in a crystal of silver, 4 atoms effectively occupy the volume of a cube 4.06 Å on an edge. Taking the density of silver to be 10.64 g/cc, calculate the number of atoms in one gram atomic weight of silver (Avogadro's number).

3   CHEMICAL FORMULAS AND EQUATIONS

Many of the essential features of a reaction can be summarized by means of a balanced equation which expresses, among other things, the relationship between the amounts of reactants and products. One of our objectives in this chapter will be to learn how to write and interpret chemical equations. In order to do this, we must first represent the compositions of reactants and products in terms of their chemical formulas. We will consider two different types of formulas — the simplest formula and the molecular formula.

1. The **simplest** or **empirical** formula of a compound gives the simplest whole number ratio between the numbers of atoms of the different elements making up the compound. An example is the simplest formula of water, H_2O, which tells us that there are twice as many hydrogen atoms as oxygen atoms. In potassium chlorate, it has been established by experiment that the three elements potassium, chlorine, and oxygen are present in an atom ratio of $1:1:3$; the simplest formula of potassium chlorate is $KClO_3$.

The simplest formula of a compound can always be obtained by the techniques of *quantitative analysis*, in which the proportions by weight of the elements making up the compound are determined. Whenever a new compound is prepared in the laboratory or discovered in nature, it is analyzed and, on that basis, is assigned a simplest formula. Even if the compound has been prepared previously, it is often analyzed to determine its purity. The importance of analysis as a research tool is indicated by the fact that in one recent issue (Sept. 20, 1972) of the *Journal of the American Chemical Society*, elementary analyses are reported for some 80 different compounds.

2. The **molecular** formula indicates the number of atoms in a molecule of a molecular substance. The molecular formula may be identical with the simplest formula, as is the case with water, H_2O: there are two atoms of hydrogen and one atom of oxygen per molecule of water. Alternatively the molecular formula may be an integral multiple of the simplest formula. This is the case with hydrogen peroxide, where we write the molecular formula H_2O_2 to indicate that two hydrogen atoms are combined with two oxygen

atoms to form a single structural unit, the hydrogen peroxide molecule. The simplest formula of this compound would be HO. Where there is a choice between representing a substance by either its molecular formula or its simplest formula, the molecular formula is invariably used, since it is more informative. Some substances, namely those which do not contain molecules, have only an empirical formula, e.g., NaCl.

The chemical formula for a metal is the symbol of the element, e.g., iron is represented by Fe, sodium by Na

3.1 SIMPLEST (EMPIRICAL) FORMULAS FROM ANALYSIS

In order to determine experimentally the simplest formula of a compound, we must establish by chemical analysis the proportions by weight of the elements making up the compound. Combining this information with the known atomic weights of the elements, we can readily calculate the simplest formula. The reasoning involved is indicated in Example 3.1.

Example 3.1 Vitamin C, which is reputed to be effective in preventing colds, is an organic compound containing the three elements carbon, hydrogen, and oxygen. Analysis of a carefully purified sample of vitamin C gives the following results.

$$\%C = 40.9; \quad \%H = 4.58; \quad \%O = 54.5$$

From these data, determine the empirical formula.

Solution. We need to know the relative numbers of atoms of C, H, and O. To obtain this information, let us calculate the relative numbers of gram atomic weights of C, H, and O in a given weight of vitamin C, for convenience, 100 g.

In 100 g, we have 40.9 g of C, 4.58 g of H, and 54.5 g of O. Converting to numbers of gram atomic weights, we obtain

$$\text{no. GAW C} = 40.9 \text{ g C} \times \frac{1 \text{ GAW C}}{12.0 \text{ g C}} = 3.41 \text{ GAW C}$$

$$\text{no. GAW H} = 4.58 \text{ g H} \times \frac{1 \text{ GAW H}}{1.01 \text{ g H}} = 4.53 \text{ GAW H}$$

$$\text{no. GAW O} = 54.5 \text{ g O} \times \frac{1 \text{ GAW O}}{16.0 \text{ g O}} = 3.41 \text{ GAW O}$$

These quantities represent not only the relative numbers of gram atomic weights of C, H, and O, *but also the relative numbers of atoms of these three elements*, since the conversion factor relating gram atomic weights to atoms (1 GAW = 6.02×10^{23} atoms) is the same for all elements. In other words, the numbers of atoms of C, H, and O in vitamin C are in the ratio 3.41:4.53:3.41.

To deduce the simplest formula, we need to know the simplest, whole number ratio between these three numbers. To obtain this, we divide each number by the smallest, 3.41:

$$\text{C:} \frac{3.41}{3.41} = 1.00; \quad \text{H:} \frac{4.53}{3.41} = 1.33; \quad \text{O:} \frac{3.41}{3.41} = 1.00$$

We see that for every C atom, there are 1.33 = 4/3 H atoms and one oxygen atom. Multiplying by three to obtain the simplest whole number ratio, we have

$$3 \text{ C:4 H:3 O;} \quad \text{simplest formula} = C_3H_4O_3$$

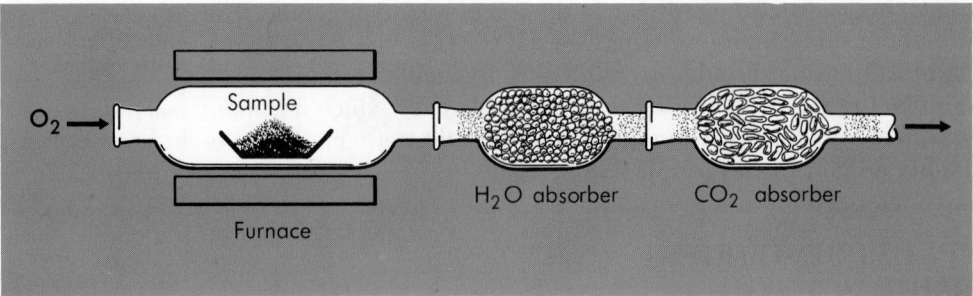

Figure 3.1 Combustion train used for carbon-hydrogen analysis. The absorbent for water is magnesium perchlorate, $Mg(ClO_4)_2$. Carbon dioxide is absorbed by finely divided sodium hydroxide supported on asbestos.

Once the percentages by weight of the constituent elements have been determined, the calculation of the simplest formula of a compound requires only a knowledge of atomic weights and a little arithmetic. The determination of the percentages, however, is not necessarily a simple matter. We must first obtain a sample of the pure substance, since even small amounts of contaminants can alter percentages and the derived formulas in unpredictable and unfortunate ways. Given a pure sample, we can proceed to determine its composition by any of several experimental methods. In a beginning chemistry course we cannot hope to become acquainted with all the techniques of elemental analysis that are used by analytical chemists. It may be helpful, however, to describe a few of the experiments that have led to data of the sort used in Example 3.1.

The percentages of carbon and hydrogen in an organic compound can be determined with the aid of the apparatus shown in Figure 3.1. A sample, which may weigh as little as a few milligrams, is burned in oxygen to form carbon dioxide and water. The amounts of CO_2 and H_2O produced are determined by measuring the increase in weight of the two absorption tubes. From these data, it is possible to calculate the percentages of carbon and hydrogen. If only one other element (e.g., oxygen) is present, its percentage can be calculated by difference (Example 3.2).

Example 3.2 The percentages of C, H, and O in vitamin C (Example 3.1) are determined by burning a sample weighing 2.00 mg; the weights of CO_2 and H_2O formed are 3.00 mg and 0.816 mg, respectively. What are the percentages of C, H, and O?

Solution. Let us first calculate the weight of carbon in 3.00 mg of CO_2. Since one gram molecular weight of CO_2, 44.0 g, contains one gram atomic weight of carbon, 12.0 g: 12.0 g C $\simeq$ 44.0 g CO_2

$$\text{no. mg C} = 3.00 \text{ mg CO}_2 \times \frac{12.0 \text{ g C}}{44.0 \text{ g CO}_2} = 0.818 \text{ mg C}$$

Similarly, since there are 2.02 g of hydrogen in 18.0 g of H_2O:

$$\text{no. mg H} = 0.816 \text{ mg H}_2\text{O} \times \frac{2.02 \text{ g H}}{18.0 \text{ g H}_2\text{O}} = 0.0916 \text{ mg H}$$

The samples are weighed on a microbalance

Noting that the sample weighed 2.00 mg, we have

$$\%C = \frac{0.818}{2.00} \times 100 = 40.9\% \qquad \%H = \frac{0.0916}{2.00} \times 100 = 4.58\%$$

We obtain the percentage of oxygen by difference:

$$\%O = 100.0 - (40.9 + 4.6) = 54.5\%$$

We see from Figure 3.1 and Example 3.2 that the determination of the percentage of carbon in an organic compound depends upon converting a weighed sample to a compound of known composition, CO_2. Similarly, to find the percentage of hydrogen, we need to know the formula of the compound to which it is converted, H_2O.

The same principle can be applied to determine the percentages by weight of the elements in inorganic compounds. For example, the percentage of chlorine in a water-soluble metal chloride can be determined by precipitating a weighed sample with silver nitrate, which forms insoluble silver chloride, $AgCl$. The silver chloride is filtered off and weighed; knowing the weight of the sample and that of the $AgCl$ to which it is converted, we can calculate the percentage of chlorine. Once again the percentage of the metal can be obtained by difference if it is the only other element present (Example 3.3).

Example 3.3   A sample of scandium chloride weighing 2.159 g is dissolved in water and treated with Ag^+ to precipitate the Cl^- ions as $AgCl$, which weighs 6.134 g. Calculate the percentage composition and the simplest formula of scandium chloride.

Solution.   Let us first calculate the weight of chlorine in the precipitated silver chloride. Since one gram formula weight of $AgCl$, 143.32 g, contains one gram atomic weight of Cl, 35.45 g, it follows that 35.45/143.32 of the silver chloride is chlorine. Hence,

$$\text{weight Cl in 6.134 g AgCl} = \frac{35.45}{143.32} \times 6.134 \text{ g} = 1.518 \text{ g}$$

It is preferable, where possible, to analyze for all the elements in a substance. Can you suggest why?

This is the weight of chlorine not only in the precipitated silver chloride but also in the original sample of scandium chloride. The percentage composition of scandium chloride must then be:

$$\%Cl = \frac{\text{wt. Cl}}{\text{wt. sample}} \times 100 = \frac{1.518 \text{ g}}{2.159 \text{ g}} \times 100 = 70.31\%$$

$$\%Sc = 100.00 - 70.31 = 29.69\%$$

To obtain the simplest formula of scandium chloride

$$29.69 \text{ g Sc} \simeq 70.31 \text{ g Cl}$$

$$\text{no. of GAW Sc in 29.69 g} = 29.69 \text{ g Sc} \times \frac{1 \text{ GAW Sc}}{44.96 \text{ g Sc}} = 0.660$$

$$\text{no. of GAW Cl in 70.31 g} = 70.31 \text{ g Cl} \times \frac{1 \text{ GAW Cl}}{35.45 \text{ g Cl}} = 1.98$$

Hence
$$0.660 \text{ GAW Sc} \simeq 1.98 \text{ GAW Cl}$$
$$1 \text{ GAW Sc} \simeq 3 \text{ GAW Cl} \quad \text{Simplest formula} = ScCl_3$$

Before proceeding further, it is worthwhile to comment on two points suggested by the experiment described in Example 3.3.

1. The procedure is readily adapted to checking the purity of a compound. If analysis of a sample believed to be $ScCl_3$ showed the percentage of chlorine to be less than the theoretical value of 70.31%, we would suspect the presence of an impurity such as Sc_2O_3. If the percentage of chlorine were too high, let us say 72%, we might be dealing with an impurity such as $AlCl_3$, containing a higher percentage of chlorine (79.7%). (It is only fair to point out that on occasion a chemist has ascribed an incorrect analysis to the presence of an impurity when, in fact, he had actually isolated a substance which was not the one he expected.)

2. It is not really necessary to calculate the *percentages* of the elements in a compound to determine its simplest formula. So long as the relative weights of the elements are known from experiment, we can obtain the simplest formula. For example, knowing that in scandium chloride, 1.518 g of Cl is combined with $(2.159 - 1.518)$ g $= 0.641$ g of Sc we have

$$\text{no. GAW Sc} = 0.641 \text{ g Sc} \times \frac{1 \text{ GAW Sc}}{44.96 \text{ g Sc}} = 0.0143 \text{ GAW Sc}$$

$$\text{no. GAW Cl} = 1.518 \text{ g Cl} \times \frac{1 \text{ GAW Cl}}{35.45 \text{ g Cl}} = 0.0428 \text{ GAW Cl}$$

The symbol $\simeq$ means "is chemically equivalent to"

Hence $0.0143 \text{ GAW Sc} \simeq 0.0428 \text{ GAW Cl}$
$1 \text{ GAW Sc} \simeq 3 \text{ GAW Cl}$ Simplest formula $= ScCl_3$

3.2 MOLECULAR FORMULAS

The **molecular formula** of a substance can be derived from the simplest formula if the molecular weight of the substance is known. We write the molecular formula O_2 for oxygen because it has been shown experimentally that oxygen has a molecular weight of 32. The molecular weight of water vapor is known to be 18; consequently, water must have the molecular formula H_2O rather than H_4O_2, H_6O_3, or some other multiple of the simplest formula.

Example 3.4 The empirical formula of vitamin C was determined in Example 3.1 to be $C_3H_4O_3$. From another experiment, its molecular weight is found to be about 180. Determine its molecular formula.

Solution. The formula weight corresponding to the simplest formula, $C_3H_4O_3$, is

$$(3 \times 12) + (4 \times 1) + (3 \times 16) = 88$$

Clearly the approximate molecular weight, 180, is twice the formula weight. It follows that the simplest formula must be multiplied by two to obtain the molecular formula.

Molecular formula vitamin C $= C_6H_8O_6$

Notice that an approximate value of the molecular weight is sufficient to obtain the molecular formula from the simplest formula. An experiment that establishes the molecular weight of vitamin C to be in the range from 160 to 200 would demonstrate that the simplest formula should be multiplied by 2 rather than 1, 3, or some other integer to obtain the molecular formula.

3.3 THE MOLE

In considering mass relations in reactions, chemists make frequent use of a quantity known as the mole. Always associated with a formula, **a mole contains Avogadro's number of formula units.** To understand what this statement means, consider the substance water, whose molecular formula is H_2O. One mole of H_2O contains Avogadro's number of molecules; it weighs 18.0 g (the gram formula weight of H_2O). In a similar manner:

For atoms,
  1 mole = 1 GAW
For molecules,
  1 mole = 1 GMW

1 mole of C contains 6.02×10^{23} carbon atoms and weighs 12.0 g
1 mole of CO_2 contains 6.02×10^{23} CO_2 molecules and weighs 44.0 g
1 mole of Na^+ contains 6.02×10^{23} Na^+ ions and weighs 23.0 g

The mole is frequently used in connection with an empirical formula. Consider, for example, the ionic compound sodium chloride in which there is one Na^+ ion for every Cl^- ion. We can, if we wish, say that one mole of NaCl contains Avogadro's number of formula units of NaCl. A "formula unit" of NaCl consists of one Na^+ ion and one Cl^- ion. A mole of NaCl must then contain Avogadro's number of Na^+ ions and Avogadro's number of Cl^- ions. The mass of one mole of NaCl is readily calculated.

1 mole of NaCl weighs 23.0 g + 35.5 g = 58.5 g

Similarly,   1 mole of $CaCl_2$ weighs 40.0 g + 71.0 g = 111.0 g

---

Example 3.5

    a. Determine the mass in grams of 2.60 moles of baking soda, $NaHCO_3$.
    b. How many moles of penicillin, $C_{16}H_{18}O_4N_2S$, are there in 218 g?
    c. How many molecules are there in 0.0820 mole of CO_2? 0.0820 mole of penicillin?

Solution

    a. One mole of $NaHCO_3$ weighs

$$23.0 \text{ g} + 1.0 \text{ g} + 12.0 \text{ g} + (3 \times 16.0 \text{ g}) = 84.0 \text{ g}$$

$$\text{no. g } NaHCO_3 = 2.60 \text{ moles } NaHCO_3 \times \frac{84.0 \text{ g } NaHCO_3}{1 \text{ mole } NaHCO_3}$$

$$= 218 \text{ g } NaHCO_3$$

    b. Mass of one mole of penicillin:

$$16(12.0 \text{ g}) + 18(1.0 \text{ g}) + 4(16.0 \text{ g}) + 2(14.0 \text{ g}) + 32.0 \text{ g} = 334 \text{ g}$$

$$\text{no. moles penicillin} = 218 \text{ g penicillin} \times \frac{1 \text{ mole penicillin}}{334 \text{ g penicillin}}$$

$$= 0.653 \text{ mole penicillin}$$

    c. 6.02×10^{23} molecules = 1 mole

$$\text{no. molecules} = 0.0820 \text{ mole} \times \frac{6.02 \times 10^{23} \text{ molecules}}{1 \text{ mole}}$$

$$= 4.94 \times 10^{22} \text{ molecules}$$

Here, it makes no difference whether we refer to CO_2, penicillin, or any other molecular substance. The conversion factor relating molecules to moles is the same for all species.

3.4 TRANSLATION OF REACTIONS INTO EQUATIONS

Once the nature of the reactants and products has been established, it is possible to describe a reaction by means of an equation that indicates the nature and amounts of the substances that participate in it. Writing a balanced chemical equation to describe a given reaction is by no means as simple a process as many beginning students have been led to believe. To illustrate the principles involved, let us consider the reaction that takes place in a rocket motor which uses hydrazine as a fuel and dinitrogen tetroxide as an oxidizer. It has been established experimentally that the molecular formulas of these two species, both of which are liquids, are N_2H_4 and N_2O_4, respectively. Analysis of the mixture coming out of the rocket exhaust shows that it consists of gaseous elementary nitrogen, N_2, and liquid water. To write a balanced equation for this reaction, we proceed as follows:

1. We first write an unbalanced equation in which the formulas of the reactants appear on the left and those of the products on the right. In this case, we have

$$N_2H_4 + N_2O_4 \rightarrow N_2 + H_2O$$

2. We balance the equation by taking into account the Law of Conservation of Mass, which requires that there be the same number of atoms of each element on the two sides of the equation. To accomplish this, we might begin by writing a coefficient of 4 for the H_2O, thereby obtaining 4 oxygen atoms on both sides.

$$N_2H_4 + N_2O_4 \rightarrow N_2 + 4\ H_2O$$

Looking now at the hydrogen atoms, we notice that we have $4 \times 2 = 8$ H atoms on the right. To obtain 8 H atoms on the left, we write a coefficient of 2 for N_2H_4.

$$2\ N_2H_4 + N_2O_4 \rightarrow N_2 + 4\ H_2O$$

Finally, we consider the number of atoms of nitrogen; there are a total of $(2 \times 2) + 2 = 6$ N atoms on the left. We balance the nitrogen by writing a coefficient of 3 for N_2.

$$2\ N_2H_4 + N_2O_4 \rightarrow 3\ N_2 + 4\ H_2O \tag{3.1}$$

Atoms are always conserved in a chemical reaction

The procedure that we have followed is essentially an intuitive one. We balance an equation one element at a time until the Law of Conservation of Mass is satisfied for each element. It is simplest to start with an element such as oxygen which appears in only one species on both sides of the equation. We could equally well have started with hydrogen, but it would have been awkward to have worked first with nitrogen, since both reactants, N_2H_4 and N_2O_4, contain this element.

3. Throughout this text, the physical states of reactants and products will be indicated in the equation. We will use the letters (g), (l) and (s) to represent gases, *pure* liquids, and solids, respectively. For the reaction between hydrazine and dinitrogen tetroxide, we write

$$2\ N_2H_4(l) + N_2O_4(l) \rightarrow 3\ N_2(g) + 4\ H_2O(l) \qquad (3.2)$$

When we deal with reactions taking place in water solutions, we will use the symbol (aq) to designate a dissolved species (ion or molecule). Thus, the equation for the reaction that takes place when a water solution containing Ag^+ ions is mixed with one containing Cl^- ions is written as

Chemical equations describe reactions in a very meaningful, although symbolic, way

$$Ag^+(aq) + Cl^-(aq) \rightarrow AgCl(s) \qquad (3.3)$$

Again, the equation

$$Zn(s) + 2\ H^+(aq) \rightarrow Zn^{2+}(aq) + H_2(g) \qquad (3.4)$$

is written to represent the reaction observed when solid zinc is added to an acidic water solution containing H^+ ions; the products are Zn^{2+} ions in aqueous solution and molecules of H_2 in the gas state.

3.5 INTERPRETATION OF BALANCED EQUATIONS

The coefficients of a balanced equation tell us the relative numbers of formula units of reactants and products participating in a reaction. For example, the equation

$$2\ N_2H_4(l) + N_2O_4(l) \rightarrow 3\ N_2(g) + 4\ H_2O(l)$$

indicates that 2 formula units (molecules) of N_2H_4 react with 1 formula unit (molecule) of N_2O_4 to give 3 formula units (molecules) of N_2 and 4 formula units (molecules) of H_2O.

It is important to keep in mind that a balanced equation remains valid if each of the coefficients is multiplied by the same number, x. Thus, for the reaction between hydrazine and dinitrogen tetroxide, we could write

2x molecules N_2H_4 + x molecules $N_2O_4 \rightarrow$
$$3x \text{ molecules } N_2 + 4x \text{ molecules } H_2O$$

where x can be any positive number, including 1, 1000, 2.0×10^5, ... A situation which is of particular interest is that in which x is Avogadro's number, N, (6.02×10^{23}).

2 N molecules N_2H_4 + N molecules $N_2O_4 \rightarrow$
$$3 \text{ N molecules } N_2 + 4 \text{ N molecules } H_2O$$

Recalling from Section 3.3 that one mole contains Avogadro's number, N, of molecules, we see that

$$2 \text{ moles } N_2H_4 + 1 \text{ mole } N_2O_4 \rightarrow 3 \text{ moles } N_2 + 4 \text{ moles } H_2O$$

The relationship which we have just derived is a general one, valid for any chemical equation. **The coefficients of a balanced equation represent the relative numbers of moles of reactants and products.** For example, the equation

$$4\ Al(s) + 3\ O_2(g) \rightarrow 2\ Al_2O_3(s) \qquad (3.5)$$

tells us that 4 moles of Al combine with 3 moles of O_2 to give 2 moles of Al_2O_3 or that, in this reaction

$$4 \text{ moles Al} \simeq 3 \text{ moles O}_2 \simeq 2 \text{ moles Al}_2\text{O}_3$$

where the symbol $\simeq$ is taken, as usual, to mean "chemically equivalent to."

We have seen that a mole of a substance has a certain mass in grams. Thus, 1 mole of Al weighs 27.0 g (1 GAW Al); 1 mole of O_2 weighs 32.0 g (2 × GAW O); 1 mole of Al_2O_3 weighs 102 g (2 × GAW Al + 3 × GAW O). Consequently, for the reaction of aluminum with oxygen, not only do we have

$$4 \text{ moles Al} \simeq 3 \text{ moles O}_2 \simeq 2 \text{ moles Al}_2\text{O}_3$$

but also
$$4 \,(27.0 \text{ g Al}) \simeq 3 \,(32.0 \text{ g O}_2) \simeq 2 \,(102 \text{ g Al}_2\text{O}_3)$$

or
$$108 \text{ g Al} \simeq 96.0 \text{ g O}_2 \simeq 204 \text{ g Al}_2\text{O}_3$$

In summary, the coefficients of a balanced equation give us directly the relative numbers of **formula units** or the relative numbers of **moles** of reactants and products. They also make it possible to calculate the relative numbers of grams of reactants or products. Stated in a slightly different way, the coefficients of a balanced equation give us, directly or indirectly, the **conversion factors** which allow us to relate the quantities of reactants and products in a chemical reaction. The use of these conversion factors in calculations involving balanced equations is illustrated in Examples 3.6 and 3.7.

In Reaction 3.5,
$$\frac{4 \text{ moles Al}}{3 \text{ moles O}_2} \simeq 1$$

$$\frac{4 \text{ moles Al}}{2 \text{ moles Al}_2\text{O}_3} \simeq 1$$

Example 3.6   For the reaction
$$2 \text{ N}_2\text{H}_4(l) + \text{N}_2\text{O}_4(l) \rightarrow 3 \text{ N}_2(g) + 4 \text{ H}_2\text{O}(l)$$
determine:

a. the number of moles of N_2O_4 required to react with 2.72 moles of N_2H_4.

b. the number of grams of N_2 produced when 2.72 moles of N_2H_4 are consumed.

c. the mass in grams of H_2O formed when 1.00 g of N_2O_4 reacts.

Solution.   In each case, we use the coefficients of the balanced equation to obtain the conversion factor required.

Chemical formulas are important in part because they facilitate an understanding of the mass relationships that exist between substances which participate in a chemical reaction

a. Here the conversion factor follows directly from the coefficients
$$2 \text{ moles N}_2\text{H}_4 \simeq 1 \text{ mole N}_2\text{O}_4$$

no. moles N_2O_4 = 2.72 moles $N_2H_4 \times \dfrac{1 \text{ mole N}_2\text{O}_4}{2 \text{ moles N}_2\text{H}_4} = 1.36$ mole N_2O_4

b. In this case, we need a conversion factor relating moles of N_2H_4 to grams of N_2. From the coefficients of the balanced equation
$$2 \text{ moles N}_2\text{H}_4 \simeq 3 \text{ moles N}_2$$

To obtain the desired conversion factor, we need only translate 3 moles of N_2 into grams. Since 1 mole of N_2 weighs 28.0 g we have
$$2 \text{ moles N}_2\text{H}_4 \simeq 3 \,(28.0 \text{ g N}_2) = 84.0 \text{ g N}_2$$

Hence, no. g N_2 formed = 2.72 moles $N_2H_4 \times \dfrac{84.0 \text{ g N}_2}{2 \text{ moles N}_2\text{H}_4} = 114$ g N_2

c. Again, we start with the mole relationship

$$1 \text{ mole } N_2O_4 \simeq 4 \text{ moles } H_2O$$

This time we translate both quantities into grams (1 mole $N_2O_4 = 92.0$ g; 1 mole $H_2O = 18.0$ g)

$$92.0 \text{ g } N_2O_4 \simeq 4 \,(18.0 \text{ g}) \, H_2O = 72.0 \text{ g } H_2O$$

$$\text{no. g } H_2O \text{ formed} = 1.00 \text{ g } N_2O_4 \times \frac{72.0 \text{ g } H_2O}{92.0 \text{ g } N_2O_4} = 0.783 \text{ g } H_2O$$

Example 3.7 When coal containing the mineral pyrites, FeS_2, is burned, the atmosphere is polluted by sulfur dioxide, formed by the reaction

$$4 \text{ FeS}_2(s) + 11 \text{ O}_2(g) \rightarrow 2 \text{ Fe}_2O_3(s) + 8 \text{ SO}_2(g)$$

What weight of sulfur dioxide is formed from the combustion of a kilogram of coal containing 2.0% of FeS_2?

Solution. This problem is similar to part (c) of Example 3.6 except that we must first calculate the weight of FeS_2 present

$$\text{weight } FeS_2 = 0.020 \times 1.0 \text{ kg} = 0.020 \text{ kg} = 20 \text{ g}$$

To find the weight of SO_2 produced, we start by relating moles of FeS_2 to moles of SO_2.

$$4 \text{ moles } FeS_2 \simeq 8 \text{ moles } SO_2$$

or $$1 \text{ mole } FeS_2 \simeq 2 \text{ moles } SO_2$$

Converting moles to grams:

$$1 \text{ mole } FeS_2 = 1 \text{ GAW Fe} + 2 \text{ GAW S} = 55.8 \text{ g} + 64.0 \text{ g} = 119.8 \text{ g}$$

$$1 \text{ mole } SO_2 = 1 \text{ GAW S} + 2 \text{ GAW O} = 32.0 \text{ g} + 32.0 \text{ g} = 64.0 \text{ g}$$

Thus $$120 \text{ g } FeS_2 \simeq 2 \,(64.0 \text{ g}) \, SO_2 = 128 \text{ g } SO_2$$

and hence $$\text{no. g } SO_2 = 20 \text{ g } FeS_2 \times \frac{128 \text{ g } SO_2}{120 \text{ g } FeS_2} = 21 \text{ g } SO_2$$

3.6 THEORETICAL YIELDS. PERCENTAGE YIELDS

When a chemist carries out a reaction in the laboratory, he seldom uses exactly equivalent quantities of reactants. Instead, he usually works with an excess of one reactant, hoping in this way to completely convert the other reactant, which may be more expensive or more difficult to obtain, to products. Consider, for example, the reaction of benzene with nitric acid:

$$C_6H_6(l) + HNO_3(l) \rightarrow C_6H_5NO_2(l) + H_2O(l) \qquad (3.6)$$

Let us suppose that we wish to form one mole of nitrobenzene, $C_6H_5NO_2$, from one mole of benzene, C_6H_6. In principle, one mole of nitric acid could be used to accomplish this purpose. In practice, if we wish to convert as much of the benzene as possible to nitrobenzene, it is advisable to use a considerable excess of HNO_3.

In calculating the **theoretical yield,** i.e., the maximum amount of product that can be produced in the reaction, we must be sure to base the calculation on the "limiting" reagent, the one that is not in excess. If, for example, a student were to allow one mole of benzene to react with five moles of nitric acid, he could not hope to obtain more than one mole of nitrobenzene, for he would run out of benzene to be nitrated. The theoretical yield of nitrobenzene, using these quantities of reagents, would be one mole (123 g).

The **actual yield** of product in a reaction is ordinarily less than the theoretical yield. If we react one mole of benzene with excess nitric acid, the actual yield of nitrobenzene will not be one mole. Instead, we may obtain 0.90, 0.80, or even as little as 0.50 mole of product. There are many reasons for this. In the first place, the reaction may not go to completion. Frequently an equilibrium is set up whose position is such that significant amounts of starting material remain unreacted (Chapter 13). Another factor which reduces the yield of product below that to be expected theoretically is the possibility of side reactions. For example, in the reaction of benzene with nitric acid, small quantities of dinitrobenzene, $C_6H_4(NO_2)_2$, are likely to be formed, thereby reducing the yield of the desired product, $C_6H_5NO_2$. Finally, even if the desired product is obtained in nearly the theoretical yield, part of it is likely to be lost in the separation and purification processes that follow the preparation.

Clearly, it is the goal of a chemist carrying out a reaction to make the **percentage yield,** defined as

$$\% \text{ yield} = \frac{\text{actual yield}}{\text{theoretical yield}} \times 100$$

as large as possible. To do this, he chooses conditions such that the desired reaction will occur to as great an extent as possible without significant interference from side reactions. He can often do this by changing the temperature at which the reaction is carried out or the time during which it is allowed to proceed.

If it takes a sequence of 10 reactions to produce a desired antibiotic and each reaction has a 75% yield, only 6% of the starting material will appear in the final product

---

Example 3.8 Aspirin is made commercially by adding acetic anhydride to a water solution of salicylic acid. The equation for the reaction is

$$2\ C_7H_6O_3(aq) + C_4H_6O_3(l) \rightarrow 2\ C_9H_8O_4(aq) + H_2O(aq)$$
$$\text{salicylic acid} \qquad \text{acetic anhydride} \qquad \text{aspirin}$$

If 2.00 kg of acetic anhydride are added to 1.00 kg of salicylic acid, calculate

    a. the theoretical yield of aspirin in g.

    b. the percentage yield if 1.12 kg of aspirin is actually isolated.

Solution

    a. To obtain the theoretical yield, we must first decide on which reagent we should base our calculations. It will be convenient to start by determining the numbers of moles of both reagents available (MW salicylic acid = 138, acetic anhydride = 102)

$$\text{no. moles } C_7H_6O_3 = \frac{1000\ g}{138\ g/\text{mole}} = 7.25$$

$$\text{no. moles } C_4H_6O_3 = \frac{2000\ g}{102\ g/\text{mole}} = 19.6$$

We see from the equation that only 1/2 mole of acetic anhydride is required to react with one mole of salicylic acid. Clearly we have a tremendous excess of acetic anhydride; only 3.63 moles would be *required* to react with 7.25 moles of salicylic acid. Our calculations must therefore be based on the amount of salicylic acid, the limiting reagent in the reaction.

This line of reasoning *must* be carried out in finding the limiting reagent

From the coefficients of the equation, it is clear that one mole of salicylic acid yields one mole of aspirin, $C_9H_8O_4$. Since the molecular weight of aspirin is 180, it follows that

$$\text{Theor. yield} = 7.25 \text{ moles } C_9H_8O_4 \times \frac{180 \text{ g } C_9H_8O_4}{1 \text{ mole } C_9H_8O_4} = 1.30 \times 10^3 \text{ g } C_9H_8O_4$$
$$= 1.30 \text{ kg}$$

b. $\% \text{ yield} = \dfrac{\text{actual yield}}{\text{theor. yield}} \times 100 = \dfrac{1.12 \text{ kg}}{1.30 \text{ kg}} \times 100 = 86.2\%$

PROBLEMS

3.1 Nicotine has the molecular formula $C_{10}H_{14}N_2$.

   a. What is the simplest formula of nicotine?

   b. What are the percentages by weight of C, H, and N in nicotine?

   c. How many grams of combined nitrogen are there in 242 g of nicotine?

3.2 What are the simplest formulas of compounds with the following compositions?

   a. 22.9% Ca, 40.5% Cl, 36.6% O.

   b. 29.1% Na, 40.5% S, 30.4% O.

   c. 25.4% Cu, 12.8% S, 57.7% O, 4.0% H.

3.3 A sample of powdered iron weighing 1.23 g is heated in a stream of chlorine gas to give a compound of the two elements weighing 3.59 g.

   a. What is the simplest formula of the chloride of iron formed?

   b. Write a balanced equation for the reaction that took place.

3.4 A certain zinc ore contains 95.0% by weight of ZnS with the remainder being cadmium sulfide, CdS. The ore is treated to convert the sulfides to the free metals, zinc and cadmium. What is the percentage of cadmium in the mixture that is produced?

3.5 Xylene is an organic compound containing only carbon and hydrogen. A sample weighing 2.00 mg is burned to form 6.64 mg of CO_2. What is the simplest formula of xylene? In another experiment the molecular weight of xylene is found to be 110 (two significant figures). What is its molecular formula?

3.19 The amino acid cystine has the molecular formula $C_6H_{12}N_2O_4S_2$.

   a. What is the simplest formula of cystine?

   b. What are the percentages by weight of the elements in cystine?

   c. How many grams of nitrogen are there in 18.6 g of cystine?

3.20 What are the simplest formulas of compounds with the following compositions?

   a. 73.4% Co, 26.6% O.

   b. 55.2% Cs, 21.6% Cr, 23.2% O.

   c. 32.0% C, 6.7% H, 18.7% N, 42.6%O.

3.21 A sample of a certain oxide of scandium weighing 1.423 g is heated in a stream of hydrogen gas to give a residue of 0.929 g of scandium.

   a. What is the simplest formula of scandium oxide?

   b. How much water vapor is formed in the reaction?

3.22 A certain drug has, as its major ingredient, aspirin, $C_9H_8O_4$. None of the other substances present in the drug contain carbon. Analysis shows the percentage of carbon in the drug to be 45.0. What is the percentage of aspirin?

3.23 Naphthalene, the principal component of ordinary mothballs, is an organic compound containing no elements other than carbon and hydrogen. A 1.000 mg sample of naphthalene burns to form 3.44 mg of CO_2. The molecular weight of naphthalene is found to be about 130.

   a. What is the simplest formula of naphthalene?

   b. What is the molecular formula?

3.6 A sample of benzoic acid, containing only C, H, and O, which weighs 3.66 mg, burns to form 9.24 mg of CO_2 and 1.62 mg of H_2O. Determine the percentages by weight of the elements in benzoic acid and its simplest formula.

3.7 Upon reaction with silver nitrate, a sample of indium bromide weighing 0.500 g precipitates 0.795 g of AgBr. Determine the simplest formula of indium bromide.

3.8 A sample of barium hypochlorite (Ba, Cl, O) weighing 0.850 g is heated in air to form 0.737 g of barium chloride which is then converted to 1.014 g of AgCl. Determine the simplest formulas of barium hypochlorite and barium chloride.

3.9 The following quantities are placed on the left pan of a two-pan balance. How many moles of H_2O must be placed on the other pan to exactly balance

    a. one mole of dry ice (CO_2)?
    b. 3.01×10^{23} molecules of CO_2?
    c. 3.01×10^{23} molecules of H_2O?

3.10 Determine

    a. the number of grams in 1.48 moles of quartz, SiO_2.
    b. the number of moles of camphor, $C_{10}H_{16}O$, in a sample weighing ten g.
    c. the number of molecules in 2.42 moles of boric acid.

3.11 Estimate the mass in grams of one mole of

    a. electrons (mass = 1/1837 that of a hydrogen atom).
    b. protons.
    c. Cl^- ions.

3.12 Write balanced equations to represent

    a. the combustion of acetylene gas, C_2H_2, to form carbon dioxide and water vapor.
    b. the precipitation reaction that occurs when aqueous solutions containing Ca^{2+} and PO_4^{3-} ions are mixed.
    c. the decomposition of cobalt carbonate in air, for which the unbalanced equation is $CoCO_3(s) + O_2(g) \rightarrow Co_3O_4(s) + CO_2(g)$.

3.24 Aniline is an organic compound containing only C, H, and N. A 3.00 mg sample burns to form 8.52 mg of CO_2 and 2.03 mg of H_2O.

    a. What are the percentages of C, H, and N in aniline?
    b. What is the simplest formula of aniline?

3.25 Bismuth chloride is treated with silver nitrate to give a precipitate of AgCl which weighs 1.364 times as much as the original sample. Determine the simplest formula of bismuth chloride.

3.26 A sample of bismuth oxychloride weighing 0.620 g forms a precipitate of AgCl weighing 0.341 g. Another sample of equal mass precipitates 0.612 g of Bi_2S_3. Calculate the percentages of Bi, Cl, and O in bismuth oxychloride and the simplest formula of the compound.

3.27 Suppose the mole were defined to contain 1.00×10^{24} particles. On this basis, what would be

    a. the mass of a single O_2 molecule?
    b. The mass of one mole of O_2?

3.28 Determine

    a. the number of grams in one mole of Epsom salts, $MgSO_4 \cdot 7 H_2O$.
    b. the number of grams of ethyl alcohol, C_2H_6O, in 3.90 moles.
    c. the number of moles in 1.80×10^{22} molecules of dinitrobenzene.

3.29 Estimate the mass of a mole of spherical fog droplets 10 Å in diameter, taking the density of water to be 1.0 g/ml.

3.30 Write balanced equations to represent

    a. the combustion of the rocket fuel $B_5H_9(l)$, which burns to form $B_2O_3(s)$ and $H_2O(l)$.
    b. the precipitation reaction that occurs when solutions containing Al^{3+} and OH^- ions are mixed.
    c. the detonation of TNT, for which the unbalanced equation is $C_7H_5N_3O_6(s) \rightarrow N_2(g) + H_2O(g) + CO(g) + C(s)$.

3.13 In a rocket propellant composed of hydrazine, N_2H_4, and hydrogen peroxide, H_2O_2, the reaction is: $N_2H_4(l) + 2\ H_2O_2(l) \rightarrow N_2(g) + 4\ H_2O(l)$.

    a. What is the total number of moles of products (nitrogen and water) formed when 1.81 moles of N_2H_4 react?

    b. How many moles of water are formed from 15.0 g of H_2O_2?

    c. What should be the mass ratio of H_2O_2 to N_2H_4 to achieve complete reaction?

    d. What should be the volume ratio of H_2O_2 to N_2H_4 (density $H_2O_2 = 1.44$ g/ml, $N_2H_4 = 1.01$ g/ml)?

3.14 The principal constituent of natural gas is methane, CH_4, which burns in air to form carbon dioxide and water vapor.

    a. Write a balanced equation for the reaction.

    b. How many grams of oxygen are required to burn one gram of methane?

    c. How many grams of air (23% by weight O_2) are required to burn one gram of methane?

3.15 One way to remove carbon dioxide from the air in a spacecraft is to react it with sodium hydroxide

$$CO_2(g) + 2\ NaOH(s) \rightarrow Na_2CO_3(s) + H_2O(l)$$

It is estimated that over a 24-hour period, a person exhales about 1.0 kg of CO_2. How many grams of sodium hydroxide are required to remove the carbon dioxide that is formed in a six-day lunar expedition involving three astronauts?

3.16 Carbon dioxide can be prepared by allowing concentrated sulfuric acid to drop on sodium hydrogen carbonate:

$$2\ NaHCO_3(s) + H_2SO_4(l) \rightarrow$$
$$2\ CO_2(g) + Na_2SO_4(s) + 2\ H_2O(l)$$

    a. If the $NaHCO_3$ used in this preparation is 94% pure, how many grams of it will be required to form 13.6 g of CO_2?

    b. If the sulfuric acid is used in the form of a water solution which is 52% H_2SO_4 by weight and has a density of 1.26 g/ml, what volume of this solution must be used to produce 13.6 g of carbon dioxide?

3.31 Hydrogen sulfide, given off by decaying organic matter (e.g., rotten eggs), is converted to sulfur dioxide in the atmosphere by the reaction

$$2\ H_2S(g) + 3\ O_2(g) \rightarrow 2\ SO_2(g) + 2\ H_2O(l)$$

    a. How many moles of H_2S are required to form 6.19 moles of SO_2?

    b. How many moles of O_2 are required to react with one gram of H_2S?

    c. How many grams of H_2S are required to give an average concentration of SO_2 of 4.0×10^{-8} mole/liter in a column of air one km high over a city 100 km² in area?

3.32 Sulfide minerals, associated with coal deposits, are responsible for "acid mine drainage," the contamination of streams in coal mining areas by sulfuric acid. A typical reaction of such minerals is

$$4\ FeS(s) + 9\ O_2(g) + 4\ H_2O(l) \rightarrow$$
$$2\ Fe_2O_3(s) + 4\ H_2SO_4(l)$$

How many grams of FeS are required to form 100 liters of a solution containing 2.0×10^{-3} mole/liter of H_2SO_4?

3.33 At an altitude of about 20 km in the atmosphere, there is a layer of very tiny ammonium sulfate particles, presumably formed by the reaction

$$2\ NH_3(g) + H_2O(g) + SO_3(g) \rightarrow (NH_4)_2SO_4(s)$$

How many grams of ammonium sulfate can be formed from a cubic meter of air in which the concentration of SO_3 is 2.0×10^{-8} mole/liter?

3.34 Ammonia can be produced by heating ammonium chloride with calcium oxide:

$$2\ NH_4Cl(s) + CaO(s) \rightarrow$$
$$2\ NH_3(g) + CaCl_2(s) + H_2O(g)$$

    a. How many grams of NH_4Cl are needed to produce 0.316 mole of NH_3?

    b. If it is desired to use a 50% excess of calcium oxide, how many grams of CaO should be used to produce 0.316 mole of NH_3?

3.17 In the Haber process for the synthesis of ammonia

$$N_2(g) + 3 H_2(g) \rightarrow 2 NH_3(g)$$

10 g of N_2 are reacted with 1.0 g of H_2.

a. Which reactant is in excess?
b. What is the theoretical yield of NH_3?
c. What is the percentage yield if 2.12 g of NH_3 is formed?
d. Can you suggest why the yield of ammonia is so low?

3.18 A student in the organic chemistry laboratory prepares aspirin by the reaction described in Example 3.8. He is told to use a 50% excess of acetic anhydride and to expect a 74% yield of aspirin. If he needs to obtain 20 g of aspirin, how many grams of the two starting materials should he use?

3.35 A student in the organic chemistry laboratory prepares ethyl bromide, C_2H_5Br, by reacting ethyl alcohol with phosphorus tribromide:

$$3 C_2H_5OH(l) + PBr_3(l) \rightarrow 3 C_2H_5Br(l) + H_3PO_3(s)$$

He is told to react 24.0 g of ethyl alcohol with 59.0 g of phosphorus tribromide.

a. What is the theoretical yield of ethyl bromide?
b. If the student actually obtains 36.0 g of C_2H_5Br, what is the percentage yield?

3.36 A student in the inorganic chemistry laboratory wishes to prepare 20 g of the compound $[Co(NH_3)_5SCN]Cl_2$, from $[Co(NH_3)_5Cl]Cl_2$:

$$[Co(NH_3)_5Cl]Cl_2(s) + KSCN(s) \rightarrow$$
$$[Co(NH_3)_5SCN]Cl_2(s) + KCl(s)$$

He is instructed to use a 60% excess of potassium thiocyanate, KSCN, and is told that he can expect to get a 63% yield in the reaction. How many grams of each starting material should he use?

*3.37 A sample of LSD (D-lysergic acid diethylamide, $C_{24}H_{30}N_3O$) is diluted with sugar, $C_{12}H_{22}O_{11}$. When a 1.00 mg sample is burned, 2.00 mg of CO_2 is formed. What is the percentage of LSD in the sample?

*3.38 When water containing Ca^{2+} and HCO_3^- ions is heated, a precipitate of calcium carbonate is formed.

$$Ca^{2+}(aq) + 2 HCO_3^-(aq) \rightarrow CaCO_3(s) + CO_2(g) + H_2O(l)$$

Water in which the concentrations of Ca^{2+} and HCO_3^- are 1.4×10^{-3} and 1.0×10^{-3} mole/liter, respectively, is admitted to a boiler at the rate of 400 liters per day. Assuming a "percentage yield" of $CaCO_3$ of 25%, how long will it take to build up a deposit of $CaCO_3$ 0.10 mm thick if the boiler has an inside surface area of 100 ft²? (d $CaCO_3 = 2.9$ g/ml)

*3.39 A student determines the formula of copper sulfide by heating a piece of copper wire weighing 0.510 g with 0.682 g of sulfur. The excess sulfur burns off as SO_2. What weight of SO_2 is formed

a. if the formula of the sulfide formed is Cu_2S?
b. if the formula of the sulfide formed is CuS?

*3.40 A student is asked to prepare 0.250 mole of a pure compound C by the reaction sequence

$$2 A \rightarrow B \quad \text{and} \quad 3 B \rightarrow 2 C$$

in which A and B represent two other compounds. He is told to expect a 76% yield in the first step of this sequence and a 63% yield in the second step. He is also required to purify compound C by recrystallizing it from hot water; it is estimated that 19% of it will be lost in the recrystallization. How many moles of A should he start with?

4   THERMOCHEMISTRY

When a chemical reaction occurs, it is accompanied by an energy change which may take any of several different forms. Consider, for example, the reactions

$$CH_4(g) + 2\ O_2(g) \rightarrow CO_2(g) + 2\ H_2O(l) \tag{4.1}$$

$$(CH_3)_2N_2H_2(g) + 2\ N_2O_4(l) \rightarrow 2\ CO_2(g) + 4\ H_2O(l) + 3\ N_2(g) \tag{4.2}$$

$$6\ CO_2(g) + 6\ H_2O(l) \rightarrow C_6H_{12}O_6(s) + 6\ O_2(g) \tag{4.3}$$

Reaction 4.1, which takes place when a mixture of natural gas and air is ignited, supplies the heat required to cook a steak on a gas range or decompose potassium chlorate with a Bunsen burner. The reaction of dimethylhydrazine with dinitrogen tetroxide (4.2) furnishes the mechanical energy that enables an Apollo spacecraft to take off from the surface of the moon. The formation of glucose, $C_6H_{12}O_6$, by the process of photosynthesis requires the absorption of light energy from the sun.

In this chapter we shall concentrate upon a particular type of energy change, namely, the heat flow associated with a reaction when it is carried out directly in the laboratory or in the world around us. We will be interested both in reactions which proceed with the *evolution* of heat (**exothermic**) and those which require the *absorption* of heat (**endothermic**). Later, in Chapter 12, we shall consider other forms of energy that may be evolved or absorbed in a chemical reaction.

4.1   THERMOCHEMICAL EQUATIONS

A thermochemical equation specifies the magnitude and direction of the heat flow accompanying a reaction. To illustrate, consider the combustion of methane, Reaction 4.1. When one mole of methane burns in a Bunsen burner, 212.8 kcal * of heat is evolved. We express this information by writing the thermochemical equation:

$$CH_4(g) + 2\ O_2(g) \rightarrow CO_2(g) + 2\ \overset{\cdot}{H_2}O(l);\ Q = -212.8\ kcal \tag{4.1'}$$

where Q is the general symbol used for heat flow.

---

* 1 kcal = 1000 cal, where one calorie is the heat which must be absorbed to raise the temperature of one gram of water from 14.5 to 15.5°C. Alternatively, 1 cal = 4.184 joule, where the joule is taken to be the fundamental unit of energy.

There are some specific conventions ("ground rules") that are used in writing thermochemical equations such as 4.1'. In particular:

1. Q is given a negative sign if, as in 4.1', a reaction evolves heat. A positive sign for Q indicates an endothermic reaction.

2. The coefficients in a thermochemical equation always refer to numbers of moles. That is, −212.8 kcal is the heat flow when *one mole* of CH_4 reacts with *two moles* of O_2 to form *one mole* of CO_2 and *two moles* of H_2O.

3. Since the heat flow in a reaction depends upon the state of aggregation of reactants and products, those states must be specified in the thermochemical equation. If in 4.1' water vapor, i.e., $H_2O(g)$, were the product rather than liquid water, Q would be −191.8 kcal rather than −212.8 kcal. Less heat would be evolved because some of it (21 kcal) would be absorbed in vaporizing 2 moles of water.

4. In thermochemical equations it is always assumed that reactants and products are at the same temperature; in addition, unless specifically stated otherwise, that temperature is taken to be 25°C.

A reaction is exothermic when heat must flow out of the products to cool them down to the temperature of the reactants

THE ENTHALPY CHANGE, ΔH

Experience tells us that the amount of heat evolved or absorbed in a reaction depends upon the way in which it is carried out. When 1 mole of methane burns in an open flame, Q is −212.8 kcal. If, on the other hand, the same amount of methane is exploded in a closed container such as a bomb calorimeter (Fig. 4.9), slightly less heat is evolved (Q = −211.6 kcal). Finally, if the reaction of methane with oxygen is used to generate electrical energy in a fuel cell (Chapter 20), considerably less heat is evolved, as little as 7.2 kcal.

Most of the reactions we will consider in this chapter are those, like the combustion of methane in an open flame, that take place at *constant pressure without doing any useful work* (e.g., generating electrical energy). Indeed, reactions are usually conducted under such conditions, since we usually work with chemicals in open containers exposed to the constant pressure of the atmosphere. For reactions carried out in this way, the heat flow is directly related to a fundamental property of the pure substances involved, called **enthalpy** and given the symbol **H**. The relationship is

A mole of a pure substance in a given state will have a fixed amount of enthalpy, in much the same way as it has a fixed mass

Heat flow at constant pressure

$$= \text{Enthalpy of Products} - \text{Enthalpy of Reactants}$$

$$Q_P = H_{products} - H_{reactants} = \Delta H \qquad (4.4)$$

where we write ΔH to signify the difference between the enthalpies of products and reactants.

Applying this idea to the combustion of methane in an open flame:

$$CH_4(g) + 2\ O_2(g) \rightarrow CO_2(g) + 2\ H_2O(l);\ Q_P = \Delta H = -212.8\ \text{kcal}$$

We interpret this equation to mean that the enthalpy of the products, 1 mole of $CO_2(g)$ and 2 moles of $H_2O(l)$, is 212.8 kcal less than that of the reactants, 1 mole of $CH_4(g)$ and 2 moles of $O_2(g)$. Again, when magnesium burns in an open crucible

$$Mg(s) + \tfrac{1}{2}\ O_2(g) \rightarrow MgO(s);\ Q_P = \Delta H = -143.8\ \text{kcal} \qquad (4.5)$$

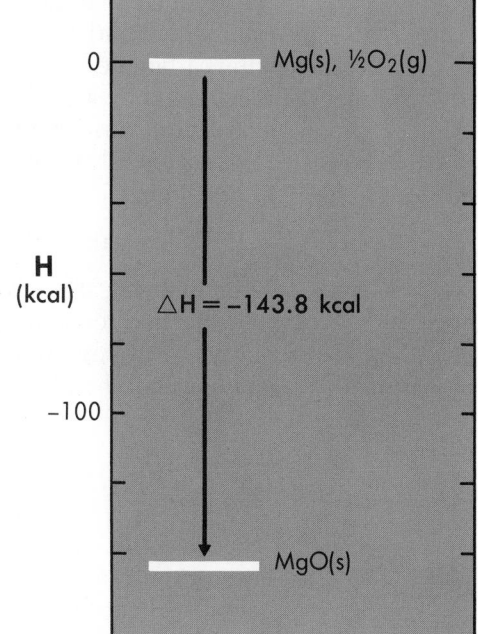

Figure 4.1 One mole of MgO has an enthalpy 143.8 kcal less than that of the elements.

which we take to mean that the enthalpy of one mole of magnesium oxide is 143.8 kcal less than that of the magnesium and oxygen which produced it (Fig. 4.1). In general, for any exothermic reaction carried out directly at constant pressure, the amount of heat evolved will simply equal the loss of enthalpy that occurs in the reaction:

Enthalpy lost by the reaction system
$$= \text{Heat gained by the surroundings} \quad (4.6)$$

We can readily extend this concept to endothermic reactions carried out under the conditions just described. An example is the thermal decomposition of calcium carbonate as carried out in a limekiln.

$$CaCO_3(s) \rightarrow CaO(s) + CO_2(g); \quad Q_P = \Delta H = +42.5 \text{ kcal} \quad (4.7)$$

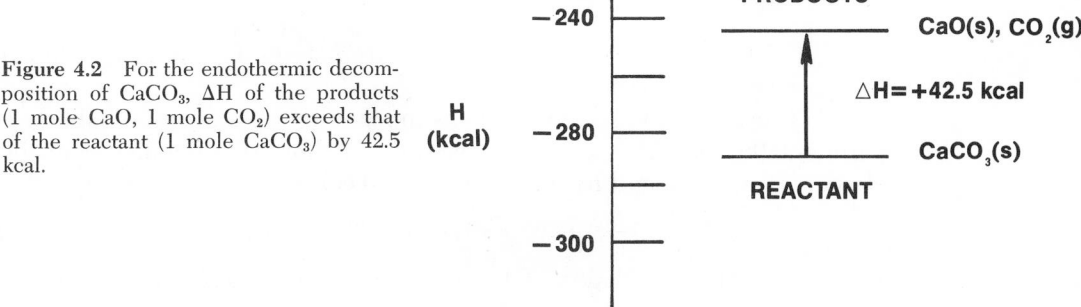

Figure 4.2 For the endothermic decomposition of $CaCO_3$, ΔH of the products (1 mole CaO, 1 mole CO_2) exceeds that of the reactant (1 mole $CaCO_3$) by 42.5 kcal.

We deduce that the enthalpy of the products, 1 mole of calcium oxide (quicklime) and 1 mole of carbon dioxide, must exceed that of 1 mole of calcium carbonate (limestone) by 42.5 kcal (Fig. 4.2). This increase in enthalpy is achieved by absorbing heat from the surroundings, which might include a coal-fired furnace or a battery of Bunsen burners.

In general, for endothermic reactions carried out in a manner similar to Reaction 4.7:

According to the First Law, a reaction mixture can change its energy only by exchanging an equal amount of energy with its surroundings

Enthalpy gained by reaction system = Heat lost by surroundings   (4.8)

Equations 4.4, 4.6, and 4.8 may be regarded as special cases of the First Law of Thermodynamics, which tells us that energy can neither be created nor be destroyed but only changed from one form (e.g., chemical energy) to another (e.g., heat).

LAWS OF THERMOCHEMISTRY

Once the existence of enthalpy is recognized, several fundamental laws of thermochemistry can be stated. These laws were originally established by heat flow experiments but are most simply expressed in terms of the enthalpy change, ΔH.

1. *ΔH is directly proportional to the amount of substance that reacts or is produced in a reaction.*

Example 4.1   Hydrogen peroxide, a common bleaching agent, decomposes as follows:

$$H_2O_2(l) \rightarrow H_2O(l) + \tfrac{1}{2} O_2(g); \quad \Delta H = -23.5 \text{ kcal}$$

How much heat is released when one g of H_2O_2, stored in an open bottle, decomposes?

Solution.   From the equation, ΔH for the decomposition of 1 mole of H_2O_2 is -23.5 kcal. Therefore

$$1 \text{ mole } H_2O_2 = 34.0 \text{ g } H_2O_2 \doteq -23.5 \text{ kcal}$$

Using the conversion factor approach:

$$1.00 \text{ g } H_2O_2 \times \frac{-23.5 \text{ kcal}}{34.0 \text{ g } H_2O_2} = -0.691 \text{ kcal}$$

We deduce that the decomposition of one g of hydrogen peroxide gives off 0.691 kcal (691 cal) of heat.

---

2. *ΔH for a reaction is equal in magnitude but opposite in sign to ΔH for the reverse reaction.*

To illustrate the use of this law, suppose that we have established experimentally that ΔH for the formation of 1 mole of ammonia from the elements, nitrogen and hydrogen, is -11.0 kcal

$$\tfrac{1}{2} N_2(g) + \tfrac{3}{2} H_2(g) \rightarrow NH_3(g); \quad \Delta H = -11.0 \text{ kcal} \tag{4.9}$$

It follows that the enthalpy of 1 mole of NH_3 is 11.0 kcal less than that of 1/2 mole of N_2 and 3/2 mole of H_2 (Fig. 4.3). Consequently, to decompose 1

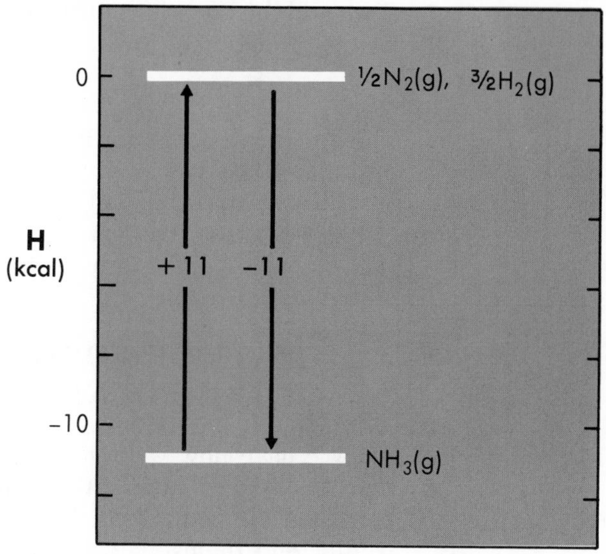

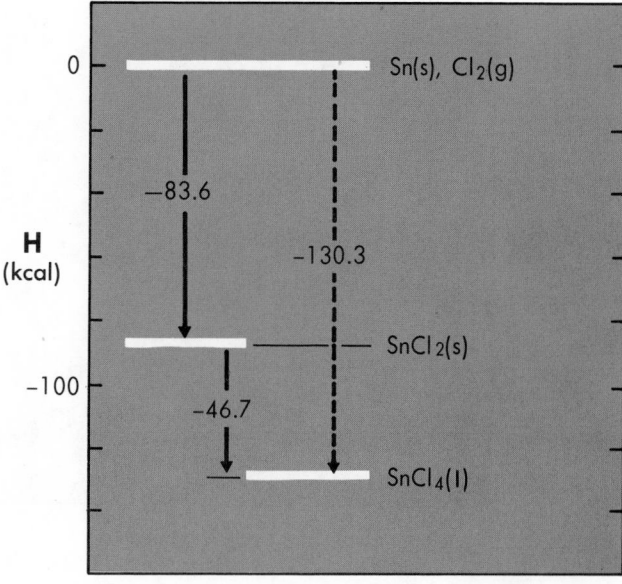

Figure 4.3 Since the enthalpy of 1 mole of NH_3 is 11.0 kcal less than that of the elements, ΔH for its decomposition is +11.0 kcal, while ΔH for its formation is −11.0 kcal.

mole of NH_3 to the elements at constant pressure would require the absorption of 11.0 kcal of heat, i.e.,

$$NH_3(g) \rightarrow \tfrac{1}{2} N_2(g) + \tfrac{3}{2} H_2(g); \ \Delta H = +11.0 \text{ kcal} \qquad (4.10)$$

3. *If a reaction can be regarded as the sum of two or more other reactions, ΔH for the overall reaction must be the sum of the enthalpy changes for the other reactions.*

Consider the following reactions:

$$Sn(s) + Cl_2(g) \rightarrow SnCl_2(s); \ \Delta H_1 = -83.6 \text{ kcal} \qquad (4.11)$$

$$SnCl_2(s) + Cl_2(g) \rightarrow SnCl_4(l); \ \Delta H_2 = -46.7 \text{ kcal} \qquad (4.12)$$

Figure 4.4 ΔH for the reaction:

$$Sn(s) + 2 Cl_2(g) \rightarrow SnCl_4(l)$$

is the same whether it is carried out in one step (−130.3 kcal) or two (−83.6 kcal − 46.7 kcal = −130.3 kcal).

If we add the equations for Reactions 4.11 and 4.12, we obtain an equation for the formation of $SnCl_4$ from the elements

$$Sn(s) + 2\ Cl_2(g) \rightarrow SnCl_4(l); \quad \Delta H = \Delta H_1 + \Delta H_2 = -130.3\ \text{kcal} \quad (4.13)$$

The enthalpy change for Reaction 4.13 is the sum of the ΔH's of Reactions 4.11 and 4.12.

This relationship between enthalpy changes in related reactions is called **Hess's Law**. It is very useful for determining the heat flow of a reaction which is difficult or impossible to carry out directly in a single step.

4.2 HEATS OF FORMATION OF COMPOUNDS

We have now written a substantial number of thermochemical equations involving changes in enthalpy, ΔH. You may wonder why we have not yet listed the actual enthalpies, H, of any substances since such numbers would lend themselves, by Equation 4.4, to a very simple calculation of heat flows and enthalpy changes. We almost blush to admit it, but it turns out that we can't find the enthalpies of individual substances by any experiment. Every reaction gives us a ΔH value, but it always involves a difference between two or more enthalpies, none of which can be determined by themselves.

Fortunately, if we think about the calculation of heat flow and enthalpy changes long enough, we conclude that we really don't need to know absolute values of enthalpies in order to establish a general method of calculating ΔH for chemical reactions. We find that quantities called heats of formation, ΔH_f, will do as well, perhaps better.

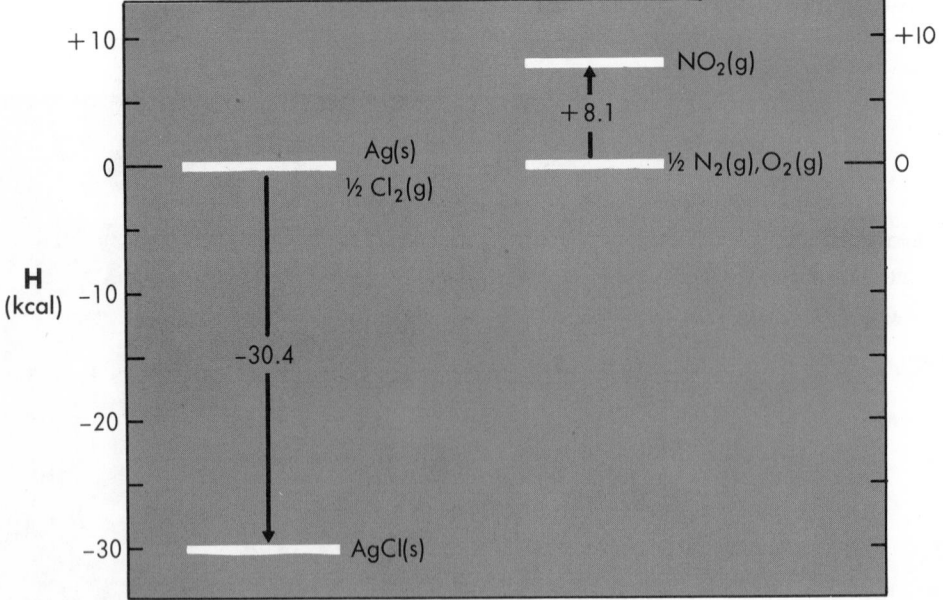

Figure 4.5 Most compounds, like AgCl, have negative heats of formation; heat is evolved when they are formed from the elements. A few compounds, like NO_2, have positive heats of formation.

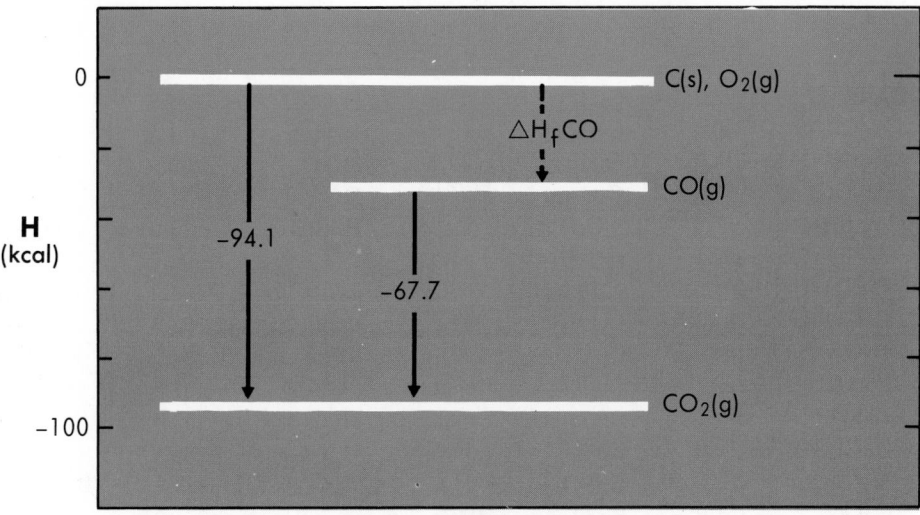

Figure 4.6 The heat of formation of CO (−26.4 kcal) can be deduced from the known values for the heat of combustion of C (−94.1 kcal) and CO (−67.7 kcal).

The heat of formation, ΔH_f, of a compound is defined as **the enthalpy change when one mole of the compound is formed from the elements in their stable forms at 25°C and 1 atm.** Thus, from the equations

$$Ag(s) + \tfrac{1}{2} Cl_2(g) \rightarrow AgCl(s); \Delta H = -30.4 \text{ kcal} \qquad (4.14)$$

$$\tfrac{1}{2} N_2(g) + O_2(g) \rightarrow NO_2(g); \Delta H = +8.1 \text{ kcal} \qquad (4.15)$$

we see that the heats of formation of silver chloride and nitrogen dioxide are −30.4 kcal and +8.1 kcal, respectively. These values are shown graphically in Figure 4.5.

For many compounds it is possible to determine heats of formation directly by measuring ΔH for the formation of the compound from the elements. This is true, for example, for silver chloride; Reaction 4.14 can be carried out directly in a calorimeter (Section 4.4). In some cases it is more convenient to measure ΔH for a reaction in which a pure compound breaks down into the elements. When we find that ΔH for the decomposition of 1 mole of nitric oxide is −21.6 kcal, i.e.,

$$NO(g) \rightarrow \tfrac{1}{2} N_2(g) + \tfrac{1}{2} O_2(g); \quad \Delta H = -21.6 \text{ kcal}$$

we deduce that ΔH for the reverse reaction and, hence, the heat of formation of NO must be +21.6 kcal. Still another approach is illustrated in Figure 4.6 for carbon monoxide. Knowing, from experiment, the values of ΔH for the reactions

$$C(s) + O_2(g) \rightarrow CO_2(g); \qquad \Delta H = -94.1 \text{ kcal}$$

$$CO(g) + \tfrac{1}{2} O_2(g) \rightarrow CO_2(g); \quad \Delta H = -67.7 \text{ kcal}$$

we deduce, using Hess's Law, that

$$-67.7 \text{ kcal} + \Delta H_f CO = -94.1 \text{ kcal}$$

and hence $\qquad \Delta H_f CO = -94.1 \text{ kcal} + 67.7 \text{ kcal} = -26.4 \text{ kcal}$

TABLE 4.1   Heats of Formation (kcal/mole) at 25°C and 1 atm

AgBr(s)	−23.8	C_2H_2(g)	+54.2	H_2O_2(l)	−44.8	NH_3(g)	−11.0
AgCl(s)	−30.4	C_2H_4(g)	+12.5	H_2S(g)	−4.8	NH_4Cl(s)	−75.4
AgI(s)	−14.9	C_2H_6(g)	−20.2	H_2SO_4(l)	−193.9	NH_4NO_3(s)	−87.3
Ag_2O(s)	−7.3	C_3H_8(g)	−24.8	HgO(s)	−21.7	NO(g)	+21.6
Ag_2S(s)	−7.6	n-C_4H_{10}(g)	−29.8	HgS(s)	−13.9	NO_2(g)	+8.1
Al_2O_3(s)	−399.1	n-C_5H_{12}(l)	−41.4	KBr(s)	−93.7	NiO(s)	−58.4
$BaCl_2$(s)	−205.6	C_2H_5OH(l)	−66.4	KCl(s)	−104.2	$PbBr_2$(s)	−66.3
$BaCO_3$(s)	−291.3	CoO(s)	−57.2	$KClO_3$(s)	−93.5	$PbCl_2$(s)	−85.9
BaO(s)	−133.4	Cr_2O_3(s)	−269.7	KF(s)	−134.5	PbO(s)	−52.1
$Ba(OH)_2$(s)	−226.2	CuO(s)	−37.1	KOH(s)	−101.8	PbO_2(s)	−66.1
$BaSO_4$(s)	−350.2	Cu_2O(s)	−39.8	$MgCl_2$(s)	−153.4	Pb_3O_4(s)	−175.6
$CaCl_2$(s)	−190.0	CuS(s)	−11.6	$MgCO_3$(s)	−266	PCl_3(g)	−73.2
$CaCO_3$(s)	−288.5	$CuSO_4$(s)	−184.0	MgO(s)	−143.8	PCl_5(g)	−95.4
CaO(s)	−151.9	Fe_2O_3(s)	−196.5	$Mg(OH)_2$(s)	−221.0	SiO_2(s)	−205.4
$Ca(OH)_2$(s)	−235.8	Fe_3O_4(s)	−267.0	$MgSO_4$(s)	−305.5	$SnCl_2$(s)	−83.6
$CaSO_4$(s)	−342.4	HBr(g)	−8.7	MnO(s)	−92.0	$SnCl_4$(l)	−130.3
CCl_4(l)	−33.3	HCl(g)	−22.1	MnO_2(s)	−124.5	SnO(s)	−68.4
CH_4(g)	−17.9	HF(g)	−64.2	NaBr(s)	−86.0	SnO_2(s)	−138.8
$CHCl_3$(l)	−31.5	HI(g)	+6.2	NaCl(s)	−98.2	SO_2(g)	−71.0
CH_3OH(l)	−57.0	HNO_3(l)	−41.4	NaF(s)	−136.0	SO_3(g)	−94.5
CO(g)	−26.4	H_2O(g)	−57.8	NaI(s)	−68.8	ZnO(s)	−83.2
CO_2(g)	−94.1	H_2O(l)	−68.3	NaOH(s)	−102.0	ZnS(s)	−48.5

Let us now consider how heats of formation can be used to calculate enthalpy changes for reactions. To illustrate the principle involved, consider the reaction

$$ZnO(s) + 2\ HCl(g) \rightarrow ZnCl_2(s) + H_2O(l)$$

According to Hess's Law, we are at liberty to imagine that this reaction takes place in two steps (Figure 4.7). In the first step, 1 mole of zinc oxide

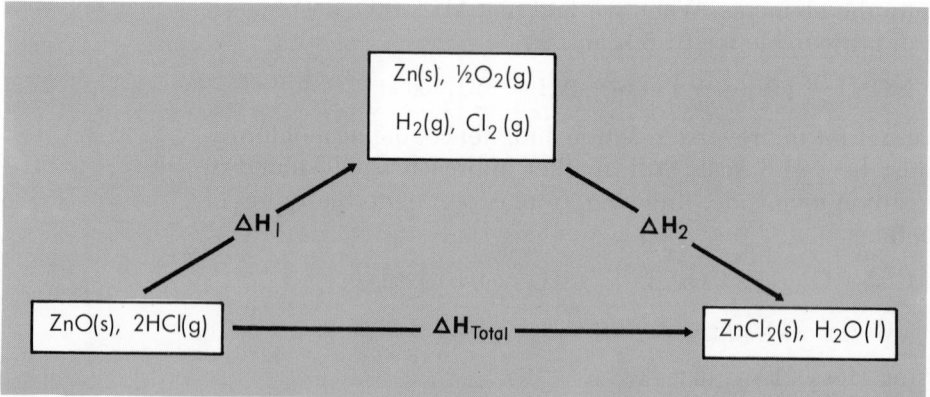

Figure 4.7 ΔH for the overall reaction is the sum of the ΔH's for the two steps. For the second step, where $ZnCl_2$ and H_2O are formed from the elements, $\Delta H_2 = \Sigma\ \Delta H_f$ products. For the first step, where ZnO and HCl decompose to the elements, $\Delta H_1 = -\Delta H_f$ reactants.

and 2 moles of hydrogen chloride decompose to the elements. Clearly, for this step

$$\Delta H_1 = -\Delta H_f ZnO - 2 \Delta H_f HCl$$

The second step involves the formation of 1 mole of zinc chloride and 1 mole of water from the elements, for which

$$\Delta H_2 = \Delta H_f ZnCl_2 + \Delta H_f H_2O$$

Adding:
$$\Delta H_{reaction} = \Delta H_1 + \Delta H_2$$
$$= \Delta H_f ZnCl_2 + \Delta H_f H_2O - \Delta H_f ZnO - 2 \Delta H_f HCl$$

In general, we can state that, for any reaction, ΔH **is equal to the sum of the heats of formation of the products minus the sum of the heats of formation of the reactants.**

$$\Delta H = \Sigma \Delta H_f \text{ products} - \Sigma \Delta H_f \text{ reactants} \qquad (4.16)$$

Any elementary substance involved in the reaction is omitted in taking this sum, since its heat of formation is zero by definition.

Equation 4.16 is, as you can see, very analogous to Equation 4.4, which involved absolute enthalpies. By taking our zero of enthalpy to be that of the pure elementary substances, we arrive at a simple equation which can be used, in conjunction with a table such as Table 4.1, to calculate enthalpy changes for a huge number of reactions.

Generally speaking, one can equate the enthalpy of a substance with its enthalpy of formation. Using this convention, the enthalpies of the elements are zero and Equations 4.16 and 4.4 become identical

---

Example 4.2 Using Table 4.1, calculate ΔH for the combustion of propane

$$C_3H_8(g) + 5 O_2(g) \rightarrow 3 CO_2(g) + 4 H_2O(l)$$

Solution. Following Equation 4.16

$$\Delta H = 4 \Delta H_f H_2O(l) + 3 \Delta H_f CO_2(g) - \Delta H_f C_3H_8(g)$$
$$= 4 (-68.3 \text{ kcal}) + 3(-94.1 \text{ kcal}) - (-24.8 \text{ kcal})$$
$$= -531 \text{ kcal}$$

We conclude that 531 kcal of heat is evolved when 1 mole of propane burns according to the above equation.

4.3 HEATS OF BOND FORMATION

At high temperatures, hydrogen molecules dissociate into atoms

$$H_2(g) \rightarrow H(g) + H(g); \quad \Delta H = +104 \text{ kcal} \qquad (4.17)$$

The heat absorbed in this reaction goes to break the chemical bonds holding hydrogen atoms together in the H_2 molecule. As we might expect, bond breaking is invariably an endothermic process. Upon cooling, the reverse of Reaction 4.17 occurs

$$H(g) + H(g) \rightarrow H_2(g); \quad \Delta H = -104 \text{ kcal} \qquad (4.18)$$

This reaction is exothermic; when gaseous atoms bond together to form a molecule, heat is evolved.

The enthalpy change in Reaction 4.18 is referred to as the heat of formation of the H—H bond (ΔH_f H—H). In general, the heat of bond formation * is defined as the enthalpy change when one mole of a particular type of bond is formed from gaseous atoms. From the enthalpy changes for the following reactions,

$$Cl(g) + Cl(g) \rightarrow Cl_2(g); \ \Delta H = -58 \text{ kcal} \tag{4.19}$$

$$HCl(g) \rightarrow H(g) + Cl(g); \ \Delta H = +103 \text{ kcal} \tag{4.20}$$

Heats of bond formation are usually found spectroscopically

we deduce that the heats of formation of the Cl—Cl and H—Cl bonds are −58 kcal and −103 kcal, respectively.

The concept of heat of bond formation helps us to understand, at the molecular level, why some reactions are exothermic and others endothermic. If the bonds in the product molecules are stronger than those in the reactants, the reaction will be exothermic.

"weak" bonds → "strong" bonds; $\Delta H < 0$

If the reverse is true, heat will have to be absorbed to bring about reaction

"strong" bonds → "weak" bonds; $\Delta H > 0$

To cite a specific example, consider the reaction between hydrogen and chlorine

$$H_2(g) + Cl_2(g) \rightarrow 2 \ HCl(g); \ \Delta H = -44 \text{ kcal}$$

The fact that this reaction evolves heat implies that the bonds in HCl are "stronger" than those in the elementary substances H_2 and Cl_2. More precisely, the amount of heat evolved when two moles of H—Cl bonds are formed is 44 kcal greater than that absorbed in breaking one mole of H—H and one mole of Cl—Cl bonds.

Hess's Law holds here because atoms as well as molecules have characteristic enthalpies

bond breaking: $H_2(g) \rightarrow 2 \ H(g); \ \Delta H = -\Delta H_f$ H—H $= +104$ kcal
$Cl_2(g) \rightarrow 2 \ Cl(g); \ \Delta H = -\Delta H_f$ Cl—Cl $= \underline{\ +58 \text{ kcal}}$
$+162$ kcal

bond forming: $2 \ H(g) + 2 \ Cl(g) \rightarrow 2 \ HCl(g); \Delta H = 2 \ \Delta H_f$ H—Cl $= -206$ kcal
$\Delta H_{total} = -206 \text{ kcal} + 162 \text{ kcal} = -44 \text{ kcal}$

---

* In many texts, the term "bond energy" is used rather than "heat of bond formation." The bond energy is defined as the enthalpy change when 1 mole of bonds is *broken* in the gas state. Bond energy is equal in magnitude but opposite in sign to heat of bond formation; H—H, Cl—Cl, and H—Cl have bond energies of +104, +58, and +103 kcal, respectively.

TABLE 4.2 Heats of Bond Formation (kcal/mole at 25°C)

H—H	−104						
C—C	−83	H—C	−99				
N—N	−38	H—N	−93	C—N	−70		
O—O	−33	H—O	−111	C—O	−84		
S—S	−51	H—S	−81	C—S	−62	N—F	−65
F—F	−37	H—F	−135	C—F	−105	N—Cl	−48
Cl—Cl	−58	H—Cl	−103	C—Cl	−79		
Br—Br	−46	H—Br	−88	C—Br	−66	O—F	−44
I—I	−36	H—I	−71	C—I	−57	O—Cl	−49

The thermal stability of molecules is directly related to the strengths of the bonds holding them together. The fluorine molecule, F_2, which contains the relatively weak F—F bond, undergoes appreciable decomposition at 1000°C

$$F_2(g) \rightarrow 2\ F(g); \quad \Delta H = +37 \text{ kcal} \qquad (4.21)$$

In contrast, the hydrogen fluoride molecule, HF, is one of the most stable known. The bond between hydrogen and fluorine is so strong that the HF molecule is stable at temperatures as high as 5000°C.

What is ΔH_f for F(g)?

$$HF(g) \rightarrow H(g) + F(g); \quad \Delta H = +135 \text{ kcal} \qquad (4.22)$$

This explains, at least in part, the interest in fluorine compounds as components of rocket propellants. The fuel-oxidizer combination is designed so that hydrogen fluoride will be a major product in the rocket exhaust.

In principle, a table of heats of bond formation can be used to calculate ΔH for chemical reactions, making use of a relationship entirely analogous to Equation 4.16.

$$\Delta H = \Sigma \Delta H_f \text{ product bonds} - \Sigma \Delta H_f \text{ reactant bonds} \qquad (4.23)$$

Example 4.3   Calculate ΔH for the reaction

$$Cl_2(g) + 2\ HF(g) \rightarrow 2\ HCl(g) + F_2(g)$$

a. Using Table 4.1 and Equation 4.16

b. Using Table 4.2 and Equation 4.23

Solution

a. Using heats of formation of *compounds*

$$\Delta H = 2\ \Delta H_f\ HCl - 2\ \Delta H_f\ HF$$

From Table 4.1: $\Delta H = 2(-22.1 \text{ kcal}) - 2(-64.2 \text{ kcal}) = +84.2 \text{ kcal}$

b. Using heats of formation of *bonds*

$$\Delta H = 2\ \Delta H_f\ H\text{—}Cl + \Delta H_f\ F\text{—}F - (\Delta H_f\ Cl\text{—}Cl + 2\ \Delta H_f\ H\text{—}F)$$

From Table 4.2: $\Delta H = 2(-103 \text{ kcal}) + (-37 \text{ kcal}) - (-58 \text{ kcal} + 2[-135 \text{ kcal}])$
$$= -243 \text{ kcal} + 328 \text{ kcal} = +85 \text{ kcal}$$

Notice that in this case the values of ΔH calculated by these two different approaches are essentially identical (+84.2 vs +85 kcal).

The calculation of enthalpy changes from data in Table 4.2 is most useful for reactions involving unstable species whose heats of formation are unknown. Consider, for example, the reaction

$$CH_4(g) + OH(g) \rightarrow CH_3(g) + H_2O(g) \qquad (4.24)$$

which is believed to be one of the steps involved in the combustion of methane. We cannot use Table 4.1 to calculate ΔH for this reaction, since values for the unstable species OH and CH_3 are not listed. However, if we look at this reaction from the standpoint of breaking and forming bonds

$$CH_4(g) \rightarrow CH_3(g) + H(g)$$
$$OH(g) + H(g) \rightarrow H_2O(g)$$

we see that all that is involved is the breaking of a C—H bond ($CH_4 \rightarrow CH_3$) and the formation of an O—H bond (OH $\rightarrow$ HOH). We can estimate ΔH from heats of bond formation listed in Table 4.2

$$\Delta H \approx \Delta H_f \text{ O—H} - \Delta H_f \text{ C—H}$$
$$= -111 \text{ kcal} - (-99 \text{ kcal}) = -12 \text{ kcal}$$

ΔH is seldom calculated by the approach outlined in Example 4.3b if heats of formation are available for all the compounds taking part in the reaction. The reason is that, in most cases, the heat of bond formation is not a fixed value; it varies to some extent with the species in which the bond is found. For example, the enthalpy changes for the two reactions

$$H—O—H(g) \rightarrow H(g) + O—H(g) \quad \Delta H = +120 \text{ kcal} \qquad (4.25)$$

$$O—H(g) \rightarrow H(g) + O(g) \qquad \Delta H = +102 \text{ kcal} \qquad (4.26)$$

Calculations of ΔH from heats of bond formation are usually approximate because of this limitation. ΔH in Example 4.3 is not limited in this way. Why?

are not exactly the same even though both involve the breaking of an O—H bond. The heat of formation of the O—H bond given in Table 4.2, −111 kcal, is actually an average value calculated from enthalpy changes for a variety of reactions such as 4.25 and 4.26 and in which an O—H bond is broken or formed.

4.4 CALORIMETRY

The amount of heat evolved or absorbed in a reaction can be measured in an apparatus known as a calorimeter. A simple type of "coffee cup" calorimeter suitable for use in the general chemistry laboratory is shown in Figure 4.8. It offers a simple, convenient way to determine enthalpy changes for reactions occurring in water solution. Suppose, for example, we want to measure the heat of solution of an ionic solid, MX.

$$MX(s) \rightarrow M^+(aq) + X^-(aq) \qquad (4.27)$$

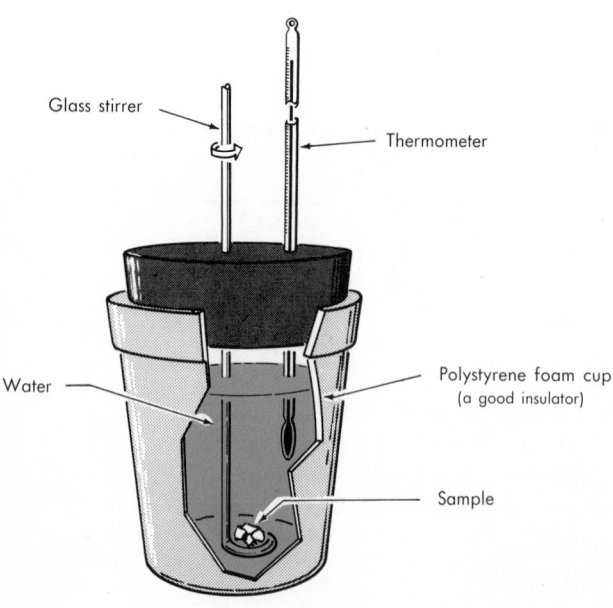

Glass stirrer

Thermometer

Water

Polystyrene foam cup (a good insulator)

Sample

Figure 4.8 Coffee cup calorimeter. Used to obtain ΔH for processes taking place in water solution.

A known amount of water is added to the cup, the cover is replaced, and the initial temperature is read on the thermometer. A weighed amount of solid is then added and dissolved by stirring. If the solution process is endothermic, heat is absorbed from the water and the temperature drops; if it is exothermic, the temperature rises. As soon as the mercury in the thermometer reaches a constant level, the temperature is read and recorded. Knowing the temperature change and the amounts of water and solid added, it is possible to calculate the heat of solution (Example 4.4).

Example 4.4   When 5.00 g of sodium hydroxide is added to 100 g of water, the temperature rises from 25.0 to 37.5°C. Calculate the molar heat of solution, i.e., ΔH for the process

$$NaOH(s) \rightarrow Na^+(aq) + OH^-(aq)$$

taking the specific heat of water to be 1.00 cal/g°C and that of NaOH to be 0.48 cal/g°C.

Solution.   We first obtain the heat given off to the components of the solution.

$$\underbrace{1.00 \frac{cal}{g°C} \times 100 \text{ g} \times (37.5 - 25.0)°C}_{\text{water}} + \underbrace{0.48 \frac{cal}{g°C} \times 5.0 \text{ g} \times (37.5 - 25.0)°C}_{\text{NaOH}}$$

sp. ht. of water × gw × Δt   +   that for other substance.

$$= 1250 \text{ cal} + 30 \text{ cal} = 1280 \text{ cal} = 1.28 \text{ kcal}$$

We conclude that the reaction gives off 1.28 kcal of heat. This means that the enthalpy of the products must be 1.28 kcal *less* than that of the reactants, or that $\Delta H = -1.28$ kcal for the solution of 5.00 g of NaOH. Noting that 1 mole of NaOH weighs 40.0 g, we obtain for the molar heat of solution

$$\Delta H = 40.0 \text{ g NaOH} \times \frac{-1.28 \text{ kcal}}{5.00 \text{ g NaOH}} = -10.2 \text{ kcal}$$

ΔH is negative because heat must be *removed* from the solution to cool it to 25°C.

Calorimeters of the "coffee cup" type are used for measuring heat flows for many reactions carried out in water solution. In order to obtain precise thermochemical data, it is necessary to take account of small heat flows between the solution and the cup or between the cup and the surroundings. In the general chemistry laboratory, these effects are usually neglected.

A more precise apparatus for measuring heat flow, known as a bomb calorimeter, is shown in Figure 4.9. It is widely used to obtain data on the heats of combustion of fuels and explosives. The sample to be burned is added to the heavy-walled steel bomb; the cover is tightened in place and enough oxygen pumped in to build up a pressure of about 20 atm. The bomb is then lowered into a metal container which fits snugly into the insulating walls of the calorimeter. A weighed amount of water sufficient to submerge the bomb is added and the entire apparatus closed.

When the temperature of the water becomes steady, it is measured precisely and the reaction is initiated, usually by electrically igniting a piece of fine iron wire passing through the sample. If all goes well, the sample burns almost instantly and the products attain a very high temperature. The product gases cool rapidly as they hit the bomb wall, which gets warmer and exchanges heat with the surrounding water. The highest temperature reached by the water is measured and recorded.

In order to calculate the amount of heat evolved in a reaction taking

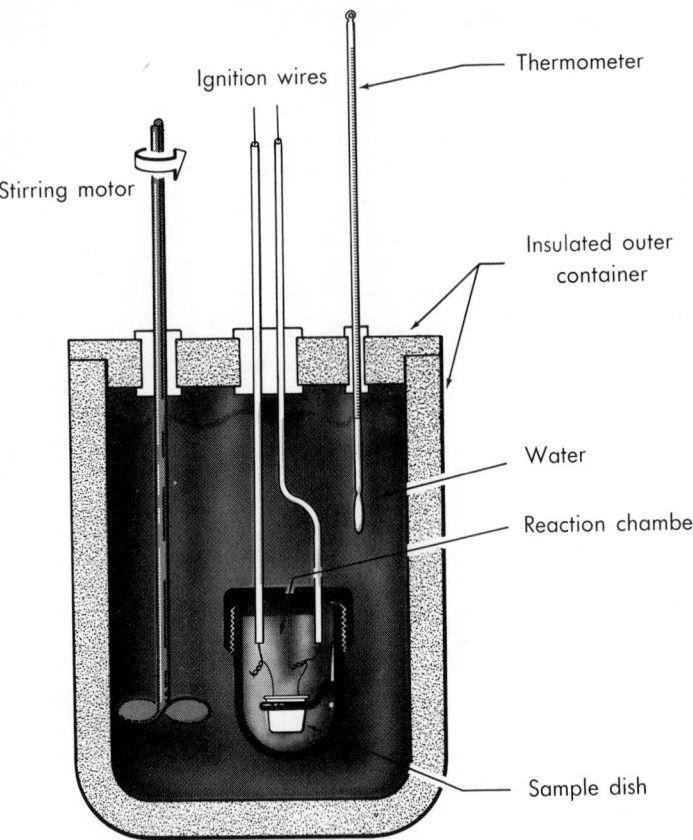

Stirring motor

Ignition wires

Thermometer

Insulated outer
container

The heats of
formation of
substances which
will burn in
oxygen are
ordinarily
measured in
devices like this

Water

Reaction chamber

Sample dish

Figure 4.9 Bomb calorimeter. Used to determine precise values of heat flows for reactions involving gases.

place in a bomb calorimeter, it is, of course, necessary to measure the amount of water present and the final and initial temperatures. Since the calorimeter (bomb + reaction products + metal container) absorbs a significant portion of the heat, we must know its *heat capacity,* i.e., the amount of heat required to raise its temperature by 1°C. In principle, the heat capacity of the calorimeter could be calculated from the masses and specific heats of all its components. In practice, it is determined experimentally, usually by carrying out a reaction which evolves a known amount of heat and measuring the temperature increase (cf. Problems 4.8 and 4.20 at the end of this chapter).

Example 4.5 A 1.00 g sample of the rocket fuel hydrazine, N_2H_4, is burned in a bomb calorimeter containing 1200 g of water. The temperature rises from 24.62 to 27.96°C. Taking the heat capacity of the calorimeter to be 200 cal/°C, calculate:

  a. Q for the combustion of the 1-g sample.

  b. the molar heat of combustion of hydrazine.

Solution. a. We proceed as in Example 4.4, first calculating the amount of heat given off to the water and the calorimeter.

$$1.00 \frac{cal}{g°C} \times 1200 \text{ g} \times (27.96 - 24.62)°C + 200 \frac{cal}{°C} (27.96 - 24.62)°C$$

$$= 4010 \text{ cal} + 670 \text{ cal} = 4680 \text{ cal} = 4.68 \text{ kcal}$$

The fact that 4.68 kcal of heat is evolved to the "surroundings" (i.e., the calorimeter) means that Q for the reaction system must be −4.68 kcal.

b. 1 mole $N_2H_4 = (28.0 + 4.0)$ g $= 32.0$ g

Hence
$$Q = 32.0 \text{ g} \times \frac{(-4.68 \text{ kcal})}{1.00 \text{ g}} = -150 \text{ kcal}$$

Strictly speaking, the heat flow measured in a bomb calorimeter is not exactly equal to the enthalpy change. You will recall that ΔH is defined so as to be equal to the heat flow at *constant pressure* (Q_P). In a bomb calorimeter, reaction takes place at *constant volume;* the pressure may, and frequently does, change if gases are produced or consumed. The difference between the heat flow at constant pressure, Q_P, and that at constant volume, Q_V, is ordinarily small compared to the heat flow itself (Table 4.3). For precise work, a correction can be made using the equation given in Problem 4.29 at the end of this chapter.

TABLE 4.3 Difference Between Q_P and Q_V at 25°C

REACTION	Q_P	Q_V	$(Q_P - Q_V)/Q_P$
$CH_4(g) + 2 O_2(g) \rightarrow CO_2(g) + 2 H_2O(l)$	−212.8 kcal	−211.6 kcal	+0.006
$N_2H_4(g) + O_2(g) \rightarrow N_2(g) + 2 H_2O(l)$	−148.7	−148.1	+0.004
$Zn(s) + S(s) \rightarrow ZnS(s)$	−48.5	−48.5	0.000
$NH_4Cl(s) \rightarrow NH_3(g) + HCl(g)$	+42.3	+41.1	+0.028

4.5 SOURCES OF ENERGY

The principles of thermochemistry which we have presented in this chapter evolved over a period of about two centuries to meet a basic human need for the development of new sources of energy. Prior to 1700 man had only one major source of mechanical energy—the power that he could develop with his own muscles and those of domesticated animals. Since the discovery of fire in prehistoric times, virtually all of the thermal energy from the combustion of fuels had been used for one purpose: domestic heating. The problem that faced our ancestors was that of converting heat into more useful forms of energy.

When Thomas Newcomen in 1705 built the first practical heat engine, man entered a new era in the utilization of energy. The development of the steam engine, the internal combustion engine, and the gas turbine have freed us from the limitations inherent in animate power. These and other devices have made it possible to use the heat evolved in the combustion of fuels to generate mechanical, electrical, and chemical energy. Modern civilization is dependent in a very real sense on this process of energy conversion.

Today we are becoming aware of some of the problems associated with the use of conventional fuels. As our reserves of petroleum and natural gas dwindle, we approach a situation not unlike that of 200 years ago. If we are to continue to raise our standard of living in the face of an expanding population, we must develop new sources of energy, probably by the end of this century. In the remainder of this chapter, we shall explore some of the aspects of this problem. Before doing so, it will be helpful to examine the properties of the solid, liquid, and gaseous fuels that today supply more than 95 per cent of our energy requirements.

FOSSIL FUELS

All our natural fuels are derived from plants that grew on the earth in recent or ancient times. All these substances contain carbon and hydrogen along with varying amounts of oxygen. While the plant grew, it absorbed energy from the sun and stored that energy chemically. The composition of the dead plant changed to some extent over a period of time, but its remains can serve as a fuel because the chemical energy stored within is released as heat when the fuel burns.

Solid fuels of relatively recent photosynthetic origin (wood, peat, and lignite coal) contain appreciable percentages of combined oxygen. Wood consists largely of cellulose, a polymeric substance of empirical formula $C_6H_{10}O_5$. As cellulose was converted to coal over millions of years, oxygen was driven off, presumably by a process of incomplete combustion in which water vapor was a major product. The end result was anthracite coal, which contains the highest percentage of carbon of all natural fuels. Most of the carbon in coal is combined with hydrogen in the form of solid hydrocarbons of very high molecular weight. Petroleum is a complex mixture of liquid hydrocarbons. The gas phase above petroleum deposits, referred to as natural gas, consists primarily of a single hydrocarbon, methane.

Sulfur is a relatively minor component of natural fuels, but an important one from the standpoint of air pollution. Much of the sulfur dioxide that contaminates the atmosphere in our cities comes from the combustion of fuels. In coal, sulfur occurs mainly in the form of the minerals pyrites, FeS_2, and gypsum, $CaSO_4 \cdot 2\ H_2O$. Relatively small amounts of organic sulfur compounds are a major source of the foul odors that we associate with oil

How would you explain the existence of oil in northern Alaska?

TABLE 4.4   Properties of Natural Fuels

	APPROXIMATE ATOM PERCENTAGES			HEATING VALUE (kcal/g)
	C	H	O	
Wood	32	46	21	4.5
Peat (dry)	40	44	16	3.8
Lignite coal	49	37	14	3.8
Bituminous coal	56	40	4	7.0
Anthracite coal	70	27	2	7.3
Petroleum	36	63	0.4	11.3
Natural gas	20	76	0.4	11.6

refineries. Natural gas, as it comes from the well, is often contaminated with hydrogen sulfide. Virtually all of the H_2S is removed and converted to elementary sulfur in gas treatment plants.

In the last column of Table 4.4 are listed the heating values of natural fuels. These represent the amount of heat evolved when 1 g of the fuel burns. The increase in heating value as we go from wood to natural gas is readily explained in terms of chemical composition. If we take wood to be pure cellulose, we calculate a heating value of about 4.2 kcal/g. *Wood has less heating value than methane because it is already partially oxidized*

$$C_6H_{10}O_5(s) + 6\ O_2(g) \rightarrow 6\ CO_2(g) + 5\ H_2O(l);\quad \Delta H = -680\ \text{kcal}$$

$$\Delta H \text{ per gram } C_6H_{10}O_5 = \frac{-680\ \text{kcal}}{162\ \text{g}} = -4.2\ \text{kcal}$$

The value for natural gas, 11.6 kcal/g, is somewhat less than that of its principal constituent, methane, due to the presence of inert components such as nitrogen and carbon dioxide.

$$CH_4(g) + 2\ O_2(g) \rightarrow CO_2(g) + 2\ H_2O(l);\quad \Delta H = -213\ \text{kcal}$$

$$\Delta H \text{ per gram } CH_4 = \frac{-213\ \text{kcal}}{16.0\ \text{g}} = -13.3\ \text{kcal}$$

CONSUMPTION OF FUELS

Until about a century ago a single fuel, wood, supplied virtually the entire energy requirement of the United States. At about the time of the Civil War, coal began to replace wood as a fuel for use in industry and transportation. The production of coal in the United States reached 500 million tons in 1910 and has fluctuated around that figure ever since. Our increasing demand for energy during the past 60 years has been met almost entirely by increased consumption of the hydrocarbon fuels, petroleum and natural gas. Today we are using 10 times as much petroleum and 20 times as much natural gas as we were in 1910.

From the standpoint of conservation of resources, there is a disturbing trend evident from Table 4.5. Over the past century our major energy source has shifted from wood, which takes a few years to grow, first to coal

TABLE 4.5 Consumption of Energy in the United States

	Wood	Coal	Petro-leum	Natural Gas	Water Power	Nuclear	TOTAL (kcal)
1800	99.3	0.5	0.0	0.0	0.2	0.0	0.015×10^{16}
1850	91.1	8.6	0.0	0.2	0.1	0.0	0.075
1900	28.3	66.8	2.4	2.4	0.1	0.0	0.26
1920	10.9	71.8	13.1	3.8	0.4	0.0	0.54
1940	7.6	50.6	31.3	9.8	0.7	0.0	0.64
1950	5.6	39.7	36.2	17.5	1.0	0.0	0.93
1960		25.0	43.8	30.0	1.2	0.0	1.2
1970		22.8	41.4	33.0	2.6	0.2	1.8

PER CENT OF TOTAL

and then to oil and natural gas, all of which are fossil fuels, formed millions of years ago and essentially irreplaceable. Clearly we are no longer living on our income from stored solar energy; instead we are dipping into our principal at an ever increasing rate. We have already consumed nearly 40 per cent of our reserves of petroleum and natural gas; it appears that they will be virtually exhausted within 50 years.

To gain some idea of how long the world's total supply of fossil fuels can be expected to last, we need to consider two factors: the rate at which fuels are being consumed, and the total reserves available in the earth's crust. The first figure is known quite accurately; in 1971 it amounted to an energy equivalent of 5.4×10^{16} kcal/year (one third of this was consumed in the United States). The energy equivalent of fossil fuel reserves is more difficult to come by. A perhaps optimistic estimate of proved and potential reserves would be in the neighborhood of 6000×10^{16} kcal. Over 80 per cent of this is in the form of coal; the remainder is made up of deposits of petroleum and natural gas. (Included in this total, incidentally, are oil shales and tar sands, which cannot be extracted profitably at present.)

It is tempting to carry out a simple division to show that our deposits of fossil fuels could last more than 1000 years.

$$\frac{6000 \times 10^{16} \text{ kcal}}{5.4 \times 10^{16} \text{ kcal/year}} = 1100 \text{ years}$$

If this calculation were realistic, we could stop worrying about new sources of energy. There is, however, a fallacy in this line of reasoning. We have not taken into account the fact that energy requirements increase each year as population grows and industrial development accelerates. Over the past 50 years, the world's consumption of energy has increased, on the average, about 3 per cent per year. Using this as a growth rate, we can anticipate a doubling of our energy requirements every 20 to 25 years. At this rate, if fossil fuels remain our only major source of energy, they will last, not for 1100 years, but *only a little more than 100 years.* (The details of this calculation are outlined in Problem 4.30.)

One way to stretch out our supply of fossil fuels is to reduce the annual increase in energy consumption, perhaps from its present value of 3 per cent to a figure close to 1 per cent. To do this would require controls on the growth of both population and per capita consumption of energy. Historically these two factors have been of about equal importance, at least in highly industrialized countries such as the United States (Problem 4.12). Controls on per capita consumption of energy would pose serious economic problems, particularly in underdeveloped countries striving to raise their standard of living. The only alternative is to supplement our supply of fossil fuels from energy sources that are largely untapped at the moment.

PROSPECTS FOR THE FUTURE

Nuclear fuels (Chapter 22), which now contribute less than 1 per cent of the world's energy, may assume a much larger role in future years. Reserves of uranium and other fissionable elements provide an energy potential perhaps 100 times greater than that available from fossil fuels. The technology of nuclear power plants is reasonably well developed and, although they can create pollution problems of various kinds, they are likely to offer at least a short-term source of electrical energy.

Essentially unlimited amounts of energy will become available if we learn to control the process of nuclear fusion (Chapter 22), in which nuclei of relatively abundant light elements such as hydrogen combine to form heavier nuclei. Such reactions evolve tremendous amounts of energy and the products appear to be much less objectionable from an environmental standpoint than those of fission. So far, the enormous temperatures required to bring about fission reactions have prevented them from becoming a commercial source of power.

Today there is a growing awareness that nuclear fuels, in themselves, are not the ultimate answer to the energy problem. The difficulty is that the energy of nuclear fuels, like that of fossil fuels, is given off as heat. To make use of this heat, it must ultimately be converted into useful work by some type of heat engine. Unfortunately this conversion can never be complete. No matter how cleverly we design and operate a heat engine, it will have a maximum, or theoretical, efficiency less than 100 per cent. The limitation is imposed by the Second Law of Thermodynamics, which tells us that

$$\text{Maximum efficiency} = \frac{T_a - T_r}{T_a} \qquad (4.28)$$

where T_a is the temperature in °K at which the engine absorbs heat and T_r the temperature at which it rejects heat to the surroundings.

To illustrate the restriction imposed by Equation 4.28, suppose we operate a steam engine using boiling water at atmospheric pressure as the source of heat. This sets T_a at 373°K (i.e., 100°C). The temperature at which the engine discharges heat can hardly be less than that of the surroundings, which we will take to be 298°K (i.e., 25°C). Under these conditions the maximum efficiency will be

The conversion of heat into useful work is inexorably subject to this limitation

$$\frac{373 - 298}{373} = 0.20 = 20\%$$

In other words, if we burn enough coal, oil, or gas to supply an engine with 100 kcal of heat, we will recover at best 20 kcal of mechanical work; the other 80 kcal will be discharged as heat.

One can readily show with use of Equation 4.28 that the theoretical efficiency of a heat engine goes up when T_a is increased (see Problem 4.11). This explains why superheated steam is ordinarily used in steam engines and why automobile engines are adjusted to run at as high a temperature as is feasible. However, the actual efficiency of a heat engine is always considerably less than the maximum permitted by the Second Law. Anyone who can build an engine and power transmission system that will convert appreciably more than 30 per cent of the heat it absorbs into useful work is doing very well indeed.

It is somewhat disconcerting to realize that at least three fourths and probably 90 per cent of all the chemical energy associated with the fossil fuels burned over the past 200 years has been wasted as useless heat. What is even more disturbing, at least in its implications for the future, is the effect of this wasted heat upon our environment. The subject of thermal pollution will be considered in more detail in Chapters 11 and 15, but it may be well to comment briefly at this point on the magnitude of the problem. If we continue to depend upon heat from fossil or nuclear fuels to meet our increasing demands for energy, it seems likely that within 50 years we will be raising the temperature of the earth at the rate of 0.1°C per year or 10°C per century. At the very least, a temperature increase of this magnitude would have a profound effect on our climate and would threaten the survival of many cold-blooded organisms, including most species of fish. These considerations alone make it imperative that we find alternative sources of energy which will not raise the temperature of the environment.

While this situation is potentially important, it seems probable that fuel limitations will pose more immediate problems

WATER POWER AND SOLAR ENERGY

All of us are familiar with one source of energy in use today which does not require the conversion of heat. *Water power* has been used for centuries, first to drive water wheels in small mills and then, in this century, to propel electric generators. The contribution from this source has been increasing steadily throughout this century. Indeed, hydroelectric plants built in the past 20 years in this country alone are producing more electrical energy than all the nuclear power plants in the world. Unfortunately the prospects for the future are not so encouraging. Engineers have estimated that harnessing the water power of all the major rivers of the world would yield at most 0.6×10^{16} kcal of energy per year, which is only a little more than 10 per cent of our present energy requirement.

Another approach to both the energy problem and the heat problem is to make more effective use of *solar energy*. Each year we receive from the sun the enormous total of 5×10^{20} kcal of energy. If we could trap as little as 0.01 per cent of this energy, we could meet the world's annual energy requirements (5.4×10^{16} kcal) without disturbing the earth's heat balance.

The problem of trapping and utilizing solar energy is one that has intrigued scientists for centuries. Joseph Priestley, 200 years ago, used a simple magnifying glass to focus the sun's rays on a sample of mercuric oxide and thereby bring about the endothermic reaction

$$HgO(s) \rightarrow Hg(l) + \tfrac{1}{2} O_2(g); \quad \Delta H = +21.7 \text{ kcal} \qquad (4.29)$$

Devices which make more effective use of solar energy are perhaps the most important scientific need of modern civilization

Since the time of Priestley a variety of ingenious devices have been proposed to make direct use of the sun's energy. Solar furnaces, in which sunlight is focused through circular parabolic mirrors, can achieve temperatures as high as 3500°C. Such furnaces have been used for melting and sintering ceramic materials such as aluminum oxide (mpt = 2000°C). Nitric oxide can be made by passing air through the focus of a solar furnace. The reaction

$$N_2(g) + O_2(g) \rightarrow 2 \text{ NO}(g); \quad \Delta H = +43.2 \text{ kcal} \qquad (4.30)$$

is spontaneous only at temperatures above 2000°C.

In recent years a great deal of research has been devoted to the problem of using solar energy for space heating. "Solar houses" which derive all their heat from the sun are now in operation in some of the more temperate areas of the world. Installed on the roof or walls of these houses is a device which collects heat from the sun. The simple collector shown in Figure 4.11 consists of a shallow water tank with a glass cover (a sheet of black plastic can be substituted for the glass). The cover exerts a "green-

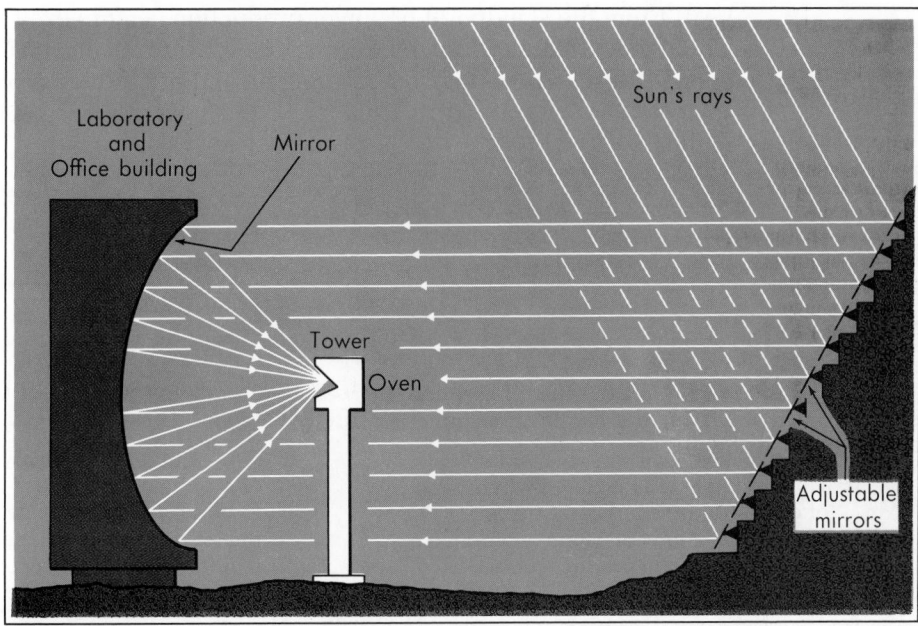

Figure 4.10 Solar furnace. By focusing the sun's rays, it is possible to reach temperatures as high as 3500°C. The furnace shown is one in operation in France.

house" effect, allowing sunlight to penetrate but preventing infrared radiation from escaping. In this way it is possible on a sunny day to raise the temperature of the water in the tank by as much as 20°C. The warm water is fed into a well-insulated storage tank. At night, when the temperature drops, this water is circulated through heating units inside the house.

The devices that we have considered for utilizing solar energy are essentially heat storage systems. At best, they would preserve our dwindling deposits of fossil fuels by providing an alternative source of domestic

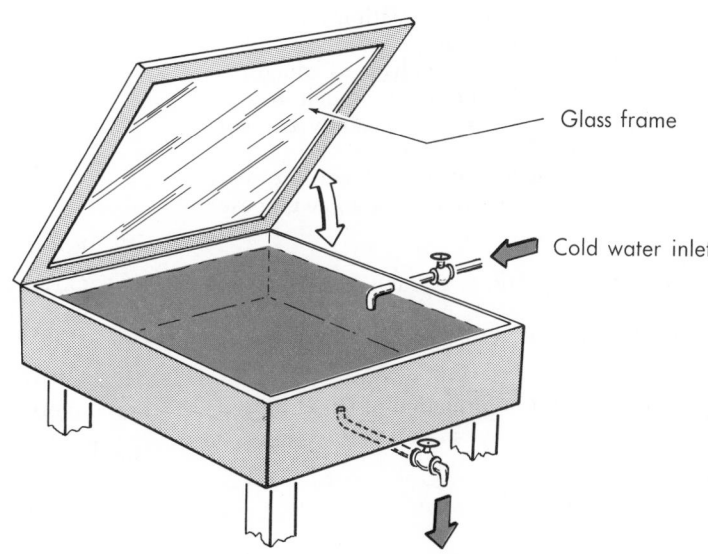

Figure 4.11 Solar collector. Heat from the sun warms the water which is then circulated through radiators to heat a house.

heat. From the standpoint of power production, a more exciting possibility would be to convert solar energy directly into mechanical or electrical energy. Research in this area is going on and has indeed produced a small solar battery, discussed in Chapter 9, but progress has been disappointingly slow.

PROBLEMS

4.1   One of the major constituents of gasoline is octane, C_8H_{18}. ΔH for the combustion of octane is -1300 kcal/mole.

   a. How much heat is evolved when one g of octane burns?
   b. How many grams of octane would have to be burned to produce one kcal of heat?

4.2   When iron rusts in air, the following reaction occurs

$$4\ Fe(s) + 3\ O_2(g) \rightarrow 2\ Fe_2O_3(s)$$
$$\Delta H = -393\ kcal$$

   a. How much heat is evolved when one g of iron reacts?
   b. Calculate ΔH for the formation of one g of Fe_2O_3.
   c. Calculate the heat of formation of one mole of Fe_2O_3.

4.3   Glucose, $C_6H_{12}O_6$, is sometimes eaten by athletes to give them quick energy. It is metabolized in the body to give $CO_2(g)$ and $H_2O(l)$.

   a. Write a balanced equation for the metabolism of glucose.
   b. ΔH for the metabolism of glucose is -673 kcal/mole. How much heat is produced by eating a candy bar containing 40.0 g of glucose?
   c. How much heat is produced for every gram of O_2 used to metabolize glucose?

4.4   Given:

$$Cu_2O(s) + \tfrac{1}{2} O_2(g) \rightarrow 2\ CuO(s);\ \Delta H = -34.4\ kcal$$
$$CuO(s) + Cu(s) \rightarrow Cu_2O(s);\quad \Delta H = -2.7\ kcal$$

Calculate the heat of formation of Cu_2O.

4.13   The enthalpy change for the combustion of ethyl alcohol, C_2H_5OH, is -227 kcal/mole.

   a. How much heat is evolved when one g of ethyl alcohol burns?
   b. If ethyl alcohol and octane were selling for the same price per pound, which would be the more economical fuel to use in an automobile (cf. Problem 4.1)?

4.14   The decomposition of mercuric oxide, carried out by Priestley, is represented by the equation

$$2\ HgO(s) \rightarrow 2\ Hg(l) + O_2(g);\ \Delta H = 43.4\ kcal$$

   a. How much heat must be absorbed to decompose 1.00 g of HgO?
   b. How much heat must be absorbed to form 0.100 g of oxygen?
   c. Calculate the heat of formation of HgO.

4.15   When one g of nitroglycerine, $C_3H_5(NO_3)_3$, decomposes to form $N_2(g)$, $O_2(g)$, $CO_2(g)$ and $H_2O(l)$, 1.90 kcal of heat is evolved.

   a. Write a balanced equation for the reaction.
   b. Calculate ΔH for the decomposition of one mole of nitroglycerine.
   c. How much heat is evolved per mole of gas formed in the decomposition of nitroglycerine?

4.16   For the following reactions

$$C_2H_2(g) + \tfrac{5}{2} O_2(g) \rightarrow 2\ CO_2(g) + H_2O(l)$$
$$C(s) + O_2(g) \rightarrow CO_2(g)$$
$$H_2(g) + \tfrac{1}{2} O_2(g) \rightarrow H_2O(l)$$

ΔH is -310.7 kcal, -94.1 kcal and -68.3 kcal respectively.

Calculate ΔH for the reaction

$$2\ C(s) + H_2(g) \rightarrow C_2H_2(g)$$

4.5 Using the heats of formation listed in Table 4.1, calculate ΔH for the following reactions.

a. $C_2H_6(g) + \frac{7}{2} O_2(g) \rightarrow 2 CO_2(g) + 3 H_2O(l)$

b. $MgSO_4(s) \rightarrow MgO(s) + SO_2(g) + \frac{1}{2} O_2(g)$

4.6 Estimate ΔH for each of the following reactions, using heats of bond formation given in Table 4.2. Where possible, check your answers by using the data in Table 4.1.

a. $Cl_2(g) + 2 HBr(g) \rightarrow 2 HCl(g) + Br_2(g)$
ΔH_{vap} $Br_2(l) = 7.3$ kcal

b. $H_3C—O(g) + HOH(g) \rightarrow$
$H_3C—OH(g) + OH(g)$

c. $NH_3(g) + Cl_2(g) \rightarrow NH_2Cl(g) + HCl(g)$

4.7 When 1.23 g of KF (specific heat = 0.20 cal/g°C) dissolves in 10.0 g of water (specific heat = 1.00 cal/g°C), the temperature rises from 23.2 to 33.1°C. Calculate the heat of solution of

a. 1.23 g of KF
b. one mole of KF

4.8 When one g of benzoic acid is burned in a bomb calorimeter of the type shown in Figure 4.9 containing 3.62 kg of water, the temperature rises from 24.33 to 25.99°C. The heat of combustion of benzoic acid under these conditions is 6.315 kcal/g. What is the heat capacity of the calorimeter, in cal/°C?

4.9 A sample of methanol, CH_3OH, weighing two g is burned in a bomb calorimeter, which has a heat capacity of 300 cal/°C and contains 3.56 kg of water (specific heat = 1.00 cal/g°C). The temperature rises from 25.27 to 28.10°C. Determine

a. the amount of heat evolved in the reaction.
b. the heat evolved in the combustion of one mole of methanol. Compare with the value calculated from heat of formation data (CO_2 and $H_2O(l)$ are formed).

4.10 Consider the data in Table 4.4.

a. Calculate the percentages by weight of carbon and hydrogen in wood and anthracite coal.
b. If petroleum were pure octane, C_8H_{18}, what would be its heating value? (cf. Problem 4.1)?

4.17 Using the data in Table 4.1, calculate ΔH for:

a. $CH_3OH(l) + O_2(g) \rightarrow CO_2(g) + 2 H_2O(g)$
b. the combustion of one g of acetylene to give CO_2 and $H_2O(l)$.

4.18 Estimate ΔH for each of the following, using the data in Table 4.2:

a. $CH_4(g) + 3 F_2(g) \rightarrow$
$CHF_3(g) + 3 HF(g)$
b. the first step in the thermal decomposition of hydrazine, $HN\overset{\text{H H}}{—}NH$, assuming two NH_2 radicals are formed.
c. repeat (b) if the fragments formed are H and N_2H_3.

4.19 The molar heat of solution of K_2SO_4 is +5.8 kcal; its specific heat is 0.19 cal/g°C. If 2.00 g of K_2SO_4 dissolves in 10.0 g of water originally at 20.0°C, what will be the final temperature?

4.20 To determine the heat capacity of a bomb calorimeter, a student adds 125 g of water at 50.0°C to the calorimeter, which is at an initial temperature of 25.0°C. When equilibrium is reached, the temperature is 33.0°C. What is the heat capacity of the calorimeter in cal/°C?

4.21 When one g of the rocket fuel dimethyl hydrazine, $(CH_3)_2N_2H_2$, is burned in a bomb calorimeter (heat capacity = 440 cal/°C) containing 5.00 kg of water, the temperature rises from 24.62 to 26.07°C. Calculate the amount of heat that would be given off by the combustion of one mole of dimethyl hydrazine.

4.22 Calculate the heating value (kcal/g) of a sample of bottled gas which consists of 50 mole % propane and 50 mole % butane (heats of combustion = −588 kcal/mole and −531 kcal/mole, respectively). Propane: C_3H_8, butane: C_4H_{10}.

4.11 Using Equation 4.28, calculate the theoretical efficiency of a heat engine using superheated steam at 200°C and 16 atm pressure. Assume the discharge temperature is 30°C. If the engine absorbs 200 kcal of heat, how much heat would it discharge to the environment?

4.12 The population of the United States in 1900 was 7.6×10^7; in 1970, it was 20.4×10^7. Using the data in Table 4.5, find

   a. the per capita consumption of energy in 1900; in 1970.
   b. what the total consumption would have been in 1970 if the per capita value had remained at its 1900 value.
   c. what the total consumption would have been in 1970 if the population had remained constant, assuming the per capita figure for 1970.

4.23 Assume that the absorption of solar energy raises the temperature of a kilogram of water 20°C.

   a. How many kcal of heat are absorbed?
   b. If the heat is used to operate an engine calculate its efficiency using Equation 4.29 and taking $Tr = 25°C$.

4.24 Using the data in Table 4.5, determine

   a. the number of kcal of energy obtained by burning coal in 1900; in 1970.
   b. the number of tons of coal burned in 1900 and in 1970, assuming that 7.0 kcal are obtained per gram of coal.
   c. the per capita consumption of coal (in tons) in 1900 and in 1970 (cf. Problem 4.12).

°4.25 Equation 4.28 can be applied to a refrigerator in the form: $W/Q_a = (T_r - T_a)/T_a$, where W is the work required to remove an amount of heat, Q_a, from the refrigerator.

   a. How much work must be done to remove one kcal of heat from inside a refrigerator at 5°C to a room at 30°C?
   b. If electrical work costs 5.0×10^{-3} cents/kcal, how much will it cost to freeze a kilogram of ice cubes ($\Delta H_{fus} = 80$ cal/g)?

°4.26 Someone has said that the First Law of Thermodynamics means that "you can't win" (i.e., you can't create energy). Again, the Second Law has been interpreted to mean that "you can't break even, except at absolute zero." Using Equation 4.28, explain what this witticism means. (The Third Law, incidentally, "says" that you can't reach absolute zero.)

°4.27 Calculate the heating value of the following foods, taking the heating values of fats, proteins, and carbohydrates to be 9.0 kcal/g, 4.0 kcal/g and 4.0 kcal/g, respectively.

	% fat	% protein	% carbo- hydrates	% water
beef	12.0	23.0	0.0	65.0
cheese	36.8	27.7	4.1	14.0
cornflakes	1.5	5.5	81.0	11.8
potato chips	40.0	9.0	46.5	4.5

How many "calories" would be given off when one ounce of each of these foods is metabolized in the body? (The calorie referred to in nutrition is really a kilocalorie.)

°4.28 Consider the thermite reaction, which was once used to weld iron rails.

$$2 \, Al(s) + Fe_2O_3(s) \rightarrow Al_2O_3(s) + 2 \, Fe(s)$$

   a. Calculate ΔH for this reaction from heat of formation data.
   b. Taking the specific heats of Al_2O_3 and Fe to be 0.19 cal/g°C and 0.10 cal/g°C, calculate the temperature to which the products of the reaction will be raised, starting at room temperature, assuming no heat is lost.
   c. Will the reaction produce molten iron (mpt Fe = 1275°C, $\Delta H_{fus} = 65$ cal/g)?

°4.29 The relation between Q_P and Q_V in a chemical reaction is

$$Q_P = Q_V + \Delta n_g \, R \, T$$

where Q_P is the heat flow at constant pressure, Q_V is the heat flow at constant volume, Δn_g is the change in the number of moles of gas in the reaction, R is the gas constant, 2.0 cal/mole°K, and T is the temperature in °K. Using this relation, calculate the values of Q_P in Table 4.3 from the values given for Q_V, and compare with those listed.

*4.30 Starting with a consumption of 5.4×10^{16} kcal of energy in 1971 and assuming an increase of 3.0 per cent per year,

a. Set up expressions for the consumption of energy in 1972; in 1973; a general expression for the consumption "n years" after 1971.

b. Estimate the annual consumption of energy for each year of the decade 1971–1980. What would be the total consumption during this decade?

c. Using the procedure in (b), estimate the total consumption in successive decades (i.e., 1981–1990, 1991–2000, . . .). Note that the ratio of total consumption per decade to annual consumption in the first year of the decade stays constant.

d. Assuming a total energy equivalent for fossil fuel reserves to be 6000×10^{16} kcal, when will these reserves be exhausted if fossil fuels serve as the only source of energy?

5 THE PHYSICAL BEHAVIOR OF GASES

Gases have been involved, in one way or another, with many of the fundamental developments of chemistry. Because of their invisibility, gases contributed, at least in part, to the confusion which reigned for centuries with regard to the kinds of chemical substances that exist and the nature of the combustion reaction. They were used by Cannizzaro in establishing for the first time a rational method for the determination of atomic weights. In more recent years the spectra of gases have been important in showing the validity of some of the results obtained from quantum mechanics. Although at present we know much more about the basic properties of gases than we do about liquids and solids, much remains to be learned, especially in connection with the details of the changes that gas molecules experience during chemical reactions.

5.1 SOME GENERAL PROPERTIES OF GASES

When a liquid is heated sufficiently, it begins to evaporate or boil. In this process the substance is said to make a transition from the liquid to the gaseous state. During the change in state the particles in the liquid become free from one another and pass into space as molecules of gas. In general this process is accompanied by a great change in volume. If half a cupful of water is evaporated, the resulting water vapor (water in the gas phase) at one atmosphere and 100°C occupies a volume roughly equal to that of a 50-gallon oil drum. Since molecules in the gas phase are the same size as they are in the liquid state, it follows that the distances between them in the gas are much greater than they are in the liquid.

In view of the rather large distances between gas molecules, we might expect that it would be fairly easy to compress a gas, at least as compared to a liquid. Experimentally we find that this is so, and that in general the volume of a gas varies inversely with the applied pressure; that is, doubling

In the air you breathe the molecules are roughly 10 diameters apart

88

the pressure reduces the volume to about half its previous value. Similarly, if we double the amount (mass) of gas in a container we find that the pressure approximately doubles. Increasing the temperature of a gas in a closed container will increase the pressure of the gas.

Gases can be expanded indefinitely and will always tend to occupy their containers completely and uniformly. If one milliliter of hydrogen gas at one atmosphere pressure is let into an evacuated 10,000-liter container, the hydrogen almost instantly diffuses to give a constant density and pressure throughout the container. The pressure of the gas would be very low under such conditions, about 10^{-7} atm, but could be measured easily. You might wonder whether the situation would be the same if there had already been another gas at one atmosphere in the container. Under such circumstances the amount of hydrogen in any part of the container would ultimately be the same as if the container were initially evacuated, but the time required to attain a homogeneous mixture would be of the order of hours rather than a fraction of a second; the diffusion of the hydrogen molecules would be hindered greatly by collisions with the other gas molecules.

The H_2 molecule would encounter the same difficulties as a commuter trying to get on a New York city subway at rush hour

All gases mix readily with one another to form completely homogeneous solutions. Ordinary air is such a solution. No one has ever been able to prepare a mixture of gases which tended to settle out into two or more regions of different composition. This situation is very different from that with mixtures of liquids and solids, in which the solubility of one substance in another is usually limited.

5.2 ATMOSPHERIC PRESSURE AND THE BAROMETER

The most common gas we encounter, and the only one known until about 1750, is the air about us. This gas lies over the earth in a blanket about 50 miles thick. Like all earthly matter, the air is subject to the gravitational pull of the earth. The air near the earth is compressed by the weight of the air above it. The pressure of the air at the earth's surface is by no means negligible, amounting to about 14.7 pounds per square inch; this means that our bodies are at all times subject to a rather gigantic force. The pressure of the atmosphere varies with the height above sea level and weather conditions. In Denver, Colorado (altitude, 5000 ft above sea level), the atmospheric pressure is about 13.5 pounds per square inch. During a hurricane the pressure may become that low at sea level. Above 10,000 feet, breathing becomes uncomfortable for people not accustomed to such altitudes, and for that reason modern aircraft have pressurized cabins. Spacecraft, which operate at much higher altitudes, are also pressurized; on landing and takeoff, however, astronauts wear pressurized suits and oxygen masks. (Why?) At a height of 10 miles the atmospheric pressure is only about 10 per cent of that at sea level. Uncomfortable is not the word to describe the sensations of an unpressurized human being at such an altitude.

Although the facts of atmospheric pressure are really very simple, they were not clearly understood until about 1650. Men had learned earlier in a very practical way that they could not lift water more than about 33 feet

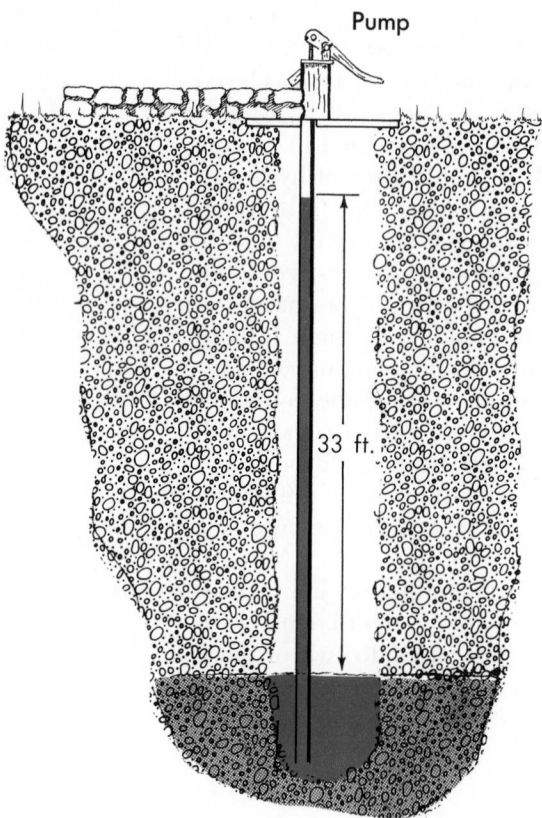

Pump

Figure 5.1 The suction pump.

33 ft.

Question of the year in 1600

Why can't we pump water out of
a well more than 33 feet deep?

with a suction pump; the reason for this was unknown, and the explanation
given by philosophers was that "nature abhors a vacuum." Admittedly, that
wasn't much of an answer; in fact, it wasn't an answer at all. The solution
to this ancient problem was given by Torricelli, an Italian scientist. He
applied the natural law that in a liquid equal pressures exist at equal
heights (Fig. 5.1). At the surface of the water in the pipe, the pressure is
essentially zero, since the surface there is in contact with a vacuum. The
pressure inside the water pipe at the same height as the lower surface is
equal to atmospheric pressure and also has to be equal to the pressure
exerted by the water column, 33 feet high, above that surface. Once the
pump lowers the air pressure inside the well pipe to zero the water in the
pipe will stop rising, no matter how fast the pump works.

How could you
prove the validity
of this explana-
tion to a non-
believer?

Torricelli recognized that the maximum height of the liquid column
was essentially a measure of the atmospheric pressure, and made a simple
device for making the measurement conveniently. His instrument, called a
barometer, is shown in Figure 5.2; it consists of a closed glass tube filled
with mercury and inverted over a pool of mercury. Provided the tube is
sufficiently long, liquid mercury flows into the reservoir when the tube is
first inverted, leaving a nearly perfect vacuum above the liquid; the height
of the liquid column remaining in the tube is then a measure of the atmospheric pressure. Since the density of mercury is high, the height of the column is much less than with water and is, at one standard atmosphere pres-

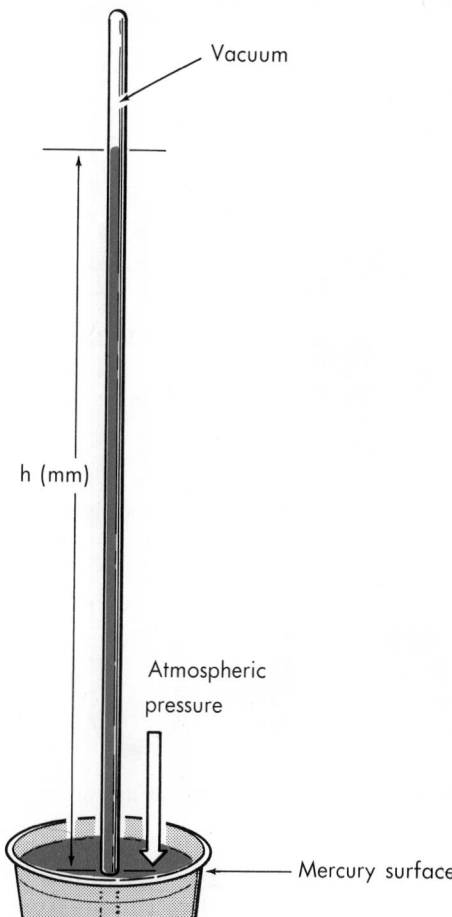

Figure 5.2   The barometer.

Could a barometer be used to
measure the air pressure in a
submarine?

sure, 760 mm if the mercury is at 0°C. It was several years before Torricelli's
contemporaries accepted his reasoning and recognized the barometer as
the scientific instrument it is.

THE MANOMETER AND THE MEASUREMENT OF GAS PRESSURE

The measurement of the pressure exerted by a gas in a closed container
is ordinarily accomplished by a device called a manometer, which is similar but not identical to a barometer (Fig. 5.3).

A manometer consists of a glass U-tube partially filled with mercury.
One side of the U-tube is connected to the container in which the pressure
is to be measured, and the other side is connected to a region of known
pressure. The gas in the container will exert a force on the mercury column
and will tend to make it go down. This force will be opposed by that created
by the gas over the other surface. The difference between the mercury

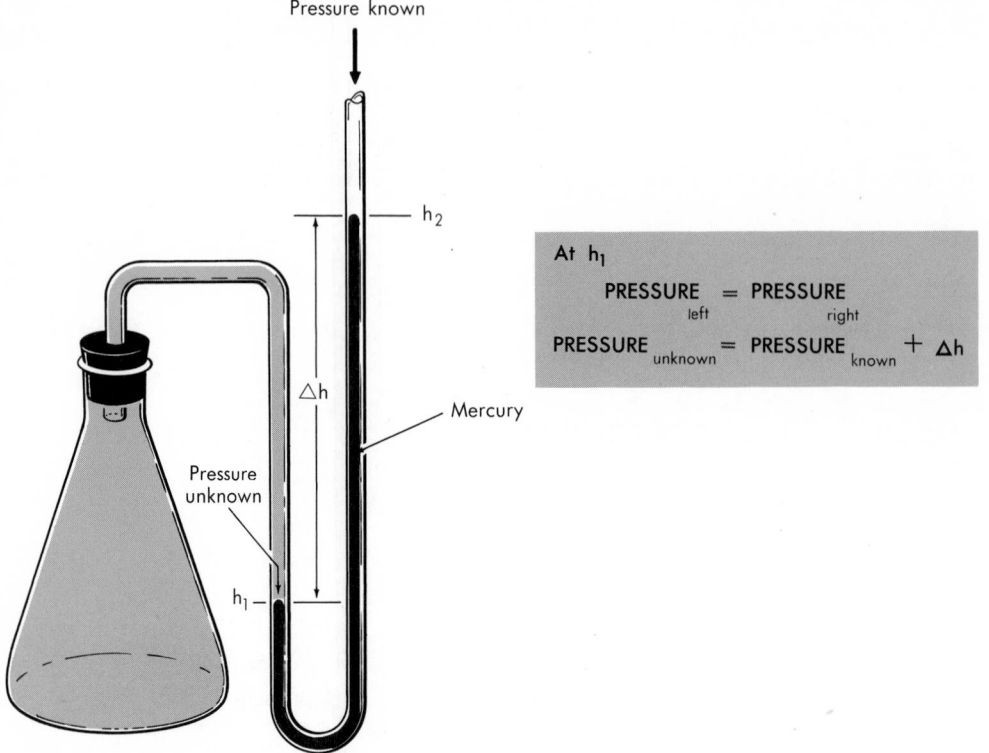

Figure 5.3 Manometer and the measurement of gas pressure.

levels at equilibrium is directly proportional to the difference between the two gas pressures.

Noting Figure 5.3, we can say mathematically that at level h_1

Pressure$_{unknown}$ = Pressure$_{known}$ + Pressure due to Δh mm Hg

Ordinarily the known pressure is expressed in terms of the height of a column of mercury to which it is equivalent; usually the known pressure is either that of the air in the laboratory or that of a vacuum. In the former case, the known pressure would be barometric pressure; in the latter case, it would be zero. Because of the way in which gas pressures are determined, they are often reported in mm Hg. (In recent years, scientists have begun using the word torr for mm Hg; stating a pressure as being, say, 340 torr means that it is 340 mm Hg. The authors feel that this terminology is unfortunate; we are not against honoring Torricelli, but we do oppose changing a unit with a clearly implied physical meaning, like mm Hg, to one with no such meaning.) Pressures can be expressed in many other units, such as lb/in², or absolute units, dynes/cm²; high pressures are frequently given in atmospheres.

5.3 THE IDEAL GAS LAW

When one compresses a sample of gas at constant temperature, he finds that the volume change which occurs does not depend on the kind of gas used, but is determined only by the initial volume and the ratio of the initial and final pressures. Similarly, heating a gas sample at constant pressure results in an increase in volume which is not dependent on the nature of the gas in the sample. Gases, unlike liquids and solids, show volume-pressure-temperature behavior that can be described in terms of general relations, which are as valid for methane as for helium, for air as for fluorine. The most important of these relations, and the one with which this chapter is mainly concerned, is called the Ideal Gas Law.

The Ideal Gas Law is an equation which describes the relationship among four of the fundamental properties of a gas. The Law is usually stated in the following form:

This Law is more profound than it appears to be

$$P V = n R T \tag{5.1}$$

where P is the pressure, V is the volume and n is the number of moles of gas. In the equation T is the absolute temperature of the gas and is related to the Celsius temperature t in the following way:

$$T = t + 273° \tag{5.2}$$

The quantity R is called the gas constant; it has the same value for all gases and, in the usual dimensions for pressure, volume and temperature, is equal to 0.0821 lit-atm/mole°K.

As you can see, the Ideal Gas Law is an expression in four variables, P, V, n and T. It tells us how these properties of a gas depend on each other and is really a rather remarkable equation in that it is one of the very few natural laws that involves four variables. Usually scientists perform experiments in which, hopefully, there are only two variables, since then it is easiest to recognize how one depends on the other. Actually the Ideal Gas Law is based on several experiments of this sort and, as we shall see, took its final form when it was realized that it was possible to summarize the results from such experiments in a general equation.

BOYLE'S LAW

Probably the simplest experiment that can be done on a gas is to trap a sample of air and measure its volume at several different pressures, holding the temperature constant. This experiment will determine the relationship between the two variables, pressure and volume, under conditions

TABLE 5.1   Compression of a Sample of Air at 25°C

STATE	PRESSURE mm Hg	VOLUME ml	PRESSURE × VOLUME (mm Hg) ml
I	400	100	4.0×10^4
II	670	60	4.0×10^4
III	800	50	4.0×10^4
IV	1000	40	4.0×10^4
V	1200	33	4.0×10^4
VI	1600	25	4.0×10^4

in which the amount and temperature are fixed. When we perform this experiment we find that as the pressure on the sample is increased, the volume decreases in such a way that the product of pressure and volume remains essentially constant. In Table 5.1 we have listed some experimental data for the compression of air at 25°C.

As you can see, although the pressure and volume change a great deal, their product remains constant within the accuracy of the experiment. Mathematically, we can express the results of this kind of experiment in the equation:

$$P V = B \quad \text{or} \quad V = B/P \quad \text{at constant n and T} \tag{5.3}$$

This was one of the first physical laws to be discovered

where B is a constant; in the experiment cited, B would equal 4.0×10^4 (mm Hg) ml.

In Figure 5.4 we have plotted the data in Table 5.1 and connected the data points with a smooth curve on which we would expect any other data in this experiment to fall.

Equation 5.3 is called Boyle's Law, since it was Robert Boyle who first discovered that a gas behaves this way when compressed at constant temperature. Boyle, a British natural philosopher, did his experiments on air in 1660, in a glass U-tube like that shown in Figure 5.5, using mercury to compress the entrapped gas. By assuming that the diameter of the tube was constant, Boyle was able to obtain all the data he needed to establish his law with just a meter stick and a barometer. Any schoolboy these days could perform Boyle's experiment in a few hours, but even now he might have a bit of trouble interpreting his data properly. (See Problem 5.6.)

If you refer back to Equation 5.1, the Ideal Gas Law, you will see that if we hold the amount of gas, n, and the temperature, T, fixed, the right side of the equation becomes fixed, and will remain constant as P and V are

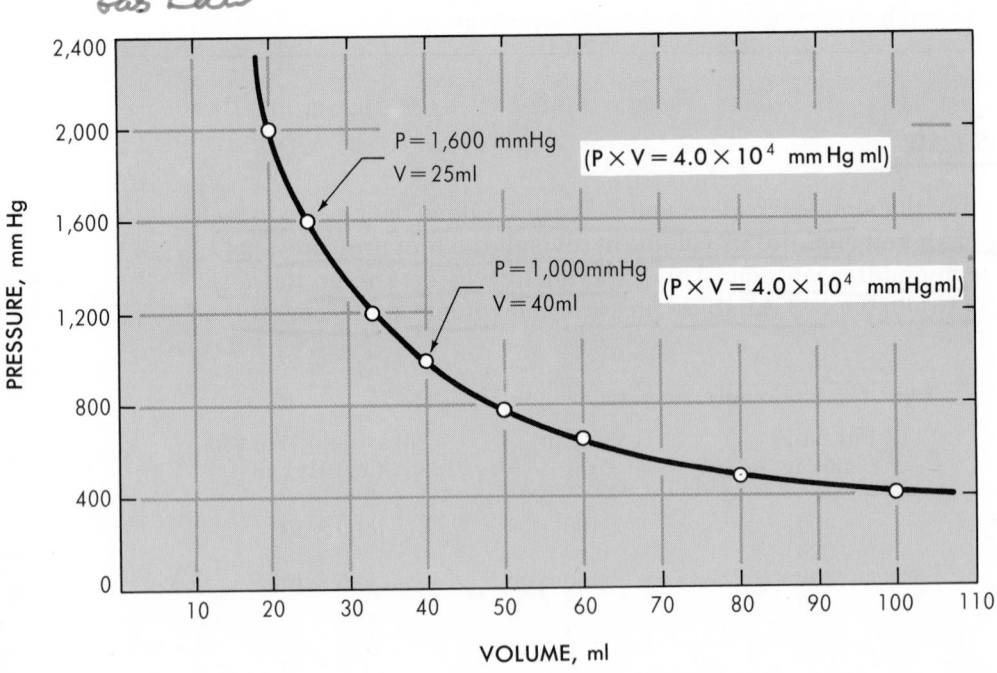

Figure 5.4 Boyle's Law.

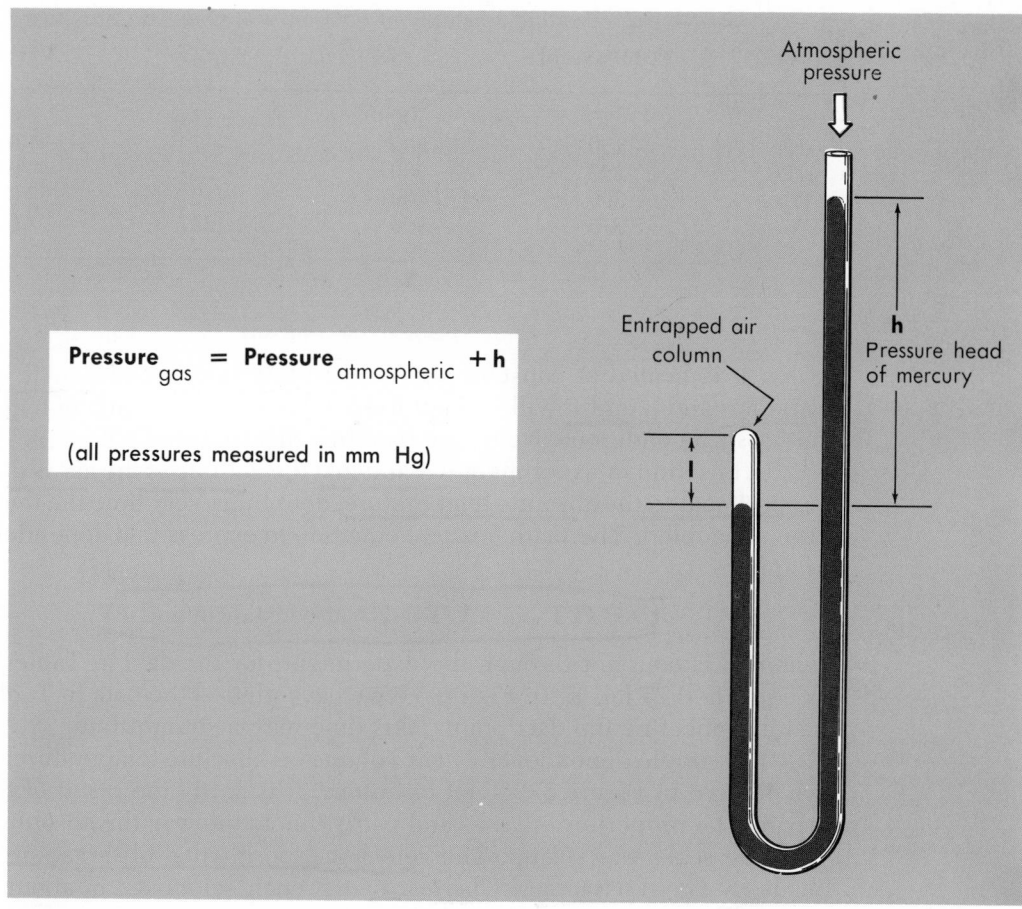

Figure 5.5 Boyle's experiment.

varied. Under such conditions the Ideal Gas Law becomes identical to Boyle's Law, as indeed it should.

Robert Boyle was one of the first experimental scientists. In addition to his discovery of the law which bears his name, he made many contributions to scientific thought. One of the most important of these was a book entitled *The Sceptical Chemist* in which he challenged, on the basis of his experiments, the then prevailing notion that salt, sulfur and mercury were the true principles of nature. In this regard he proposed that matter was ultimately composed of particles of various sorts which could arrange themselves into groups, and that groups of one kind constituted a chemical substance. Boyle thus used concepts of atomic and molecular theory similar to those we have today, and by some has been called one of the fathers of modern chemistry. Since his ideas on chemistry had of necessity to be extremely primitive and were not in any sense quantitative, we would prefer to consider the first modern chemists to be men like Dalton, Lavoisier and Priestley, who actually performed experiments to establish fundamental chemical relationships.

CHARLES AND GAY-LUSSAC'S LAW

We can readily extend the approach just described to another pair of variables. This time let us look at how the volume of a gas sample changes

TABLE 5.2   Heating a Sample of Hydrogen at One Atm Pressure

STATE	VOLUME, ml	T, °K	t, °C	V/T, ml/°K
I	75	100	−173	0.75
II	150	200	−73	0.75
III	225	300	27	0.75
IV	300	400	127	0.75
V	375	500	227	0.75

when it is heated at constant pressure. In this case V and T will be the variables and n and P will be kept fixed. If we study a sample of hydrogen under such conditions, we obtain data like that in Table 5.2.

In this kind of experiment we find that the volume of the gas is directly proportional to the absolute temperature, doubling every time the temperature is doubled. The mathematical equation to express this dependence is simply

$$V = C\,T \quad \text{or} \quad V/T = C \quad \text{at constant n and P} \tag{5.4}$$

where C is constant through the experiment; for the data in Table 5.2, C is equal to 0.75 ml/°K. In Figure 5.6 we have plotted the data in Table 5.2, again connecting the data points, this time with a straight line.

The simple dependence of gas volume on absolute temperature which we observe in Figure 5.6 is not fortuitous. It is partly the result of the experimental properties of gases and partly due to the way the absolute temperature scale was set up. The relevant experimental observations were made by Gay-Lussac and Charles, two French scientists, in about 1780.

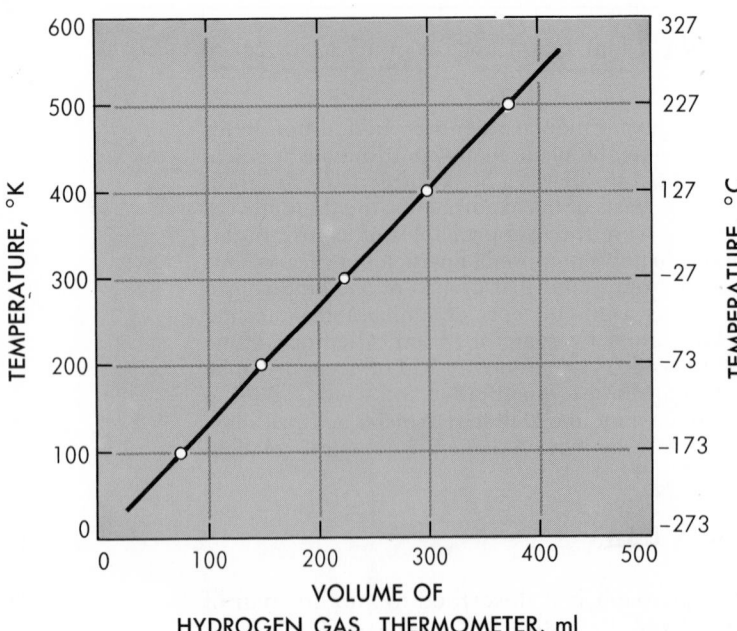

Figure 5.6 The hydrogen gas thermometer.

They studied the expansion of gases with temperature, and found that the rate of expansion with Celsius temperature was constant and had the same value for all gases. In modern terms we would say that any two gases, when each was heated at constant pressure from one given temperature to another, would have the same percentage increase in volume. If the two temperatures involved are 0°C and 100°C, any gas will increase in volume by about 37 per cent. (Actually, Charles and Gay-Lussac were not so interested in the fundamental laws of gases as they were in the use of hot, or light, gases in lighter-than-air balloons. Charles was on board the second balloon ever to lift man off the face of the earth. This happened on December 1st, 1783, with a balloon filled with hydrogen gas; in the first ascent, which occurred only 10 days earlier, a hot-air balloon was used.)

Much of our science had its origin in practical research

It was nearly a century after Charles and Gay-Lussac did their experiments on gas expansion that scientists first realized that gas volume could be used to set up the absolute temperature scale shown in Figure 5.6. Essentially what is done is to define absolute temperature in terms of the volume of a gas heated at constant pressure; that is, Equation 5.4 is *defined* to be true. In order to retain a simple relationship between the absolute scale and the Celsius scale of temperature, the size of the degree is kept the same, so that 100° separate the freezing and boiling points of water on both scales. The actual numerical relation between volume and temperatures on the two scales is determined by the ratio of the volume of a gas at 100°C to that of the same sample at 0°C and the same, low, pressure; that ratio is found in accurate experiments to be 1.3661. Rewriting Equation 5.4 to obtain volume ratios, we have

Both the Celsius and absolute temperature scales are ultimately based on the gas volume thermometer

$$\frac{V_{100}}{V_0} = 1.3661 = \frac{T_{100}}{T_0} = \frac{T_0 + 100°}{T_0}$$

Solving this equation, we see that the T_0, the temperature on the absolute scale corresponding to 0°C, must be:

$$\frac{100}{0.3661} = 273.16°$$

In general,

$$T = t + 273.16°$$

Equation 5.4 is, as you may have suspected, called Charles and Gay-Lussac's Law, although perhaps it should have been named Kelvin's Law, since he had far more than they to do with its derivation. It is, indeed, a rather remarkable relation, and far less simple than it looks. One thing that might bother you, if you worry about matters of this sort, is the extrapolation of the volume of the gas to zero in Figure 5.6. Obviously this cannot go all the way, since at some finite temperature every gas will condense to a liquid. This is not a serious objection, however, to the concept of absolute temperature, since the fixing of the absolute zero is established from data obtained at 0 and 100°C, where the intermolecular attractions responsible for condensation are of negligible importance. It is true, though, that a very tight theoretical argument, borne out by experiment, can be made for the impossibility of attaining the absolute zero of temperature. The best we have been able to do so far is to get to about 0.001°K.

AVOGADRO'S LAW

Thus far we have dealt with the behavior of samples containing a fixed amount of gas. However, one of the most useful features of the Ideal Gas Law is that it allows us to relate the amount of gas, in moles, to its pressure, volume and temperature. There are two relations involving the amount of gas and its dependence on other properties that we should mention.

The first of these is concerned with the relation between the volume of a gas and the amount, as measured under conditions of constant temperature and pressure. As you might expect, we find that the volume is directly proportional to amount, doubling each time the amount is doubled. The equation relating volume and amount is

$$V = nD \quad \text{at constant P and T} \tag{5.5}$$

Here D is the proportionality constant between the volume and the amount in moles of different samples of the same gas, all measured at the same pressure and temperature.

The other relation concerning amount of gas was first stated in 1813 by Amadeo Avogadro, an Italian physicist. In modern terminology we would state his Law as follows:

Equal volumes of different gases under the same conditions of temperature and pressure contain the same number of moles.

The mathematical equivalent of the above statement is:

$$n_x = n_y \quad \text{for two gases x and y having the same values of V, T and P} \tag{5.6}$$

Although Avogadro's Law is now considered to be a fundamental tenet of chemistry, it was denounced on faulty grounds by John Dalton when it was proposed. Following its vindication by Cannizzaro in about 1860, Avogadro's Law furnished for many years the most reliable basis for the determination of accurate molecular weights.

The Ideal Gas Law incorporates in one equation the relationships in Boyle's Law, Charles' Law and Avogadro's Law. Once these laws, plus Equation 5.5, are established, it is relatively easy to formulate the Ideal Gas Law. We simply write the volume as it occurs in each law:

C and D depend on P, but B doesn't. Therefore, V must vary as 1/P

$$V = \frac{1}{P} B = TC = nD$$

Boyle's Law   Charles' Law   Equation 5.5
(n, T fixed)   (n, P fixed)   (P, T fixed)

The dependence of volume on *pressure, temperature* and *amount* must be as given in the three equations, since all other variables are constant in these expressions. We therefore can combine the three equations, obtaining

$$V = \frac{nRT}{P} \tag{5.1}$$

which is the Ideal Gas Law. Since, by Avogadro's Law, the amount of gas, in moles, is determined only by the pressure, volume and temperature, the value of R, the gas constant, must be the same for all gases.

EVALUATION OF R, THE GAS CONSTANT

To establish the magnitude of R one merely needs to determine under known conditions the volume of a mole of gas of known molecular weight. For example, oxygen gas at 0°C and 1 atm pressure is known to have a density of 1.429 g/lit. Since oxygen is known, or can be defined, to weigh 32.00 grams per mole, its molar volume under these conditions is

For many years the mole was defined this way

$$V_{molar} = \frac{32.00 \text{ g } O_2}{1 \text{ mole}} \times \frac{1.00 \text{ lit}}{1.429 \text{ g } O_2} = 22.4 \text{ liters}$$

R follows directly by substitution into the Ideal Gas Law:

$$PV = nRT \quad \text{or} \quad R = \frac{PV}{nT}$$

where $P = 1.00$ atm, $V = 22.4$ lit, $n = 1.00$ moles, $T = 0° + 273° = 273°K$

$$R = \frac{1.00 \text{ atm} \times 22.4 \text{ lit}}{1.00 \text{ mole} \times 273°K} = 0.0821 \text{ lit-atm/mole°K}$$

The value of R as obtained under the most precise conditions, using oxygen at low pressure, is 0.082054 lit-atm/mole°K. Since R involves the units of atm, lit, moles and °K, those are the units that must be used for pressure, volume, amount and temperature in any problems where this value of R is employed.

5.4 USING THE IDEAL GAS LAW

The Ideal Gas Law can be used to solve a wide variety of problems dealing with the experimental behavior of gases. It can, for example, be used to determine the effect of a change in conditions upon a variable such as volume (Example 5.1) or temperature (Example 5.2).

Example 5.1 In a McLeod gauge a gas at a low pressure, to be determined, is compressed at room temperature to a much smaller volume, where the pressure is increased sufficiently to be measured directly on a manometer. In a certain experiment a sample of helium in a vacuum system was compressed at 25°C from a volume of 200 ml to a volume of 0.240 ml, where its pressure was found to be 3.00 cm Hg. What was the original pressure of the helium?

Solution. For the gas in both states, $PV = nRT$

Initially: $P_1 = ?$, $V_1 = 200$ ml, $n_1 = n$, $T_1 = T$

$$P_1V_1 = nRT$$

Finally: $P_2 = 3.00$ cm Hg, $V_2 = 0.240$ ml, $n_2 = n$, $T_2 = T$

$$P_2V_2 = nRT$$

Since both the temperature and amount of gas are held constant in this problem, the right sides of the two final equations are equal, and so

$$P_1V_1 = P_2V_2 \quad P_1 = \frac{P_2V_2}{V_1}$$

$$P_1 = \frac{3.00 \text{ cm Hg} \times 0.240 \text{ ml}}{200 \text{ ml}} = 3.60 \times 10^{-3} \text{ cm Hg}$$

This problem is typical of many encountered with gases. The state of the gas
is changed, with one or more of its variables remaining fixed; in this case, at
constant n and T, we have essentially a Boyle's Law problem. By writing the
Gas Law for the two states, one can usually quickly recognize which terms in
the two equations are equal and use that information and the given data to
find the unknown quantity. Note that in the problem here the units of volume
(ml) must be consistent if they are to cancel properly. The pressure is obtained in cm Hg, but can be readily converted to mm Hg or to atm if necessary:

$$3.60 \times 10^{-3} \text{ cm Hg} \times \frac{10 \text{ mm}}{1 \text{ cm}} = 3.60 \times 10^{-2} \text{ mm Hg}$$

$$3.60 \times 10^{-3} \text{ cm Hg} \times \frac{1 \text{ atm}}{76.0 \text{ cm Hg}} = 4.74 \times 10^{-5} \text{ atm}$$

$P_1 V_1 = P_2 V_2$ if amt of gas + T are constants
$P_1 = \frac{P_2 V_2}{V_1}$

Example 5.2 A hydrogen gas volume thermometer has a volume of 100.0 ml
when immersed in an ice-water bath at 0°C. When immersed in boiling liquid
chlorine, the volume of the hydrogen at the same pressure is 87.2 ml. Find the
temperature of the boiling point of chlorine in °K and in °C.

Solution. For the hydrogen in the two states, $PV = nRT$

Initially: $P_1 = P$, $V_1 = 100.0$ ml, $n_1 = n$, $T_1 = 0° + 273° = 273°K$

$$PV_1 = nRT_1$$

Finally: $P_2 = P$ $V_2 = 87.2$ ml $n_2 = n$ $T_2 = ?$

$$PV_2 = nRT_2$$

In this problem, P, n and R are the same in the two states; we collect those
variables on the left side of the two equations, obtaining

$$\frac{P}{nR} = \frac{T_1}{V_1} = \frac{T_2}{V_2}$$

Therefore

$$T_2 = \frac{V_2T_1}{V_1}$$

if n, P, R are constant.

$$T_2 = \frac{87.2 \text{ ml} \times 273°K}{100.0 \text{ ml}} = 238°K \qquad t_2 = T_2 - 273° = -35°C$$

In the previous examples the value of the gas constant R was not
needed, since R cancelled from the calculations. Cases in which explicit
evaluation or use of the amount of gas is involved require the full use of
the Ideal Gas Law, including the value of R. The following problems are
illustrative and show the large amount of information that the law allows
one to obtain about gases.

Example 5.3 2.50 g of XeF_4 gas are introduced into an evacuated 3.00-lit container at 80°C. Find the pressure in atmospheres in the container.

Solution. In this problem only one state is involved, and for it, $PV = nRT$.

$$P = ?; \quad V = 3.00 \text{ lit}; \quad n = 2.50 \text{ g } XeF_4 \times \frac{1 \text{ mole}}{207.3 \text{ g } XeF_4} = 0.0121 \text{ moles}$$

$$R = 0.0821 \text{ lit atm/mole°K}; \quad T = 273 + 80 = 353°K$$

Substituting: $P = \dfrac{nRT}{V} = \dfrac{0.0121 \text{ mole} \times 0.0821 \dfrac{\text{lit atm}}{\text{mole°K}} \times 353°K}{3.00 \text{ lit}} = 0.117 \text{ atm}$

Here all the elements in the Ideal Gas Law enter the calculation directly; if we use 0.0821 lit atm/mole°K for R, the units of all the terms are of necessity those that appear in R, and any quantities which do not have those units must be converted before substituting in the Gas Law.

Example 5.4 A lighter-than-air balloon is designed to rise to a height of 6 miles, at which point it will be fully inflated. At that altitude the atmospheric pressure is 210 mm Hg and the temperature is −40°C. If the full volume of the balloon is 100,000 liters, how many pounds of helium will be needed to inflate the balloon?

Solution. To calculate the number of pounds of helium required, we shall first use the Ideal Gas Law to obtain the number of moles, n, and then convert from moles to grams and finally to pounds.

About how much could this balloon lift?

Since balloons have flexible walls, the pressure in the balloon is substantially equal to the outside atmospheric pressure. Hence, for the helium at a height of 6 miles,

$$P = 210 \text{ mm Hg} \times \frac{1 \text{ atm}}{760 \text{ mm Hg}} = 0.276 \text{ atm}$$

$$V = 1.000 \times 10^5 \text{ lit}; \quad n = ?; \quad R = 0.0821 \text{ lit atm/mole°K}; \quad T = 273 - 40 = 233°K$$

$$n = \frac{PV}{RT} = \frac{0.276 \text{ atm} \times 1.000 \times 10^5 \text{ lit}}{\dfrac{0.0821 \text{ lit atm}}{\text{mole°K}} \times 233°K} = 1440 \text{ moles}$$

Since 1 mole of helium weighs 4.00 grams, the mass of the helium is

$$1440 \text{ moles} \times \frac{4.00 \text{ g}}{1 \text{ mole}} = 5760 \text{ g} \times \frac{1 \text{ lb}}{453.6 \text{ g}} = 12.7 \text{ lb}$$

There are some problems in which the use of the Ideal Gas Law is not quite so direct as in the previous examples. Two important cases involve the determination of the density of a gas under given conditions and the evaluation of the molecular weight from experimental data. These problems are most readily treated by rewriting the Ideal Gas Law so that it contains explicitly the number of grams of gas, g. Recalling that *must be rewritten*

$$\text{number of moles} = \frac{\text{number of grams}}{\text{gram molecular weight}} \qquad n = \frac{g}{M}$$

we can write

$$PV = nRT = \frac{gRT}{M} \tag{5.7}$$

The use of this equation is illustrated in Examples 5.5 and 5.6.

Example 5.5 Uranium hexafluoride, UF_6, is perhaps the heaviest of all gases. What is the density in g/lit of UF_6 at 100°C and 1 atm?

Solution. The problem essentially is to find the number of grams per liter of UF_6 which will be present in a container at 100°C and 1 atm. We solve Equation 5.7 for the density, d, of the gas

$$\text{density} = \frac{\text{mass}}{\text{volume}} \qquad \boxed{d = \frac{g}{V} = \frac{MP}{RT}}$$

$$M_{UF_6} = 238 + 6(19) = 352 \text{ g/mole; } P = 1.00 \text{ atm}$$

$$R = 0.0821 \text{ lit-atm/mole°K; } T = 273° + 100° = 373°K$$

$$d = \frac{352 \text{ g/mole} \times 1.00 \text{ atm}}{0.0821 \text{ lit-atm/mole°K} \times 373°K} = 11.5 \text{ g/lit}$$

Example 5.6 A sample of chloroform weighing 0.495 g is collected as a vapor (gas) in a flask having a volume of 127 ml. At 98°C the pressure of the vapor in the flask is 754 mm Hg. Calculate the molecular weight of chloroform.

Solution. Rewriting Equation 5.7, this time to obtain an explicit expression for M,

$$M = \frac{gRT}{PV}$$

This is probably the easiest way to find the molecular weight of a volatile substance

We need merely to express the variables in the equation in the proper units and solve for the gram molecular weight by substitution.

$$g = 0.495 \text{ grams; } R = 0.0821 \text{ lit-atm/mole°K; } T = 98° + 273° = 371°K$$

$$P = 754 \text{ mm Hg} \times \frac{1 \text{ atm}}{760 \text{ mm Hg}} = 0.992 \text{ atm; } V = 127 \text{ ml} \times \frac{1 \text{ lit}}{1000 \text{ ml}} = 0.127 \text{ lit}$$

$$M = \frac{0.495 \text{ g} \times 0.0821 \text{ lit-atm/mole°K} \times 371°K}{0.992 \text{ atm} \times 0.127 \text{ lit}} = 119 \text{ g/mole}$$

The calculated molecular weight of chloroform is, therefore, 119.

The data in Example 5.6 are typical of those obtained in one of the simplest experimental methods for the determination of molecular weights. In the experiment a few milliliters of a volatile liquid are placed in a flask fitted with a stopper in which there is a fine orifice (Fig. 5.7). The flask is then heated in a water bath to a temperature somewhat above the boiling point of the liquid. The liquid evaporates, and its vapor replaces the air in the flask. After all the liquid has evaporated and the flask is filled with vapor, the flask is removed from the bath and allowed to cool. The vapor condenses and air re-enters the flask. The mass of the vapor is taken to be the difference between the mass of the flask containing the condensed vapor and its mass when it contains just air. The method is an approximate one for several reasons, but when used properly it gives results accurate to within a few per cent.

Another application of the gas laws arises in connection with our interpretation of chemical reactions. We have seen that, given any chemical

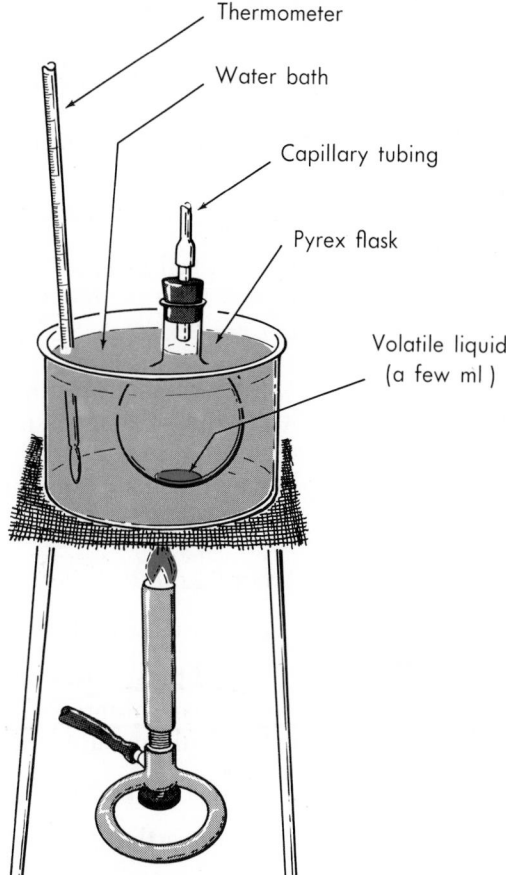

Figure 5.7 Determination of molecular weight by vapor density.

equation, it is possible to calculate the relative masses and numbers of moles of reactants and products. Where gases are present it is possible to extend the interpretation of chemical equations to include the volumes of the gases that would be involved under given conditions. The procedure is shown in the following example.

Example 5.7 In a Saab 99E engine the maximum cylinder volume is about 500 ml. If air enters the cylinder at 60°C and one atm pressure, how many grams of gasoline, C_8H_{18}, should the fuel injection system send into the cylinder if the gasoline is to be able to burn completely in that air when the spark plug fires at the end of the compression stroke? Assume air is 20 mole per cent O_2. The reaction is: $C_8H_{18}(g) + \frac{25}{2} O_2(g) \rightarrow 8 CO_2(g) + 9 H_2O(g)$

Solution. From the equation for the reaction we see that

$$1 \text{ mole } C_8H_{18} \simeq \tfrac{25}{2} \text{ moles } O_2$$

so that, if we find the number of moles of O_2 in the air in the cylinder, we can calculate the number of moles, and grams, of gasoline needed.

By the Gas Law: $n_{O_2} = \dfrac{P_{O_2}V}{RT}$

$$P_{O_2} = 0.20 P_{tot} = 0.20 \text{ atm;} \quad V = 500 \text{ ml} = 0.500 \text{ lit;} \quad T = 273° + 60° = 333°K$$

$$n_{O_2} = \frac{0.20 \text{ atm} \times 0.500 \text{ lit}}{0.0821 \text{ lit atm/mole}°K \times 333°K} = 0.0037 \text{ moles O}_2$$

$$0.0037 \text{ moles O}_2 \times \frac{1 \text{ mole C}_8H_{18}}{\frac{25}{2} \text{ moles O}_2} = 0.00029 \text{ moles C}_8H_{18}$$

$$\text{MW C}_8H_{18} = 114; \quad 0.00029 \text{ moles C}_8H_{18} \times \frac{114 \text{ g}}{1 \text{ mole}} = 0.033 \text{ g C}_8H_{18}$$

(In actual practice, to ensure an excess of air, a somewhat smaller amount of gasoline would be injected.)

A rather interesting interpretation of chemical reactions can be made when the substances involved are all gases whose volumes are measured under the same conditions of temperature and pressure. Consider the following chemical reaction:

$$4 \text{ NH}_3(g) + 5 \text{ O}_2(g) \rightarrow 4 \text{ NO}(g) + 6 \text{ H}_2O(g)$$

According to the usual interpretation, we would say

$$4 \text{ moles NH}_3 + 5 \text{ moles O}_2 \rightarrow 4 \text{ moles NO} + 6 \text{ moles H}_2O$$

If all these gases are measured at the same temperature and pressure, their molar volumes will all be equal to some fixed value, V_m liters. Under such conditions it would be true that

$$4 \text{ } V_m \text{ lit NH}_3 + 5 \text{ } V_m \text{ lit O}_2 \rightarrow 4 \text{ } V_m \text{ lit NO} + 6 \text{ } V_m \text{ lit H}_2O$$

or, dividing through by V_m, simply

$$4 \text{ lit NH}_3 + 5 \text{ lit O}_2 \rightarrow 4 \text{ lit NO} + 6 \text{ lit H}_2O$$

Can you recall a simple experiment that implies this kind of relationship?

Volumes of gases, then, when measured under the same conditions, have the same simple numerical relationships that exist between moles of substances in chemical reactions. The remarkable fact is that, whereas the

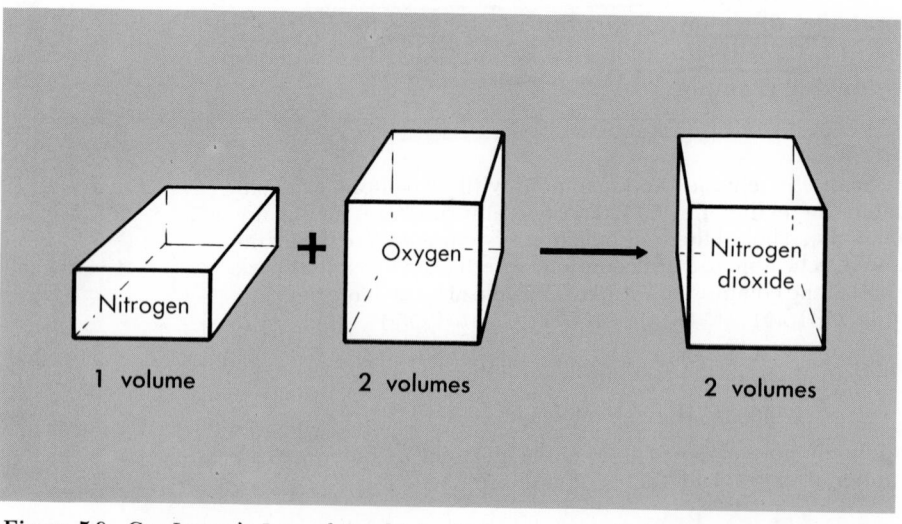

Figure 5.8 Gay-Lussac's Law of Combining Volumes as applied to the reaction: $N_2(g) + 2 O_2(g) \rightarrow 2 NO_2(g)$.

relation between moles is deduced from theory, that between volumes can be found experimentally. Indeed, in 1808 Gay-Lussac discovered the relationship and stated his Law of Combining Volumes: **In any chemical reaction involving gaseous substances the volumes of the various gases reacting or produced are in the ratios of small whole numbers.** (The gases are measured at the same temperature and pressure.) It was this Law that Dalton felt contradicted his atomic theory. Since Avogadro had used Gay-Lussac's Law to deduce his Law, Dalton didn't like Avogadro's Law either.

Can you guess Avogadro's line of reasoning?

5.5 MIXTURES OF GASES: DALTON'S LAW OF PARTIAL PRESSURES

So far we have considered the behavior of gaseous systems in which only one component is present. If several substances are present in a gaseous solution, we can still use the gas laws, but must take proper account of the presence of the different substances. Suppose we have a gaseous mixture of A, B and C, confined in a container of volume V. The Ideal Gas Law will apply to the mixture, provided we let n equal the total number of moles present. That is,

$$PV = nRT \quad \text{or} \quad P = \frac{nRT}{V}$$

in which P is the total pressure in the container and n equals $n_A + n_B + n_C$.

The three substances, A, B and C, each contribute to the total pressure in the system. Their individual contributions, P_A, P_B and P_C, may be obtained by the Gas Law. We can say that

$$P_A = \frac{n_A RT}{V}; \quad P_B = \frac{n_B RT}{V}; \quad P_C = \frac{n_C RT}{V}$$

P_A, P_B and P_C are called the partial pressures of A, B and C in the container. The partial pressure of a gas is the pressure the gas would exert if it alone were present in the container at the same temperature as the mixture. It is clear that since

$$P = \frac{nRT}{V} = (n_A + n_B + n_C)\frac{RT}{V} = \frac{n_A RT}{V} + \frac{n_B RT}{V} + \frac{n_C RT}{V}$$

it follows that

$$P = P_A + P_B + P_C \tag{5.8}$$

This equation is a mathematical statement of Dalton's Law of Partial Pressures, first proposed by John Dalton in 1807. In words the law is: *The total pressure in a container is equal to the sum of the partial pressures of the component gases.*

Dalton's Law makes it possible to easily handle problems in which gaseous solutions are involved. The following example illustrates a typical application.

Example 5.8 Air from the prairies of North Dakota in winter contains essentially only nitrogen, oxygen and argon. A sample of air collected at Bismarck at $-22°C$ and 742 mm Hg was analyzed and found to contain 78 mol per cent N_2, 21 mol per cent O_2 and 1 mol per cent argon. Find the partial pressures of each of these gases in the sample when it was collected.

Solution. The analysis tells us that 78 per cent of the molecules, or moles, of gas in the air are nitrogen, 21 per cent are oxygen and 1 per cent are argon. For the whole sample of air,

$$P = \frac{nRT}{V}$$

where
$$n = n_{N_2} + n_{O_2} + n_{Ar}$$

For the nitrogen in the sample,

$$P_{N_2} = \frac{n_{N_2}RT}{V}$$

Dividing P_{N_2} by P, we obtain

$$\frac{P_{N_2}}{P} = \frac{n_{N_2}}{n} = 0.78$$

Therefore, $P_{N_2} = 0.78P = 0.78 \times 742 \text{ mm Hg} = 5.8 \times 10^2 \text{ mm Hg}$

By similar reasoning,

$$P_{O_2} = 0.21 \times 742 \text{ mm Hg} = 1.6 \times 10^2 \text{ mm Hg}$$

$$P_{Ar} = 0.01 \times 742 \text{ mm Hg} = 7 \text{ mm Hg}$$

Dalton's Law often finds practical use in experiments involving gases. In order to measure the amount of a gas produced in a chemical reaction we must collect the gas under known conditions. Probably the easiest way of doing this is to let the gas displace water in a system such as that shown

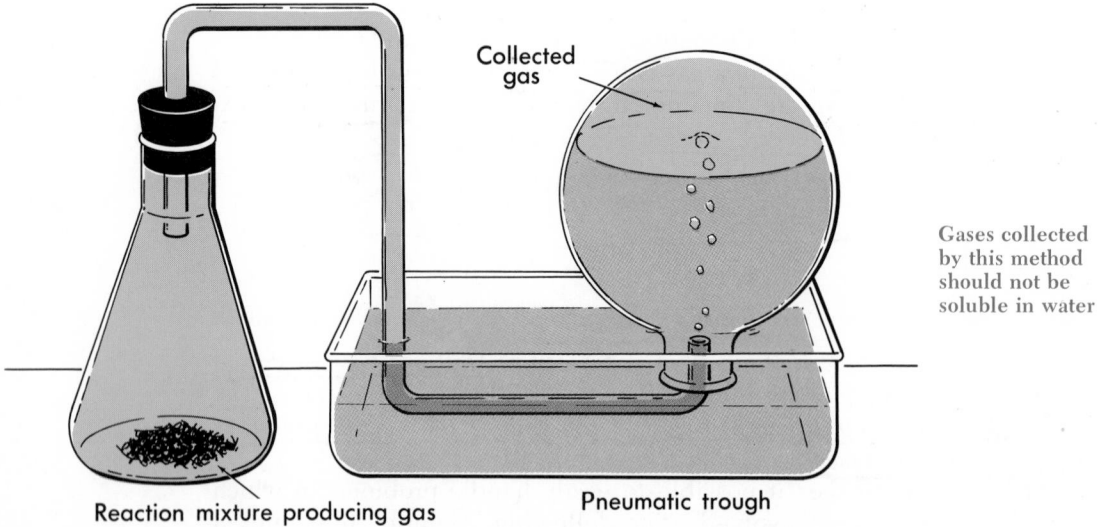

Collected gas

Gases collected by this method should not be soluble in water

Reaction mixture producing gas

Pneumatic trough

Figure 5.9 The collection of gases over water.

in Figure 5.9. In this way we can measure the volume of gas at atmospheric pressure and known temperature. If the gas were pure, we could immediately use the Gas Law to calculate the number of moles produced by the reaction. However, under the conditions of the experiment the gas collected contains water vapor in addition to the gas of interest. The true pressure of the gas produced is, therefore, by Dalton's Law, equal to the total pressure minus the partial pressure of the water vapor. It is found experimentally that the pressure of water vapor in the presence of liquid water is a constant at a given temperature; its value can be obtained from a table of vapor pressures (Appendix 1). The following example is illustrative.

Example 5.9 In a laboratory experiment, concentrated hydrochloric acid was reacted with aluminum. Hydrogen gas evolved and was collected over water at 25°C; it had a volume of 355 ml at a total pressure of 750 mm Hg. How many moles of hydrogen were collected? At 25°C the vapor pressure of water is known to be about 24 mm Hg.

Solution We must first obtain the partial pressure of the hydrogen, P_{H_2}. By Dalton's Law: $P = P_{H_2} + P_{H_2O}$

$$P_{H_2O} = 24 \text{ mm Hg, the vapor pressure of } H_2O \text{ at } 25°C$$

Therefore

$$P_{H_2} = P - P_{H_2O} = (750 - 24) \text{ mm Hg} = 726 \text{ mm Hg}$$

$$= 726 \text{ mm Hg} \times \frac{1 \text{ atm}}{760 \text{ mm Hg}} = 0.955 \text{ atm}$$

By the Gas Law:

$$P_{H_2} = \frac{n_{H_2}RT}{V}; \; n_{H_2} = \frac{P_{H_2}V}{RT} = \frac{0.955 \text{ atm} \times 0.355 \text{ lit}}{0.0821 \dfrac{\text{lit atm}}{\text{mole°K}} \times 298°K} = 0.0139 \text{ mole}$$

5.6 REAL GASES

In this chapter we have applied the various gas laws to all gases, tacitly assuming that the laws were obeyed exactly. Under ordinary conditions (and in nearly all problems involving gases in this text) the assumption is a very good one. Actually, however, all gases deviate to some extent from the ideal laws by amounts that depend on the gas, its temperature and its pressure. For gases in the vicinity of room temperature and 1 atm pressure the deviation is small, at most a few per cent. Gases like oxygen and hydrogen, which boil far below 25°C, have molar volumes that are within 0.1 per cent of the value calculated by the Ideal Gas Law. Sulfur dioxide and chlorine, which boil at −10° and −35°C, respectively, are, at 25°C and 1 atm, not so nearly ideal and have molar volumes that are 2.4 and 1.6 per cent lower than the ideal value.

We can illustrate the behavior of real gases graphically by plotting PV/RT for a mole of gas as a function of pressure. Figure 5.10 is such a graph for several gases at 0°C. For a mole of ideal gas PV/RT would be a

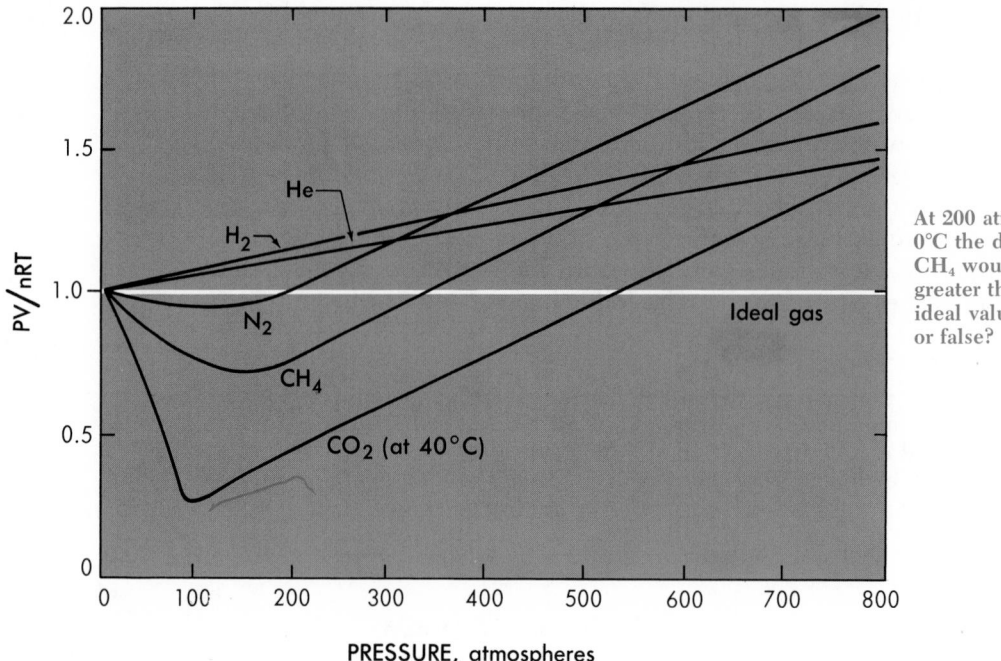

At 200 atm and 0°C the density of CH_4 would be greater than the ideal value. True or false?

Figure 5.10 The behavior of real gases at 0°C.

At low pressure all gases become essentially ideal

constant equal to 1.00 at all pressures. Actually, at relatively low pressures (less than 100 atm) PV/RT is less than 1 in the vicinity of 25°C for all gases except hydrogen and helium, and approaches 1 as the pressure approaches zero. The pressure of most gases at a given volume and temperature is, then, somewhat less than would be predicted by the Gas Law. This behavior is caused, we believe, by the small attractions that exist between molecules even in the gas phase. (See Chapter 8.) As the temperature of the gas is lowered, the energy of motion of the molecules is lowered, and the effect of the attractive forces becomes more important, until, at the boiling point of the substance, the forces are large enough to cause the molecules to condense to a liquid. These forces in all cases tend to reduce the value of PV/RT and, in view of the above line of reasoning, are important at temperatures near the boiling point of the substance. This qualitatively explains why SO_2 is less nearly ideal in its behavior than oxygen at 25°C.

The PV/RT behavior of gases at very high pressures (500 atm and above) is shown at the right of Figure 5.11. For all gases, as the pressure is increased, PV/RT ultimately begins to increase, finally becoming much greater than the ideal value. Such an effect is best understood by a consideration of the volumes of gas molecules. As we noted, in general the gas volume is much larger than the actual volume of the constituent gas molecules. However, the volume of the molecules is not zero (being roughly equal to the volume of the liquid made by condensing the molecules); as the pressure increases, the volume of the molecules becomes a larger and larger fraction of the gas volume. The volume in which the gas molecules can actually move about is thus decreased by the volume taken up by the

TABLE 5.3   van der Waals Constants for Some Typical Gases

SUBSTANCE	a in $\frac{lit^2\ atm}{mole^2}$	b in lit/mole	bp (°C) at 1 atm	V_{molar} (lit) of liquid
H_2	0.244	0.027	−253	0.022
CH_4	2.25	0.043	−161	0.039
H_2O	5.46	0.030	100	0.018
SO_2	6.71	0.056	−10	0.045
CCl_4	20.39	0.138	77	0.096

molecules; this causes the pressure of the gas to be somewhat larger than that predicted by the Gas Law.

It is possible to write equations of state for gases which take into account both intermolecular attractions and finite molecular volumes. Perhaps the best known of these relations is the van der Waals equation, which for one mole of gas takes the form

$$P = \frac{RT}{V-b} - \frac{a}{V^2}$$ (5.9)

where a and b are constants selected for each gas to give the best possible agreement between the equation and actual experimental behavior. The term a/V^2, which is associated with the attractive forces between molecules, serves to decrease P below the ideal gas value. The van der Waals b corrects for the effect of molecular volumes and, as you can see, serves to increase the observed pressure. In Table 5.3 we have listed values of van der Waals a and b for a few gases, along with their boiling points and molar volumes of the liquid. The van der Waals b is about equal to the molar volume of the liquid, as it should be if the effective volume of the gas is decreased by the volume of the molecules. The van der Waals a is not so readily accounted for, but its increase with increasing boiling point seems very reasonable, since high boiling point would imply strong intermolecular attractive forces. In spite of its relative simplicity the van der Waals equation gives good qualitative prediction of real gas behavior. It is used in treating problems in which the nonideality of the gas is important and a more exact analytical relation between P, V and T is required than that given by the Ideal Gas equation.

The van der Waals equation explains many properties of real gases including, surprisingly enough, their condensation to liquids at low temperatures

5.7   THE KINETIC THEORY OF GASES

The fact that the Ideal Gas Law can be used to summarize reasonably accurately the physical behavior of all gases, whatever their degree of molecular complexity, is a clear indication that the gaseous state of matter is a relatively simple one to attempt to treat from a theoretical point of view. There must be certain properties common to all gases which cause them to follow the same natural law and exhibit so many generally similar characteristics.

With the experimental work on gases by Charles, Gay-Lussac, Graham, and others, enough information on gas behavior became available to allow the development of a theory of gases. Between 1850 and 1880, Maxwell, Boltzmann, Clausius, and others, using the notion that the properties of gases are the result of molecular motions, developed the kinetic theory of gases. The theory was shown to be consistent with the known laws of gas behavior and implied much about gases that was then unknown. Since the time of its creators, who, incidentally, rank among the very best theoreticians the world has known, the theory has had to be modified to only a small extent to make it consistent with quantum theoretical principles. In its present form it is one of the most successful of scientific theories, ranking in stature with the Copernican theory for the motion of the planets and the atomic theory of the nature of matter.

POSTULATES OF THE KINETIC THEORY OF GASES

1. **Gases consist of molecules in continuous, random motion.** The molecules undergo collisions with one another and with the container walls. The pressure of a gas arises from the forces associated with wall collisions.

2. **Molecular collisions are elastic.** During collisions there are no frictional losses which result in loss of energy of motion. The temperature of a gas insulated from its surroundings does not change.

The energy of a rocket is mainly translational. That of a spinning top is rotational

3. **The average energy of translational motion of a gas molecule is proportional only to the absolute temperature.** The energy associated with the motion of a molecule from one place to another depends on the temperature, but not on the pressure or the nature of the molecule. Mathematically, we can state this postulate in the following way:

$$\epsilon = (\tfrac{1}{2}) mu^2 = cT \tag{5.10}$$

where ϵ is the average molecular energy of translation, m is the mass of the molecule, u is its average speed, and c is a proportionality constant which has the same value for all gases.

In addition to these postulates, it is often assumed that the volumes of the molecules are negligible as compared to container volume and that molecules do not exert forces on each other except by collisions.

The postulates of the kinetic theory are easily stated. Their implications, however, are by no means obvious, and mathematical derivations based on the postulates tend to be very complex. The difficulty arises essentially from the fact that gas molecules move about in a completely random manner, with velocities that are constantly changing both in magnitude and direction because of molecule-molecule and molecule-wall collisions. To take account of this kind of motion in anything resembling a rigorous way requires, as you can well imagine, mathematics of a very high level of sophistication. In this text we shall not attempt even a simplified mathematical development using the kinetic theory, but shall proceed directly to some of the results which the theory has produced.

PRESSURE OF A GAS

One of the important relations which can be derived from the kinetic theory and the laws of mechanics relates the pressure of a gas to its volume and the number, mass and speed of its molecules. The equation which one obtains from the theory is really quite simple:

$$PV = (\tfrac{1}{3}) Mnu^2 = nRT \qquad (5.11)$$

where M is the gram molecular weight of the gas, n is the number of moles in the volume V, and u is the average molecular speed. The first and last terms comprise the Ideal Gas Law and confirm that the Law indeed follows from the postulates of the kinetic theory. The middle term relates the speed of the molecules, first, to the pressure they exert in a container and, second, to the temperature of the gas. Much of the rest of this chapter will be devoted to exploring some of the many implications of Equation 5.11.

GRAHAM'S LAW

Characteristic of a successful theory is the fact that it can predict the experimental properties of the system to which it is applied. At the heart of the kinetic theory is the idea that gas properties arise from molecular motions. In testing the theory it is important to perform experiments in which the speeds of molecules are directly involved. Probably the simplest of these involves the phenomena of diffusion and effusion.

If two different gases in separate containers at the same temperature and pressure are separated by stopcocks from a third gas confined at the same temperature and pressure, they will, when the stopcocks are opened, tend to diffuse into the third gas (Fig. 5.11). The rate of diffusion of each

At a given temperature light molecules move faster than heavy ones

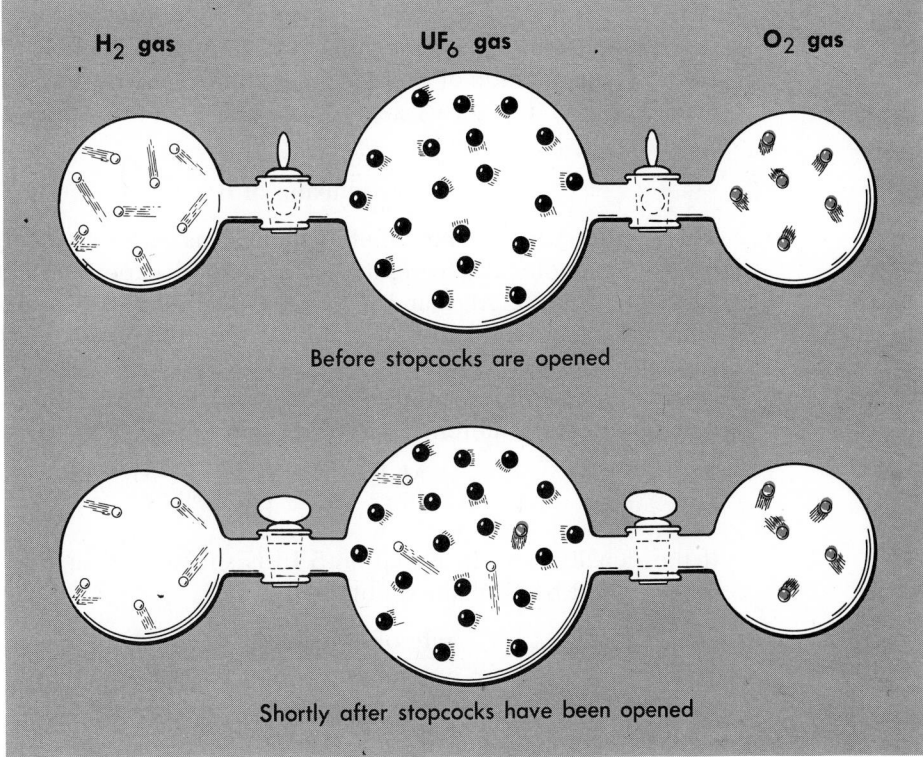

Before stopcocks are opened

Shortly after stopcocks have been opened

Figure 5.11 Diffusion of gases.

gas will be proportional to the average speed of its molecules. The less massive molecules will, according to Equation 5.11, tend to move more rapidly than the heavier ones and, hence, will diffuse more quickly. Hydrogen gas diffuses more quickly than does oxygen, as would be expected in light of the above line of reasoning. Uranium hexafluoride, UF_6, with its massive molecules, would diffuse much more slowly than would hydrogen or oxygen.

We can make a quantitative test of the kinetic theory of gases if we measure the rate at which a gas flows into a vacuum through a small opening in its container. This phenomenon is called **effusion,** and here too we would predict that the rate of effusion of a gas would be proportional to the average speed of its molecules. In this case, however, we could measure the rate very simply by noting how rapidly the pressure in the container drops and relating this to the number of molecules leaving the container in a given time interval. If the rates of effusion of two gases, A and B, were measured in the same container under the same initial conditions, we would predict that

A toy balloon gradually deflates because the gas molecules effuse slowly through the walls

$$\frac{\text{rate of effusion of A}}{\text{rate of effusion of B}} = \frac{\text{average speed of molecule A}}{\text{average speed of molecule B}} = \frac{u_A}{u_B} \quad (5.12)$$

By Equation 5.11, if gases A and B are studied at the same temperature,

$$\tfrac{1}{3} M_A u_A^2 = RT = \tfrac{1}{3} M_B u_B^2$$

or

$$\frac{u_A^2}{u_B^2} = \frac{M_B}{M_A} \quad (5.13)$$

in which M_B and M_A are the molecular weights of B and A. If Equation 5.13 is solved for the ratio of average molecular speeds and this is substituted in Equation 5.12, it is easily seen that

$$\frac{\text{rate of effusion of A}}{\text{rate of effusion of B}} = \left(\frac{M_B}{M_A}\right)^{1/2} \quad (5.14)$$

This relation, that the rate of effusion of a gas varies inversely as the square root of its molecular weight, was observed experimentally by Thomas Graham in 1828. At that time molecular weights were not thoroughly understood, and Graham stated the relation in terms of the densities of the gases he studied. As we saw in Example 5.5, the density of a gas and its molecular weight are related by the equation $d = MP/RT$. Applying this equation to two different gases at the same T and P:

$$d_B = \frac{M_B P}{RT}; \quad d_A = \frac{M_A P}{RT}; \quad \text{so} \quad \frac{M_B}{M_A} = \frac{d_B}{d_A} \quad (5.15)$$

If the density ratio in Equation 5.15 is substituted in Equation 5.14, the law reported by Graham is obtained:

$$\frac{\text{rate of effusion of A}}{\text{rate of effusion of B}} = \left(\frac{d_B}{d_A}\right)^{1/2} \quad (5.16)$$

when the densities of gases A and B are measured under the same conditions of temperature and pressure.

Graham's Law gives us an alternate method for measuring the molecular weights of gases, and one of its main uses is in this area. One needs merely to compare the rate of effusion of the unknown gas or vapor to that of a known reference gas. The procedure is straightforward and has found some practical application in industry in cases in which the purity of a gas is clearly reflected by its molecular weight. The following example is illustrative.

Example 5.10   In an effusion experiment it required 45 seconds for a certain number of moles of an unknown gas to pass through a small orifice into a vacuum. Under the same conditions it required 18 seconds for the same number of moles of oxygen to effuse. Find the molecular weight of the unknown gas.

Solution.   The rates of effusion are *inversely* proportional to the times needed for the flow. Hence, by Equation 5.14

$$\frac{\text{rate}_x}{\text{rate}_{O_2}} = \frac{t_{O_2}}{t_x} = \frac{18}{45} = \left(\frac{32}{M_x}\right)^{1/2}$$

Squaring both sides, $\dfrac{32}{M_x} = \left(\dfrac{18}{45}\right)^2 = 0.16$; $M_x = \dfrac{32}{0.16} = 200$

During World War II Graham's Law had a rather unexpected application in connection with a very complicated chemical problem. It had been found that the isotope of uranium having a mass of 235, ^{235}U, has a nucleus unstable to collisions with neutrons. Such collisions result in a splitting of the uranium nucleus into lighter fragments (fission) and the liberation of large amounts of energy in the form of heat and γ-rays (see Chapter 22). It became necessary to separate ^{235}U from the much more plentiful (but not fissionable) isotope, ^{238}U. Because of the great chemical similarity of the isotopes of an element, the chemical resolution of uranium into its isotopes was not feasible, and some physical method was sought. Since the rate of diffusion of a gas varies with its molecular weight, the composition of a gas mixture coming through an orifice will not be quite the same as that in the original sample and, hence, the resolution of a gas mixture by successive diffusions is possible, at least in principle. Preliminary diffusion experiments with uranium hexafluoride, UF_6, a volatile uranium compound, indicated that $^{235}UF_6$ could indeed be separated from $^{238}UF_6$ by diffusion, and an enormous plant was built for the purpose in Oak Ridge, Tennessee. In the process, UF_6 diffuses many thousands of times through porous barriers, with the lighter fractions moving on to the next stage and the heavier fractions being recycled through earlier stages.

If UF_6 were not volatile we might never have developed nuclear energy

MOLECULAR SPEEDS AND ENERGIES

Of greater theoretical interest than the relative molecular speeds treated by Graham's Law are the actual speeds of molecules. Referring back again to Equation 5.11, it can be seen that a simple manipulation allows one to obtain an explicit expression for the average molecular speed u:

$$u^2 = \frac{3RT}{M}; \quad u = \left(\frac{3RT}{M}\right)^{1/2} \qquad (5.17)$$

where R is the gas constant and M is the gram molecular weight. It is clear from Equation 5.17 that molecular speeds increase with increasing temperature, decrease with increasing molecular weight and, rather surprisingly, are not directly influenced by gas pressure.

If we wish to actually calculate an average molecular speed, we can do so by substitution into Equation 5.17. The only possible difficulty is in selecting a proper value for R, the gas constant, which so far has always been taken to be 0.0821 lit-atm/mole°K. It is obvious that this value is unsuitable for use in this kind of problem, since lit-atm do not involve speed dimensions in any simple way. This difficulty can be resolved by carefully examining the dimensions of R. We find that a lit-atm is a unit of work, or energy, and would indeed be the amount of work done by a gas in a container expanding in volume by one liter against a pressure of one atmosphere. Recognizing that R must always contain the dimensions of energy, it remains only to select units for R which will provide speed in cm/sec in Equation 5.17. It turns out that for that purpose R must be expressed in *absolute units*, where it has the value *8.31 × 10⁷ ergs/mole°K*. It is also possible to determine R in terms of heat units, where it has the value 1.987 cal/mole°K. In every case, R involves the same amount of energy; we simply use the appropriate conversion factors to change its units.

It is good to know the following values for the gas constant:
$R = 0.0821$ lit-atm/mole°K
$= 8.31 \times 10^7$ erg/mole°K
$= 1.987$ cal/mole°K

Example 5.11 Find the average speed of an oxygen molecule in air at room temperature (25°C).

Solution. Remembering that R must be given in absolute units to get the proper units for the speed, we substitute directly into Equation 5.17:

$$u = (3RT/M)^{1/2} = \left(\frac{3 \times 8.31 \times 10^7 \text{ ergs/mole°K} \times (25 + 273)°K}{32.0 \text{ g/mole}}\right)^{1/2}$$

$$= 4.82 \times 10^4 \text{ cm/sec} = 482 \text{ m/sec}$$

By a simple conversion (1 m/sec = 2.24 miles/hr) we can show that, in more familiar units, the average speed is about 1000 miles/hr!

According to the kinetic theory, molecular speeds are, on the average, very high by ordinary standards. It has been possible to test this prediction by several direct experiments, and the quantitative data agree very well with the theoretical values. Qualitatively, the very high average speed of molecules appears reasonable when one considers the speed of sound in air. Since sound is propagated by molecular motion, one would expect that the speed of sound and that of the molecules in the gas through which it passes would be roughly equal. The speed of sound in air is about 800 miles an hour, which is indeed very close to the average speed of an oxygen or nitrogen molecule at 25°C.

It is possible to use Equation 5.17 in another manner to determine

Propeller-driven airplanes never went faster than sound, but jet aircraft do. Can you rationalize these observations in terms of the kinetic theory?

molecular energy instead of speed. A single molecule moving through space will have a translational energy given by Equation 5.10:

$$\epsilon = (\tfrac{1}{2})\, mu^2$$

A mole of such molecules will have an energy of motion E_{trans} which is equal to $N\epsilon$, where N is Avogadro's number:

$$E_{trans} = N\epsilon = (\tfrac{1}{2})\, mNu^2 = (\tfrac{1}{2})\, Mu^2$$

where M is the gram molecular weight. By manipulation of Equation 5.17 we obtain

$$E_{trans} = (\tfrac{1}{2})\, Mu^2 = \tfrac{3}{2}\, RT \tag{5.18}$$

Consistent with a postulate of the theory, the energy in translation of a mole of gas is directly proportional to temperature and has the same value for all gases. At 25°C, using R with units of heat energy, 1.987 cal/mole°K, we find the molar energy,

$$E_{trans} = \tfrac{3}{2}\, RT = (\tfrac{3}{2}) \times (1.987 \text{ cal/mole°K}) \times (25 + 273)°K = 888 \text{ cal/mole}$$

This amount of energy, about 1 kcal/mole, is rather small by comparison with the energy changes typically observed in chemical reactions. E_{trans}, however, is by no means of negligible importance. The fact that E_{trans} turns out to have the same value at any given temperature for every gas, from He to UF_6, is, more than any other single factor, responsible for the common physical behavior we observe for gases.

Although it is useful to be able to calculate an average speed or energy for a molecule, one must remember that not all the molecules will have that speed and energy. The motion of molecules in a gas is utterly chaotic. In the course of a second a molecule will typically undergo millions of collisions with other molecules, with each collision resulting in a change of the molecule's speed and direction of motion. In view of this situation you

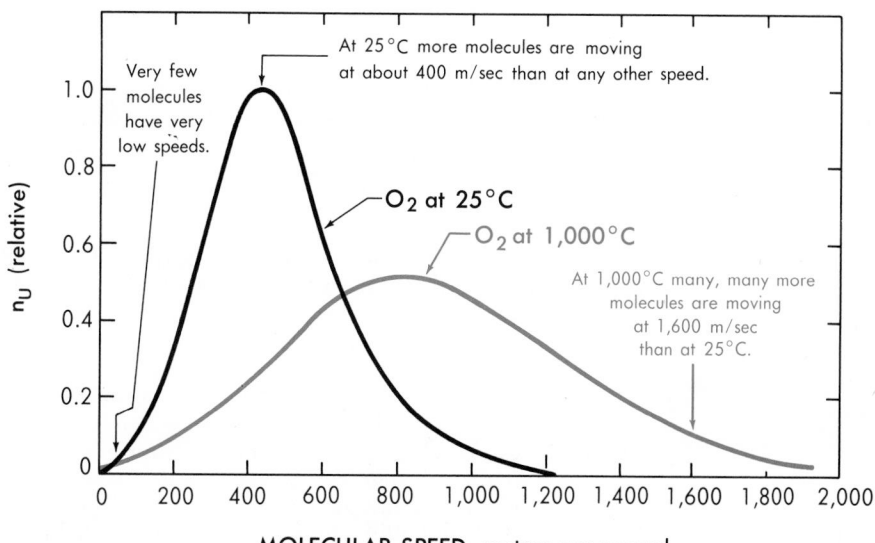

Figure 5.12 The Maxwellian distribution function for molecular speeds.

might well wonder whether one could hope to determine anything more than average molecular properties; in fact, the calculation of average speeds and molar energies seems a rather remarkable achievement.

Surprisingly, we do know considerably more about molecular motion in a gas than would first appear. In 1860 James Clerk Maxwell showed that different possible speeds were distributed among molecules in a definite way, and by a careful analysis of the gaseous system he derived a mathematical expression for this distribution. Figure 5.12 illustrates graphically the relation he obtained for oxygen gas at 25°C and 1000°C. On the graph the relative number of oxygen molecules having the speed u is plotted as a function of the speed. We see that there are very few molecules having speeds near zero, and that the number having a given speed increases rapidly with speed, up to a maximum at about 400 m/sec at 25°C. This is the most probable speed of an oxygen molecule at 25°C, with the majority of the molecules having speeds between 200 and 600 m/sec. Above about 400 m/sec, the number of molecules moving at any particular speed decreases, so that the likelihood of finding a molecule moving at about 800 m/sec is only about one tenth as large as that of finding a molecule moving at about 400 m/sec. For speeds in excess of about 1200 m/sec the fraction of molecules drops off rapidly.

As the temperature of the gas increases, the speeds of the molecules increase, and the distribution curve for molecular speeds is displaced to the right and becomes broader. From the curves at 25 and 1000°C it is clear that the probability of a molecule having a given high speed is enormously larger at high temperatures than it is at low.

Since translational energy for a molecule is simply related to the speed u, it is true that one can also speak of a distribution of energy among molecules which is analogous to that for speed among molecules. At high molecular energies we find very few molecules; the fraction of molecules having a given energy E or greater per mole drops off exponentially with increasing energy and to a good degree of approximation is given by the equation

$$f = e^{-E/RT} \qquad (5.19)$$

where f is the fraction of molecules having molar energy E or greater in a gas at temperature T, and small e is the base of natural logarithms. Since molecules having high energies are the ones which tend to participate in chemical reactions, Equation 5.19 is important in studies of reaction rates (Chapter 14).

The mathematical relation obtained by Maxwell for the distribution of molecular speeds is one of the most formidable in all of physical science. It is

$$n_u \text{ (rel)} = 4\pi u^2 \left(\frac{M}{2\pi RT}\right)^{3/2} e^{-Mu^2/2RT}$$

The fact that Maxwell was able to develop this equation from first principles is indicative of the intellectual ability of this remarkable man.

Maxwell was born in 1831 in Scotland. After education at the University of Edinburgh and at Cambridge he became a professor of natural philosophy, first in Scotland and later at Cambridge. His mathematical abilities became apparent at an early age and were applied in many areas. In addition to his accomplishments with gases, which laid the foundation of the science now known as statistical mechanics, Maxwell worked extensively in thermodynamics, developing several fundamental equations which bear his name. His greatest successes, however, were in connection with the theory of light and electricity, where he discovered and formulated the general equations of the electromagnetic field. He was first to recognize that light is a form of electromagnetic radiation and anticipated the development of what we now call radio waves. For several years during his life Maxwell was forced to be inactive because of illness; he was 48 years old when he died, having completed much of his work by the time he was 30.

The number of truly outstanding theoreticians the world has known is very small, certainly numbering less than one hundred. James Clerk Maxwell belongs among the elite of this group. He was truly an intellectual giant, to be ranked with Newton, Einstein, and J. Willard Gibbs (Chapter 12).

PROBLEMS

5.1   Cite three simple experiments you could do to prove that you are surrounded by a gas and not by a vacuum.

5.2   What three laws are special cases of the Ideal Gas Law? Show how the Ideal Gas Law takes the form of each of the simpler laws under the conditions where that law is valid.

5.3   What properties of a gas determine

    a. the average speed of its molecules?
    b. the average energy of its molecules?
    c. the pressure exerted by the gas?

5.4   In a McLeod gauge, operating at room temperature, a 250 ml sample of neon was compressed to a volume of 0.083 ml, where it exerted a pressure of 3.14 cm Hg. What was the original pressure of the neon?

5.5   Describe an experiment by which you might test at least qualitatively the validity of Equation 5.17.

5.6   In Boyle's original experiment he measured the length (l) (directly proportional to volume) of a gas column as a function of the pressure head (h) of mercury (Fig. 5.4). The following table of data was obtained by his procedure

l (cm)	40	35	30	25	20
h (cm Hg)	0.0	11.0	25.3	45.5	75.6

The barometric pressure during the experiment was 76.0 cm Hg and the temperature was 25°C. Make calculations from these data which demonstrate the validity of Boyle's Law.

5.20   What properties of air give the automobile tire its desirable characteristics? Can you think of any other devices which use the physical behavior of gases to good advantage?

5.21   What are the variables in the Ideal Gas Law? What simpler law is obtained when pressure and amount are held constant? Show that the volume of a gas is a linear function of the Celsius temperature if Charles' Law and Equation 5.2 are obeyed; i.e., show that $V = a + bt$.

5.22   What would you do to a sample of carbon dioxide gas in order to

    a. increase its pressure?
    b. increase the average speed of its molecules?
    c. make it behave more nearly ideally?

5.23   The argon gas in a 0.35 ml bulb at one atm is vented into an evacuated light bulb with a volume of 210 ml to establish a low gas pressure so as to inhibit vaporization of the tungsten filament. If temperature remains constant, what pressure is exerted by the argon in the light bulb?

5.24   Cite three common observations that would lead you to conclude that the molecules in a gas have very high speeds.

5.25   A student measures the pressure of a gas, using an apparatus like that in Figure 5.3. He finds that the mercury level on the side of the manometer in contact with air is 15.0 cm lower than the side in contact with the gas being studied. If the barometric pressure is 753 mm Hg, what is the pressure in the container in mm Hg?

5.7 The gauge pressure in an automobile tire was 24 lbs/in² one August morning when the air was at 25°C. Gauge pressure is the difference between total pressure and atmospheric pressure, which is 15 lb/in². After driving on the Pennsylvania Turnpike for 4 hours, the driver measured the pressure again and found it to be 32 lb/in². Assuming the volume of the tire hadn't changed, what was the temperature of the air in the tire?

5.8 The natural gas in a storage reservoir in St. Paul, Minn., under atmospheric pressure has a volume of 15.0×10^8 lit at 25°C. The temperature on the next day falls to −25°C, but the pressure remains constant. What does the volume of the gas become?

5.9 A 10.0 lit sample of steam at 100°C and one atm is cooled to 30°C and expanded until the pressure is 25 mm Hg. If no water condenses, find the final volume of the water vapor.

5.10 Calculate the density of H_2 at 75°C and 1.00 atm pressure. What is the density of CCl_4 gas under the same conditions? Compare the number of molecules in one liter of each of these gases under the given conditions.

5.11 Five ml of a volatile organic liquid were vaporized at 99°C in a 225 ml flask with a fine-holed stopper. After all the liquid was vaporized, the flask was cooled and weighed with the condensed liquid. The liquid was found to weigh 0.563 g. The barometric pressure in the laboratory was 735 mm Hg. Find the molecular weight of the liquid from these data.

5.12 One way to prepare pure oxygen is to heat solid potassium chlorate:

$$KClO_3(s) \rightarrow KCl(s) + \tfrac{3}{2} O_2(g)$$

If the oxygen is collected over water at 22°C and a total pressure of 755 mm Hg, how many grams of $KClO_3$ would have to react to produce 2.00 lit of wet gas? VP $H_2O = 20$ mm Hg at 22°C.

5.13 Carbon monoxide slowly oxidizes in the air according to the following equation:

$$2\ CO(g) + O_2(g) \rightarrow CO_2(g)$$

How many liters of O_2 would it take to react with 3.0 liters of CO, if each gas is measured at 28°C and 0.050 atm?

5.14 Air contains about 21% oxygen molecules and 79% nitrogen molecules, so is said to be about 21 mole per cent oxygen. In a commercial plant producing liquid O_2, how many liters of air at 20°C and 750 mm Hg would be needed to produce 100 lit of $O_2(1)$, whose density is 1.14 g/ml?

5.26 Houses with well water systems have ballast tanks to maintain reasonably constant water pressure. In a house with a 300-gallon ballast tank, the tank is 0.75 full of water when the gauge pressure is 50 lb/in². How much water will be delivered before the gauge, which registers the pressure of air in excess of 15 lb/in² in the tank, drops to 20 lb/in², at which point the pump is set to turn on?

5.27 A helium gas thermometer has a volume of 125 ml of 0°C and 1.00 atm pressure. It is immersed in some boiling liquid hydrogen, and its volume at one atm drops to 9.2 ml. Find the temperature of the liquid hydrogen.

5.28 How many grams will the steam (H_2O) filling a 50-gallon oil drum at 100°C and one atm weigh? (1 gal = 3.88 lit.)

5.29 Given that one mole of air contains 0.21 mole O_2 and 0.79 mole N_2, calculate the density of air in g/lit at 20°C and 750 mm Hg. What is the average "molecular weight" of air?

5.30 A student carries out the experiment described in Problem 5.11. He finds that the vapor filling a 235-ml flask at 100°C and 745 mm Hg pressure weighs 1.332 g. Estimate the molecular weight of the liquid.

5.31 A sample of wet oxygen, saturated with water vapor at 22°C, exerts a total pressure of 748 mm Hg.

  a. What is the partial pressure of O_2 in the sample? (VP $H_2O = 20$ mm Hg.)
  b. If the volume of the sample is 320 ml, how many grams of O_2 does it contain?

5.32 $NaHCO_3(s)$, sodium hydrogen carbonate, sometimes called sodium bicarbonate or baking soda, is readily decomposed by heat:

$$2\ NaHCO_3(s) \rightarrow Na_2CO_3(s) + H_2O(g) + CO_2(g)$$

If the H_2O and CO_2 produced in the reaction are separated and collected at 1 atm and 200°C, how many liters of $H_2O(g)$ would be formed during a period in which 16 liters of CO_2 were produced at the same temperature and pressure?

5.33 Using the data in Problem 5.14, find the total mass of air in pounds in a room 5.0 m × 3.0 m × 4.0 m at 23°C and 0.96 atm.

5.15 A 0.150 g sample of an Al-Zn alloy when treated with HCl evolves 138 ml of hydrogen, measured dry at 27°C and 1.00 atm. Calculate the per cent Al in the alloy (1 mole Al → 3/2 mole H_2; 1 mole Zn → 1 mole H_2).

5.16 Hydrogen has two naturally occurring isotopes, normal hydrogen with at. wt. = 1, and deuterium, at. wt. 2. A mixture containing 50 mole per cent H_2 and 50 mole per cent D_2 is allowed to effuse down a capillary. Which species would predominate in the first effused gas? What per cent of the molecules effusing first would be D_2?

5.17 How hot would a sample of oxygen have to be if its molecules were to have the same average speed as He molecules at 27°C? What would the temperature have to be if they were to have the same average energy of motion?

5.18 The density of liquid hydrogen is about 0.070 g/ml. Assuming that hydrogen behaved ideally at high pressures, calculate the pressure required to compress H_2 to that density at 25°C. Referring to Figure 5.10, would you expect that the pressure actually required to effect such a compression would be higher or lower than the ideal value?

5.19 It took 105 seconds for a sample of carbon dioxide to effuse through a porous plug and 126 seconds for the same number of moles of an unknown gas to effuse under the same conditions. Estimate the molecular weight of the gas.

5.34 A 0.356 g sample of $XH_2(s)$ reacts with water according to the following equation:

$$XH_2(s) + 2\ H_2O(l) \rightarrow X(OH)_2(s) + 2\ H_2(g)$$

The hydrogen evolved is collected over water at 23°C and occupies a volume of 431 ml at 746 mm Hg total pressure. Find the number of moles of H_2 produced and the atomic weight of X(VP $H_2O = 21$ mm Hg).

5.35 Calculate the ratio of the rates of diffusion of UF_6 molecules derived from uranium-235 and uranium-238 isotopes, respectively. A 50-50 mixture of the two kinds of UF_6 is allowed to diffuse through a porous plug. What would be the composition of the gas which effuses first? Why was the isolation of enriched U-235 so difficult?

5.36 On the average, how fast will H_2 molecules in a mixture of H_2 and UF_6 be moving at 25°C and 1 atm? What will be the speed of the UF_6 molecules? Compare the average energies of these two kinds of molecules in the mixture.

5.37 The molecular weight of the liquid as found by the method outlined in Problem 5.11 will be inherently incorrect for several reasons. Name as many reasons as you can, and indicate the one you feel would be most important.

5.38 Arrange the following gases in order of increasing speeds of their molecules at 100°C:

CO_2 $CHCl_3$
Cl_2 N_2
  Ar

*5.39 The escape velocity, v_e, from the surface of the moon is about 2.4×10^5 cm/sec. Unless $u \leqslant 0.20v_e$, essentially all of a gas will escape in 10^9 years or less. Would you look for H_2 in the lunar atmosphere? O_2? (Take T = 300°K.)

*5.40 In a classic experiment the density of carbon dioxide was carefully measured at 0°C at several low pressures, with the following results:

Pressure (atm)	Density (g/lit)
1.00000	1.97676
0.66667	1.31485
0.33333	0.65596

Use the Ideal Gas Law to calculate the apparent molecular weight of carbon dioxide at each pressure. Use a desk calculator or five-place logarithms for the calculations, since the accuracy of the data exceeds that possible with a slide rule. R = 0.082054 lit-atm/mole°K and 0°C = 273.16°K. The differences between values of MW obtained are due to deviations of the gas from the ideal. Make a graph of $MW_{apparent}$ vs. pressure, and extrapolate the (straight) line to zero pressure, where the gas behaves ideally. The limiting value of $MW_{apparent}$ at zero P is the best value for the molecular weight of CO_2. Assuming the molecular formula for carbon dioxide and the atomic weight of oxygen, calculate the atomic weight of carbon from the molecular weight of CO_2 you obtained.

   The approach employed in this problem is called the limiting density method for molecular weight determination. It was used extensively in the early part of this century to find molecular and atomic weights.

°5.41 Arrange the following gases in order of increasing magnitude of the van der Waals constant b; do the same for the constant a for these gases: C_2H_2, O_2, He, C_6H_6. You may use any physical properties of these gases that seem reasonable as a basis for your ranking. Consult a handbook to find the values of the properties.

°5.42 It takes 3 cal to raise the temperature of one mole of any monatomic gas by 1°C when the gas is heated at constant volume. Can you suggest how this experimental fact could be used to support the validity of Equation 5.18?

°5.43 A thermometer on the dark side of a satellite on the way to the moon records a temperature of −270°C. The gas molecules in the vicinity of the thermometer have speeds corresponding to a temperature of over 1000°C. How can this possibly happen?

°5.44 For a mole of ideal gas, plot graphs of

    a. P vs V at 77°C
    b. P vs T at V = 25 lit
    c. PV vs V at T = 77°C
    d. u vs T for N_2
    e. u vs M at 77°C
    f. E vs M at 77°C

°5.45 The buoyant force on a balloon is equal to the weight of air it displaces. The gravitational force on the balloon is equal to the sum of the weights of the balloon, the gas it contains and the balloonist. If the balloon and balloonist together weigh 400 lbs, what would the diameter of a hydrogen-filled balloon have to be in meters if the rig was to get off the ground at 20°C and one atm? If the balloon were filled with hot air at 250°C, how big would it have to be? (Take the average MW of air to be 28.8).

6 THE ELECTRONIC STRUCTURE OF ATOMS

The kinetic theory of gases attempts to explain observed properties of gases in terms of the physical properties of their component molecules. The development of the theory not only allowed us to better understand why gases behave as they do but also led to other ideas about gases which proved to be valid and important in studies of their heat capacities, spectra, and reaction rates. Theories of the electronic structure of atoms attempt in a similar way to deal with the chemical properties of matter through understanding of the manner in which electrons are arranged in atoms and molecules.

In Chapter 2 we presented some experimental and theoretical information about the general nature of atoms. You will recall that we believe each atom consists of a central, positively charged nucleus, surrounded by a group of negatively charged electrons. The nucleus is very small, about 10^{-12} cm in diameter, and includes just about all the mass of the atom. According to our model, the nucleus is made up of protons and neutrons in sufficient number to account for its mass and charge. The number of protons in the nucleus of an atom is characteristic of the element to which that atom belongs and is called the atomic number of that element.

Since electrons have charges equal in magnitude but opposite in sign to those of protons, a neutral atom of atomic number Z will have Z electrons outside its nucleus. These electrons move about in a spherical region roughly 2 or 3 Å in diameter. Atoms may gain or lose one or two electrons relatively easily, becoming charged particles called ions with chemical properties which differ greatly from those of the atoms from which they were derived.

6.1 SOME PROPERTIES OF ELECTRONS IN ATOMS AND MOLECULES

The behavior of electrons in atoms and molecules has been the subject of extensive research, both experimental and theoretical, during all this century. It must, however, be admitted that at present our knowledge of the detailed electronic structure of all but the simplest atoms is still incomplete. Much progress has been made on this problem, but much more work remains to be accomplished. In this book we shall present some of the common current models for electron arrangements in atoms and molecules.

The main obstacle to our understanding of the properties of chemical substances in terms of the electrons and nuclei of which they are composed is that small particles, such as atoms, molecules, nuclei, and particularly electrons, appear to obey different laws regarding energy and motion than do larger objects such as billiard balls and rotating bicycle wheels. Systems with which we are ordinarily concerned, with masses many, many times those of atoms and molecules, follow the laws of motion first formulated by Isaac Newton. These laws constitute that part of physics called classical mechanics. Small particles obey the laws of a somewhat different kind of mechanics, called quantum mechanics. (Actually it turns out that classical mechanics is, in a very real sense, a special case of quantum mechanics and is valid for all but those particles which have exceedingly small masses.)

Quantum mechanics is part of a general theory, called the quantum theory, which had its beginnings early in this century. Like the atomic theory, the quantum theory has evolved considerably during its development; some of its original postulates have been retained, while others have been modified or discarded. It is at present the fundamental theory used to explain the behavior of electrons and other small particles. Some experiments which led to the quantum theory will be considered in the next section after we have stated and discussed three of the underlying principles of the theory that are of chemical interest.

A flying flea obeys classical mechanics

POSTULATES OF THE QUANTUM THEORY

1. **Atoms and molecules can only exist in certain states, characterized by definite amounts of energy. When an atom or molecule changes its state, it must absorb or emit an amount of energy just sufficient to bring it to another state.**

Atoms and molecules can, as we have seen, possess various kinds of energy. One form of energy of particular importance when considering atomic structure arises from the motion of electrons about the atomic nucleus and from the charge interactions among the electrons and between the electrons and the nucleus. This kind of energy is called *electronic energy*. Only certain values of electronic energy are allowed to an atom. When an atom goes from one allowed electronic state to another, it must absorb or emit just enough energy to bring its own energy to that of the final state.

Analogous considerations apply to the other forms of energy possessed by atoms, molecules, and other small particles. Translational energy of motion, rotational and vibrational energy, in addition to electronic energy, are subject to the limitations of the quantum theory.

The energy of systems that can exist only in discrete states is said to be *quantized*. A change in the energy level of such a system involves the absorption or emission of a quantum of energy.

2. **When atoms or molecules absorb or emit light in the process of changing their energies, the wavelength λ of the light is related to the change in energy |ΔE| by the equation**

$$|\Delta E| = hc/\lambda \tag{6.1}$$

where h is a physical constant, called Planck's constant, and c is the speed of light.

Max Planck was one of the first men to use quantum ideas

A ray of light can be considered to consist of photons, which appear to have some properties characteristic of particles. In particular, each photon of wavelength λ has an associated energy equal to hc/λ. The energy of a photon in ergs can be easily found, given the values of Planck's constant, 6.626×10^{-27} ergs sec, the speed of light, 3.00×10^{10} cm/sec, and the wavelength of the photon in cm (see Example 6.1).

The energy lost by an atom or molecule in going from a higher to a lower energy state will equal the energy of the photon which may be emitted during the transition. Similarly, to absorb a photon, an atom or molecule must, during the absorption, be able to make a jump between two energy levels separated by an amount equal to the energy of the absorbed photon.

3. **The allowed energy states of atoms and molecules can be described by sets of numbers called quantum numbers.**

The mathematical solutions to problems regarding the energies of atoms and molecules as obtained by quantum mechanics usually result in sets of integral numbers, which serve to denote the allowed states and permit calculation of their energies. These numbers are called *quantum numbers*.

Associated with each electronic state of an atom is a group of quantum numbers that identify the state (see Section 6.5). In the usual model, the quantum numbers are associated with the individual electrons in the atom. Each electron is assigned a set of quantum numbers according to a set of rules. A statement of all the quantum numbers of all the electrons in an atom is used to designate the energy level, or quantum state, of the atom.

6.2 EXPERIMENTAL BASIS OF THE QUANTUM THEORY

Probably the best way to get some measure of understanding of the quantum theory is to examine some of the experiments and theoretical relations on which it is based. There are actually a great many experiments which are best explained by the theory; among them should be mentioned black body radiation, the photoelectric effect, atomic spectra, and the several kinds of molecular spectra. Though these are all important phenomena, we shall restrict our attention here to atomic spectra, postponing a discussion of the other experiments to more advanced courses of study in physics and chemistry. Atomic spectra bear directly on the problem of atomic structure and nicely illustrate an area of experiment in relation to the general quantum theory.

Classical mechanics failed completely to explain these phenomena

ATOMIC SPECTRA

When atoms are exposed to high energy from nearly any source, they tend to become excited and to give off energy in the form of radiation. The nature of the radiation emitted depends on the excitation which is used.

If we heat a metal in a furnace or in a flame, it will, depending on its temperature, give off visible light; at 1000°C it will look red, at 1500°C it will appear white. If the emitted light is examined with a spectroscope, a device which breaks up light into its component colors, the light is found to contain essentially all colors. More precisely, we would say that, over the region in which the metal radiates, its *spectrum* is *continuous*, containing light at all wavelengths.

Not all emitters of light radiate at all wavelengths. If we observe in a spectroscope the light emitted by sodium chloride when it is placed in a flame, say from a Bunsen burner, we see only a few bright lines, which indicate the few wavelengths at which sodium atoms, excited by the flame, are emitting light. In this case we are seeing the *atomic spectrum* of sodium, which, since it contains light at only specific wavelengths, is said to be *discrete*.

Atomic spectra are emitted when atoms are mildly excited; this can be accomplished by a flame, by a spark, or by electrons which have been accelerated by falling through a few volts of potential. Our common fluorescent lights and mercury vapor highway lights give off light of this sort, with well-resolved components (or lines) which can be observed with a spectroscope (Fig. 6.1).

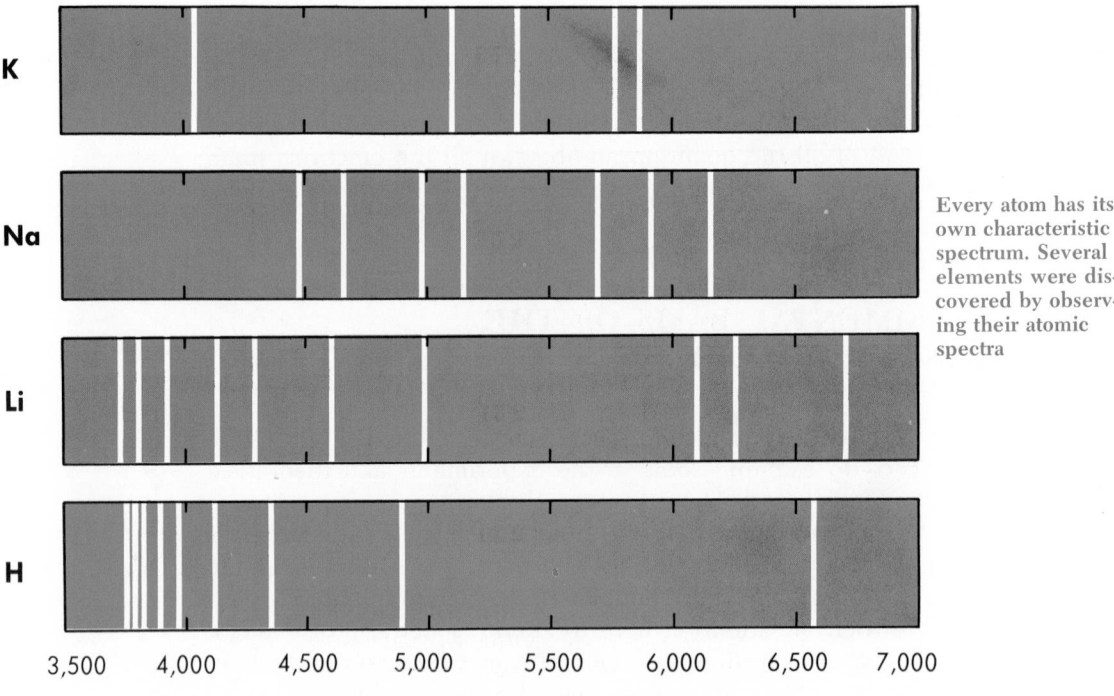

Every atom has its own characteristic spectrum. Several elements were discovered by observing their atomic spectra

WAVELENGTH, angstroms

Figure 6.1 Atomic spectra of hydrogen and some alkali metals in the visible region.

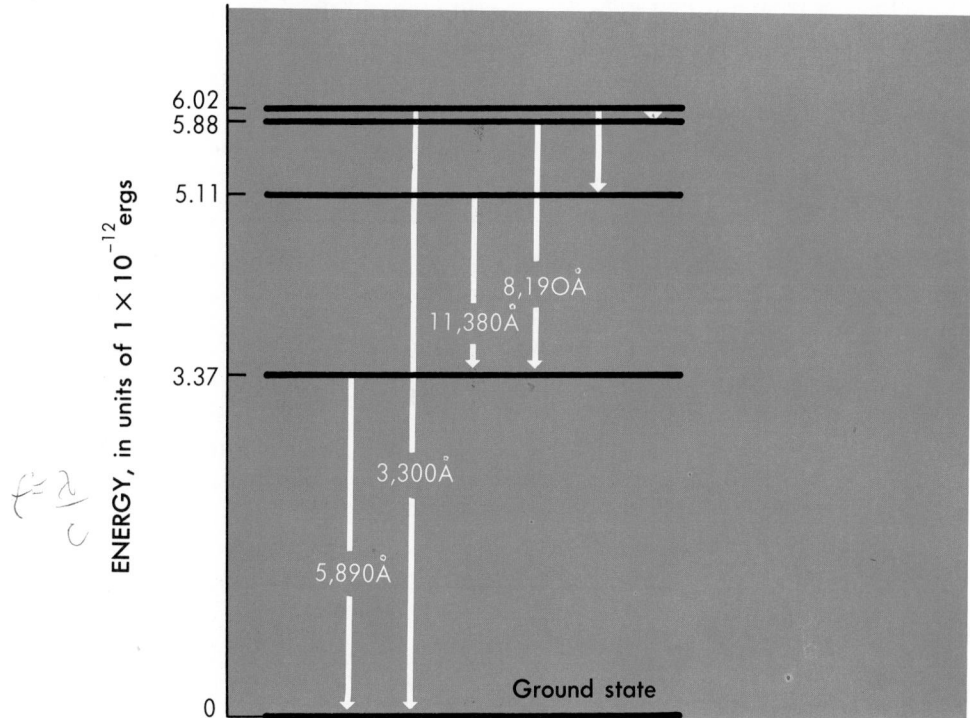

Figure 6.2 Some transitions occurring between low lying energy levels of the sodium atom.

The existence and character of atomic spectra are readily treated by the quantum theory. When a gas containing sodium atoms is heated to a high temperature or is exposed to fast-moving electrons, some of the sodium atoms will, through collisions, absorb energy and become excited, moving in the process to higher allowed electronic energy levels. Once excited, the atoms will be unstable and will tend to return to lower electronic energy levels, ultimately reaching the lowest energy level, the ground state of the atom. Under excitation the sodium gas will quickly reach a state of dynamic equilibrium in which some atoms are releasing energy at the same rate that it is being absorbed by others.

One of the ways in which electronically excited sodium atoms can lose energy is by radiating it as light. In a process in which an excited atom makes a transition to a lower energy level, a photon of light may be emitted. The wavelength λ of the photon is related to the magnitude of the change in energy $|\Delta E|$ of the atom by Postulate 2 of the quantum theory:

$$|\Delta E| = \frac{hc}{\lambda} = E \text{ photon} \qquad (6.2)$$

This equation was first proposed by Einstein in 1905, in the very early days of the quantum theory. It was based on the notion that interchanges of energy between radiation and matter take place in discrete units, called quanta, with energies given by Equation 6.2.

Atomic spectra, then, are believed to consist of the radiation resulting from transitions made by electronically excited atoms from higher to lower allowed energy levels. In Figure 6.2 we have arranged, in somewhat simplified form, some of the lower electronic energy levels of the sodium atom, along with the transitions that are observed between those levels. Each transition gives rise to a line in the atomic spectrum of sodium, with a wavelength given by Equation 6.2. One of the lines in the sodium spectrum in Figure 6.1 is the result of a transition shown in Figure 6.2.

Example 6.1 Excited sodium atoms may emit radiation having a wavelength of 5890 Å. What is the energy in ergs of the photons in this radiation?

Solution. The energy is obtained by substitution into Equation 6.2:

$$|\Delta E_{atom}| = E_{photon} = \frac{hc}{\lambda}$$

Given that Planck's constant, h, equals 6.626×10^{-27} erg sec and the speed of light, c, is 3.00×10^{10} cm/sec, the wavelength, λ, must be expressed in cm to obtain the energy of the photon in ergs.

Since $1 \text{ Å} = 1 \times 10^{-8}$ cm

$$\lambda = 5890 \text{ Å} \times \frac{1 \times 10^{-8} \text{ cm}}{1 \text{ Å}} = 5.890 \times 10^{-5} \text{ cm}$$

$$E_{photon} = \frac{6.626 \times 10^{-27} \text{ erg sec} \times 3.00 \times 10^{10} \text{ cm/sec}}{5.890 \times 10^{-5} \text{ cm}} = 3.37 \times 10^{-12} \text{ erg}$$

In emitting the photon, the atom will lose 3.37×10^{-12} ergs of energy since in the overall process energy is conserved. This means that there must be two levels in the atom which are separated by this amount of energy. See Figure 6.2.

It is of interest to compare the energy changes which occur when atoms emit their characteristic spectra to the energy changes involved in chemical reactions. To make such a comparison, let us calculate the energy evolved, in *kcal,* when *one mole* of sodium atoms undergoes the transition referred to in Example 6.1. Using the appropriate conversion factors, we have,

$$|\Delta E| = \frac{3.37 \times 10^{-12} \text{ erg}}{1 \text{ atom}} \times \frac{6.02 \times 10^{23} \text{ atoms}}{1 \text{ mole}} \times \frac{1 \text{ kcal}}{4.184 \times 10^{10} \text{ ergs}}$$

$$= 48.5 \text{ kcal/mole}$$

or about 50 kcal per mole. This is about half the amount of energy that would be evolved if a mole of sodium metal reacted with chlorine to form NaCl. (See Table 4.1.) Energy changes involved in the emission of atomic spectra are, mole for mole, of the same order of magnitude as energy changes observed in chemical reactions.

REGULARITIES IN ATOMIC SPECTRA. Beginning about 1880 it was recognized that there is a certain amount of order in atomic spectra. It was found possible in some cases to sort out the lines in a spectrum into series; lines were assigned to each series on the basis of wavelength, intensity, and breadth. In the atomic spectrum of sodium we talk about the so-called principal (intense), sharp, and diffuse series.

In 1885 Balmer found a very remarkable relationship between the wavelengths of the lines in the atomic spectrum of hydrogen (Fig. 6.1).

TABLE 6.1   Calculated and Observed Wavelengths (in Å) in
the Balmer Series for Hydrogen

$$\lambda_{calc} = 3646.00 \, [n^2/(n^2 - 4)]$$

LINE	n	λ_{obs}	λ_{calc}	LINE	n	λ_{obs}	λ_{calc}
1	3	6562.79	6562.80	6	8	3889.06	3889.07
2	4	4861.33	4861.33	7	9	3835.40	3835.40
3	5	4340.47	4340.48	8	10	3797.91	3797.92
4	6	4101.74	4101.75	9	11	3770.06	3770.65
5	7	3970.07	3970.09				

He showed that one could mathematically express the wavelengths of the nine then known lines in the atomic spectrum of hydrogen by the equation

$$\lambda = 3646.00[n^2/(n^2 - 4)] \tag{6.3}$$

Balmer was a high school math teacher when he made his discovery

in which λ is the wavelength in Å and n is an integer which has the values 3, 4, 5, and so on, for the first, second, third, . . . , lines in the hydrogen spectrum.

Balmer's formula for the series of lines which now bears his name predicts the wavelengths exceedingly well. In Table 6.1 we have tabulated the calculated and observed wavelengths of the first nine lines in the series. Balmer's equation undoubtedly gives the most nearly exact prediction of experimental quantities in all physical science, and its discovery gave others great incentive to look for other such relations. To this day no equations have been found for atoms other than hydrogen that approach the Balmer formula in exactness. For hydrogen there are several known series, all of which are given by relations similar to that found by Balmer. Probably the simplest of these is the equation for the lines found in the ultraviolet region of the hydrogen spectrum, the so-called Lyman series:

The agreement between λ_{obs} and λ_{calc} in Table 6.1 is within about 2 parts in 1 million

$$\lambda = 911.5 \, n^2/(n^2 - 1) \tag{6.4}$$

where n is any integer larger than one and λ is in Å.

6.3   THEORIES OF ELECTRONIC STRUCTURE

By the beginning of this century knowledge of atomic structure had advanced to the point at which scientists could begin to speculate on the way in which positive and negative charges were arranged in atoms. Part of the problem was solved when Rutherford demonstrated the existence of atomic nuclei (Chapter 2). It was only two years later, in 1913, that Niels Bohr presented a theory for the structure of the hydrogen atom that added greatly to our ideas regarding the behavior of the electrons in atoms.

Bohr based his approach on Rutherford's nuclear atom and on Planck's suggestion that atoms and other small particles can possess only certain definite amounts of energy. He assumed that a hydrogen atom consisted of a central proton, around which an electron moved in a circular orbit (Fig. 6.3). By balancing the force of attraction of the proton for the electron

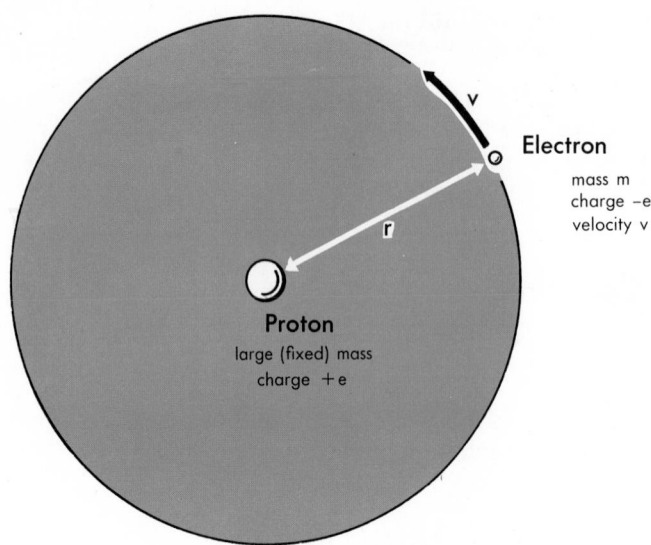

Electron

mass m
charge −e
velocity v

Proton

large (fixed) mass
charge +e

Figure 6.3 Bohr's model of the H atom.

by the centrifugal force due to the motion of the electron, he was able to express the energy of the atom in terms of the radius of the electron's orbit. He then, arbitrarily and boldly, imposed a quantizing condition on the angular momentum, mvr, of the electron, namely, that it was given by the equation

$$mvr = nh/2\pi$$

where m is the electronic mass, v its speed, r the radius of the orbit, n a quantum number which can take on any positive integral value, (i.e., 1, 2, 3, . . .), and h is Planck's constant.

Bohr showed that his quantum condition restricted the energies of the hydrogen atom to those values given by the equation

$$E = -B/n^2 \tag{6.5}$$

where B was found by the theory to be a constant equal to 2.179×10^{-11} ergs.

In Figure 6.4 we have shown the energy levels of the hydrogen atom as predicted by Bohr's theory. In setting up the model, Bohr took the zero of energy to be the ionized atom; since, in general, energy is required to ionize the atom, the allowed energies are all negative. The lowest energy, at −B, occurs when n equals one; this level is called the ground state of the atom, and is the state in which the atom is ordinarily found.

When n equals two, the energy, by Equation 6.5, is −B/4; an H atom in this state is said to be excited, and will tend to quickly make a transition to the ground state. As the quantum number n increases, the energy of the atom increases (i.e., becomes less negative); as n gets large, the atom becomes less and less stable and is more easily ionized.

By Bohr's theory the size of the H atom is directly proportional to n^2

The triumph of Bohr's theory was its ability to predict the observed spectral series of the hydrogen atom. Transitions would be expected to occur from all excited levels to lower levels, each transition being associated with a photon having a particular wavelength and energy. Mathemat-

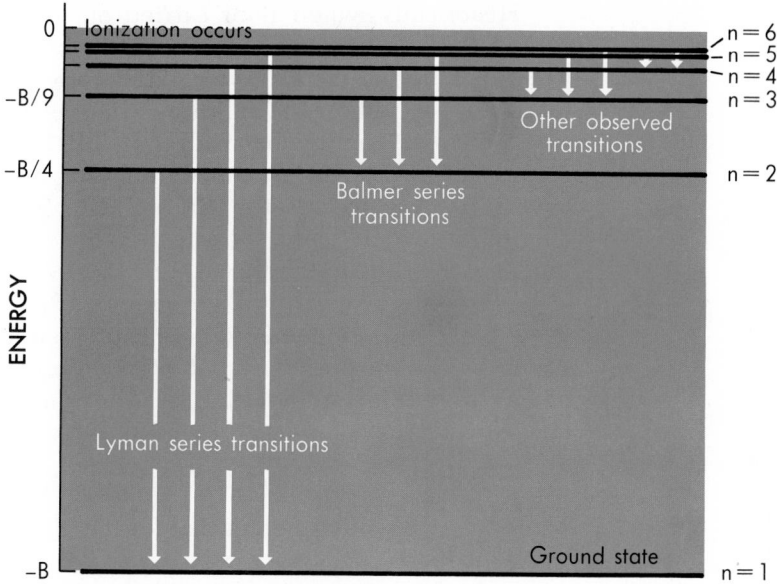

Figure 6.4 Some energy levels of the hydrogen atom.

ically, for any transition from an upper level, E_{hi}, to a lower level, E_{lo}, involving an energy change of magnitude $|\Delta E|$,

$$|\Delta E| = E_{hi} - E_{lo} = \frac{-B}{n_{hi}^2} - \left(\frac{-B}{n_{lo}^2}\right) = -B\left(\frac{1}{n_{hi}^2} - \frac{1}{n_{lo}^2}\right)$$

By Equation 6.2 the wavelength of an emitted photon is related to the energy change in the atom as follows:

$$E_{photon} = \frac{hc}{\lambda} = |\Delta E| = -B\left(\frac{1}{n_{hi}^2} - \frac{1}{n_{lo}^2}\right) = B\left(\frac{n_{hi}^2 - n_{lo}^2}{n_{hi}^2 n_{lo}^2}\right)$$

$$\lambda = \frac{hc}{|\Delta E|} = \frac{hc}{B}\left(\frac{n_{hi}^2 n_{lo}^2}{n_{hi}^2 - n_{lo}^2}\right) \tag{6.6}$$

Substituting known values of h, c, and B, we obtain

$$\lambda = \frac{6.626 \times 10^{-27} \text{ erg sec} \times 3.00 \times 10^{10} \text{ cm/sec}}{2.179 \times 10^{-11} \text{ erg}}\left(\frac{n_{hi}^2 n_{lo}^2}{n_{hi}^2 - n_{lo}^2}\right)$$

$$\lambda = 912 \times 10^{-8} \frac{n_{hi}^2 n_{lo}^2}{n_{hi}^2 - n_{lo}^2} \text{ cm} = 912\left(\frac{n_{hi}^2 n_{lo}^2}{n_{hi}^2 - n_{lo}^2}\right) \text{ Å} \tag{6.7}$$

If we compare this equation with Equation 6.4, we see that Bohr's relation is essentially identical with that obtained for the Lyman series if n_{lo} is set equal to one, and n_{hi} equals n, the quantum number of the excited state. Bohr's theory, then, indicates that the Lyman series in hydrogen arises from transitions from any excited level to the ground state (n = 1) of the atom, and, amazingly enough, is able to predict quantitatively the wavelengths of all the lines in that series. Similarly, if n_{lo} equals two and n_{hi} equals n, the Bohr Equation (6.7) and the equation for the Balmer series (6.3) become identical.

Historically, when Bohr carried out the analysis we have just presented, the Lyman series, which is in the ultraviolet region, was unknown. The prediction of the existence of this series and its subsequent discovery lent great support for the validity of the theory.

A corollary prediction made possible by the theory is the ionization energy of the H atom. To ionize the normal atom one must give it enough energy to raise it from the ground state (n = 1, E = −B) to the zero of energy, at which point the electron is completely removed from the proton. According to the theory, this means that the atom must receive an amount of energy equal to B, or about 2.179×10^{-11} ergs, if it is to ionize. Frequently ionization energies are expressed in electron volts; an electron which has fallen through a potential drop of one volt is said to have an energy of one electron volt, 1 ev. The conversion from electron volts to ergs is readily accomplished by using the conversion factor:

$$1 \text{ electron volt} = 1.60 \times 10^{-12} \text{ erg}$$

Therefore, we would predict that for the H atom,

The ionization energies of atoms range from 3.87 ev for cesium to 24.5 ev for helium

$$\text{ionization energy} = 2.179 \times 10^{-11} \text{ erg} \times \frac{1 \text{ ev}}{1.60 \times 10^{-12} \text{ erg}} = 13.6 \text{ ev}$$

The experimentally observed ionization energy is 13.5 electron volts.

6.4 MODERN QUANTUM THEORY

Bohr's theory for the structure of the hydrogen atom (and all other one-electron species like He^+ and Li^{2+}) was highly successful, and scientists of the day must have thought that they were on the verge of being able to predict the allowed energy levels of all the atoms. However, it soon became obvious that the extension of Bohr's ideas to atoms with two or more electrons gave, at best, only qualitative agreement with experiment. The failure of the Bohr theory in such cases ultimately led to an extensive search, beginning about 1920, for other approaches to the problem of energies of electrons in atoms and molecules.

In 1924 a young French physicist, Louis de Broglie, reasoning as physicists do, sometimes with strikingly successful results, suggested that if light rays have particle properties, then perhaps particles may, under some circumstances, exhibit wave properties. By an argument which is beyond the level of this text, de Broglie predicted the wavelength associated with a particle of mass m moving at velocity v to be given by the relation

$$\lambda = h/mv \tag{6.8}$$

where h equals, as you might now expect, Planck's constant.

An electron with an energy of 20 ev has a wavelength of about 3 Å

Within a few years Davisson and Germer, working at the Bell Telephone Laboratories, tested de Broglie's prediction by diffracting electrons of known energy from crystals. They established that a beam of electrons did indeed have wave properties and that the wavelength to be associated with electrons of known mv value was exactly that predicted by the de Broglie equation.

SOME WAVE PROPERTIES OF PARTICLES

The de Broglie relation is a key to the wave properties of small particles, much as the Einstein equation is a key to the particle properties of light. We shall now consider one problem involving de Broglie-type waves, so as to show you the kinds of waves that appear to be associated with small particles, how quantum numbers arise in the wave treatment, and how the allowed energy levels of a system can be found. We shall then apply the solution obtained to several examples to illustrate how the observed properties of small particles are correlated with their masses and locations.

The system we shall work with consists simply of a particle confined to move in one dimension between two impenetrable walls. We assume that the particle has only kinetic energy and moves across the space allowed to it at a constant speed, v, bouncing elastically every time it hits a wall (Fig. 6.5). By the classical relations of mechanics, the energy ϵ and momentum are given by the relations

$$\epsilon = \tfrac{1}{2} mv^2 \quad \text{and} \quad \text{momentum} = mv$$

The de Broglie relation states that there must be a wave associated with the particle. Our theory tells us that the wave must just fit into the space allowed to the particle. The wave can have finite amplitudes only inside the system, and the wave amplitude at the walls must be zero. Examples of some waves which meet these conditions are shown in Figure 6.6.

Mathematical expressions called wave functions can be written which give the amplitude of each wave as a function of position; the waves of Figure 6.6 result from plotting the amplitude in the wave function against position in the container. Wave functions are often given the symbol ψ, and for the wave in Figure 6.6a ψ obeys the following equation

The waves are similar in appearance to those obtained with vibrating strings

$$\psi = A \sin \frac{\pi x}{d} \tag{6.9}$$

where x is the distance from the left-hand wall and d is the total distance between walls.

According to current theory, the square of the value of ψ at any point is equal to the relative probability of finding the particle at that point. If we could examine experimentally a system which had a wave function like that in Figure 6.6a, we would be most likely to find the particle in the middle of the box and would essentially never be able to detect it at the wall! A strange situation indeed, but one which, insofar as it can be tested, is consistent with the behavior of such systems.

When we treat this problem by classical methods, we find that we can, if we wish, predict the path a single particle would follow, given constant speed and perfectly reflecting walls. When we use the wave approach, all we can hope to do in this regard is to determine the waves to be associated with a particle, and so predict the likelihood of finding it at any given point in space. We cannot predict its path as Bohr attempted to do in his approach to the hydrogen atom problem. This difference between classical and quantum systems appears to be fundamental and makes our "understanding" of quantum systems more difficult.

Although the paths of particles are not obtainable by the wave ap-

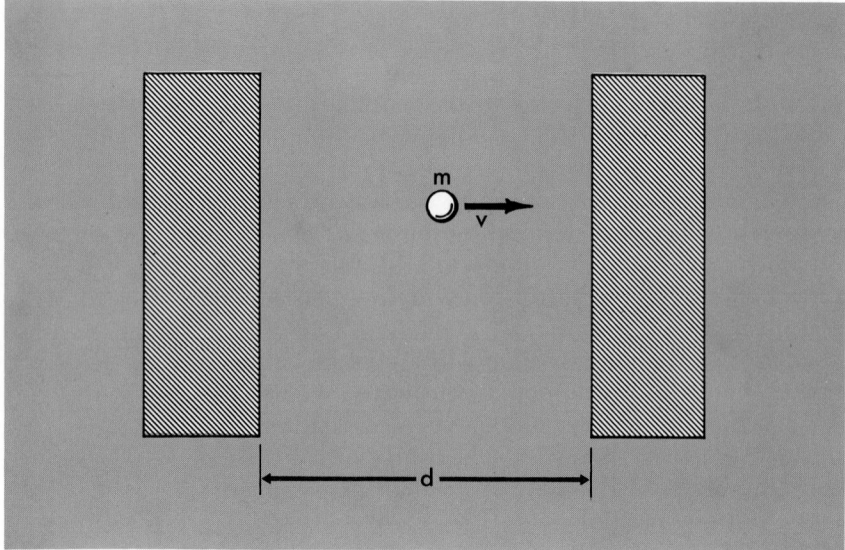

Figure 6.5 Particle in a one-dimensional box.

proach, their energies are. From the properties of the waves in Figure 6.6, we can see that for this system there are many waves that satisfy the condition that the amplitude be zero at the walls. The wavelengths, λ, of allowed waves all can be seen to satisfy the relation

$$\lambda = \frac{2d}{n} \qquad (6.10)$$

where d is the distance between the walls and n is an integer equal to 1, 2, 3, ... The number n is the **quantum number** for this system and arises here naturally as a result of the condition that the waves must fit properly in the space allowed to the particle.

By combining de Broglie's Equation 6.8 with the values of λ in Equation 6.10, we obtain

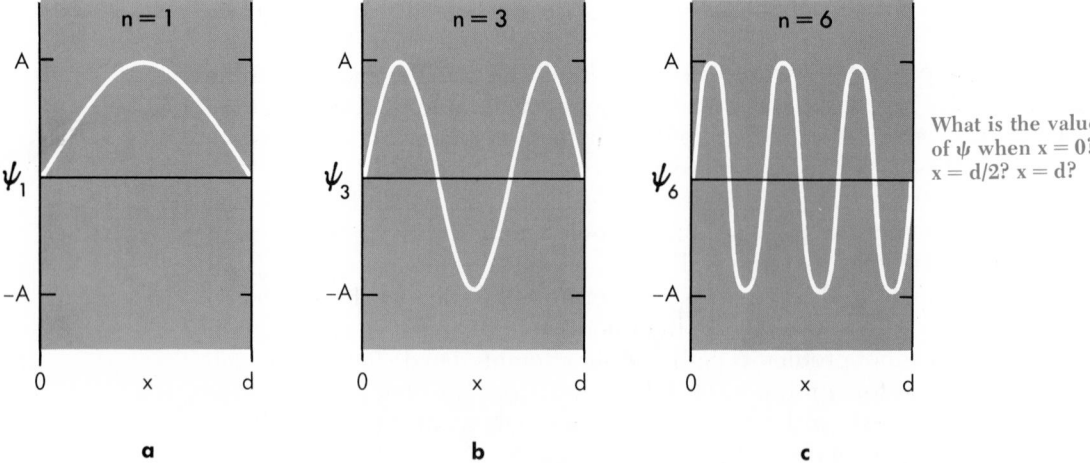

What is the value of ψ when x = 0? x = d/2? x = d?

Figure 6.6 Some wave functions of a particle in a one-dimensional box.

Figure 6.7 Low lying energy levels of a particle in a box.

$$\lambda = \frac{h}{mv} = \frac{2d}{n} \quad \text{or} \quad mv = \frac{nh}{2d} \tag{6.11}$$

Equation 6.11 effectively imposes a quantum condition on the momentum mv of the particle, which limits its energy ϵ to certain allowed values:

$$\epsilon = \tfrac{1}{2}\,mv^2 = \frac{1}{2m}\,(mv)^2 = \frac{1}{2m}\left(\frac{nh}{2d}\right)^2 = \frac{n^2h^2}{8md^2} \tag{6.12}$$

This equation tells us that a particle, of any kind, confined to move in a region of any length, will have only certain energies allowed to it. Those energies depend, as indicated by Equation 6.12, on the mass m of the particle, the interval d through which it can move, and the quantum number n. The energy levels available to such a particle are shown in Figure 6.7. For each allowed energy there will be an associated value of the quantum number, n, and a wave describing a wave function, ψ_n, with n loops, analogous to ψ_3 in Figure 6.6. The implications of Equation 6.12 with regard to the behavior of electrons, atoms, molecules, and matter in general are surprisingly far-reaching and we shall briefly explore a few of them.

IMPLICATIONS OF THE
de BROGLIE RELATION

Using Equation 6.12 one can calculate the energies available to essentially any particle confined to a region of any length. There are several features of the equation that should be recognized. The first is that a confined particle cannot have zero energy, since n cannot be zero if a wave is to be present. Since n can equal 1, 2, 3 . . . , the particle may take on many energies; of these, the lowest value will be for n = 1.

Secondly, the lowest energy available to a particle depends markedly on its mass and on the size of the space to which it is confined. This energy varies inversely with the mass; every time the mass is decreased by a factor of two, the energy is doubled. The minimum energy also varies inversely

Small particles tend to be in continual motion

with the square of the length through which the particle can move; cutting this length in half will always quadruple the lowest energy the particle can have. This means that particles of low mass such as electrons confined to small regions, like atoms or molecules, have very much higher minimum energies than relatively heavy particles such as gas molecules kept in a rather large space like a room. It turns out that these two particular systems are important to the argument, so let us pursue their properties further.

We first apply Equation 6.12 to obtain the minimum energy (n = 1) of an electron (m = 9.1 × 10⁻²⁸g) in a hydrogen atom. The force holding the electron near the nucleus is coulombic rather than a wall, and the electron moves in a sphere rather than along a line; but as a very approximate model, we can assume that an electron in a hydrogen atom behaves as a particle confined to a box of the order of 1 Å long (d = 1 × 10⁻⁸ cm). Substituting into Equation 6.12:

$$\epsilon \min = \frac{n^2h^2}{8md^2} = \frac{1^2(6.626 \times 10^{-27})^2}{8(9.1 \times 10^{-28})(1 \times 10^{-8})^2} = 6.0 \times 10^{-11} \text{ ergs}$$

Note that this is about the ionization energy of the H atom

For a nitrogen molecule moving about in a room, the situation is quite different. In the first place, the N_2 molecule (m = 4.7 × 10⁻²³ g) is about 50,000 times as heavy as an electron. Even more important, taking a typical room to be about 3 meters across (3 × 10¹⁰ Å), we see that the space available to the N_2 molecule in a room is 3 × 10¹⁰ as great as that for an electron restricted to a hydrogen atom. From Equation 6.12, taking an energy ratio:

$$\frac{\text{Lowest energy } N_2 \text{ molecule}}{\text{Lowest energy electron in H atom}} = \frac{\text{mass electron}}{\text{mass } N_2} \times \left(\frac{\text{diam. H atom}}{\text{size room}}\right)^2$$

$$= \frac{1}{50,000} \times \left(\frac{1}{3 \times 10^{10}}\right)^2 = 2 \times 10^{-26}$$

Clearly, a N_2 molecule in the translational ground state has an energy which is only a tiny fraction of that for an electron in its lowest energy state in a hydrogen atom or, indeed, in any atom or simple molecule.

In Figure 6.8 the minimum energies of several systems, including the two just discussed, are indicated. Shown as a heavy horizontal line in the figure is a very important quantity, the *average kinetic energy* of a gas molecule at 25°C. You will recall from Chapter 5 (Equation 5.18) that this amounts to about 900 cal/mole or:

$$\frac{900 \text{ cal}}{1 \text{ mole}} \times \frac{4.184 \times 10^7 \text{ ergs}}{1 \text{ cal}} \times \frac{1 \text{ mole}}{6.02 \times 10^{23} \text{ molecules}}$$

$$= 6.2 \times 10^{-14} \text{ ergs/molecule}$$

Notice from the figure that the minimum translational energy of a N_2 molecule is much *smaller*, by a factor of about 1 × 10⁻²³, than that of an average gas molecule. This means that a N_2 molecule in its ground state would gain energy by collision with other molecules. As its energy increased, its quantum number would go from 1 to a very large number. In a sample of nitrogen gas our chances of finding a molecule with the minimum energy (n = 1) would be very small indeed.

In contrast, we see from Figure 6.8 that the minimum energy of an electron in a hydrogen atom is about 1000 times *larger* than that of an average gas molecule. This means that an electron, unlike a N_2 molecule, can-

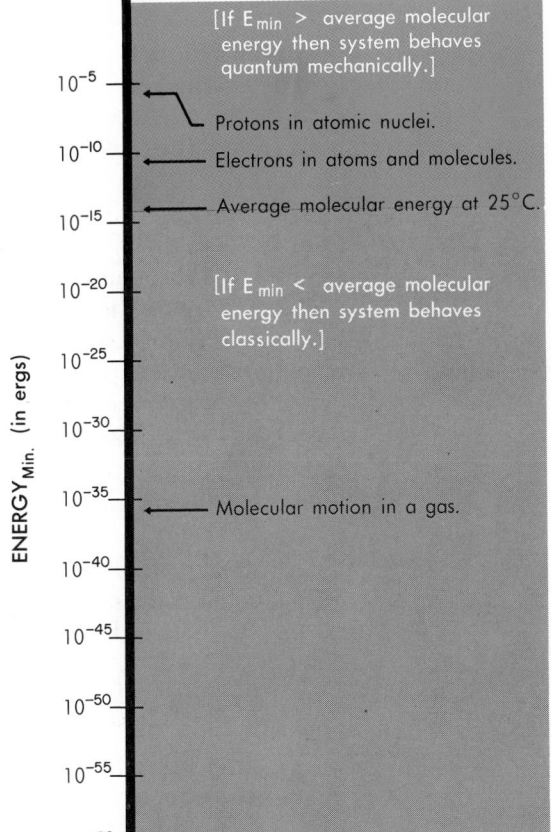

Figure 6.8 Minimum energies allowed to some simple systems.

To decide if a particular atomic or molecular system will behave classically, it is necessary to compare its lowest allowed energy with the average energy of gas molecules which might interact with it

not gain energy as a result of molecular collisions; gas molecules have energies which are much too small to excite electrons to higher energy states. Consequently, *electrons in atoms and molecules,* at ordinary temperatures, *are found in their lowest possible energy states.* This means that the behavior of electrons in atoms and molecules must be deduced from quantum mechanics; classical mechanics fails here because we do not have a statistical distribution of particles among a huge number of energy states. As we see from Figure 6.8, protons in atomic nuclei resemble electrons in their behavior, while billiard balls follow the same laws as nitrogen molecules. In general, relatively heavy particles that have plenty of room to move around in can be treated by classical mechanics; quantum mechanical principles are required to predict the behavior of light particles confined to tiny, submicroscopic volumes.

The de Broglie relation can be applied rather easily to small systems in which the energy of the particle can be described, as in the case of the particle in a box, in terms only of its speed. There are many systems, including those involving electrons in atoms and molecules, in which the energy of a particle is partly due to its speed and partly due to its position. For such systems it is not possible to apply the de Broglie equation except as an approximation, and we are forced to resort to a more sophisticated approach if we are to obtain exact solutions.

THE SCHRÖDINGER WAVE EQUATION

Even before the experimental proof of de Broglie's theory was established, Erwin Schrödinger had applied de Broglie's ideas in a somewhat different way than we have done. He developed in 1926 another equation, now called the *Schrödinger wave equation,* which allowed him to determine the energy levels and wave properties of the hydrogen atom, the particle in the box, and a few other relatively simple systems. Unlike the Bohr theory, the wave equation appears to be applicable both to atoms other than hydrogen and to molecules. Unfortunately the extension of the wave equation to these other systems has proved to be much more difficult in fact than in principle, since the form of the equation for such systems is, in general, so complicated mathematically as to make it insoluble. However, in the relatively few cases in which a satisfactory solution has been possible, the results have agreed essentially perfectly with experiment. This has led scientists to believe that the approach to atomic and molecular properties through the Schrödinger wave equation is a correct although complex one.

Even with the largest computers we have been unable to solve the wave equation exactly for any system containing 3 or more electrons

Although we shall not deal with any mathematical problems using the Schrödinger wave equation, we write it down so that you may have some idea of its nature. For a one-particle system, the equation takes the form

$$\frac{\partial^2\psi}{\partial x^2} + \frac{\partial^2\psi}{\partial y^2} + \frac{\partial^2\psi}{\partial z^2} + \frac{8\pi^2 m}{h^2}(E - V)\psi = 0 \qquad (6.13)$$

where m is the mass of the particle, E and V are its total and potential energies, respectively, h is Planck's constant, and the first three terms are partial derivatives of ψ with respect to the coordinates (x, y, z) of the particle, the wave function to be associated with the particle.

The wave equation is a formidable mathematical relation, as you will probably agree. The quantum mechanical problem is to solve the equation for the wave functions, ψ, allowed to the system. In general, for a given problem there are a great many wave functions that satisfy Equation 6.13. In Figure 6.6 we sketched some of these functions for the one-dimensional particle in a box problem. Those functions were obtained by the rigorous

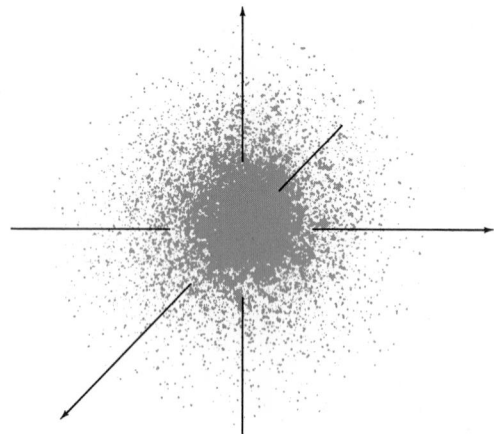

Figure 6.9 Electron cloud surrounding an atomic nucleus.

solution of the wave equation for that problem. Similar wave functions are found in the exact solution to the H atom problem and to the other problems which have been treated exactly.

As we noted, the interpretation of the wave function in a given state is that ψ^2 at any point in space is proportional to the probability of finding the particle at that point. In problems involving electrons, we can interpret the value of ψ^2 as being proportional to the electric charge density at the point. "Electron cloud" diagrams showing charge density in atoms and molecules are frequently drawn on this basis. In Figure 6.9 we have drawn schematically the "electron cloud" surrounding the nucleus of a typical atom.

6.5 ELECTRON ARRANGEMENTS IN ATOMS

QUANTUM NUMBERS OF ELECTRONS

The Schrödinger wave equation can readily be set up for the problem of the hydrogen atom. In that problem we consider the hydrogen nucleus to be fixed in position and examine the energy, E, of the electron in its allowed energy states. The potential energy, V, of the electron is obtained from the laws of electrostatics and is simply $-e^2r$, where e is the electronic charge and r is the distance from the nucleus to the electron. Knowing this, we can immediately write the wave equation

$$\frac{\partial^2 \psi}{\partial x^2} + \frac{\partial^2 \psi}{\partial y^2} + \frac{\partial^2 \psi}{\partial z^2} + \frac{8\pi^2 m}{h^2}\left(E + \frac{e^2}{r}\right)\psi = 0 \qquad (6.14)$$

Schrödinger solved this equation for the energy levels of the hydrogen atom, obtaining the identical relation obtained by Bohr with his theory 15 years earlier. Schrödinger also obtained the wave functions and quantum numbers associated with each of the allowed states of the atom.

Schrödinger found that the electron in the H atom could be described by three quantum numbers, which are now called n, l, and m_l. There are three such numbers because the electron requires three coordinates to determine its motion. The quantum numbers of the electron establish its state, fixing both its energy and wave function. Quantum numbers are also used in describing other atoms, where they furnish similar but less quantitative information about the electrons. We shall now discuss the quantum numbers of electrons as they are used with atoms in general.

The first or *principal* quantum number, given the symbol n, is of primary importance in determining the energy of an electron. For the H atom, the energy is completely fixed by the value of n:

$$E = -\frac{B}{n^2}$$

In other atoms, in which there are more electrons, the energy of each electron is dependent mainly (but not completely) on the value of the principal quantum number. As the quantum number increases, the energy of the electron increases, and the average distance of the electron cloud from the nucleus also increases. The principal quantum number is always integral and can take on the values

$$n = 1, 2, 3, 4, 5, \ldots, \quad \text{but not } 0$$

In an atom, electrons having the same value of **n** move about in roughly the same region and are said to be in the same *level* or *shell*.

Each level of electrons in an atom includes one or more sublevels or subshells. The sublevels are denoted by the second quantum number, **l**, called the *angular momentum quantum number*. The general geometric contour of the electron cloud associated with the electron is determined by **l**. The quantum number **l** for an electron is related to the quantum number **n** for the electron in that state; for a given value of **n**, **l** is limited to the values:

$$l = 0, 1, 2, \ldots (n - 1)$$

How many sub-levels are there in the n = 5 level? What are their l quantum num-bers?

An electron with a quantum number **n** equal to 1 will, by the rules, have of necessity an **l** quantum number equal to 0; this means that in the **n** = 1 level, there will be only one sublevel, in which **l** equals 0. Electrons in the second level, where **n** equals 2, can have an **l** quantum number of either 0 or 1, so that there are two sublevels in the **n** = 2 level, denoted by **l** = 0 and **l** = 1. Similarly, in the **n** = 3 level, electrons can have **l** values of 0, 1, or 2. In general, in the nth level, there will be **n** sublevels, with **l** quantum numbers equal to 0, 1, 2, . . . , **n** − 1. We find from experience that electrons in sublevels within a given principal level will have energies which increase slightly with increasing **l** value. In the **n** = 3 level, an electron with **l** equals 0 will have less energy than one with an **l** value of 1. The most energetic electrons in the third level will be those in the sublevel in which **l** equals 2.

The third quantum number m_l is often called the *magnetic orbital quantum number*. This quantum number is associated with the orientation of the electron cloud with respect to a given direction, usually one which is imposed on the atom by a strong magnetic field. The quantum number m_l has very little effect on the energy of an electron in any atom. Its importance has to do with its influence on the orientation of the electron cloud and its relation to the angular momentum quantum number **l**. For a given value of **l**, m_l can have any integral value between **l** and −**l**. That is:

$$m_l = l, l - 1, l - 2, \ldots, 0, -1, -2, \ldots, -l$$

Electrons in an atom having the same values of **n, l,** and m_l are said to be in the same *orbital*. Within a given sublevel of quantum number **l** there will be $2l + 1$ orbitals.

In an **l** equals 0 sublevel, all the electrons must have quantum number m_l equal to 0; there is, therefore, only one orbital in the **l** equals 0 sublevel.

How many orbitals are there in an l = 2 sublevel?

When **l** equals 1, m_l may take on the values 1, 0, and −1; this means that in any sublevel in which **l** equals 1, an electron may be in any one of three orbitals, denoted by m_l = 1, 0, and −1.

In addition to its three quantum numbers **n, l,** and m_l, each electron appears to have a fourth quantum number m_s called its *magnetic spin quantum number*, or simply the spin quantum number. The spin quantum number can be loosely associated with the spin of the electron about its own axis. It was introduced into the theory to make the properties of atoms consistent with experiment. The quantum number m_s is not related to the values of n, l, or m_l for an electron and can have one of two possible values, $+1/2$ or $-1/2$, depending on the direction of rotation of the electron about its axis. Two electrons in the same orbital having m_s values of $+1/2$ and $-1/2$ are said to be **paired.**

The final property of electrons in atoms which we must consider relates to the possible sets of quantum numbers which the electrons in an atom can possess. As we have said, each electron in an atom requires four quantum numbers to describe it; an electron in an atom will be in a state determined by its quantum numbers n, l, m_l, and m_s. For instance, according to the rules we have cited, we might have an electron in a sodium atom with the quantum numbers 2, 0, 0, +1/2. The question we might ask is, How many electrons in the atom can be assigned this particular set of quantum numbers? The Pauli exclusion principle answers this question by telling us: **No two electrons in any atom can have the same set of four quantum numbers.** This condition was first stated by Pauli in 1925, again to make the theory of atomic structure consistent with experimental observations.

By the Pauli principle, within a given *orbital* there can be only two electrons, one with a spin quantum number of $+1/2$, and the other with a spin quantum number of $-1/2$. If there were more than two electrons in an orbital, they could not all have a different set of quantum numbers, and the Pauli principle would be violated. Since the population of each electronic orbital is thus limited to two electrons, the population of each sublevel, and of each level, is also limited.

Table 6.2 includes all allowed quantum number combinations up to $n = 3$. Each sublevel contains one or more orbitals, each of which has a capacity of two electrons. An $l = 0$ sublevel includes only one orbital ($m = 0$) and hence can contain *two* electrons. An $l = 1$ sublevel has three associated orbitals ($m_l = 1$, 0, -1), and so can hold *six* electrons. An $l = 2$ sublevel can contain *ten* electrons. Similarly each level has one or more associated sublevels and will have a capacity equal to the sum of those of its sublevels. The $n = 1$ level, containing only an $l = 0$ sublevel and one orbital, has a capacity of two electrons. The $n = 3$ level, with three sublevels which include a total of nine orbitals, can hold up to 18 electrons. The capacities of the electronic sublevels and levels up to $n = 5$ are summarized in Table 6.3. It is interesting to note that the total capacity of each electron level is equal to $2n^2$, where n is the principal quantum number for the level.

If $m_S = +\frac{1}{2}$, this is often indicated with a vertical arrow, ↑. If $m_S = -\frac{1}{2}$, ↓ is used

TABLE 6.2 Allowed Sets of Quantum Numbers for Electrons in Atoms

Level n	1	2				3								
Sublevel l	0	0	1			0	1			2				
Orbital m_l	0	0	1	0	−1	0	1	0	−1	2	1	0	−1	−2
Spin m_s	⥮	⥮	⥮	⥮	⥮	⥮	⥮	⥮	⥮	⥮	⥮	⥮	⥮	⥮

TABLE 6.3 Capacities of Electronic Levels and Sublevels in Atoms

LEVEL n	TOTAL NUMBER OF ELECTRONS IN LEVEL	NUMBER OF ELECTRONS IN SUBLEVELS 1 = 0	1	2	3	4	5
1	2	2	–	–	–	–	–
2	8	2	6	–	–	–	–
3	18	2	6	10	–	–	–
4	32	2	6	10	14	–	–
5	50	2	6	10	14	18	–

ELECTRON CONFIGURATIONS OF ATOMS

In order to describe the electronic state of an atom we need to specify as completely as we can the sets of quantum numbers of all its electrons. This specification is called the **electron configuration** of the atom and is of considerable use in connection with its chemical bonding properties. We have considered the rules governing the four quantum numbers of electrons in atoms in order that we might be able to predict the electron configuration of any atom.

Once the allowed sets of quantum numbers for electrons have been determined, as in Table 6.2, all that is required to establish the electron configuration of an atom is to assign its electrons to orbitals in the proper order. Electrons will enter the possible orbitals in order of their increasing energy, filling all the lower lying orbital positions before entering any of higher energy.

You will recall that the energy of an electron is dependent almost completely on the values of its **n** and **l** quantum numbers, increasing sharply with increasing **n** value and only slightly as the **l** value goes up. The m_l and m_s quantum numbers of an electron have only a very small effect on its energy. It is, then, the electronic sublevels, characterized by different **n, l** values, which have appreciably different energies in atoms.

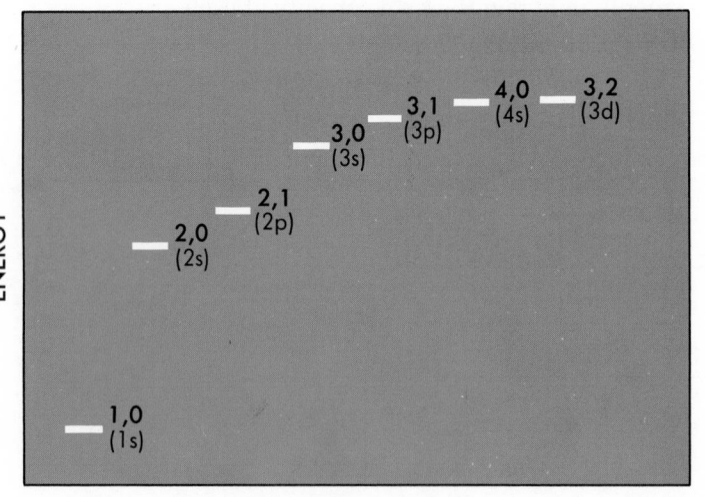

Figure 6.10 Relative energies of electronic sublevels.

Although the energies of the different sublevels in atoms cannot be calculated exactly, their relative magnitudes are obtainable from experiments and are qualitatively quite similar in different atoms. In Figure 6.10 we have shown the relative energies of some of the lower lying sublevels for electrons in atoms.

On this basis we can say that the two following sets of electron quantum numbers are associated with the lowest lying energy level in an atom:

$$1, 0, 0, +\tfrac{1}{2}; \quad 1, 0, 0, -\tfrac{1}{2}$$

These are the only two sets which have $n = 1$, and both necessarily have an l value of 0.

A hydrogen atom with its one electron will ordinarily be in one or the other of the two states described by the above sets of quantum numbers. The complete electron configuration of the H atom would then be simply

$$1, 0, 0, +\tfrac{1}{2} \quad \text{or} \quad 1, 0, 0, -\tfrac{1}{2}$$

Since only the values of n and l affect electron energies appreciably, the above description is often abbreviated to omit reference to the values of the m_l and m_s quantum numbers. This leaves the configuration specification of hydrogen as

$$1, 0$$

For some reason, probably traditional, we do not express electron configurations in quite this way. The notation for the principal quantum number, **n**, is as we have written it, but a new symbol is used for the second quantum number, l. Electrons for which $l = 0$ are called s electrons; if $l = 1$ they are called p electrons; for $l = 2$ they are called d electrons; if their l value is 3, they are called f electrons. The symbols s, p, d, and f come from the sharp, principal, diffuse, and fundamental series observed in atomic spectra early in this century. Spectroscopists were among the first to describe atoms in this way, and their notation has persisted. On this basis the electron in the ordinary H atom is a 1s electron, and the electron configuration of the H atom is simply 1s.

In the helium atom, with two electrons, we have an electron in each of the two states in the lowest lying level, with quantum number descrptions

$$1, 0, 0, +\tfrac{1}{2}: \ 1, 0, 0, -\tfrac{1}{2}$$

Since the two electrons are in the same sublevel, the electron configuration is written 1s², pronounced wun ess too, meaning there are two electrons with **n**, l values of 1, 0, respectively.

The three electrons in the lithium atom cannot all have 1s quantum numbers, since as we have seen there are only two sets of four quantum numbers having $n = 1$ and $l = 0$. We see from Figure 6.9 that the third electron would be assigned to the 2s sublevel, since that is the one having the next higher energy. With the first two electrons in the 1s level, we have an electron configuration for the Li atom of 1s²2s. *Why can't the electron configuration of the Li atom be 1s³?*

Since the 2s sublevel contains only one orbital ($m_l = 0$), it can hold only two electrons, and so will be filled at beryllium, atomic number 4, with electron configuration 1s²2s².

Each of the first two electronic sublevels contains only one orbital, so that the assignment of electrons to orbitals in atoms up to Be is unequivocal. When we reach boron, with five electrons per atom, we begin to fill orbitals in the 2p sublevel. In this sublevel there are three orbitals, with m_l values of 1, 0, and −1, all of which have the same energy. The manner in which these orbitals are populated by electrons has an effect on the magnetic and chemical properties of atoms and must be considered.

A 2p electron may be described by any one of the six sets of quantum numbers listed below:

$$2, 1, 1, +\tfrac{1}{2} \qquad 2, 1, 0, +\tfrac{1}{2} \qquad 2, 1, -1, +\tfrac{1}{2}$$

$$2, 1, 1, -\tfrac{1}{2} \qquad 2, 1, 0, -\tfrac{1}{2} \qquad 2, 1, -1, -\tfrac{1}{2}$$

$$m_l = 1 \qquad\qquad m_l = 0 \qquad\qquad m_l = -1$$

The fifth electron in the B atom will assume one of the six sets of quantum numbers listed above. Each set, being of equal energy, will be equally probable, and since we ordinarily cannot distinguish between the sets, we would say simply that the electron is in a 2p orbital and write the electron configuration of the B atom as $1s^2 2s^2 2p$.

When we come to the carbon atom, with its six electrons, the situation is somewhat different. The two 2p electrons in the C atom could both go into the same orbital, with paired spins (spin quantum numbers of $+1/2$ and $-1/2$), or they could go into different orbitals with either unpaired parallel spins (both spin quantum numbers having the same sign) or with paired opposed spins (spin quantum numbers having opposite signs). The experimental properties of the carbon atom indicate that there is an energy difference between the different possible arrangements and that the most stable configuration is the one in which the two electrons are unpaired, with parallel spins, and so are in different orbitals.

The condition we observe on the electron arrangement in the carbon atom can be generalized to include other atoms for which similar situations exist. There is a general principle governing all such cases, Hund's rule, which states that

In an atom in which orbitals of equal energy are to be filled by electrons, the order of filling is such that as many electrons remain unpaired as possible.

In order to keep the electron arrangement notation determined by Hund's rule as simple as possible, a system which avoids listing all four quantum numbers of the electrons is commonly employed. We simply indicate by means of a drawing the number of electrons in each orbital and the relative directions of their spins. In the case of the carbon atom, there would be five orbitals to be considered, a 1s, a 2s, and three 2p. These might be denoted by boxes, as in the following diagram. The populations and electron spins in each orbital are indicated by arrows: ↑ means an electron is present with spin quantum number equal to $+1/2$; ↓ means there is an electron in the orbital with spin quantum number equal to $-1/2$. By this system, the electron arrangement in the carbon atom would be

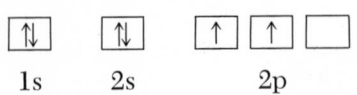

1s   2s   2p

The interpretation of the above representation, which we will call an **orbital diagram,** is that the 1s orbital contains two electrons which are paired (opposed spins), the 2s orbital also contains two electrons with opposed spins, and the two 2p electrons are in different orbitals and are unpaired (parallel spins). Since the other possible arrangements in the 2p orbitals which are consistent with Hund's rule are equivalent to the one given, they are not stated.

The orbital diagrams for atoms beyond carbon which contain 2p electrons in the outermost sublevel follow immediately by the approach we used for carbon. In Figure 6.11 we have shown the orbital diagrams for the atoms from boron to neon. You can see that the neon atom has no unpaired electrons, boron and fluorine both have one, carbon and oxygen two, and in the nitrogen atom the three 2p electrons can all remain unpaired. As we shall see in succeeding chapters, the orbital diagram notation for electronic arrangements in atoms is useful in discussions of chemical bonding in many chemical substances. Hund's rule applies to electronic structures in any atoms, ions, or molecules in which there are several orbitals with equal, or nearly equal, energies. Those arrangements will be energetically preferred where the maximum number of electrons are unpaired.

How many unpaired electrons are there in an Al atom? a P atom?

Using the building up, or Aufbau, principle, adding electrons one by one as atomic number increases (taking due regard of the Pauli principle and Hund's rule), one can readily determine the electron configurations and orbital diagrams of any atom. The electrons are assigned to the sublevels in order of increasing energy, in general filling each sublevel to capacity before beginning the next. The order of sublevels is that in Figure 6.10; a complete list in proper order is given as follows:

1s, 2s, 2p, 3s, 3p, 4s, 3d, 4p, 5s, 4d, 5p, 6s, 4f, 5d, 6p, 7s, 5f, 6d

You will note several cases where there is an "overlap" between principal energy levels. For example, the 4s sublevel ($n = 4$, $l = 0$) is filled

Atom	Orbital diagram			Electron configuration
B	↑↓ ↑↓ ↑ □ □			$1s^2 2s^2 2p$
C	↑↓ ↑↓ ↑ ↑ □			$1s^2 2s^2 2p^2$
N	↑↓ ↑↓ ↑ ↑ ↑			$1s^2 2s^2 2p^3$
O	↑↓ ↑↓ ↑↓ ↑ ↑			$1s^2 2s^2 2p^4$
F	↑↓ ↑↓ ↑↓ ↑↓ ↑			$1s^2 2s^2 2p^5$
Ne	↑↓ ↑↓ ↑↓ ↑↓ ↑↓			$1s^2 2s^2 2p^6$
	1s 2s 2p			

Figure 6.11 Orbital diagrams showing electron arrangements in some second row elements.

TABLE 6.4 The Electron Configurations of the Atoms of the Elements

Element	Atomic Number	1s	2s	2p	3s	3p	3d	4s	4p	4d	4f	5s
H	1	1										
He	2	2										
Li	3	2	1									
Be	4	2	2									
B	5	2	2	1								
C	6	2	2	2								
N	7	2	2	3								
O	8	2	2	4								
F	9	2	2	5								
Ne	10	2	2	6								
Na	11				1							
Mg	12				2							
Al	13		Neon core		2	1						
Si	14				2	2						
P	15				2	3						
S	16				2	4						
Cl	17				2	5						
Ar	18	2	2	6	2	6						
K	19							1				
Ca	20							2				
Sc	21						1	2				
Ti	22						2	2				
V	23						3	2				
Cr	24						5	1				
Mn	25		Argon core				5	2				
Fe	26						6	2				
Co	27						7	2				
Ni	28						8	2				
Cu	29						10	1				
Zn	30						10	2				
Ga	31						10	2	1			
Ge	32						10	2	2			
As	33						10	2	3			
Se	34						10	2	4			
Br	35						10	2	5			
Kr	36	2	2	6	2	6	10	2	6			
Rb	37											1
Sr	38											2
Y	39									1		2
Zr	40									2		2
Nb	41		Krypton core							4		1
Mo	42									5		1
Tc	43									6		1
Ru	44									7		1
Rh	45									8		1
Pd	46									10		
Ag	47									10		1
Cd	48									10		2

Table 6.4 *Continued.*

ELEMENT	ATOMIC NUMBER	POPULATIONS OF SUBSHELLS									
		4d	4f	5s	5p	5d	5f	6s	6p	6d	7s
In	49	10		2	1						
Sn	50	10		2	2						
Sb	51	10		2	3						
Te	52	10		2	4						
I	53	10		2	5						
Xe	54	10		2	6						
Cs	55	10		2	6			1			
Ba	56	10		2	6			2			
La	57	10		2	6	1		2			
Ce	58	10	2	2	6			2			
Pr	59	10	3	2	6			2			
Nd	60	10	4	2	6			2			
Pm	61	10	5	2	6			2			
Sm	62	10	6	2	6			2			
Eu	63	10	7	2	6			2			
Gd	64	10	7	2	6	1		2			
Tb	65	10	9	2	6			2			
Dy	66	10	10	2	6			2			
Ho	67	10	11	2	6			2			
Er	68	10	12	2	6			2			
Tm	69	10	13	2	6			2			
Yb	70	10	14	2	6			2			
Lu	71	10	14	2	6	1		2			
Hf	72	10	14	2	6	2		2			
Ta	73	10	14	2	6	3		2			
W	74	10	14	2	6	4		2			
Re	75	10	14	2	6	5		2			
Os	76	10	14	2	6	6		2			
Ir	77	10	14	2	6	9					
Pt	78	10	14	2	6	9		1			
Au	79	10	14	2	6	10		1			
Hg	80	10	14	2	6	10		2			
Tl	81	10	14	2	6	10		2	1		
Pb	82	10	14	2	6	10		2	2		
Bi	83	10	14	2	6	10		2	3		
Po	84	10	14	2	6	10		2	4		
At	85	10	14	2	6	10		2	5		
Rn	86	10	14	2	6	10		2	6		
Fr	87	10	14	2	6	10		2	6		1
Ra	88	10	14	2	6	10		2	6		2
Ac	89	10	14	2	6	10		2	6	1	2
Th	90	10	14	2	6	10		2	6	2	2
Pa	91	10	14	2	6	10	2	2	6	1	2
U	92	10	14	2	6	10	3	2	6	1	2
Np	93	10	14	2	6	10	5	2	6		2
Pu	94	10	14	2	6	10	6	2	6		2
Am	95	10	14	2	6	10	7	2	6		2
Cm	96	10	14	2	6	10	7	2	6	1	2
Bk	97	10	14	2	6	10	9	2	6		2
Cf	98	10	14	2	6	10	10	2	6		2
Es	99	10	14	2	6	10	11	2	6		2
Fm	100	10	14	2	6	10	12	2	6		2
Md	101	10	14	2	6	10	13	2	6		2

Krypton core

[handwritten annotation: $1s^2 2s^2 2p^6 3s^2 3p^6 3d^{10} 4s^2 4p^3$]

[handwritten annotation: $1s^2 2s^2 2p^6 3s^2 3p^6 3d^5 4s^1$]

before the 3d sublevel ($n = 3$, $l = 2$). This means that in potassium and calcium, electrons enter the 4s sublevel in preference to the 3d. The electronic configuration of potassium is $1s^2 2s^2 2p^6 3s^2 3p^6 4s^1$ rather than $1s^2 2s^2 2p^6 3s^2 3p^6 3d^1$.

Let us determine the electron configuration and the orbital diagram for the iron atom, atomic number 26. There are two 1s electrons, two 2s, six 2p, two 3s, six 3p, and two 4s electrons, filling the first six sublevels and requiring 20 electrons; these electrons will of necessity be paired in the orbitals in which they are found (why?). The remaining six electrons will go into the 3d sublevel, with its five equivalent orbitals, according to Hund's rule. The electron configuration of iron, Fe, atoms is, therefore, $1s^2 2s^2 2p^6 3s^2 3p^6 4s^2 3d^6$. The complete orbital diagram for the Fe atom is

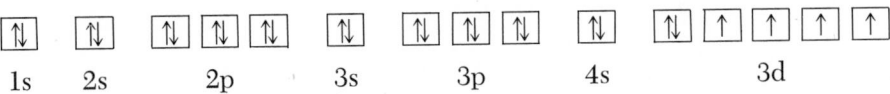

In applying Hund's rule we place the 3d electrons one by one in the orbitals available, keeping the electron spins all parallel (spin quantum number the same, $+1/2$ chosen arbitrarily). Having filled the orbitals with the first five such electrons, we put the sixth electron, with opposite spin, in the first orbital. In the atom we would anticipate that there would be four unpaired electrons.

In Table 6.4 we have listed the electron configurations of the elements of atomic numbers 1 through 101. The configurations generally follow the rules we have given. The major exceptions occur for those elements which are close to filling or half-filling sublevels. By Hund's rule we might surmise that half-full sublevels would be relatively stable; it is also true that a full sublevel, be it s, or p, or d, confers a measure of stability on the electronic structure of an atom. Where 3d sublevels are involved, with energies near those of the 4s sublevel, the effect is large enough in chromium (atomic number 24) to allow promotion of a 4s electron to the 3d sublevel, giving the Cr atom five 3d electrons, all with parallel spins. In the copper atom, with 29 electrons, a 4s electron is promoted to the 3d sublevel, thereby filling that sublevel.

Orbital diagrams are useful for determining the number of unpaired electrons in an atom

GEOMETRIC REPRESENTATIONS OF ELECTRON CHARGE CLOUDS

You will recall that the value at any point of the square of the wave function, ψ, associated with an electron is proportional to the probability of finding the electron at that point, and hence to the charge density at that point. The electron cloud sketched in Figure 6.9 indicates that the charge density in the atom is largest near the nucleus and drops off rapidly as the distance from the nucleus is increased.

The electron charge cloud associated with a single electron is dependent, as you might expect, on the values of n, l, and m_l for the state under consideration. The dependence on n can be expressed in a function that includes only r, the distance of the electron from the nucleus, and so is

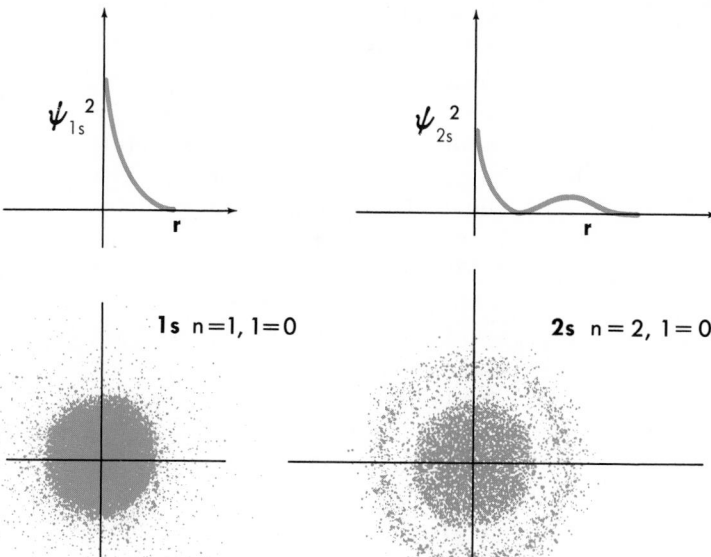

Figure 6.12 Charge clouds associated with s orbitals.

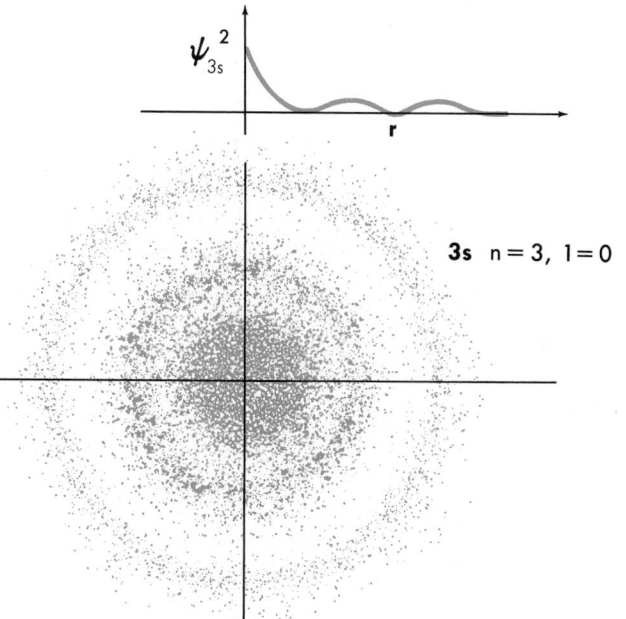

A 3s electron ranges over a much larger volume than does a 1s electron. Note, however, that it too is allowed to be near the nucleus

always spherically symmetric. As **n** increases, the charge cloud moves on the average farther from the nucleus and takes on a very roughly shell-like structure, with maxima and minima that depend in their position and magnitude on the value of **n**. The charge clouds for an electron with different **n** values and a value of $l = 0$ (an s electron) are shown in Figure 6.12. The charge cloud for any s electron is spherically symmetric.

For states in which **l** is not equal to zero the associated electron clouds are more complicated in their geometric structure. If we are working with p electrons, where $l = 1$, the value of m_l is no longer restricted to zero, but may take on the values +1, 0, and −1. We would anticipate that there would be three different electron clouds possible for $l = 1$ states, one for each possible value of m_l. This is indeed the case. The three clouds are identical in their overall structure but differ in their orientation with respect to a given set of axes, whose position might be established, for instance, by a strong magnetic field. The three clouds may be considered to be concentrated along the x, y, and z axes in the manner indicated in Figure 6.13. There are three possible p orbitals; they are often called p_x, p_y, and p_z orbitals. (The actual structure of the charge cloud for a p orbital will depend on the value of **n** as well as that of m_l, and in Figure 6.13 **n** has been taken to be 2.)

It is possible to draw charge clouds for orbitals associated with a d electron. In this case there are five different orbitals, corresponding to the five possible values for m_l. We shall consider d orbitals in some detail in Chapter 19.

The charge cloud associated with an np^6 configuration, like that for an ns^2 configuration, is spherically symmetrical

The charge clouds we have drawn so far have been for single electrons. Qualitatively one can assume that the charge cloud for an atom containing several electrons would be equal to the superposition of charge clouds associated with the electrons in each occupied orbital in the atom. In succeeding chapters we shall make extensive use of electron configurations, orbital diagrams, and geometric properties of atomic orbitals in our discussions of chemical bonding.

SOME COMMENTS ON THE ELECTRON CONFIGURATION THEORY. The theory of electron configurations associates the electrons in an atom with quantum states appropriate to their energies and space properties. The electron configuration and orbital diagram of an atom describe, as best we can at present, the properties of an atom in terms of the properties of its electrons. Associated with each electron quantum state is its set of quantum numbers and a cloud of negative charge, whose density at any point is proportional to the probability of finding the electron at that point in space. An atom, according to the model, consists of a positive nucleus surrounded by a cloud of negative charge resulting from the superposition of the charge clouds associated with each of its electrons.

Since the charge clouds for different sets of quantum numbers, or different quantum states, may differ in their properties, particularly in the distance from the atomic nucleus at which they have appreciable densities, our notion of the elec-

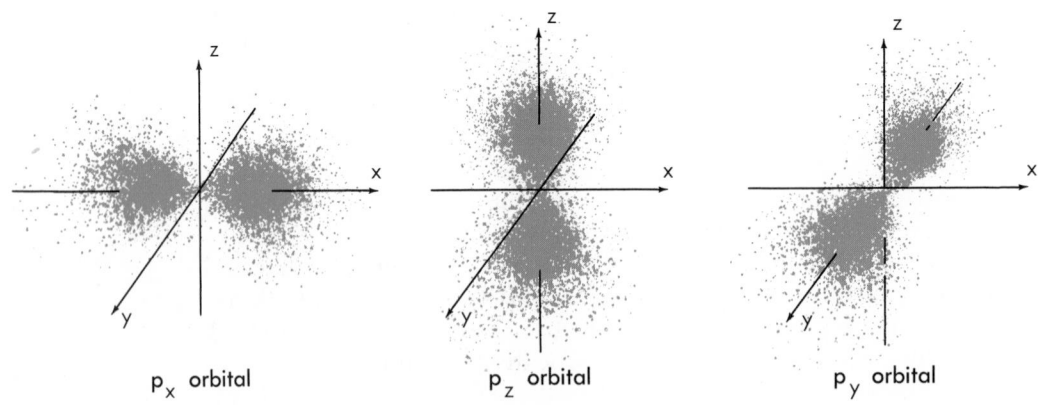

Figure 6.13   Electron clouds associated with p orbitals.

tronic structure of the atom is not quite so nebulous as it might first appear. Theory tells us that electrons with high principal quantum number **n** have high energy and, on the average, are much more likely to be found in regions relatively distant from the nucleus than are electrons with low **n** values. On this basis we can assign electrons to regions, which are usually called "shells" or "levels," one shell or level for each value of **n**, with shells associated with low **n** values near the nucleus and those with higher **n** values more distant. To discriminate between electrons in the same level (same **n** value) but different values of the quantum number **l**, we use the term "same sublevel" or "same subshell" to denote those electrons with the same values of both **n** and **l**. Since the regions over which the charge clouds for electrons have appreciable density are not at all sharply defined, "shell" terminology is somewhat unfortunate, since to the novice it implies more than it should. We use the shell notation occasionally because it is convenient, and we must simply remember the meaning of the term in light of this discussion. In this and subsequent chapters we frequently use the terms "level" and "sublevel" in place of "shell" and "subshell"; this terminology is perhaps better in principle in view of the fact that electron configurations describe energy levels, but it has the disadvantage that when used in this context "level" must be taken to mean that group of states having a given **n** value. Similarly, "sublevel" will mean that group of states with the same values of both **n** and **l**.

There is one difficulty with the electron configuration theory that should be mentioned. The theory is based on the assumption that the electrons in an atom can be described by assigning them quantum numbers. This means that in the theory we consider the electrons to have properties as individuals. (Indeed the quantum numbers we use and the interpretations we give them are based on the wave mechanical solution of the one-electron problem and, hence, involve only the electron-nucleus interactions.) In such an atomic model the electronic properties of the atom will simply be equal to the sum of the properties of all the electrons in the atom, with individual electron properties determined by solutions to one-electron problems. Such a model is clearly incorrect, since in an atom there are important electron-electron charge interactions and electron-electron spin interactions, as well as the electron-nucleus charge interaction covered by the model. So far, the theory of atomic structure has been unable to properly treat electron-electron interactions in atoms. Such interactions are usually considered to be "averaged in" when one interprets electron configurations. The qualitative structure of atoms, involving the existence of shells and their populations, does not seem to be appreciably altered by such an averaging; this explains the usefulness of electron configurations. Clearly, however, the necessity for such an averaging renders impossible the use of electron configurations in quantitative calculations of atomic energy levels and atomic dimensions.

6.6 ELECTRON CONFIGURATIONS AND IONIZATION POTENTIALS

The electronic configurations of atoms are based on theoretical evidence, particularly the wave mechanical solution to the hydrogen atom problem, and on experimental evidence, mainly atomic spectra. The configurations are also closely related, as we shall see in many later sections, to the chemical properties of the elements.

It would take us too far afield to show the detailed relations between atomic spectra and electron configurations. However, similar relations are obtained in investigations of the ionization energies of atoms and ions. The ionization energy of a particle is a direct measure of the energy required to remove an electron from that particle. A given element will have several ionization energies, one for each electron in its atom. The first ionization energy of sodium ($1s^2 2s^2 2p^6 3s^1$) is that required to remove the outermost,

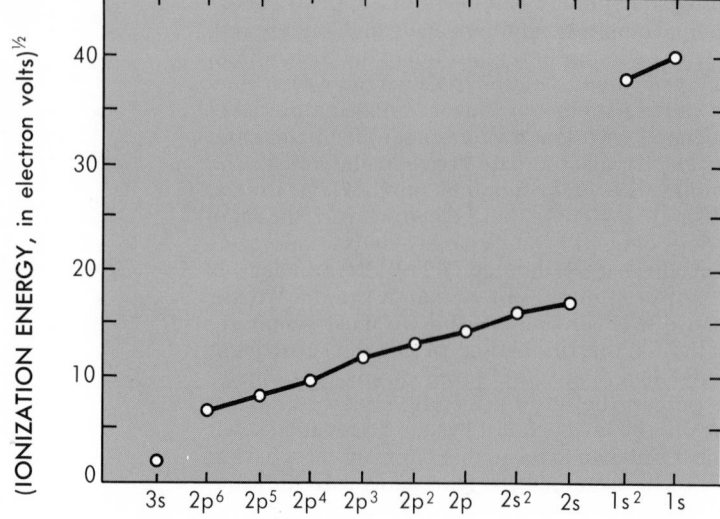

Figure 6.14 Ionization energies of the sodium atom.

The electrons in the Na atom are found experimentally in three sets of energy levels in accord with the electron configuration theory

3s, electron from the atom; the second ionization energy measures the energy required to remove a 2p electron from Na^+; the third, the energy needed to remove a second 2p electron from Na^{2+}; and so on. As the charge on the ion increases, the ionization energy increases. In Figure 6.14 we have plotted the square roots of the ionization energies of sodium as a function of the electron being removed.

The first electron to be removed from sodium, being a 3s electron, would be expected to come off relatively easily. Then we would predict that there would be a group of eight electrons, in the 2s and 2p sublevels, each of which would be somewhat less easily removed than the one before it, but which would not be grossly different in character from one another as far as energy of removal is concerned. The last two electrons, in the $n = 1$ level, would be very tightly bound and could be removed only with great difficulty.

Figure 6.14 clearly supports the theory. The ionization energies are grouped as expected, indicating the three sets of energy levels and their electron populations. The eight electrons for which $n = 2$ clearly fall in a single group, with no sharp break between the six 2p and the two 2s electrons.

6.7 THE ELECTRONIC STRUCTURES OF ATOMS AND THE PERIODIC TABLE

In the previous section we associated the electronic configurations of atoms with one of their physical properties, their ionization energies. Probably the most striking support of such configurations, however, comes from a study of the chemical properties of the elements.

We saw in Chapter 4 that when substances undergo a chemical reaction there is typically an energy change observed, usually in the form of heat which is evolved from the reaction mixture. Perhaps the simplest rational explanation of the origin of the energy change is that, during the reaction, a rearrangement of the electrons in the reacting atoms or molecules occurs, resulting in a final electronic structure which has a somewhat lower energy than that of the reactants. In chemical changes the energy liberated is of the order of 100 kcal per mole of reactant, or about 4 ev per molecule. This energy is roughly of the same order of magnitude as the ionization energy of the electrons in the outermost shell of an atom, but much smaller than the ionization energy of any of the inner electrons (Fig. 6.14).

From this rather simple observation one can conclude with some confidence that when atoms take part in chemical reactions the electron rearrangement which occurs involves only the outermost electrons. The electrons in the inner shells simply are energetically too stable to be appreciably affected by only a few electron volts of energy. This line of reasoning leads us to believe that, since only the outermost shells of electrons participate in chemical reactions, *the chemical properties of an atom are to be associated primarily with the electron configuration in its outermost shell.* *This principle will furnish the basis of much of our chemical bonding theory*

On examining the structures of atoms from this point of view, it is at once apparent that similar outermost electron configurations recur periodically as one scans the atoms in order of increasing atomic number. For example, as seen in Table 6.4, a single outermost s electron exists in the following elements:

Element	H	Li	Na	K	Rb	Cs
Atomic number	1	3	11	19	37	55
Outermost electron	1s	2s	3s	4s	5s	6s

If this approach has validity and chemical properties of atoms are indeed closely related to outermost electron configurations, we would deduce that the above elements should have similar physical and chemical properties. Experimentally we find that this is the case. Except for hydrogen, which turns out to be rather exceptional, all these elements are soft, low melting metals. Each of the metals, M, reacts rather violently with water, evolving hydrogen gas and forming an alkaline (basic) solution containing a hydroxide of the formula MOH. Essentially all the compounds formed by these elements are soluble in water. The formulas of these compounds are, almost without exception, clearly related to one another; some examples are

$$LiCl \quad NaCl \quad KCl \quad RbCl \quad CsCl$$
$$Li_2SO_4 \quad Na_2SO_4 \quad K_2SO_4 \quad Rb_2SO_4 \quad Cs_2SO_4$$

We can cite many other properties of these elements, but they all support the fact that these elements are very similar; because of their similarities they are called the group of alkali metals.

On studying the properties of other sets of elements in which the atoms have the same number of outermost electrons, one finds that they too can be assigned to a group whose members exhibit strong similarities in chemical reactivity, physical properties, and formulas of compounds. Within any group, trends in properties tend to occur gradually and smoothly as one proceeds from lower to higher atomic numbers (Fig. 6.15). The classifica-

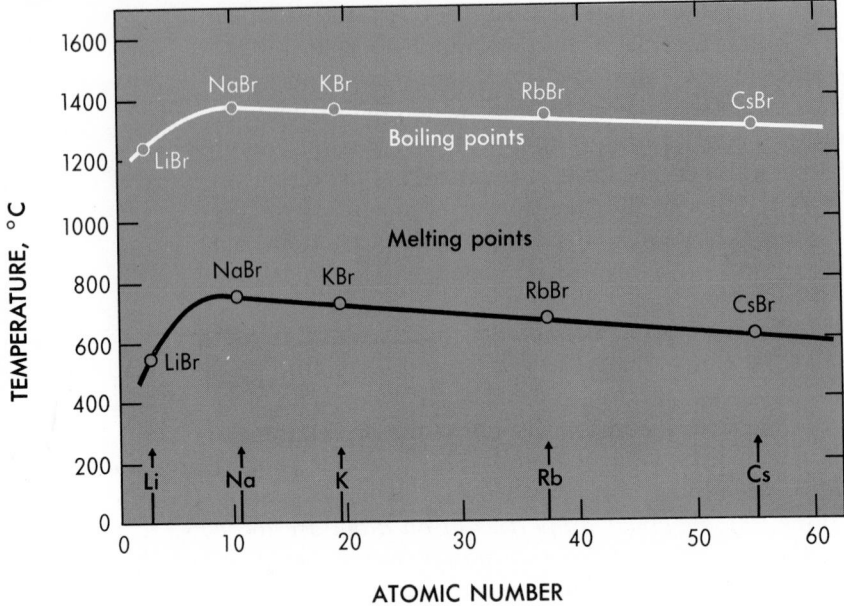

Figure 6.15 Trends in some physical properties of the bromides of the alkali metals.

tion of essentially all the known elements in this way is possible and results in a sorting of the elements into families, or groups, each of which has
its own characteristic properties and outermost electron configuration.

The correlation of the properties of the elements with the configurations of the outermost electrons in their atoms is evidence of the strongest
sort in support of the theory of electron configurations. You must not, however, conclude from this that the correlation was predicted by theory.
Historically, long before the theory was available, it was recognized
that there was a periodic recurrence of similar properties when the
elements were ordered according to atomic weight. The resulting classification of the elements into families was indeed one of the experimental facts which guided scientists in the development of the atomic
theory, and although the theory may well explain why the elements are
related this way, it was not involved at all in predicting the existence of
the relationship. Ordinarily the essential features of the periodic classification are summarized by arranging the elements with similar properties
in an array called the *Periodic Table*, a classic form of which is printed inside the front cover of this book.

The tale of the development of the Periodic Table is an interesting one, and
probably familiar to many of you who have studied some chemistry, so we shall
make it very brief. Early in the nineteenth century some chemists recognized that
there were triads of elements with similar properties. In particular, for calcium, strontium, and barium, it was shown that the atomic weight and other properties of the middle element were very close to the mean of the properties of the
other two. Somewhat later, about 1860, the periodic, or cyclically repeating, character of the elements when arranged in order of increasing atomic weight was
suggested by Newlands, an English chemist, but his "octaves," as he called them,
had such a resemblance to the musical scale that the idea appeared preposterous. In
those days the notion that the elements were related in some way must have

seemed revolutionary, to say the least. In 1870, however, Mendeleev, a Russian chemist, classified, much as we do in the modern Periodic Table, the 70 elements which were then known and boldly asserted that the gaps in his arrangement corresponded to elements yet to be discovered. He then went on to predict the properties of those elements on the basis that they would lie at the average of the properties of the elements above and below them in the Periodic Table. This must have appeared to be a fantastic idea at the time, but when, over the next 20 years, the elements he predicted were discovered and had the properties he had said they would, the Periodic Table was confirmed as a scientific fact and the name of Mendeleev was entered on the list of the great men of science.

Although the noble gases were unknown at the time of Mendeleev, they fit readily into his Periodic Table

STRUCTURE OF THE PERIODIC TABLE

Although the Periodic Table was developed on the basis of experiment, it is probably easier to understand its structure if one approaches it through the electron configurations of atoms, relating the chemical properties of elements to the electronic structures of their outermost shells. The Table lists elements, in order of increasing atomic number, in horizontal rows of such length that elements with the same outermost electron configuration fall directly beneath one another. To illustrate this point, consider the elements of the second and third periods (Table 6.5).

You can see that elements 3 to 10 all differ in their outer structures; starting with element 11 we begin a new cycle in which the elements have the same configuration as those directly above them (ns for the 1A group, ns^2 for 2A, and so on). There are eight elements in each of these two rows or periods because eight electrons are required to completely fill the s and p sublevels corresponding to a given value of n (i.e., ns^2p^6).

In the 4th period, ten additional elements, often referred to as transition metals, appear. In these elements, starting with scandium (AN = 21) and going across to zinc (AN = 30), the 3d sublevel is being filled. Since there are 10 3d electrons, 10 elements are required to fill that sublevel. As one might expect, since these elements differ in structure only in an inner sublevel, the transition metals resemble one another in some of their properties; they are classified as belonging to the B groups. In the 5th row there are again 18 elements, eight Group A elements and 10 in the B Groups. Elements like Cr and Mo in Group 6B, having similar structures, resemble one another as do elements in a given A Group. In the 6th and 7th rows the general pattern continues, with the addition of 14 more elements in each

The cycles, or periods, in the Periodic Table are not all of equal length

TABLE 6.5   The Second and Third Rows of the Periodic Table

GROUP	1A	2A	3A	4A	5A	6A	7A	8A
Element	Li	Be	B	C	N	O	F	Ne
At. No.	3	4	5	6	7	8	9	10
Outermost Configuration	2s	$2s^2$	$2s^22p$	$2s^22p^2$	$2s^22p^3$	$2s^22p^4$	$2s^22p^5$	$2s^22p^6$
Element	Na	Mg	Al	Si	P	S	Cl	Ar
At. No.	11	12	13	14	15	16	17	18
Outermost Configuration	3s	$3s^2$	$3s^23p$	$3s^23p^2$	$3s^23p^3$	$3s^23p^4$	$3s^23p^5$	$3s^23p^6$

row, due to the filling of yet another inner subshell, 4f in the 6th row and 5f in the 7th. The series of 14 elements in the 6th row are called the lanthanides, or rare earths; they resemble one another even more than do the transition metals, since in this case the 4f electrons are very deeply buried in the electronic structure of the atoms. In the 7th period the 14 elements are called the actinides, many of which are radioactive in addition to having very similar properties. The lanthanides and actinides are listed separately at the bottom of the Table to keep the width manageable.

Hydrogen is the most difficult element to place in the Periodic Table. We have put it in Group 7A, but it has some similarity to the elements in Group 1A

Among the A group elements, we find a general decrease in metallic character as we move from left to right in the Table. Looking at the third period, for example, we find that the first three elements, Na, Mg, and Al, have typical metallic properties. Of these, sodium gives up electrons most readily and, hence, is the most reactive. The next four elements, Si, P, S, and Cl, show to an increasing extent the characteristic properties of nonmetals. Of these elements, chlorine, which has the greatest tendency to acquire electrons, is the most reactive toward metals. The last element in the period, argon, is extremely unreactive. This suggests that the electronic structure ns^2p^6 is particularly stable, an idea whose consequences we will examine in Chapter 7.

Metallic character ordinarily increases for A group elements as one moves down in the Periodic Table. This is particularly evident in group 4A, which includes carbon and silicon, rather typical nonmetals, followed by germanium, which is on the borderline between metals and nonmetals, and then by tin and lead, both of which are clearly metals. Groups 5A and 6A show a similar trend; nitrogen and oxygen are both gaseous nonmetals, while the elements at the bottom of these groups, bismuth and polonium, are primarily metallic. In Group 7A we move from the most reactive of nonmetals, fluorine, to a solid element, iodine, which shows the beginning of metallic properties, in particular metallic luster.

The elements along the diagonal line are called metalloids and include all the useful transistor materials

The broad diagonal line in the table represents an approximate division between metals and nonmetals. We see that most of the elements in the Periodic Table are metals; the relatively few nonmetals are clustered at the upper right corner. Since most compounds contain at least one of the nonmetals, their importance in chemistry is much greater than their number might imply.

USING THE PERIODIC TABLE

Given an understanding of the nature of the Periodic Table one can, on the basis of relatively little information, make predictions of the properties of many substances. The principle to be used is very simple, namely,

TABLE 6.6 Germanium in the Periodic Table

3A	4A	5A
Al	Si	P
Ga	Ge	As
In	Sn	Sb

that the properties of the elements within a Group vary gradually and smoothly, so that an unknown element will have properties that are a properly chosen average of the adjacent elements in the same Group. This principle was the basis on which Mendeleev made his successful predictions and is a very useful one. Mendeleev made several predictions regarding the element germanium, and we might profitably follow his line of reasoning.

Example 6.2 Predict the formula of germanium chloride and its boiling point.

Solution. Germanium is in Group 4A in the same family as silicon and tin and, by our general principle, will tend to form compounds having formulas similar to those of silicon and tin. Two chlorides of each element are known:

$$SiCl_4 \quad Si_2Cl_6 \qquad SnCl_2 \quad SnCl_4$$

Since $SiCl_4$ and $SnCl_4$ both exist, a likely formula for the germanium compound would be $GeCl_4$, and this was Mendeleev's prediction. The possibility that Ge_2Cl_6 and $GeCl_2$ also exist is suggested by the other formulas. Actually, the ordinary chloride observed is $GeCl_4$. Ge_2Cl_6 and $GeCl_2$ have been reported but are less stable.

We would expect the boiling point of $GeCl_4$ to be between those of $SiCl_4$ and $SnCl_4$, which boil at 58° and 114°C, respectively. Since germanium has an atomic number halfway between those of silicon and tin, we should take the average of the two boiling points as the predicted boiling point of $GeCl_4$:

$$BP \; GeCl_4 = \frac{BP \; SiCl_4 + BP \; SnCl_4}{2} = \frac{58° + 114°}{2} = 86°C$$

The actual boiling point of $GeCl_4$ is 83°C. Mendeleev, being somewhat conservative in this case, predicted a boiling point below 100°C.

The condition which allowed us to predict in Example 6.2 that $GeCl_4$ is the formula of germanium chloride is obeyed by the formulas of many inorganic compounds. This condition allows the correlation of the chemical formulas in a very powerful way, and in its most general form is embodied in the following rule:

Given the formula of any inorganic compound, the replacement of an element in the compound by another from the same group will often give the correct formula for another compound.

Given, for example, the formula for magnesium chloride, $MgCl_2$, one would correctly predict on the basis of this rule that the other 2A chlorides would have the formulas $CaCl_2$, $SrCl_2$, and $BaCl_2$, and that the 7A binary magnesium compounds would be MgF_2, $MgBr_2$, and MgI_2. The general formula for 2A to 7A binary compounds would be predicted to be MX_2, as is observed.

The general relation is really far broader than this. It would imply, for instance, that since calcium molybdate has the formula $CaMoO_4$, that of barium molybdate would be $BaMoO_4$, that calcium chromate would be $CaCrO_4$, and that strontium tungstate would have the formula $SrWO_4$. These formulas are all correct and follow directly on the substitutions of elements in the same family in the basic formula.

Like many general rules, the relation we have cited has some exceptions. The difficulty is that in some cases we would predict the existence of substances that do not occur or are extremely unstable. Indeed, as we noted, $GeCl_2$ and Ge_2Cl_6 are not ordinarily observed. Another example of such a failure would arise if carbon monoxide, CO, were used as a reference substance. The rule would predict the existence of a sulfide of carbon with the formula CS; this compound, if it exists at all, appears to be highly unstable. The fact that the rule would also predict the existence of substances with the formulas PbS and SnO, both of which are well known, may increase our confidence in the relation, but does not remove the fact that an exception occurs. Such exceptions occur most frequently when one or more of the elements involved is the first member of a group of nonmetals, 4A through 7A, or the first member of a group in the transition series. With practice and due caution, however, the student will find the general relation for systematizing chemical formulas to be a very useful one.

6.8 THE CORRELATION OF ATOMIC RADII WITH ELECTRON CONFIGURATIONS

As we have just seen, a knowledge of the electron configurations of atoms allows us to correlate and predict many of their physical and chemical properties in a truly remarkable way. In the next few chapters we shall discuss the applicability of the theory to problems of chemical bonding. Here we shall show how electron configurations can be used to aid our understanding of atomic sizes.

According to our model, the electrons in an atom are arranged in diffuse shells and subshells around the nucleus, with the average radius of a shell increasing with increasing principal quantum number of its component electrons. The electrons and the nucleus obey, within quantum-mechanical limits, the laws of electrostatics. Electrons, being of like negative charge, repel each other; electrons and nuclei, on the other hand, having negative and positive charges, tend to attract each other. Mathematically these forces are given by Coulomb's Law:

The sizes of atoms are the result of a balance between electron-electron repulsions and electron-nucleus attractions

$$\text{for electron-electron repulsion} \quad F_R = \frac{e^2}{r^2}$$

$$\text{for electron-nucleus attraction} \quad F_A = \frac{Ze^2}{r^2}$$

where e is the magnitude of the electronic charge, Ze is the nuclear charge, and r is the interparticle distance. It is also true that the force between a set of concentric spherical charges and a distant charge is proportional to the algebraic sum of all the charges included in the set of spheres (Fig. 6.16). In our model (admitted|y very simplified), Figure 6.16 can be considered to represent an atom, with its nuclear charge +Ze "shielded" from the outermost electron by the charges $-q_1$ and $-q_2$ of the electrons in the two inner electron shells. The force of attraction pulling the outermost electrons toward the nucleus is:

$$F = \frac{e(Ze - q_1 - q_2)}{r^2}$$

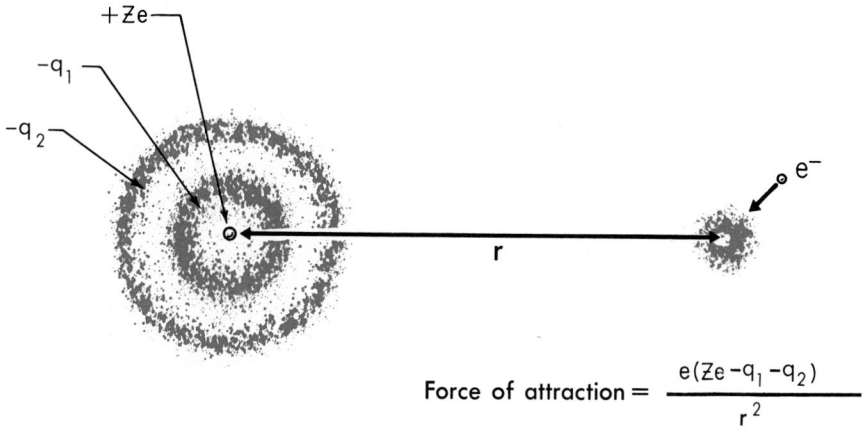

$$\text{Force of attraction} = \frac{e(\bar{Z}e - q_1 - q_2)}{r^2}$$

Figure 6.16 Shielding of nuclear charge by inner electronic subshells.

On the basis of the model let us now consider sizes of atoms. As a first example, examine the electronic structures of the elements in Group 1A (Table 6.7). The atoms of the elements in this Group all have a single s electron outside a closed shell or a closed p subshell. The full subshells are relatively much closer to the nucleus than is the outermost s electron and form essentially concentric spheres of negative charge around the nucleus, thereby decreasing the effective positive charge of the nucleus very markedly and so decreasing the net attractive force on the outermost electron. If the shielding were perfect, the outermost electron would experience attraction due to a single positive charge and would have the properties of an electron with the same quantum number in a hydrogen atom. Since the average distance of the electron from the hydrogen nucleus increases rapidly with increasing value of **n,** the size of the alkali metal atoms should increase with increasing principal quantum number of the outer electron and so increase as one proceeds through the family from Li to Cs. The shielding is not perfect, but nevertheless the general line of reasoning is valid enough to properly predict the increase of size with period for the atoms within any given family.

A similar argument can be applied to atoms of elements in the same period, that is, with the same number of shells but differing configurations in the outermost shell. A good illustration here would be the third period elements shown in Table 6.8.

In each of the atoms in this period the electron configuration in the inner shells is the same, namely $1s^2 2s^2 2p^6$. If shielding were perfect, the outer electrons, in the third shell, would each be attracted by a charge equal to $Z - 10$ for an atom of atomic number Z; electrons in the third shell,

TABLE 6.7 Radii of the Alkali Metals

	Li	Na	K	Rb	Cs
Configuration in outermost shell	2s	3s	4s	5s	6s
Atomic radius in Å	1.52	1.86	2.31	2.44	2.62

TABLE 6.8   Radii of Elements in the Third Period

	Na	Mg	Al	Si	P	S	Cl
Outer configuration	3s	$3s^2$	$3s^23p$	$3s^23p^2$	$3s^23p^3$	$3s^23p^4$	$3s^23p^5$
Atomic radius in Å	1.86	1.60	1.43	1.17	1.10	1.04	0.99

all being at about the same distance from the nucleus, are much less effective "shields" than are the inner electrons. In view of the fact that Z increases as we proceed across the period from left to right, increasing the attractive force on the outer electrons as we move in that direction, we can readily understand the observed decrease in atomic size on going from Na on the left to Ar on the right end of the third period.

PROBLEMS

6.1   Atoms are very small particles, yet we feel that we know a great deal about them. Why do we believe, for example, that

    a. atoms contain nuclei?
    b. atoms are about 1 Å in diameter?
    c. there are electrons outside atomic nuclei?
    d. the nature of an atom is fixed by its nuclear charge and not by its mass?

6.2   Cite some experiments involving the properties of atoms and molecules which cannot be explained by classical mechanics.

6.3   What theoretical basis does the electron configuration theory have? What additions need to be made to explain experimental properties of atoms?

6.4   An electron in an atom has the quantum numbers 3, 1, 0, $+\frac{1}{2}$. Compare its energy and position with those of an electron with quantum numbers 2, 0, 0, $-\frac{1}{2}$.

6.5   One of the strong lines in the atomic spectrum of mercury occurs at 4358 Å. What is the energy in ergs of photons having this wavelength? What is the energy in calories per mole of these photons?

6.21   Describe each of the following components of atoms as to size, mass, and charge:

    a. electrons
    b. nuclei
    c. protons
    d. neutrons

6.22   How do the laws of motion governing the electrons in a fluorine atom differ from the laws governing the motion of planets in the solar system?

6.23   State the role that each of the following has in establishing the theory for electron configurations:

    a. Pauli principle.
    b. Hund's rule.

6.24   An electron has four quantum numbers, n, l, m_l, and m_s. What relationship does each of the quantum numbers have to the energy and position of the electron?

6.25   The first line in the Balmer series for hydrogen is found at 6562 Å. What change in energy occurs in an H atom when it emits a photon with this wavelength? Compare the energy in calories of a mole of these photons with the heat evolved when a mole of water vapor is formed from the elements.

6.6   The chemical and physical properties of sulfur resemble those of tellurium but not those of strontium. How does atomic theory explain these experimental observations? Is the fact that the atomic number of tellurium is 36 greater than that of sulfur fortuitous, or does the theory explain that too?

6.7   The electron configuration of the S atom is $1s^2 2s^2 2p^6 3s^2 3p^4$. What does this statement mean?

6.8   What is the electron configuration of each of the following atoms in the ground state?

$$_{15}P, \; _{29}Cu, \; _{23}V, \; _{18}Ar$$

6.9   Which of the following electron configurations would correspond to atoms in excited states? Which are ground state configurations? Which are incorrect?

$1s^2 2p$	$1s^2 2s^2 2p$
$1s^2 2s^2 2d$	$1s^2 2s^2 2p^3 3d$
$1s^2 2s^3$	$1s^2 2s^2 3s$

6.10   Draw orbital diagrams for atoms with the following electron configurations:

a. $1s^2 2s^2 2p^6 3s^2 3p^4$ c. $1s^2 2s$
b. $1s^2 2s^2 2p^2$ d. $1s^2 2s^2 2p^5$

How many unpaired electrons would each of the above atoms have?

6.11   An atom has the following orbital diagram. What is its electron configuration? Name the element to which the atom belongs.

1s   2s   2p
[↑↓] [↑↓] [↑↓][↑↓][↑↓]
3s   3p
[↑↓] [↑↓][↑][↑]

6.12   Given electrons with the following sets of quantum numbers, arrange them in order of increasing energy. If any have the same energy, put them together.

$2, 1, 0, \frac{1}{2}$
$2, 0, 0, -\frac{1}{2}$
$3, 1, 1, \frac{1}{2}$
$3, 1, -1, -\frac{1}{2}$
$3, 2, 1, \frac{1}{2}$
$3, 1, 0, \frac{1}{2}$

6.26   Give the atomic number of an element which

a. has both metallic and nonmetallic character.
b. is in Group 5A and least resembles arsenic.
c. cannot easily be assigned a position in the Periodic Table.

6.27   How would you go about determining the electron configuration of the Si atom? What are some of the things it would tell you about the atom?

6.28   State the electron configurations of each of the following species in their ground states:

$$Al \quad Ca \quad Na^+ \quad O^- \quad F^+ \quad Mn$$

6.29   Which of the following electron configurations would correspond to atoms in excited states? Which are ground state configurations? Which are just plain wrong?

$1s^2 2s^2 2p^6 3s$	$1s^2 3d$
$1s^2 2s^2 2p^5 3p$	$1s^2 2s^2 3p^7$
$1s^2 2s^2$	$1s^2 2s^2 2p^6 2d^2$

6.30   Draw orbital diagrams for atoms with the following electron configurations:

a. $1s^2 2s^2 2p^6 3s^2$ b. $1s^2 2s^2 2p^6 3s^2 3p^6 3d^4$
c. $1s^2 2s^2 2p^3$

Name the atom that has each configuration and state the number of unpaired electrons it would have.

6.31   An ion with charge +1 has the following orbital diagram. Name the ion and state its electron configuration

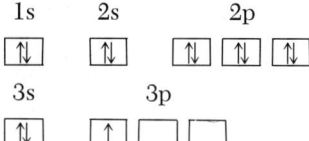

1s   2s   2p
[↑↓] [↑↓] [↑↓][↑↓][↑↓]
3s   3p
[↑↓] [↑][ ][ ]

6.32   Given the following possible sets of quantum numbers for electrons, indicate those sets which could not occur and, starting from the right of each set, correct it by changing the numbers in the smallest possible way in order to obtain an allowed set:

$3, 1, \frac{1}{2}, 0$
$3, 1, 0, \frac{1}{2}$
$3, 2, 2, -\frac{1}{2}$
$4, 0, 1, \frac{1}{2}$
$2, 1, 1, \frac{1}{2}$

6.13 Consider the element Sr. From the properties of its neighbors in the Periodic Table, estimate its atomic weight and atomic radius. Compare the observed values, available in this text.

6.14 Given the formulas KI, $BiCl_3$, La_2O_3, WF_6, $Na_2Sn(OH)_6$, and $RbClO_3$ predict the formula of

a. lithium chloride.
b. potassium hydroxo-plumbate.
c. arsenic bromide.
d. molybdenum fluoride.
e. sodium bromate.
f. scandium sulfide.

6.15 Name three ions which could have the following electron configuration, remembering that ions usually have the same configuration as atoms with the same number of electrons:

$$1s^2 2s^2 2p^6 3s^2$$

6.16 Select the species in each of the following pairs which you would expect to be larger:

a. Ca   Ba   d. Mg   P
b. C   N   e. Sc   Mn
c. Sr   Rh   f. Na Na^+

6.17 The energy of any one-electron species, like He^+ or Li^{2+}, in its nth state is found by the Bohr theory to be equal to $-Z^2B/n^2$, where Z is the charge on the nucleus and B is 2.179×10^{-11} ergs. On this basis, find the ionization energy of the He^+ ion, first in ergs and then in electron volts. (1 ev = 1.60×10^{-12} ergs.)

6.18 Referring to Problem 6.17, find the wavelength in Å of the light emitted when He^+ ions make a transition from the n = 2 to the n = 1 state. What would it be for n = 5 to n = 4 transition?

6.19 What would be the minimum energy in ergs of an electron confined to a 2.0 Å box? Why would the electron obey only quantum mechanical laws?

6.20 Calculate the minimum energy of an electron confined to an atomic nucleus of diameter equal to 1.0×10^{-12} cm. Compare this energy with that available for binding nuclear particles, which is about 8×10^6 ev. Suggest why nuclear theoreticians do not feel that there are electrons in atomic nuclei.

6.33 Predict the boiling point of Br_2, using values for its neighbors in the Periodic Table, available in a handbook such as the Handbook of Chemistry and Physics.

6.34 Given the chemical formulas H_2S, HBr, Na_3PO_4, Sb_2O_3, $FeCl_3$, and K_2CrO_4, predict likely formulas for the following substances:

a. osmium bromide.
b. iodine chloride.
c. hydrogen selenide.
d. potassium phosphate.
e. arsenic sulfide.
f. sodium molybdate.

6.35 When atoms are excited, usually only one electron is affected. What would be the electron configuration of the O atom in its first three excited states?

6.36 Select the species in each of the following pairs which you would expect to be larger:

a. O   S   d. Na^+ F^-
b. O   F   e. Nd   Eu
c. Mg Mg^{2+} f. Cl Cl^-

6.37 Referring to Problem 6.17 for information, find the energy in ergs of the Li^{2+} ion in the ground state. What would be the ionization energy of the ion in electron volts?

6.38 Referring to Problem 6.17 for information, find the wavelength of the light emitted when H atoms make a transition from the n = 4 to the n = 2 state. What would it be for n = 3 to n = 1 transitions? Which wavelength belongs to the Balmer series? Which line in the series is it?

6.39 Calculate the minimum energy allowed to a 100-kg man confined to a closet one meter long. What would be his quantum number if he moved at a speed of one mm/sec (Equation 6.12)? Would his movements in the closet obey classical laws?

6.40 Calculate the minimum energy of a proton confined to an atomic nucleus. Compare this energy with that available for binding nuclear particles. Would this binding energy be enough to hold a proton? a neutron? an alpha particle? See Problem 6.20 for data.

°6.41 A rotating O_2 molecule can be represented in quantum mechanics by a particle of atomic weight equal to 8 moving on a circle of radius r equal to 1.2 Å. The energies allowed to the particle are identical to those it would have if confined to a box of length equal to $2\pi r$. Determine whether this system will behave classically. What would be the wavelength of photons which could excite this system from the n = 1 to the n = 2 state? In what region of the spectrum would light of this wavelength lie?

*6.42   If the electronic charge were only one tenth as large as it is, atoms would be about 100 times as large as they now are. Under such conditions would the electrons in atoms behave quantum mechanically? Hint: Consider the discussion on p. 134.

*6.43   X-ray spectra originate from transitions which occur in an atom when electrons in innermost shells are removed by bombardment with high-energy electrons. A very common transition involves an n = 2 electron moving down in energy to replace an ejected n = 1 electron. For this transition, the system behaves almost exactly like a one-electron species of atomic number Z-1 undergoing an n = 2 to n = 1 transition. On this basis, and using the data in Problem 6.17, find the wavelength in the x-ray spectrum of the Mo atom which is associated with that transition.

*6.44   Sodium atoms can be made to emit yellow light at a wavelength of 5890 Å if they are heated sufficiently in a Bunsen flame. This light is the basis of the test for sodium in qualitative analysis. In the flame the atoms are excited by fast-moving molecules, having an average energy of 3RT/2 per mole. Account as best you can for the fact that a sodium-containing substance will give off yellow light when in a flame at about 1500°C.

*6.45   Suppose that the spin quantum number could have the values $-\frac{1}{2}$, 0, and $+\frac{1}{2}$. Assuming that the rules governing the possible values of the other quantum numbers and the order of filling sublevels remain unchanged,

    a. what would be the capacity for electrons of an s sublevel? a p sublevel?
    b. what would be the capacity of the n = 2 level?
    c. what would be the electron configuration of the element with atomic number 10? 18?
    d. what would be atomic numbers of two elements which would be expected to most resemble the one with atomic number 10?

7

CHEMICAL BONDING

In Chapter 6 we discussed the electron distribution in isolated atoms. In this chapter we will consider how these distributions change when atoms combine with one another. Specifically, we will be concerned with the nature and properties of three different kinds of interatomic forces which are strong enough to be called chemical bonds.

1. *Ionic bonds*, the electrostatic forces that hold oppositely charged ions together in such familiar compounds as table salt (NaCl), fluorite (CaF$_2$) and limestone (CaCO$_3$).

2. *Covalent bonds*, which operate between atoms in nonmetallic elementary substances such as hydrogen, chlorine and carbon and in molecular compounds such as water, ammonia and carbon dioxide.

3. *Metallic bonds*, found in metals such as copper, zinc, tin, and lead and in alloys such as brass and pewter.

7.1 IONIC BONDING

When the metal sodium and the nonmetal fluorine are brought together, a vigorous exothermic reaction takes place.

$$Na(s) + \tfrac{1}{2} F_2(g) \rightarrow NaF(s); \quad \Delta H = -136 \text{ kcal} \qquad (7.1)$$

The product of this reaction is a crystalline, high melting solid (mpt = 1000°C) which is a good conductor of electricity either in the molten state or in water solution. These properties are characteristic of ionic compounds which consist of positively and negatively charged ions held together by strong electrostatic forces. In this particular compound, there are equal numbers of Na$^+$ and F$^-$ ions; the way in which these ions are packed in the crystal is shown in Figure 7.1.

CHARGES AND ELECTRONIC STRUCTURES OF IONS

When an ionic compound is formed by the transfer of electrons from metal to nonmetal atoms, there is a tendency for ions that have particularly

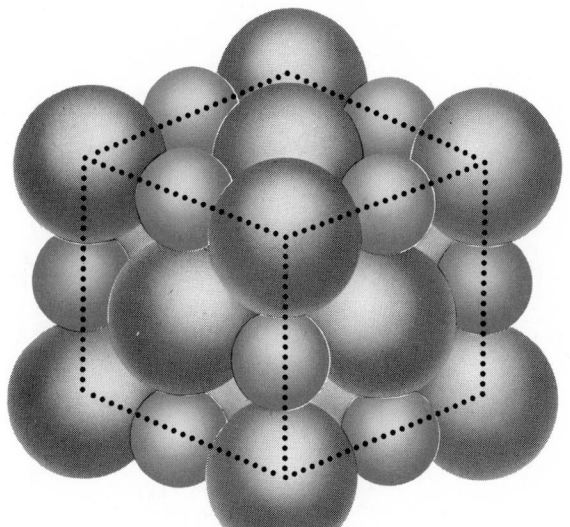

Figure 7.1 Arrangement of ions in NaF crystal. The small spheres represent Na^+ ions, the large ones F^- ions.

This is one of the most common structures for ionic solids of formula MX

stable electronic structures to be produced. One such structure is that of the noble gas atoms (ns^2np^6). The fluoride ion, formed when fluorine reacts with metals, has the neon structure

$$_9F(1s^22s^22p^5) + e^- \rightarrow {}_9F^-(1s^22s^22p^6)$$

The other halogens (Cl, Br, I) all form -1 ions (Cl^-, Br^-, I^-), thereby acquiring the electronic structure of atoms of the noble (inert) gas (Ar, Kr, Xe) which immediately follows them in the periodic table.

The nonmetals in Group 6A of the periodic table (O, S, Se, Te) form -2 ions when they react with metals (O^{2-}, S^{2-}, Se^{2-}, Te^{2-}). These ions, like those formed from the halogens, have noble gas electronic structures.

$$_8O(1s^22s^22p^4) + 2\ e^- \rightarrow {}_8O^{2-}(1s^22s^22p^6)$$

The argument which we have just gone through can be extended to predict, correctly, that metals with 1 or 2 more electrons than the preceding noble gas would, in reacting with nonmetals, lose electrons to form positive ions with charges of $+1$ and $+2$, respectively. Examples include

$$_{11}Na(1s^22s^22p^63s^1) \rightarrow {}_{11}Na^+(1s^22s^22p^6) + e^-$$

$$_{12}Mg(1s^22s^22p^63s^2) \rightarrow {}_{12}Mg^{2+}(1s^22s^22p^6) + 2\ e^-$$

Among the metals which have three more electrons than the preceding noble gas, aluminum in group 3A and all the metals in group 3B are capable of forming $+3$ ions (Al^{3+}, Sc^{3+}, . . .) when they react with nonmetals. Boron, the first member of group 3A, shows no tendency to do so.

The metals located to the right of group 3B in the periodic table cannot form positive monatomic ions with noble gas electronic structures. In order to do so, they would have to lose four or more electrons, for which the energy requirement is prohibitively large. Yet the properties of many of the compounds that these elements form with nonmetals indicate them to be ionic. For example, silver fluoride, AgF, is a high melting solid (mpt =

TABLE 7.1 Outer Electronic Structures of Some Transition and
Post-Transition Metal Cations

	1ST SERIES		2ND SERIES			3RD SERIES		
	Cr^{3+}	$3d^3$	Mo^{2+}	$4d^4$				
	Mn^{2+}	$3d^5$						
	Fe^{3+}	$3d^5$	Ru^{3+}	$4d^5$				
	Fe^{2+}	$3d^6$						
	Co^{2+}	$3d^7$	Rh^{3+}	$4d^6$				
	Ni^{2+}	$3d^8$	Pd^{2+}	$4d^8$	Pt^{2+}	$5d^8$		
	Cu^{2+}	$3d^9$						
	Cu^+	$3d^{10}$	Ag^+	$4d^{10}$	Au^+	$5d^{10}$		
	Zn^{2+}	$3d^{10}$	Cd^{2+}	$4d^{10}$	Hg^{2+}	$5d^{10}$		
	Ga^{3+}	$3d^{10}$	In^{3+}	$4d^{10}$	Tl^+	$5d^{10}$	$6s^2$	
			Sn^{2+}	$4d^{10}$	$5s^2$	Pb^{2+}	$5d^{10}$	$6s^2$

435°C) with a high electrical conductivity in the molten state. This suggests the presence of Ag^+ ions combined with an equal number of F^- ions. The same sort of evidence leads us to believe that Cu^+ and Mn^{2+} ions are present in the compounds CuCl and $MnCl_2$.

The electronic structures of some of the more common ions of the transition and post-transition metals are given in Table 7.1. **Notice that when these ions are formed, the outer s electrons are lost first.** For example, the Mn^{2+} ion is formed from a manganese atom (electronic structure: $Ar3d^54s^2$) by the loss of two 4s electrons.

Most of the cations in Table 7.1 form stable complex ions (Chapter 19)

Perhaps the most striking feature of Table 7.1 is the frequency with which ions with completed subshells (d^{10} or $d^{10}s^2$) show up. Metals which can acquire such a structure by losing 1, 2 or 3 electrons commonly do so. It would appear that these structures, like that of the noble gases (s^2p^6), have associated with them a certain "extra" stability.

We should emphasize that although many of the transition and post-transition metals are capable of forming positive ions such as those listed above, the bonding in many of their compounds is not primarily ionic. Consider, for example, the compound $SnCl_4$, where we might be tempted to postulate the existence of Sn^{4+} ions. Experimental observations show

TABLE 7.2 Some Common Polyatomic Ions

+1	−1	−2	−3
Ammonium, NH_4^+	Hydrogen sulfate, HSO_4^-	Dichromate, $Cr_2O_7^{2-}$	Phosphate, PO_4^{3-}
Mercurous, Hg_2^{2+}	Hydrogen carbonate, HCO_3^-	Chromate, CrO_4^{2-}	
	Hydroxide, OH^-	Sulfate, SO_4^{2-}	
	Nitrate, NO_3^-	Sulfite, SO_3^{2-}	
	Chlorate, ClO_3^-	Carbonate, CO_3^{2-}	
	Perchlorate, ClO_4^-	Peroxide, O_2^{2-}	
	Acetate, $C_2H_3O_2^-$		
	Cyanide, CN^-		
	Permanganate, MnO_4^-		

clearly that $SnCl_4$ is molecular rather than ionic in nature. It is a liquid at room temperature (mpt $= -30°C$), soluble in nonpolar organic solvents such as carbon tetrachloride, and a poor conductor of electricity.

The ions discussed up to this point have all been derived from a single atom by the loss or gain of electrons. Such ions are referred to as *monatomic*. In contrast, many of the ions most frequently encountered in general chemistry contain more than one atom. The electronic structures of *polyatomic* ions are discussed in Section 7.4; the names and charges of a few of the more common cations and anions of this type are given in Table 7.2.

Example 7.1   Deduce the formulas of the following ionic compounds: calcium fluoride, aluminum oxide, ammonium sulfate, sodium peroxide.

Solution.   In each case we obtain the formula from the charges of the ions involved, taking into account the fact that the compound must be electrically neutral. Thus, for calcium fluoride (Ca^{2+}, F^- ions) we obtain the formula CaF_2. For aluminum oxide, where two Al^{3+} ions are required to balance three O^{2-} ions, the formula is Al_2O_3. Applying this principle to ammonium sulfate (NH_4^+, SO_4^{2-}) gives $(NH_4)_2SO_4$. In the case of sodium peroxide, two Na^+ ions are required to balance one O_2^{2-} ion. The formula is usually written as Na_2O_2 to suggest its structure: the empirical formula, NaO, is misleading.

SIZES OF IONS

The fact that it is impossible to obtain a sample of matter made up of only one kind of ion makes it difficult to determine ionic radii. From x-ray diffraction studies (Chapter 9), we obtain the distances between the centers of touching ions. For most ionic crystals, where the contact is between positive and negative ions, this distance is the sum of the two ionic radii, i.e.,

$$d = \text{radius cation} + \text{radius anion}$$

Without further information, it is impossible to decide how the internuclear distance should be split up among the two ions.

One equation with two variables can't be solved

One way out of this dilemma is to make x-ray measurements on crystals in which anions are in contact with each other. One such crystal is lithium iodide, where the small Li^+ ions fit into holes in the lattice between adjacent I^- ions. In this case the measured internuclear distance (Fig. 7.2) is twice the radius of the anion, i.e.,

$$\text{radius } I^- = 4.32 \text{ Å}/2 = 2.16 \text{ Å}$$

Once the radius of the I^- ion is known, it is possible to choose cationic radii which fit the observed internuclear distances in such compounds as RbI and CsI, where the cations are large enough to prevent anion-anion contact. From the radii of Rb^+ and Cs^+, one can establish radii for F^-, Cl^-, Br^-, and so on. The ionic radii listed in Appendix 2, p. 660, were obtained by a somewhat different approach than the one we have described, but the values obtained are very similar.

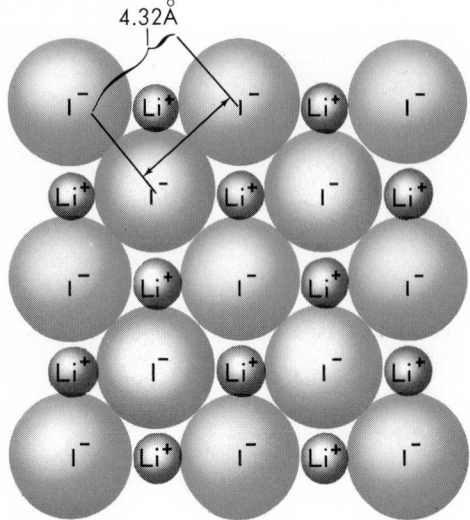

Figure 7.2 In LiI, the anions touch each other. One half the distance between the centers of adjacent anions can be taken to be the radius of the I⁻ ion.

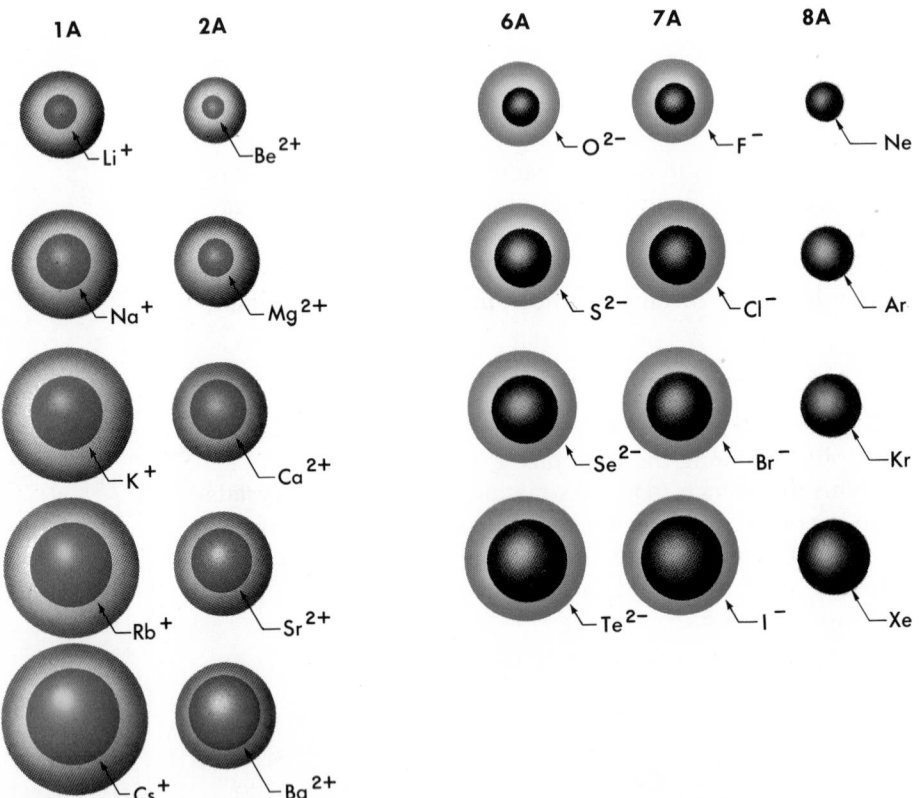

Figure 7.3 Relative sizes of atoms and ions of the A group elements. Notice that + ions are generally smaller than − ions.

In Figure 7.3 we show the sizes of the monatomic ions (and atoms) of the Group A elements. From the figure, several trends are apparent.

1. A series of species such as O^{2-}, F^-, Ne, Na^+ and Mg^{2+}, all of which have the same number of electrons, shows a steady decrease in radius with increasing nuclear charge. In these *isoelectronic* species, an increase in nuclear charge (from 8 in O^{2-} to 12 in Mg^{2+}) pulls the outer electrons closer to the nucleus (radius $O^{2-} = 1.40$ Å, radius $Mg^{2+} = 0.65$ Å). All these ions have the same configuration, that of Ne

2. When an atom loses electrons to form a positive ion (e.g., $Na \rightarrow Na^+ + e^-$), there is a decrease in radius (1.86 to 0.95 Å). The formation of a negative ion from an atom (e.g., $F + e^- \rightarrow F^-$) leads to an increase in radius (0.64 to 1.36 Å).

3. Within a given group in the periodic table, ionic radius ordinarily increases with increasing atomic number (e.g., $Li^+ < Na^+ < K^+ \ldots$).

7.2 FORMATION OF THE COVALENT BOND

We can rationalize the existence of oppositely charged ions in compounds such as sodium fluoride or magnesium oxide by postulating the transfer of electrons from metal (Na, Mg) to nonmetal (F, O) atoms. However, electron transfer cannot explain the formation of a molecule of an elementary substance such as H_2, where we are dealing with two identical atoms.

The atoms in the H_2 molecule are held together by what is referred to as a *covalent* or *electron-pair* bond. To emphasize the fact that a pair of electrons is shared between two atoms, the bond is often represented as

$$H : H \quad \text{or} \quad H—H$$

with the understanding that the two dots or the straight line drawn between the two hydrogen atoms represent a covalent bond. This simple picture can be misleading if it is taken to imply that the two electrons are located at a fixed position between the two nuclei. A more accurate map of the electron density would resemble that shown in Figure 7.4. At any instant the two electrons may be located at any of various points around the two nuclei. There is, however, a somewhat greater probability of finding the electrons between the two nuclei than at the far ends of the molecule.

A question that has long intrigued chemists is: why should the sharing of two electrons between two nuclei result in increased stability? Why, for example, should the H_2 molecule be more stable, by about 100 kcal/mole,

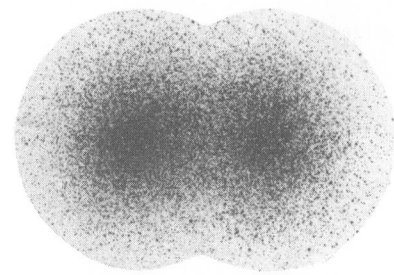

Figure 7.4 Electron density in H_2. The depth of shading is proportional to the probability of finding an electron in a particular region.

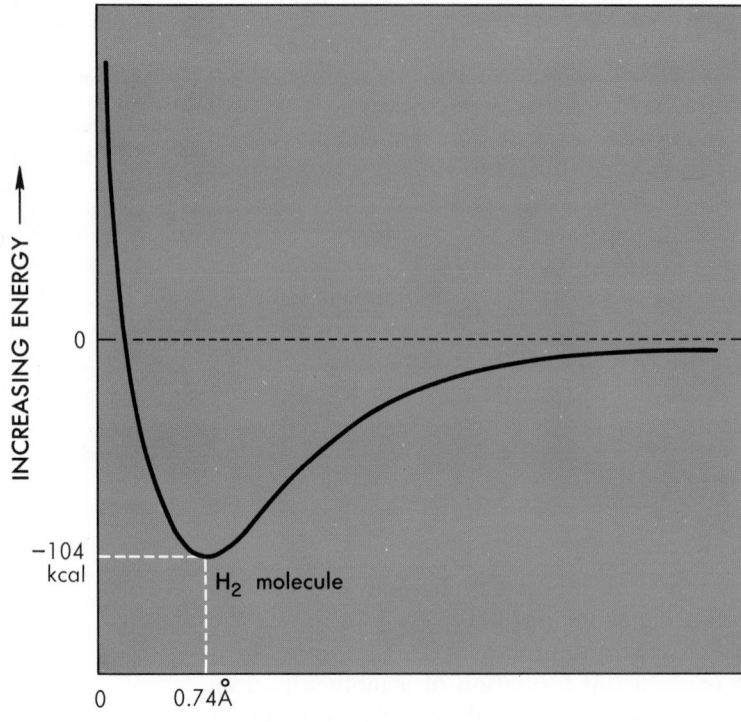

Figure 7.5 Energy of a system of two H atoms as a function of the distance between them. The minimum represents the equilibrium distance of separation in the H_2 molecule.

than two isolated hydrogen atoms? The first plausible answer to this question was put forth in 1927 by two physicists, W. A. Heitler and T. London. By combining the wave functions of two hydrogen atoms, using the principles of quantum mechanics, they were able to calculate the interaction energy of two hydrogen atoms as a function of the distance between them.

At large distances of separation (far right, Figure 7.5) there is essentially no interaction between the two hydrogen atoms. As the atoms come closer together (moving to the left in Figure 7.5), they experience an attraction which leads gradually to an energy minimum. This minimum occurs at an internuclear distance of 0.74 Å; the attractive energy at this point is 104 kcal/mole, the bond energy of the H_2 molecule. At this distance of separation the molecule is in its most stable state. If we attempt to bring the atoms closer together, repulsive forces become increasingly important and the energy curve rises steeply.

All stable chemical bonds must possess minima of this sort

The existence of the energy minimum shown in Figure 7.5 is responsible for the stability of the hydrogen molecule. Two factors are responsible for this minimum; the second of these, based on quantum mechanical considerations, turns out to be the more important.

1. At distances in the vicinity of 0.74 Å, the electrostatic attraction between the electron of one atom and the nucleus of the other (electron 1 and nucleus 2, electron 2 and nucleus 1) exceeds the electrostatic repulsion between particles of like charge (electron 1 and electron 2, nucleus 1 and nucleus 2). This factor accounts for about 10 per cent of the total bond energy.

2. When two hydrogen atoms approach each other closely, the elec-

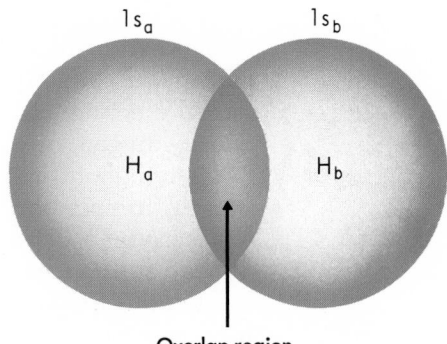

Figure 7.6 Overlap of two 1s orbitals in H_2.

tron density is spread out over the entire volume of the molecule instead of being confined to a particular atom. This amounts in effect to an increase in the volume available for electrons to move around in. You will recall from our discussion of the "particle in a box" model of Chapter 6 that the energy of an electron is inversely related to the dimension of the box. For a three-dimensional box

$$E \propto \frac{1}{V^{2/3}}; \quad V = \text{volume} \tag{7.2}$$

Admittedly, the particle in a box model is a crude oversimplification of the situation that exists in the H_2 molecule. Nevertheless, the prediction that we would make on the basis of Equation 7.2 is qualitatively correct. The increased volume available to an electron when it is allowed to move around two nuclei rather than one lowers its energy and, hence, contributes to the minimum shown in Figure 7.5. In the jargon of quantum mechanics, this principle is often described in terms of an *overlap* of the charge clouds of the electrons. We say that in the H_2 molecule, the 1s orbitals of the atoms overlap in such a way that an electron originally confined to a single orbital can spread itself over both orbitals.

The mathematical equations upon which the Heitler-London model is based cannot be solved exactly for any but the simplest of molecules. However, by certain approximation methods, it is possible to extend the model to rationalize the stability of covalent bonds in a variety of elementary and compound substances. One method of doing this was developed in the 1930's by Linus Pauling and J. C. Slater, among others. The general approach, which emphasizes the overlap between electron orbitals in the bonded atoms, is referred to as the **atomic orbital** or **valence-bond** method.

According to the **valence-bond** approach, if two atoms are to form a covalent bond, they must each have an unshared electron to contribute. Looking at the electronic configurations of the hydrogen and fluorine atoms,

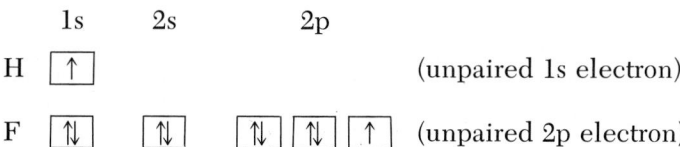

we would predict, correctly, that two hydrogen atoms, two fluorine atoms, or one hydrogen and one fluorine atom could combine to form the molecules H_2, F_2 and HF. In contrast, helium atoms, in which there are no unpaired electrons, would not be expected to form bonds with themselves, with hydrogen, or with fluorine.

Throughout most of this chapter, we shall use the atomic orbital approach, in part because of its simplicity, but more important because it can account satisfactorily for the properties of most of the covalently bonded substances with which we deal in general chemistry. It does, however, suffer from certain rather serious deficiencies. In particular, it tends to underemphasize the extent to which the energy levels of electrons in atoms are modified by covalent bond formation.

Within the past decade an alternative approach to covalent bonding, developed originally by F. Hund and R. S. Mulliken around 1930, has become increasingly popular. This approach is known as the **molecular orbital** method. In the derivation of the electronic structure of a molecule by this method, all the electrons are considered to belong to the molecule as a whole. These electrons are then distributed among a set of energy levels, called molecular orbitals, which are analogous to the atomic orbitals used to describe the energies of electrons in isolated atoms. We shall present an introduction to molecular orbital theory in Section 7.7.

7.3 PROPERTIES OF THE COVALENT BOND

Three of the most important characteristics of a covalent bond are its *polarity*, which tells us how the bonding electrons are distributed between the two atoms, its *enthalpy of formation*, which is a measure of the heat evolved when it is formed, and the *length* of the bond, which measures the distances between the centers of the two bonded atoms. In this section, we shall examine how these properties vary with the nature of the atoms which share the electron pair.

POLAR AND NONPOLAR BONDS, ELECTRONEGATIVITY

As we might expect, the two electrons joining the atoms in the H_2 molecule are equally shared by the two nuclei. In quantum mechanical terms, we say that the electron density is symmetrical about a line joining the two nuclei: an electron is as likely to be found in the vicinity of one nucleus as the other. Bonds of this type are described as **nonpolar.** We expect to find nonpolar bonds whenever the two atoms joined are identical, as is the case in H_2 or F_2.

In the HF molecule, the distribution of the bonding electrons is somewhat different from that in H_2 or F_2. Here, the density of the electron cloud is concentrated around the fluorine atom; the bonding electrons, on the average, are shifted toward fluorine and away from hydrogen (Fig. 7.7).

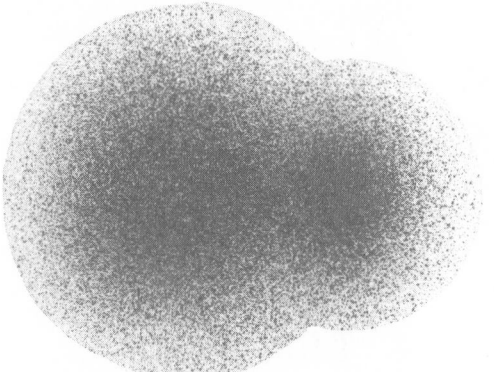

Figure 7.7 Electron density in HF. Probability of finding a bonding electron is greater in the vicinity of F atom.

Bonds in which the electron distribution is unsymmetrical are referred to as **polar.** The hydrogen fluoride molecule can be described as a dipole, with the fluorine atom acting as a negative pole and the hydrogen atom as a positive pole.

Since atoms of two different elements always differ in their affinity for electrons, covalent bonds between unlike atoms are always polar. The degree of polarity can be interpreted in terms of the relative **electronegativity** (attraction for electrons) of the two atoms joined by a covalent bond. In the ClF molecule, the bonding electrons are only slightly displaced toward fluorine, which is slightly more electronegative than chlorine. On the other hand, fluorine has a much stronger attraction for electrons than does hydrogen, so the polarity in the HF molecule is much more pronounced than in ClF.

The difference in electronegativity between atoms of two different elements can be estimated in various ways. One method, based upon bond energy calculations, leads to the scale of electronegativities listed in Table 7.3. On this scale, each element is assigned a number ranging from 4.0 for the most strongly electronegative element, fluorine, to 0.7 for the element having the least attraction for electrons, cesium. Among the A-group elements, electronegativity values increase from left to right in the periodic table and ordinarily decrease as one moves down in a given group.

TABLE 7.3   Electronegativity Values

H						
2.1						
Li	Be	B	C	N	O	F
1.0	1.5	2.0	2.5	3.0	3.5	4.0
Na	Mg	Al	Si	P	S	Cl
0.9	1.2	1.5	1.8	2.1	2.5	3.0
K	Ca	Sc	Ge	As	Se	Br
0.8	1.0	1.3	1.8	2.0	2.4	2.8
Rb	Sr	Y	Sn	Sb	Te	I
0.8	1.0	1.2	1.8	1.9	2.1	2.5
Cs	Ba	La–Lu	Pb	Bi	Po	At
0.7	0.9	1.0–1.2	1.9	1.9	2.0	2.2

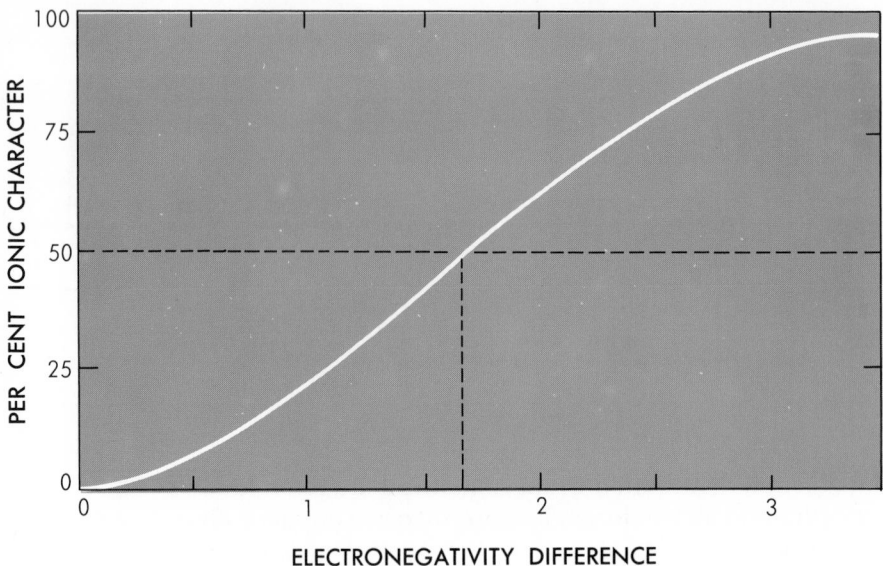

ELECTRONEGATIVITY DIFFERENCE

Figure 7.8 Relation between ionic character of bond and difference in electronegativity of atoms joined by the bond.

A polar covalent bond may be thought of as being intermediate between a pure (nonpolar) covalent bond, in which the electrons are equally shared, and a pure ionic bond, in which there has been a complete transfer of an electron from one atom to the other. In this sense, we sometimes express bond polarity in terms of *partial ionic character*. The greater the difference in electronegativity between two elements, the more ionic will be the bond between them. The relationship between these two variables is shown graphically in Figure 7.8. Notice that a difference of 1.7 electronegativity units corresponds to a bond with approximately 50 per cent ionic character. Such a bond might be described as being halfway between a pure covalent and a pure ionic bond.

It is clearly an oversimplification to refer to a bond between two elements as being "ionic" or "covalent." Consider, for example, the bonding in compounds formed by a 1A or 2A metal with a nonmetal in group 6A or 7A. The difference in electronegativity ranges from a minimum of 0.6 for the beryllium-tellurium pair to a maximum of 3.3 for cesium and fluorine. The percentage of ionic character in the bonds formed between these pairs of elements shows a corresponding variation, from about 10 per cent for BeTe to 95 per cent for CsF.

Ionic and co-valent bonds should be interpreted in light of these considerations

The difference in electronegativity between oxygen or fluorine on the one hand and a 1A or 2A metal on the other always exceeds 1.7 units, the value corresponding to 50 per cent ionic character. In this sense the bonding in the oxides and fluorides of these metals is predominantly ionic. The same statement applies to the oxide and fluoride of aluminum in group 3A, where electronegativity differences of 2.0 and 2.5 units correspond to about 65 per cent and 80 per cent ionic character in Al_2O_3 and AlF_3, respectively. The situation is quite different with the chloride, bromide and iodide of aluminum; in each of these three compounds, electronegativity differences less than 1.7 units imply that the bonding is predominantly covalent.

How ionic is the bond in NaCl; in H_2O?

BOND DISTANCES AND ENTHALPIES

It is possible to estimate the distance between atoms joined by a covalent bond by taking the sum of their atomic radii (Appendix 2, p. 660). For the F_2 molecule, the value obtained in this way

$$\text{F—F distance} = 0.64 \text{ Å} + 0.64 \text{ Å} = 1.28 \text{ Å}$$

is exactly equal to the observed internuclear distance, for the very simple reason that the atomic or, more properly, the covalent radius of a nonmetal) is taken to be one half the distance between centers of bonded atoms in the elementary substance! In other words, the atomic radius of a nonmetal X is *defined* to be

$$\text{radius X} = (\text{X—X distance})/2$$

When two unlike atoms, differing in electronegativity, are joined by a covalent bond, the observed internuclear distance is ordinarily smaller than that calculated by taking the sum of the atomic radii. Thus, for HF, we would calculate, using covalent radii,

The radius of atom X is *not* determined from the size of the free atom

$$\text{H—F distance} = \text{radius H} + \text{radius F} = 0.37 \text{ Å} + 0.64 \text{ Å} = 1.01 \text{ Å}$$

as compared to an observed distance of 0.92 Å. The simplest way to explain the bond shortening in HF and similar molecules is to assume that the introduction of partial ionic character into a covalent bond strengthens it and, hence, tends to pull the bonded atoms closer together.

A more direct measure of bond strength is the amount of heat evolved when the bond is formed from individual gaseous atoms. This quantity, known as the heat or enthalpy of bond formation, was introduced in Chapter 4. Values of ΔH_f for various bonds are listed in Table 4.2, Chapter 4. Here, again, we have an indication that polarity contributes to bond strength. Compare, for example, the observed enthalpy of bond formation of H—F

$$\Delta H_f \text{ H—F} = -135 \text{ kcal}$$

to the value we would calculate by taking the average for the two nonpolar bonds H—H and F—F,

$$\frac{\Delta H_f \text{ H—H} + \Delta H_f \text{ F—F}}{2} = \frac{-104 \text{ kcal} - 37 \text{ kcal}}{2} = -71 \text{ kcal}$$

We see that the H—Cl bond is some 64 kcal/mole stronger than we might expect if it were nonpolar.

TABLE 7.4 Calculated and Observed Enthalpies of Bond Formation of Hydrogen Halides (kcal/mole)

	X = F	X = Cl	X = Br	X = I
ΔH_f H—H	−104	−104	−104	−104
ΔH_f X—X	−37	−58	−46	−36
ΔH_f H—X (calc)	−71	−81	−75	−70
ΔH_f H—X (obs)	−135	−103	−88	−71
Y (calc − obs)	64	22	13	1

In Table 7.4 we list the "extra bond enthalpy," Y, associated with bond polarity for the hydrogen halide molecules. Note that the discrepancy between calculated and observed values as expressed by Y decreases as the bond becomes less polar. It has its maximum value, 64 kcal/mole, in the strongly polar HF molecule. In HI, where the two atoms have nearly equal electronegativities, the difference has nearly disappeared; Y is only 1 kcal/mole.

Observations of this sort led Linus Pauling to suggest the following relationship between Y and the difference in electronegativity

$$Y = 23(\Delta EN)^2 \tag{7.3}$$

Electronegativity is mainly defined by this equation

The electronegativities listed in Table 7.3 were calculated with the aid of this equation. Thus, for H—Br, where Y = 13, we obtain

$$\Delta EN = \left(\frac{13}{23}\right)^{1/2} = 0.75 \approx (2.8 - 2.1)(\text{values taken from Table 7.3})$$

MULTIPLE BONDS

Ordinarily it is found that the distance between two atoms joined by a covalent bond and the amount of heat evolved when it is formed are nearly constant in different molecules. Consider, for example, the first three compounds listed in Table 7.5, ethane, propane and butane. The measured C—C bond distance and the calculated value of ΔH_f C—C in these three compounds are identical.

When we find that one or the other of these properties changes significantly, we suspect a change in bond type. This is the case with the last two compounds listed in Table 7.5 (ethylene, C_2H_4, and acetylene, C_2H_2). The distance between the carbon atoms in ethylene (1.33 Å) is significantly shorter than the C—C bond distance (1.54 Å). At the same time, the bond holding the carbon atoms together in this compound is significantly stronger than the C—C bond. These effects become even more pronounced in acetylene.

This evidence is interpreted to mean that there are **multiple** bonds between the carbon atoms in ethylene and acetylene. Specifically, we say that there is a **double bond** in ethylene, consisting of two pairs of electrons joining the two carbon atoms,

TABLE 7.5   Carbon-Carbon Bond Distances and Enthalpies

	BOND DISTANCE (Å)	ΔH_f C—C (kcal/mole)
C_2H_6	1.54	−83
C_3H_8	1.54	−83
C_4H_{10}	1.54	−83
C_2H_4	1.33	−143
C_2H_2	1.20	−196

$$
\begin{array}{cc}
\text{H} & \text{H} \\
| & | \\
\text{H—C} & = & \text{C—H}
\end{array}
$$

and a **triple bond** in acetylene (three pairs of electrons)

$$\text{H—C} \equiv \text{C—H}$$

We always find that double or triple bonds are stronger than single bonds between the same two atoms. Compare, for example, the enthalpy of bond formation of the triple bond in the N_2 molecule to that of the single bond in the N_2H_4 molecule

:N≡N:

$$
\begin{array}{cc}
\text{H} & \text{H} \\
\diagdown & \diagup \\
\ddot{\text{N}}\text{—}\ddot{\text{N}} \\
\diagup & \diagdown \\
\text{H} & \text{H}
\end{array}
$$

The N≡N bond is the most stable bond known

$\Delta H_f \ \text{N} \equiv \text{N} = -225 \ \text{kcal/mole}$ $\Delta H_f \ \text{N—N} = -38 \ \text{kcal/mole}$

Hydrazine, N_2H_4, is used as a rocket fuel, in part because its combustion produces the very stable N_2 molecule and hence liberates a large amount of energy.

7.4 LEWIS STRUCTURES; THE OCTET RULE

The concept of the covalent or electron-pair bond was first introduced by the American physical chemist G. N. Lewis in 1916. To rationalize the stability of this bond, he pointed out that *atoms, by sharing electrons, can acquire a stable, noble-gas configuration.* For example, when two hydrogen atoms, each with one electron, combine to form an H_2 molecule,

$$\text{H} \cdot + \text{H} \cdot \rightarrow \text{(H} \colon \text{H)}$$

each hydrogen atom acquires a share in the two electrons and, in that sense, attains the electronic configuration of the noble-gas helium (atomic number = 2). Similarly, when two fluorine atoms, each with seven electrons in its outermost principal energy level (n = 2), combine to form the F_2 molecule,

$$:\ddot{\text{F}} \cdot + \cdot \ddot{\text{F}}: \rightarrow (:\ddot{\text{F}} \colon \ddot{\text{F}}:)$$

each atom attains the neon structure with eight electrons in the outermost level.

The structures written above are referred to as **Lewis structures** (or, in the vernacular, "flyspeck formulas"). In writing a Lewis structure for an atom or molecule, we include only the electrons in the *outermost principal energy level,* the so-called **valence electrons.** For the A group elements, the number of valence electrons is given by the group number, i.e., 4 for a 4A element, 5 for a 5A element, and so on. Lewis structures for the simple molecules formed by hydrogen with the nonmetals of the third period are:

$$SiH_4 \qquad PH_3 \qquad H_2S \qquad HCl$$

H—Si—H with H above and H below (SiH₄)
H—P̈—H with H below (PH₃)
H—S̈—H (H₂S)
H—C̈l: (HCl)

Many of the polyatomic ions listed in Table 7.2 can be assigned simple Lewis structures. For example, the OH^- and NH_4^+ ions can be represented as

$$(:\ddot{O}{-}H)^- \quad \text{and} \quad \left[H{-}\underset{\underset{H}{|}}{\overset{\overset{H}{|}}{N}}{-}H \right]^+$$

In both these ions, hydrogen atoms are joined by covalent bonds to non-metal atoms (O, N). In both ions there are eight valence electrons. In the case of the hydroxide ion, this is one more than the number associated with the neutral atoms (6 + 1 = 7), in agreement with the −1 charge of the ion. The +1 charge of the NH_4^+ ion is accounted for when one realizes that four hydrogen atoms and one nitrogen atom would supply 9 valence electrons (4 + 5 = 9), one more than the number present in the ion.

The rule that molecules or ions in which each atom has a noble-gas configuration are particularly stable is often referred to as the **octet rule.** Nonmetals, with the exception of hydrogen, achieve a noble-gas structure by acquiring an "octet" of electrons. As we shall see later, many stable, covalently bonded species "violate" the octet rule. Nevertheless, we shall find it very useful in suggesting plausible electronic structures for molecules and polyatomic ions.

The octet rule is a good first approximation to the electronic structures of many molecules and ions

WRITING LEWIS STRUCTURES

Lewis structures for many molecules and ions are readily written by inspection. However, it may be helpful to follow the general procedure outlined below.

1. *Draw a skeleton structure for the molecule or ion, joining atoms by single bonds.* In some cases, only one arrangement of atoms is possible; in others, experimental evidence must be used to decide between two or more alternative structures.

2. *Count the number of valence electrons.* For a molecule, we simply sum up the valence electrons of the atoms present. For a *polyatomic anion*, electrons are *added* to take into account the negative charge. For a polyatomic *cation*, a number of electrons equal to the positive charge must be *subtracted*.

3. *Deduct two valence electrons for each single bond written in step 1. Distribute the remaining electrons as unshared pairs so as to give each atom eight electrons if possible.* If you find that there are "too few electrons to go around," convert single to multiple bonds. The formation of a double bond compensates for a deficiency of two electrons; a triple bond, a deficiency of four electrons.

The application of these rules and some further guiding principles are illustrated in Example 7.2.

Example 7.2 Draw Lewis structures for:

   a. ClO^- b. SO_4^{2-} c. CH_2O

Solution

   a. Only one skeleton structure is possible for the hypochlorite ion

$$(Cl—O)^-$$

   To obtain the total number of valence electrons we add one (the charge of the ion) to the number contributed by chlorine in group 7A (7) and oxygen in group 6A (6).

   no. of valence electrons $= 7 + 6 + 1 = 14$

   Deducting the two electrons used to make the covalent bond leaves 12. Putting 6 electrons around each atom, we arrive at a reasonable structure for the ClO^- ion which satisfies the octet rule, since both the Cl and O atoms have a share in eight electrons. (The bonding electrons are counted for both atoms.)

$$(:\overset{..}{\underset{..}{Cl}}—\overset{..}{\underset{..}{O}}:)^-$$

   b. Various skeletons could be written for the sulfate ion. However, in *oxyanions* such as SO_4^{2-}, NO_3^- and CO_3^{2-} we ordinarily find that *each oxygen atom is bonded to the central, nonmetal atom.* Following this general rule, we write

$$\begin{bmatrix} & O & \\ & | & \\ O{-}&S&{-}O \\ & | & \\ & O & \end{bmatrix}^{2-}$$

   The number of valence electrons is found by adding the charge of the ion to the total contributed by the sulfur and oxygen atoms:

   no. of valence electrons $= 6 + 4(6) + 2 = 32$

   Deducting 8 electrons for the four covalent bonds in the skeleton structure leaves 24. Putting 6 electrons around each oxygen atom gives us a plausible Lewis structure for the sulfate ion.

$$\begin{bmatrix} & :\overset{..}{O}: & \\ & | & \\ :\overset{..}{\underset{..}{O}}{-}&S&{-}\overset{..}{\underset{..}{O}}: \\ & | & \\ & :\underset{..}{O}: & \end{bmatrix}^{2-}$$

   c. A reasonable skeleton structure for formaldehyde would be

$$\begin{array}{c} H \\ | \\ H{-}C{-}O \end{array}$$

   The number of valence electrons in this neutral species is simply the total of those contributed by the carbon, hydrogen and oxygen atoms.

   no. of valence electrons $= 4 + 2(1) + 6 = 12$

   Deducting 6 electrons for the three covalent bonds in the skeleton structure leaves us only 6 to work with. We might spend these remaining electrons in either of two ways:

$$H-\overset{\overset{\displaystyle H}{|}}{\underset{..}{C}}-\overset{..}{\underset{..}{O}} \quad \text{or} \quad H-\overset{\overset{\displaystyle H}{|}}{\underset{..}{C}}-\overset{..}{\underset{..}{O}}:$$

But, whatever we do, we are 2 electrons shy of the number required to give both carbon and oxygen an octet. To remedy this deficiency, we form a double bond between the carbon and oxygen atoms

$$H-\overset{\overset{\displaystyle H}{|}}{C}=\overset{..}{O}:$$

to give a reasonable and, as it happens, the correct structure for the formaldehyde molecule.

There are other structures that could be written for CH_2O which would put 8 electrons around each atom. One such structure is

$$:\overset{\overset{\displaystyle H}{|}}{C}=\overset{..}{O}-H$$

Even though this structure might seem plausible, we are able to dismiss it by taking account of the general principle that *carbon*, in virtually all its stable compounds, *forms four covalent bonds*.

When drawing Lewis structures one needs to make use of a few empirical rules, such as the fact that nonbonding electron pairs are rarely found around carbon atoms

DEFICIENCIES OF OCTET RULE STRUCTURES: RESONANCE FORMS

In certain cases the Lewis structure does not adequately describe the properties of the ion or molecule which it is written to represent. Consider, for example, the SO_2 molecule, for which we would derive the following structure

$$\overset{\displaystyle \overset{..}{O}:}{\underset{\displaystyle :\overset{}{\underset{..}{O}}:}{\underset{\displaystyle |}{\overset{\displaystyle \diagup\!\!\diagup}{:S}}}}$$

This structure implies that there are two different kinds of sulfur-to-oxygen bonds in SO_2, double and single, yet experiment tells us that there is only one kind of bond in this molecule. In particular, we find the two bond distances are equal, 1.43 Å.

We might "explain" this situation by assuming that each of the bonds in SO_2 is a "hybrid" intermediate between a single and double bond. The fact that the observed bond distance is intermediate between those expected for a single and double bond lends some support to this idea. To express this concept within the framework of Lewis structures, we sometimes write two structures

$$\overset{\displaystyle \overset{..}{O}:}{\underset{\displaystyle :\overset{}{\underset{..}{O}}:}{\underset{\displaystyle |}{\overset{\displaystyle \diagup\!\!\diagup}{:S}}}} \quad \longleftrightarrow \quad \overset{\displaystyle \overset{..}{O}:}{\underset{\displaystyle :\overset{}{\underset{..}{O}}:}{\underset{\displaystyle \|}{\overset{\displaystyle \diagup}{:S}}}}$$

with the understanding that the true structure is intermediate between them. In valence bond terminology, these are referred to as *resonance* forms. The concept of resonance is introduced to rationalize the fact that a single Lewis structure does not adequately describe the properties of substances such as sulfur dioxide.

Another species for which it is necessary to invoke the idea of resonance is the nitrate ion. Here three equivalent structures can be written

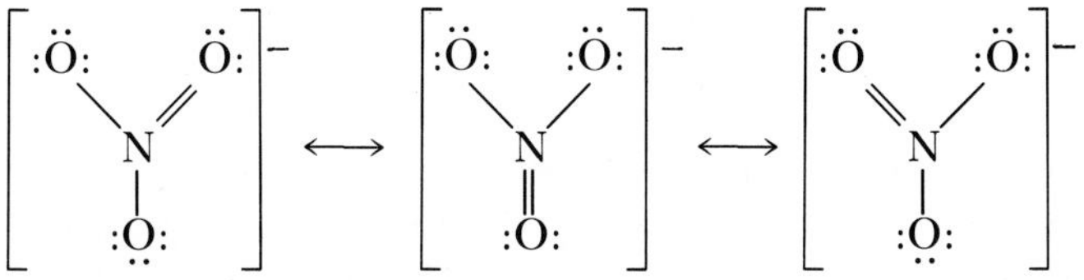

Valence bond theory assigns equal weight to each of these structures in describing the structure of the ion

to explain the experimental observation that the three nitrogen to oxygen bonds in the NO_3^- ion are identical in all respects.

We will encounter other examples of molecules and ions whose properties can be interpreted by invoking the concept of resonance. In this connection it may be well to point out that:

1. Resonance forms do not imply different kinds of molecules. Sulfur dioxide is built up of only one type of molecule, whose structure is assumed to be between those of the two resonance forms.

2. Resonance can be anticipated when it is possible to write, for the same atomic geometry, two or more Lewis structures which are about equally plausible. In the case of the nitrate ion, the three structures we have written are equivalent. One could, in principle, write many other structures, including:

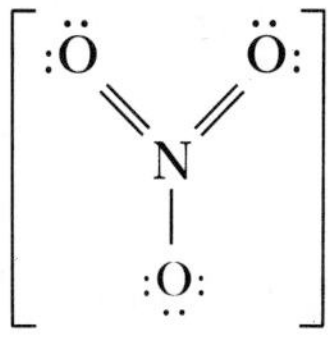

However, since this structure involves putting 10 electrons around the nitrogen atom, it seems unlikely that it would make a major contribution to the true structure of the nitrate ion.

3. Molecules or ions which can be represented as resonance hybrids are found to be more stable than we might predict on the basis of their Lewis structures. For example, the heat of formation of SO_2 is a larger negative number than we would calculate, using enthalpies of bond formation, for a molecule in which there is one single and one double bond. In this particular case, the difference between observed and calculated values, often referred to as the "resonance energy," amounts to about 60 kcal/mole.

FAILURE OF THE OCTET RULE

A few molecules, of which nitric oxide, NO, and nitrogen dioxide, NO_2, are perhaps the most common examples, contain an odd number of valence electrons.

$$NO: \text{ number of valence electrons} = 5 + 6 = 11$$

$$NO_2: \text{ number of valence electrons} = 5 + 6(2) = 17$$

For such *odd electron* species (sometimes called free radicals) it is impossible to write Lewis structures in which each atom obeys the octet rule. In valence-bond terminology, the NO molecule is considered to be a resonance hybrid with the two contributing structures

$$\cdot \ddot{N}{=}\ddot{O}: \ \leftrightarrow \ :\ddot{N}{=}\ddot{O}\cdot$$

Of the two resonance structures, the one on the left is probably the more important since it assigns more electrons to the more electronegative atom, oxygen. Several different resonance structures can be written for NO_2, of which the more plausible are of the type

$$\begin{array}{c} \dot{N}{=}\ddot{O}: \\ {\Large/} \\ :\ddot{O}: \end{array}$$

Species such as NO and NO_2, in which there are unpaired electrons, are **paramagnetic**; that is, they show a weak attraction toward a magnetic field. Elementary oxygen is also paramagnetic, which suggests that the conventional Lewis structure

$$:\ddot{O}{=}\ddot{O}:$$

is incorrect, since it requires that all the electrons be paired. The paramagnetism of oxygen could be explained by the structure

$$:\dot{\ddot{O}}{-}\dot{\ddot{O}}:$$

in which there are two unpaired electrons. However, this structure, like the one written previously, is unsatisfactory. In the first place, it does not conform to the octet rule; much more important, it does not agree with experimental evidence. The distance between the two oxygen atoms in O_2 (1.21 Å) is considerably smaller than that ordinarily observed with an O—O single bond (1.48 Å). These properties of oxygen are difficult to explain in terms of valence-bond theory. As we shall see in Section 7.7, the molecular orbital approach leads to a more satisfactory picture of the electron distribution in the O_2 molecule.

There are many other species for which Lewis structures written to conform to the octet rule are unsatisfactory. Examples include the fluorides of beryllium and boron. Although one could write multiple bonded structures for these molecules in accordance with the octet rule, experimental evidence strongly supports the structures

$$:\ddot{F}{-}Be{-}\ddot{F}: \quad \text{and} \quad \begin{array}{c} :\ddot{F}: \quad \ :\ddot{F}: \\ {\Large\diagdown} \ \ {\Large\diagup} \\ B \\ | \\ :\ddot{F}: \end{array}$$

in which the central atom is surrounded by 4 and 6 valence electrons, respectively, rather than 8.

At the opposite extreme, certain of the halides of the 5A and 6A elements have structures in which the central atom is surrounded by more than 8 valence electrons. In PF_5 and PCl_5, the phosphorus atom is joined by single bonds to each of five halogen atoms and consequently must be surrounded by 10 bonding electrons. An analogous structure for SF_6 requires that a sulfur atom have 12 valence electrons around it.

7.5 MOLECULAR GEOMETRY

The "geometry" of a simple diatomic molecule can be specified by giving the internuclear distance. With molecules containing more than two atoms, another factor is involved: the angles between the bonds. In such molecules as H_2O and CO_2, it is important to know whether the three atoms are in a straight line, as would be the case if the angle between the bonds were 180 degrees

or arranged in a triangular pattern

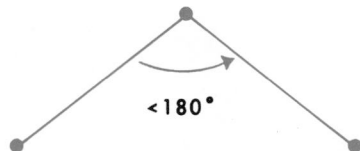

corresponding to a bond angle of less than 180 degrees.

ELECTRON PAIR REPULSION

The major features of the geometry of molecules and polyatomic ions can be predicted by a simple principle first suggested by Sidgwick and Powell in 1940 and developed more recently by R. J. Gillespie of McMaster University in Canada.

The electron pairs surrounding an atom, because of electrostatic repulsion, are oriented so as to be as far apart as possible.

The application of this principle is illustrated in Figure 7.9, in which we show the orientations to be expected for 2, 3, 4, 5 and 6 electron pairs. The molecules cited as examples at the far right of the figure are ones in which all the electron pairs are used by a central atom to form covalent bonds (i.e., there are no unshared pairs of electrons). In such species, the geometry of the molecule is fixed by the orientation of the electron pairs. In the methane molecule, for example, the carbon atom is located at the center of a regular tetrahedron with the four hydrogen atoms arranged symmetrically around it at the corners of the tetrahedron. In sulfur hexafluoride, the

Number of electron pairs	Orientation of electron pairs	Predicted bond angles	Examples	
2	Straight line	180°	BeF$_2$	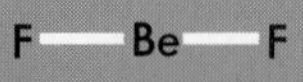 F—Be—F
3	Equilateral triangle	120°	BF$_3$	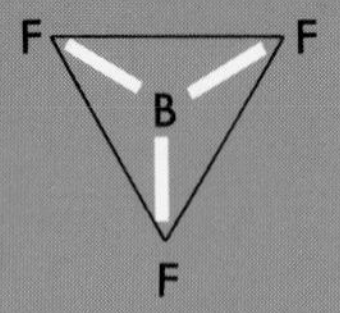 (trigonal planar diagram)
4	Tetrahedron	109.5°	CH$_4$	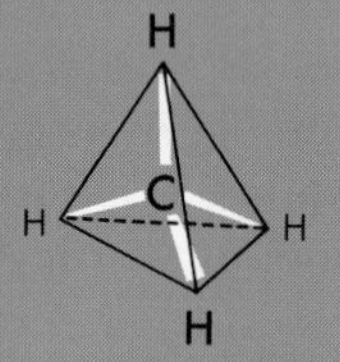 (tetrahedron diagram)
5	Trigonal bipyramid	120°, 90°	PCl$_5$	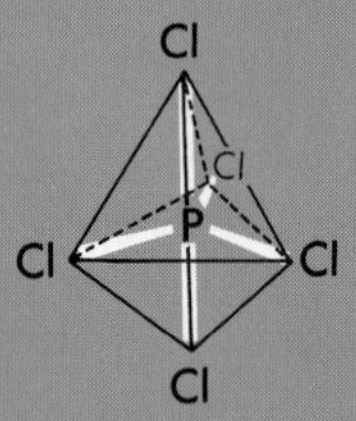 (trigonal bipyramid diagram)
6	Octahedron	90°	SF$_6$	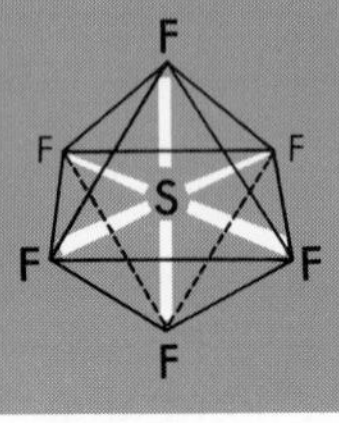 (octahedron diagram)

Figure 7.9

six fluorine atoms are found at the corners of a regular octahedron with a sulfur atom at the center.

The same principle can be used to deduce the geometries of molecules in which one or more of the electron pairs surrounding the central atom are unshared. In Figure 7.10 we show the geometries of the NH$_3$ and H$_2$O molecules. The positions of the unshared pairs, shown as charge clouds, cannot be determined experimentally. We can visualize the ammonia molecule as a "tetrahedron with one corner missing" or, more properly, as a *trigonal pyramid* with the nitrogen atom located above the center of the equilateral triangle formed by the three hydrogen atoms. When two corners of the tetrahedron are removed, as in H$_2$O, we are left with a bent molecule with an oxygen atom at the center.

Experimentally, we find that bond angles in molecules where the central atom has at least one unshared pair of electrons tend to be somewhat smaller than those listed in Figure 7.9. For example, in NH$_3$ the measured bond angle is about 107 degrees, a little less than the predicted tetrahedral

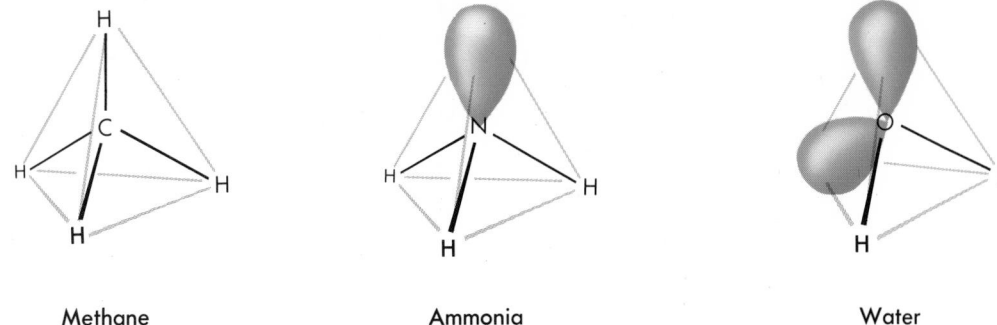

Methane   Ammonia   Water

Figure 7.10 Geometries of CH_4, NH_3, H_2O. When the four pairs of electrons are all shared, the molecule is tetrahedral. When one pair is unshared (NH_3), we obtain a trigonal pyramid. Two unshared pairs, as in H_2O, lead to a bent molecule.

angle of 109.5 degrees. The effect is slightly greater in H_2O, where the bond angle is 105 degrees. These discrepancies have been attributed to the influence of the unshared pair(s). We might expect the electron cloud formed by the unshared pair in NH_3 to spread out over a greater volume than that of the three pairs connected to hydrogen atoms. This would tend to force the bonding pairs closer to one another and thereby reduce the bond angle. Where there are two pairs of unshared electrons, as in H_2O, this effect is more pronounced.

The description of the structure of a molecule reflects the position of the nuclei rather than that of the valence electrons

SPECIES CONTAINING MULTIPLE BONDS

The idea of electron pair repulsion can be extended to predict the geometries of molecules or ions containing double or triple bonds if we assume that *so far as molecular geometry is concerned, a multiple bond behaves as if it were a single electron pair.* Thus, we find that the CO_2 molecule, like BeF_2, is linear

$$F—Be—F, \quad O{=}C{=}O$$

while the SO_3 molecule and the NO_3^- ion, like the BF_3 molecule, are triangular

$$
\underset{\overset{\|}{O}}{\overset{O}{\underset{}{S}}}{\diagdown}\diagup O
\qquad
\left[\underset{\overset{\|}{O}}{\overset{O}{\underset{}{N}}}{\diagdown}\diagup O\right]^{-}
\qquad
\underset{\underset{F}{|}}{\overset{F}{\underset{}{B}}}{\diagdown}\diagup F
$$

Applying this principle to the acetylene molecule, where there is a triple bond between the carbon atoms, we predict that each of the bond angles must be 180 degrees, which gives the observed linear structure

$$H—C{\equiv}C—H$$

The ethylene molecule, with a double bond between the two carbon atoms, has the geometry to be expected if each carbon atom had only three pairs of electrons around it.

$$\underset{H}{\overset{H}{\diagdown}}C=C\underset{H}{\overset{H}{\diagup}}$$

The six atoms are located in a plane with bond angles of 120°.

Example 7.3 Predict the geometries of the following species: SO_4^{2-}, SO_2, XeF_4.

Solution. Clearly, in order to use the electron pair repulsion principle, we have to know how many pairs of electrons there are around the central atom. Perhaps the easiest way to find this out is to derive the Lewis structure for the molecule or ion. For the SO_4^{2-} ion, this is available from Example 7.2. Referring back, we see that the sulfur atom in the sulfate ion is surrounded by four electron pairs. As we would predict, the SO_4^{2-} ion is tetrahedral; i.e., it has the same geometry as the CH_4 molecule shown in Figure 7.9.

Looking at the Lewis structure of SO_2 on p. 178, we count around the central sulfur atom one double bond, one single bond and one unshared pair of electrons. Pretending that the double bond is a single electron pair, we obtain a bent molecule with a 120-degree bond angle (the observed angle is 119.5°)

$$\underset{\underset{\cdot\cdot}{S}}{\overset{O\qquad O}{\diagdown\quad\diagup}}$$

In compounds like BF₃, PCl₅ and SF₆ each halogen atom shares one of its electrons with the central atom

With XeF_4 we have to start from scratch by counting the number of valence electrons. Noting that Xe is in group 8A and fluorine in group 7A we have: $8 + 4(7) = 36$. To find the number of electron pairs around the Xe atom, we might work out the Lewis structure of the molecule. We can save time by recalling that each of the four fluorine atoms will have three pairs of nonbonding electrons around it, giving a total of $4 \times 3 \times 2 = 24$ unshared electrons. The remaining 12 electrons (6 electron pairs) must be located around the Xe atom. From Figure 7.9, we find that these six electron pairs will be directed toward the corners of a regular octahedron. To decide how the four fluorine atoms are located relative to xenon, we lop off the top and bottom of the octahedron, where the unshared electron pairs are presumably located. We are left with a square with a xenon atom at the center. Experimental studies carried out in 1963 showed that this was indeed the structure of XeF_4.

$$\underset{\underset{F}{\diagup}\quad\underset{F}{\diagdown}}{\overset{F}{\diagdown}\quad\overset{F}{\diagup}}{Xe}$$

POLARITY OF MOLECULES

A polar molecule is one in which there is a separation of positive and negative charge, that is, + and − poles. Any diatomic molecule in which the atoms differ from each other (HF, HCl, . . .) is polar, with a negative pole

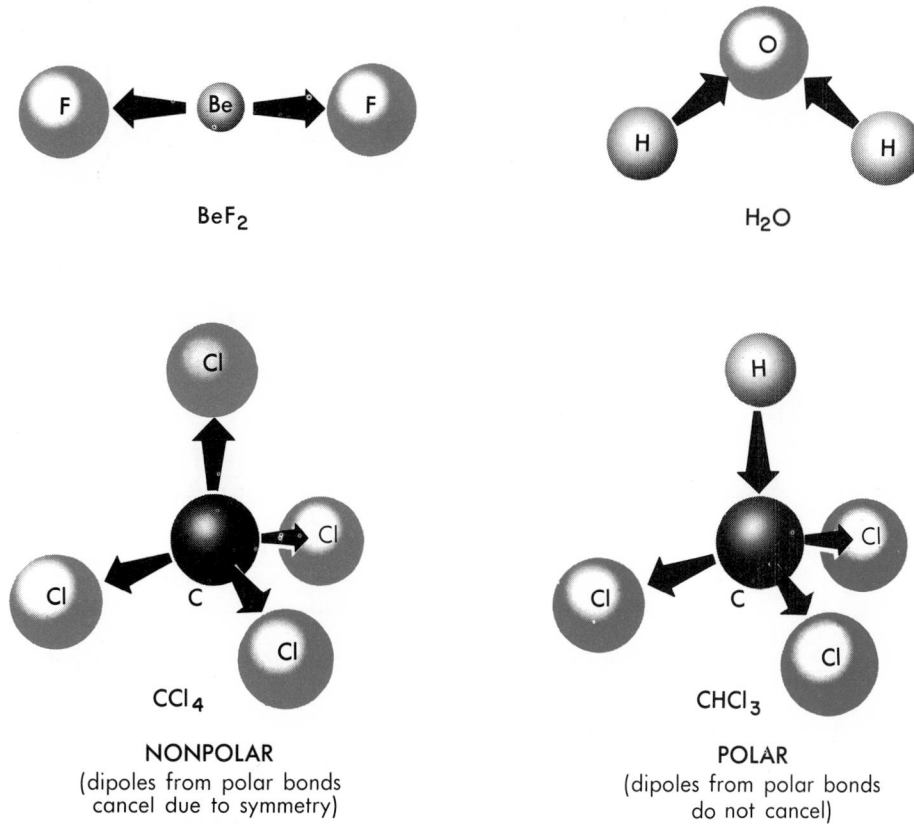

Figure 7.11 The molecules BeF_2 and CCl_4 are nonpolar because the polar bonds cancel each other. This does not happen in the polar H_2O and $CHCl_3$ molecules. (The arrows point to the atom of higher electronegativity.)

located at the more electronegative atom. A diatomic molecule such as H_2, in which the two atoms are identical, is nonpolar.

In molecules containing more than two atoms, we must know the bond angles to decide whether or not the molecule is polar. Consider, for example, the two triatomic molecules shown at the top of Figure 7.11, BeF_2 and H_2O. The linear BeF_2 molecule is nonpolar; the two polar bonds cancel each other. From a slightly different point of view, we might say that the centers of negative and positive charge coincide in this molecule. In contrast, the bent water molecule is polar: the charge centers do not coincide. One can visualize a negative pole located at the oxygen atom with the compensating positive pole located midway between the two hydrogen atoms.

Another molecule which is nonpolar despite the presence of polar bonds is CCl_4. The four C—Cl bonds are polar, with the bonding electrons slightly displaced toward the chlorine atoms. However, because of the symmetrical pattern in which the four chlorines are arranged around the carbon atom, the polar bonds cancel each other. If one of the chlorine atoms in CCl_4 is replaced by hydrogen, the symmetry is destroyed and we obtain a polar species, the chloroform molecule, $CHCl_3$.

Would CH_2Cl_2 be polar?

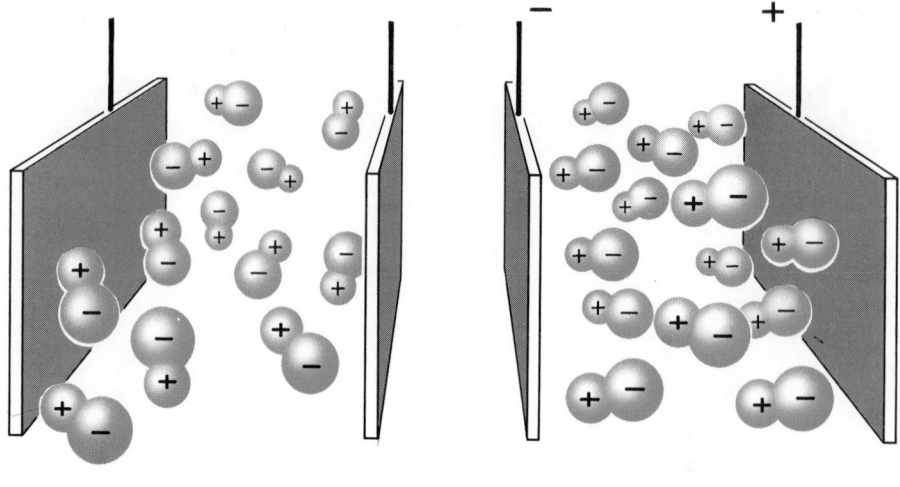

Field off   Field on

Figure 7.12   Orientation of polar molecules in an electrical field.

Example 7.4   Would you expect the SO_2 molecule to be polar? the XeF_4 molecule?

Solution.   In Example 7.3, we decided that the SO_2 molecule is bent, with a bond angle of 120 degrees. We would expect it to be slightly polar, with a + pole at the less electronegative sulfur atom and a − pole midway between the two oxygens. Again, we recall from Example 7.3 that in XeF_4 the four fluorine atoms are located at the corners of a square with the Xe atom in the center. The four polar bonds cancel each other to give a nonpolar molecule.

A possible alternate polar structure of XeF_4 is ruled out because the molecule has no dipole movement

We can test these predictions by studying the behavior of molecules in an electrical field, such as that between the plates of a condenser (Fig. 7.12. Polar molecules tend to line up with their + poles oriented toward the negative plate. We describe this behavior by saying that the molecule has a *dipole moment* which causes it to turn into a preferred position in the field. Nonpolar molecules have a dipole moment of zero.

7.6   HYBRID ATOMIC ORBITALS

Simple valence-bond theory explains covalent bond formation in terms of the sharing of unpaired electrons. From this point of view, it would seem that the number of bonds formed by a given atom would be governed by the number of unpaired electrons in the valence shell. In the case of carbon, whose orbital diagram is

	1s	2s	2p		
$_6$C	↑↓	↑↓	↑	↑	

we might predict the formation of two covalent bonds. Yet, we know from experiment that carbon invariably forms four bonds rather than two. In order to explain this and other discrepancies within the framework of the valence-bond model, it is necessary to invoke a new kind of atomic orbital, the so-called hybrid bond orbital.

sp^3 HYBRID BONDS

The fact that a carbon atom forms four covalent bonds could be explained by assuming that, prior to reaction, one of the 2s electrons is promoted to the 2p level

$$\text{1s} \qquad \text{2s} \qquad \text{2p}$$

$$_6\text{C} \quad \boxed{\uparrow\downarrow} \quad \boxed{\uparrow} \quad \boxed{\uparrow}\ \boxed{\uparrow}\ \boxed{\uparrow}$$

Now that the carbon atom has four unpaired electrons, it can form four covalent bonds by sharing electrons with another atom such as hydrogen.

$$\text{1s} \qquad \text{2s} \qquad \text{2p}$$

$$_6\text{C} \quad \boxed{\uparrow\downarrow} \quad \boxed{\uparrow\downarrow} \quad \boxed{\uparrow\downarrow}\ \boxed{\uparrow\downarrow}\ \boxed{\uparrow\downarrow} \quad (\text{CH}_4 \text{ molecule})$$

(The colored arrows indicate electrons supplied by hydrogen: the horizontal lines are drawn to enclose the orbitals involved in bond formation.)

There is one fundamental objection to the model we have just described. It implies that two different kinds of bonds are formed. One of these would be an "s" bond while the other three would be "p" bonds. Experimentally it is found that the four bonds formed are identical in all respects. This leads us to believe that the four orbitals used for bond formation by carbon must be equivalent. In valence-bond terminology, we rationalize this situation by saying that an s and three p orbitals have been *mixed*, or *hybridized*, to give four new bonding orbitals described as **sp³ hybrids.**

It should be clearly understood that sp³ hybrid orbitals have their own unique properties, quite different from those of the orbitals from which they are formed. It can be shown, by arguments based on quantum mechanics, that the four equivalent orbitals are directed toward the corners of a regular tetrahedron. This is, as we have seen, the orientation that keeps four pairs of electrons as far apart as possible.

The fact that the bond angles in ammonia and water are very nearly tetrahedral suggests that the four electron pairs surrounding the central atom in these molecules also occupy sp³ hybrid orbitals. If the bonds formed by nitrogen in NH₃ or oxygen in H₂O were pure "p" bonds, we would expect them to be oriented at right angles to each other. In PH₃ and H₂S the bond angles *are* very close to 90 degrees (93 and 92 degrees, respectively), implying that in these molecules hybridization may not occur.

OTHER TYPES OF HYBRID ORBITALS, sp AND sp^2 HYBRIDS

The various types of hybrid orbitals that have been postulated to explain the geometries of molecules and complex ions are listed in Table 7.6.

TABLE 7.6 Types of Hybrid Orbitals

No. of Electron Pairs	Atomic Orbitals		Hybrid Orbitals	Orientation	Example
2	s,p	→	sp	linear (180°)	BeF_2
3	s,p,p	→	sp²	triangular (120°)	BF_3
4	s,p,p,p	→	sp³	tetrahedral (109.5°)	CH_4
	°d,s,p,p	→	dsp²	square, planar (90°)	$Ni(CN)_4^{2-}$
5	s,p,p,p,d	→	sp³d	† trigonal bipyramid (90°, 120°)	PCl_5
6	s,p,p,p,d,d	→	sp³d²	octahedral (90°)	SF_6
	°d,°d,s,p,p,p	→	d²sp³	octahedral (90°)	$Co(NH_3)_6^{3+}$

° The d orbitals involved here are inner orbitals, e.g., 3d hybridized with 4s and 4p in $Ni(CN)_4^{2-}$ (Chapter 19).

† See Figure 7.9 for geometry.

Several of these involve d orbitals; we shall have more to say on this topic when we discuss coordination chemistry in Chapter 19. The only kinds of hybrid orbitals which need concern us here, in addition to **sp³**, are **sp²** and **sp** hybrids, found in the compounds BF_3 and BeF_2, respectively.

The sp² hybrid orbitals are formed by hybridizing an s and two p orbitals in the same energy principal energy levels, to form three new, equivalent, hybrid orbitals. These orbitals are directed toward the corners of an equilateral triangle, in agreement with the observed bond angles in BF_3 (120 degrees).

Perhaps the simplest type of hybrid orbital is the sp hybrid found in BeF_2. Two such orbitals are formed by mixing together an s and a p orbital. The two sp hybrid orbitals in BeF_2 and other compounds are directed at 180° angles to each other, giving a linear molecule.

BONDING ORBITALS IN COMPOUNDS CONTAINING MULTIPLE BONDS

We saw in Section 7.5 that the bond angles in C_2H_4 are the same as those in BF_3. This suggests that the bonding orbitals in the two molecules are similar. Recalling that in BF_3 boron is joined to three fluorines by sp² hybrid bonds, it seems reasonable to suppose that a carbon atom in C_2H_4 forms three such bonds, two to hydrogen, one to the other carbon atom. That is, the geometry of the C_2H_4 molecule is fixed by a framework of sp² bonds, three of which extend from each carbon atom. This framework is shown in Figure 7.13a. You will notice that in each of these bonds, the electron density is completely symmetrical about a line joining the two bonded atoms. Bonds meeting this criterion are often described as **sigma (σ) bonds**; sp^3, sp^2 and sp hybrid bonds all fall in this category.

We have now accounted, through sp^2 hybridization, for three of the four valence electrons on each carbon atom in ethylene. We have left the fourth valence electron sitting all by itself in the lonely confines of an unhybridized p orbital. This situation is shown pictorially in Figure 7.13b. Realizing that we have two such unpaired electrons, one from each carbon atom, the obvious solution is to bring them together through p orbital overlap to form a bonding orbital of the type shown in Figure 7.13c. Notice that the electron density in this orbital is not completely symmetrical about the C—C axis: instead it is concentrated in two lobes, one above the axis, the other below. Bonds such as this, in which the electron density is not symmetrical about a line joining the bonded atoms, are referred to as **pi (π) bonds**.

The fact that the acetylene molecule, C_2H_2, has the linear geometry of beryllium fluoride, BeF_2, implies that its backbone consists of sp hybrid bonds, two per carbon atom. One of these is directed to a hydrogen atom, the other to the neighboring carbon. We are left with two unhybridized p electrons on each carbon atom. By overlap of p orbitals, we can then form two π bonds between the carbon atoms. In other words, the "triple bond" in acetylene, according to this picture, is made up of one σ and two π bonds.

The planar structure of C_2H_4 is required if stabilization of the molecule by maximum overlap of p orbitals is to occur

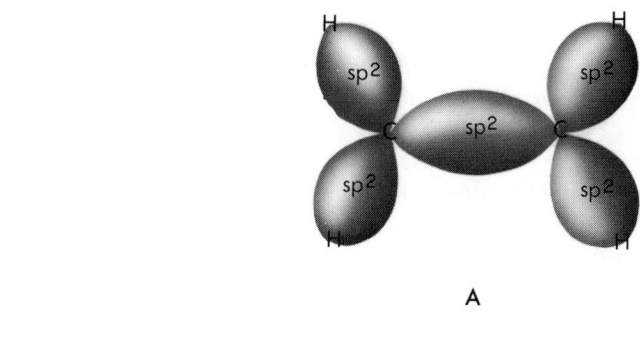

A

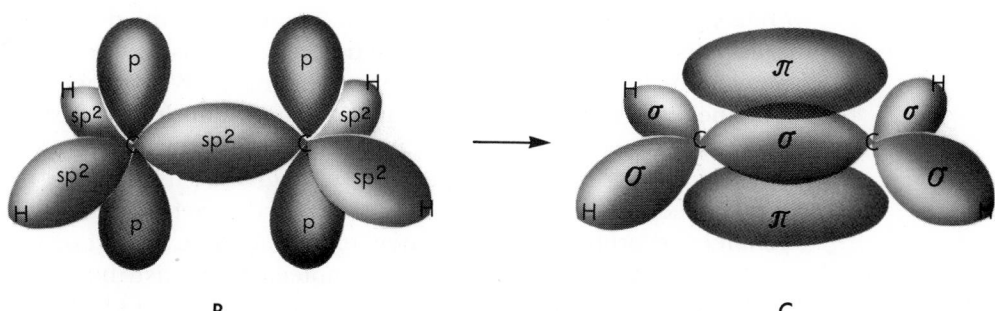

B   C

Figure 7.13 Bonding in C_2H_4. In *A* the sigma-bond framework is shown; in *B* the unoccupied p orbitals, one per carbon atom, are shown. Overlap between p orbitals of the two atoms gives the final structure shown in *C*.

7.7 MOLECULAR ORBITALS

The valence-bond model presented in this chapter explains a great many of the structural features of molecules and polyatomic ions. Most important, it accounts, at least qualitatively, for the stability of the covalent bond in terms of an overlap of atomic orbitals. By making use of the concept of hybridization, valence-bond theory succeeds splendidly in accounting for the principles of molecular geometry. By introducing the idea of resonance, it is even possible to rationalize the properties of molecules such as SO_2 for which a conventional Lewis structure is inadequate.

A major weakness of the valence-bond approach has been its inability to predict correctly the magnetic properties of many simple molecules. An example previously cited is molecular oxygen. The same problem arises with the B_2 molecule found in boron vapor at high temperatures. This molecule, like O_2, is paramagnetic even though it has an even number, 6, of valence electrons.

The deficiencies of valence-bond theory arise from an oversimplification inherent in its approach. It assumes that the electrons in a molecule occupy atomic orbitals of the individual atoms. For example, in the CH_4 molecule, the bases for consideration of the electrons in a bonding pair are the 1s orbital of the H atom and one of the four sp^3 hybrid orbitals of the C atom. Clearly this is an approximation, since each electron is in an orbital characteristic of the molecule as a whole. In line with this idea, the molecular orbital theory attempts, again very approximately, to treat electron arrangements in molecules in terms of orbitals involving the whole molecule.

The molecular orbital approach to the electronic structures of molecules, referred to briefly in Section 7.2, involves three fundamental operations:

A molecular orbital is a linear combination of atomic orbitals of the atoms making up the molecule. For that reason MO theory is sometimes called LCAO theory

1. The atomic orbitals associated with isolated atoms are combined to give a new set of molecular orbitals characteristic of the molecule as a whole. In doing this we always find that *the number of molecular orbitals formed is equal to the number of atomic orbitals combined.* For example, when two atoms combine to form a diatomic molecule, two s orbitals, one from each atom, yield two molecular orbitals. Again, six p orbitals, three from each atom, give a total of six different molecular orbitals.

2. Having arrived at a set of molecular orbitals, we attempt to arrange them in order of increasing energy. In principle, the energies of molecular orbitals, like those of atomic orbitals, can be derived by solving the Schrödinger wave equation. In practice, it is impossible to make precise calculations for any but the simplest of molecules. What is actually done is to deduce the relative energies of molecular orbitals from experimental observations on molecules or ions in which those orbitals are utilized. The spectra and magnetic properties of these species are useful here, but it must be admitted that we sometimes reason after the fact, reshuffling the order of molecular orbitals to explain new experimental evidence.

3. The valence electrons in the molecule are distributed among the available molecular orbitals in much the same way that electrons in atoms are fed into atomic orbitals. In particular, we ordinarily find that:

a. *Each molecular orbital can hold a maximum of two electrons.*

b. *Electrons go into the lowest molecular orbital available.* A higher orbital starts to fill only when each orbital below it has its quota of two electrons.

c. *Hund's rule is obeyed.* When two orbitals of equal energy are available, one electron goes into each, giving two half-filled orbitals.

If you read between the lines in the above description of the molecular orbital theory, you can see that it attempts to do for molecules what the electron configuration theory does for atoms. By semi-empirical means, we arrive at a description of the orbitals available to the electrons in a molecule. We then fill these orbitals, using much the same rules as were used to fill sublevels in atoms.

To illustrate molecular orbital theory let us apply it to the diatomic molecules of the elementary substances in the first two periods of the periodic table. Our choice of these elements as examples is dictated by the fact that MO theory is simplest to apply here and gives results for these molecules which are in many ways superior to those of the atomic orbital approach. In short, we are presenting molecular orbital theory in its most favorable light.

FIRST PERIOD ELEMENTS. COMBINATION OF 1s ATOMIC ORBITALS

Molecular orbital (MO) calculations show that the combination of two 1s atomic orbitals leads to the formation of two molecular orbitals, one of which has an energy lower than that of the atomic orbitals from which it is formed (Figure 7.14); placing electrons in this orbital gives a species which is more stable than the combination of two isolated atoms. For this reason the lower molecular orbital in Figure 7.14 is referred to as a **bonding orbital**. The other molecular orbital has a higher energy than the corresponding atomic orbitals; electrons entering it find themselves in an unstable, higher energy state. It is referred to as an **antibonding orbital**.

The electron density corresponding to these two molecular orbitals is shown at the bottom of Figure 7.14. It will be observed that the bonding orbital has a high electron density in the region between the nuclei, which accounts for its stability. In the antibonding orbital, the probability of finding an electron between the nuclei is very small; the electron density is concentrated at the far ends of the "molecule." Since the nuclei are less shielded from each other than they are in the isolated atoms, the antibonding orbital is unstable with respect to the individual atomic orbitals. The electron density in both the bonding and antibonding orbitals is symmetrical with respect to the internuclear axis; both of these are sigma orbitals. In MO notation, these molecular orbitals are designated as σ_{1s}^b and σ_{1s}^*, respectively; the asterisk is used to represent the antibonding orbital.

Let us now consider the order in which electrons enter these molecular orbitals.

1. *1 electron.* In the H_2^+ ion we would expect to find the single electron in the bonding orbital. This corresponds to one half of an electron pair bond with one unpaired electron.

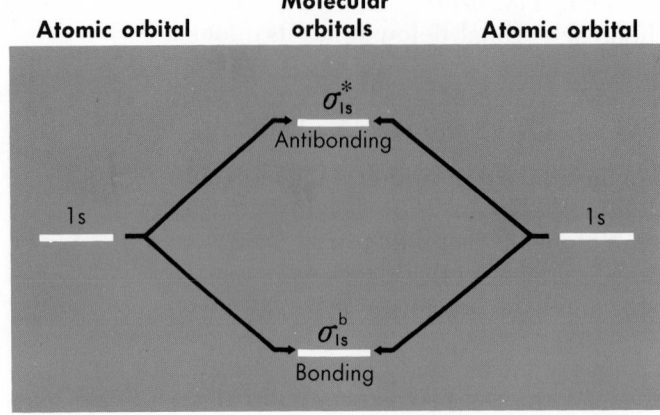

Figure 7.14 Molecular orbitals formed by combining two 1s orbitals. Upper diagram shows the relative energies, lower diagram the electron density clouds associated with bonding and antibonding orbitals.

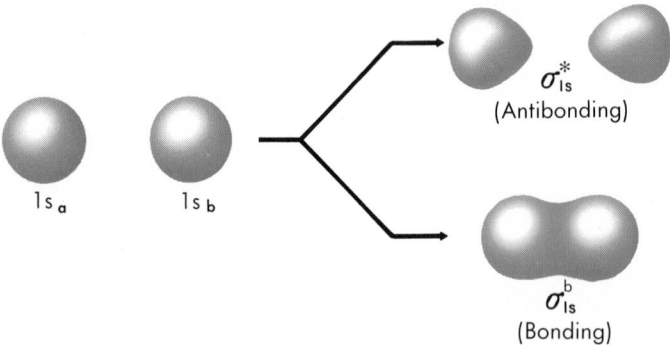

2. *2 electrons.* In the H_2 molecule the bonding orbital, σ_{1s}^b, is filled with two electrons. We expect a single bond with no unpaired electrons (so did G. N. Lewis and everyone else!).

3. *3 electrons.* The He_2^+ ion should have two electrons in the bonding orbital and 1 in the antibonding σ_{1s}^* orbital. This is equivalent to

$$\frac{2-1}{2} = \frac{1}{2} \text{ electron pair bond}$$

There would be one unpaired electron.

4. *4 electrons.* The He_2 molecule would fill both the bonding and antibonding orbitals. These would cancel each other and the net number of bonds would be zero. The stability of He_2 would be no greater than that of two He atoms, and so we would not expect this molecule to exist.

Of these four species, H_2 is, of course, well known. H_2^+ and He_2^+ have been detected in discharge tubes and their properties have been established. No one has ever been able to detect the He_2 molecule in the ground state. As Table 7.7 indicates, our predictions for H_2^+, H_2 and He_2^+ are in good agreement with experiment. Note particularly the correlation between the enthalpy of bond formation and the number of bonds.

TABLE 7.7 Properties of H_2^+, H_2, He_2^+

PREDICTED MO THEORY			OBSERVED	
Total Electrons	No. Bonds	No. Unpaired Electrons	ΔH_f Bond	Magnetic Behavior
H_2^+ 1	$\frac{1}{2}$	1	−61 kcal/mole	paramagnetic
H_2 2	1	0	−103 kcal/mole	diamagnetic
He_2^+ 3	$\frac{1}{2}$	1	−60 kcal/mole	paramagnetic

SECOND PERIOD ELEMENTS, COMBINATION OF 2s AND 2p ORBITALS

The fact that molecular orbital theory correctly predicts the properties of species in which there are 1, 2, 3 or 4 electrons encourages us to extend it to cases where orbitals from the second principal energy level are involved. Here, we will limit our discussion to the diatomic molecules of the second period. Some of these, including N_2, O_2 and F_2, are familiar to all of us. The molecules Li_2, B_2 and C_2 are less common but have been observed and studied in the gas phase. The molecules Be_2 and Ne_2 in the ground state have never been detected: hopefully MO theory can explain why this should be the case.

Combining two 2s atomic orbitals, one from each atom, gives two molecular orbitals completely analogous to those previously discussed. These orbitals are designated as σ_{2s}^b (sigma, bonding, 2s) and σ_{2s}^* (sigma, antibonding, 2s).

When the three 2p orbitals associated with each of two atoms are combined, six molecular orbitals are formed. Three of these are bonding orbitals, since their energies lie below those of the individual atomic orbitals (Fig. 7.15). Of this set of bonding orbitals, one is higher in energy than the other two. This orbital, in which the electron density is symmetrical about a line joining the nuclei, is described as a σ_{2p}^b orbital. The other two bonding orbitals, which have equal energies, are referred to as π_{2p}^b orbitals. You will note from the diagram at the base of Figure 7.15 that the electron density in these orbitals is not symmetrical with respect to the axis joining the nuclei; hence, their designation as π orbitals.

The three antibonding orbitals in this set have energies that are higher than those of the atomic orbitals from which they are derived. In these orbitals, the electron density is concentrated away from the region between the two nuclei (Figure 7.15). As a result, putting electrons in these orbitals reduces the stability of a species. Of the three antibonding orbitals one, designated σ_{2p}^* (sigma, antibonding, 2p), is higher in energy than the other two π_{2p}^* (pi, antibonding, 2p) orbitals.

Let us now consider how these molecular orbitals are filled as we put in electrons to form the diatomic molecules of the elementary substances in the second period. In Li_2 we have two valence electrons to locate (a 2s electron from each atom). These will fill the σ_{2s}^b orbital to give a stable molecule with one electron pair bond and no unpaired electrons. Experimentally we find that the Li_2 molecule is indeed diamagnetic. The enthalpy of bond for-

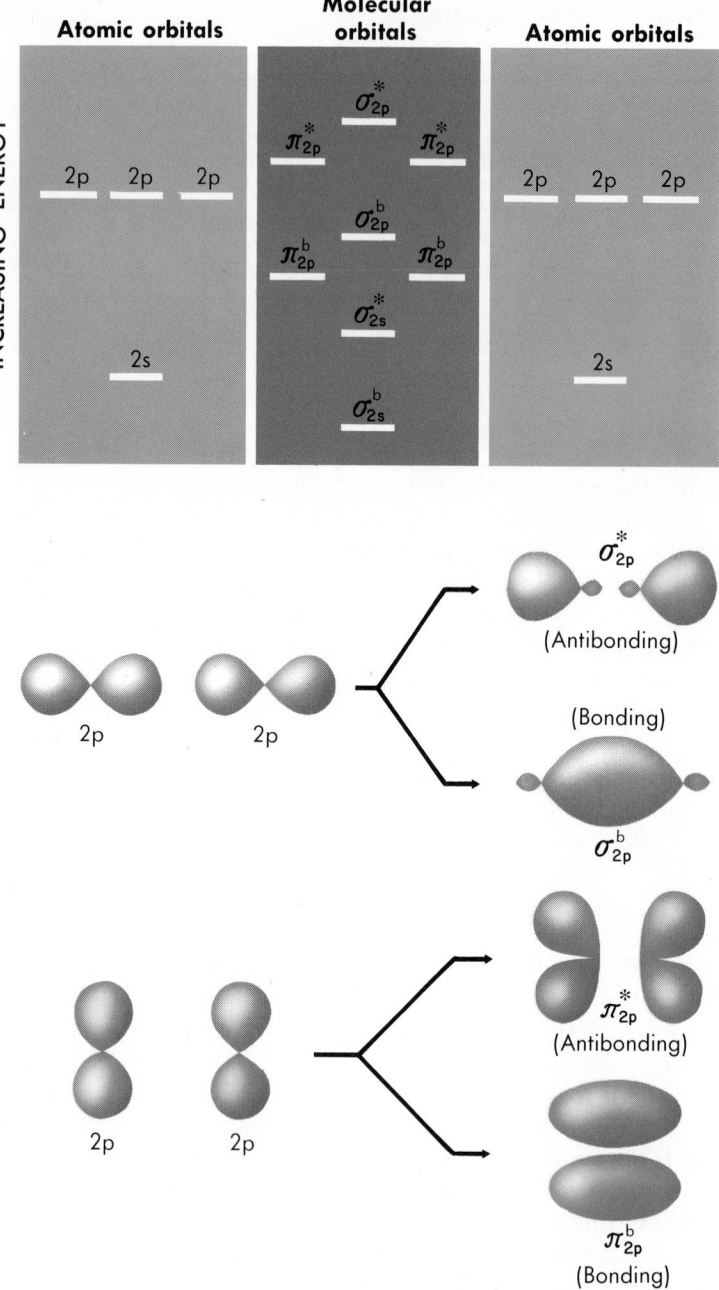

Figure 7.15 Molecular orbitals formed by combining 2p atomic orbitals. The order of energy levels applies at least through N_2. There is evidence to suggest that, starting with O_2, the σ_{2p}^b orbital lies below the π_{2p}^b orbitals.

mation, -25 kcal/mole, is actually somewhat smaller than we might expect for a single bond. In the hypothetical molecule Be_2, both the σ_{2s}^b and σ_{2s}^* orbitals would be filled. This would give an unstable species with no net bonding; small wonder that it does not exist.

With the diatomic molecule of boron, B_2, we start to fill the p orbitals.

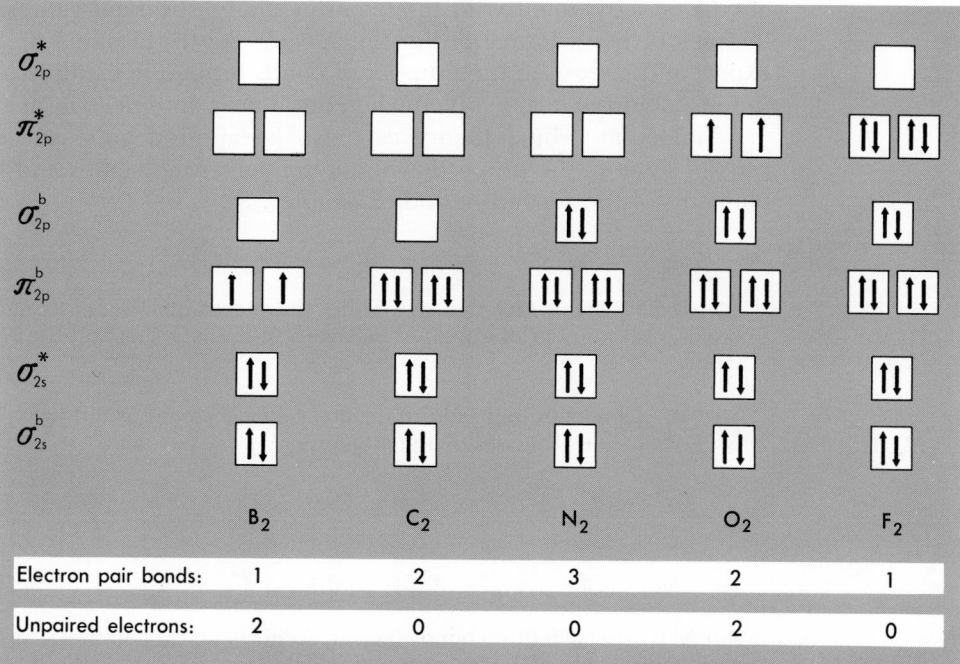

Figure 7.16   Molecular orbitals occupied in the diatomic molecules B_2 to F_2.

The electronic structures of the species B_2, C_2, N_2 and O_2 are shown in Figure 7.16 along with the calculated number of electron pair bonds and unpaired electrons. In the last member of this series, Ne_2, all the 2s and 2p orbitals, both bonding and antibonding, would be filled. This would give no net bonding, leading to an unstable species.

A major triumph of MO theory is its ability to predict correctly that the O_2 molecule should contain both a double bond and two unpaired electrons; simple valence-bond theory is unable to rationalize this observation. Again, the molecular orbital approach is successful where valence-bond theory fails in explaining the paramagnetism of the B_2 molecule. For the N_2 and F_2 molecules, both theories agree in predicting a triple and single bond, respectively, with no unpaired electrons.

The number of covalent bonds predicted by MO theory is in reason-

In O_2 the two electrons of highest energy remain unpaired because of Hund's rule

TABLE 7.8   Enthalpies of Bond Formation of Diatomic Molecules

MOLECULE	ΔH_f BOND	NO. OF BONDS (MO THEORY)
B_2	−69 kcal/mole	1
C_2	−150	2
N_2	−225	3
O_2	−118	2
F_2	−37	1

ably good agreement with the experimentally observed values of the enthalpies of bond formation for the species $B_2 \rightarrow F_2$ (Table 7.8). We would expect an increase in the number of bonds formed to result in an increase in the amount of heat evolved when the atoms combine to form a molecule.

Molecular orbital theory can also be applied quite successfully to predict some of the properties of heteronuclear molecules and ions of the elements of the second period (Example 7.5).

Example 7.5 Using MO theory, predict the electronic structures, number of bonds, and number of unpaired electrons in the NO^+ ion and the NO molecule.

Solution. Concentrating only upon the valence electrons (10 in NO^+, 11 in NO), we arrive at the following structures:

	σ_{2s}^{b}	σ_{2s}^{*}	π_{2p}^{b}	π_{2p}^{b}	σ_{2p}^{b}	π_{2p}^{*}	π_{2p}^{*}	σ_{2p}^{*}
NO^+	⇅	⇅	⇅	⇅	⇅	☐	☐	☐
NO	⇅	⇅	⇅	⇅	⇅	↑	☐	☐

For NO^+, we have three bonds (i.e., an excess of $6e^-$ in bonding as opposed to antibonding orbitals) and no unpaired electrons. For NO, there is one unpaired electron and

$$\frac{6-1}{2} = \frac{5}{2} \text{ bonds}$$

As we might expect, NO^+ is a particularly stable ion; the ionization energy of NO is relatively small. At high levels in the atmosphere, NO^+ is among the more abundant species (see Chapter 15).

7.8 THE METALLIC BOND

Any model of the electronic structure of metals must be able to explain the unique properties of the metallic elements which distinguish them from nonmetals. These include:

1. *High Electrical Conductivity.* Metals as a group have electrical conductivities several orders of magnitude greater than those of typical nonmetals. Silver is the best metallic conductor but is too expensive to be widely used for this purpose. Copper, with a conductivity only slightly less than that of silver, is used extensively in electrical wiring. Lead, one of the poorer metallic conductors, still has a conductivity nearly 4000 times that of silicon or germanium.

2. *High Thermal Conductivity.* Among solids, metals are by far the best conductors of heat. Here again, silver is the best conductor as you may have guessed if you have ever stirred a hot cup of coffee with a sterling silver spoon. Plated silverware (silver over brass) is a somewhat poorer conductor; aluminum and stainless steel are farther down the list. The fact that the thermal conductivity of metals parallels their electrical conductivity (Table 7.9) suggests that the mechanism is the same in the two cases.

TABLE 7.9   Relative Electrical and Thermal Conductivities (Pb = 1)

| | METALS | | | NONMETALS |
	Electrical	Thermal		Electrical
Ag	13.6	12.2	C (graphite)	0.016
Cu	13.0	11.1	Si	2.6×10^{-4}
Al	8.4	6.1	Ge	2.5×10^{-4}
Zn	3.7	3.2	I	2×10^{-14}
Fe	2.2	2.0	C (diamond)	4×10^{-20}
Pb	1.0	1.0	S	1×10^{-22}

3. *Luster.* The pleasing appearance of polished metal surfaces is due to their ability to reflect light. Most metals have a silvery white color, indicating that light of all wavelengths is reflected. Gold and copper absorb some light in the blue region of the spectrum and, hence, have a yellow or red color.

4. *Ductility, Malleability.* Most metals are ductile (capable of being drawn out into wire) and malleable (capable of being hammered into thin sheets). Nonmetallic solids tend to shatter if drawn out or hammered.

5. *Emission of Electrons.* Metals can emit electrons if heated (thermionic effect) or exposed to light of the proper wavelength (photoelectric effect.)

So far as electronic structure is concerned, the most obvious difference between a metal and a nonmetal atom is the number of valence electrons that it can use in bonding. Nonmetal atoms, with four or more valence electrons, are able to form strong covalent bonds with neighboring atoms. In some cases (N_2, O_2, F_2, . . .), this results in the formation of stable molecules; with other nonmetals, notably carbon, it leads to macromolecular solids in which each atom is bonded to several other atoms. The situation is quite different with metal atoms, which have very few valence electrons, typically 1, 2 or 3, and a large number of neighbors (8 or 12 in most metals). Clearly, in a metal we do not have enough valence electrons available to use them in ordinary covalent bond formation between adjacent atoms.

Considerations of this sort led to a simple model of the metallic bond, usually referred to as the **electron-sea model.** Here, the metallic lattice is visualized as a regular array of positive ions (i.e., metal atoms minus their valence electrons) anchored in position like bell buoys in a mobile "sea" of electrons. At least to a first approximation, the valence electrons are supposed to be able to wander through the lattice in much the same way that gas molecules move freely throughout their container. Presumably they would tarry a while in the vicinity of a positively charged metal ion, but they would not be permanently imprisoned between two metal ions. In Figure 7.17 we have attempted to show what a tiny portion of a metal crystal might look like according to the electron-sea model.

This simple picture of metallic bonding, involving relatively mobile electrons, offers an obvious explanation of the high electrical conductivity of metals and their ability to give off electrons readily when heated or exposed to ultraviolet light. High thermal conductivity can be explained by assuming that heat is transferred through the metal by "collisions" between

In metals the electrons are freer to move than they are in any other substances

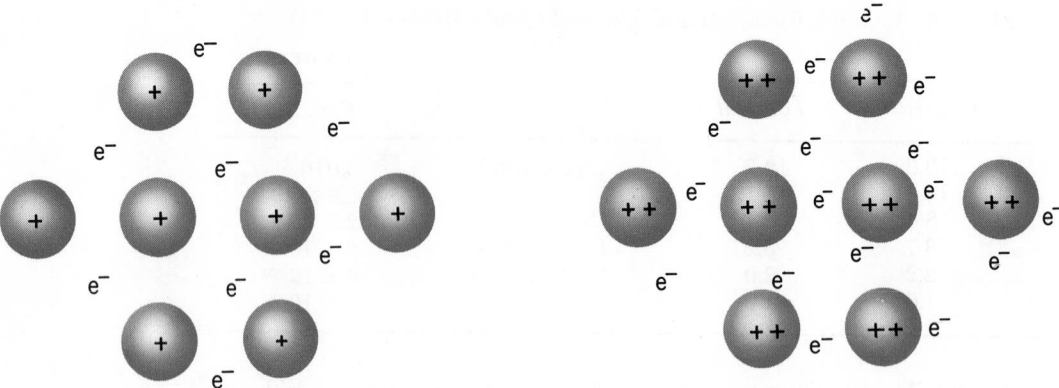

Figure 7.17 Electron-sea model of metallic bonding in Li (left diagram) and Mg (right diagram).

electrons, which must occur frequently. If electrons are responsible for carrying heat as well as electrical energy, it is hardly surprising that metals fall in the same order when arranged according to decreasing thermal or electrical conductivity.

Since the electrons are not tied down to a particular bond with a characteristic energy, they are able to absorb and re-emit light over a wide range of wavelengths, thereby explaining why metal surfaces are excellent reflectors. The electron-sea model also offers a plausible explanation for the ductility and malleability of metals (Fig. 7.18). Under an applied stress, a layer of positive ions can move relative to one another with little electrostatic resistance. In contrast, if we attempt to slide one layer across another in a crystal such as NaCl, where the negative charges are fixed in position,

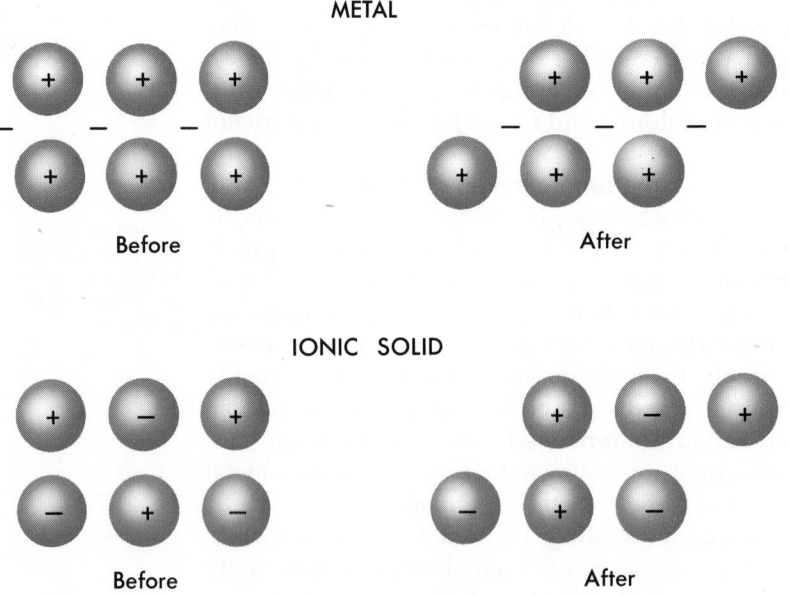

Figure 7.18 In an ionic crystal, movement of crystal planes relative to one another tends to bring ions of like charge into contact. In a metal, the structure remains essentially unchanged because of the mobility of electrons.

we create the unstable situation shown in Figure 7.18 (bottom, right) and the crystal lattice.

Even though the electron-sea model qualitatively explains many of the characteristic properties of metals, it is inadequate in certain respects. In particular, it is difficult to explain, in terms of this model, why the molar heat capacity of metals is not considerably higher than that of nonmetals. One would suppose that if the electrons in a metal were all free to move throughout the crystal, they would make a significant contribution to the heat capacity. In the case of calcium, for example, one can calculate that the heat capacity on the basis of a free-electron model should be nearly twice the observed value of 6.2 cal/mole.

Deficiencies of this type led to the development of a more sophisticated model of metallic bonding, known as the *band theory* of metals. This approach, based upon molecular orbital theory, distributes the valence electrons of a metal among a series of closely spaced energy levels which "belong" to the entire metal sample rather than a particular atom or group of atoms. A huge number of such levels, as many as 10^{23}, are grouped in a "band" which may stretch over an energy range of as much as 100 kcal/mole. Only those electrons near the top of the band are believed to contribute to such properties as conduction and heat capacity.

PROBLEMS

7.1 Certain of the following statements are valid; others are open to criticism. Which ones would you criticize and why?

    a. Carbon, in all its stable compounds, obeys the octet rule.

    b. The number of covalent bonds formed by a nonmetal atom is determined by the number of unpaired electrons.

    c. The reaction of a metal with a nonmetal leads to the formation of a compound containing ionic bonds.

    d. Molecules containing polar covalent bonds have dipole moments.

7.2 Explain, in your own words, why

    a. positive ions are smaller than the atoms from which they are formed.

    b. the bonds in methane are directed at tetrahedral angles.

    c. molecules with an odd number of valence electrons do not obey the octet rule.

    d. sulfur forms compounds involving sp^3d^2 hybrid bonds but oxygen does not.

7.21 Criticize each of the following statements:

    a. Ionic bonds are stronger than covalent bonds.

    b. Cations are smaller than anions.

    c. In CH_2Cl_2, the hydrogen and chlorine atoms are located at the corners of a regular tetrahedron.

    d. Electron repulsion can never account for the formation of a square, planar molecule.

7.22 Explain, in your own words, why

    a. linings of high temperature furnaces often are made of ionic compounds such as MgO or Al_2O_3.

    b. MO theory gives a better explanation of the properties of O_2 than does valence bond theory.

    c. some molecules of the type AB_2 are polar while others are not.

    d. a single Lewis structure does not adequately describe the SO_2 molecule.

7.3 Predict the formulas of the following compounds.

    a. lithium sulfide
    b. aluminum fluoride
    c. ammonium nitrate
    d. calcium phosphate
    e. silver carbonate
    f. zinc sulfide
    g. nickel hydroxide

7.4 Complete and balance the following equations

    a. $Ag(s) + O_2(g) \rightarrow$
    b. $Cr(s) + Br_2(l) \rightarrow$
    c. $Al(s) + O_2(g) \rightarrow$
    d. $Sc(s) + Cl_2(g) \rightarrow$
    e. $Mg(s) + S(s) \rightarrow$
    f. $Co(s) + Se(s) \rightarrow$

7.5 Give the electronic configurations of the following ions

    a. Na^+ d. Tl^{3+}
    b. Se^{2-} e. H^-
    c. Cr^{2+}

7.6 Consider the bonds formed by Be, Mg and Ca with S, Se and Te. Arrange these bonds in order of increasing per cent of ionic character.

7.7 Using the table of covalent radii given on p. 660, estimate the internuclear distance in HF, HCl, HBr and HI. The observed distances are 0.92 Å, 1.27 Å, 1.41 Å and 1.61 Å. What, if anything, does this reveal about the relative electronegativities of the halogens?

7.8 Draw Lewis structures for

    a. ClO_2^- d. C_2H_6O (two structures)
    b. PH_3 e. CN^-
    c. PO_4^{3-}

7.9 Draw Lewis structures for the following species. (The skeleton structure is indicated by the way the molecule is written.)

    a. SCN^-
    b. $ClNO_2$
    c. $O_2N—NO_2$
    d. $FN—NF$

7.10 The compound $Na_2S_2O_3$ is used as a fixing agent in photography.

    a. What are the charges of the ions present?
    b. Write a Lewis structure for the polyatomic anion.
    c. Describe the geometry of the polyatomic anion.

7.23 Give the number of moles of ions in one mole of

    a. Na_2CO_3.
    b. silver phosphate.
    c. calcium oxide.
    d. nickel chromate.
    e. aluminum oxide.
    f. chromium (III) sulfide.

7.24 Complete and balance the following equations

    a. $K(s) + Te(s) \rightarrow$
    b. $Al(s) + S(s) \rightarrow$
    c. $Li(s) + O_2(g) \rightarrow$
    d. $Sr(s) + O_2(g) \rightarrow$
    e. $Y(s) + Br_2(l) \rightarrow$

7.25 Give the formulas of two different ions which have the same electronic structure as

    a. Ar   c. Au^+
    b. Mn^{2+} d. O^{2-}

7.26 If we arbitrarily call a bond "ionic" when the electronegativity difference between the two atoms joined by the bond is greater than 1.7, with what elements does boron form ionic bonds?

7.27 Using Equation 7.3 and Table 7.4, calculate the difference in electronegativity between hydrogen and each of the halogens.

7.28 Draw Lewis structures for

    a. CO   d. NO^+
    b. NF_3 e. ClO_3^-
    c. NO^-

7.29 Draw Lewis structures for

    a. Cl_2SO
    b. $HClC—CClH$
    c. $F_2N—NF_2$
    d. $SO_2(OH)_2$
    e. $CO_2(OH)^-$

7.30 One of the important components of the upper atmosphere is ozone, O_3. Three possible structures for ozone are

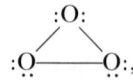 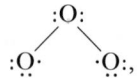 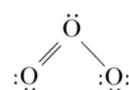

How could you determine experimentally which of these structures is correct?

7.11 For which of the following species would it be necessary, in valence-bond theory, to postulate resonance?

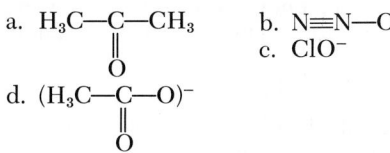

a. $H_3C—C—CH_3$ (with $\parallel$ O below the central C)   b. $N\equiv N—O$

c. ClO^-

d. $(H_3C—C—O)^-$ (with $\parallel$ O below the central C)

7.12 Give the formula of a polyatomic ion which you would expect to have the same Lewis structure as

a. HF   c. F_2O
b. Cl_2 d. CCl_4

7.13 A major eye irritant in smog is acrolein, which has the skeleton structure

$$\begin{array}{ccc} H & H & H \\ | & | & | \\ H—C—C—C—O \end{array}$$

Draw the Lewis structure for this molecule and indicate all the bond angles.

7.14 For each of the following species, draw Lewis structures and describe the geometry

a. NO_2^- c. SO_3
b. CO_2 d. SO_3^{2-}

7.15 Which of the *molecules* listed in Problems 7.12–7.14 are polar?

7.16 For each of the following species, determine the number of pairs of electrons around the central atoms, their orientation in space, and the geometry of the species.

a. SF_2 d. PCl_5
b. SF_4 e. PCl_4^+
c. SF_6 f. PCl_4^-

7.17 Give the formula of a molecule or ion in which an atom of

a. carbon forms 2 sp hybrid bonds.
b. boron forms 3 sp² hybrid bonds.
c. nitrogen forms 3 sp² hybrid bonds.
d. nitrogen forms 2 π bonds.
e. nitrogen forms 4 sp³ bonds.

7.18 Consider the +1 ions formed by removing an electron from the diatomic molecules of the elements of atomic numbers 5 through 9. Using the MO approach, list

a. the number of electrons in each of the 2s and 2p MO's.
b. the number of bonds in each of these ions.

7.31 Draw all the major resonance forms for

a. CO_2 b. CO_3^{2-} c. NO_2

7.32 Give the formula of a molecule which you would expect to have the same Lewis structure as

a. OH^- c. NO_2^-
b. CO_3^{2-} d. PO_4^{3-}

7.33 One of the most objectionable compounds in photochemical smog is peroxyacetyl nitrate, PAN. Its skeleton is

$$\begin{array}{ccccccc} H_3C—C—O—O—N—O \\ | \qquad\qquad | \\ O \qquad\qquad O \end{array}$$

Write the Lewis structure for this molecule and give all the bond angles.

7.34 Predict the geometry of each of the species given in Problem 7.28.

7.35 Which of the following molecules would you expect to have a dipole moment?

ClF   CO PCl_3 PCl_5 XeF_4

7.36 For each of the following species, determine the number of pairs of electrons around the central atom and their orientation in space.

a. XeF_2 c. ClF_3
b. SF_4^{2-} d. IF_5

7.37 Indicate the hybrid orbitals used by carbon in bond formation in

a. C_2H_6 d. H_2CO
b. C_2H_4 e. CO
c. CO_3^{2-}

How many π bonds would you expect to find in each species?

7.38 Suppose that in building up molecular orbitals, the σ_{2p}^b were placed below the π_{2p}^b (this was believed to be the case until quite recently). Prepare a diagram similar to Figure 7.16 based on this assignment. For which species would this change in relative energies affect your prediction of number of bonds? number of unpaired electrons?

7.19 Using the MO approach, decide which of the following species would be paramagnetic

   a. NO
   b. NO⁻
   c. N_2^+
   d. O_2^-
   e. O_2^{2-}

7.20 Explain in terms of the electron-sea model why

   a. metals are good conductors of heat.
   b. metals are malleable and ductile.
   c. magnesium has a higher melting point than sodium.

7.39 Consider the O_2 molecule, whose MO structure is given in Figure 7.16. If successive electrons are removed to give the 1st, 2nd, 3rd, . . . ionization energies, where would you expect to find the largest jump in ionization energy?

7.40 Explain why

   a. Mg is a better conductor of electricity than silicon.
   b. a freshly cut piece of sodium has a shiny appearance.
   c. it is relatively easy to grind large crystals of sodium chloride into a fine powder, but difficult to do this with iron.

*7.41 Assume that an electron is confined to move about within an atom which has a radius of one angstrom. Two such atoms are brought together to form a molecule; the total volume of the system does not change but each electron can move in both atoms. Using the equation $E = n^2h^2/8MV^{2/3}$ with $n = 1$, calculate ΔE in ergs for each electron for this process. Using the appropriate conversion factor, calculate ΔE in kcal/mole of bonds. Compare to ΔH for bond formation, which is of the order of 100 kcal.

*7.42 Using the argument of Problem 7.41,

   a. what would you predict to be the ratio of the heat of formation of a double bond to a single bond, neglecting differences in bond distances?
   b. taking into account the fact that a double bond is shorter than a single bond, would you expect the actual ratio to be larger or smaller than the number predicted in (a)? (The observed ratio for carbon is about 1.7.)
   c. remembering that the I atom and the I_2 molecule are, of course, larger than the Cl atom and the Cl_2 molecule, would you expect the I—I bond to be stronger or weaker than the Cl—Cl bond? (The observed values are ΔH_f I—I $= -36$ kcal, ΔH_f Cl—Cl $= -58$ kcal.) Hint: Obtain an expression relating ΔE for the formation of a bond to the volume of the atom, assuming the volume to the molecule to be twice that of the atom.

*7.43 In the molecule IF_7, there are seven fluorine atoms around iodine. By extending the electron repulsion approach for five and six pairs of electrons, predict the geometry for IF_7.

*7.44 It was stated on p. 198 that ionic crystals tend to shatter under stress. Yet, in some instances, planes of ions are known to slide over one another if a stress is applied in the right direction. For example, geologists have attributed changes in the shoreline along the Gulf of Mexico to flow of salt deposits beneath the surface of the earth. Looking at the diagram of the NaCl structure on p. 33, can you suggest how such might occur? (It may help to turn the page sideways.)

*7.45 It was pointed out in the text that the Xe atom in XeF_4 is surrounded by six pairs of electrons directed toward the corners of a regular octahedron. The square planar geometry of XeF_4 was derived by removing two corners of the octahedron. Such a process could give another, quite different structure. Describe this structure and suggest why it is not observed.

8 PHYSICAL PROPERTIES OF MOLECULAR SUBSTANCES; NATURE OF ORGANIC COMPOUNDS

In Chapter 7 we discussed the nature of the covalent bonds that hold atoms together in molecular substances. The chemical properties of these substances are dependent in large part on the strength of these bonds, since they must be broken for reaction to occur. In contrast, the physical properties of molecular substances (e.g., melting point, boiling point, solubility) are determined primarily by the extent to which molecules interact with one another. When a substance such as ethyl alcohol boils or dissolves in water, no covalent bonds are broken; instead, intermolecular forces are overcome.

In this chapter we shall consider the various types of intermolecular forces that govern the physical behavior of molecular substances. The principles developed, along with those introduced in Chapter 7, will then be applied to interpret the physical and chemical properties of a very important class of molecular substances, the organic compounds.

8.1 GENERAL CHARACTERISTICS OF MOLECULAR SUBSTANCES

There are several properties of molecular substances that serve to distinguish them from those which are ionic or metallic. As a class, molecular substances are more volatile than either metals or ionic substances, which are typically solids with high melting points and very low vapor pressures at room temperature. Copper melts at 1083°C and boils at 2336°C; common salt, NaCl, melts at 801°C and begins to boil at 1413°C. In contrast, many molecular substances such as H_2, O_2, and CH_4 are gaseous at 25°C; many others at that temperature are liquids (e.g., water, carbon

203

tetrachloride) or solids (iodine, naphthalene) with appreciable vapor pressures. Most molecular substances have melting and boiling points below 300°C.

Molecular species are ordinarily poor conductors of electricity. To be an electrical conductor a substance either must contain relatively mobile electrons, as do metals, or it must in some way furnish freely moving ions. Ionic substances, when melted or dissolved in water, yield ions which conduct electricity quite well. Most molecular substances, either as pure liquids or as solutes in water solution, do not ionize appreciably and so have low electrical conductivity. There are a few polar molecular species, like HCl and HNO_3, which react with water to form ions and, hence, are good conductors in water solution; there is another larger group of polar substances which in water solution ionize to a small extent, forming slightly conducting solutions. The large bulk of molecular solutes, however, are poor conductors in water.

INTERATOMIC VS INTERMOLECULAR FORCES

The generally low melting and boiling points of molecular substances are a direct consequence of the weak forces between molecules. They imply nothing about the forces within a molecule, which are ordinarily quite strong. Consider, for example, elementary hydrogen, which is made up of diatomic molecules. To melt or boil hydrogen, it is necessary only to overcome the attractive forces holding the molecules together—rigidly in the solid, loosely in the liquid state. The low melting point (−259°C) and boiling point (−253°C) of hydrogen reflect the weakness of these forces. The fact that the covalent bond between hydrogen atoms in the H_2 molecule is extremely strong is immaterial; this bond remains intact when hydrogen melts or boils. The greater strength of interatomic as opposed to intermolecular forces is implied by the fact that at a temperature of 2400°C only about 1 per cent of the H_2 molecules are dissociated into atoms. A comparison of the energy of dissociation of hydrogen molecules, 104 kcal/mole, with the energy of sublimation, 0.122 kcal/mole, gives another indication of the relative magnitudes of these forces.

One must distinguish between forces which hold molecules together in a liquid or solid and the forces which hold the atoms together in a molecule

TRENDS IN MELTING AND BOILING POINTS

Among molecular substances of similar nature we find that the melting point, boiling point, heat of fusion, and heat of vaporization all tend to increase with molecular weight. Consider, for example, the four halogens listed at the top of Table 8.1; there is a steady increase in each of these properties as we go from F_2 (MW = 38) to I_2 (MW = 254).

Close examination of Table 8.1 reveals that although there is a general correlation between molecular weight and physical properties such as boiling point, other factors must be operating as well. Notice, for example, that SF_6 has a much lower boiling point than Br_2, even though the molecular weights of these two species are similar. Perhaps most striking of all,

TABLE 8.1   Physical Properties of Some Molecular Substances

SUBSTANCE	MOLEC-ULAR WEIGHT	MELTING POINT (°C)	BOILING POINT (°C)	HEAT OF FUSION (kcal/mole)	HEAT OF VAPOR-IZATION (kcal/mole)
F_2	38	−223	−187	0.38	2.80
Cl_2	71	−102	−35	1.63	4.79
Br_2	160	−7	59	2.59	7.17
I_2	254	113	184	4.01	10.50
CO_2	44	−56	−78 subl	2.00	3.84
CS_2	76	−109	46	1.05	6.55
SO_2	64	−75	−10	2.06	6.08
SO_3	80	17	45	1.92	9.48
SF_6	146	−56	−64 subl		
HF	20	−92	19	1.09	1.79
HCN	27	−14	26	1.72	6.03

we see that HF and HCN, which have molecular weights lower than F_2, boil about 200° higher!

One of the factors in addition to molecular weight which affects the volatility of molecular substances is polarity. Compounds built up of polar molecules melt and boil at slightly higher temperatures than nonpolar substances of comparable molecular weight (Table 8.2). The effect of polarity on melting or boiling point is ordinarily small enough to be obscured by differences in molecular weight. For example, in the series HCl → HBr → HI, the boiling point increases steadily with molecular weight despite decreasing polarity. However, in molecular compounds in which hydrogen is bonded to a small, highly electronegative atom (N, O, F), polarity has a much greater effect on volatility. Hydrogen fluoride (bp 19°C), despite its low molecular weight, has the highest boiling point of all the hydrogen halides. Water (bp 100°C) and ammonia (bp −33°C) also have abnormally high boiling points compared to those of the other hydrides of the 6A and 5A elements. In these three cases at least, the effect of polarity reverses the normal trend to be expected on the basis of molecular weight alone.

As one might expect, the substances that are most difficult to condense to liquids or solids are those in which the basic structural unit is both light in weight and nonpolar. Such substances include the nonmetallic elements

TABLE 8.2   Boiling Points of Polar vs Nonpolar Substances

NONPOLAR			POLAR		
Formula	Molecular Weight	Boiling Point (°C)	Formula	Molecular Weight	Boiling Point (°C)
N_2	28	−196	CO	28	−192
SiH_4	32	−112	PH_3	34	−85
GeH_4	77	−90	AsH_3	78	−55
Br_2	160	59	ICl	162	97

of low molecular weight (H_2, bp = $-253°C$; N_2, bp = $-196°C$; O_2, bp = $-183°C$; F_2, bp = $-187°C$) and the lower members of the noble gas group (He, bp = $-269°C$; Ne, bp = $-246°C$; Ar, bp = $-186°C$).

An explanation of these trends in volatility lies in the nature of the forces holding molecular substances together. These forces, for reasons which we shall now consider, increase in magnitude with the size and polarity of the molecule.

TYPES OF INTERMOLECULAR FORCES

DIPOLE FORCES. The effect of polarity on the physical properties of molecular substances is readily explained in terms of the dipole forces existing between polar molecules. It has been pointed out (Section 7.5) that such molecules tend to line up in an electrical field. A similar orientation exists in a crystal composed of polar molecules. In solid iodine chloride, for example, the ICl molecules are aligned in such a way that the iodine atom (+pole) of one molecule is adjacent to the chlorine atom (−pole) of the next molecule (Fig. 8.1).

The electrostatic attraction holding neighboring molecules together in an iodine chloride crystal is similar in origin to that between adjacent ions in solid sodium chloride. However, the dipole forces between polar molecules in ICl are an order of magnitude weaker than the ionic bonds in NaCl. In the former case the unequal electronegativities of iodine and chlorine produce only partial + and − charges within the molecule. In NaCl, on the other hand, a complete transfer of electrons leads to ions with full + and − charges.

When iodine chloride is heated to 27°C the comparatively weak dipole forces are no longer able to hold the molecules in rigid alignment and the solid melts. Dipole forces remain significant in the liquid state, in which the polar molecules are still relatively close to each other. Only in the gas, where the molecules are very far apart, do the electrical forces become negligible. Consequently the boiling points as well as the melting points

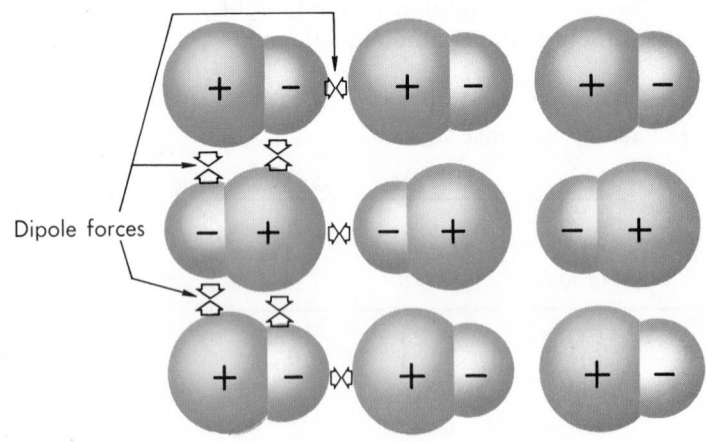

Dipole forces

Figure 8.1 Dipole forces in an ICl crystal.

of polar compounds such as ICl are higher than those of nonpolar substances of comparable molecular weight (Table 8.2).

HYDROGEN BONDS. The abnormal properties of hydrogen fluoride, water, and ammonia result from the presence in these substances of an unusually strong type of intermolecular force. This attractive force, exerted between the hydrogen atom of one molecule and the fluorine, oxygen, or nitrogen atom of another, is sufficiently unique to be given a special name, the hydrogen bond. There are two reasons why hydrogen bonds are stronger than ordinary dipole forces:

1. The difference in electronegativity between hydrogen (2.1) and fluorine (4.0), oxygen (3.5), or nitrogen (3.0) is great enough to cause the bonding electrons in HF, H_2O, and NH_3 to be markedly displaced from the hydrogen atom. Consequently, the hydrogen atoms in these molecules, insofar as their interaction with adjacent molecules is concerned, behave almost like bare protons. The hydrogen bond is strongest in HF, in which the difference in electronegativity is greatest, and weakest in NH_3, where the difference in electronegativity is relatively small.

2. The small size of hydrogen allows the fluorine, oxygen, or nitrogen atom of one molecule to approach the hydrogen atom in another very closely. It is significant that hydrogen bonding appears to be limited primarily to compounds containing these three elements, all of which have comparatively small atomic radii. The larger chlorine and sulfur atoms, with electronegativities (3.0, 2.8) similar to that of nitrogen, show little or no tendency to form hydrogen bonds in such compounds as HCl and H_2S.

What properties of H_2S might be used to show that it does not exhibit hydrogen bonding?

The energy required to break a hydrogen bond varies from about 1 to 10 kcal; even though the hydrogen bond is much weaker than a covalent bond, it is the strongest type of intermolecular force. Since many molecules of biological importance contain O—H and N—H bonds, hydrogen bonding is very common in such substances and often has an important influence on their properties. Protein molecules, which are long chains, sometimes take on a helical coil structure stabilized by hydrogen bonds between N—H groups and C≡O groups, which occur at corresponding points on successive turns in the coil (Chapter 23).

Although hydrogen bonding is most commonly encountered in organic molecules, it does occur in several inorganic substances, of which we have mentioned HF, H_2O, and NH_3. (The effect of hydrogen bonding on the properties of liquid water and ice will be discussed in Chapter 11.) Hydro-

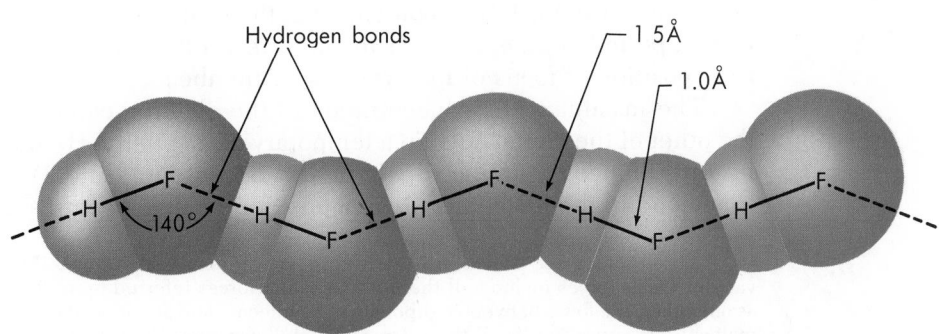

Figure 8.2 Hydrogen bonding in HF.

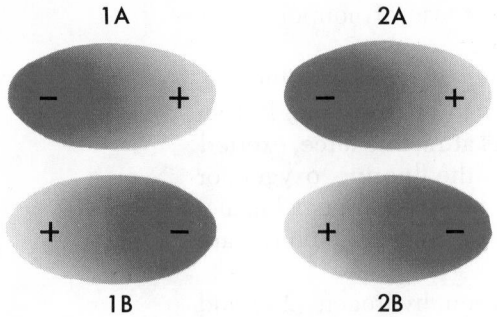

Dipoles like these exist only momentarily and may quickly change magnitude and direction

Figure 8.3   Temporary dipoles in H_2 molecules.

gen bonding is responsible for the high boiling point of hydrogen cyanide, HCN (Table 8.1).

DISPERSION (LONDON) FORCES.*   The two types of intermolecular forces already discussed can exist only between polar molecules. A different type of attractive force must be postulated to explain the existence of the liquid and solid states of such nonpolar substances as bromine and iodine. Since the melting and boiling points of nonpolar substances tend to increase with molecular weight, we deduce that the magnitude of this force increases with the mass or size of the molecule. The fact that the volatility of polar as well as nonpolar substances decreases with increasing molecular weight suggests that this third type of intermolecular force must be common to all molecular substances.

The intermolecular force whose characteristics we have speculated upon is known as a dispersion force. Its origin is more difficult to visualize than that of the dipole force or hydrogen bond. Like them, it is basically electrical in nature. However, while hydrogen bonds and dipole forces arise from an attraction between permanent dipoles, dispersion forces are due to what might be called temporary or instantaneous charge separations.

It has been pointed out that, on the average, the two bonding electrons in a nonpolar molecule such as H_2 are as close to one nucleus as to the other; the molecule has no permanent dipole moment. However, at any given instant, the electron cloud may be concentrated to some extent at one end of the molecule (position 1A in Figure 8.3). A fraction of a second later it may be located at the opposite end of the molecule (position 1B). The situation is analogous to that of a person watching a tennis match from a position directly in line with the net. At one instant his eyes are focused on the player to his left; a moment later, they shift to the player on his right. Over a period of time, he looks to one side as often as the other; the "average" position of focus of his eyes is straight ahead.

The instantaneous concentration of the electron cloud on one side or the other of the center sets up a temporary dipole in the H_2 molecule. This, in turn, induces a similar dipole in an adjacent molecule. When the electron cloud in the first molecule is at position 1A, the electrons in the second

---

* These forces are sometimes referred to as van der Waals forces. Properly speaking, van der Waals forces include all the intermolecular forces referred to in this section, as well as one other, the force between a dipole in one molecule and an induced dipole in an adjacent molecule. It is the sum of all these forces which determines the magnitude of the deviation of real gases from ideal behavior (cf. van der Waals equation, Chapter 5).

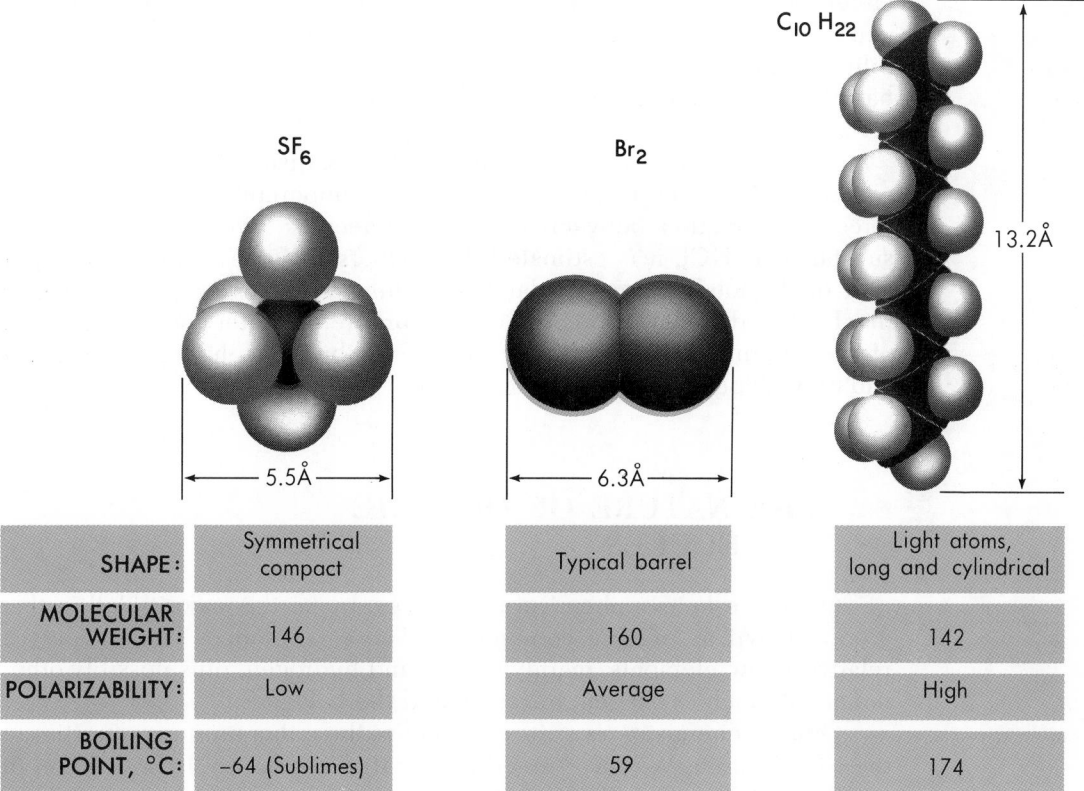

	SF$_6$	Br$_2$	C$_{10}$H$_{22}$
	5.5Å	6.3Å	13.2Å
SHAPE:	Symmetrical compact	Typical barrel	Light atoms, long and cylindrical
MOLECULAR WEIGHT:	146	160	142
POLARIZABILITY:	Low	Average	High
BOILING POINT, °C:	−64 (Sublimes)	59	174

Figure 8.4 Effect of molecular size and shape on boiling points of substances with similar molecular weights.

molecule are attracted to 2A. As the first electron cloud shifts to 1B, the electrons of the second molecule are pulled back to 2B. These temporary dipoles, both oriented in the same direction, lead to a small attractive force between the molecules; this is the dispersion force.

The strength of dispersion forces depends upon the ease with which the electronic distribution in a molecule can be distorted or "polarized" by a temporary dipole set up in an adjacent molecule. As one might expect, the ease of polarization depends primarily upon the size of the molecule. Large molecules where the electrons are far removed from the nuclei are more readily polarized than small, compact molecules in which the nuclei exert greater control over the position of the electrons. In general, molecular size and molecular weight parallel each other; hence, the observation that dispersion forces increase in magnitude with molecular weight.

The fact that dispersion forces are more directly related to the size and shape of a molecule than to its mass enables us to explain some of the anomalies previously noted in Table 8.1. Given two molecules of the same weight but of different shape or size, the electrons in the smaller or more compact molecule will have less freedom to move about, giving the molecule a smaller polarizability and the substance a lower boiling point. The effect can be quite substantial; SF$_6$, with small, compact, symmetrical molecules, has a relatively low polarizability and a low boiling point, −64°C, whereas bromine, Br$_2$, of about the same weight but with a somewhat larger volume, boils at 59°C. Molecules containing light atoms like hydrogen will tend to be rather bulky and will have relatively high polarizabilities and boiling points. Decane, $C_{10}H_{22}$, a relatively large molecule, has about the same molecular weight as bromine, but boils at 174°C (Fig. 8.4).

Of the three types of intermolecular forces, dispersion forces are most common and, in the majority of cases, most important. They are the only forces of attraction between nonpolar molecules; even with as polar a substance as HCl, it is estimated that dispersion forces contribute 85 per cent of the total intermolecular force. Only where hydrogen bonding is involved do dispersion forces play a minor role. In water about 80 per cent of the intermolecular attraction can be attributed to hydrogen bonding and only 20 per cent to dispersion forces.

8.2   THE NATURE OF ORGANIC MOLECULES

Of all the known chemical substances, by far the majority fall in that group known as organic compounds. These substances contain only a relatively few elements, mainly carbon and hydrogen, plus possibly other nonmetals such as oxygen, nitrogen, and the halogens.

Organic chemistry owes its existence to the rather unique properties of the carbon atom. We have noted earlier that carbon, by hybridization of atomic orbitals, ordinarily forms four covalent bonds. What is unique about carbon and distinguishes it from other elements in Group 4A, such as silicon and germanium which also can form four bonds, is that the C—C bond is strong ($\Delta H_f = -83$ kcal). In contrast, Si—Si bonds ($\Delta H_f = -42$ kcal) and Ge—Ge bonds are relatively weak. This enables carbon atoms to bind together in long chains, whereas silicon and germanium atoms cannot. Carbon also forms strong bonds with hydrogen ($\Delta H_f = -99$ kcal) and with other nonmetals (ΔH_f C—O $= -84$ kcal, ΔH_f C—N $= -70$ kcal). The huge number of organic compounds is the result of the fact that C—C, C—H, C—O, and C—N bonds can occur in molecules in an enormous number of different combinations, each of which gives rise to a different substance with different properties. In this section we will describe some of the general classes of combinations which are found. The rather remarkable fact is that, if you draw almost *any* Lewis structure in which each carbon atom forms four bonds with C, H, O, or N, it is extremely likely that your structure will correspond to that of a known substance!

It was once thought that organic molecules could be made only by living organisms

HYDROCARBONS

Since all organic substances contain carbon and hydrogen, we might reasonably begin a study of organic chemistry by considering the structures of some molecules containing only C and H atoms; substances of this sort are given the not unlikely name of *hydrocarbons*. In Figure 8.5 we have drawn Lewis structures of some hydrocarbons containing only single bonds. There are many more, and given the examples, you should be able to think

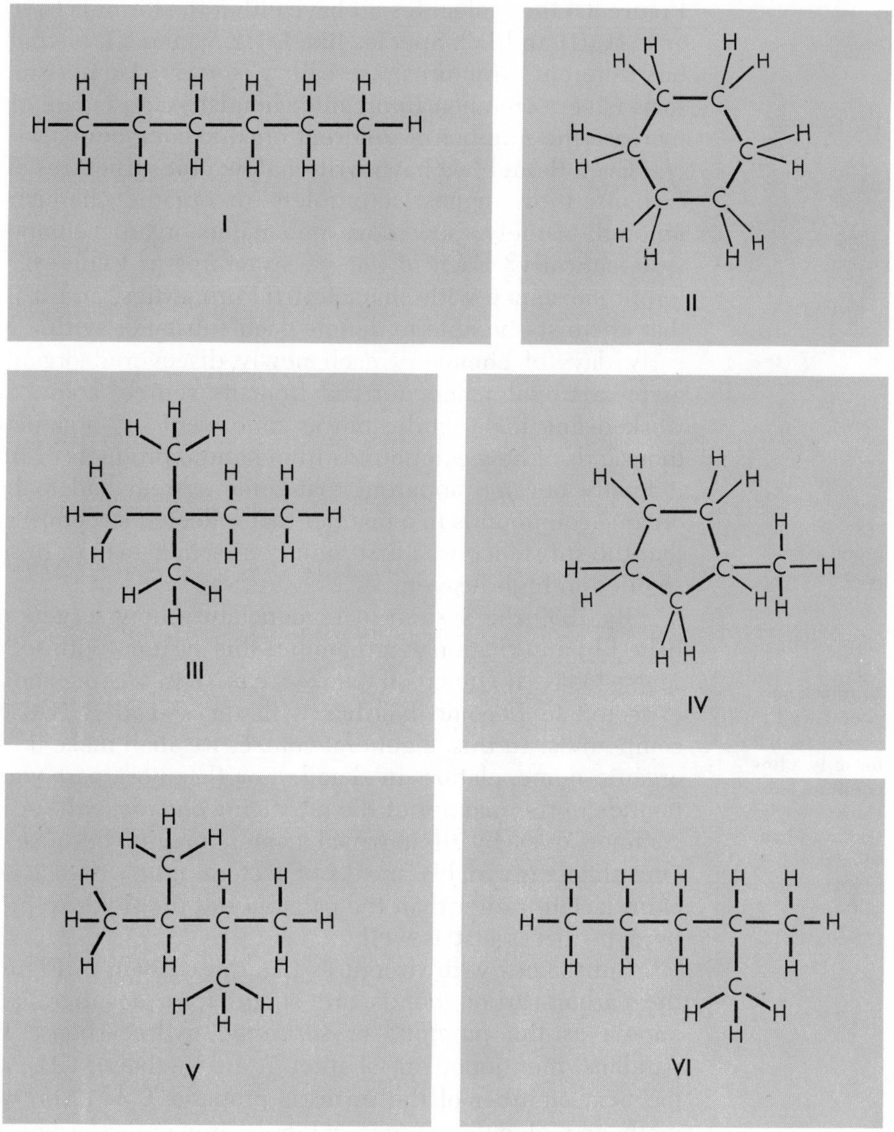

Figure 8.5   Structures of some hydrocarbons containing six carbon atoms.

of several others. Each of the structures corresponds to that of a known, reasonably common substance.

Hydrocarbons, and indeed all organic molecules, contain carbon chains, which may be short or as long as a hundred or even a thousand carbon atoms. The main chain may be branched one or more times and may contain rings of various sizes. In Figure 8.5 we limited ourselves to using only six carbon atoms, linked in different ways. When you consider the possible molecular structures containing 10, 20, or 50 carbon atoms, you can easily see why there are far more organic than there are inorganic substances.

In organic chemistry, we commonly find that a given molecular formula has several different molecular structures associated with it. In

Figure 8.5 the molecules all have either the formula C_6H_{14}(I, III, V, and VI) or C_6H_{12}(II and IV). Species like I, III, V, and VI, having the same formula but different structures, are called *isomers*. Isomerism of this and other sorts is very common in organic chemistry and is one of the factors which increase the number of different organic compounds.

Even though we have written down the structures of only a few hydrocarbons, they suggest a problem in organic chemistry that cannot be ignored, namely, how does one name organic compounds simply and systematically? Each of the six structures in Figure 8.5 represents a different substance with characteristic properties, and it is highly desirable that chemists be able to denote each substance with a word name. In the early days of chemistry each newly discovered organic compound was given a trivial name derived from its source, color, odor, or use. This worked fine for a while, but as time went by, organic chemists isolated thousands of new compounds from natural products or from syntheses, and it finally became apparent that some system had to be set up to name organic compounds in a manner that reflected structure rather than the fact that the substance was first found on or in a goat, or in sour milk, or that it smelled to high heaven.

By 1930 the system of nomenclature now in general use was established by international agreement; this system, with some modification, is in use today. If you go on to take a course in organic chemistry, you will be expected to become familiar with the so-called IUC names of organic compounds. In this, a general course, we shall make no attempt to discuss organic nomenclature in detail. For the most part we shall name compounds in the manner of the practicing chemist, who still calls most of the common ones by their trivial names, mainly because they are simpler. One almost invariably speaks of acetone rather than 2-propanone and calcium lactate rather than the calcium salt of 2-hydroxypropanoic acid. And perhaps that is just as well.

Substances with structures like those given in Figure 8.5 in which all the carbon-carbon bonds are single form a subgroup of hydrocarbons known as the *paraffins* or *saturated* hydrocarbons. The two simplest paraffins, mentioned in Chapter 7, are methane, CH_4, and ethane, C_2H_6; the next member of the series is propane, C_3H_8. You will recall that the methane molecule is a perfect tetrahedron with a carbon atom at the center. The bonding around each carbon atom in ethane is also tetrahedral, as it indeed is around carbon in *all* of the paraffins. Because of the resulting zigzag nature of chains of carbon atoms, paraffinic molecules are more compact than the planar diagrams would imply.

As you can see in Figure 8.6, the outer regions of paraffin molecules consist mainly of hydrogen atoms, and as far as the surface is concerned, one molecule looks much like another. Since the electronegativities of carbon and hydrogen are of about the same magnitude, C—H bonds are virtually nonpolar, so molecular interactions in paraffins are limited to dispersion forces. We would expect these forces to increase with molecular size, producing a reasonably smooth increase in melting and boiling points of the paraffins as one goes to longer and longer chains. These general ideas are borne out quite well by the data in Table 8.3.

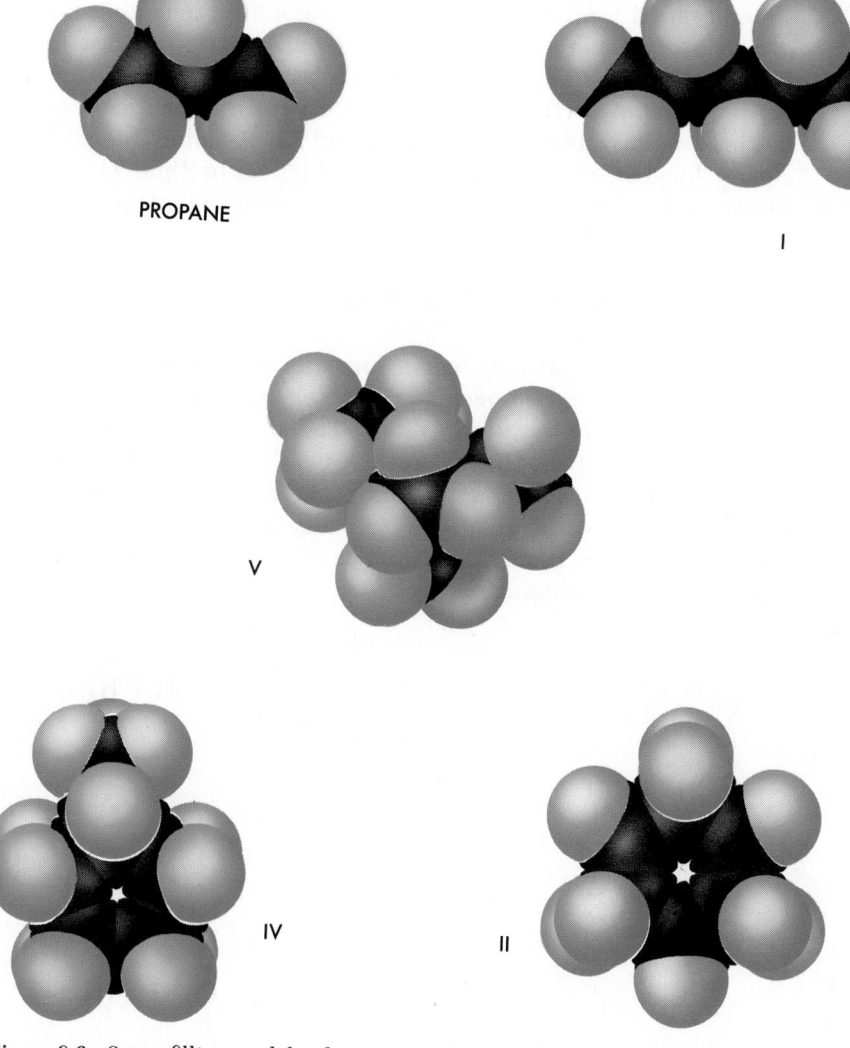

PROPANE

I

V

IV

II

Figure 8.6 Space-filling models of propane and some of the molecules in Figure 8.5.

TABLE 8.3 Physical Properties of Some Saturated Hydrocarbons

FORMULA	MOL WT	MP, °C	BP, °C
CH_4	16	−184	−161
C_2H_6	30	−172	−88
C_3H_8	44	−190	−42
C_4H_{10}	58	−135	−1
C_5H_{12}	72	−132	36
$C_6H_{14}(I)$ *	86	−94	69
$C_6H_{14}(III)$ *	86	−98	50
$C_6H_{14}(V)$ *	86	−135	58
$C_6H_{14}(VI)$ *	86		60

* Refer to Figure 8.5 for the structures of these isomers.

In Table 8.3 the boiling points of the first six compounds (all are the straight chain forms) do increase very smoothly, reflecting the gradual increase in dispersion forces with molecular size. The last four isomeric compounds in the Table have structures as shown in Figure 8.5. The lowest boiling isomer is III, which is the most nearly spherical and, therefore, the least polarizable. Compound (I), which is called "normal hexane," is clearly longest, most polarizable, and highest boiling. These compounds illustrate the trends and range of boiling points of typical isomeric species.

UNSATURATED HYDROCARBONS

It is possible, as you recall, for carbon to form double and triple as well as single bonds. Species containing carbon-carbon double or triple bonds are said to be *unsaturated*. The double-bonded compounds are sometimes called *olefins* or *alkenes*, while those with triple bonds are called *alkynes* or *acetylenes*, after C_2H_2, the first member of that series. Any saturated hydrocarbon can, formally at least, be converted to an associated olefin by removing a molecule of hydrogen; with butane, for example, we could obtain

1-butene

or (8.1)

2-butene

"Cracking" also serves to break the carbon chain of large molecules, producing more volatile substances from high-boiling oils and tars

This kind of reaction can be made to go by simply heating butane in the presence of an Fe catalyst to about 450°C. The reaction is widely used in the petroleum industry, where it is called "cracking"; it is one of the steps in the remodeling of the many hydrocarbons naturally found in petroleum that is necessary to make modern gasolines. (Unfortunately "cracking" also occurs in an automobile engine to some extent; any unburned olefins in the exhaust tend to oxidize in the air and form some of the components of smog.)

The atom geometry around a double-bonded carbon atom is always planar, with the bond angles around the atom all about equal to 120°; this configuration is due to the three sp^2 hybrid bonds formed by double-bonded carbon. Since the C=C bond consists of a π bond plus a σ bond from the sp^2 hybrid, there is no free rotation about a double bond; a molecule like 2-butene can therefore exist in two isomeric forms, called *geometric* isomers, in which the two hydrogen atoms attached to double-

bonded carbon atoms can be either on the same side (**cis** form) or on opposite sides (**trans** form) of the double bond:

CH₃   H   H   H
C=C   C=C
H   CH₃   CH₃   CH₃

trans 2-butene   cis 2-butene

The olefins have a higher chemical reactivity, by virtue of the double
bond, and so are more directly useful in industry and in the laboratory. A
common reaction of olefins is one of addition, essentially the reverse of
Reaction 8.1; it is possible to add molecules like HCl or Br_2, as well as H_2,
directly to the double bond. Another very important reaction is that of
polymerization, in which an olefinic molecule adds to itself to form a long
chain. Since polymers as a group have rather characteristic properties and
can be made from a variety of organic compounds, we shall treat them
separately after we have discussed the general classes of organic substances.

Substances containing triple-bonded carbon atoms are less common
than are the paraffins and the olefins. Acetylene (bp, −84°C), which is the
simplest of these compounds, is very reactive chemically. The best known
use of acetylene is in welding torches, where it is burned in pure oxygen
to furnish a very high temperature flame. The acetylenes readily undergo
addition reactions with hydrogen or the halogens to form olefins or substituted paraffins, many of which are important industrially. The acetylenes,
in general, are unstable with respect to decomposition to the elements;
acetylene itself is prone to spontaneous detonation with little provocation;
special procedures are used to compress the gas in tanks for use in industry.

Bonding theory correctly predicts that the geometric arrangement of
atoms around a carbon atom on which there is a triple bond is linear. All
acetylenes therefore contain four colinear atoms; the C≡C bond is made
up of one σ and two π bonds.

Acetylene can be
made by reacting
calcium carbide
with water:
$$CaC_2(s) + 2\ H_2O \rightarrow$$
$$C_2H_2(g) + Ca(OH)_2(s)$$

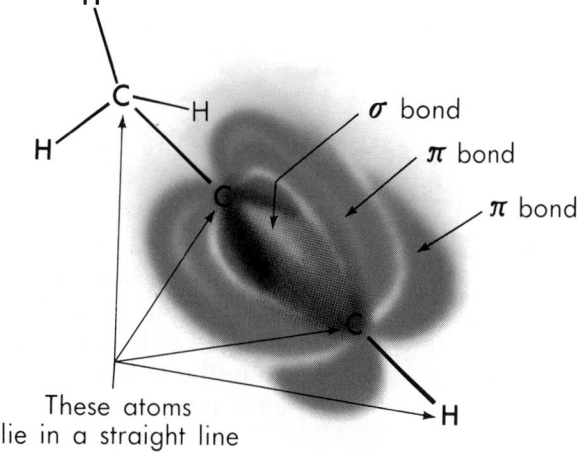

Figure 8.7 Structure of methyl acetylene.

AROMATIC HYDROCARBONS

There is one group of unsaturated hydrocarbons that is so important as to be classified separately. Substances in this class are called *aromatics*, probably because many of them have rather strong, and often pleasant, odors. All aromatic substances can be considered to derive from one compound, benzene, C_6H_6:

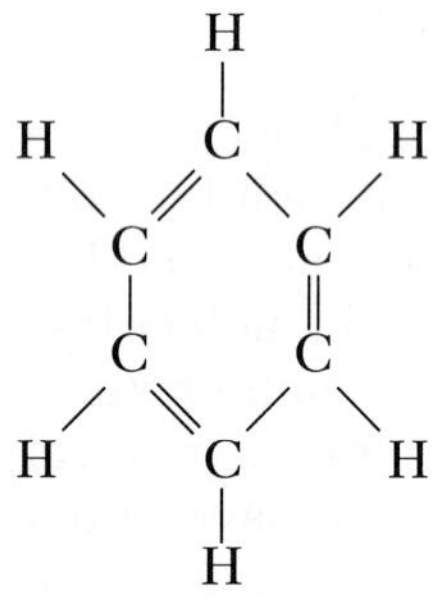

Benzene is a most interesting substance, from both a theoretical and a practical point of view. Knowing that the bonds around a double-bonded carbon atom tend to form three 120 degree angles, we would expect that a ring such as that present in benzene could form without strain and so might be reasonably stable. Theory would also predict that the molecule would be planar, since the atomic geometry around a double-bonded carbon atom is always planar.

Benzene is indeed a planar molecule, and in addition is actually a regular hexagon, with all C—C bonds of equal length. The molecule is energetically much more stable than its high degree of unsaturation would imply; it does not readily add halogens or hydrogen and shows no tendency to polymerize.

The hexagonal symmetry and high stability of benzene are thought to be the result of resonance forms analogous to those previously used to explain the properties of molecules like SO_2 and the NO_3^- ion:

These are called
Kekulé structures
after the man who
first proposed that
benzene was hexag-
onal

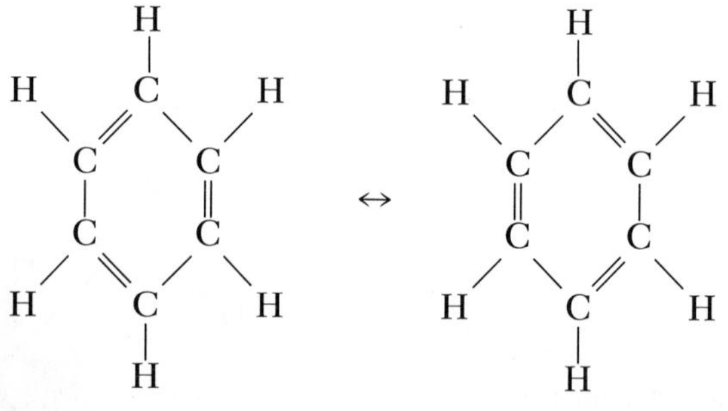

The π electrons in benzene appear to be quite mobile, so that as a result of resonance π bonding occurs uniformly above and below the plane of the molecule (Fig. 8.8). Molecular orbital theory has been rather successfully applied to benzene and indicates that the π electronic structure is as shown in the figure.

There are many aromatic substances, all of which contain one or more

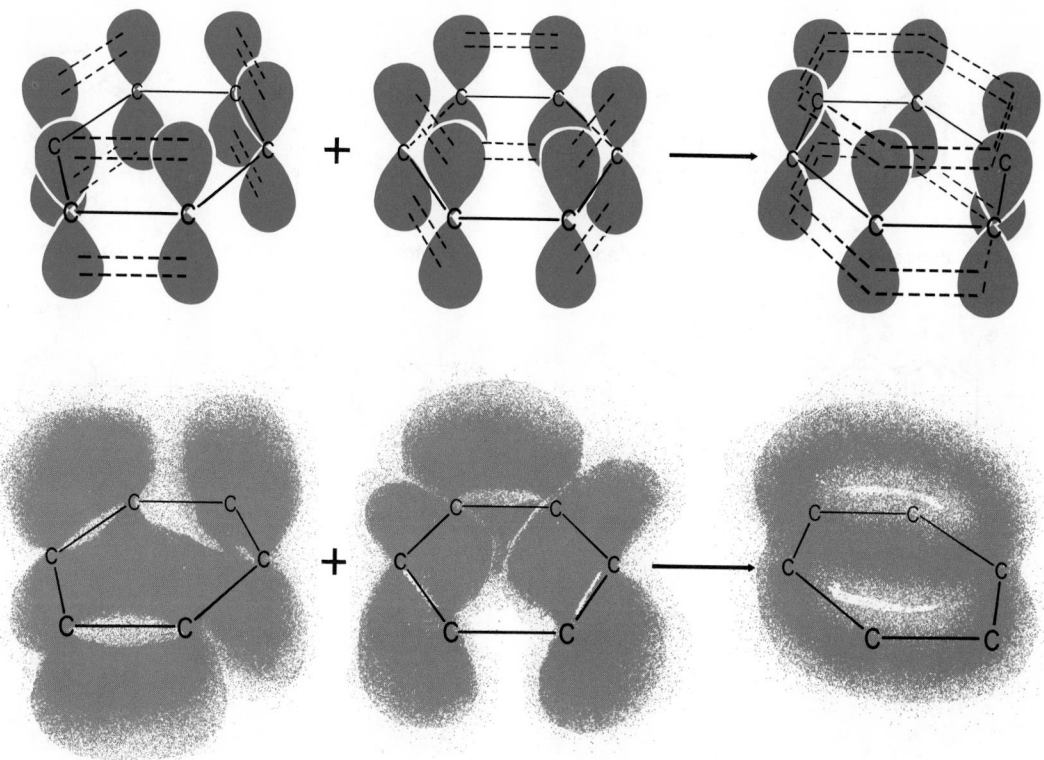

Figure 8.8 Pi bonding in benzene. Resonance of π bonds produces doughnut-shaped charge clouds above and below the plane of the molecule.

benzene rings. In Figure 8.9 we indicate the structures of some of the simpler ones, all of which can be found in coal tar, a gummy black material obtained when soft coal is heated to about 1200°C in the absence of air. (In the drawings the ring carbon and hydrogen atoms are omitted.) These compounds are all of commercial importance. Benzene, toluene, and the xylenes are used in gasoline and are industrial intermediates. Naphthalene is still found to some extent in mothballs but is mainly used in the manufacture of those polymers known as alkyd resins, which form the film in many automobile enamels. Phenol, sometimes called carbolic acid, was the first antiseptic; it is effective but unfortunately is also corrosive and causes severe skin burns. These days phenol is used in enormous amounts in the manufacture of Bakelite plastics and glues for plywood. The cresols are the main components of the wood preservative known as creosote.

In addition to the substances in Figure 8.9, coal tar contains a good many polynuclear hydrocarbons composed of various numbers of benzene rings fused together. Some of these are very powerful carcinogenic (cancer-producing) substances. This property was discovered early in this century after coal tar workers were seen to have an abnormally high incidence of skin cancer. One of the most active carcinogens in coal tar is 1,2 benzopyrene, a yellow crystalline substance, identified as the culprit about 1933; this material is also a component of cigarette smoke and is thought to be a

Although the aromatics are found in coal tar they are produced industrially mainly from petroleum in reactions where paraffins are catalytically cyclized and dehydrogenated

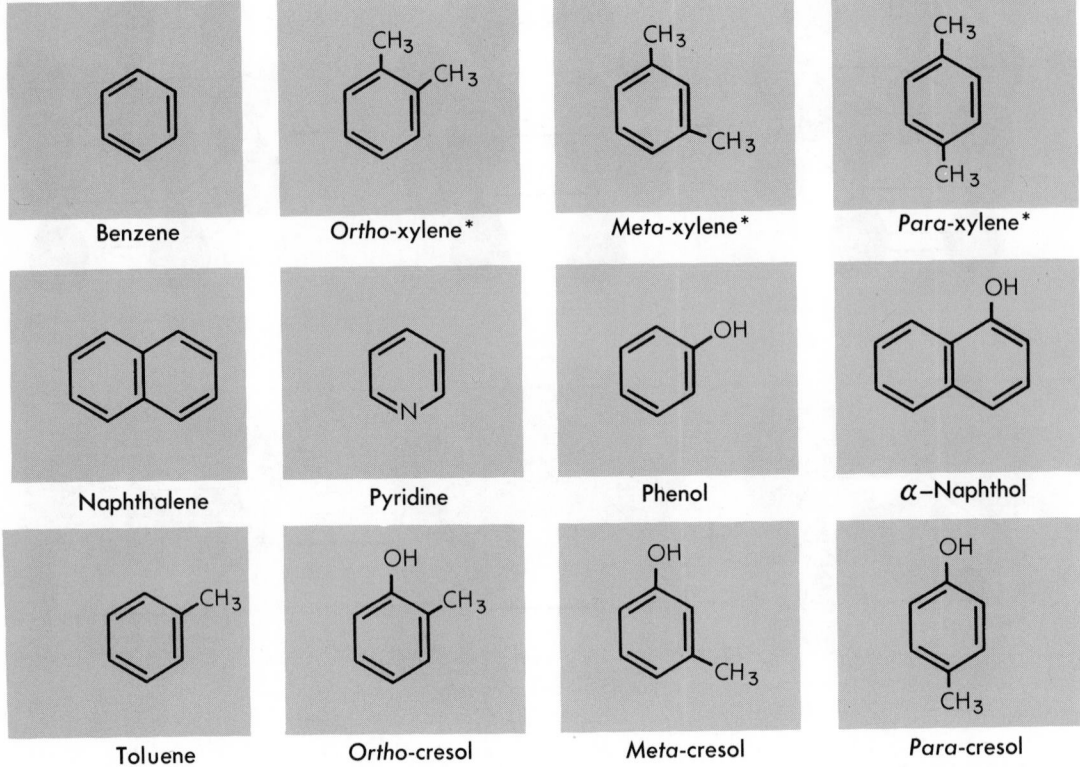

Figure 8.9 Some aromatic substances found in coal tar.

° Ortho, meta and para are used as prefixes to indicate where disubstituted species exist on the benzene ring.

cause of the lung cancer so prevalent among heavy smokers. There are many other recognized carcinogens, one of the most potent being methyl-cholanthrene, which can be produced by a simple, although unlikely, reaction of cholesterol, a steroid present in all animal blood. Common to many carcinogens, and to these two examples, are three coplanar, linear, benzene rings.

1,2-Benzopyrene   Methylcholanthrene

HALOGEN-CONTAINING ORGANIC COMPOUNDS

Although there are many very common organic compounds which contain halogen atoms, almost none of these occur naturally. The synthesis of chlorinated compounds is usually by direct reaction of chlorine with the hydrocarbon and occurs quite readily. Although preparation of organic bromides and iodides is feasible by various methods, chlorine is both cheaper and more reactive than bromine or iodine, so most of the important halogenated hydrocarbons are chlorides. Fluorine reacts so violently with hydrocarbons that fluoro derivatives must be prepared indirectly, usually by reaction of an inorganic fluoride of silver, mercury, antimony, or cobalt with an alkyl bromide or chloride.

The possibilities for isomerism when one or more hydrogen atoms in a hydrocarbon are replaced by halogen atoms are enormous. (Just consider the number of isomers obtainable from chlorine substitution in hexane, Structure I in Figure 8.5, recognizing that anywhere from one to 14 Cl atoms may substitute.) Since industrial halogenation processes usually produce many of the possible isomers, and since those of similar molecular weight are not easily separated, the standards of purity of many of the commercially available halogenated hydrocarbons are well below those of common inorganic substances. The halogenated hydrocarbons are much more reactive chemically than are the hydrocarbons themselves and are extremely useful intermediates in organic chemical manufacture. More than five million tons of halogenated hydrocarbons are manufactured annually in this country. You may be familiar with some of the simpler ones (Fig. 8.10).

Many, indeed most, halogenated compounds are toxic to animal and plant life. (Freon, CCl_2F_2, used in aerosol spray cans and refrigerators because of its low boiling point, is, along with other highly fluorinated species, inert.) Most of the common insecticides and herbicides are chlorinated compounds; these will be considered further in Chapter 11.

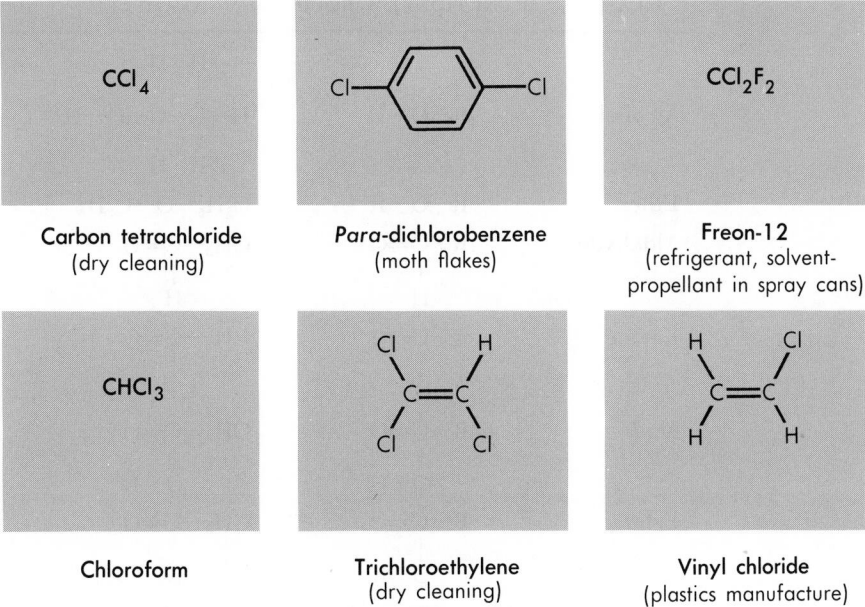

CCl_4	Cl—⬡—Cl	CCl_2F_2
Carbon tetrachloride (dry cleaning)	**Para-dichlorobenzene** (moth flakes)	**Freon-12** (refrigerant, solvent-propellant in spray cans)
$CHCl_3$	Cl,H / C=C \ Cl,Cl	H,Cl / C=C \ H,H
Chloroform	**Trichloroethylene** (dry cleaning)	**Vinyl chloride** (plastics manufacture)

Figure 8.10 Some common halogenated hydrocarbons.

OXYGEN-CONTAINING ORGANIC COMPOUNDS

When one considers organic compounds containing oxygen as well as carbon and hydrogen, the number and kinds of possible substances increase enormously. Whereas hydrogen and the halogen atoms form only one covalent bond, oxygen typically forms two, and so is not restricted to a substitutional role as are the halogens.

Oxygen occurs in organic molecules in several different bonding arrangements, each of which is given a class name. Each class is characterized by the presence of a particular molecular fragment, called a **functional group**. Some of the most common classes of organic compounds are listed in Table 8.4, along with their associated functional groups and a typical representative of each class.

You may recognize some of the names in Table 8.4, since they are among the most common organic compounds. The oxygen-containing substances have some common properties, such as high polarity and appreciable water solubility; each class also has its own characteristic properties, which often make its members more suitable for particular purposes than those in another class.

Ethyl alcohol, C_2H_5OH, like alcohols in general, exhibits strong hydrogen bonding, which gives it a much higher boiling point, 78°C, than, say, ethyl chloride, C_2H_5Cl, 12°C. Ethyl alcohol, sometimes called grain alcohol or ethanol, is usually made by fermentation of sugar or starch solutions in the presence of yeasts and the absence of air. There are several other very common alcohols, including methyl alcohol, CH_3OH, and glyc-

TABLE 8.4   Classes of Oxygen-containing Organic Substances

CLASS	FUNCTIONAL GROUP	EXAMPLE	NAME
Alcohol	R—O—H	H—C—C—O—H (with H, H above and H, H below the two carbons)	Ethyl alcohol
Ether	R—O—R′	C_2H_5—O—C_2H_5	Diethyl ether
Aldehyde	R—C=O with H below	CH_3—C=O with H below	Acetaldehyde
Ketone	R—C=O with R′ below	CH_3—C=O with CH_3 below	Acetone
Acid	R—C=O with O—H below	CH_3—C=O with O—H below	Acetic acid
Ester	R—C=O with O—R′ below	CH_3—C=O with O—CH_3 below	Methyl acetate

R and R′ = CH_3, C_2H_5 or more complex hydrocarbon radicals

erol, a trihydroxyalcohol, $CH_2OHCHOHCH_2OH$, which is a sweet, viscous liquid used in various foods, cosmetics, and drugs.

Diethyl ether, sometimes called ether, is best known as an anesthetic, although in recent years it has been replaced almost completely by less flammable chemicals. The ethers as a group are not as polar as the other oxygen-containing classes and resemble the saturated hydrocarbons with the same chain length. For example, diethyl ether, C_2H_5—O—C_2H_5, has a boiling point (35°C) closer to that of pentane, C_5H_{12} (36°C) than to butyl alcohol, C_4H_9OH (117°C).

Ketones differ from aldehydes only in the substitution of an R′ group, such as CH_3 or C_2H_5, for an H atom. As we might expect, the physical and chemical properties of these two classes of oxygen compounds are quite similar. The carbonyl group, C=O, is quite reactive; it can, for example, be converted to either the C—O—H group, to form alcohols, or the methylene group, CH_2, to make saturated hydrocarbons. Aldehydes and ketones of relatively high molecular weight have pleasant odors and are found in flowers and spices. The C_8 to C_{14} straight chain aldehydes are used in the formulation of perfumes. The most common ketone, acetone, has a penetrating odor, but seems to give some satisfaction to glue sniffers; its vapor is about as toxic as that of gasoline. Formaldehyde, CH_2O, marketed as formalin in a 40 per cent water solution, is familiar to zoology students as a preservative.

An organic molecule containing both a carbonyl and a hydroxyl group on the same carbon atom behaves as a weak acid in water solution, producing hydrogen ions by the following reaction:

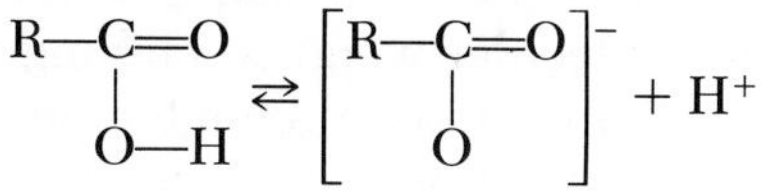

$$\begin{matrix} R-C=O \\ | \\ O-H \end{matrix} \rightleftarrows \left[\begin{matrix} R-C=O \\ | \\ O \end{matrix}\right]^- + H^+$$

Organic acids are present in many foods. Acetic acid gives vinegar its characteristic taste; citric acid is present in oranges and lemons; malic acid is present in several fruit juices and was first isolated from apples:

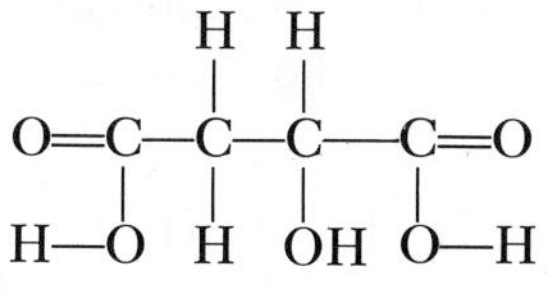

malic acid 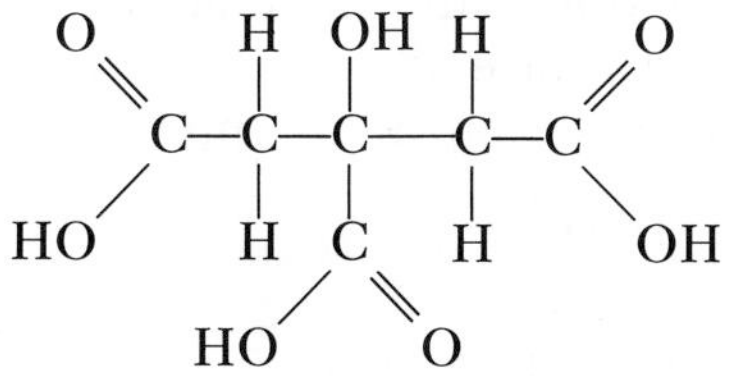   citric acid

The organic acids readily react with alcohols to form esters; methyl acetate can be made by reacting methyl alcohol and acetic acid under such conditions that water, the other product, is removed:

The organic acids readily react with alcohols to form esters; methyl acetate can be made by reacting methyl alcohol and acetic acid under such conditions that water, the other product, is removed:

$$CH_3-OH + HO-\overset{\overset{\textstyle O}{\|}}{C}-CH_3 \rightarrow CH_3-O-\overset{\overset{\textstyle O}{\|}}{C}-CH_3 + H_2O$$

methyl alcohol   acetic acid   methyl acetate

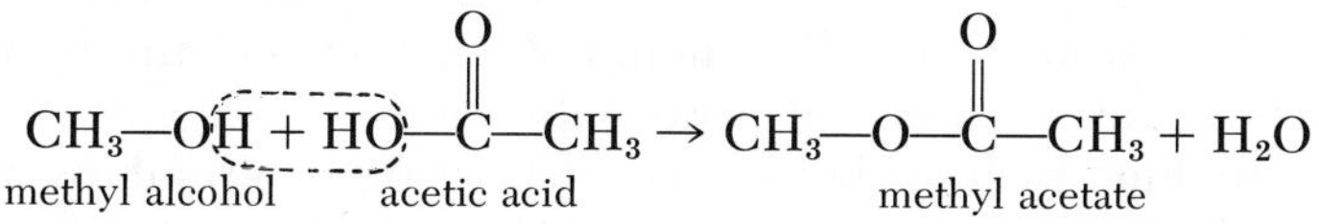

Fats are esters of long chain organic acids and various alcohols, usually glycerol. A very important industrial reaction of fats is that with boiling sodium hydroxide solution, which produces substances known as soaps:

Would you expect ethers to exhibit hydrogen bonding?

$$CH_2OC\overset{O}{\overset{\|}{-}}(CH_2)_{14}CH_3$$

$$CHOC\overset{O}{\overset{\|}{-}}(CH_2)_{14}CH_3 + 3OH^- \rightarrow CHOH + 3\ (CH_3(CH_2)_{14}C\overset{O}{\overset{\|}{-}}O)^-$$

$$CH_2OC\overset{O}{\overset{\|}{-}}(CH_2)_{14}CH_3$$

$$CH_2OH$$
$$CH_2OH$$

a fat   glycerol   a soap anion

A soap is a sodium or potassium salt of a fatty acid. Its useful cleaning properties are related to its structure; in a single molecule, a soap combines a long chain hydrocarbon, which has good solvent action on other hydrocarbons, with the polar COO^- group, which has high water solubility.

SYNTHETIC ORGANIC POLYMERS

Some of the most important organic substances differ from those we have considered up to now in that their molecules contain very long chains, consisting for the most part of carbon atoms. Within the chains there are repeating, usually rather simple, units. Molecules of this sort are called *polymers*, in that they contain many *monomer* units. Polymers by their nature have high molecular weights, ranging from about 1×10^4 to more than 1×10^6, and all are solids at room temperature. There are many naturally occurring polymeric substances, but we shall limit our discussion here to the simpler ones, all of which are synthetic.

Probably the simplest organic polymer is polyethylene, which is made by simple end-to-end addition of ethylene molecules. The polymerization reaction will occur at high temperatures and pressures in the presence of a trace of oxygen, but can also be carried out under much more moderate conditions in the presence of a catalyst:

A polyethylene with a molecular weight of 50,000 would contain carbon chains about 4000 atoms long

ethylene   section of a polyethylene molecule

Since all the carbon-carbon bonds are single, there is essentially tetrahedral bonding around each carbon atom, making the chain zigzag. The chains are typically several thousand units long and usually have a few short branches, so that a certain amount of tangling of chains occurs. Polyethylene is a waxy, rather soft solid with chemical properties similar to those of paraffinic hydrocarbons, whose structure it resembles. Since the molecular chains do not interact except through dispersion forces, polyethylene can be melted and so is a thermoplastic. Beginning at about 120°C, the polymer softens to a viscous fluid and can be cast into film or molded into articles which solidify on cooling. Because of its high molecu-

TABLE 8.5 Some Common Addition Polymers

MONOMER	NAME	POLYMER	USES	AMT PRODUCED IN 1970 (millions of lbs)
$H_2C{=}CH_2$	ethylene	polyethylene	bags, coatings, toys	6000
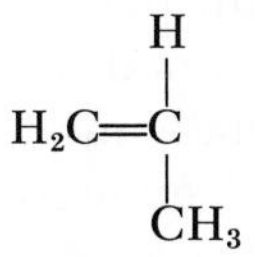 $H_2C{=}\overset{\displaystyle H}{\underset{\displaystyle CH_3}{C}}$	propylene	polypropylene	beakers, milk cartons	1000
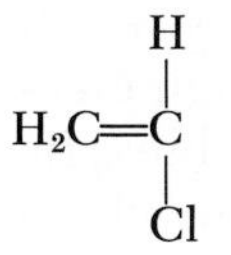 $H_2C{=}\overset{\displaystyle H}{\underset{\displaystyle Cl}{C}}$	vinyl chloride	polyvinyl chloride, PVC	floor tile, raincoats	3000
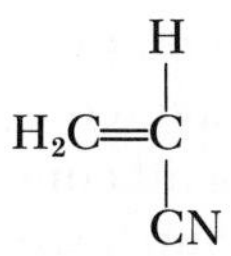 $H_2C{=}\overset{\displaystyle H}{\underset{\displaystyle CN}{C}}$	acrylonitrile	polyacrylonitrile, PAN	rugs; Orlon, Acrilan, and Dynel are copolymers with vinyl chloride or other polar monomers	500
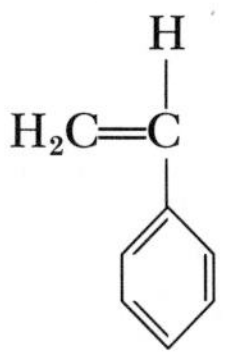 $H_2C{=}\overset{\displaystyle H}{C}$	styrene	polystyrene	cast articles requiring a transparent plastic	2500
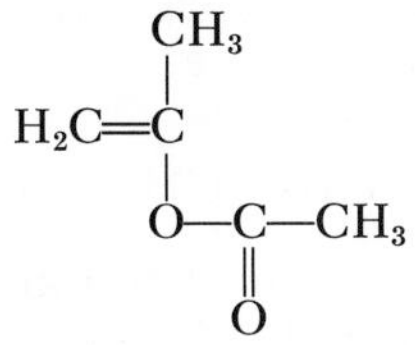 $H_2C{=}\overset{\displaystyle CH_3}{\underset{\displaystyle O{-}C{-}CH_3}{C}}$	methyl methacrylate	polymethyl methacrylate, Plexiglas, Lucite	high quality transparent cast objects	50
$F_2C{=}CF_2$	tetrafluoroethylene	Teflon	gaskets, insulation, bearings, pan coatings	20

lar weight, polyethylene, like all polymers, is not volatile and at high temperatures will decompose before it vaporizes.

Several derivatives of ethylene, most of which have the general formula $H_2C{=}CHR$ can be polymerized to form useful plastics. Like polyethylene, these are *addition* polymers, being made by direct addition of the monomer units, and contain chains having the following structure:

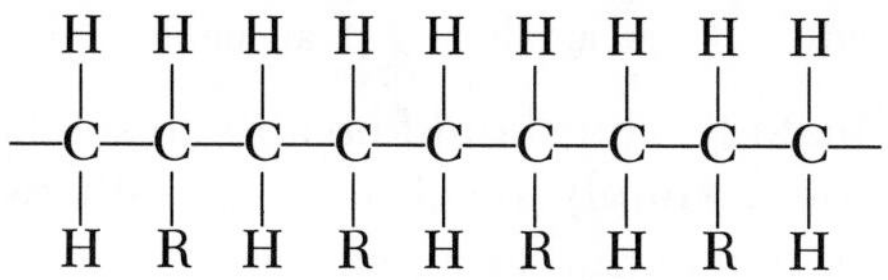

$$-\overset{\displaystyle H}{\underset{\displaystyle H}{C}}-\overset{\displaystyle H}{\underset{\displaystyle R}{C}}-\overset{\displaystyle H}{\underset{\displaystyle H}{C}}-\overset{\displaystyle H}{\underset{\displaystyle R}{C}}-\overset{\displaystyle H}{\underset{\displaystyle H}{C}}-\overset{\displaystyle H}{\underset{\displaystyle R}{C}}-\overset{\displaystyle H}{\underset{\displaystyle H}{C}}-\overset{\displaystyle H}{\underset{\displaystyle R}{C}}-\overset{\displaystyle H}{\underset{\displaystyle H}{C}}-$$

These polymers differ to some extent from polyethylene in their properties, depending on the nature of the R group they contain, and so have different industrial applications. Some of the more common ones are listed in Table 8.5.

Teflon, the last polymer listed in Table 8.5, is unique in that it is completely substituted, with all hydrogen atoms replaced by fluorine. Whereas all the other polymers will burn to some degree and can be degraded chemically, teflon is extremely inert, resisting chemical attack by all known reagents except molten alkali metals. It is a useful plastic from about −70° to 250°C, a much larger range than can be claimed for other polymers, which tend to become brittle when cold and usually melt below 200°C. Teflon is a moderately hard, white, waxy solid, tough enough to be machined to close tolerances. Its inertness and temperature stability make it *the* substance of choice for making gaskets and seals which are subject either to high temperatures or caustic reagents. Its high electrical resistance and flexibility make it ideal for use in coaxial cables. Its most mundane use is as a nonstick coating for frying pans and pots, where its desirable properties are put to the test daily by thousands of housewives.

Having noted that polymerization reactions are common among organic olefins, one might wonder whether other substances we have considered might form polymers. If, for example, we started with an acid containing two carboxyl groups and an alcohol with two hydroxyl groups, we might hope to cause them to react, under proper conditions, in such a way as to form a linear polyester:

$$HO-\overset{\overset{O}{\|}}{C}-R-\overset{\overset{O}{\|}}{C}-\overbrace{OH+H}O-R'-O\overbrace{H+HO}-\overset{\overset{O}{\|}}{C}-R-\overset{\overset{O}{\|}}{C}-\overbrace{OH+H}O-R'-OH \rightarrow$$

a dicarboxylic acid   a dihydroxy alcohol

$$-O-\overset{\overset{O}{\|}}{C}-R-\overset{\overset{O}{\|}}{C}-O-R'-O-\overset{\overset{O}{\|}}{C}-R-\overset{\overset{O}{\|}}{C}-O-R'- + x\ H_2O$$

section of a polyester molecule

When —R— is a benzene ring, ⟨◯⟩, and —R'— is —CH₂—CH₂—, a

Finding R and R'
groups which yield
polymers with
optimum proper-
ties was probably
accomplished by
trial and error
polymer which forms very strong fibers and films is obtained. This material is now produced in large amounts commercially and is called Dacron or Mylar, depending on whether it is spun into thread or rolled into a film.

The polymer known as nylon is made by a reaction quite similar to that used to produce Dacron. If instead of reacting an acid with an alcohol, one uses an analogous substance called an amine, which contains —NH₂ rather than an —OH group, one can form, on elimination of water, a substance known as an amide:

$$R-\overset{\overset{O}{\|}}{C}-OH + H_2N-R' \rightarrow R-\overset{\overset{O}{\|}}{C}-\underset{\underset{H}{|}}{N}-R' + H_2O$$

an acid   an amine   an amide

Now, reasoning by analogy, you should be able to see how we might make a polyamide. We would simply react an acid with two carboxyl groups with an amine having two amino groups:

$$\underset{\text{a dicarboxylic acid}}{HO-\overset{\overset{O}{\|}}{C}-R-\overset{\overset{O}{\|}}{C}-OH} + \underset{\text{a diamine}}{H_2N-R'-NH_2} \rightarrow$$

$$-\overset{\overset{O}{\|}}{C}-R-\overset{\overset{O}{\|}}{C}-\underset{\underset{H}{|}}{N}-R'-\underset{\underset{H}{|}}{N}-\overset{\overset{O}{\|}}{C}-R-\overset{\overset{O}{\|}}{C}-\underset{\underset{H}{|}}{N}-R'-\underset{\underset{H}{|}}{N}- + x\ H_2O$$

section of a polyamide molecule

Dr. Wallace Carothers, working at the du Pont Company laboratories in Wilmington, Delaware, in the early 1930's, found that the polymer formed when —R— is —$(CH_2)_4$— and —R'— is —$(CH_2)_6$— had very promising properties. That material has proved to be one of the most important of the synthetic polymers and is known throughout the world as nylon.

Nylon is an ideal polymer in many ways. Perhaps its main virtue is in its ability to form strong but moderately elastic fibers, which are essentially perfect for the manufacture of women's stockings. The reason nylon makes good fibers appears to be, in part at least, that between polymer chains hydrogen bonding can occur, with the oxygen atom from a carbonyl group on one chain interacting with an H atom from a nearby N—H group on another. A typical fiber will contain a large number of parallel molecular chains, held together by hydrogen bonds in such a way as to produce a strong, flexible, slightly elastic filament. In solid form nylon is also a useful polymer; it is tough and suitable for use in gears and other high quality parts requiring resistance to wear.

The successful development of nylon was at least partly responsible for the tremendous expansion of industrial chemical research following World War II

Organic polymers of the type we have discussed here are important components in fabrics, nearly all paints, rugs, and floor tiles, films for use in packaging, photography, and recording, all sorts of small cast articles, glues and adhesives, even shoe soles and uppers. Polymers appear destined to play an ever-increasing role as materials from which the world's goods are made. At present there are more chemists working on the development and production of polymers than in any other area of the chemical industry.

Many organic compounds are even more complex than any of those we have mentioned. Included among such substances are many that are important to life processes. These materials are typically polymeric, with long chains containing carbon, plus nitrogen or oxygen atoms; to the carbon atoms are attached smaller groups, ranging in size from H atoms to fragments similar in size and structure to the molecules we have discussed. These substances, whose properties and reactions form that part of chemistry known as biochemistry, will be discussed in Chapter 23.

8.3 MACROMOLECULAR SUBSTANCES

We have seen that in the case of polymers it is possible for a molecule to contain thousands of atoms; the molecular weight of a polyethylene polymer can easily be 50,000 or greater. There are some covalently bonded inorganic substances with molecular weights even larger than this in

The MW's of
macromolecular
substances ap-
proach infinity

which simply enormous numbers of atoms are linked together in two- or three-dimensional networks. These materials contain no small discrete molecules; instead, a given crystal may consist of one "molecule" of macroscopic size. Such substances, whether organic or inorganic, are called *macromolecular*. They typically have very high melting points and are insoluble in all common solvents. There are many such substances, but among the most common ones are the crystalline forms of carbon and silicon dioxide, the main component of sand.

ALLOTROPIC FORMS OF CARBON

Many elements, of which carbon is one example, can occur in more than one form in the same physical state. The general phenomenon is called *allotropy;* the different modifications are referred to as *allotropes.* The forms may differ in molecular formula, as is the case with the element oxygen, which can exist as O_2 or O_3 molecules in the gas state. Allotropy is most common in the solid state. Here it may result from a difference in the way the molecules are packed in the crystal, as in monoclinic and rhombic sulfur. Alternatively, allotropic forms of a solid element may differ in the way the atoms are bonded to one another. This is the case with the two crystalline forms of the element carbon, graphite and diamond. Both allotropes are macromolecular, but as we shall see, the bonding patterns are quite different.

Both graphite and diamond have very high melting points, above 3500°C. This value is higher than that of any other element; even tungsten melts at 3400°C. As soon as one examines the crystal structures of graphite and diamond he can see why they are so difficult to melt. Both contain essentially infinite networks of carbon-carbon bonds (Fig. 8.11). Many of these relatively strong bonds have to be broken to melt diamond or graphite.

The graphite crystal is planar and contains what amounts to the carbon skeletons of a multitude of hexagonal benzene rings covalently bonded

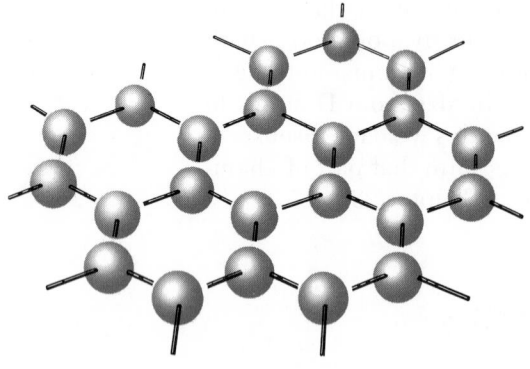

Graphite layer   Diamond crystal

Figure 8.11 The crystal structures of graphite and diamond.

together. The resonance stabilization due to π bonding observed with benzene also occurs in graphite, making for an extremely stable overall structure. The graphite crystal is two dimensional and exists as sheets one atom thick. The layers themselves are not strongly bonded together and can slide past one another, so that graphite appears to be soft and slippery to the touch. When one writes with a "lead" pencil, which is really made of graphite, thin layers of graphite rub off on the paper.

In the diamond crystal each carbon atom is covalently bonded to four other carbon atoms arranged tetrahedrally around it. The octet rule is satisfied, and the bonds are all sp^3 hybrids. In diamond the C—C bonds are single, and have just about the same length as in the paraffins, 1.54 Å. The bonds in diamond are strong and produce a rugged three-dimensional lattice. Diamond is one of the very hardest of substances, and is used industrially in cutting tools and quality grindstones.

At room temperature and atmospheric pressure, graphite is the stable form of carbon. Diamond, in principle, should slowly transform to graphite under ordinary conditions. Fortunately for the owners of diamond rings this transition occurs at zero rate unless the diamond is heated to about 1500°C, where the conversion occurs rapidly. For understandable reasons, no one has ever become very excited over the commercial possibilities of this process. The more difficult task of converting graphite to diamond has aroused much greater enthusiasm.

Since diamond has a higher density than graphite (3.51 vs 2.26 g/ml), its formation should be favored by high pressures. Theoretically, at 25°C and 8000 atm graphite should turn to diamond. However, under those conditions the reaction has a negligible rate. At higher temperatures it goes faster, but the required pressure goes up too; at 2000°C, a pressure of about 100,000 atm is needed. In 1954 scientists at the General Electric laboratories were able to achieve these high temperatures and pressures and converted graphitic carbon to diamond for the first time. The synthetic diamonds produced by this process were at first quite small, but over the last several years their size and quality has been improved so that one carat diamonds are now occasionally formed. At present, most of our industrial diamonds are synthetic.

Several more or less amorphous forms of carbon can be prepared; charcoal and carbon black are two of the more familiar. Charcoal can be made by heating wood or other materials with a high carbon content, such as sugars, to a high temperature in the absence of air, driving off water and other volatiles. Carbon black is usually made by burning natural gas under fuel rich conditions and collecting the soot on cool metal plates; this material is used for the most part in compounding rubber for automobile tires, where it adds considerable wear resistance. A high surface form of carbon, called activated carbon, is produced by much the same process as is used to make charcoal; the material is very porous and absorbent and is usually activated by heating in steam at about 1000°C. Activated carbon is used industrially to clarify sugar solutions and has had some application in public water supply systems, where it is effective in removing odors. In all the amorphous forms of carbon the atoms are arranged in irregular hexagonal patterns similar in many respects to the structure of graphite.

SILICON DIOXIDE

Silicon is in the same family as carbon and, like carbon, is able to form up to four single bonds. At one time it was thought that it might be possible to develop a whole field of chemistry around silicon-hydrogen compounds, analogous to organic chemistry. This did not turn out to be feasible since, as we mentioned previously, long chains of Si atoms cannot be readily synthesized.

However, silicon does form a strong bond with oxygen, which can then link strongly to silicon, forming chains of the type $-\overset{|}{\underset{|}{Si}}-O-\overset{|}{\underset{|}{Si}}-O-\overset{|}{\underset{|}{Si}}-$, which are stable and which cross-link through the silicon atoms to form a macromolecular structure (Fig. 8.12). In the crystal each silicon atom is at the center of a tetrahedron, bonded to *four* oxygen atoms; each oxygen atom is linked covalently to *two* silicon atoms, making the formula of the compound SiO_2.

Quartz is the most common form of silicon dioxide, or silica, and is the main component of sea sand. Since the orientation of the oxygen atoms in the lattice is not unique, there are several crystalline modifications of SiO_2, over 20 as a matter of fact, which differ in the manner in which the atoms are actually packed in the crystal. Quartz, like graphite and diamond, has a high melting point, about 1700°C. Unlike most solids, however, it does not

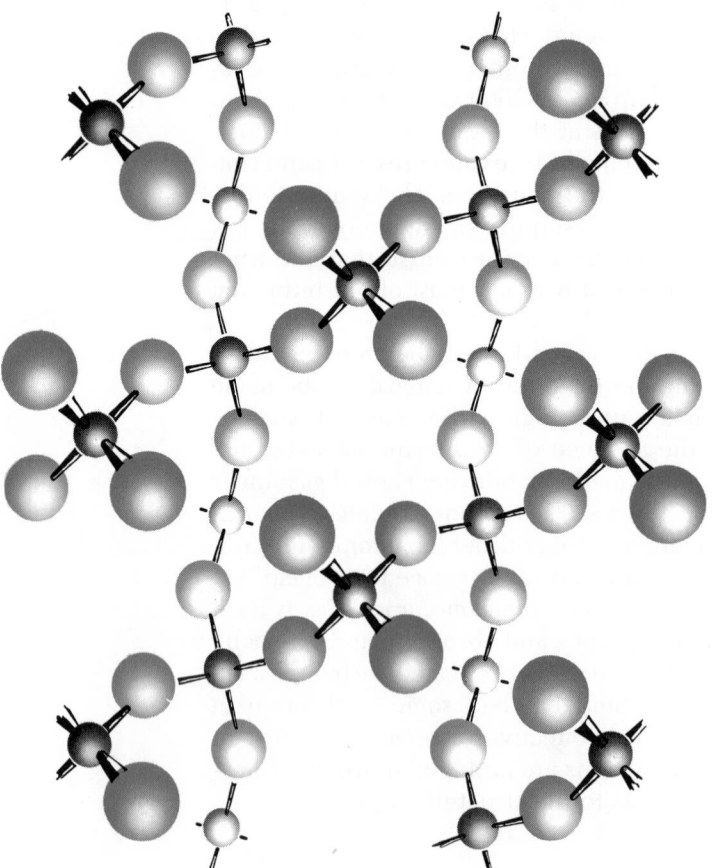

Figure 8.12 Crystal structure of quartz, SiO_2.

In SiO_2 every silicon atom (small) is bonded tetrahedrally to four oxygen atoms (large)

melt sharply to a liquid but turns to a viscous mass over a reasonably wide temperature range, first softening at about 1400°C. The viscous fluid probably contains fairly long —Si—O—Si—O— chains, with enough bonds broken so that the material can flow.

Ordinary glass is made by heating a mixture of sand, limestone ($CaCO_3$), and soda ash (Na_2CO_3) to the melting point; carbon dioxide is driven out of the hot mass, which then contains glass, a solution of the oxides of silicon, sodium, and calcium in a mole ratio of about 7-1-1. This material softens at about 600°C and has a wider temperature range of high viscosity than does quartz. Glass does not crystallize on cooling, but retains a disordered structure with many of the properties of a macromolecular substance.

Soft glass, which we have described, is produced in enormous quantities and is used in making bottles and window panes. It is not satisfactory for laboratory beakers and test tubes because it cracks fairly easily when subjected to thermal or mechanical shock. Hard, or borosilicate, glass, called Pyrex or Kimax, is made from a melt of silicon, boron, aluminum, sodium, and potassium oxides and is much more able to withstand both chemical attack and thermal shock. Hard glass softens at about 800°C and is readily worked in the flame of a propane-oxygen torch. Essentially all the glass used these days in chemical equipment is made from borosilicate glass.

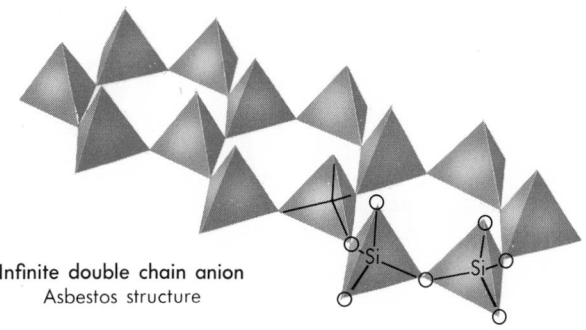

Infinite double chain anion
Asbestos structure

Figure 8.13 Silicate structures. Metal ions fit in voids in the silicate lattices to balance electrical charges.

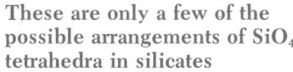

These are only a few of the possible arrangements of SiO_4 tetrahedra in silicates

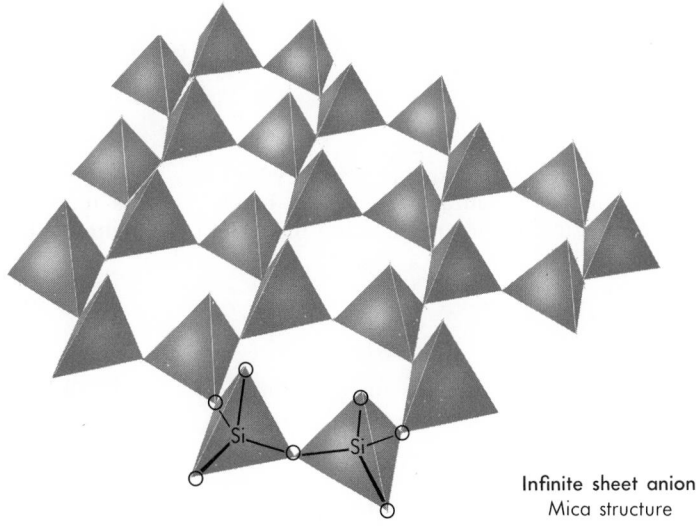

Infinite sheet anion
Mica structure

SILICATES

The bulk of the rocks, minerals, and soils found on the surface of the earth are composed of substances similar in structure to silica, in that every silicon atom in the crystal is surrounded tetrahedrally by four oxygen atoms. These materials are called *silicates* and contain silicon oxyanions of various sorts, depending on whether the SiO_4 tetrahedra are isolated or are linked together at their corners in any one of many possible arrangements. The negative charge of the silicon oxyanion is balanced by positive charges on the cations of one or more metals which fit into the crystal lattice. Strictly speaking, silicates would be classified as salts, but since the silicate lattice is frequently infinite, consisting of chains, or layers, or three-dimensional networks, many of the silicates have the properties of macromolecular substances, namely, very high melting point and insolubility in all common solvents. Some very common substances are silicates. Asbestos, $Mg_6(Si_4O_{11})$- $(OH)_6$, is fibrous, and contains long double-chain molecules. Mica, $NaAl_2$- $(Si_3AlO_{10})(OH)_2$, crystallizes in easily cleaved sheets. Ultramarine, $Na_8(Si_6Al_6O_{24})(S_2)$, is a very hard mineral with a three-dimensional silicate lattice (Fig. 8.13).

PROBLEMS

8.1   Explain in your own words the difference between each of the following:

   a. a polar bond and a nonpolar bond.
   b. a polar bond and a hydrogen bond.
   c. a molecular substance and a macromolecular substance.

8.2   Predict which member of each of the following pairs would have the higher boiling point; in each case explain your reasoning:

   a. H_2 or O_2
   b. H_2O or H_2S
   c. Na or Cl_2
   d. SiO_2 or $SiCl_4$
   e. N_2 or LiF
   f. PH_3 or AsH_3

8.3   For each of the following, predict whether the intermolecular attractions would involve polar forces, hydrogen bonds, and/or dispersion forces:

   a. CCl_4 d. CO_2
   b. HCN   e. CH_3OH
   c. C_2H_6 f. NH_3

8.4   What kinds of attractive forces or chemical bonds must be broken to

   a. boil water?
   b. melt iron?
   c. melt calcium chloride?
   d. melt quartz, SiO_2?

8.21   Distinguish between

   a. allotropic forms and resonance forms.
   b. a polymer and a monomer.
   c. a hydrogen bond and a dipole force.

8.22   Predict which member of each of the following pairs would have the higher boiling point; explain your reasoning:

   a. CH_4 or CCl_4
   b. Fe or SO_3
   c. NaCl or $C_{10}H_{22}$
   d. CO_2 or CS_2
   e. NH_3 or CH_4
   f. C or C_6H_6

8.23   For each of the following predict whether the intermolecular forces would include polar, hydrogen bonds, and/or dispersion forces.

   a. CH_2Cl_2 d. SF_6
   b. $HOCH_2$—CH_2OH e. HF
   c. SO_2 f. C_4Cl_{10}

8.24   What kinds of attractive forces or chemical bonds must be broken to

   a. melt diamond?
   b. melt potassium?
   c. boil benzene?
   d. boil HCN?

8.5 Sketch the structures of three molecules which are isomers of pentane, C_5H_{12}. Show that each structure satisfies the octet rule.

8.6 Describe the molecular geometry of polyethylene; that is, what would be the atom arrangement in space around each carbon atom in the chain making up the polymer? Would you expect that there could be rotation around each carbon-carbon bond?

8.7 Draw as many isomers as you can for C_5H_5. Can you explain why none of these are as stable as benzene, C_6H_6?

8.8 Select the member in each pair that would have the higher boiling point:

    a. C_2H_6, C_4H_{10}
    b. C_2H_5OH, CH_3OCH_3
    c. the two isomers of butane.
    d. C_4H_9OH, C_6H_{14}

8.9 Explain why graphite feels soft and diamond is very hard in spite of the fact that both materials have extremely high melting points.

8.10 How many different substances could one obtain by chlorinating propane, C_3H_8? (From one to eight Cl atoms may substitute.)

8.11 Describe each of the following molecules as belonging to an alcohol, an ether, an aldehyde, a ketone, an acid, or an ester:

    a. CH_3—CH_2—O—CH_3
    b. CH_3—C=O
            |
           C_2H_5
            H   H   H
            | | |
    c. CH_3—C——C——C—CH_3
            | | |
           OH OH OH
    d. H—C=O
            |
           OH

8.12 Draw structural formulas to correspond to the following molecular formulas:

    a. C_4H_8 c. $C_{10}H_8$
    b. C_7H_7Cl d. $C_3H_4Cl_2$

8.13 For each of the molecular formulas indicate whether the compound could be an ether, an alcohol, an aldehyde, a ketone, or an acid. Sketch the structures of the molecules which you believe to be consistent with your answers.

    a. CH_4O d. C_3H_6O
    b. C_2H_6O e. CH_2O
    c. $C_2H_4O_2$

8.25 Sketch the structures of as many isomers of C_6H_{12} as you can, not including any structures listed in Figure 8.5. Show that each structure satisfies the octet rule.

8.26 Describe the arrangement of the atoms in the CH_2=CH—CH_2—CH_3, 1-butene, molecule. Which atoms in the molecule would of necessity lie in a plane? Would this molecule exhibit cis-trans isomerism?

8.27 There are two isomeric forms of C_3H_6, one of which contains a 3-membered ring. Draw the isomers. Which would be under the most strain due to distortion of natural bond angles? Which would you expect to be the more stable?

8.28 Straight chain paraffins have lower octane numbers and, hence, make poorer gasoline than do highly branched ones. Which of the isomers of heptane, C_7H_{16}, would you expect to have the lowest octane number? the highest?

8.29 Carbon dioxide, CO_2, vaporizes at 1 atm at about $-78°C$, whereas it takes over $2000°C$ to vaporize SiO_2. Explain the difference in the properties of these two oxides of Group 4A elements in terms of the chemical bonding present in each.

8.30 Dichloroethylene, $C_2H_2Cl_2$, can occur in three isomeric forms, but dichloroethane, $C_2H_4Cl_2$, can exist in only two. Explain.

8.31 Describe each of the following molecules as belonging to an alcohol, an ether, an aldehyde, a ketone, an acid, or an ester:

               H
               |
    a. CH_3—C—O—CH_3
               |
              CH_3
    b. CH_3—CH_2—C=O
                   |
                   H
    c. CH_3—C—CH_2—CH_3
              ||
              O

8.32 Draw structural formulas to correspond to the following molecular formulas:

    a. $C_4H_6Cl_2$ c. C_3H_4
    b. C_8H_{10} d. C_4H_{10}

8.33 For each of the molecular formulas, indicate whether the compound could be an ether, an alcohol, an aldehyde, a ketone, or an acid. Sketch the structures of the molecules which you believe to be consistent with your answers.

    a. C_4H_8O c. $C_4H_{10}O$
    b. $C_3H_6O_2$ d. H_2CO_2

8.14 Classify each of the following as a paraffin, an olefin, an acetylene, or an aromatic hydrocarbon:

 a. C_3H_6 c. C_5H_8 e. C_4H_{10}
 b. C_6H_{14} d. C_7H_8

8.15 Given the atomic geometry associated with sp³ and sp² hybrid bonds, make a perspective sketch of the structure of the acetic acid, CH_3COOH, molecule. Which of the atoms would of necessity lie in a plane? Which atoms would be free to move with respect to this plane?

8.16 Sketch the structure of an alcohol, an aldehyde, and an acid, each of which is different from the examples listed in Table 8.4.

8.17 In the reaction in which an addition polymer, like polystyrene, is formed, for every styrene monomer added to the carbon chain, a carbon-carbon double bond is lost and two carbon-carbon single bonds are formed. Calculate ΔH for the polymerization of a mole of styrene ($ΔH_f$ $C{=}C = -143$ kcal, $ΔH_f$ $C{-}C = -83$ kcal). Can you suggest why, in the early days of the polymer industry, an occasional batch of polystyrene went through the roof of the plant?

8.18 In making diamonds it is necessary to have a molten metal, such as nickel or chromium, present in the press along with the graphite. Under the conditions in the press, about 2000°C and 100,000 atm, graphite is somewhat soluble in the metal. Can you suggest why the metal facilitates the conversion?

8.19 In silicates the Si:O atom ratio varies, depending on how the SiO_4 anions are linked together. Show that the Si:O ratio in asbestos is 4:11.

8.20 Sketch a section of each of the following polymeric molecules:

 a. polystyrene
 b. teflon
 c. Dacron
 d. nylon

8.34 Classify each of the following as a paraffin, an olefin, an acetylene, or an aromatic hydrocarbon:

 a. C_4H_6 c. C_5H_{10} e. C_8H_8
 b. C_8H_{18} d. $C_{11}H_{10}$

8.35 Given the atomic geometry associated with sp³ and sp² hybrid bonds, make a perspective sketch of the acetone, CH_3COCH_3, molecule, indicating bond angles and those atoms which would have to lie in the same plane.

8.36 Sketch the structures of an ether, a ketone, and an ester molecule which are different from the examples listed in Table 8.4.

8.37 Vinyl alcohol, $CH_2{=}CHOH$, exists in polymeric form in a compound which is fairly soluble in water. Sketch a section of a polyvinyl alcohol molecule. Can you suggest why it is one of the few polymers which has appreciable water solubility?

8.38 Much of our synthetic rubber is an addition polymer made from butadiene or one of its derivatives. Butadiene has the structure

$$CH_2{=}C{-}C{=}CH_2$$
$$\quad\;\; |\;\; |$$
$$\quad\;\; H\;\; H$$

In the polymerization reaction one double bond is lost per monomer unit added, and the remaining double bond ends up where the C—C single bond was. Sketch a section of a polybutadiene molecule containing three monomer units. Can you see why there might be more than one way in which the polymer could form? (The useful form has a cis structure.)

8.39 There is only one form of diamond crystal, but many kinds of SiO_2 crystals. Explain why these substances can differ in this way.

8.40 Sketch a section of each of the following molecules:

 a. Lucite
 b. polyvinyl chloride
 c. a 1:1 copolymer of vinyl chloride and acrylonitrile

°**8.41** In the development of polymers, chemists will frequently try many possible monomers in looking for a product with desirable qualities. Suggest two monomers that might make useful addition polymers. Draw the structures of a pair of monomers that might make a good polyester. In each case sketch the structure of the polymer obtained.

°**8.42** Some polymers, like Bakelite, will not melt because the polymer chains are linked into two- or three-dimensional networks, as in diamond and graphite. Can you suggest some monomer species which either by addition reactions, or by reactions in which water is eliminated, would tend to form cross-linked, nonmelting polymers? Use your imagination in finding likely monomers.

9   LIQUIDS anD SOLIDS; CHanGES In STATE

In Chapter 5, we discussed the laws governing the physical behavior of gases and the interpretation of these laws in terms of the kinetic theory. In succeeding chapters we have had frequent occasion to refer to substances in the liquid and solid states. We shall now discuss the properties of these two condensed states of matter. In doing so, we shall be particularly interested in:

1. The particle structure of liquids and solids and its influence upon their physical properties.

2. The equilibria between the gaseous, liquid and solid phases of a pure substance.

9.1   PROPERTIES OF LIQUIDS

The structure of liquids is less well established than that of gases or solids. Despite a great deal of research in this area, we still do not have a clear picture of the way in which molecules are arranged in even the most common liquid, water. We do, however, have a reasonably detailed knowledge of the average distances between atoms or molecules in a liquid. Moreover, we can estimate with considerable accuracy the magnitude of the forces between particles in a liquid. It is these two aspects of liquid structure which we shall now consider.

At ordinary temperatures and pressures, the molecules in a liquid are much closer together than they are in a gas. In the liquid there is very little free volume between molecules; in the gas only a small fraction of the total volume is occupied by the molecules themselves. The close contact between molecules in a liquid as compared to a gas explains many of the differences in physical properties between these two states of matter. In particular, we see why the volume of a liquid is much less sensitive to pressure and temperature than that of a gas.

Since the molecules in a liquid are much closer together than those in a gas, attractive forces between molecules are considerably stronger in the liquid state. Energy has to be absorbed to overcome these forces, separate the molecules, and thereby convert the liquid to its vapor. To boil water we apply heat from an outside source such as a Bunsen burner or a hot plate. When a liquid evaporates at room temperature, heat is absorbed from the surroundings. You feel cold when you come out of a shower or a swimming pool because the evaporation of moisture draws heat from the body surface and in so doing lowers its temperature.

The amount of heat required to vaporize a mole of liquid against a constant external pressure (e.g., in an open container) is referred to as the **molar heat of vaporization, ΔH_{vap}**. The heat of vaporization is a measure of the strength of the intermolecular forces in a liquid. Compare, for example, the heat of vaporization of water, approximately 10 kcal/mole, to that of methane, 2 kcal/mole. The difference between these two quantities reflects the strong hydrogen bonds in liquid water which are absent in liquid methane.

Since a molecule at the surface of a liquid has fewer neighbors than one in the interior, there is an unbalanced attractive force tending to pull it into the body of the liquid. This explains why liquids in contact with gases tend to achieve as small a surface as possible. Raindrops falling through air, or gas bubbles rising through a liquid, take on the shape of a sphere, thereby achieving the smallest possible surface area for a given mass.

The work required to expand the surface of a liquid by unit area is referred to as its **surface tension, γ**. This quantity is ordinarily expressed in ergs/cm^2 ($\equiv$ dynes/cm). The surface tension, like the heat of vaporization, is a measure of the strength of intermolecular forces. Water, in which these forces are comparatively strong, has a high surface tension, considerably greater than that of most organic liquids. This explains why benzene or ethyl alcohol tend to "wet" or spread out on solid surfaces more readily than water. The wetting ability of water can be increased by adding a soap or detergent which drastically lowers the surface tension. Mercury, with a surface tension even higher than that of water, is a notoriously poor wetting agent. When spilled on a laboratory bench or on the floor, mercury forms tiny spherical drops which are difficult to retrieve.

The ability of a liquid to spread over a solid surface depends not only upon its surface tension but also upon the nature of the surface. Water forms a thin, nearly invisible film on the inside walls of a clean buret; if the buret becomes dirty, perhaps through contact with stopcock grease, the water forms drops which stand out on the walls. Fabrics can be waterproofed by coating them with a water-repellent material such as methyl silicone, a polymer which has the following structure

Synthetic silicon polymers involve —Si—O—Si—O— chains

$$\begin{bmatrix} \begin{array}{ccc} CH_3 & CH_3 & CH_3 \\ | & | & | \\ -Si-O-Si-O-Si- \\ | & | & | \\ CH_3 & CH_3 & CH_3 \end{array} \end{bmatrix}_x$$

The nonpolar methyl groups repel water and prevent it from soaking into the fabric.

When a liquid wets a porous surface such as that of a sponge or a piece of filter paper, it tends to be soaked up into the pores of the material, rising against the force of gravity. This phenomenon, known as capillary action, is essential to the process of thin-layer or paper chromatography (Chapter 1).

TABLE 9.1   Heats of Vaporization and Surface Tensions of Liquids

LIQUID	ΔH_{vap} (kcal/mole) [*]	γ(ergs/cm^2) AT 20°C
Mercury	14.2	475
Water	9.7	72.8
Pyridine	8.5	38.0
Benzene	7.3	29.0
Ether	6.2	17.0

[*] Both the heat of vaporization and the surface tension decrease with temperature; heats of vaporization here are given at the normal boiling point (cf. Table 9.2).

9.2   LIQUID-VAPOR EQUILIBRIUM

Let us consider what happens from a molecular viewpoint when a volatile liquid is placed in a closed container (Fig. 9.1). A few fast-moving molecules have sufficient kinetic energy to overcome the intermolecular forces and escape into the vapor. Occasionally one of these molecules collides with the surface and re-enters the liquid. At first, the movement of molecules is primarily in one direction, from liquid to vapor. Gradually, however, as the concentration of molecules builds up in the vapor, the rates of vaporization and condensation approach each other. Eventually we reach a position of *dynamic equilibrium* at which these two rates become equal. From this point on, the concentration of molecules in the gas phase has a certain fixed value which does not change with time.

VAPOR PRESSURE

The process which we have just described cannot, of course, be followed visually. We cannot see molecules escape from the liquid nor

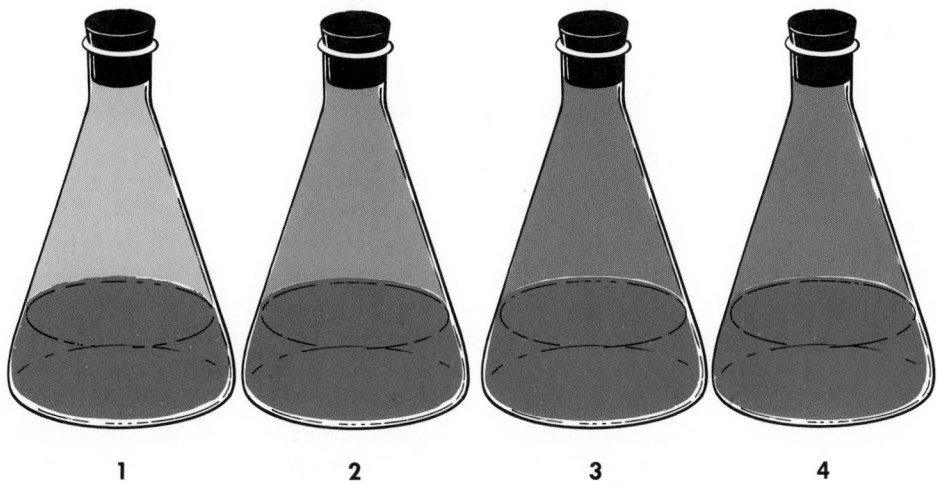

1   **2**   **3**   **4**

Figure 9.1   As time passes, the concentration of molecules in the vapor phase rises until it reaches a value (flasks 3 and 4) fixed by the equilibrium vapor pressure of the liquid.

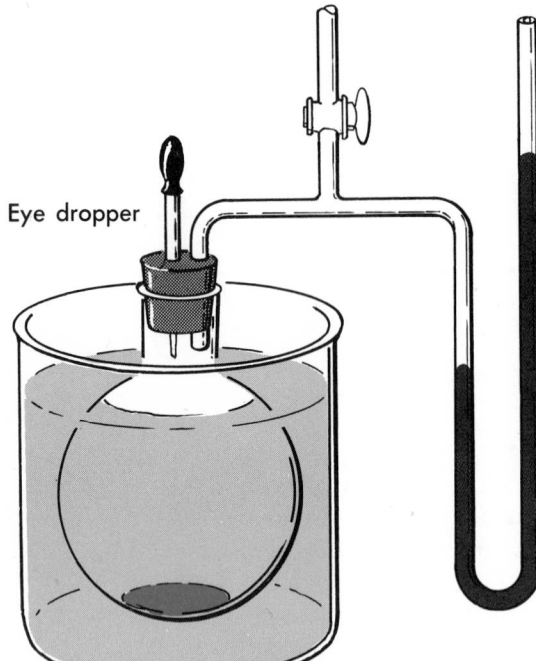

Eye dropper

Figure 9.2 Measurement of vapor pressure. The difference in the two mercury levels at equilibrium gives the vapor pressure directly in mm Hg.

In this experiment the increase in gas pressure is equal to the vapor pressure of the liquid

can we see their concentration build up to its equilibrium value. It is possible, however, to follow the process by making use of the apparatus shown in Figure 9.2. At the beginning of the experiment, the flask is empty, the stopcock is open to the atmosphere, and the mercury levels in the two arms of the manometer are at the same height. A few milliliters of liquid are squeezed out of the dropper bulb into the flask and the stopcock is closed. As vaporization occurs, molecules leaving the liquid increase the pressure in the space above it. This increase in pressure registers on the manometer; the mercury level in the left arm of the U tube falls, while that in the right arm rises. The pressure steadily increases until equilibrium is reached; at that point it becomes constant and the mercury levels stop moving.

The pressure of vapor in equilibrium with a pure liquid at a given temperature is referred to as the **vapor pressure** of the liquid. The apparatus shown in Figure 9.2 can be used to measure vapor pressure. If we place liquid benzene in the flask and allow it to come to equilibrium with its vapor at 25°C, we find that the difference in mercury levels is 92 mm Hg; it follows that the vapor pressure of liquid benzene at 25°C is 92 mm Hg. If we repeat the experiment with water, we find that the observed vapor pressure is considerably lower, about 24 mm Hg at 25°C.

As the above discussion implies, a liquid at a given temperature has a certain fixed vapor pressure that is characteristic of the substance. It is important to point out that so long as *both liquid and vapor* are present in a container, the pressure exerted by the vapor will be *independent* of the *volume*. To see what this statement means, consider Figure 9.3, which shows the vaporization of liquid benzene taking place at 25°C in a container whose volume can be varied by lowering the piston. We place a small

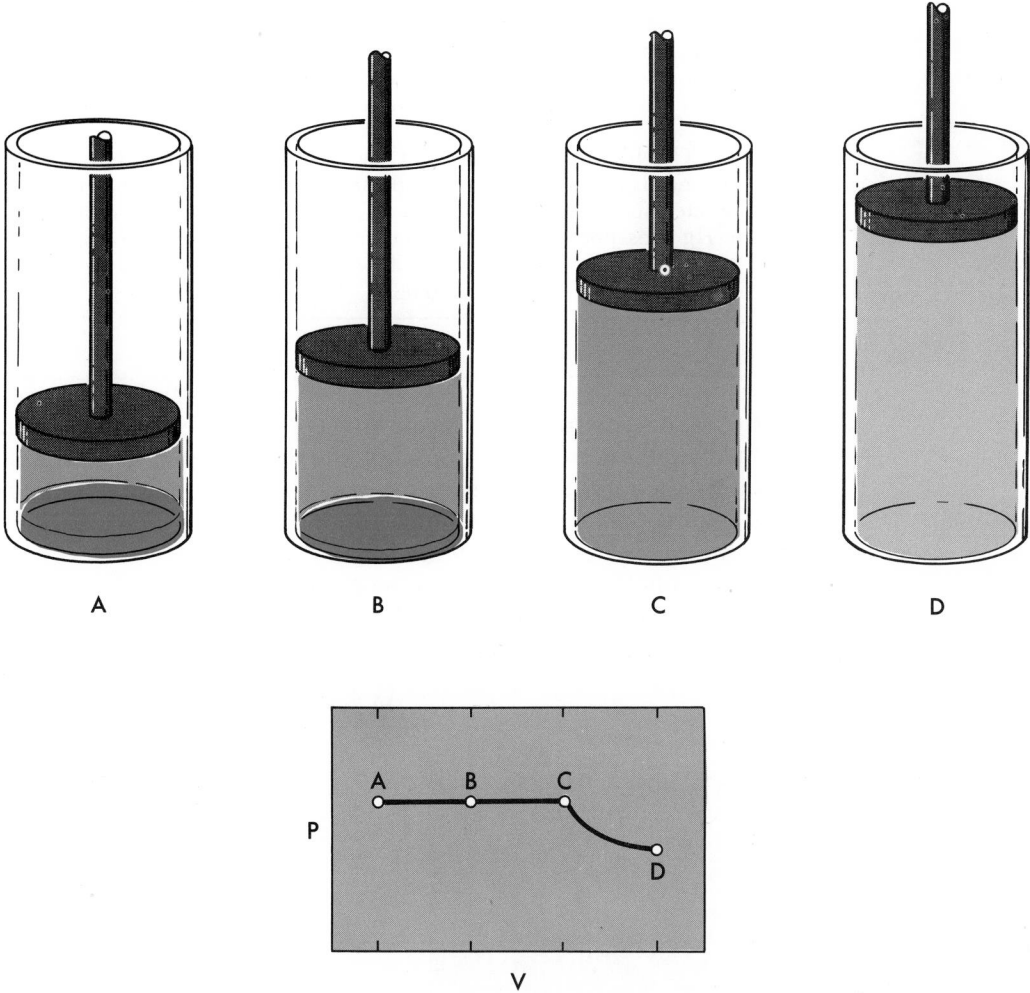

Figure 9.3 The pressure exerted by the vapor is independent of volume so long as there is liquid present (*A, B*). When all the liquid is vaporized (*C*), a further increase in volume (*D*) decreases the pressure in accordance with the Ideal Gas Law.

amount of benzene in the container, raise the piston to the level shown in A, and allow equilibrium to be established. Vaporization occurs until the concentration of molecules in the vapor becomes constant at a pressure of 92 mm Hg. If the equilibrium is disturbed by raising the piston to the level shown in B, the pressure will drop momentarily. However, liquid will quickly evaporate to establish the original concentration of molecules in the vapor and, hence, the original vapor pressure, 92 mm Hg. This process will be repeated each time we raise the piston, as more and more liquid vaporizes. Eventually, all the liquid will have been converted to vapor (Fig. 9.3*C*). If we continue to raise the piston beyond this point, no more molecules can vaporize to compensate for the increase in volume; the concentration of molecules in the vapor will drop (Fig. 9.3*D*) and the pressure will decrease in accordance with Boyle's Law.

If there were twice as much liquid present in State A, how would this affect the pressures observed in States A, B, C and D?

Example 9.1 Suppose that in the experiment just described we start with one mole of benzene at 25°C (vp = 92 mm Hg). To what value must the volume be increased in order for the liquid to just disappear (Fig. 9.3C)?

Solution. The liquid will "disappear" when the one mole of benzene originally present is completely vaporized. Rephrasing the question, we are asked to calculate the volume occupied by one mole of benzene vapor at 25°C and 92 mm Hg pressure. Neglecting deviations from ideality, we have:

$$V = \frac{n\,R\,T}{P} = \frac{1 \text{ mole} \times 0.0821 \frac{\text{lit atm}}{\text{mole °K}} \times 298°K}{(92/760) \text{ atm}} = 200 \text{ lit}$$

The vapor pressure of a liquid always increases with temperature. Water evaporates more rapidly on a hot, dry day; stoppers in bottles of volatile liquids such as ether or gasoline may pop out when the temperature rises. The effect of temperature upon vapor pressure is shown in Figure 9.4 for two liquids, water and benzene.

In studying the relationship between two variables such as vapor pressure and temperature, scientists prefer to work with linear functions which fit the general algebraic equation

$$y = Ax + B$$

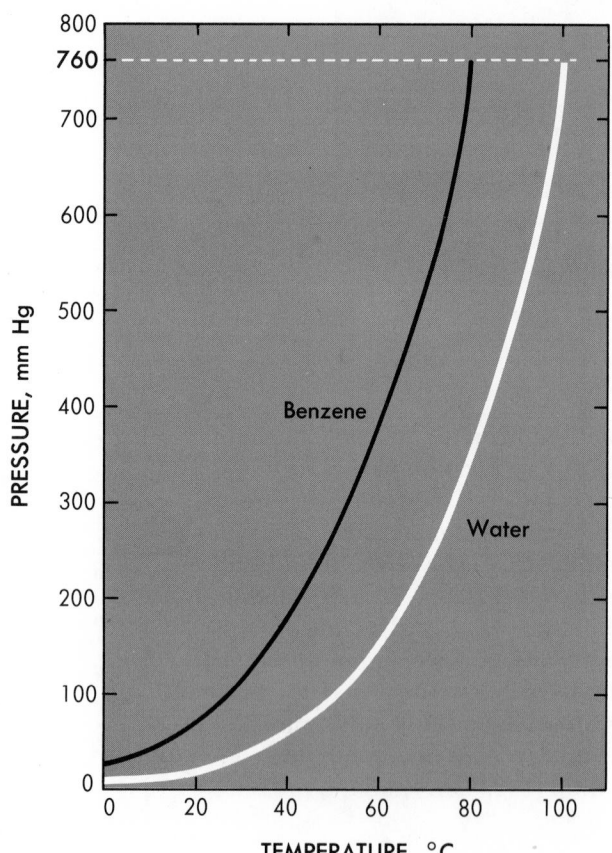

Figure 9.4 Effect of temperature on vapor pressure. Benzene has the higher vapor pressure at all temperatures and, hence, reaches a pressure of 1 atm at a lower temperature (80°C as compared to 100°C for water).

The vapor pressure of liquids increases more rapidly with increasing temperature

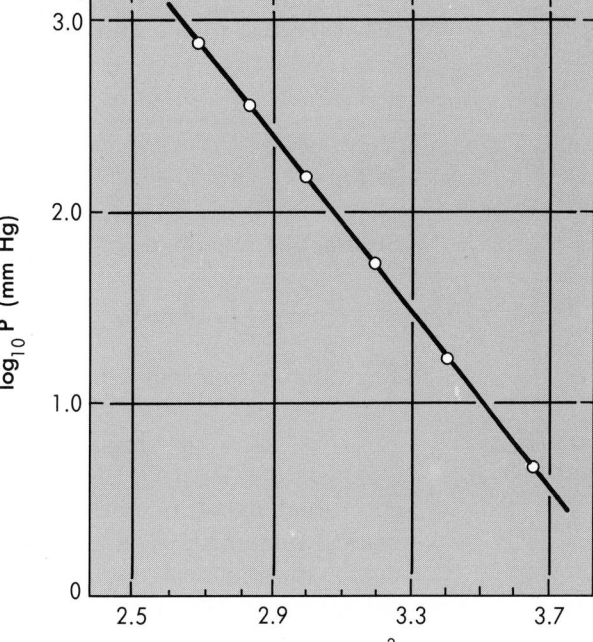

Figure 9.5 Plot of $\log_{10}P$ vs $1/T$ for water. The heat of vaporization can be calculated from the slope (Equation 9.3).

In this case we can obtain a linear relation, at least over a short temperature range, by plotting the logarithm of the vapor pressure ($\log_{10}P$) vs the reciprocal of the absolute temperature ($1/T$). Such a plot is shown in Figure 9.5 for water; the general equation is

$$\log_{10}P = \frac{A}{T} + B \tag{9.1}$$

where A is the slope of the straight line and B is the intercept.

The constant A in Equation 9.1 can be related to the heat of vaporization of the liquid. Specifically,

$$A = -\Delta H_{vap}/2.30\ R \tag{9.2}$$

where ΔH_{vap} is the molar heat of vaporization in cal/mole and R is the gas law constant $= 1.99$ cal/mole°K. Substituting for A in Equation 9.1:

$$\log_{10}P = \frac{-\Delta H_{vap}}{(2.30)(1.99)T} + B \tag{9.3}$$

A convenient method of measuring the heat of vaporization of a liquid takes advantage of the linear relationship between $\log_{10}P$ and $1/T$. The vapor pressure is measured at a series of different temperatures, using an apparatus such as that shown in Figure 9.2. At each temperature we calculate $\log_{10}P$ and $1/T$, plot these points on a graph, and then draw the best straight line through the points. By taking the slope of this line, the constant A is obtained. The heat of vaporization is then calculated using Equation 9.2 (cf. Problem 9.9 at the end of this chapter).

For many purposes it is convenient to have a two-point equation relating the vapor pressure (P_2) at one temperature (T_2) to that (P_1) at another

This is a relatively easy way to measure ΔH_{vap}. Can you think of another way to do it?

temperature (T_1). Such an equation can be obtained by applying Equation 9.3 at the two temperatures.

at T_2:
$$\log_{10}P_2 = \frac{-\Delta H_{vap}}{(2.30)(1.99)T_2} + B$$

at T_1:
$$\log_{10}P_1 = \frac{-\Delta H_{vap}}{(2.30)(1.99)T_1} + B$$

Subtracting, we obtain

$$\log_{10}P_2 - \log_{10}P_1 = \frac{-\Delta H_{vap}}{(2.30)(1.99)}\left[\frac{1}{T_2} - \frac{1}{T_1}\right]$$

Rearranging to a somewhat more convenient form

In order to use this equation you ***must*** *be able to work with logarithms*

$$\log_{10}\frac{P_2}{P_1} = \frac{\Delta H_{vap}}{(2.30)(1.99)}\left[\frac{T_2 - T_1}{T_2 T_1}\right] \tag{9.4}$$

Equation 9.4, in a more general form, is one of the most useful mathematical relationships in all of chemistry. We will meet it again when we deal with chemical equilibrium (Chapter 13) and with rate of reaction (Chapter 14). When applied to the dependence of vapor pressure upon temperature, it is known as the Clausius-Clapeyron equation, honoring the German physicist Rudolph Clausius, who was pre-eminent in the field of thermodynamics in Europe in the last half of the 19th century, and the French engineer B. P. E. Clapeyron, who first proposed it, in a somewhat different form, in 1834. The Clausius-Clapeyron equation is particularly useful in calculating the vapor pressure of a liquid at one temperature, knowing the vapor pressure at another temperature and the heat of vaporization.

Example 9.2 The vapor pressure of benzene at 50°C is 272 mm Hg. Calculate its vapor pressure at 60°C, given that the heat of vaporization is 7300 cal/mole.

Solution. Substituting in Equation 9.4, expressing temperature in °K:

$$\log_{10}\frac{P_2}{272 \text{ mm Hg}} = \frac{7300(333 - 323)}{(2.30)(1.99)(333)(323)} = +0.148$$

where P_2 is the vapor pressure at 60°C. Since the antilog of $+0.148$ is 1.41, we have:

$$P_2/272 \text{ mm Hg} = 1.41; \quad P_2 = 1.41 \times 272 \text{ mm Hg} = 384 \text{ mm Hg}$$

BOILING POINT

When we heat a liquid in an open container, bubbles form, usually at the base of the container where heat is being applied. The first bubbles that we see are often air, driven out of solution by the increase in temperature. Eventually, however, when a certain temperature is reached, vapor bubbles form throughout the liquid, rise to the surface and break. When this happens, we say that the liquid is boiling.

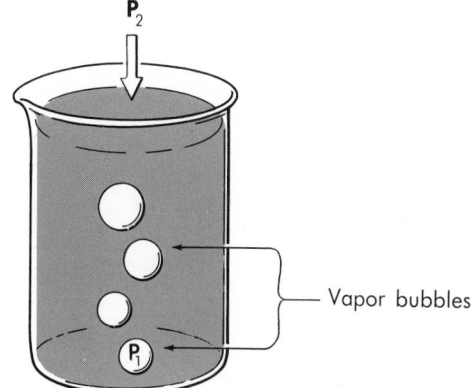

Figure 9.6 A liquid boils when its vapor pressure (P_1) exceeds the pressure above it (P_2).

Vapor bubbles

We find that the temperature at which a liquid boils depends upon the pressure above it. To understand why this is the case, let us refer to Figure 9.6, which shows vapor bubbles rising in a boiling liquid. In order for a vapor bubble to form and grow as it approaches the surface, the pressure within it, P_1, must be at least equal to the pressure above it, P_2. But P_1 is simply the vapor pressure of the liquid. From this argument we conclude that **a liquid boils at a temperature at which its vapor pressure becomes equal to the pressure above its surface.** If this pressure is one standard atmosphere, 760 mm Hg, the liquid boils at a temperature which we call the **normal boiling point.** From Figure 9.4 we note that the vapor pressure of water reaches 760 mm Hg at 100°C, while that of benzene becomes equal to one atmosphere at a somewhat lower temperature, 80°C. We refer to 100°C as the normal boiling point of water; 80°C is the normal boiling point of benzene.

As we might expect, the boiling point of a liquid can be reduced by lowering the pressure above it. It is possible to boil water at 25°C by evacuating the space above it with a vacuum pump or even an ordinary water aspirator. When we reach a pressure of 24 mm Hg, the equilibrium vapor pressure at 25°C, the water starts to boil. Chemists often purify a high-boiling compound which may decompose or oxidize at its normal boiling point by heating at a reduced temperature under vacuum and condensing the vapor.

As the water boils, it cools rapidly. Why?

If you have been fortunate enough to camp in the high Sierras or the Rockies, or live at a great distance above sea level, you may have noticed that it takes longer to boil potatoes, eggs and other foods at high altitudes. The reduced pressure lowers the boiling point of water and, hence, slows down the physical and chemical changes that take place when potatoes or eggs are cooked. In principle, this problem can be solved by using a pressure cooker, in which the pressure developed is high enough to raise the boiling point of water above 100°C. Pressure cookers are indeed used by housewives in localities such as Cheyenne, Wyoming (elevation 6100 ft), and Flagstaff, Arizona (elevation 6895 ft), but not by mountain climbers, who have to carry all their equipment on their backs.

As we see from Table 9.2, the normal boiling point of a liquid is inversely related to its vapor pressure at room temperature. A volatile liquid such as ether, where the vapor pressure has reached 442 mm Hg at 20°C, has to be heated only to 35°C to bring its vapor pressure to 760 mm Hg. In

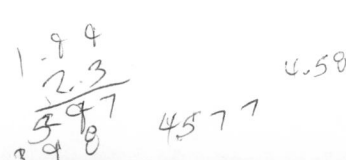

TABLE 9.2   Physical Properties of Liquids

	VAPOR PRESSURE AT 20°C		NORMAL BOILING POINT	HEAT OF VAPORIZATION	$\dfrac{\Delta H_{vap}}{T_b}$
Mercury	0.0012	mm Hg	630°K	14,200 cal/mole	22.0
Water	17.5	mm Hg	373°K	9,700 cal/mole	26.0
Benzene	75	mm Hg	353°K	7,300 cal/mole	20.7
Bromine	75	mm Hg	332°K	7,160 cal/mole	21.6
Ether	442	mm Hg	308°K	6,200 cal/mole	20.1
Ethane	27,000	mm Hg	184°K	3,700 cal/mole	20.1
Oxygen	———		90°K	1,630 cal/mole	18.2

contrast, mercury (with a very low vapor pressure at room temperature) must reach a much higher temperature before it can boil at 1 atm pressure.

There is a direct correlation between the heat of vaporization of a liquid and its normal boiling point. For many liquids we find that the heat of vaporization (cal/mole) is approximately 21 times the normal boiling point in °K.

Knowing the value of T_b for a liquid, one can use Equations 9.5 and 9.4 to predict its vapor pressure at any temperature

$$\frac{\Delta H_{vap}}{T_b} \approx 21 \qquad (9.5)$$

This generalization, known as Trouton's rule, holds to within 5 to 10% for most organic liquids. For water and other liquids in which hydrogen bonding is involved, the heat of vaporization is considerably higher than that calculated by Trouton's rule.

CRITICAL TEMPERATURE AND PRESSURE

Let us imagine an experiment in which a sample of liquid benzene is placed in an evacuated, heavy-walled glass tube (Fig. 9.7), which is then sealed and heated to higher and higher temperatures. The pressure of the vapor rises steadily; to 1 atm at 80°C, to 14 atm at 200°C, to 43 atm at 280°C . . . . Nothing spectacular happens (unless, of course, there happens to be a weak spot in the tube) until we reach 289°C, where the vapor pressure of benzene is 48 atm. Suddenly, as we pass this temperature, the meniscus separating the liquid and vapor phases disappears! The tube now contains only one phase, benzene vapor.

This phenomenon is readily reversed; if we cool the tube to 289°C, liquid benzene appears as suddenly as it disappeared upon heating. Interestingly enough, in other experiments we find that it is impossible to condense benzene vapor at temperatures above 289°C, regardless of the applied pressure. Even at pressures as high as 1000 atm, benzene vapor stubbornly refuses to liquefy at 290 or 300°C. This behavior is typical of all substances. There is a temperature, called the **critical temperature,** above

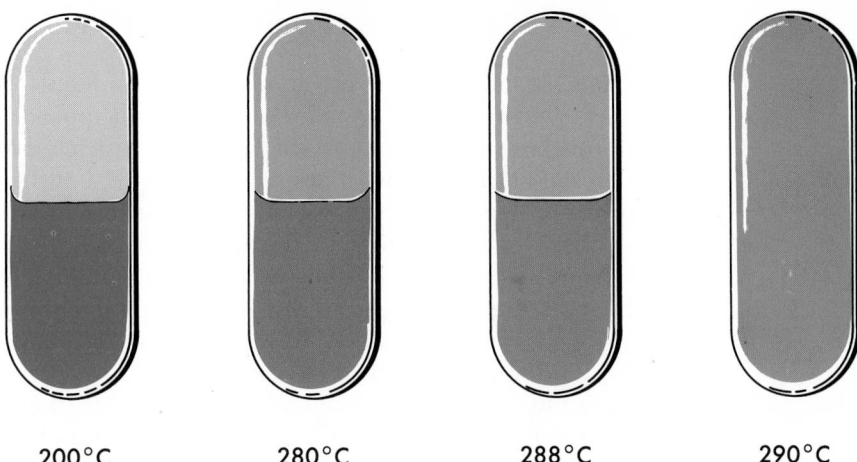

| 200°C | 280°C | 288°C | 290°C |

Figure 9.7 When a sample of benzene is heated in a sealed tube, the meniscus suddenly disappears at the critical temperature, 289°C.

which the liquid phase of a pure substance cannot exist. The pressure which must be applied to bring about condensation at that temperature is called the **critical pressure.** Alternatively, one can define the critical pressure as the vapor pressure of the liquid at its critical temperature.

Table 9.3 lists the critical temperatures of several common substances. The species in the column at the left, all of which have critical temperatures below 25°C, are often referred to as "permanent gases." Applying pressure at room temperature will not condense a permanent gas; it must be cooled as well. Helium, with a critical temperature only a few degrees above absolute zero, is the most difficult gas to liquefy. The permanent gases are stored and sold under high pressures, often 150 atm or greater; when the valve on the cylinder is opened, the pressure drops as gas escapes, in accordance with the Ideal Gas Law.

At the critical point the properties of the liquid and vapor phases become identical

The gases listed in the center column of Table 9.3 have critical temperatures above 25°C; they are available commercially as liquids in high pressure cylinders. When we open the valve on a cylinder of propane, the gas that escapes is replaced by vaporization of liquid, and the pressure returns to its original value. Only when the liquid has completely vaporized and the tank is almost empty does the gauge pressure drop.

TABLE 9.3 Critical Temperatures

"PERMANENT GASES"		"CONDENSABLE GASES"		"LIQUIDS"	
Helium	−268°C	Carbon dioxide	31°C	Ether	194°C
Hydrogen	−240	Ethane	32	Ethyl alcohol	243
Nitrogen	−147	Propane	97	Benzene	289
Argon	−122	Ammonia	132	Bromine	302
Oxygen	−119	Chlorine	146	Water	374
Methane	−82	Sulfur dioxide	158	Mercury	1460

9.3 NATURE OF THE SOLID STATE

In a crystalline solid, the atoms, ions or molecules making up the solid are arranged in a fixed geometric pattern that repeats itself over and over again in three dimensions. Each particle is restricted to a particular site in the crystal lattice. It can vibrate about that site, but the amplitude of vibration rarely becomes great enough for a particle to slip past its neighbor. The high degree of order in the microstructure often results in the formation of macroscopic crystals which have a precise geometrical form. In ordinary table salt, we can distinguish small cubic crystals of sodium chloride. Large, beautifully formed crystals of such minerals as quartz (SiO_2) and fluorite (CaF_2) are found in nature. Crystalline solids have distinct physical properties which can be used to identify them. The physical properties of a few solids are listed in Table 9.4 along with those of the corresponding liquids.

Heat is always absorbed when a solid melts

The heat which must be absorbed to break down the crystal lattice and melt one mole of a solid is referred to as its **heat of fusion.** You will notice from Table 9.4 that the heat of fusion is only a small fraction of the heat of vaporization. This implies that the intermolecular forces in the liquid are nearly as strong as they are in the solid; only when the molecules are widely separated from each other in the vapor do these forces become insignificant. The density of a substance in the solid state is ordinarily slightly greater than that of the liquid (Table 9.4), implying that the particles are somewhat more closely packed in the solid. A few substances, of which water is the most common example, actually expand upon freezing because of the formation of a relatively "open" solid structure (Chapter 11).

CRYSTAL STRUCTURES FROM X-RAY DIFFRACTION

Those of you who have had a course in physics may recall that visible light striking a glass plate ruled with a large number of closely spaced parallel lines is broken up by diffraction into a series of beams oriented at definite angles to each other. The basic requirement for diffraction is that the lines be evenly spaced and that the distance between them be about the same as the wavelength of the incident light. In 1912 Max Von Laue, an assistant lecturer in Physics at Munich, suggested that a crystal, which is

TABLE 9.4   Properties of Solids and Liquids

	m pt	SOLID		LIQUID	
		Density *	ΔH_{fus}*	*Density* *	ΔH_{vap}*
Mercury	−39°C	14.2 g/cc	0.56 kcal/mole	13.6 g/cc	13.5 kcal/mole
Water	0	0.917	1.44	1.000	10.7
Benzene	5	1.005	2.55	0.894	8.27
Naphthalene	80	1.15	4.61	0.979	9.7
Sodium chloride	800	2.0	6.9	1.55	43

° Values given at the melting point.

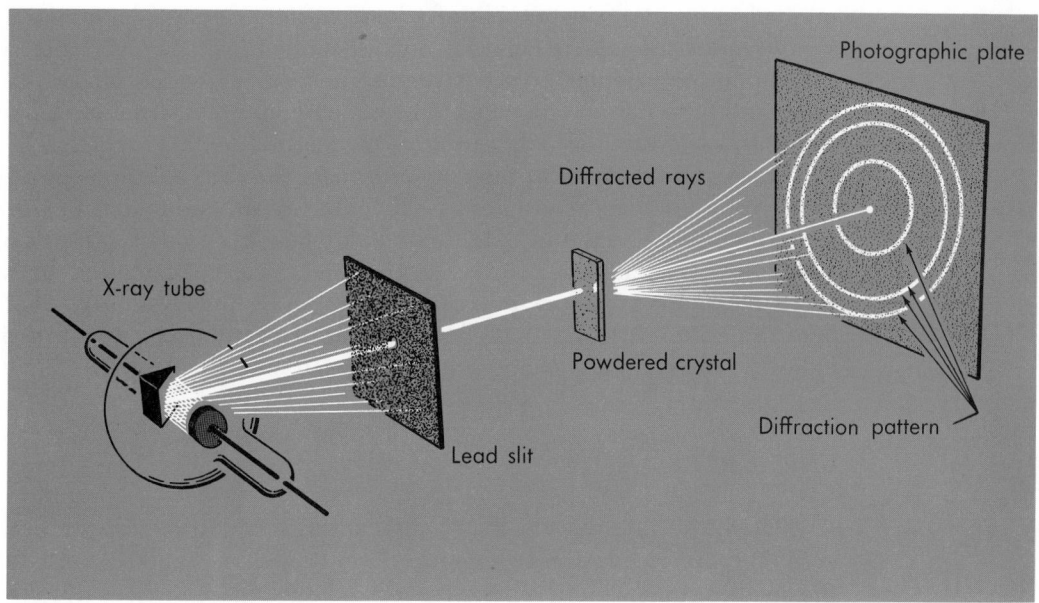

Figure 9.8 X-ray diffraction. From the position at which the diffracted rays reach the plate, the diffraction angle, θ, and hence the distance between planes of atoms, d, can be calculated (Equation 9.6).

made up of particles uniformly spaced a few angstroms apart, could act as a diffraction grating for x-rays, which have wavelengths ranging from 0.1 to 10 Å. His reasoning was confirmed by two Ph.D. students at Munich, Friedrich and Knipping, who passed a beam of x-rays through a copper sulfate crystal and observed a distinct pattern of spots on a photographic plate (Fig. 9.8).

In 1913 W. H. Bragg, at the University of Leeds, and his son W. L. Bragg (the late Sir Lawrence Bragg) adapted x-ray diffraction to determine crystal structures. The principle used is embodied in the Bragg equation, which relates the angle of diffraction to the distance between successive layers of atoms or ions in the crystal.

$$\sin \theta = \frac{n\lambda}{2d} \tag{9.6}$$

In this equation, θ is the angle between the beam and the layer of atoms, λ is the wavelength of the x-rays, d is the distance between successive layers, and n is an integer (1, 2, 3, . . .) known as the order number of the diffracted beam.

By measuring the angle at which x-rays of known wavelength are diffracted by a crystal, we can calculate from Equation 9.6 the distances between planes of atoms or ions. In this way values of atomic or ionic radii can be obtained. It is more difficult to deduce the geometric pattern in which the particles are arranged. The problem is that an x-ray beam "sees" not one but many different series of layers oriented at various angles to one another. Consequently a single beam, in passing through a crystal, is broken up into a complex pattern of diffracted beams.

Despite these difficulties, x-ray crystallographers have been able to unravel the particle structures of a wide variety of crystals, ranging from simple inorganic salts to complex organic molecules. The basic approach is to assume a particular crystal structure and calculate the angles and intensities of the diffracted beams to be expected for that structure. Comparison with experimental data can then suggest refinements which will give better agreement. Prior to the computer age, this was a tedious process. It took Dorothy Hodgkin and her associates at Oxford eight years to work out the structure of vitamin B-12 (molecular formula $C_{63}H_{84}N_{14}O_{14}PCo$). In 1964, nine years after her work was published, she was awarded the Nobel prize in Chemistry. At present, using computers, x-ray crystallographers can determine structures as complex as those of proteins, in which a single molecule may contain a thousand or more atoms (Chapter 23).

X-ray diffraction offers the only method for determining the atomic geometry in complex molecules

UNIT CELLS

The basic information that comes out of x-ray diffraction studies concerns the dimensions and geometric form of what is known as the **unit cell**, the smallest unit which, repeated over and over again in three dimensions, generates the crystal lattice. Perhaps the simplest unit cell to visualize is the **simple cubic cell**, which consists of eight atoms (or molecules, or ions)

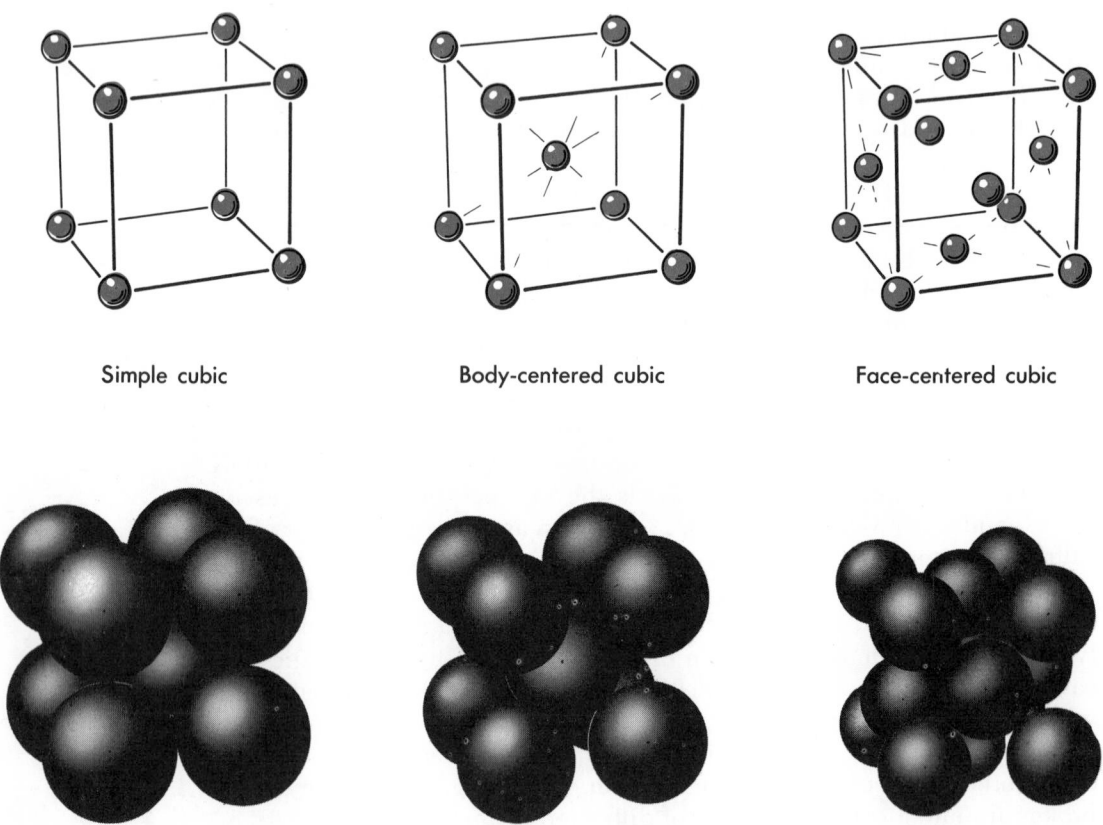

Simple cubic   Body-centered cubic   Face-centered cubic

Figure 9.9   Three types of cubic lattices.

located at the corners of a cube. Two other types of cubic cells are shown in Figure 9.9. One of these, in which there is an atom at the center of the cube, is referred to as a **body-centered cubic cell**. The third cell, in which there are atoms at the center of each face of the cube, is called a **face-centered cubic cell**.

In considering the unit cells shown in Figure 9.9, it should be kept in mind that certain of the atoms shown do not belong exclusively to a particular cell. For example, an atom in the center of the face of a cube is shared by the cube directly adjacent to it. In effect, then, only 1/2 of that atom can be assigned to one cell. An atom at the corner of a cube forms a part of eight different cubes that touch at that point. In this sense, only 1/8 of a corner atom "belongs" to a particular cell. The number of atoms to be assigned to each of the unit cells in Figure 9.9 is:

Atoms belonging to a unit cell must be counted this way

Simple cubic: 8 atoms × 1/8 = 1 atom per unit cell
Body-centered cubic: 8 atoms × 1/8 + 1 atom = 2 atoms per unit cell
Face-centered cubic: 8 atoms × 1/8 + 6 atoms × 1/2 = 4 atoms per unit cell

Example 9.3 Copper crystallizes in a structure with a face-centered cubic unit cell 3.63 Å on an edge. Calculate

    a. the atomic radius of copper.

    b. the volume of the unit cell in cc.

    c. the density of copper.

Solution

    a. It will be helpful here to draw a face of the unit cell.

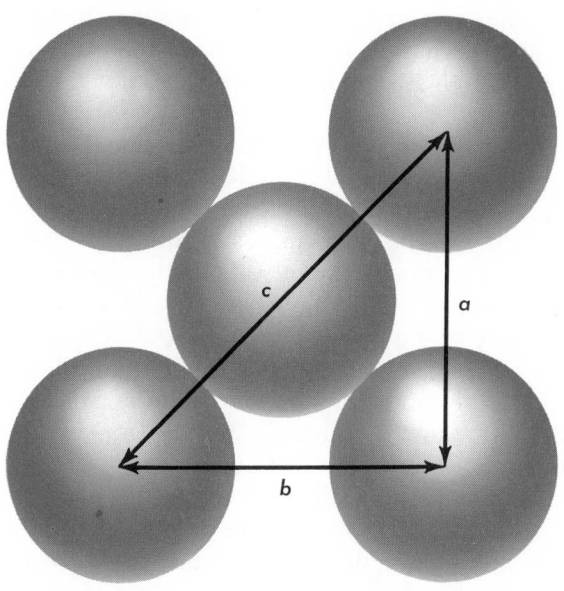

The distances a and b are each 3.63 Å. By the Pythagorean theorem,

$$c^2 = a^2 + b^2 = 2a^2$$

or $c = \sqrt{2}a = (1.41)(3.63 \text{ Å}) = 5.12 \text{ Å}$

But, from the figure, c = 4r, where r is the radius of the metal atom,

$$4r = 5.12 \text{ Å}; \; r = 1.28 \text{ Å}$$

b. $V_{cell} = a^3 = (3.63 \times 10^{-8} \text{ cm})^3 = 47.8 \times 10^{-24} \text{ cc}$

c. The density of the metal should be equal to that of the unit cell. The mass is that of the four copper atoms assigned to the face-centered cell; the mass of a single copper atom is found by dividing the gram atomic wt of copper by Avogadro's number.

$$\text{mass of unit cell} = 4 \text{ atoms} \times \frac{63.5 \text{ g}}{6.02 \times 10^{23} \text{ atoms}} = 42.2 \times 10^{-23} \text{ g}$$

$$\text{density} = \frac{\text{mass}}{\text{volume}} = \frac{42.2 \times 10^{-23} \text{ g}}{47.8 \times 10^{-24} \text{ cc}} = 8.83 \text{ g/cc}$$

(The observed density of copper is 8.92 g/cc at 20°C.)

TYPES OF PACKING

When the atoms in a crystal lattice are all the same size, as is the case with metals, they tend to pack as closely as possible. Consequently we do not expect to find metals crystallizing in a simple cubic structure, where there is a relatively large amount of empty space between atoms. A body-centered cubic structure, where each atom has eight nearest neighbors (i.e., a *coordination number* of 8), represents a more closely packed structure with less "waste space." A still more efficient way of packing spheres of the same size is the face-centered cubic structure, where the coordination number of an atom is 12.

It turns out that there are two different ways to pack atoms of equal radius most efficiently, leaving as little empty space as possible. One of these "closest-packed" arrangements is equivalent to the face-centered cubic structure just discussed; it is sometimes referred to as **cubic closest-packing.** The other, which is equally efficient, is known as **hexagonal closest-packing.**

To understand why there are two closest-packed arrangements, consider Figure 9.10. The drawing at the left represents a portion of a single layer of spheres, labeled B, packed as closely together as possible. A second closest-packed layer can be superimposed above the first if we place spheres so that their centers lie directly above the points labeled A. There are now two different ways in which we can add a third closest-packed layer below the plane of the paper.

1. In hexagonal closest-packing, the spheres are placed so that they line up exactly with the top layer, i.e., their centers fall directly below the points labeled A. In this structure, the pattern repeats itself every two layers; it is commonly described as ABABAB . . . .

2. In cubic closest-packing, the spheres in the third layer are placed so that their centers lie directly below the points marked C. Here the pattern is repeated after every three layers: the structure is described as ABCABC . . . . It can be shown that this structure gives a face-centered cubic unit cell.

Virtually all the metals crystallize in either the body-centered cubic structure (about 20 metals), or one of the closest-packed structures (about 24 metals in cubic closest-packing and 30 in hexagonal closest-packing).

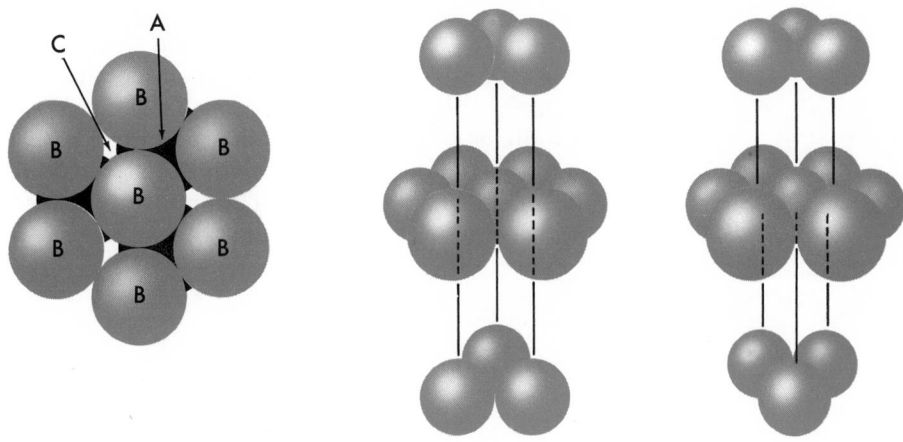

Closest-packing: hexagonal = ABA   Hexagonal closest-packed   Face-centered cubic
                 cubic   = ABC

Figure 9.10 Arrangement of atoms in hexagonal and cubic closest-packing. In HCP the pattern repeats after two layers; in CCP, after three.

The geometry of ionic crystals, where there are two different kinds of ions, is necessarily more complicated than that of metallic lattices. Nevertheless, it is possible to visualize the packing in certain ionic crystals in terms of the simple unit cells discussed above. An example is the lithium chloride crystal, where the large Cl^- ions (r = 1.81 Å) form a face-centered cubic lattice, with the small Li^+ ions (r = 0.60 Å) fitting into "holes" in the lattice, one Li^+ ion per Cl^- ion (Fig. 9.11). The structure of sodium chloride is similar to that of LiCl, except that the Na^+ ions (r = 0.95 Å) are slightly too large to fit into a close-packed lattice of Cl^- ions touching each

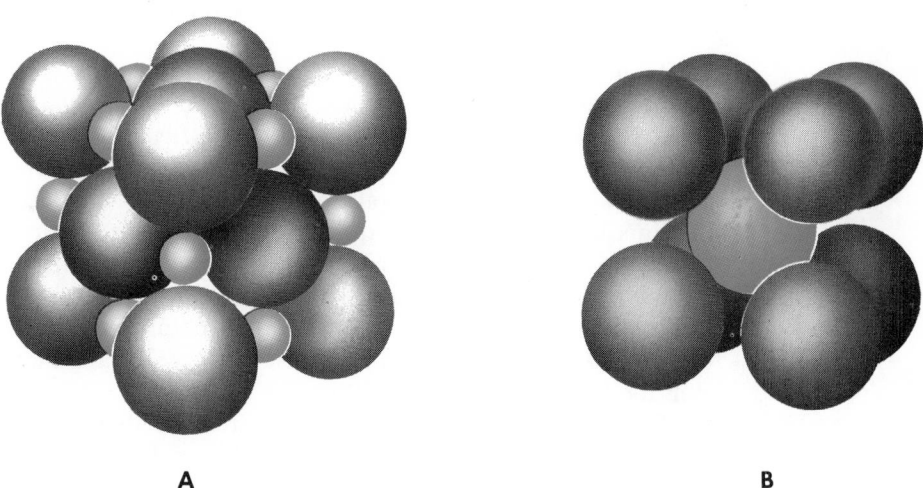

A   B

Figure 9.11 In LiCl, where the cation is much smaller than the anion, there is anion-anion contact in a face-centered cubic structure. In CsCl, the Cs^+ ion at the center of a cube touches the eight Cl^- ions at the corners.

other along the faces of a cube. Consequently the face-centered cubic array of Cl^- ions found in LiCl is slightly expanded in NaCl.

In cesium chloride, the cations are much too large (r of $Cs^+ = 1.69$ Å) to fit into a face-centered cubic array of Cl^- ions. We find that cesium chloride crystallizes in a quite different structure (Fig. 9.11) in which Cs^+ ions are located in the center of a cube outlined by Cl^- ions. The Cs^+ ion at the center touches Cl^- ions located at opposite corners of the cube; the Cl^- ions at adjacent corners of the cube almost, but not quite, touch each other.

9.4 PHASE EQUILIBRIA

Earlier in this chapter we discussed in some detail the equilibrium between a liquid and its vapor. All of us are familiar with another type of phase equilibrium, that between the liquid and the solid states of a substance. When ice cubes are added to a glass of water, the temperature drops to 0°C and an equilibrium is established between solid and liquid water. If we heat a test tube containing crystals of the organic compound naphthalene to 80°C, some of the solid melts; so long as the test tube is maintained at this temperature, liquid and solid naphthalene coexist in equilibrium with each other.

Equilibria between solids and their vapors are perhaps less familiar than liquid-vapor or liquid-solid equilibria. You may, however, have observed the faint purple color that appears in the gas phase above iodine crystals stored in a reagent bottle. The color is that of gaseous molecules of I_2 in equilibrium with the solid. If the temperature is raised, the purple color deepens, which suggests that the vapor pressure of solids, like that of liquids, increases with temperature.

Many of the important characteristics of phase equilibria can be summarized by means of what is known as a **phase diagram.** In Figure 9.12 we have drawn a portion of the phase diagram for the pure substance, water. Pressure is plotted along the vertical axis, temperature along the horizontal axis.

To understand how we make use of this diagram, let us consider first the significance of the three lines AB, AC and AD. *Each of these lines tells us the pressures and temperatures at which two phases are in equilibrium with each other.* Line AB is one that we have seen before; it is a portion of the vapor pressure-temperature curve of liquid water. At any point along this line, liquid water is in equilibrium with water vapor. Line AC represents the vapor pressure curve of ice; at any point along this line we will find the two phases, ice and vapor, in equilibrium with each other. Finally, the line AD tells us the conditions of temperature and pressure at which liquid water is in equilibrium with ice; we will have more to say a little later about the nature of this line.

The point A on the phase diagram represents the one point (temperature and pressure) at which all three phases, liquid, solid and vapor, are in equilibrium with one another. Surprisingly enough, it is called, simply, the triple point. For water the triple point temperature is 0.01°C; at this temperature liquid water and ice have the same vapor pressure, 4.56 mm Hg.

The three areas of the phase diagram, labeled "solid," "liquid" and

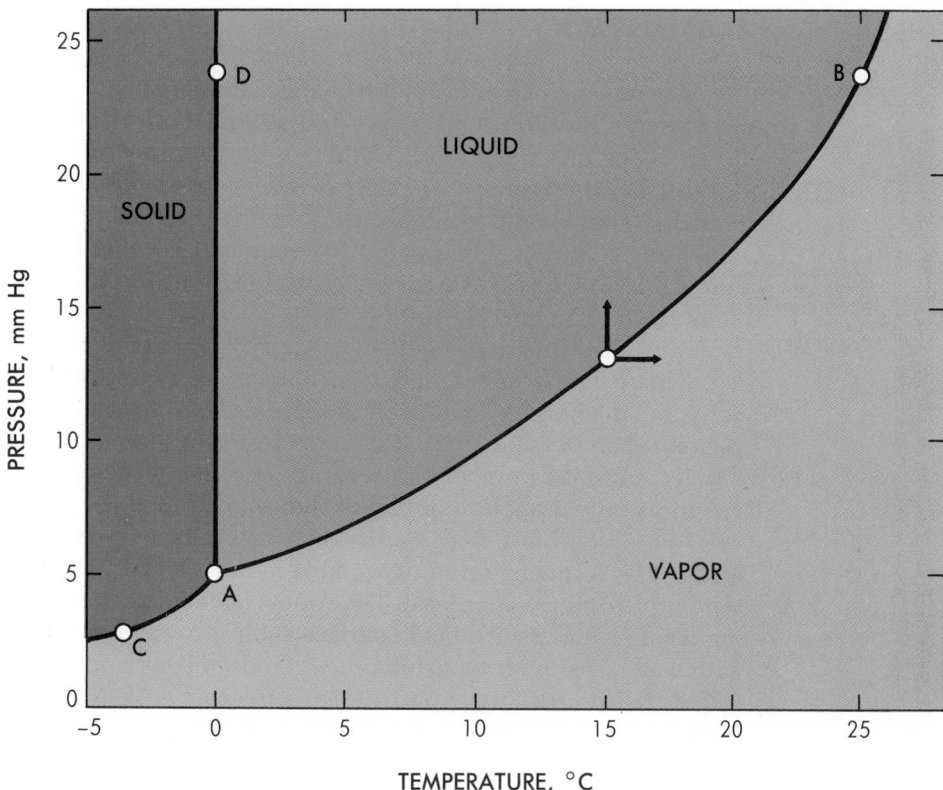

Figure 9.12 Phase diagram for water in the region −5 to 25°C.

"vapor" in Figure 9.12, comprise regions of temperature and pressure at which only one phase is present. To show that this is the case, let us consider what happens to an equilibrium mixture of two phases when the pressure or temperature is changed. Specifically, let us start at the point on line AB, indicated by an open circle, where liquid water and vapor are in equilibrium at 15°C and 13 mm Hg. Intuition tells us that if we suddenly increase the pressure on this mixture, holding the temperature at 15°C, condensation will occur. The phase diagram, based on experimental data, confirms our hunch; by increasing the pressure at 15°C we move up into the liquid region (vertical arrow). Suppose, in another experiment, we hold the pressure constant but increase the temperature. It seems reasonable that this change would cause the liquid to vaporize. The phase diagram tells us that our reasoning is correct; an increase in temperature (horizontal arrow) corresponds to a shift into the vapor region.

Superficially at least, the phase diagrams of other pure substances resemble that of water. Nevertheless, differences in the triple point temperature and in the orientation of the solid-liquid line AD cause substances such as iodine, carbon dioxide and benzene to behave quite differently from water when they undergo the phase changes of sublimation or fusion.

SUBLIMATION

The process by which a solid changes directly to its vapor without passing through the liquid state is called sublimation. Examination of Figure 9.12 tells us that a *solid will sublime at any temperature below the triple point when the pressure above it is reduced below the equilibrium vapor pressure.* To illustrate what this means, let us consider the conditions under which ice sublimes. On a cold, dry winter day, when the temperature is below 0°C and the pressure of water vapor in the air is less than the equilibrium value (4.56 mm Hg at 0°C), ice or snow "disappears" by sublimation. The rate of sublimation can be increased by evacuating the space above the ice. This is done commercially in the freeze-drying process for making dehydrated foods. A food such as eggs or lean meat is frozen, put into a vacuum chamber and evacuated to pressures of 1 mm Hg or less. Gradually the ice crystals formed on freezing sublime to give a product whose weight is only a fraction of that of the original food.

We lose a lot of our snow that way in Minnesota

Iodine sublimes more readily than ice because its triple point pressure, 90 mm Hg, is much higher. If we heat iodine crystals in a test tube, we can see purple iodine vapor form. The vapor can be made to condense by bringing it into contact with a cold surface such as a watch glass or beaker filled with water. A similar two-step process is used by organic chemists to purify volatile solids such as naphthalene or benzoic acid. A difference of a few degrees in temperature is sufficient to cause the vapor to condense to the solid. This accounts for the formation of frost or snow when water vapor at low pressures is cooled below 0°C on the surface of the earth (frost) or in the upper atmosphere (snow).

If we wish to sublime iodine by heating it in a test tube, we must be careful to stay below the triple point temperature, 115°C. If the temperature exceeds 115°C, the iodine melts. No such problem arises with solid carbon dioxide (Dry Ice), which has a triple point pressure above one atmosphere (5.2 atm at −57°C) (Figure 9.13). Liquid CO_2 cannot be prepared by heating Dry Ice in an open container; regardless of the temperature, the solid passes directly to the vapor.

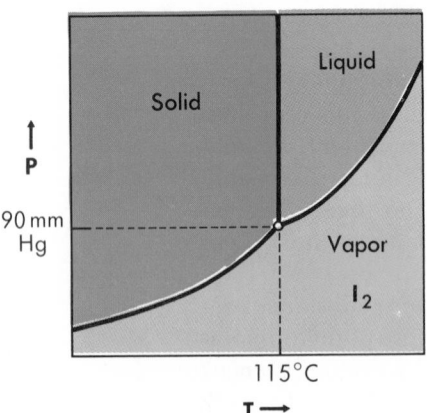
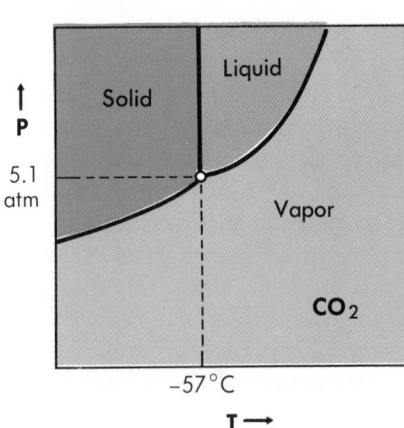

Figure 9.13 Phase diagrams of I_2 and CO_2 near their triple points. Notice that $CO_2(l)$ cannot exist at 1 atm.

Figure 9.14 An increase in pressure lowers the melting point of water (d liquid > d solid) but raises that of benzene (d solid > d liquid).

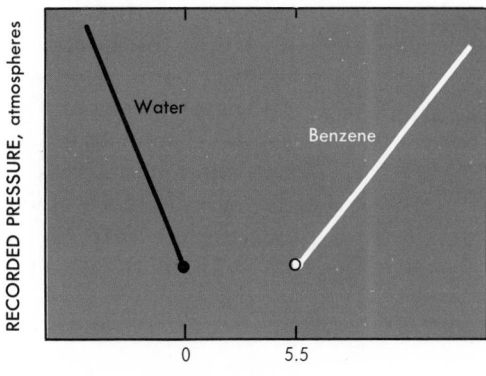

The amount of heat absorbed when one mole of a solid sublimes is referred to as the **heat of sublimation.** The heat of sublimation is often calculated by making use of Hess's Law which tells us that the enthalpy change depends only upon the final and initial states and is independent of the number of steps. Applied to sublimation, Hess's Law requires that, at a given temperature,

$$\Delta H_{subl} = \Delta H_{fus} + \Delta H_{vap} \tag{9.7}$$

For water at 0°C, these quantities are 12.1 kcal, 1.4 kcal and 10.7 kcal, respectively.

FUSION

The **melting point** of a solid or, conversely, the **freezing point** of a liquid is taken to be the temperature at which the solid and liquid phases of a pure substance are in equilibrium with each other. Melting points are ordinarily measured in an open container at atmospheric pressure. The melting points that we find recorded in the literature for various substances represent the temperature at which solid and liquid are in equilibrium with each other at an external pressure of 1 atm. For most substances the melting point under these conditions is virtually identical with the triple point. For water, the difference is only 0.01°C.

Although the effect of pressure upon melting point is very small, we are often interested in the direction of this effect. To decide whether the melting point will be increased or decreased by compressing the solid, we need only apply the principle that **an increase in pressure favors the formation of the more dense phase.** In the case of benzene, in which the solid is more dense than the liquid (Table 9.4), an increase in pressure favors the solid. This means that at higher pressures, the solid will become stable at temperatures above the normal melting point; i.e., the melting point will be raised by an increase in pressure (Figure 9.14). The behavior of benzene is typical of that of most substances, since the solid phase is usually more dense than the liquid.

We recall that water is unusual in that the liquid phase is more dense than the solid; an ice cube floats in a glass of water. We would then predict that the melting point of ice should decrease with pressure; i.e., at high pressures the liquid should be stable at temperatures below the normal melting point. This is confirmed experimentally; the melting point of ice decreases by about 1°C for every 134 atm of applied pressure. This effect can be demonstrated strikingly by suspending two heavy weights from a wire stretched across a block of ice. In time, the wire passes completely through the block of ice, which appears to be unchanged. What happens is that the pressure exerted by the weights melts a thin layer of ice around the wire. As the wire falls, the pressure above it drops and the ice re-forms.

It takes a good deal of pressure to change the MP of a solid by as much as 1°C

9.5 NONEQUILIBRIUM PHASE CHANGES

The phase changes we have considered in this chapter have been assumed to take place under equilibrium conditions. In practice, the phase changes that we observe in the laboratory or in the world around us occur at temperatures and pressures at least slightly removed from equilibrium. When experiments of the type we have described are done, we frequently find that the systems do not behave in quite the manner predicted by the phase diagram. A liquid being vaporized may not boil smoothly, particularly at low pressures. Instead, it may superheat to temperatures above the calculated boiling point and then "bump" and boil furiously. A very pure liquid on cooling does not start to freeze exactly at its freezing point; it has to be supercooled below that temperature before crystallization occurs.

The tendency of liquids to superheat or supercool reflects the difficulty of initiating a new phase within the body of the liquid. For a liquid cooled below its melting point to freeze, some centers, or nuclei, are needed on which crystallization can occur. These centers may be dust particles or small crystals of the substance itself, added to induce crystallization. It is sometimes possible to persuade a supercooled liquid to freeze by stirring vigorously, which may redistribute foreign particles through the body of the liquid. In particularly stubborn cases, scratching the sides of the container is sometimes effective, possibly because it removes tiny particles of glass from the walls and introduces them into the liquid. In a pure liquid, uncontaminated, unstirred and unscratched, crystallization nuclei are absent, and supercooling may be extensive. As an extreme example, very pure water can be cooled to −40°C without freezing.

For a liquid to boil, there must be centers at which bubbles can form. These centers may be microscopic bubbles of dissolved gas, dust particles, or sharp crystal edges or corners. When a liquid is to be distilled, a small quantity of an inert solid such as marble or porcelain is usually added. These solids have sharp edges on which vapor bubbles can form. In the absence of such solids, pronounced superheating can occur. Water that has been thoroughly purged of dust particles and gas bubbles has been heated in an open capillary tube to a temperature of 270°C without boiling. At low pressures, where the driving force for boiling is small, superheating is especially likely.

9.6 DEFECT CRYSTALS

In discussing the structures of crystalline solids, we assumed them to be perfect crystals, with all the atoms or ions lined up in a precise geometric pattern. A perfect crystal, like an ideal gas, is an abstraction; the crystals that we see in the world around us contain imperfections, no matter how carefully they are prepared. These defects, even if relatively few in number, can profoundly affect the physical and chemical properties of a solid. Three important kinds of crystal defects that we will deal with are

1. Vacant sites in the crystal lattice.

2. Interstitial atoms or ions, i.e., particles occupying positions other than those found in a perfect lattice.

3. Foreign atoms or ions incorporated into the crystal lattice.

SILVER BROMIDE. A perfect crystal of silver bromide would have a structure similar to that of lithium chloride (Fig. 9.11). A cross section of such a crystal, expanded slightly to emphasize the amount of space available, is shown at the left of Figure 9.15. Fortunately or unfortunately, the silver bromide that we make in the laboratory does not have this completely ordered structure. Instead, a few Ag^+ ions move out of lattice sites to occupy an interstitial position, creating crystal defects of the type shown at the right of Figure 9.15.

The sensitivity to visible and ultraviolet light that is characteristic of silver bromide and other silver salts has been explained in terms of crystal defects. When light strikes a silver bromide crystal, enough energy is available to remove an electron from an occasional Br^- ion.

$$Br^- \rightarrow Br + e^-$$

The electron produced is able to migrate through the crystal until it comes

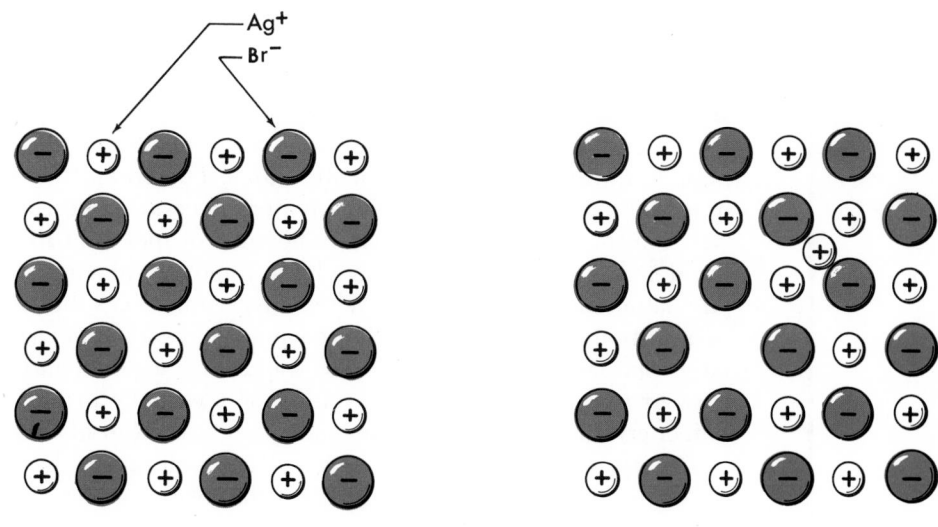

Perfect crystal   Interstitial Ag$^+$

Figure 9.15 Schematic drawing of a common type of crystal defect in AgBr.

$$Ni^{2+} \quad O^{2-} \quad Ni^{3+} \quad O^{2-} \qquad\qquad Ni^{2+} \quad O^{2-} \quad Ni^{2+} \quad O^{2-}$$

$$O^{2-} \qquad\quad O^{2-} \quad Ni^{2+} \qquad\qquad O^{2-} \quad Li^{+} \quad O^{2-} \quad Ni^{2+}$$

$$Ni^{3+} \quad O^{2-} \quad Ni^{2+} \quad O^{2-} \qquad\qquad Ni^{3+} \quad O^{2-} \quad Ni^{2+} \quad O^{2-}$$

$$O^{2-} \quad Ni^{2+} \quad O^{2-} \quad Ni^{2+} \qquad\qquad O^{2-} \quad Ni^{2+} \quad O^{2-} \quad Ni^{2+}$$

Figure 9.16 Crystal defects in NiO. The crystal at the left is nonstoichiometric (too little Ni). The one at the right contains Li^+ as an impurity. In both cases, electroneutrality is maintained by the presence of Ni^{3+}

in contact with an interstitial Ag^+ ion which it converts to an atom of silver metal.

$$Ag^+ + e^- \rightarrow Ag$$

The silver atoms formed by this mechanism apparently act as a nucleus for the formation of grains of silver, which are visible to the naked eye. At first, they are seen as a purplish discoloration at the surface of the silver bromide; on further exposure to light the entire solid turns black.

Photographic film is coated with an emulsion of silver bromide in gelatin. When the film is exposed momentarily by opening the shutter of a camera, a few interstitial silver atoms are produced, forming what is known as a latent image. This appears to consist of aggregates of from 10 to 500 silver atoms forming around an individual atom as a nucleus. The amount of silver constituting the latent image is too small to be visible under a microscope, but it can be seen as a visible negative when the film is developed chemically.

NICKEL OXIDE. When nickel is heated in air to about 800°C, it becomes coated with a green film of nickel oxide, NiO. However, when finely divided nickel is heated to 1200°C in pure oxygen, the product is black rather than green. Chemical analysis shows the black material to be "nonstoichiometric" in the sense that the atom ratio of nickel to oxygen is slightly less than one (e.g., $Ni_{0.97}O_{1.00}$).

The different properties of these two forms of nickel oxide can be explained in terms of their crystal structures. The green, stoichiometric form of nickel oxide is made up of nearly perfect crystals, in which there is one Ni^{2+} ion for every O^{2-} ion. The deficiency of nickel in the black product implies that a few Ni^{2+} ions have been removed, leaving vacancies at certain cation sites in the lattice. When a Ni^{2+} ion leaves the lattice, electrical neutrality must be maintained; we cannot have a net negative charge on a large sample of nickel oxide. One way to restore the charge balance is to convert two Ni^{2+} ions in the vicinity of a vacancy to Ni^{3+} ions, giving a defect crystal with the structure indicated at the left of Figure 9.16.

Black, nonstoichiometric nickel oxide is found to have an electrical conductivity many times greater than that of the green oxide ($Ni_{1.00}O_{1.00}$). The conductivity can be explained by the ease with which an electron can be transferred from a Ni^{2+} ion to an adjacent Ni^{3+} ion. If enough Ni^{3+} ions are present, a series of such electron transfers can progress across the crystal, amounting to a flow of current.

In practice, nonstoichiometric crystals of nickel oxide are of limited value as semiconductors because it is very difficult to produce large deviations from stoichiometry. A more effective way of increasing the electrical conductivity of an ionic crystal is to introduce certain foreign ions into it. It is possible to increase the conductivity of nickel oxide by a factor of 10^{10} by introducing Li^+ ions into the crystal lattice. The radii of Li^+ and Ni^{2+} ions are close enough (0.60 Å vs 0.74 Å) that it is possible to substitute a large fraction of the Ni^{2+} ions by Li^+ (up to 10 per cent or more). Every time a Li^+ ion is introduced into the crystal, a Ni^{2+} ion must be converted to Ni^{3+} to maintain electrical neutrality. In the resultant defect crystal, shown at the right of Figure 9.16, exchange of electrons between Ni^{2+} and Ni^{3+} is responsible for the enhanced conductivity.

GERMANIUM AND SILICON. Extremely pure samples of germanium and silicon are nonconductors because the valence electrons are localized in the four covalent bonds that each atom forms with its neighbors. The conductivity increases dramatically when small amounts of arsenic or boron (as little as 0.0001 mole per cent) are introduced into the crystal. To understand how this comes about, let us consider how the crystal structure of a 4A element might be disturbed by substituting atoms of a 5A or 3A element.

An atom of arsenic, which has five valence electrons, can fit into the crystal lattice of germanium or silicon because of the close similarity in atomic radii (1.21 Å vs 1.22 Å or 1.17 Å). In order to do so, however, it must give up its valence electron. This electron can move relatively freely through the crystal under the influence of an electrical field, giving rise to an **n-type semiconductor** (current carried by the flow of negative charge). If an atom of boron or other element with three valence electrons is introduced into the lattice, a quite different situation arises. An electron deficiency is created at the site occupied by the foreign atom; it is surrounded by seven valence electrons rather than eight. Physicists describe the introduction of such a defect in terms of the formation of a "positive hole" in the lattice. In an electrical field, an electron moves from a neighboring atom to fill this hole. In so doing, it creates an electron deficiency around the atom which it leaves. Conduction in this type of defect crystal, a **p-type semiconductor,** can be thought of as a movement of positive holes through the lattice.

A semiconductor in which there is a junction between an electron-rich and an electron-deficient region acts as a rectifier, capable of converting alternating to direct current. Such an "n-p" junction can be formed by starting with a pure silicon disc and introducing a trace of boron on one side of the disc and a trace of arsenic on the other. These impurities are allowed to diffuse into the silicon at high temperatures until the region of p-type semiconductor meets the region of n-type semiconductor. Electrons flow readily from the electron-rich region containing arsenic to the electron-deficient area created by the presence of boron atoms. Potentials as high as 1000 volts are incapable of bringing about electron flow in the opposite direction.

Perhaps the best known of all semiconductor devices is the transistor, in which n-p-n or p-n-p junctions are created by forming a sandwich of alternate electron-rich and electron-poor regions. A transistor amplifies an electric current in much the same way as a vacuum tube. The small size of transistors makes them ideal for use in hearing aids, miniature radios,

Extremely pure silicon is needed when making semiconductors

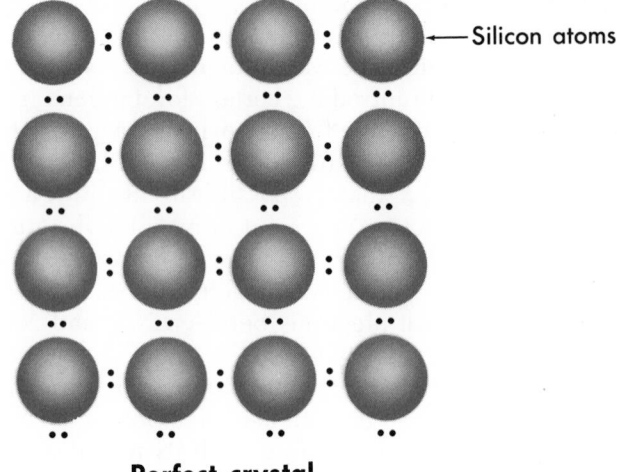

Silicon atoms

Perfect crystal

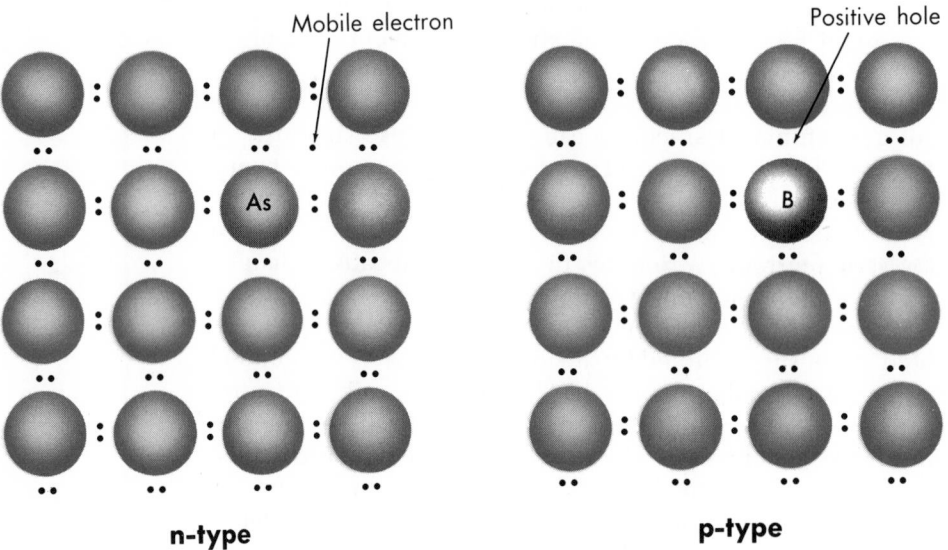

Mobile electron

Positive hole

n-type   **p-type**

Figure 9.17 Semiconductors derived from silicon.

and electronic components of missiles or satellites. Their increasing use in television and radio receivers, in which they have largely displaced vacuum tubes, reflects their greater durability and lower energy consumption.

The electrical conductance or, conversely, the resistance of all types of semiconductors is extremely sensitive to temperature. The resistance of a typical semiconductor decreases by 4 per cent per °C at room temperature; that of copper wire increases by only one tenth of that amount under the same conditions. Semiconductor devices known as thermistors have been developed to take advantage of this property. Accurate resistance readings taken in a circuit including thermistors make it possible to estimate temperature changes to 0.0001°C or better. The small size of thermistors makes them suitable for temperature measurements in restricted spaces. A thermistor inserted into the tip of a glass probe can be used to measure temperatures in various parts of the human body. Thermistors can be taped to the skin or soldered to metal surfaces to measure surface temperatures.

A semiconductor whose surface contains an n-p junction can convert radiant energy to electrical energy. This principle is used in the solar battery, which has been used as a source of power in space vehicles. In this device, light striking a very pure crystal of germanium or silicon impregnated with an electron-rich impurity such as arsenic ejects some of the loosely held electrons. These electrons are collected at the surface, which is coated with a thin, transparent layer of a p-type semiconductor. The electrons then pass through an electrical circuit where their energy is used to do useful work. At the present time a main drawback to the use of solar batteries is the economic factor. The semiconductors used in the battery are extremely expensive. They will have to be reduced in cost by a factor of at least 100 if the solar battery is to become competitive with conventional sources of electrical energy.

PROBLEMS

9.1 How would you explain to an elementary school pupil why

a. he feels cool when he comes out of a swimming pool?
b. he should use soap when he takes a bath?
c. Dry Ice disappears without leaving a residue?

9.2 How would you explain to a professor of English (or philosophy, or music)

a. why liquids rather than compressed air are used as brake fluids?
b. why a chemistry professor has a laboratory full of x-ray equipment?
c. the operation of the transistor in his radio set?

9.3 Design an experiment to

a. measure the heat of vaporization of benzene.
b. prove that ice can be melted by exerting pressure.
c. demonstrate that gases are more compressible than liquids.
d. prepare liquid carbon dioxide.

9.20 How would you explain to a teacher in an elementary school why

a. she has difficulty removing the lid on a pressure cooker when it cools?
b. she can dry clothes outdoors in the winter even though they freeze when she puts them on the line?
c. in a box packed with Ping-Pong balls, the centers of the balls in one layer do not lie directly above those of the balls underneath?

9.21 How would you explain to an outdoor enthusiast

a. why it takes longer to cook foods by boiling at high altitudes?
b. how freeze-dried foods are made?
c. why silver bromide is used in photography?

9.22 Design an experiment to

a. show that diffusion is faster in gases than in liquids.
b. measure the atomic radius of zinc.
c. determine the heat of sublimation of benzene.

9.4 The density of liquid water at its normal boiling point is 0.958 g/ml.

a. What is the volume of one mole of liquid water at this temperature?
b. Using the Ideal Gas Law, calculate the volume of one mole of water vapor at its normal boiling point.
c. Assuming that water molecules can be treated as spheres with a radius of 1.38 Å, find the volume of a mole of water molecules.
d. Using the answers to (a), (b) and (c), calculate the percentages of the total volume that are occupied by the molecules in liquid water and in water vapor at the normal boiling point.

9.5 Using data from Table 9.1, calculate the amount of heat required to vaporize one gram of pyridine (C_5H_5N).

9.6 Calculate an approximate value for the amount of heat which must be absorbed to convert 3.00 g of ice at 0°C to steam at 100°C.

9.7 One mole of liquid ether, $C_4H_{10}O$, is added to the apparatus shown in Figure 9.3. The apparatus is maintained at 20°C, at which temperature ether has a vapor pressure of 442 mm Hg.

a. What will be the volume below the piston when the liquid is just vaporized?
b. What will be the pressure of the vapor when the volume is 30 liters?
c. What will be the pressure of the vapor when the volume is 50 liters?

9.8 A sample of water vapor at 100°C and 380 mm Hg pressure is cooled in a container of constant volume.

a. What will be the pressure of the vapor at 85°C (vp water = 434 mm Hg)?
b. What will be the pressure of the vapor at 70°C (vp water = 234 mm Hg)?

9.9 A student measures the vapor pressure of a certain liquid as a function of temperature and expresses his data as follows

$\log_{10}P$	1.60	1.87	2.13	2.40
1/T	0.00330	0.00320	0.00310	0.00300

Calculate the heat of vaporization of the liquid.

9.23 The density of copper at 20°C is 8.82 g/cc.

a. What is the volume of one mole of solid copper at 20°C?
b. Taking a Cu atom to be a sphere of radius 1.28 Å, calculate the volume of a mole of copper atoms.
c. From your answers to (a) and (b), calculate the fraction of "empty space" in a copper crystal at 20°C.

9.24 Using data from Table 9.4, calculate ΔH when 12.5 g of benzene, C_6H_6, condenses from vapor to solid at its melting point.

9.25 What is the minimum amount of ice that should be added to a glass containing 100 cc of water at 25°C if you want to be sure that there will be some ice left when equilibrium is reached?

9.26 A mother uses a water vaporizer to raise the humidity in her child's bedroom, which is at 25°C and has a volume of 2.0×10^4 liters. Assuming that the air is completely dry to begin with and that no moisture leaves the room,

a. how many grams of water must she put in the vaporizer to ensure that the air becomes saturated with water vapor (v p = 24 mm Hg at 25°C)?
b. what will be the final pressure of water vapor in the room if she puts 600 g of water in the vaporizer?
c. what will be the final pressure of water vapor in the room if she puts 400 g of water in the vaporizer?

9.27 The vapor pressure of benzene at various temperatures is as follows:

P (mm Hg)	75	118	181	269
T (°C)	20	30	40	50

A sample of benzene vapor at 50°C and 150 mm Hg pressure is cooled in a stoppered flask. Will it condense when cooled to 40°C? 30°C? Estimate the temperature at which condensation will start.

9.28 Applying a graphical method to the data in Problem 9.27, calculate the heat of vaporization of benzene and compare it to the value given in Table 9.1.

9.10 The vapor pressure of n-hexane at 30°C is 189 mm Hg; its heat of vaporization is 7.14 kcal/mole. Estimate the vapor pressure of n-hexane at 50°C.

9.11 The normal boiling point of n-heptane is 98°C. Using Trouton's rule, estimate the heat of vaporization. The measured heat of vaporization is 7.6 kcal/mole. What is the per cent of error in the calculated value?

9.12 The critical point of ammonia is 133°C, 112 atm. Its normal boiling point is −33°C. Which of the following statements concerning ammonia must be true?

    a. A tank of ammonia at room temperature which has a pressure of 5 atm must contain liquid.
    b. A tank of ammonia at room temperature which has a pressure of 1 atm cannot contain liquid NH_3.

9.13 Using Table 9.4, calculate the volume change when 20.0 g of benzene (C_6H_6) freezes.

9.14 When x-rays with a wavelength of 0.400 Å are diffracted by a certain crystal, the first order beam (n = 1) is found to be displaced 10.0° from the original beam.

    a. Calculate the distance between the layers of atoms responsible for the diffraction.
    b. What will be the angle of diffraction for the second order beam (n = 2)?

9.15 Consider the following two-dimensional pattern (the lines indicate bonds; the points, atoms)

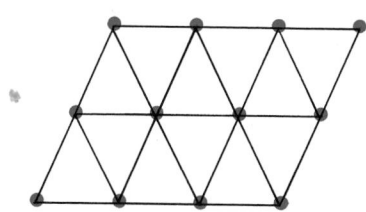

    a. What is the unit cell in this structure?
    b. How many atoms should be assigned to the unit cell?

9.29 The normal boiling point of liquid nitrogen is −196°C; its heat of vaporization is 1.34 kcal/mole. Estimate

    a. the vapor pressure of liquid nitrogen at −180°C.
    b. the temperature at which liquid nitrogen will boil under a pressure of two atmospheres.

9.30 A certain organic compound has a normal boiling point of 70°C; its heat of vaporization is measured to be 60.0 cal/g. Using Trouton's rule, estimate the molecular weight of the compound.

9.31 The critical point of sulfur dioxide is 157°C, 78 atm. Liquid sulfur dioxide has a vapor pressure of 3.8 atm at 25°C. Which of the following statements must be true?

    a. Sulfur dioxide is a gas at 25°C and atmospheric pressure.
    b. A tank of sulfur dioxide at 25°C can have a pressure of 5 atm.
    c. Sulfur dioxide gas cooled to 150°C and 80 atm pressure will condense.
    d. The normal boiling point of sulfur dioxide lies between 25 and 157°C.

9.32 A 200-cc glass contains 100 cc of water at 20°C. It is filled to the brim with ice at 0°C.

    a. How much ice must melt to bring the final temperature to 0°C?
    b. What will be the total volume of ice and water when equilibrium is reached?

9.33 The Braggs used diffraction patterns obtained with sodium chloride to determine the wavelengths of x-rays. The distance between one set of diffraction planes in NaCl is 2.76 Å. Using x-rays given off when copper is bombarded by electrons, the diffraction angles are found to be 16°10′, 33°50′, 56°40′ (n = 1, 2, 3). Calculate the wavelength of the x-rays.

9.34 Consider the structure, repeated indefinitely in two dimensions:

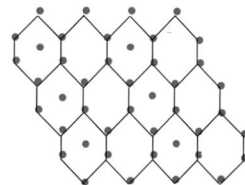

    a. How many unit cells are shown?
    b. How many atoms should be assigned to a unit cell (each dot is an atom)?

9.16 Iron (at. rad. 1.26 Å) crystallizes in a body-centered cubic structure.

    a. How many atoms are there per unit cell?
    b. What is the length of one side of the unit cell?
    c. What is the volume of the unit cell?
    d. What is the molar volume of iron?

9.17 In the LiCl structure shown in Figure 9.11, where the Cl^- ions form a face-centered cubic unit cell, what is the length of an edge of the cube? Show that it is possible to fit a Li^+ ion between Cl^- ions along an edge of the cube. Could a Na^+ ion fit?

9.18 The triple point of naphthalene is 80°C, 7.4 mm Hg. At the triple point the density of the solid is 1.15 g/cc while that of the liquid is 0.979 g/cc. Sketch the phase diagram of naphthalene in the vicinity of the triple point. Describe, in some detail, an experiment in which naphthalene could be purified by sublimation.

9.19 Suppose a small number of atoms of the following elements were introduced into a silicon crystal. Would you expect to form a semiconductor and, if so, would it be of the n- or p-type?

    a. Ga   c. Ge
    b. As   d. Se

9.35 Gold crystallizes in a face-centered cubic structure. The dimension of the unit cell is 4.08 Å. Calculate the atomic radius of gold.

9.36 In the CsCl structure shown in Figure 9.11, what is the distance between centers of the Cl^- ions at the edges of the cube?

9.37 From information given in this chapter, construct a phase diagram of benzene extending from 0°C to the critical temperature.

9.38 If you wanted to prepare a semiconductor from AgCl, what metal would you suggest as a possible substitute for Ag^+?

*9.39 It has been suggested that the pressure exerted on a skate blade is sufficient to melt the ice beneath it and form a thin film of water which makes it easier for the blade to slide over the ice. Assume that the skater weighs 150 lbs and that the blade has an area of 1/2 in². Calculate the pressure exerted on the blade (15 lbs/in² = 1 atm). From information in the text, calculate the decrease in melting point at this pressure. Comment on the plausibility of this explanation and suggest another mechanism by which the water film might be formed.

*9.40 What is the maximum radius that a cation can have if it is to fit, without distortion, into a face-centered cubic lattice consisting of anions of radius r_-, touching each other along a face diagonal?

*9.41 Describe experiments, based upon principles discussed in this chapter, by which you could

    a. estimate the molecular weight of a liquid.
    b. determine an accurate value of Avogadro's number.
    c. determine an accurate value of the atomic weight of a metal.

*9.42 Calculate the fraction of empty space or "free volume" in a crystal made up of simple cubic unit cells; body-centered cubic unit cells; face-centered cubic unit cells. Consider the volume of the atoms assigned to each cell in comparison to the volume of the cell itself.

10   SOLUTIONS

In previous chapters we have dealt almost entirely with the structures and properties of pure substances. From now on we shall be dealing increasingly with solutions, where most chemical reactions occur. In this chapter we shall develop the background for a discussion of solution chemistry by studying the structure and physical properties of solutions.

10.1   INTRODUCTION

A solution can be defined as a homogeneous mixture of two or more substances. By homogeneous we ordinarily mean uniform to visual observation by eye or microscope. Sea water, filtered if necessary to remove seaweed, surf boards and sharks, is a solution of various salts in water. "Clean air," a rare commodity these days, is a solution composed of nitrogen, oxygen, argon and small amounts of other gases.

From a structural standpoint, homogeneity implies that the particles of the different species present are of molecular size ($\sim$50 Å or less in diameter) and that they are distributed in a more or less random pattern. To illustrate, in a solution formed by adding ethyl alcohol to water, the particles are individual molecules a few Å in diameter. These molecules occur randomly throughout the liquid; there is no tendency for large clusters of alcohol or water molecules to come together in one region of the solution. In a suspension, either or both of these criteria are not met. Fog is formed by the clustering of thousands upon thousands of water molecules into droplets large enough to be visible; it is considered to be a suspension of water in air rather than a true solution.

Solutions may exist in any of the three states of matter: gas, solid or liquid. Air is the most familiar example of a gaseous solution. Solid solutions are relatively common, particularly where two metals are involved. The "nickel" coin is actually a solid solution containing 25 per cent by weight of nickel dissolved in copper. Twelve-carat gold is a solid solution containing equal parts by weight of gold and silver (pure gold is 24 carat).

Throughout this chapter we shall be dealing primarily with liquid solutions, which are often further subdivided according to the physical states of the pure components. We may have a solution in which both components are liquids (e.g., a martini). One component may be a gas (soda water) or a solid (salt water). It is even possible to have a liquid solution in which neither component, at the temperature and pressure of the solution, is a liquid. A familiar example is the solution formed when rock salt (sodium chloride) is used to melt ice below 0°C.

In a gas-liquid or solid-liquid solution, we ordinarily refer to the liquid as the "solvent" and the other component (gas or solid) as the "solute." This choice of words reflects the way in which we visualize the solution process; it is natural to think of carbon dioxide or sodium chloride as "dissolving" in water rather than the reverse. If both components are liquids, the designations solute and solvent are more ambiguous. Frequently the component present in the greater amount is called the solvent. In a solution of 1 g of ethyl alcohol in 100 g of water, we would probably speak of the water as the solvent and the alcohol as the solute. If the amounts are more nearly equal, the choice becomes less clear-cut and must be specified.

Several different adjectives are used to indicate in a qualitative way the relative amounts of the components present in a solution. We may describe a solution containing a small amount of solute as being "dilute"; another solution, containing more solute, might be referred to as "concentrated." In a few cases these terms have taken on a quantitative meaning. Dilute and concentrated solutions of certain acids and bases, labeled as such in the laboratory, have the concentrations specified in Table 10.1.

Relative concentrations of solutions are sometimes expressed in a different way by using the terms "saturated," "unsaturated" and "supersaturated." A **saturated** solution of a solute, X, is one which is in equilibrium with undissolved X. A simple way to prepare a saturated solution at 25°C of, let us say, sodium chloride in water, is to bring an excess of the solid into contact with water at that temperature and stir until no more NaCl goes into solution. The resulting solution, which contains 36.2 g of sodium chloride per 100 g of water, is said to be saturated with sodium chloride. Addition of further solid sodium chloride fails to change its concentration in solution.

An **unsaturated** solution contains a lower concentration of solute than a saturated solution. A solution at 25°C containing less than 36.2 g of sodium

TABLE 10.1   Concentrations of Laboratory Acid and Base Solutions

		SOLUTE	MOLES SOLUTE PER LITER	% SOLUTE (BY WEIGHT)	DENSITY (g/ml)
Hydrochloric acid	conc.	HCl	12	36	1.18
	dilute		6	20	1.10
Nitric acid	conc.	HNO_3	16	72	1.42
	dilute		6	32	1.19
Sulfuric acid	conc.	H_2SO_4	18	96	1.84
	dilute		3	25	1.18
Ammonia	conc.	NH_3°	15	28	0.90
	dilute		6	11	0.96

° Often labeled "NH_4OH."

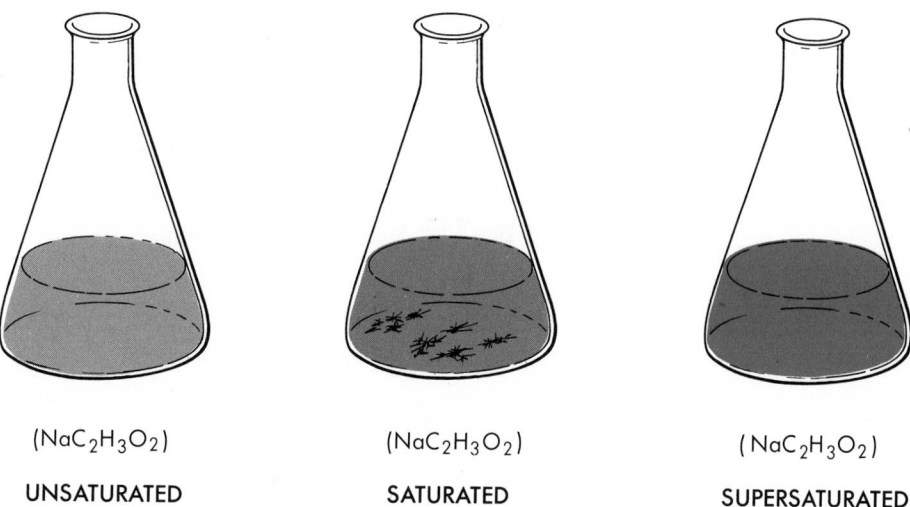

$(NaC_2H_3O_2)$ $(NaC_2H_3O_2)$ $(NaC_2H_3O_2)$

UNSATURATED   **SATURATED**   **SUPERSATURATED**

Figure 10.1 Unsaturated, saturated and supersaturated solutions of sodium acetate. Only the saturated solution is in equilibrium with undissolved solute.

chloride per 100 g of water is said to be unsaturated with respect to sodium chloride. It is not in a state of true equilibrium; if solute is added to an unsaturated solution, its concentration increases to approach saturation.

A **supersaturated** solution represents an unstable situation in which the solution actually contains more than the equilibrium concentration of solute. Supersaturated solutions most commonly arise when a solid is dissolved in a hot liquid and the solution is cooled. As an illustration of one method of preparing a supersaturated solution, consider a specific example, that of sodium acetate, $NaC_2H_3O_2$, dissolved in water. At 20°C a saturated solution contains 46.5 g of sodium acetate per 100 g of water; at higher temperatures the solubility of sodium acetate is considerably greater. If we heat 80 g of this salt with 100 g of water until it is completely dissolved (a temperature of about 50°C is required) and then cool carefully, without shaking or stirring, to 20°C, the excess solute remains in solution. This supersaturated solution can be maintained indefinitely so long as there are no nuclei upon which crystallization can start (cf. the phenomenon of supercooling, described in Chapter 9). If a small seed crystal of sodium acetate is added, crystallization quickly takes place until equilibrium is attained by the formation of a saturated solution.

Freshly opened soda pop is a supersaturated solution of CO_2 in water

10.2 CONCENTRATION UNITS

The properties of solutions depend to a large extent upon the relative amounts of solute and solvent present. For this reason, in any quantitative work involving solutions it is important to specify concentrations. This can be done by stating either the relative amounts of solute and solvent or, alternatively, the amount of one component relative to the total amount (mass or volume) of solution.

One of the simplest ways to specify solution concentrations is to give the **weight percentages** of the different components. We might, for example, describe "concentrated hydrochloric acid" (Table 10.1) by saying that it contains 36% by weight of hydrogen chloride. This tells us that in 100 g of concentrated hydrochloric acid, we can expect to find 36 g of HCl and 64 g of H_2O; in 1 g of the acid there will be 0.36 g HCl and 0.64 g H_2O, and so on.

In general, the weight percentage of component A in a solution is given by

$$\% \text{ of A} = \frac{\text{mass A}}{\text{total mass of solution}} \times 100 \tag{10.1}$$

Example 10.1   Using Table 10.1, calculate
    a. the number of grams of NH_3 in 240 g of concentrated ammonia.
    b. the number of grams of HNO_3 in 10.0 ml of dilute nitric acid.

Solution
    a. From the table, we note that the weight percentage of NH_3 is 28%. Rearranging Equation 10.1, we have

$$\text{mass } NH_3 = \frac{\% \text{ of } NH_3}{100} \times \text{total mass of solution}$$
$$= 0.28 \times 240 \text{ g} = 67 \text{ g } NH_3$$

    b. Before we can calculate the mass of HNO_3, we must obtain the total mass of 10.0 ml of dilute nitric acid. Noting from Table 10.1 that the density of dilute HNO_3 is 1.19 g/ml, we have

$$\text{total mass of solution} = 10.0 \text{ ml} \times 1.19 \frac{\text{g}}{\text{ml}} = 11.9 \text{ g}$$

Now we proceed as in (a) to find the mass of HNO_3:

$$\text{mass } HNO_3 = \frac{\% \text{ } HNO_3}{100} \times \text{total mass of solution}$$
$$= 0.32 \times 11.9 \text{ g} = 3.8 \text{ g } HNO_3$$

In Chapter 3 we stressed the usefulness of the mole concept in discussing the properties of pure substances. In this chapter we shall have occasion to use three different concentration units in which the amount of solute is expressed in moles. The first of these is the **mole fraction**; the mole fraction of component A, which we shall designate as X_A, is given by the expression

$$X_A = \frac{\text{no. of moles of A}}{\text{total no. moles of all components}} \tag{10.2}$$

Example 10.2   What are the mole fractions of benzene, C_6H_6, and toluene, C_7H_8, in a solution prepared by adding 500 g of benzene to 500 g of toluene?

Solution.   Since 1 mole $C_6H_6 = 6(12.0 \text{ g}) + 6(1.0 \text{ g}) = 78.0 \text{ g}$
      and 1 mole $C_7H_8 = 7(12.0 \text{ g}) + 8(1.0 \text{ g}) = 92.0 \text{ g}$

$$\text{no. of moles of } C_6H_6 = 500 \text{ g} \times \frac{1 \text{ mole}}{78.0 \text{ g}} = 6.41 \text{ moles}$$

$$\text{no. of moles of C}_7\text{H}_8 = 500 \text{ g} \times \frac{1 \text{ mole}}{92.0 \text{ g}} = 5.43 \text{ moles}$$

Using Equation 10.2,

$$X_{C_6H_6} = \frac{6.41}{6.41 + 5.43} = 0.541$$

$$X_{C_7H_8} = \frac{5.43}{5.43 + 6.41} = 0.459$$

Molality (m) is defined as the number of moles of solute per kilogram of solvent

$$m = \frac{\text{no. of moles solute}}{\text{no. kg solvent}} \qquad (10.3)$$

Example 10.3 Calculate the molality of a solution prepared by dissolving 1.00 g of urea, $CO(NH_2)_2$, in 48.0 g of water.

Solution. To obtain the number of moles of solute, we note that

$$1 \text{ mole } CO(NH_2)_2 = 12.0 \text{ g} + 16.0 \text{ g} + 2(14.0 \text{ g}) + 4(1.0 \text{ g}) = 60.0 \text{ g}$$

$$\text{no. moles solute} = 1.00 \text{ g urea} \times \frac{1 \text{ mole urea}}{60.0 \text{ g urea}} = 0.0167 \text{ mole urea}$$

$$\text{no. kg solvent} = 0.0480$$

$$m = \frac{0.0167}{0.0480} = 0.348$$

Students sometimes forget that a solution will have a definite concentration only after solute and solvent have been *well mixed*

As Example 10.3 implies, we can prepare a solution of a given molality by dissolving a known weight of solute in a predetermined weight of solvent. The precision with which the concentration is known is limited only by that of the balance used to make the weighings. An advantage of molality as a concentration unit is that it is independent of temperature (and pressure as well); a one molal solution prepared at 20°C will retain the same molality at 100°C, provided there is no loss of solute or solvent on heating.

Concentrations of reagents in the general chemistry laboratory are most often specified in terms of **molarity** (M), which is defined as the number of moles of solute per liter of solution, or

$$M = \frac{\text{no. moles solute}}{\text{no. liters solution}} \qquad (10.4)$$

Example 10.4
    a. How would you prepare 25 liters of 0.10 M $BaCl_2$ solution, starting with solid $BaCl_2$?
    b. What volume of the solution in (a) would you take to get 0.020 mole of $BaCl_2$?

Solution
    a. Using Equation 10.4, we can calculate the number of moles of $BaCl_2$ required.

$$\text{no. moles } BaCl_2 = (M \ BaCl_2) \times (\text{no. liters solution})$$

$$= 0.10 \ \frac{mole}{lit} \times 25 \ lit = 2.5 \ moles$$

To find out how many grams of $BaCl_2$ we should weigh out, we note that one mole of $BaCl_2$ weighs $137 \ g + 2(35.5) = 208 \ g$.

$$\text{no. g } BaCl_2 = 2.5 \ mole \times \frac{208 \ g}{1 \ mole} = 520 \ g$$

It follows that we should weigh out 520 g of $BaCl_2$ and add sufficient water to give a final volume of 25 liters, mixing well to make a homogeneous solution.

b. Solving Equation 10.4 for volume, we have

$$\text{no. liters solution} = \frac{\text{no. moles } BaCl_2}{M \ BaCl_2}$$

$$= \frac{0.020 \ mole}{0.10 \ mole/lit} = 0.20 \ lit \ (200 \ cc)$$

In dilute aqueous solution the molarity and molality of a solute are essentially identical. Why?

Since the volume of a liquid is more easily measured than its mass, laboratory reagents are usually made up to a specified molarity rather than a given molality. One way to make up a solution of the desired molarity is to start with the calculated weight of pure solute and add enough water to give the required volume, as illustrated by Example 10.4a. Another method, which is often more convenient, is to start with a concentrated solution and dilute it with water. The calculations involved are illustrated in Example 10.5.

Example 10.5 How would you prepare 100 ml of 2.0 M HCl, starting with concentrated hydrochloric acid (12 M)?

Solution. Clearly, the more concentrated solution (12 M) should be diluted with water to give a 2.0 M solution. The question is: What volume of 12 M HCl should we start with to prepare 100 ml of 2.0 HCl? The key to answering this question is to realize that *the number of moles of solute is not changed by dilution.* In the final solution, we have

$$2.0 \ \frac{mole}{lit} \times 0.100 \ lit = 0.20 \ mole$$

We must then take a sufficient volume of 12 M HCl to give us 0.20 mole. Proceeding as in Example 10.4b:

$$\text{no. liters 12 M HCl} = \frac{\text{no. moles HCl}}{M \ HCl} = \frac{0.20 \ mole}{12 \ mole/lit} = 0.017 \ lit \ (17 \ cc)$$

We deduce that we should start with 17 cc of 12 M HCl and dilute with water to a final volume of 100 ml to give a 2.0 M solution.

Still another concentration unit, which happens to be particularly useful in quantitative analysis, is **normality** (N), defined as the number of gram equivalent weights of solute per liter of solution.

$$N = \frac{\text{no. of GEW solute}}{\text{no. of liters solution}} \tag{10.5}$$

We shall postpone further discussion of the concepts of normality and gram equivalent weight to Chapter 18.

10.3 PRINCIPLES OF SOLUBILITY

Theory has not yet progressed to the point where we can predict the solubilities of even the simplest solutes in common solvents. The best that we can do is to apply general principles, based on structural considerations, to predict the relative solubilities of different solutes in a common solvent or, vice versa, the relative solubilities of a given solute in a series of solvents. The following brief discussion of solubility principles deals with solutions in the liquid state and is subdivided according to whether the solute is a liquid, a gas or a solid. Solubilities of ionic solutes in water are not included here; that topic is discussed in Chapter 16.

LIQUID-LIQUID

In discussing the solubility of two liquids in each other, it is sometimes stated that "like dissolves like." A more meaningful way of expressing this idea is to say that substances with similar molecular structures and, consequently, intermolecular forces of about the same magnitude will be soluble in each other in all proportions. To illustrate, consider the liquid aliphatic hydrocarbons of general formula C_nH_{2n+2} (C_5H_{12}, C_6H_{14} . . . $C_{18}H_{38}$), all of which are completely miscible with one another. Molecules of these nonpolar substances are held together by dispersion forces of about the same magnitude. The forces between C_5H_{12} molecules in pure liquid pentane are about as strong as those between C_5H_{12} and C_6H_{14} molecules in a solution of pentane in hexane. A pentane molecule readily passes into solution in hexane because it undergoes no significant change in environment in the solution process.

Moderate differences in polarity between solute and solvent seem to have little effect on solubility. Chloroform ($CHCl_3$, dipole moment = 1.15 Debye units) and carbon tetrachloride (CCl_4, dipole moment = 0) are soluble in each other in all proportions. Moreover, chloroform and carbon tetrachloride show similar solvent properties; both have been used in dry cleaning to remove grease, food and other organic stains from clothing. This implies that the intermolecular forces in $CHCl_3$ and CCl_4 are nearly equal, despite the considerable difference in polarity. Apparently dipole forces make only a minor contribution in $CHCl_3$; it is about as easy to "break into" the liquid structure of chloroform as it is in carbon tetrachloride.

We commonly observe that nonpolar substances have very small water solubilities. Petroleum, a complex mixture of hydrocarbons, spreads out in a thin film on the surface of a lake or an ocean rather than dissolving in the water. A typical nonpolar hydrocarbon, pentane, has a mole fraction solubility in water of only 0.00003. This is readily understood in terms of the structure of liquid water. In order to dissolve appreciable quantities of pentane in water, it would be necessary to break the hydrogen bonds holding water molecules together. There is no compensating attractive force between C_5H_{12} and H_2O to supply the energy required to break into the water structure.

Of the relatively few organic liquids which dissolve readily in water, the majority are oxygen-containing compounds of low molecular weight.

Two familiar examples are the alcohols containing one and two carbon atoms,

$$
\begin{array}{ccc}
& \text{H} & \\
& | & \\
\text{H} & - \text{C} - \text{O} - \text{H} & \\
& | & \\
& \text{H} &
\end{array}
\qquad\qquad
\begin{array}{ccc}
\text{H} & \text{H} & \\
| & | & \\
\text{H} - \text{C} & - \text{C} - \text{O} - \text{H} \\
| & | & \\
\text{H} & \text{H} &
\end{array}
$$

Methyl alcohol   Ethyl alcohol

both of which are soluble in water in all proportions. Methyl and ethyl alcohol each contain an —OH group, as does water. Even more important, both these compounds are known to be hydrogen-bonded in the liquid state. Consequently it is hardly surprising that they dissolve readily in water. One would expect the intermolecular forces between alcohols and water in solution to be roughly comparable to those in the pure liquids.

As the number of carbon atoms in the alcohol molecule increases, we find that the solubility in water decreases. The compound n-butyl alcohol, C_4H_9OH, has a limited solubility in water; its mole fraction in a saturated water solution at 20°C is only about 0.02. Octyl alcohol, $C_8H_{17}OH$, is extremely insoluble ($X = 0.0008$ in a saturated water solution at 20°C). The same trend is observed with many other types of organic compounds; there is a general tendency for solubility to decrease with an increase in chain length. This effect can be explained most simply in terms of the large number of hydrogen bonds that must be broken if a long-chain molecule is to be inserted into the water structure. A considerable amount of energy must be absorbed to break these bonds, making the solution process less spontaneous and decreasing the solubility.

Would you expect octyl alcohol to be soluble in octane, C_8H_{18}?

SOLID-LIQUID

In contrast to liquid-liquid pairs, where complete miscibility is common, solids are always found to have limited solubilities in liquids. If we add iodine (mp = 115°C) to carbon tetrachloride at 25°C, a saturated solution is formed when the mole fraction of I_2 is only 0.011. On the other hand, bromine, Br_2, which is a liquid at room temperature, is infinitely soluble in carbon tetrachloride; it is impossible to form a saturated solution of these two liquids.

To understand why solids should be less soluble in a given solvent than liquids of similar structure, it is helpful to visualize the process by which a solid A goes into solution in a solvent B as taking place in two steps:

1. Solid A melts to form pure liquid A:

$$A(s) \rightarrow A(l)$$

2. Liquid A dissolves in B to form a solution:

$$A(l) + B(l) \rightarrow \text{solution}$$

The first step is clearly nonspontaneous, since the dissolving solid (e.g., I_2) is at a temperature (25°C) below its melting point (115°C). It follows that the combination of Steps 1 and 2, which represent the overall solution

TABLE 10.2   Solubilities of Solid Hydrocarbons
in Benzene at 25°C *

SOLUTE	MELTING POINT (°C)	X SOLUTE
Anthracene	218	0.008
Phenanthrene	100	0.21
Naphthalene	80	0.26
Biphenyl	69	0.39

* Mole fraction of solid in saturated solution.

process for a solid, must be less spontaneous than Step 2 alone, the only process involved when a liquid dissolves.

Following this reasoning, we can deduce that the closer a solid is to its melting point, the more soluble it will be in a particular solvent. Putting it another way, low melting solids should be more soluble than high melting solids of similar structure. This argument is confirmed by data such as those in Table 10.2. Of the four hydrocarbon solutes, the most soluble is biphenyl, which is only 44° below its melting point at the solution temperature, 25°C. The least soluble is the high melting solid anthracene, which is nearly 200°C from its melting point at room temperature.

To estimate the relative solubilities of a given solid in different solvents, we take into account the principles discussed earlier in connection with liquid-liquid solubilities. Nonpolar solids are most soluble in solvents of low polarity and least soluble in hydrogen-bonded solvents such as water. An example is the well known pesticide DDT, which has a structure similar to that of carbon tetrachloride or chloroform:

$$Cl-\text{⟨benzene⟩}-\overset{\overset{\displaystyle H}{|}}{\underset{\underset{\displaystyle Cl}{|}}{\underset{Cl-C-Cl}{C}}}-\text{⟨benzene⟩}-Cl$$

DDT, like carbon tetrachloride or chloroform, is quite soluble in nonpolar or slightly polar organic solvents. This explains why it concentrates in the fatty tissue of fish, birds and game, often with lethal effects. In contrast, DDT is almost quantitatively insoluble in water; only about 10^{-6} g of DDT dissolves in a liter of water. Its low water solubility contributes to the persistence of this pesticide in the environment; DDT is not washed out of contaminated soil even by repeated rainfalls.

Suggest a good solvent for DDT

GAS-LIQUID

Following the reasoning outlined for solid-liquid and liquid-liquid solutions, we arrive at the following conclusions regarding the solubilities of gases in liquids:

1. The higher the boiling point of the gas (i.e., the closer it is to the liquid state), the more soluble it will be in a given solvent.

TABLE 10.3 Solubility of the Noble Gases in Benzene
and Water at 25°C and 1 Atm *

GAS	BOILING POINT (°C)	SOLVENT	
		Benzene	Water
He	−269	0.76×10^{-4}	0.069×10^{-4}
Ne	−246	1.14×10^{-4}	0.082×10^{-4}
Ar	−186	8.9×10^{-4}	0.25×10^{-4}
Kr	−152	27.3×10^{-4}	0.45×10^{-4}
Xe	−109	110×10^{-4}	0.86×10^{-4}
Rn	−62	310×10^{-4}	1.63×10^{-4}

* Mole fraction of gas in saturated solution.

2. The best solvent for a given gas will be the one whose inter-molecular forces are most similar to those of the gaseous solute.

Both of these principles are illustrated by the solubility data for the noble gases given in Table 10.3. Note the steady increase in solubility with boiling point (He < Ne < Ar < Kr < Xe < Rn) for both solvents. The reduced solubility of all these gases in water as compared to benzene reflects the strong intermolecular hydrogen bonding in water; in the nonpolar solvent benzene, as in the noble gases themselves, the attractive forces between particles are of the dispersion type.

10.4 EFFECT OF TEMPERATURE AND PRESSURE ON SOLUBILITY

The mutual solubilities of two substances A and B depend not only upon their structures and physical properties, but also upon the temperature and pressure. The effects of these two variables upon solubility are readily explained if we regard the solution process as an equilibrium:

$$A + B \rightleftarrows \text{solution}$$

and apply the principles discussed in Chapter 9 concerning the influence of temperature and pressure upon physical equilibria.

TEMPERATURE

In any equilibrium, an increase in temperature favors the endothermic process. This means that if heat is absorbed when A dissolves in B

$$A + B \rightleftarrows \text{solution} \quad \Delta H > 0$$

an increase in temperature will increase the solubility. Conversely, if dissolving A in B evolves heat

$$A + B \rightleftarrows \text{solution} \quad \Delta H < 0$$

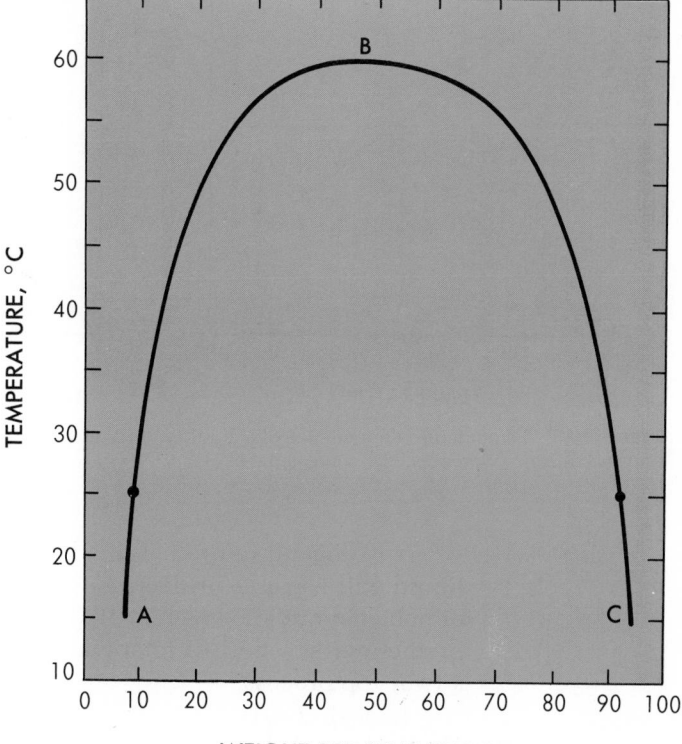

Figure 10.2 Solubility of hexane in aniline (Curve AB) and aniline in hexane (curve BC). Above 60°C (point B), the two liquids are miscible in all proportions.

an increase in temperature will favor the reverse process; i.e., it will reduce the solubility.

When two liquids with different intermolecular forces are mixed, the heat of solution is ordinarily a positive quantity. This explains why many liquid pairs which are only partially miscible at room temperature become infinitely soluble in each other when the temperature is raised. An example is the aniline-hexane system. When these two organic liquids are mixed at 25°C, two layers form. The aniline-rich layer contains only about 9 per cent by weight of hexane, while the other layer contains even less aniline, about 8 per cent by weight. As the temperature is raised, the mutual solubilities of the two liquids increase and the compositions of the two layers approach each other (Fig. 10.2). At 60°C, the two layers reach the same composition. Above this temperature only one layer forms, regardless of the relative amounts of aniline and hexane present. In other words, above 60°C, aniline and hexane are soluble in each other in all proportions.

It is possible to have a one phase aniline-hexane system at 25°C. How?

The solubility of a solid in a liquid solvent usually increases with temperature. An example is naphthalene, whose mole fraction solubility in benzene increases from 0.26 at 25°C to 0.54 at 50°C and to 1.0 (i.e., infinite solubility), at the melting point, 80°C. To explain this effect, it is convenient once again to imagine a two-step solution process:

    1. $A(s) \rightarrow A(l)$
    2. $A(l) + B(l) \rightarrow$ liquid solution

Step 1, the melting of the solid, is always endothermic ($\Delta H > 0$). If step 2 is also endothermic, as is ordinarily the case, ΔH will be positive for the overall solution process, and solubility will increase with temperature.

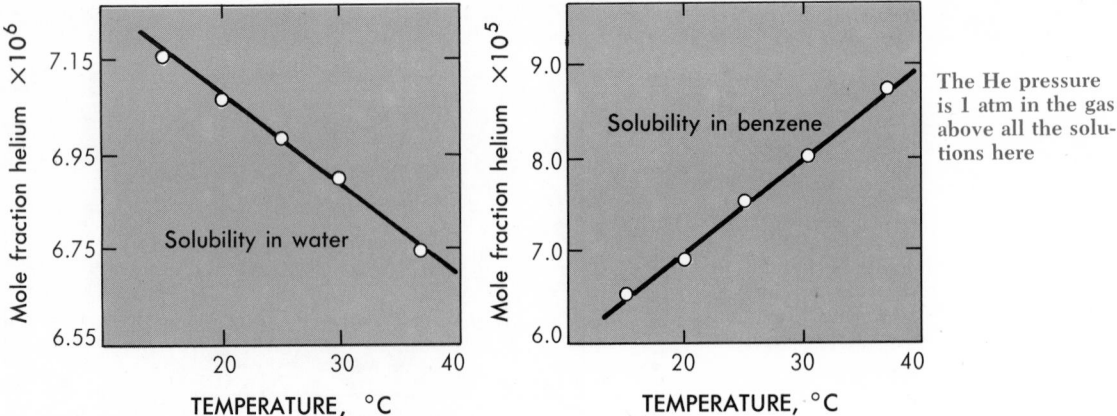

Figure 10.3 Temperature dependence of solubility of He in water (decreases with increasing T) and in benzene (increases with increasing T).

It is difficult to predict in advance whether the solubility of a gas in a liquid will increase or decrease with temperature. Examples of both types of behavior are known (Fig. 10.3). To appreciate the problems involved in predicting the effect of temperature on gas solubility, imagine, as before, a two-step process:

1. $A(g) \rightarrow A(l)$
2. $A(l) + B(l) \rightarrow$ liquid solution

Step 1, the condensation of a gas to a liquid, is exothermic. If Step 2 is endothermic, the sign of the overall heat of solution will depend upon the relative amounts of heat evolved in Step 1 and absorbed in Step 2. In the helium-benzene system, the heat absorbed in forming a cavity in the solvent large enough to accommodate a helium atom (Step 2) is numerically greater than the heat evolved when the gas condenses (Step 1). The overall solution process is endothermic and solubility increases with temperature. This behavior is characteristic of gases with relatively small heats of condensation (e.g., He, H_2, Ne) dissolving in nonpolar organic solvents.

In the helium-water system, ΔH for Step 2 is actually a negative quantity, for reasons which we shall examine in Chapter 11. Consequently the overall process is exothermic and solubility decreases with temperature. The aqueous solubility of most gases, like that of helium, decreases as the temperature is raised. This explains why bubbles of air form when a sample of water is heated. It is also a major factor in "thermal pollution," which occurs when surface waters are heated above their normal temperature. The increased temperature lowers the oxygen concentration in the water, thereby making it difficult for fish and other aquatic life to survive.

PRESSURE

Pressure has a major effect on solubility only for gas-liquid systems. We find that at moderate pressures the solubility of a gas is directly proportional to its partial pressure in the gas phase over the solution,

$$C_a = kP_a \tag{10.6}$$

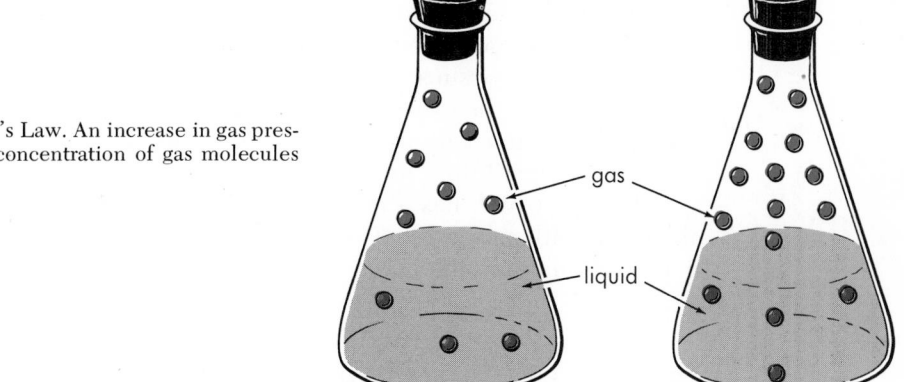

Figure 10.4 Henry's Law. An increase in gas pressure increases the concentration of gas molecules in both phases.

where P_a = partial pressure of gas, C_a = concentration of gas in solution, and k is a constant characteristic of the particular gas-liquid system. A simple kinetic explanation of Equation 10.6 (Henry's Law) is suggested in Figure 10.4.

Example 10.6 The solubility of pure oxygen in water at 20°C and one atmosphere pressure is 1.38×10^{-3} mole/liter. Calculate the concentration of O_2 in water saturated with air (mole fraction $O_2 = 0.210$) at 740 mm Hg and 20°C.

Solution. One way to solve this problem is to use the solubility data for pure oxygen to calculate k in Equation 10.6. Knowing k and the partial pressure of O_2 in air, we can then obtain the concentration of O_2 in water saturated with air.

$$k = \frac{\text{conc pure } O_2}{\text{pressure pure } O_2} = \frac{1.38 \times 10^{-3} \text{ mole/lit}}{1.00 \text{ atm}} = 1.38 \times 10^{-3} \frac{\text{mole}}{\text{lit atm}}$$

To obtain the partial pressure of O_2 in air, we recall from Dalton's Law (Chapter 5) that:

$$P_{O_2} = X_{O_2} \times P_{tot}$$
$$= 0.210 \times \frac{740}{760} \text{ atm} = 0.204 \text{ atm}$$

Applying Henry's Law to obtain the concentration of O_2 at this partial pressure

$$C_{O_2} = k \times P_{O_2}$$

$$= 1.38 \times 10^{-3} \frac{\text{mole}}{\text{lit atm}} \times 0.204 \text{ atm} = 2.82 \times 10^{-4} \frac{\text{mole}}{\text{lit}}$$

In the air the concentration of O_2 is 85×10^{-4} moles/lit

The influence of partial pressure upon gas solubility is utilized in the commercial production of carbonated beverages. Beer, certain wines and many soft drinks are bottled under a carbon dioxide pressure slightly greater than 1 atm. When the container is open, exposing the liquid to air, in which the partial pressure of CO_2 is very small (0 to .001 atm), the carbon dioxide bubbles out of solution, giving a froth or "head." Pressurized containers for whipped cream work on a similar principle. Opening a valve causes dissolved gas to come out of solution, carrying the liquid with it as a foam.

The excruciatingly painful and sometimes fatal affliction known as the "bends" is another consequence of the effect of pressure on gas solubility. Compressed air breathed by divers working underwater dissolves in the body fluids and tissues. If the individual is suddenly exposed to atmospheric pressure, the excess air comes out of solution as tiny bubbles which impair circulation and affect nerve impulses. One way to minimize this effect is to substitute a mixture of helium and oxygen for the compressed air. Helium, which has a much lower boiling point than nitrogen, is only about one fifth as soluble. Consequently much less gas comes out of solution on decompression.

10.5 COLLIGATIVE PROPERTIES OF NONELECTROLYTE SOLUTIONS

Certain properties of solutions are found to depend primarily upon the concentrations of solute particles rather than their nature. These are called **colligative properties**; they include vapor pressure lowering, boiling point elevation, freezing point depression and osmotic pressure. In this chapter we shall consider colligative properties of nonelectrolyte solutions, where the solute particles are molecules. Properties of electrolyte solutions (ionic solutes) are discussed in Chapter 11.

The laws relating colligative properties to solute concentration are presented below as Equations 10.7 to 10.11. They are best regarded as limiting laws, which are approached more and more closely as the solution becomes more dilute. In practice these equations will ordinarily be valid to within, at most, a few per cent at concentrations as high as 1 molal.

VAPOR PRESSURE LOWERING

Experience tells us that the equilibrium vapor pressure of solvent above a solution is lower than that of the pure solvent. Concentrated water solutions of substances such as sugar or urea evaporate more slowly than pure water, reflecting the lowering of the water vapor pressure by the presence of solute. Indeed, if the solute concentration is high enough, water vapor from the atmosphere may condense into the solution, thereby diluting it.

A quantitative study of vapor pressure lowering shows that it is a true colligative property; it is directly proportional to the concentration of solute but independent of the nature of the solute molecule. For example, the vapor pressure of water above a 0.10 molal solution of either sugar or urea is the same, about 0.043 mm Hg less than that of pure water. In a 0.20 molal solution, the vapor pressure lowering is twice as great, 0.086 mm Hg.

The relationship between solvent vapor pressure and concentration can be expressed as

$$P_1 = X_1 P_1^0 \qquad (10.7)$$

where P_1 is the vapor pressure of solvent over the solution, X_1 is the mole fraction of solvent in solution and P_1^0 is the vapor pressure of pure solvent at the same temperature. Note that since X_1 in a solution must be less than 1, P_1 must be less than P_1^0.

We can obtain a direct expression for the vapor pressure lowering by making the substitution $X_2 = 1 - X_1$, where X_2 is the mole fraction of solute.

$$P_1 = (1 - X_2)P_1^0$$

Rearranging
$$P_1^0 - P_1 = X_2 P_1^0$$

The quantity $(P_1^0 - P_1)$, the difference between the solvent vapor pressure in the pure liquid and in solution, is, by definition, the vapor pressure lowering, so

$$VPL = X_2 P_1^0 \qquad\qquad (10.8)$$

Example 10.7 Calculate the vapor pressure lowering of a solution containing 100 g of sugar, $C_{12}H_{22}O_{11}$, in 500 g of water at 25°C.

Solution. In order to use Equation 10.8, we need to know the vapor pressure of pure water (23.76 mm Hg at 25°C) and, X_2, the mole fraction of sugar. Since the molecular weights of sugar and water are 342 and 18.0, respectively, we have

$$X_2 = \frac{\text{no. moles } C_{12}H_{22}O_{11}}{\text{no. moles } C_{12}H_{22}O_{11} + \text{no. moles } H_2O} = \frac{100/342}{100/342 + 500/18.0} = 0.0104$$

From Equation 10.8, we have

$$VPL = 0.0104 \times 23.76 \text{ mm Hg} = 0.247 \text{ mm Hg}$$

(The vapor pressure of water over the solution would be: $(23.76 - 0.25)$ mm Hg = 23.51 mm Hg.)

Equation 10.7, known as Raoult's Law, can, in principle at least, be applied to solutions in which both components are volatile. If *both* components follow Raoult's Law, we have

$$P_{tot} = P_1 + P_2 = X_1 P_1^0 + X_2 P_2^0$$

Solutions which follow this relation are said to be **ideal**; an example of such a solution is that formed by benzene and toluene, two organic liquids of very similar structures. For an equimolar mixture ($X_1 = X_2 = 1/2$) at 20°C, where the vapor pressures of the pure liquids are 76 mm Hg (benzene) and 24 mm Hg (toluene), we calculate a total pressure of

Benzene   Toluene

$$P_{tot} = \tfrac{1}{2}(76 \text{ mm Hg}) + \tfrac{1}{2}(24 \text{ mm Hg}) = 50 \text{ mm Hg}$$

Experimentally we find that the total vapor pressure over the solution is 51 mm Hg, within 2 per cent of the calculated value.

In practice very few solutions in which both components are volatile behave ideally. Although the solvent ordinarily follows Raoult's Law in dilute solution, the solute seldom does so. It is not difficult to see why this should be the case. When a small amount of solute is dissolved in a large amount of solvent, the solute molecules find themselves in an environment quite different from that of the pure solute. The escaping tendency of the solute, which is reflected in its vapor pressure, is determined by the attractive forces between a solute molecule and the solvent particles

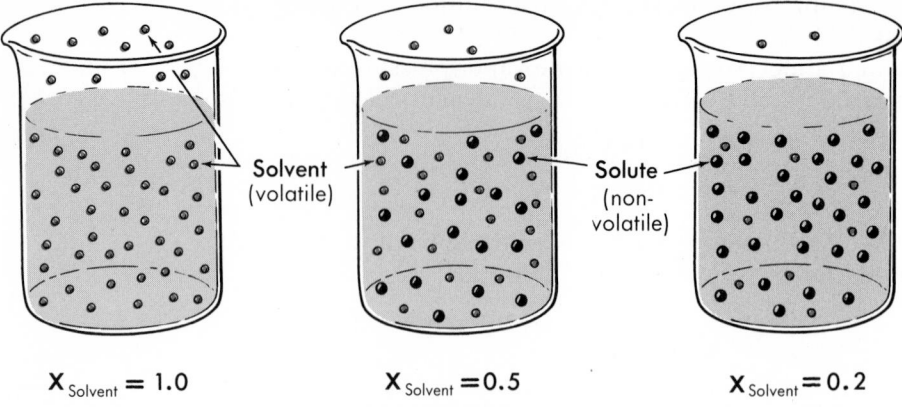

Figure 10.5 Raoult's Law. Adding solute decreases the concentration of solvent molecules in both phases.

which surround it. These forces will rarely be exactly equal to the solute-solute attractive forces, as they would have to be if Raoult's Law were to apply to the solute.

BOILING POINT ELEVATION

A solution of a *nonvolatile* solute invariably boils at a higher temperature than the pure solvent. In dilute solutions the boiling point elevation is found to be directly proportional to solute concentration. The relationship between these two variables can be written in the form

$$\Delta T_b = k_b \times m \qquad (10.9)$$

Volatile solutes, such as alcohol, might actually depress the boiling point

where ΔT_b is the boiling point elevation in °C, m is the molality, and k_b is a constant characteristic of the particular solvent. For water, k_b is 0.52°C. In other words, a 1 molal water solution of a *nonvolatile* solute (sugar, urea, ethylene glycol, but *not* methyl alcohol) has a boiling point 0.52°C higher than that of pure water (100.52°C at 760 mm Hg). Organic solvents ordinarily have values of k_b which are considerably larger than that of water (Table 10.4).

The elevation of the boiling point associated with solutions of nonvolatile solutes is readily explained in terms of a lowering of the vapor

TABLE 10.4   Molal Freezing Point and Boiling Point Constants

SOLVENT	FREEZING POINT	k_f	BOILING POINT	k_b
Water	0°C	1.86	100°C	0.52
Acetic acid	17	3.90	118	2.93
Benzene	5.50	5.10	80	2.53
Cyclohexane	6.5	20.2	81	2.79
Camphor	178	40.0	208	5.95

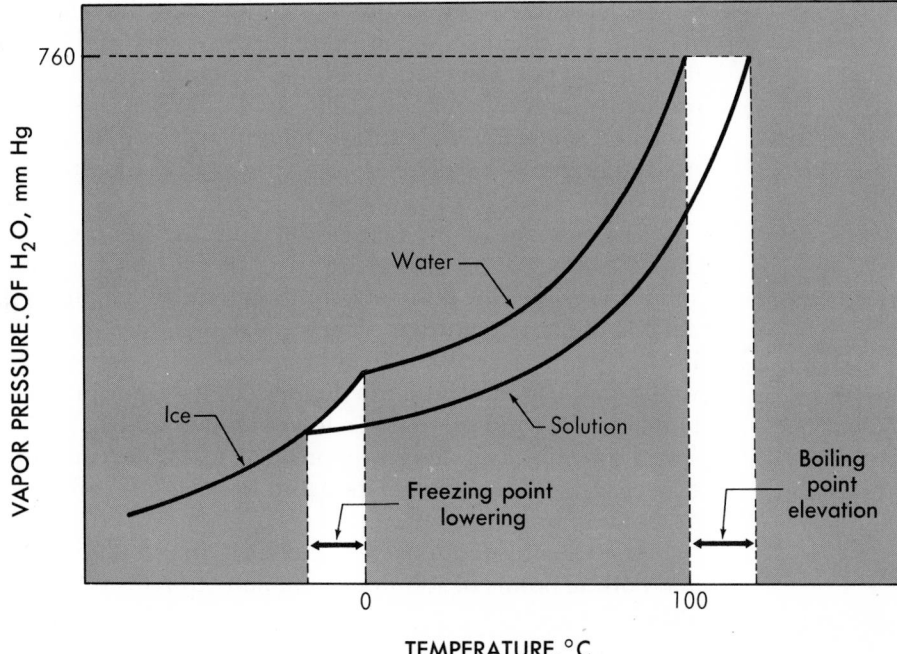

Figure 10.6 The lowering of solvent vapor pressure by addition of solute lowers the freezing point and raises the boiling point.

pressure. Since the solution, at any given temperature, has a vapor pressure lower than that of the pure solvent, a higher temperature must be reached before the solution boils, that is, before its vapor pressure becomes equal to the external pressure. Moreover, since the extent of vapor pressure lowering is directly proportional to solute concentration, we might expect to find the same type of linear relationship between boiling point elevation and solute concentration. Figure 10.6 illustrates this reasoning graphically.

FREEZING POINT LOWERING

Solutions containing small amounts of solute freeze or melt at temperatures below that of the pure solvent. For example, a mixture of naphthalene and biphenyl containing only 5 per cent by weight of biphenyl is completely melted at 77°C, some 3°C below the melting point of pure naphthalene (80°C). Organic chemists frequently make use of this behavior to check the purity of a solid which they have prepared in the laboratory. By comparing the melting point of their sample to that of an authentic sample of the pure solid, they get a rough estimate of the amount of impurity present.

The lowering of the freezing point in dilute solution, like the lowering of the vapor pressure and the elevation of the boiling point, is directly proportional to solute concentration. The relationship may be expressed as

$$\Delta T_f = k_f \times m \qquad (10.10)$$

where ΔT_f is the freezing point lowering, m is the molality, and k_f is a constant for a particular solvent. Note the resemblance between the equations for freezing point lowering and boiling point elevation. The constant k_f is ordinarily larger than k_b (cf. Table 10.4). For water, k_f is 1.86°C; for organic solvents, it is commonly a larger number. Camphor, a complex ketone of molecular formula $C_{10}H_{16}O$, has an unusually large freezing point constant, 40.0°C.

The freezing point depression, like the boiling point elevation, is a direct consequence of the lowering of the solvent vapor pressure by a solute. The freezing point of a solution from which pure solvent crystallizes out * is the temperature at which the solvent has the same vapor pressure in the two phases, liquid solution and solid solvent. Since the solvent vapor pressure in solution is depressed, its vapor pressure becomes equal to that of the solid solvent at a lower temperature. Again, since vapor pressure lowering is directly proportional to solute concentration, we might expect to find a similar relationship between freezing point lowering and concentration (Fig. 10.6).

The use of Equations 10.9 and 10.10 in calculations involving freezing and boiling points of solutions of nonelectrolytes is illustrated by Example 10.8.

Freezing point depression experiments can be done with volatile solutes

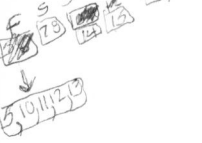

Example 10.8 Calculate the freezing point and boiling point, at 760 mm Hg, of a solution of 2.60 g of urea, $CO(NH_2)_2$ in 50.0 g of water.

Solution. To use Equations 10.9 and 10.10, we must first calculate the molality. Noting that the molecular weight of urea is 60.0, we have

$$\text{no. moles urea} = 2.60/60.0 = 0.0433$$

$$m = \frac{\text{no. moles urea}}{\text{no. kg water}} = \frac{0.0433}{0.0500} = 0.866$$

Hence:

$$\Delta T_b = 0.52°C \times 0.866 = 0.45°C; \quad bp = 100.00°C + 0.45°C = 100.45°C$$

$$\Delta T_f = 1.86°C \times 0.866 = 1.61°C; \quad fp = 0.00°C - 1.61°C = -1.61°C$$

We take advantage of the phenomenon of freezing point lowering when we add antifreeze to a car radiator to prevent the water from freezing. So-called permanent antifreeze uses ethylene glycol, an organic dialcohol which has the structure

$$\begin{array}{ccc} H & & H \\ | & & | \\ H-C & - & C-H \\ | & & | \\ OH & & OH \end{array}$$

Ethylene glycol can be used year round as a radiator coolant since it is relatively nonvolatile (b pt = 197°C) and raises the boiling point of water.

Another organic compound sometimes used as an antifreeze is methyl alcohol, CH_3OH. Methyl alcohol is cheaper than ethylene glycol. Moreover, since its molecular weight is only a little more than half that of ethy-

---

* If a solid solution separates on freezing, this argument will not be valid. In practice, it is almost always pure ice that separates when dilute aqueous solutions freeze.

TABLE 10.5   Freezing Points of Ethylene Glycol-Water Solutions °

WEIGHT % $C_2H_6O_2$	VOLUME % $C_2H_6O_2$	MOLALITY $C_2H_6O_2$	OBSERVED fp	CALC. fp (Eqn. 10.10)
10	9.2	1.79	−3.6°C	−3.3°C
20	18.3	4.03	−7.9	−7.5
30	28.0	6.91	−14.0	−12.9
40	37.8	10.8	−22.3	−20.0
50	47.8	16.1	−33.8	−30.0
60	58.1	24.2	−49.3	−45.0

° The ethylene glycol-water system reaches a minimum freezing point of −65°C at about 70 weight per cent of ethylene glycol; at higher concentrations the freezing point *rises* toward that of pure ethylene glycol, −16°C.

lene glycol (32 vs 62), its freezing point lowering per gram is nearly twice as great. The principal disadvantage of methyl alcohol is its volatility (b pt = 65°C); it lowers the boiling point of water and also tends to distill out of solution.

Table 10.5 compares the observed freezing points of ethylene glycol-water solutions with those calculated using Equation 10.10. Notice that up to 60 weight per cent, the measured freezing point is lower than the calculated; at high concentrations of ethylene glycol, the difference amounts to several degrees. The deviations occur because Equation 10.10 is a limiting law, valid only at low concentrations.

OSMOTIC PRESSURE

Imagine an experiment in which two beakers, one containing pure water, the other a sugar solution, are placed under a bell jar (Figure 10.7). As time passes, it is found that the liquid level in the beaker containing the solution rises, while the level of pure water in the other beaker falls. Eventually, by evaporation and condensation, all the water is transferred to the solution; at the end of the experiment, the beaker that contained pure water is empty.

The driving force behind the process just described is the difference in vapor pressure of water in the two beakers. Water moves spontaneously from a region in which its vapor pressure is high (pure water) to a region in which its vapor pressure is low (sugar solution). The air in the bell jar is permeable only to water molecules; the nonvolatile solute is unable to move from one beaker to the other.

The apparatus shown at the bottom of Figure 10.7 can be used to achieve a result similar to that found in the bell jar experiment. Here, the sugar solution is separated from the pure water by a semipermeable membrane, which may be an animal bladder, a slice of vegetable tissue (a hollow carrot works quite well) or a piece of parchment. This membrane, by a mechanism which is poorly understood, allows solvent but not solute molecules to pass through it. As before, water moves from a region in which its vapor pressure is high to a region in which its vapor pressure is low. This process, taking place through a membrane permeable only to the solvent, is referred to as **osmosis**.

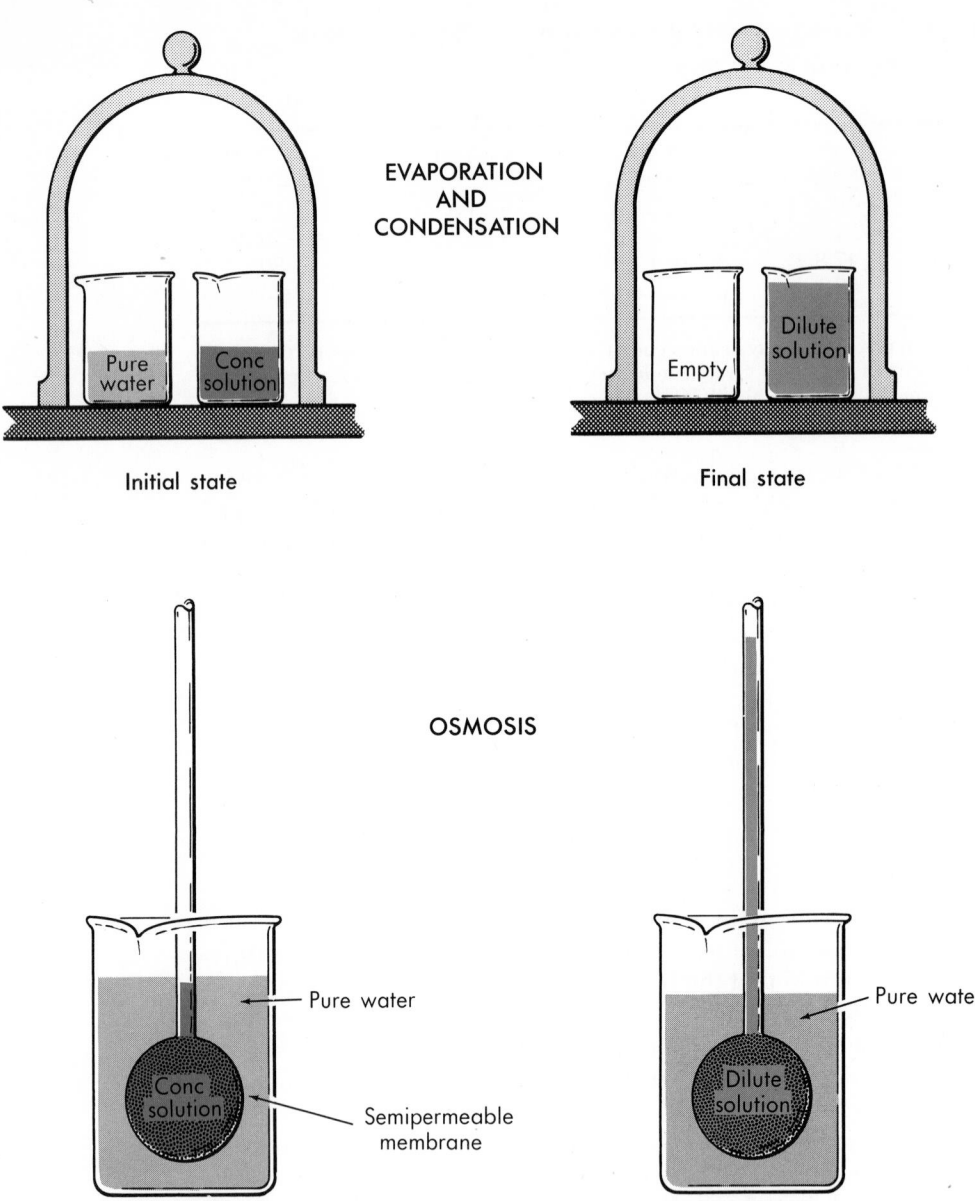

Figure 10.7 Water moves toward the region of lower vapor pressure, either by evaporation and condensation (upper diagram) or osmosis (lower diagram).

The passage of water molecules through a membrane into a solution may be prevented by applying pressure to the solution. The external pressure which is just sufficient to prevent osmosis is referred to as the **osmotic pressure** of a solution. Osmotic pressure may be measured in an apparatus such as that shown in Figure 10.8. The inner, porous tube A contains within it a strong, semipermeable membrane consisting of a film of copper ferrocyanide, $Cu_2Fe(CN)_6$. This insoluble film is formed by allowing solutions containing Cu^{2+} and $Fe(CN)_6^{4-}$ ions to diffuse into each other through the walls of the tube. The tube is filled with pure water; the compartment B surrounding the tube is filled with the solution whose osmotic pressure is to be measured. Pressure is applied to the solution at C so as to maintain a constant level D in the capillary attached to the tube containing pure water. This pressure is, by definition, the osmotic pressure.

The **osmotic pressure,** π, of a dilute solution is a colligative property in that it is directly proportional to the concentration of solute and independent of its nature. The equation relating osmotic pressure to solute concentration may be written in the form

$$\pi = MRT \qquad (10.11)$$

where π = osmotic pressure, M = molarity, R = gas constant (0.0821 lit atm/mole °K), and T = temperature in °K. (Notice the similarity to the Ideal Gas Law, with molarity, M, equal to number of moles per liter, n/V.)

Substitution into Equation 10.11 shows that the osmotic pressure, even in dilute solution, is relatively large. Consider, for example, a 0.10 M solution at 25°C, where we calculate

$$\pi = 0.10 \,\frac{\text{mole}}{\text{lit}} \times 0.0821 \,\frac{\text{lit atm}}{\text{mole °K}} \times 298°K = 2.45 \text{ atm}$$

The osmotic pressure is the *difference* between the pressure on the solution and the pressure on the pure solvent; the latter is usually 1 atm

A pressure of 2.45 atm is equivalent to a column of water 83 feet high! This gives some indication of the driving force behind osmosis and the difficulties of measuring osmotic pressures accurately using ordinary membranes.

Osmosis plays a vital role in many biological processes. Nutrient and waste materials pass by osmosis through the cell walls of animal tissues, which show varying degrees of permeability to different solutes. A striking

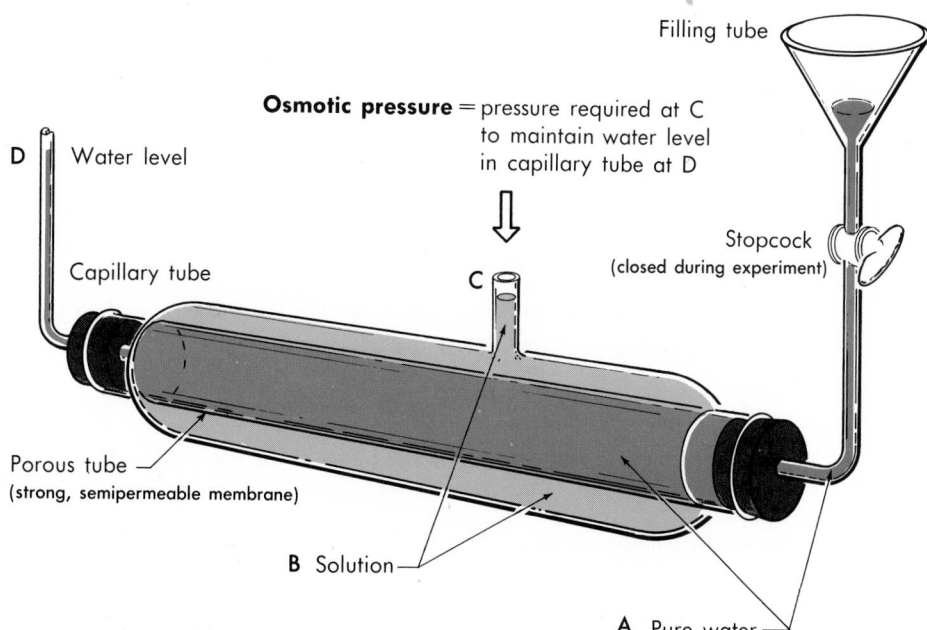

Figure 10.8 Measurement of osmotic pressure.

example of a natural osmotic process can be followed by watching under a microscope what happens when red blood cells are placed in pure water. Water passes through the walls to dilute the solution inside the cell, which swells and eventually bursts, releasing its red pigment. If the blood cells are placed in a concentrated sugar solution, water moves in the reverse direction; the cells shrink in size and shrivel up. To avoid these effects, nutrient solutions used in intravenous feeding are carefully adjusted so as to have the same osmotic pressure as blood (about 7.7 atm).

The walls of plant as well as animal cells act as semipermeable membranes. Flowers immersed in sugar or salt solution wilt as they are dehydrated by osmosis; if transferred to pure water, they regain their freshness as water moves back into the cells. It is generally believed that the flow of nutrients up a plant or tree is due to osmosis. The movement of water through the roots into the nutrient solution inside the plant cells can develop an osmotic pressure sufficient to push fluid to the top of the highest tree (cf. Problem 10.16).

DETERMINATION OF MOLECULAR WEIGHTS FROM COLLIGATIVE PROPERTIES

In Chapter 5 we discussed the determination of molecular weights from gas density data. This method works quite well for many gases and volatile liquids but is useless for substances which decompose on heating, such as sugar or urea. An alternative approach, applicable to a wide variety of nonelectrolytes, involves measuring one of the colligative properties of their solutions. The calculations involved are illustrated in Example 10.9.

Example 10.9 A student determines the molecular weight of a certain nonelectrolyte by weighing out 1.100 g of the solid and dissolving it in 20.0 g of benzene. He measures the freezing point of the solution to be 4.38°C. Knowing that pure benzene has a freezing point of 5.50°C and a k_f of 5.10, calculate the molecular weight of the solute.

Solution. Let us first calculate from the data the observed freezing point lowering. Then we can use Equation 10.10 to obtain the molality of the solution. Finally, from the defining equation for molality (10.3), we can calculate the molecular weight of the solute.

$$\Delta T_f = \text{fp pure benzene} - \text{fp solution} = 5.50°C - 4.38°C = 1.12°C$$

Applying Equation 10.10, we have

$$\Delta T_f = k_f \times m; \quad m = \frac{\Delta T_f}{k_f} = \frac{1.12°C}{5.10°C} = 0.220$$

But
$$m = \frac{\text{no. moles solute}}{\text{no. kg solvent}} = \frac{\text{no. g solute/GMW solute}}{\text{no. kg solvent}}$$

Rearranging
$$\text{GMW solute} = \frac{\text{no. g solute}}{(m)(\text{no. kg solvent})}$$

Substituting
$$\text{GMW} = \frac{1.100 \text{ g}}{(0.220)(0.0200)} = 250 \text{ g}$$

It is, of course, possible to determine molecular weights by measuring the vapor pressure lowering, boiling point elevation or osmotic pressure. Freezing point lowering is perhaps most commonly used because the effect is comparatively large (compare, for example, the constants for the freezing point lowering, 1.86°C, and the boiling point elevation, 0.52°C, for water solutions). Organic solvents are generally preferred to water because their k_f values are larger. Camphor, with $k_f = 40°C$, is a popular choice, although its high melting point (178°C) complicates the temperature measurement. Another organic solvent which has been used effectively is p-dichlorobenzene, which has a comparatively large freezing point constant (7.1°C) and a convenient melting point (53°C).

Osmotic pressure measurements are frequently used to determine the molecular weights of polymeric materials, in which the molar concentration of solute is ordinarily extremely low. Consider, for example, a solution containing 10 g of a polymer of molecular weight 10,000 dissolved in 1 kg of water. Such a solution would have a molality of 0.0010; its osmotic pressure would be approximately 0.025 atm (19 mm Hg) as compared to a freezing point lowering of only 0.0019°C. The principal difficulty associated with osmotic pressure measurements has always been that of finding a membrane which is both semipermeable and strong enough to withstand pressure. Instruments are now on the market which are capable of measuring osmotic pressures of solutions of polymers covering a wide range of molecular weights.

Osmotic pressure measurements probably give the best values for the MW's of polymers

PROBLEMS

10.1 How would you explain to a high school chemistry student why

    a. supersaturated solutions form on cooling rather than heating?

    b. gases are more soluble at high pressures?

    c. the freezing point of a substance is depressed by impurities?

10.2 Write out instructions for an untrained laboratory technician telling him how to

    a. prepare a supersaturated solution of sodium thiosulfate, $Na_2S_2O_3$ (its solubility at 20°C is 70 g/100 ml water).

    b. prepare 10 liters of 0.50 M NaOH, starting with the solid.

    c. set up a lecture demonstration to illustrate osmosis.

10.20 How would you explain to a nurse

    a. why she must not inject pure water into a patient's blood?

    b. why water is a poor solvent for iodine?

    c. why salads made with salt and vinegar wilt in a few hours?

10.21 Describe in some detail how you would

    a. prepare one liter of 0.10 M NaOH, starting with 6.0 M NaOH.

    b. determine the molecular weight of insulin, a high molecular weight protein.

    c. determine which of two aqueous solutions has the lower vapor pressure.

10.3 Using the data in Table 10.1, calculate

    a. the weight of HCl in 100 g of concentrated hydrochloric acid.

    b. the weight of HCl in 100 ml of concentrated hydrochloric acid.

    c. the volume of water which must be added to a liter of concentrated HNO_3 to form dilute HNO_3 (neglect the volume change on mixing).

10.4 A water solution contains 12.0 g of sugar, $C_{12}H_{22}O_{11}$, in 200 ml. The density of this solution is 1.022 g/ml. Calculate

    a. the molarity of sugar.

    b. the molality of sugar.

    c. the weight per cent of sugar.

    d. the mole fraction of sugar.

10.5 A solution is made up by dissolving 12.0 g of sulfuric acid, H_2SO_4, to form 800 ml of solution. Determine

    a. the molarity of H_2SO_4.

    b. the number of grams of H_2SO_4 in 120 ml of this solution.

    c. the volume of solution which must be taken to give 0.100 mole of H_2SO_4.

10.6 Assuming no volume change on mixing, how much water must be added to a solution containing 10.0 g of ethyl alcohol, C_2H_5OH, in 50 ml of water to give a one molar solution?

10.7 The "proof" of an alcoholic beverage is defined as twice the percentage by volume of alcohol, C_2H_5OH. How many grams of alcohol are there in one quart of 90 proof Scotch whiskey? (d alcohol = 0.80 g/ml; 1 liter = 1.057 qt.)

10.8 The solubility of carbon dioxide in sea water is 3.0×10^{-2} mole/liter at 25°C and one atmosphere pressure. Calculate the molarity of CO_2 in sea water in equilibrium with air in which the partial pressure of CO_2 is 0.50 mm Hg.

10.9 The mole fractions of Cl_2, Br_2 and I_2 in a saturated solution in carbon tetrachloride at 20°C and one atmosphere are 0.187, 1.00 and 0.010, respectively. Explain the relative magnitudes of these numbers in terms of solubility principles.

10.22 The solution of hydrogen peroxide sold as a bleach and disinfectant contains 3.0% by weight of H_2O_2 and has a density of 1.0 g/ml.

    a. What is the weight of H_2O_2 in one ml of this solution?

    b. What volume of oxygen, at 25°C and one atm, is liberated when one ml of this solution decomposes? The reaction is $H_2O_2(aq) \rightarrow H_2O(aq) + \frac{1}{2} O_2(g)$. The label on the bottle says that it forms ten times its own volume of oxygen.

10.23 A solution prepared by dissolving 20.0 g of ethyl alcohol, C_2H_5OH, in 60.0 g of carbon tetrachloride, CCl_4, has a density of 1.28 g/ml. The density of pure CCl_4 is 1.51 g/ml. Calculate

    a. the weight per cent of ethyl alcohol.

    b. the mole fraction of ethyl alcohol.

    c. the molality of ethyl alcohol.

    d. the molarity of ethyl alcohol.

10.24 A saturated solution of hydrogen sulfide in water is approximately 0.10 M in H_2S.

    a. How many grams of H_2S are there in one liter of the solution?

    b. What volume of this solution must be taken to give 0.0025 mole of H_2S?

    c. How many milliliters of $H_2S(g)$, at 25°C and one atm, are required to saturate one liter of solution?

10.25 Derive a formula for the volume of water, V_w, which must be added to V_c ml of concentrated solution of molarity M_c to give a dilute solution of molarity M_d. Assume no change in volume on dilution.

10.26 The solubility of the pesticide lindane, $C_6H_6Cl_6$, in water is about 7.5 parts per million (i.e., 7.5 g of lindane dissolves in 10^6 g of water). What is the weight per cent of lindane in a saturated aqueous solution? the molality?

10.27 The solubility of nitrogen in water at 25°C and 1 atm is 6.4×10^{-4} mole/liter. How many grams of N_2 are there in 10 liters of rainwater saturated with air ($X_{N_2} = 0.78$) at 25°C?

10.28 Arrange each of the following sets of compounds in order of increasing water solubility.

    a. H_3C—O—CH_3, H_3C—OH, H_3C—CH_2—O—CH_2—CH_3 (all liquids)

    b. CO_2, H_2, H_2O_2

10.10 For each of the following pairs of substances, predict which will be the more soluble, first in benzene and then in water:

    a. C_2H_6 or C_3H_8.
    b. Chloroform ($CHCl_3$) or iodoform (CHI_3, mp = 119°C).
    c. KCl or CCl_4.

    d. Formic acid, $H-\overset{\overset{O}{\|}}{C}-OH$, or butyric acid, $H_3C-CH_2-\overset{\overset{O}{\|}}{C}-OH$, both of which are liquids.

10.11 Calculate the vapor pressure of water over a solution containing 50.0 g of sugar, $C_{12}H_{22}O_{11}$, in 50.0 g of water at 100°C.

10.12 The two liquids n-hexane and n-heptane form an ideal solution. Their vapor pressures at 30°C are 189 and 58 mm Hg, respectively. What is the total vapor pressure over a solution of these two liquids at 30°C in which the mole fraction of n-hexane is 0.48?

10.13 Calculate the boiling point at 760 mm Hg and the freezing point of the following solutions:

    a. 10.0 g of citric acid, $C_6H_8O_7$, in 50.0 g of water.
    b. 1.96 g of iodoform, CHI_3, in 10.0 g of benzene.

10.14 Tetrahydrofuran, C_4H_8O, has been suggested as an antifreeze. How many grams of tetrahydrofuran would have to be added to water to give the same freezing point lowering as one g of ethylene glycol, $C_2H_6O_2$? If the prices of tetrahydrofuran and ethylene glycol are $1.40/lb and $0.83/lb, respectively, which would be the cheaper antifreeze?

10.15 What is the osmotic pressure of a solution containing 3.80 g of sugar, $C_{12}H_{22}O_{11}$, in one liter of solution at 20°C?

10.16 If osmosis is responsible for sap rising in a tree, estimate the height in feet to which sap can rise at 25°C if it is 0.15 M in sugar and the water outside the sap-bearing tubule contains nonelectrolytes at a concentration of 0.01 M (1 cm Hg exerts the same pressure as 13.6 cm of water).

10.29 You are asked to prepare pure phenanthrene ($C_{14}H_{10}$, mp = 100°C) by recrystallization from a 50-50 mole % mixture with naphthalene ($C_{10}H_8$, mp = 80°C).

    a. Explain why toluene, (C_7H_8), would be a better choice as a solvent for recrystallization than water.
    b. To what temperature would you heat the solvent to make sure all the solid dissolved?
    c. When the hot solution cools, which solid will crystallize out first?
    d. From the data in Table 10.2, would you expect a single recrystallization to give a good yield of pure phenanthrene?

10.30 How many grams of urea, $CO(NH_2)_2$, must be added to 100 g of water to give a solution with a vapor pressure one mm Hg less than that of pure water at 25°C? at 50°C? (vp water at 25°C = 23.8 mm Hg, at 50°C = 92.5 mm Hg.)

10.31 The vapor pressures of benzene and toluene at 90°C are 1016 mm Hg and 405 mm Hg, respectively. The two liquids form an ideal solution. What must be the mole fraction of benzene in a toluene solution which boils at 90°C at 760 mm Hg?

10.32 How many grams of the following substances would have to be dissolved in 100 g of water to give a solution freezing at −1.00°C? What would be the boiling points of these solutions at 760 mm Hg?

    a. urea, $CO(NH_2)_2$
    b. glucose, $C_6H_{12}O_6$

10.33 Isopropyl alcohol, C_3H_8O, is used as an antifreeze in windshield washers. What volume of isopropyl alcohol (d = 0.79 g/ml) should be added to one quart of water to give a solution which would not freeze above 0°F?

10.34 The osmotic pressure of blood at body temperature, 98.6°F, is about 7.7 atm. Taking the solutes in blood to be nonelectrolytes, estimate their total concentration.

10.35 Of the three solutions: 1.0 M sugar in water, 1.0 M methyl alcohol in water, 1.0 M methyl alcohol in benzene, which will have the same

    a. freezing point?
    b. boiling point?
    c. osmotic pressure?

10.17   A student finds that a solution of 1.26 g of a certain nonelectrolyte in 15.0 g of water freezes at −0.46°C. Calculate the molecular weight of the nonelectrolyte.

10.18   The freezing point of p-dichlorobenzene is 53.1°C; its k_f value is 7.10. A solution of 1.00 g of sulfanilamide (one of the sulfa drugs) in 15.0 g of p-dichlorobenzene freezes at 50.2°C. What is the molecular weight of sulfanilamide?

10.19   The osmotic pressure of a solution prepared by dissolving 1.0 g of hemoglobin per 100 ml of solution is 2.75 mm Hg at 25°C. Estimate the molecular weight of hemoglobin.

10.36   Progesterone, a female hormone, is found on analysis to contain 9.5% H, 10.2% O and 80.3% C. A solution of 1.00 g of progesterone in 10.0 g of benzene freezes at 3.88°C. What is the molecular formula of progesterone?

10.37   Nicotine has the empirical formula C_5H_7N. A solution of 0.50 g of nicotine in 12.0 g of water boils at 100.14°C at 760 mm Hg. What is the molecular formula of nicotine?

10.38   A saturated solution of a certain protein in water contains 4.60 g of solute per liter. The solution has an osmotic pressure of 3.10 mm Hg at 20°C. What is the molecular weight of the protein? Estimate the freezing point of the solution.

*10.39   Fifty kilograms of lindane, an agricultural pesticide, is spread in several applications on a field 50 meters × 50 meters. An inch of rain falls on the field; half of the water runs off into a nearby river. Assume the run-off water is saturated with lindane (solubility = 7.5 parts per million). What fraction of the lindane has been washed off the field? If the average annual rainfall is 30 in, how long will it take to wash off the 50 kg of lindane?

*10.40   The concentration of gold in sea water is estimated to be 4×10^{-9} g/liter. The total volume of the North Sea is estimated to be 5.4×10^4 km³. With gold selling at $35 per ounce, what is the total value of the gold in the North Sea? (The famous chemist Fritz Haber proposed to pay off Germany's World War I debt by extracting gold from the ocean; the process proved not to be economically feasible.)

*10.41   A beaker containing 1.68 g of sugar, $C_{12}H_{22}O_{11}$, in 20.0 g of water and another containing 0.815 g of vitamin C in 24.0 g of water are placed in an evacuated container. At equilibrium the total mass of the sugar solution is 24.2 g. What is the molecular weight of vitamin C?

*10.42   The water-soluble form of the drug heroin has a molecular weight of 423. A 0.100 g sample of this drug mixed with sugar (MW = 342) is added to 1.00 g of water to give a solution freezing at −0.528°C. Estimate the weight per cent of heroin in the sample.

*10.43   A martini, weighing about 5.0 ounces, contains 30% by weight of alcohol. About 15% of the alcohol in the martini passes directly into the blood stream, which, for an adult, has a volume of about 7.0 liters. Estimate the concentration of alcohol in the blood, in g/ml, of a person who drinks two martinis before dinner. (A concentration of 0.0030 g/ml is frequently taken as indicative of intoxication in a "normal" adult.)

11 WATER, PURE AND OTHERWISE

Life as we know it depends on water, which is by far the most abundant substance in plant and animal tissues. Water accounts for about 70% of the weight of a human being; fresh fruits and vegetables contain from 70 to 95% water. The chemical reactions that sustain life take place in aqueous solution; many of them involve water as a reactant. An important example of such a reaction is photosynthesis, the process by which green plants synthesize complex organic molecules from carbon dioxide and water.

$$x\ CO_2(g) + y\ H_2O(l) \rightarrow C_xH_{2y}O_y(s) + x\ O_2(g) \tag{11.1}$$

Man has always been aware of his urgent need for fresh water. When this country was settled in the 17th century, centers of population grew up at the mouths of rivers: the Charles in Massachusetts, the Hudson in New York, the Delaware in Pennsylvania, and the James and the Potomac in Virginia. At that time, in a predominantly agricultural society, man needed relatively small amounts of water for his crops, his livestock, and his own life processes. Today, much of the demand for water comes from industry, where 250 kg of water are required to produce a kilogram of paper, 300 kg of water for every kilogram of steel, and 600 kg for every kilogram of fertilizer.

Fortunately our planet is well endowed with water in its solid, liquid and gaseous forms. Over 80% of the earth's surface is covered by water—as a relatively pure liquid in lakes and rivers, as a dilute salt solution in the

TABLE 11.1 Distribution of the Earth's Water

	MASS	% OF TOTAL	% OF FRESH WATER
Oceans	$13{,}000 \times 10^{20}$ g	97.3	—
Ice and snow	250×10^{20}	1.9	83.0
Underground	50×10^{20}	0.4	16.6
Lakes	1.2×10^{20}	0.009	0.40
Atmosphere	0.17×10^{20}	0.0013	0.056
Rivers and streams	0.02×10^{20}	0.0002	0.007

ocean, or as a nearly pure solid in snowfields, glaciers and the polar ice caps. Huge quantities of water are available in underground deposits; smaller but still appreciable quantities appear as clouds or vapor in the atmosphere. The total amount of water above, below, and on the surface of the earth is estimated to be 1.33×10^{24} kg, about 5% of the earth's total mass.

Despite the enormous quantities potentially available, water pure enough to meet man's needs is in short supply in many parts of the world. In part, this is a consequence of changing patterns of industrial and agricultural development; the concentration of industry in urban areas and the increasing use of irrigation to grow crops in arid regions create a demand for fresh water that cannot be met from local supplies. Even more critical is the fact that most surface waters are too impure to meet the needs of society. Dissolved salts in the oceans make sea water unsuitable for human consumption or for most industrial and agricultural purposes. To an increasing extent, we are compounding this problem by contaminating our lakes and rivers with a variety of pollutants which make their waters unfit for our use.

Lack of water is the main reason many otherwise attractive regions in this country remain unsettled

A principal objective of this chapter will be to discuss how water, contaminated by natural processes or by man, can be purified to meet our needs. Before doing this, however, it will be helpful to consider the properties and structure of water itself and of aqueous salt solutions.

11.1 PHYSICAL PROPERTIES OF WATER

Water is so familiar that we sometimes fail to appreciate its unusual properties. Its boiling point, for example, is some 260° higher than that of methane, even though these two compounds have comparable molecular weights (18 or 16). Another way to emphasize the peculiar behavior of water is to compare its properties to those of a "normal" liquid, n-heptane, which happens to have a boiling point, 98°C, very close to that of water.

The rather exceptional properties of water have a profound effect upon our environment. Its high specific heat helps to prevent large variations in the surface temperature of the earth. Oceans and lakes absorb solar heat during the day and release it to the atmosphere at night without undergoing an appreciable temperature change. On the moon, whose surface

TABLE 11.2 Physical Properties of Water vs n-Heptane

	WATER	n-HEPTANE
Specific heat (cal/g°C)*	1.00	0.49
Surface tension (erg/cm²)	72.8	19.2
ΔH of vaporization (cal/g)	540	76
ΔH of fusion (cal/g)	79.7	34
ΔV of fusion (ml/g)	−0.093	+0.079

* Specific heats and surface tensions are at 20°C, heats of vaporization at the boiling point, heats and volume changes of fusion at the freezing point.

consists of rocky material with a specific heat only about one fifth that of water, the temperature can vary from 120°C at noon to −150°C at midnight.

Another factor holding the earth's temperature relatively constant is the high heat of vaporization of water which, on a cal/g basis, is greater than that of any other liquid. About one third of the solar energy which reaches the surface of the earth is dissipated by vaporizing water from oceans, lakes, rivers and ice fields. The same mechanism is in large part responsible for maintaining the temperature of our bodies within very narrow limits. Much of the heat generated by metabolism is consumed by the evaporation of water through the pores of the skin. If water had the same heat of vaporization per gram as n-heptane, we might have to consume seven times as much of it to keep cool on a hot summer day.

The water in clouds acts as an insulating blanket at night by absorbing energy radiated from the earth

Water is one of the very few substances whose solid form, ice, is less dense than the liquid. An ice cube floats upon a glass of water; more important, the ice that forms on lakes in winter remains at the surface. If ice were the denser phase, it would sink to the bottom where it would melt very slowly, perhaps incompletely, during the spring and summer. Even more unique is the volumetric behavior of liquid water, which *contracts* when heated above 0°C, reaching a maximum density at 4°C. Figure 11.1 shows the changes in volume of one gram of water in the range −10 to +10°C. Normal behavior is indicated in the volume-temperature plot for n-heptane.

In Chapter 8 we rationalized the abnormally high melting and boiling points of water in terms of hydrogen bonding. Certainly this unusually

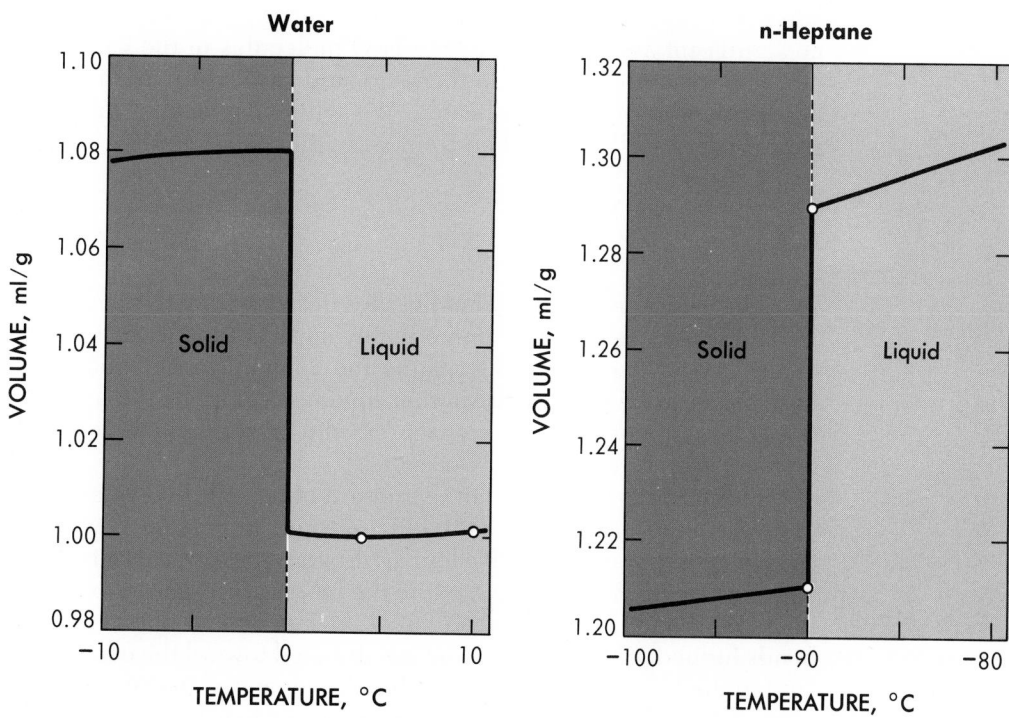

Figure 11.1 Changes in volume of one gram of water (left) and one gram of n-heptane (right) near the freezing point. Water decreases in volume upon melting; liquid water continues to contract from 0 to 4°C.

TABLE 11.3   Properties of Ammonia, Water and Hydrogen Fluoride

	NH₃	H₂O	HF
Boiling point (°C)	−33	+100	+19
Freezing point (°C)	−78	0	−92
Heat of vaporization (cal/g)	330	540	360
ΔV of fusion (ml/g)	+0.14	−0.093	+0.22

strong type of intermolecular force has a pronounced effect upon the properties we have been discussing. However, it is worth pointing out that the *properties of water are out of line even when compared to other hydrogen-bonded liquids.* From Table 11.3 we see that the boiling point, freezing point and heat of vaporization of water are significantly higher than those of ammonia or hydrogen fluoride, two substances which have molecular weights comparable to water and form strong hydrogen bonds. Moreover, neither ammonia nor hydrogen fluoride contracts upon melting as ice does. Clearly, although hydrogen bonding helps to explain the properties of water, it is not sufficient to stop there without inquiring further into the structure of liquid water and ice.

11.2   THE STRUCTURE OF WATER

Of the three physical states in which water can exist, the vapor, as we might expect, has the simplest structure. From gas density data, we calculate a molecular weight of 18 for water vapor. This implies that there is no significant association between H_2O molecules in the gas.

The structure of water in the solid, and particularly in the liquid, state is considerably more complicated. We will consider first the structure of ice, which is well established from x-ray diffraction studies.

ICE

The arrangement of H_2O molecules in an ice crystal is shown in Figure 11.2. One feature which is immediately apparent is the hexagonal pattern of the crystal itself, which explains the common observation that snowflakes tend to be six-sided. Another obvious feature of the ice structure is the large amount of "empty space" in the crystal, which accounts for its low density.

The stability of the "open" structure of ice can be explained in terms of the geometry around an individual oxygen atom (Fig. 11.3). Notice that the oxygen is surrounded by four hydrogens: two of these are covalently bonded at a distance of 0.99 Å while the others are joined to oxygen through hydrogen bonds at a distance of 1.77 Å. The crucial point is that the four bonds formed by an oxygen atom are directed toward the corners of a regular tetrahedron, which is precisely the geometry preferred by an atom surrounded by an octet of electrons. In other words, the open, hydrogen-bonded structure of ice is one which satisfies the requirement of fourfold, tetrahedral coordination around oxygen.

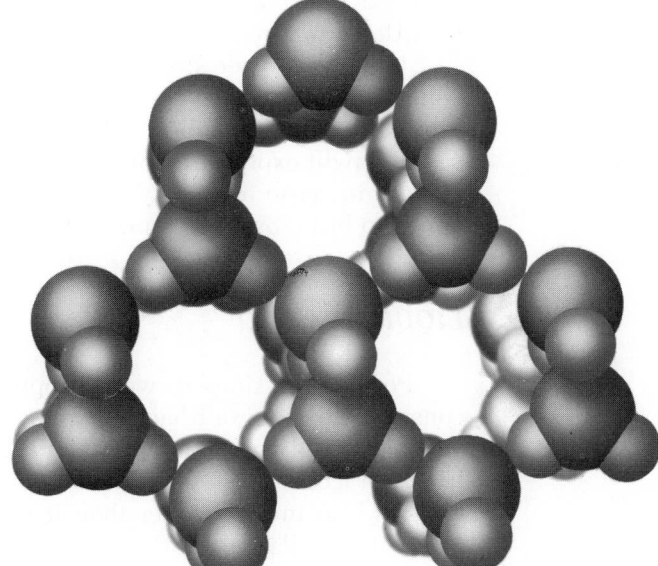

Figure 11.2 Crystal structure of ice. The large circles represent oxygen atoms, the small ones hydrogen. Note the hexagonal pattern and the "open" structure.

The crystal structure of ice is quite different from that of solid NH_3 or HF, in which the atoms are more closely packed. The H_2O molecule, with two hydrogen atoms and two unshared pairs of electrons, is unique in its ability to form three-dimensional networks held together by hydrogen bonds. The HF molecule, with three unshared pairs of electrons and only one hydrogen atom, cannot do this; at most, hydrogen bonding in HF can only lead to chains (one dimension) or rings (two dimensions). The same restriction applies to NH_3, which has only one unshared pair of electrons. Now we can begin to understand why hydrogen bonding has a more pro-

Figure 11.3 Tetrahedral arrangement of four hydrogens around a central oxygen atom in ice.

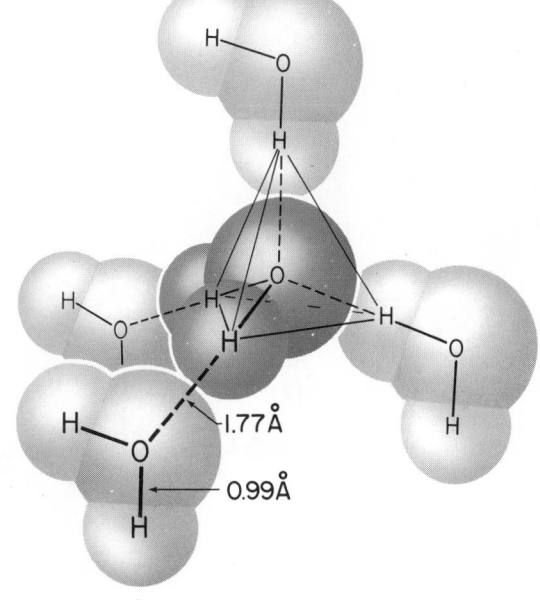

nounced effect on the properties of water than it does with hydrogen fluoride or ammonia.

The ordinary ice structure shown in Figure 11.3 is not the only way in which H_2O molecules can pack to give a stable, three-dimensional structure. A total of eight other forms of ice, stable at high pressures, are known. As one might expect, the high-pressure forms are more dense than ordinary ice; the increase in density is achieved by slightly distorting the hydrogen bonds to bring oxygen atoms closer to one another.

These forms become stable at pressures between 2000 and 20,000 atm

LIQUID WATER

Perhaps the simplest way to approach the structure of liquid water is to consider what doesn't happen when ice melts! The heat of fusion of ice, 80 cal/g, is only a fraction of the energy needed to break all the hydrogen bonds in ice, estimated to be about 500 cal/g. The volume change, −0.093 ml/g, is also much smaller than it would be if the "open" ice structure completely collapsed. If the molecules in water were close-packed, as they are in most liquids, its density would be nearly twice the observed value.

These observations imply that the structure of liquid water may resemble that of ice. W. K. Roentgen suggested in 1892 that water is a mixture of "ice-like" and "vapor-like" molecules. He does not seem to have pursued this idea further, perhaps because his discovery of x-rays three years later distracted him. For many years, however, it was believed that liquid water, at least near 0°C, consists largely of clusters of H_2O molecules with a geometry identical to that of a small portion of an ice crystal. Such a model has one serious weakness; it seems inconsistent with the fact that very pure water can be supercooled to −40°C without freezing. It is difficult to believe that this could happen if liquid water contains what would amount to microcrystals of ice. Nowadays people who propose models for

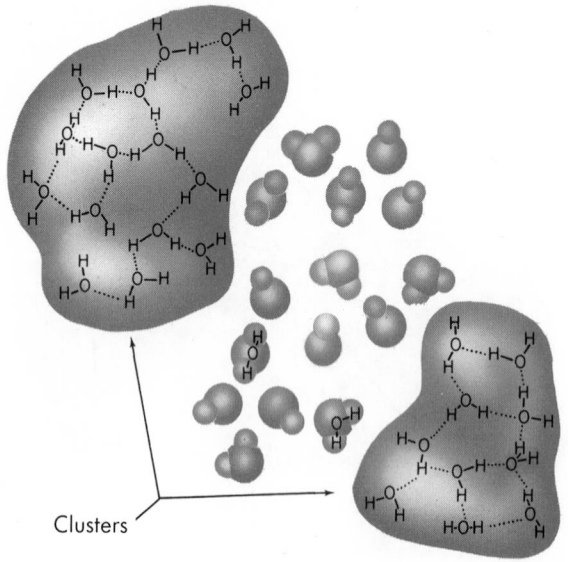

Figure 11.4 "Flickering clusters" of hydrogen-bonded H_2O molecules in liquid water. Relatively "free" H_2O molecules are shown between the clusters.

Clusters

water structure usually hedge their bets by referring to "ice-like" clusters. The implication is that the H_2O molecules in these clusters are bonded together in a pattern similar to but not identical with that of ordinary ice.

Perhaps the most popular model of this type is that developed by Henry Frank at the University of Pittsburgh in the 1950's. The model pictures open clusters of H_2O molecules, held together by hydrogen bonds, swimming in a sea of relatively "free" water molecules (Fig. 11.4). The clusters come in a variety of shapes and sizes with a geometry which may be quite different from that of ice. They are supposed to be constantly collapsing and re-forming, giving rise to the picturesque description "flickering clusters."

One attractive feature of this model is its ability to explain the changes in volume undergone by water near the freezing point. The gradual collapse of the open clusters upon heating is balanced by a slight increase in the distance between H_2O molecules, leading to a minimum in the volume-temperature curve at about 4°C (Fig. 11.1). The model also explains the peculiar observation, noted in Chapter 10, that heat is evolved when nonpolar molecules such as Ar and CH_4 mix with water. Such molecules stabilize the "flickering clusters" by occupying the space within them. In this way nonpolar species promote the formation of clusters at the expense of free H_2O molecules. This process would be expected to be exothermic, since it involves the formation of hydrogen bonds.

This model also helps to explain the high specific heat of water. How?

There is, to be sure, a disturbing vagueness associated with this model. We have no clear picture of the geometry of a "flickering cluster," nor do we know precisely what is meant by a "free" water molecule. Many other models, some of them much more specific in nature, have been proposed for the structure of liquid water. The fact that we cannot settle upon one particular model which explains all the properties of water reflects our basic ignorance of the nature of the liquid state in general and of that most peculiar liquid, water, in particular.

11.3 ELECTROLYTE SOLUTIONS

Solutes are often classified into three categories according to the electrical conductivities of their water solutions. Substances which dissolve as molecules and, hence, give nonconducting solutions are classed as *nonelectrolytes*. Examples include methyl alcohol, CH_3OH, and sugar, $C_{12}H_{22}O_{11}$. The properties of nonelectrolyte solutions were discussed in Chapter 10.

Substances which exist in water solution as an equilibrium mixture of ions and molecules are called *weak electrolytes*. An example is hydrogen fluoride; a 0.1 M water solution of this compound contains a few H^+ and F^- ions in equilibrium with a large number of HF molecules (cf. Chapter 17).

Strong electrolytes exist almost exclusively as ions in water solution. Their electrical conductivities, at concentrations as low as 0.1 M, are at least 100,000 times that of pure water. Compounds which are ionic in the solid state act as strong electrolytes; examples include NaCl (Na^+, Cl^-) and $Ba(OH)_2$ (Ba^{2+}, 2 OH^-). A few molecular species ionize almost completely when added to water. An example is hydrogen chloride, which exists as

HCl molecules in the pure gas but is converted almost entirely to ions (H^+, Cl^-) in water. The remainder of this section will be devoted to a discussion of the properties and structures of water solutions of strong electrolytes.

COLLIGATIVE PROPERTIES

We saw in Chapter 10 that in dilute solution the vapor pressure lowering, boiling point elevation, freezing point depression and osmotic pressure are directly proportional to the concentration of solute *particles*. On this basis we would predict that, at a given molality, an electrolyte such as sodium chloride would have a greater effect on colligative properties than a nonelectrolyte such as sugar. When a mole of sugar dissolves, *one* mole of solute molecules is obtained; a mole of NaCl, on the other hand, yields *two* moles of ions. By the same argument, a 1 molal solution of $CaCl_2$ (three moles of ions per mole of solute) should have a lower vapor pressure, a higher boiling point, a lower freezing point and a greater osmotic pressure than 1 molal NaCl.

These predictions are confirmed experimentally. At a given molality the vapor pressures of solutions of sugar, sodium chloride and calcium chloride decrease in that order (i.e., sugar > NaCl > $CaCl_2$). Many electrolytes form saturated solutions whose vapor pressures are so low that the solids pick up water when exposed to moist air. This phenomenon, called *deliquescence*, is shown by calcium chloride, whose saturated solution has a vapor pressure only 20% that of pure water. If dry $CaCl_2$ is exposed to air in which the vapor pressure of water is greater than 20% of the equilibrium value, it deliquesces to form a concentrated aqueous solution. Calcium chloride is sometimes spread on dirt roads in summer to prevent them from becoming dusty and developing a rough, "washboard" surface.

These salts also unfortunately increase the rate of corrosion in cars which use those highways

The freezing points of electrolyte solutions, like their vapor pressures, are lower than those of nonelectrolytes at the same concentration. Sodium chloride and calcium chloride are used to melt ice on highways; their aqueous solutions have freezing points as low as $-21°C$ and $-55°C$, respectively.

The equations for the freezing point lowering, boiling point elevation and osmotic pressure of electrolytes are similar to those given in Chapter 10 for nonelectrolytes except for the introduction of a multiplier i.

$$\Delta T_f = i(1.86°C)m \qquad (11.2)$$

$$\Delta T_b = i(0.52°C)m \qquad (11.3)$$

$$\pi = iMRT \qquad (11.4)$$

If we regard an electrolyte simply as a source of solute particles which act independently of one another in determining colligative properties, we would predict that i should be equal to *the number of moles of ions per mole of electrolyte;* that is, i would equal 2 for NaCl, KCl and $MgSO_4$, 3 for $CaCl_2$ and Na_2SO_4, and so on.

Looking at the data in Table 11.4, we see that the situation is not so simple as the above discussion implies. The observed freezing point lowerings of KCl and $MgSO_4$ are smaller than we would predict from Equation

TABLE 11.4   Freezing Point Lowerings of Solutions of KCl and MgSO$_4$

| Molality | ΔT_f OBSERVED | | i CALCULATED (EQUATION 11.2) | |
	KCl	MgSO$_4$	KCl	MgSO$_4$
0.005	0.0182	0.0158	1.96	1.70
0.01	0.0361	0.0301	1.94	1.62
0.02	0.0714	0.0573	1.92	1.54
0.05	0.175	0.132	1.88	1.42
0.10	0.346	0.246	1.86	1.32
0.20	0.682	0.454	1.83	1.22
0.50	1.67	1.00	1.80	1.08

11.2 with i = 2. In other words, at any finite concentration, the multiplier i is less than 2; it approaches a limiting value of 2 as the solution becomes more and more dilute.

Comparing the values in the last two columns of Table 11.4, we see that i depends on both the nature of the electrolyte (i is smaller for MgSO$_4$ than for KCl) and its concentration (i decreases as concentration increases). If one makes a careful study of the dependence of i upon concentration, it turns out that for KCl and many other strong electrolytes in very dilute solution, *i is a linear function of the square root of the molality* (Fig. 11.5). This observation, which intrigued solution chemists for many years, is one that has to be explained by any structural model of electrolyte solutions.

Despite the complications just noted, we can use measurements on colligative properties to decide upon the nature or extent of ionization of an electrolyte in water (Example 11.1).

Example 11.1   The freezing points of 0.010 m solutions of NaHCO$_3$ and HF are −0.038°C and −0.024°C. Use this information to determine
  a. whether NaHCO$_3$ ionizes predominantly as

$$NaHCO_3(s) \rightarrow Na^+(aq) + HCO_3^-(aq); \quad i \rightarrow 2$$

  or $$NaHCO_3(s) \rightarrow Na^+(aq) + H^+(aq) + CO_3^{2-}(aq); \quad i \rightarrow 3$$

  b. the extent to which HF, in 0.010 m solution, ionizes to H$^+$ and F$^-$.

Solution
  a. Let us calculate the value of i in Equation 11.2

$$0.038°C = i \, (1.86°C)(0.010)$$

$$i = \frac{0.038}{0.0186} = 2.0$$

  Clearly, the first mode of ionization is indicated.
  b. Again, we start by calculating i.

$$i = \frac{0.024}{0.0186} = 1.3$$

If HF were not ionized at all, i would be 1. If it were completely ionized to H$^+$ and F$^-$, i would be 2. Since the observed value of i is 30% of the way from 1 to 2, we conclude that HF is about 30% ionized in 0.010 m solution.

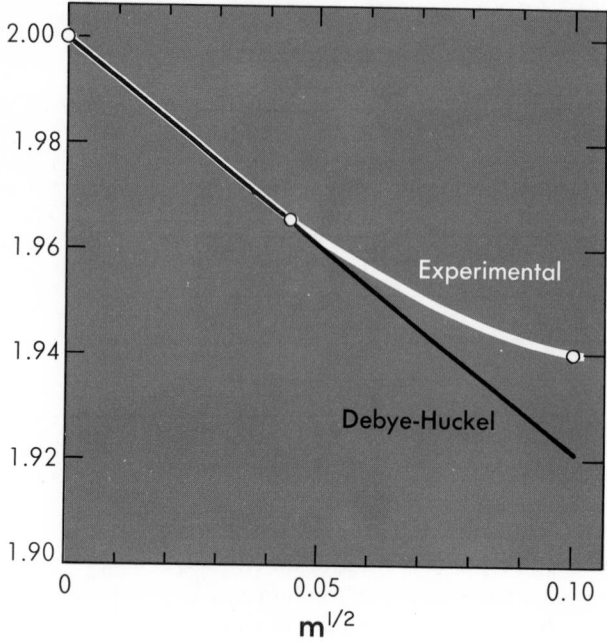

Figure 11.5 Freezing points of dilute KCl solutions. At concentrations above about 0.01 m ($m^{1/2} = 0.10$), deviations from the Debye-Hückel relation (dotted line) are pronounced.

The Debye-Hückel theory works well only with very dilute solutions

STRUCTURE

The deviations between the observed properties of electrolyte solutions and those predicted on the basis of completely independent solute ions have been explained in various ways. The structural models involved to explain these properties have led, slowly and painfully, to a better understanding of the forces between ions and molecules in water solution. Many prominent chemists have worked in this area, starting with Svandte Arrhenius, who in 1903 won a Nobel Prize for his theory of ionization in water solution. He believed that in a solution of KCl, as in a solution of HF, there was an equilibrium between free ions (K^+, Cl^-) and molecules (KCl). According to Arrhenius, the ions were the predominant species but there were enough molecules present to give values of i less than 2 (Table 11.4). This simple picture fell into disfavor when x-ray studies gave no evidence for "molecules" of KCl in the solid state. If they do not exist as a solid, it hardly seems reasonable that they would be present in solution.

In 1923 Peter Debye, then at Zurich, later Nobel laureate at Cornell, and a student of his, Erich Hückel, pointed out that, because of electrostatic attraction, there will be, near a given ion in solution, more ions of opposite than of like charge. In other words, an ion in solution will have associated with it an *ionic atmosphere,* containing an excess of oppositely charged ions (Fig. 11.6). The existence of such an atmosphere prevents ions from acting as completely independent solute particles. One result is to make an ion somewhat less effective than a nonelectrolyte molecule in its influence on colligative properties.

Qualitatively at least, the Debye-Hückel approach offers an attractive

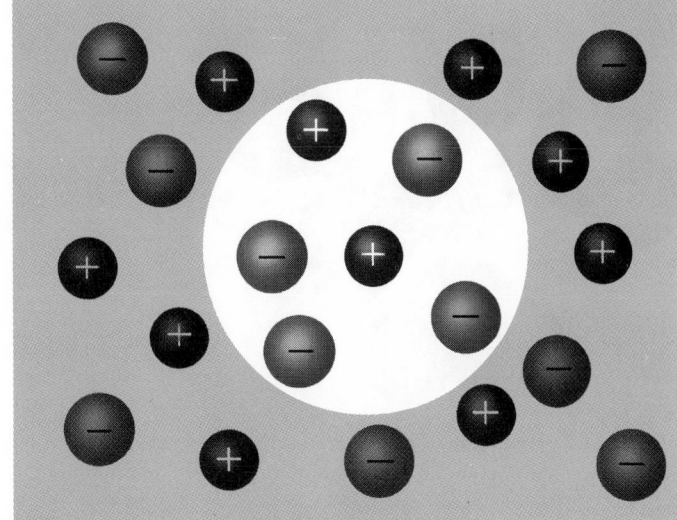

Figure 11.6 Ion atmosphere. An ion is surrounded by more ions of opposite than of like charge.

explanation for the colligative properties of an electrolyte solution. Since ion-atmosphere effects become more pronounced at high concentrations, we would predict, in agreement with experiment, that i in Equations 11.2 to 11.4 should decrease with increasing concentration. Moreover, the Debye-Hückel model explains why, at a given concentration, i should be smaller for $MgSO_4$ than KCl; the ion atmosphere should be more extensive around ions of high charge (+2, −2 in $MgSO_4$) than near ions of low charge (+1, −1 in KCl).

The ion atmosphere imposes a loose structure on the ions which lowers their effective concentrations

The ion atmosphere model can be used to derive equations for the concentration dependence of many of the properties of electrolyte solutions. In particular, Debye arrived at the following equation for the multiplier i in solutions of 1:1 electrolytes (+1 and −1 ions) such as KCl:

$$i = 2 - 0.78 \ m^{1/2} \qquad (11.5)$$

Notice that the equation predicts, in agreement with experiment, that i will approach the theoretical value of 2 in very dilute solution (i.e., $i \to 2$ as $m \to 0$), that it will decrease with increasing concentration and, most important, that i will be a *linear function of the square root of the molality*.

As you can well imagine, the Debye-Hückel model of electrolyte solutions was received initially with a great deal of enthusiasm. It was applied to many different properties of electrolyte solutions, notably to electrical conductivities by Lars Onsager (Nobel laureate, 1968) at Yale. However, careful comparison of the experimental values of i with those predicted by Equation 11.5 reveals a fundamental weakness in the simple Debye-Hückel model. Referring back to Figure 11.5, we note that below 0.01 m ($m^{1/2} = 0.1$), theory (dashed curve) and experiment (solid curve) are in reasonably good agreement. However, at higher concentrations, the two curves diverge more and more from each other. In general, Equation 11.5 predicts smaller values of i than those found experimentally.

The essential weakness of the ion-atmosphere model is that it neglects at least three other structural factors that operate even in relatively dilute solution.

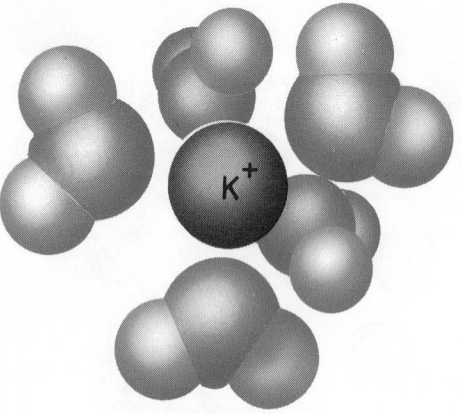 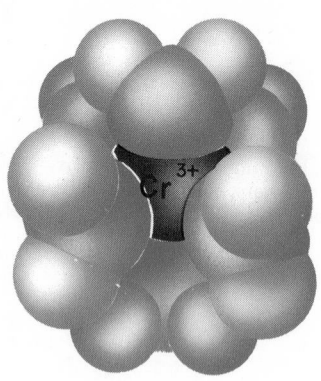

Figure 11.7 Hydration of K^+ (weak ion-dipole forces) and Cr^{3+} (discrete hydrated ion).

1. *Electrostatic forces between ions and polar water molecules* (Fig. 11.7). These forces are particularly strong for small cations of high charge, such as Cr^{3+}, where they lead to the formation of a discrete hydrated species, $Cr(H_2O)_6^{3+}$. With a large cation of low charge such as K^+, the ion-dipole forces are weaker, but become significant at high concentrations.

2. *Changes in the water structure due to the presence of ions.* Ions can shift the equilibrium between "flickering clusters" and "free" water molecules. It appears that most simple ions act as "structure breakers" in the sense that they favor the formation of free water molecules at the expense of hydrogen-bonded clusters.

The fundamental difficulty in developing a good theory of electrolytes is that so many factors need to be considered

3. *Short-range association between ions of opposite charge, to form "ion pairs."* This idea, first advanced by Niels Bjerrum in 1926, was largely ignored at the time because of the popularity of the Debye-Hückel theory, but is now coming back into favor. Ion pair formation appears to be of minor importance in solutions of 1:1 salts such as KCl, except perhaps at high concentrations. However, there is general agreement that with magnesium sulfate, where we are dealing with +2 and −2 ions, the principal species, even at concentration as low as 0.10 m, is the $MgSO_4$ ion pair rather than "free" Mg^{2+} and SO_4^{2-} ions.

Unfortunately, despite many years of research, we still have no quantitative theory of electrolyte solutions which incorporates the structural factors just pointed out. We cannot, for example, write a general equation analogous to 11.5 to predict the colligative behavior of simple salts at concentrations above, at best, 0.1 molal. To be candid about it, our knowledge of the structure of electrolyte solutions has not increased substantially in the past 20 years. We have become more aware of the factors that must be taken into account to develop a comprehensive theory of salt solutions but, so far at least, no one has been able to put them all together.

11.4 WATER POLLUTION

During the past decade we have been made increasingly aware of the problem of water pollution. Fish kills caused by indiscriminate dumping

Thanks, but I think I'll go
swimming somewhere else

Figure 11.8 (Photo by Donald Chandler, courtesy of Dr. Arthur D. Hasler, Professor of Zoology, University of Wisconsin.)

of human and industrial wastes have been reported in rivers as far apart as the Androscoggin in northern New England and the Columbia in the Pacific Northwest. Oil slicks along our beaches and pesticide residues leaching into inland waters have been held responsible for reductions in the population of waterfowl and songbirds. Many of our lakes and rivers are so badly contaminated that they cannot even be used for recreational purposes.

Perhaps we should start our discussion of water pollution by explaining what it means. "Polluted water" is, quite simply, water that is unfit for the purpose for which it is intended. It is entirely possible for water to be "polluted" for one purpose yet pure enough for other uses. Water containing 0.05 part per million of lead ion (i.e., 0.05 g of Pb^{2+} per million g of water) is considered unfit for drinking but is perfectly satisfactory for most industrial purposes. Water contaminated with 10% alcohol along with smaller amounts of sugar and other organic compounds would not be used for washing dishes or irrigating asparagus, yet, as a beverage, many people prefer wine to distilled water.

SOURCES OF POLLUTION

Many of the inorganic cations that we find objectionable in surface and ground waters come from natural sources. Water coming into contact with limestone ($CaCO_3$) or dolomite ($CaCO_3$, $MgCO_3$) picks up Ca^{2+} and Mg^{2+} ions which make it too "hard" for many household and industrial uses. Underground deposits of iron compounds are responsible for the presence of Fe^{3+} in certain ground waters. This ion reacts with hot water to form rust (hydrated ferric hydroxide) which deposits as a brown stain on bathtubs, clothing, and so on.

$$Fe^{3+}(aq) + 3\ H_2O(l) \rightarrow Fe(OH)_3(s) + 3\ H^+(aq) \qquad (11.6)$$

Another natural pollutant that we see in many of our rivers is silt, which consists of suspended mineral particles resulting from land erosion. One billion tons of silt are carried into the oceans each year by our rivers; four times as much settles out along the way, gradually filling lakes and reservoirs. Hoover Dam, built only 40 years ago, has already lost one half of its capacity because of precipitation of sediment.

Perhaps the most obvious man-made pollutant is human sewage, which began to contaminate water supplies when the flush toilet was introduced around 1800. A hundred years later the city of London was still discharging raw sewage into the Thames and withdrawing untreated drinking water from wells along its banks. Among the catastrophes that resulted were the cholera epidemics of 1849 and 1853 that killed 20,000 Londoners. Typhoid fever, another water-borne bacterial disease, was common in the United States as late as 1900. Even today, microorganisms in human waste can be a serious health hazard. Infectious hepatitis is carried in blood and feces by a virus that is highly resistant to ordinary methods of water treatment.

Another development occurred around 1800 that greatly increased the extent of water pollution. This was the Industrial Revolution. Today, it is estimated that in this country industry is the source of two thirds of the total amount of man-made pollutants. Traditionally the worst offenders have been food processors, the pulp and paper industry, metal producers and chemical manufacturers, not necessarily in that order. For years many industries dragged their feet in the area of pollution abatement. Now the situation is changing; many larger corporations are rapidly increasing their efforts in this area.

Much, much more water is used in industry than in homes

One of the most serious effects of human and industrial wastes is to increase the **biochemical oxygen demand (BOD)** of natural water supplies. Aerobic bacteria use oxygen to degrade the complex organic compounds in sewage to simpler species, ultimately to carbon dioxide and water. This process lowers the amount of dissolved oxygen in the water, sometimes to the point where fish and other aquatic life cannot survive.

The BOD of polluted water is determined by measuring the amount of oxygen consumed by a sample of known volume. The water sample is first diluted with air-saturated distilled water to ensure an excess of oxygen. The concentration of dissolved oxygen in the diluted sample is determined immediately and again after a period of five days. From the decrease in oxygen concentration, one can calculate the BOD, which is usually expressed as

$$BOD = \frac{\text{no. mg } O_2}{\text{no. liter sample}} \equiv \text{parts } O_2/\text{million parts sample} \qquad (11.7)$$

Example 11.2 5.0 ml of a sewage sample is diluted to 100 ml with distilled water. The concentration of dissolved oxygen immediately after dilution is found to be 8.3×10^{-3} g/liter; after 5 days, it is 2.0×10^{-3} g/liter. Estimate the BOD of the sample.

Solution. Let us first calculate the amount of oxygen consumed. The decrease in concentration of dissolved oxygen is

$$8.3 \times 10^{-3} \text{ g/lit} - 2.0 \times 10^{-3} \text{ g/lit} = 6.3 \times 10^{-3} \text{ g/lit}$$

Since the sample after dilution had a volume of 100 ml (0.10 lit), the total amount of oxygen consumed must be

$$6.3 \times 10^{-3} \frac{g}{liter} \times 0.10 \text{ liter} = 6.3 \times 10^{-4} \text{ g}$$

The BOD, in mg O_2 per liter of sample, must be

$$\frac{6.3 \times 10^{-4} \text{ g}}{5.0 \text{ ml sample}} \times \frac{1000 \text{ ml}}{1 \text{ liter}} \times \frac{1000 \text{ mg}}{1 \text{ g}} = 130 \text{ mg/liter} = 130 \text{ ppm}$$

TYPES OF POLLUTANTS

The U.S. Public Health Service classifies pollutants into eight major categories (Table 11.5). Rather than attempting to discuss all these types of pollutants, we will consider only a few examples of current interest.

DETERGENTS. The structure of a type of detergent widely used in this country prior to 1965 is shown at the top of Figure 11.9. This detergent had one serious shortcoming from an ecological standpoint—it is not biodegradable; that is, bacteria are unable to chew up the organic anion, breaking it down into simpler species. Consequently the detergent stayed around in water supplies more or less indefinitely, causing some rivers and sewage treatment ponds to become covered with mountains of foam. Fortunately, chemists were able to modify the structure of the anion to give a detergent which is degradable by aerobic (oxygen-consuming) bacteria. The modification was a simple one; the branched side chain was replaced by a straight chain to give the structure shown at the bottom of Figure 11.9. For reasons which we do not understand, aerobic bacteria seem to find straight chains more palatable than branched chains.

In 1965 the use of branched chain detergents was prohibited in this

TABLE 11.5 Classification of Water Pollutants

TYPE	EXAMPLES
1. Oxygen-demanding wastes	1. Human, animal waste; decaying vegetation
2. Infectious agents	2. Bacteria and viruses
3. Plant nutrients	3. Nitrates, phosphates
4. Organic molecules	4. Detergents, pesticides
5. Other minerals and chemicals	5. CN^-, Pb^{2+}, mercury
6. Suspended material	6. Silt from land erosion
7. Radioactive substances	7. Fallout products, radioactive waste
8. Heat	8. Water used for cooling in industry

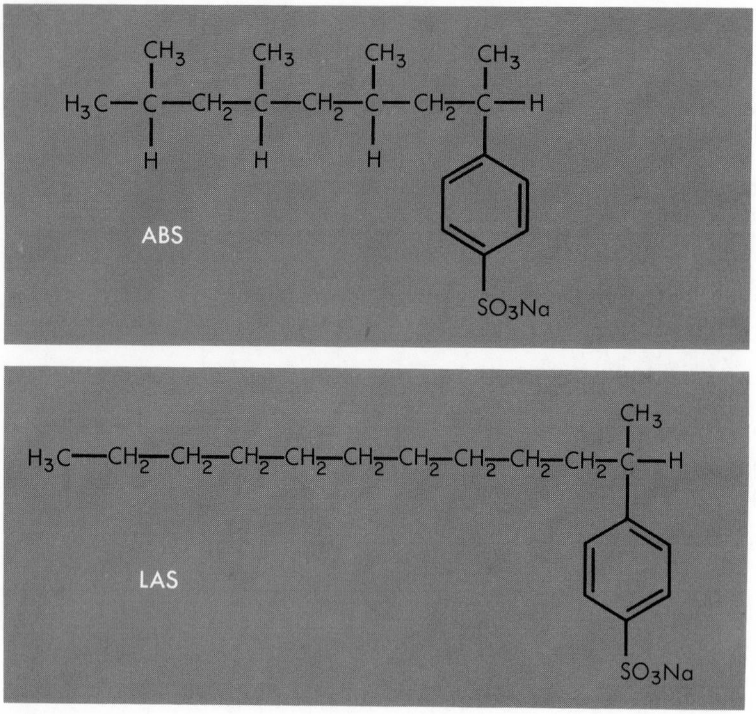

Figure 11.9 Molecules of a branched chain detergent (alkylbenzene sulfonate) and straight chain detergent (linear alkylsulfonate).

For the difference in structure between a detergent and a soap, see page 222

country, and the foam problem was solved. Commercial detergent products can, however, contribute to pollution in another way. In some brands of detergents, as much as one half of the total weight consists of inorganic phosphates (e.g., sodium tripolyphosphate, $Na_3P_3O_9$), which form very stable complexes (ion pairs) with cations such as Ca^{2+}, Mg^{2+} and Fe^{3+} that would otherwise interfere with the cleansing action of the detergent. Phosphates are important plant nutrients and as such tend to promote the growth of algae. There is evidence to indicate that detergent phosphates are a factor in the blooms of algae that are choking many of our lakes, notably Lake Erie. There is considerable controversy on this point; it has been shown that in many water systems the critical element in algae growth is carbon or nitrogen rather than phosphorus.

For a time the best substitute for phosphates in detergents appeared to be NTA (nitrilotriacetic acid)

$$N \begin{cases} CH_2COOH \\ CH_2COOH \\ CH_2COOH \end{cases}$$

which is a good cleansing agent, a poor fertilizer and biodegradable. However, evidence accumulated that NTA, under certain conditions, can be toxic to animal life; in December 1970, its use in detergents was halted at the request of the U.S. Surgeon General.

Most of the "low-phosphate" and "non-phosphate" detergents on the market contain carbonate or silicate salts as a substitute for phosphates. These products are generally less effective cleansing agents; moreover some of them give strongly alkaline solutions that can be hazardous for household use. The best long-term solution to the "phosphate" problem probably involves installing advanced sewage treatment systems, which are capable of removing phosphates and other plant nutrients as well.

INSECTICIDES. Prior to World War II the effective ingredients of most insecticides were inorganic compounds of copper, lead and arsenic. These compounds were toxic to nearly all forms of animal life, including man. In 1939 a "safe" organic insecticide was found, *di*chloro*di*phenyl-*tri*chlorethane (DDT)

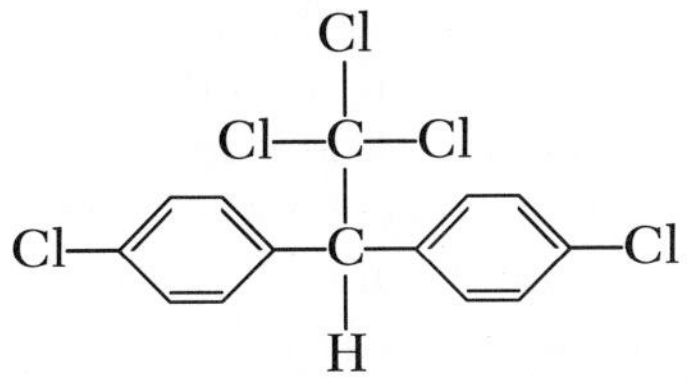

DDT was found to be effective against a wide variety of disease-bearing and agricultural insects. It was used during World War II to prevent epidemics of typhus and malaria carried by lice and mosquitoes, respectively. After the war, its use in agriculture and forestry increased to the point where 180 million pounds were manufactured in 1963. Its production has been dropping ever since.

In many communities DDT was sprayed routinely during the summer for mosquito control

As early as 1946 a species of housefly was found to be resistant to DDT. This phenomenon has now become quite common; over 100 insect pests are known to have developed immunity in a classic example of Darwinian evolution. To meet this problem a variety of other insecticides were developed, many of which, like DDT, were chlorinated hydrocarbons. Some of the more popular insecticides of this type are shown in Figure 11.10. Most of these compounds are nontoxic to humans, at least over a short time span; an exception is endrin, which is a deadly poison.

It has long been known that DDT and its relatives are persistent insec-

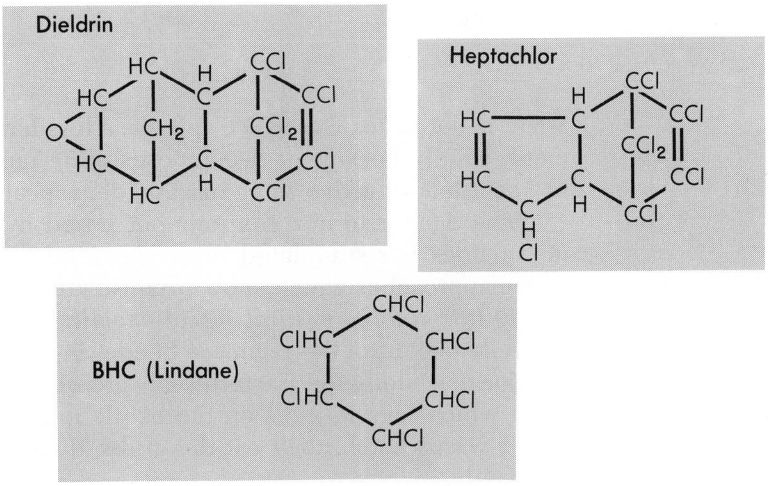

Figure 11.10 Molecules of chlorinated insecticides.

ticides in that they are not readily degraded in the environment. At one time this was considered to be an advantage since these insecticides stayed around long enough to control successive generations of insects. Today, the disadvantages are all too apparent. The stability of chlorinated organic pesticides, coupled with their high solubility in fatty tissues, causes them to concentrate in the food chain. A study of aquatic life in the Long Island marshes showed that minnows eating plankton containing 0.04 part per million of DDT accumulated 1 part per million; in sea gulls feeding on the minnows, the concentration jumped to 75 ppm.

The toxicity to wildlife of DDT, lindane and other insecticides of this type is well documented. Insecticides are at least partially responsible for the near extinction of such predatory birds as the bald eagle, the golden eagle and the peregrine falcon. The reproductive capacities of these and other birds are inversely related to DDT uptake. Ospreys, which normally average 2.5 young per year, produce 1.1 offspring at a DDT level of 3 ppm and only 0.5 when the concentration of DDT reaches 5 ppm.

Several insecticides are now available which decompose rather rapidly in water and, hence, pose less of a long-term threat to wildlife than DDT. Most of these compounds are organic phosphorus derivatives: parathion is one example.

Parathion

Parathion decomposes readily in water; within 20 days, half of the insecticide is gone. Unfortunately parathion and many related phosphorus insecticides are far more toxic to humans than DDT. In one year, 1958, 100 fatalities in India and over 300 in Japan were attributed to the careless use of parathion.

An insecticide which was widely used in the summer of 1972 to control gypsy moth infestations along the Eastern seaboard is Sevin:

Sevin

Sevin is not so toxic as DDT and has a less harmful effect on the environment, largely because it decomposes quite rapidly. This factor, however, makes it less effective as an insecticide; repeated sprayings are required.

The danger to our environment posed by DDT and other chemical insecticides has stimulated research on other methods of insect control. Three approaches which show promise are:

1. *Introducing natural insect enemies.* This technique was used successfully to control the Japanese beetle, an import from Asia which became a major pest along the Eastern seaboard of the United States. An Oriental wasp which lays its eggs on the beetle grubs was introduced. When the young wasps hatch, they eat the grubs, thereby breaking the life cycle of the beetle.

2. *Sterilization.* The screwfly worm was eradicated in the southeastern

part of the United States and the state of Texas by sterilizing male flies through irradiation and releasing them in infested areas. This program, carried out several years ago by the United States Department of Agriculture, brought the population of this insect down drastically within a single year.

3. *Use of sex attractants.* This approach was applied with some success on a small scale to control gypsy moths. The female gypsy moth secretes a tiny amount of a chemical sex attractant during the mating period. By baiting traps with this material or with equally potent synthetic compounds, it is possible to attract and destroy the male moth.

We can hope that these and other nonchemical methods of insect control will ultimately free us from our dependence on insecticides. We cannot forget, however, that in many parts of the world today chemical insecticides are essential to the prevention of disease and famine. The United Nations' World Health Organization has warned that banning DDT before effective substitutes are found would be "a disaster to world health." The dilemma that we face here may serve to remind us that man tampers with the environment at his peril; the greater the perturbation that he creates, the more difficult it is to bring under control.

Modern agriculture in this country requires the use of many chemicals for insect and weed control

TRACE ELEMENTS. The presence in water supplies of low concentrations of elements whose compounds are known to be toxic has become a major concern in environmental pollution. Some of these elements are most familiar to us as air pollutants, introduced by the combustion of coal and petroleum products. Lead compounds, for example, enter the atmosphere when gasoline containing lead tetraethyl, $Pb(C_2H_5)_4$, is burned. A major source of arsenic, cadmium, mercury and selenium compounds is the combustion of coal, which contains traces of all these elements.

The problem of evaluating the effects of low concentrations of these elements on human health is complicated by the fact that toxic limits have yet to be established for most of them. The acute effects listed in Table 11.6 are those shown by individuals unfortunate enough to have taken in large quantities of a particular element over a short period of time. We do not know whether these disorders will result from continued exposure to low concentrations over long periods. However, several trace elements, notably arsenic and mercury, are known to be cumulative poisons; that is, they are eliminated very slowly by the body. This suggests that we take every pre-

TABLE 11.6   Trace Elements in the Water Environment

ELEMENT	SOURCE	EFFECT OF LARGE DOSES
Arsenic	Detergents, pesticides, smelter wastes	Attacks tissues of internal organs, nerve tissues
Cadmium	Water mains, plating wastes, mining wastes	Anemia, hypertension, heart disorders
Lead	Water pipes, paints, pesticides	Brain damage, behavioral disorders
Mercury	Industrial wastes, natural deposits	Nerve and brain damage
Selenium	Mining wastes	Poisonous to cattle, suspected carcinogen

caution to avoid raising their concentrations in the environment to levels above those found naturally.

The recent development of instrumental methods capable of detecting metals in the part per million or even part per billion range prompted a systematic search for trace elements in water supplies and the food chains that they support. The results have been disturbing. You may recall that in March of 1970 Norwald Finreite, a graduate student at the University of Western Ontario, reported mercury concentrations in fish caught in Lake St. Clair to be as high as 7 parts per million. This is to be compared to the "safe" limit set for fish by the Food and Drug Administration of 0.5 ppm, and the concentration of 0.2 ppm later found to be typical of fish in unpolluted waters.

Most of the mercury found in fish is present as an organic derivative, dimethyl mercury, $(CH_3)_2Hg$, a toxic substance which is known to concentrate in the food chain in much the same way as DDT. This compound is synthesized by certain aquatic microorganisms, starting with elementary mercury. Some of this mercury undoubtedly comes from natural sources; much of it, however, can be attributed to the careless disposal of industrial wastes. In particular, the mercury in Lake St. Clair was traced to electrochemical plants which had "lost" large quantities of mercury in the process of making chlorine and sodium hydroxide (Chapter 20).

Since many elements in trace amounts are beneficial to living organisms, it is often hard to decide how much of a given element is too much

HEAT. In Chapter 4 it was pointed out that whenever we attempt to convert heat into useful work, a considerable portion of that heat is discharged into the surroundings. The "surroundings" may be the atmosphere; alternatively, the heat may be discharged to a neighboring body of water. This can create a problem of *thermal pollution* quite as serious as the types of material pollution discussed up to this point. The electrical power industry requires large quantities of cooling water to condense steam that has been used to generate electrical energy. In a conventional power plant, at least 2 kcal of heat are discharged to the water circulated through condensers for every kcal of electrical energy produced. In a nuclear power plant, this ratio is even higher, as high as 3 to 1.

The discharge of hot water into a river or lake raises its temperature, typically by 5 to 10°C at the source. Occasionally temperatures as high as 60°C (140°F) have been recorded. Even a relatively small temperature increase can have an adverse effect on aquatic life; salmon and trout cannot live in water much above 25°C. At higher temperatures the rate of metabolism increases, creating a greater demand for oxygen. Less oxygen is available, since its solubility decreases with increasing temperature.

One way to minimize the problem of thermal pollution is to use a cooling tower to reduce the temperature of waste water before it is discharged into a stream. In one type of tower, air is blown through the water to take advantage of the cooling effect of evaporation. The disadvantage of this design is that large quantities of water vapor are introduced into the atmosphere. Not only does this waste water; it may even form clouds or fog banks in the area around the tower. This problem is avoided in towers in which the warm water does not come into direct contact with air. By circulating the water through a maze of tubes over which compressed air is blown, its temperature can be reduced in much the same way as in an ordinary automobile radiator. Unfortunately this type of tower is more expensive to build and maintain than one of the evaporative type.

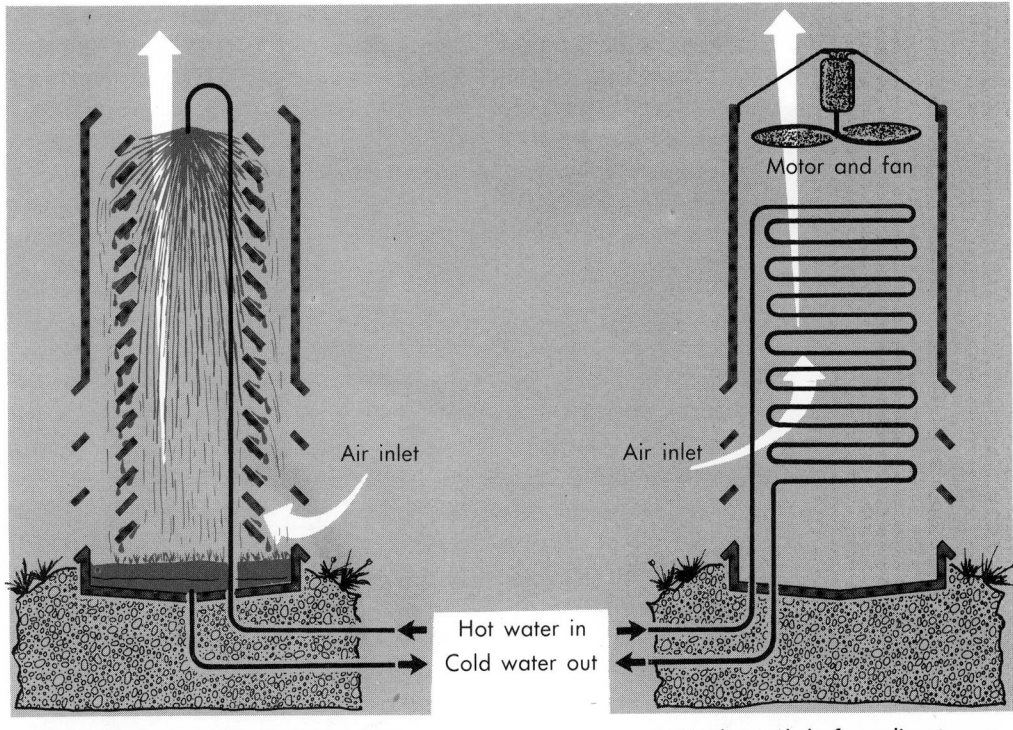

Air inlet

Motor and fan

Air inlet

Hot water in
Cold water out

Evaporative cooling tower

Mechanical draft cooling tower

Figure 11.11 Cooling of water by evaporation or heat transfer to compressed air.

11.5 WATER PURIFICATION

Perhaps you have heard the statement that "running water purifies it-self." To some extent this is true for an isolated mountain stream. However, the method is not very practical for purifying the water used by a city the size of Chicago. Methods capable of treating enormous quantities of slightly or moderately contaminated water are necessary and are operating in virtually every metropolitan area in this country. In this section we shall consider some of the processes that are used to purify the water we use in our homes and in industry.

REMOVAL OF SUSPENDED MATTER

The simplest way to clarify muddy water is to allow the suspended material in it to settle. This process, called *sedimentation*, takes place when water is allowed to stand in a reservoir or to flow through a series of settling tanks. Sedimentation alone is often insufficient to remove all the suspended matter; silt particles small enough to be in colloidal suspension cannot be removed in this manner.

The process of *coagulation* is used to remove colloidal particles from water. It involves adding a compound to water which forms a gelatinous precipitate that carries suspended particles down with it. Aluminum salts

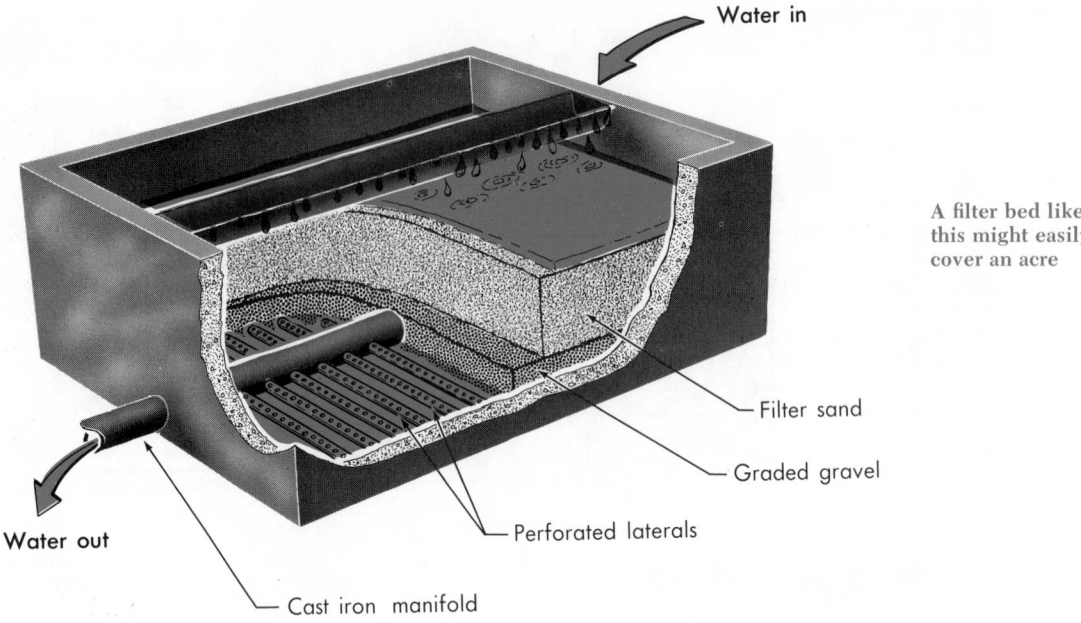

Water in

A filter bed like
this might easily
cover an acre

Filter sand

Graded gravel

Perforated laterals

Water out

Cast iron manifold

Figure 11.12 Rapid sand filter bed.

are among the more popular coagulants; the Al^{3+} ion reacts with water to give a precipitate which can be represented most simply as $Al(OH)_3$.

$$Al^{3+}(aq) + 3\ H_2O \rightarrow Al(OH)_3(s) + 3\ H^+(aq) \qquad (11.8)$$

Millions of pounds of hydrated aluminum sulfate, $Al_2(SO_4)_3 \cdot 18\ H_2O$, are added annually to clarify municipal and industrial water supplies.

Water which has passed through a coagulation plant is often filtered to remove any remaining suspended material. *Filtration* can be accomplished by allowing the water to trickle through a layer of sand supported on several layers of successively coarser gravel. A thin layer of precipitated material forms on the sand. Up to a point at least, this added layer of finely divided particles improves the efficiency of the filter. When it becomes too thick, it is removed by back-flushing with water.

DISINFECTION

City water which is to be used for drinking must be treated chemically to remove disease-causing bacteria. Disinfection is also essential for water in municipal swimming pools, although the tolerable bacteria count there can be somewhat higher. The most common disinfectant is chlorine, which kills bacteria apparently by inhibiting the activity of certain enzymes essential to their metabolism.

Large water treatment plants use liquid chlorine from high pressure tanks as a disinfectant. Enough is added to give a residual Cl_2 concentration of 0.2 to 1.0 part per million. This concentration is ordinarily high enough to destroy any bacteria entering the water after it leaves the treatment plant. To ensure that the water remains safe to drink, ammonia is often added; it reacts with dissolved chlorine to give a mixture of mono-chloramine, NH_2Cl, and dichloramine, $NHCl_2$.

$$2\ NH_3(aq) + Cl_2(aq) \rightarrow NH_2Cl(aq) + NH_4^+(aq) + Cl^-(aq) \qquad (11.9)$$

$$3\ NH_3(aq) + 2\ Cl_2(aq) \rightarrow NHCl_2(aq) + 2\ NH_4^+(aq) + 2\ Cl^-(aq) \qquad (11.10)$$

The two chloramines are less effective disinfectants than elementary chlorine, but they are more stable in water and, hence, afford better long-term protection against infection.

Many chemicals other than chlorine can be used as disinfectants. Ozone, O_3, is actually more effective in killing bacteria, but is more expensive and less stable than chlorine. Iodine is receiving increased attention as a possible substitute for chlorine; it has a less objectionable effect on the taste and odor of water and also attacks certain microorganisms which are immune to chlorine.

REMOVAL OF TASTE AND ODOR

No matter how safe water may be to drink, we object to it if it has an unpleasant taste or odor. City dwellers accustomed to chlorinated water enjoy, for a while at least, drinking from a clear, cold mountain stream even though it may be teeming with bacteria. Most of us are turned off by the odor of hydrogen sulfide associated with water from hot mineral springs, despite its supposed medicinal effect.

One way to remove objectionable tastes and odors from water is to pass it through a filter bed containing activated charcoal, a finely divided form of carbon. The enormous surface area of this material (as much as 6 million cm^2/g) enables it to adsorb large quantities of noxious impurities. For the most part, this is a physical process; in the case of chlorine, however, a chemical reaction occurs at the surface.

$$2\ Cl_2(aq) + C(s) + 2\ H_2O(aq) \rightarrow CO_2(aq) + 4\ H^+(aq) + 4\ Cl^-(aq) \qquad (11.11)$$

Through this reaction the taste and odor associated with chlorinated drinking water can be eliminated.

A process used to remove hydrogen sulfide and certain other gaseous impurities from water is that of *aeration*. Water can be aerated by spraying it into the air in a fine film. This saturates the water with oxygen, which converts hydrogen sulfide to sulfur

$$H_2S(aq) + \tfrac{1}{2}\ O_2(aq) \rightarrow S(s) + H_2O(l) \qquad (11.12)$$

WATER SOFTENING

"Hard" water, caused by the presence of Ca^{2+} and Mg^{2+} ions, has several undesirable properties. For one thing, it forms a precipitate with soaps, which are sodium salts of long chain organic acids (Chapter 8). A typical soap is sodium stearate, $NaC_{18}H_{35}O_2$. Since the calcium and magnesium salts of stearic acid are insoluble, the following reaction occurs when a sodium stearate soap is used in hard water:

$$M^{2+}(aq) + 2\ C_{18}H_{35}O_2{}^-(aq) \rightarrow M(C_{13}H_{35}O_2)_2(s);\ M = Ca\ or\ Mg \quad (11.13)$$

This reaction continues until nearly all the Ca^{2+} and Mg^{2+} ions are removed; only then does the soap become effective as a cleaning agent.

From an industrial standpoint, a more serious drawback of hard water is the tendency of calcium and magnesium salts to precipitate when water is heated or partially evaporated, as in a steam boiler. The nature of the precipitate depends upon the anion present. A common anion in surface water is the $HCO_3{}^-$ ion, formed by the reaction of atmospheric CO_2 with water. When water containing Ca^{2+} and $HCO_3{}^-$ ions is heated, a precipitate of calcium carbonate forms as a result of the reaction sequence.

$$2\ HCO_3{}^-(aq) \rightarrow CO_3{}^{2-}(aq) + CO_2(g) + H_2O(aq)$$
$$\underline{Ca^{2+}(aq) + CO_3{}^{2-}(aq) \rightarrow CaCO_3(s)}$$
$$2\ HCO_3{}^-(aq) + Ca^{2+}(aq) \rightarrow CaCO_3(s) + CO_2(g) + H_2O(aq) \quad (11.14)$$

The scale on teakettles and coffeepots is usually calcium carbonate; it can be removed by treating with a weakly acidic solution such as vinegar.

The problem becomes more serious if there are $SO_4{}^{2-}$ ions in hard water. In this case, heating yields a precipitate of calcium sulfate, one of the few salts which is less soluble at high than at low temperatures.

$$Ca^{2+}(aq) + SO_4{}^{2-}(aq) \rightarrow CaSO_4(s) \quad (11.15)$$

CaSO₄, once formed, is not easily dissolved. It often deposits as plaster of Paris $(CaSO_4)_2 \cdot H_2O$

Calcium sulfate is particularly likely to form in a boiler from which water is being vaporized, thereby increasing the concentration of Ca^{2+} and $SO_4{}^{2-}$ ions. It deposits as a tightly adherent scale which lowers the heat conductivity of the boiler and may even block the flow of water through the tubes.

We will discuss two of the more common methods used to soften water, the lime-soda and cation exchange processes. Our discussion will deal only with the removal of Ca^{2+} ions, although both processes will remove Mg^{2+} ions as well.

LIME-SODA METHOD. This process, introduced into the United States over 100 years ago, is still used in most municipal water-softening plants. It involves adding two chemicals, slaked lime, $Ca(OH)_2$, and soda ash, Na_2CO_3, to the hard water. The purpose of adding sodium carbonate is obvious; it acts as a source of $CO_3{}^{2-}$ ions to precipitate Ca^{2+}

$$Ca^{2+}(aq) + CO_3{}^{2-}(aq) \rightarrow CaCO_3(s) \quad (11.16)$$

The function served by the lime is less apparent. Indeed, it might seem that the addition of $Ca(OH)_2$ to water in an attempt to remove Ca^{2+} ions would be a step in the wrong direction. As a matter of fact, lime is an effective water softener only when there are $HCO_3{}^-$ ions in the hard water. To understand why this is the case, consider what happens when one mole of $Ca(OH)_2$ is added to water containing two moles of $HCO_3{}^-$ ions:

dissolving of Ca(OH)₂: $Ca(OH)_2(s) \rightarrow Ca^{2+}(aq) + 2\ OH^-(aq)$

conversion of
$HCO_3{}^- \rightarrow CO_3{}^{2-}$: $2\ HCO_3{}^-(aq) + 2\ OH^-(aq) \rightarrow 2\ CO_3{}^{2-}(aq) + 2\ H_2O$

precipitation of
added Ca²⁺: $\underline{Ca^{2+}(aq) + CO_3{}^{2-}(aq) \rightarrow CaCO_3(s)}$
overall reaction: $Ca(OH)_2(s) + 2\ HCO_3{}^-(aq) \rightarrow CaCO_3(s) + CO_3{}^{2-}(aq) + 2\ H_2O \quad (11.17)$

We note from this reaction sequence that the lime furnishes the OH^- ions required to convert HCO_3^- ions in the water to CO_3^{2-} ions. The net effect of the addition of the lime, as represented by Equation 11.17, is to form an extra mole of free CO_3^{2-} ions. These ions are then capable of removing a mole of Ca^{2+} ions originally present in the hard water. Thus, for every mole of $Ca(OH)_2$ added, an extra mole of Ca^{2+} ions is removed from the hard water. It is, of course, extremely important not to add too much $Ca(OH)_2$ to the water; any excess over that required for Reaction 11.17 will tend to increase the Ca^{2+} ion concentration in the water.

The sequence of steps involved in softening hard water by the lime-soda process may be summarized as follows:

1. The water is first analyzed for Ca^{2+} and HCO_3^-.

2. Sufficient lime is added to give one mole of $Ca(OH)_2$ for every two moles of HCO_3^-. Every mole of $Ca(OH)_2$ added removes one mole of Ca^{2+} from the hard water.

3. Any Ca^{2+} remaining in the hard water is removed by adding soda ash, Na_2CO_3, on a 1:1 mole basis.

In practice, the slaked lime and soda ash are added simultaneously. In a water treatment plant designed to remove suspended solids it is convenient to add these chemicals in the coagulation step. The calcium carbonate that precipitates comes down with the suspended material.

Example 11.3 A sample of water is found on analysis to contain 0.0030 mole/lit of Ca^{2+}, 0.0040 mole/lit of HCO_3^- and 0.0010 mole/lit of SO_4^{2-}. Calculate the number of moles of $Ca(OH)_2$ and Na_2CO_3 that should be added to one liter of this water to soften it.

Solution. The first step is to add 1 mole of $Ca(OH)_2$ for every 2 moles of HCO_3^- present:

$$\text{no. moles } Ca(OH)_2 = 0.0040/2 = 0.0020 \text{ mole } Ca(OH)_2$$

This removes an equal amount, 0.0020 mole, of Ca^{2+} from the water, leaving in one liter of the water

$$(0.0030 - 0.0020) \text{ mole} = 0.0010 \text{ mole of } Ca^{2+}$$

Consequently, it is necessary to add 0.0010 mole of Na_2CO_3 to the water.

CATION EXCHANGE. The lime-soda method of water softening is impractical for household use; one can hardly expect a housewife to analyze water, calculate how much of each chemical to add, and then filter off the precipitates. For many industrial purposes, the lime-soda process is also inappropriate, because it usually leaves water supersaturated with $CaCO_3$. This can be particularly serious in a commercial laundry, where the calcium carbonate may end up on the customer's clothing.

In situations such as these, the preferred method of water softening is the cation exchange process, in which Ca^{2+} ions in the hard water are replaced by Na^+ ions. This exchange takes place when hard water is passed through a column containing either a certain type of mineral known as a zeolite, which occurs widely in nature, or a man-made "synthetic zeolite."

To understand the principle of cation exchange, let us examine the structure of a natural zeolite, which has the empirical formula $NaAlSiO_4$

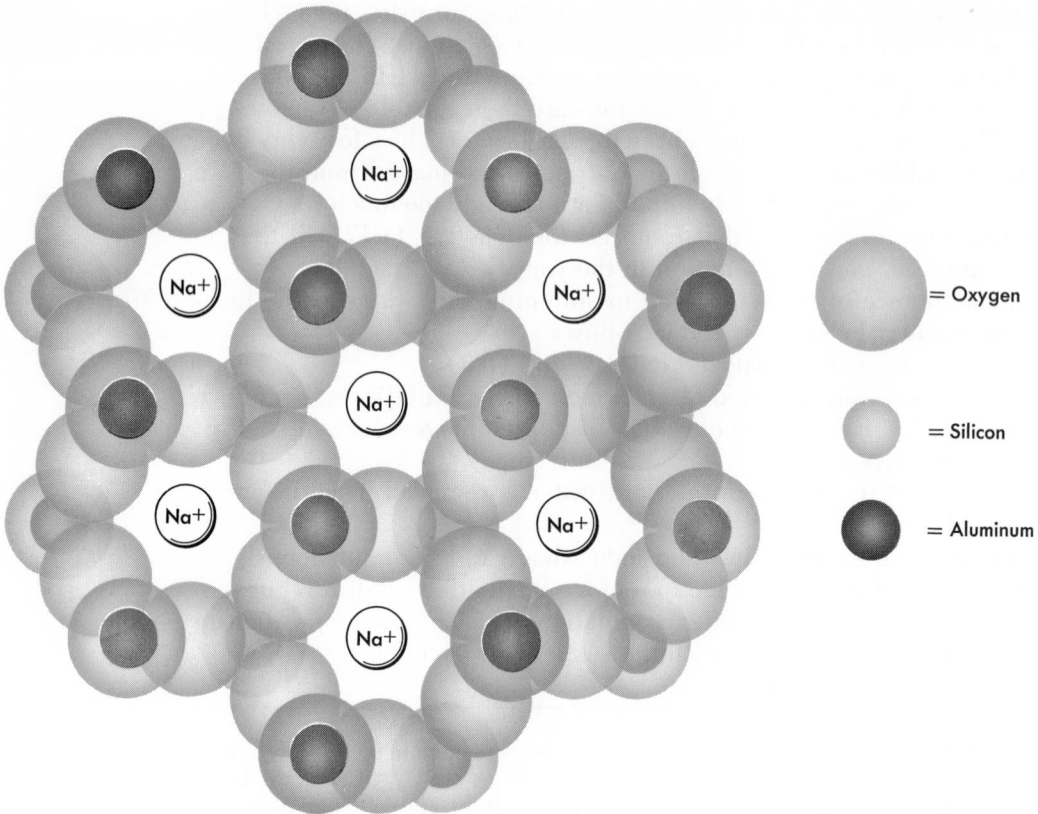

Figure 11.13 Structure of a natural zeolite, $NaAlSiO_4$.

(Fig. 11.13). Notice that atoms of aluminum, silicon and oxygen are bonded together to form a huge "macroanion" similar in structure to a macromolecule such as quartz. The negative charge of the anion is balanced by Na^+ ions trapped in "holes" in the anionic lattice. When pure water flows through the zeolite, nothing spectacular happens; the Na^+ ions cannot leave the crystal since this would create an unbalance of charge. However, if hard water comes into contact with the zeolite, some of the Na^+ ions migrate out of the lattice, being replaced by an equivalent number of Ca^{2+} ions. This process may be represented by the equation

Most home water softening systems use zeolites

$$Ca^{2+}(aq) + 2\ NaZ(s) \rightarrow CaZ_2(s) + 2\ Na^+(aq) \qquad (11.18)$$

where Z stands for a small portion of the macroanion. The effect is to exchange a Ca^{2+} ion responsible for hardness for two less objectionable Na^+ ions.

After a zeolite column has been used for some time, more and more of the vacancies in the lattice become filled with Ca^{2+} ions. Eventually, an equilibrium is reached beyond which no further exchange of cations will occur. The column can, however, be returned to its original state by flushing with a concentrated solution of sodium chloride. Reaction 11.18 is reversed and the zeolite is ready for reuse.

Modern cation exchangers use synthetic organic resins which have a greater capacity for picking up Ca^{2+} ions than do natural zeolites. One type

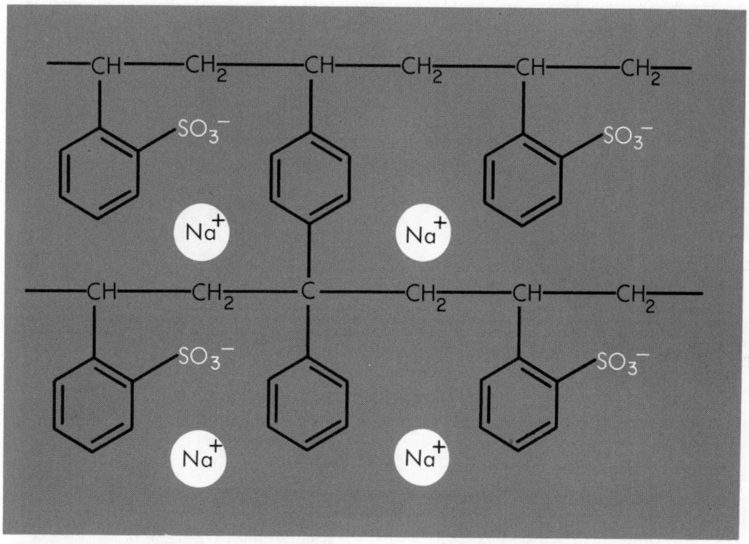

Figure 11.14 Structure of a synthetic cation exchange resin.

of resin, known as a "sulfonated polystyrene," has the structure indicated in Figure 11.14. It operates on somewhat the same principle as a zeolite; Na^+ ions associated with the resin are exchanged for Ca^{2+} ions in the hard water. A major advantage of this type of cation exchanger is that it can be modified to form part of a unit which will remove *all* the ions from water (see later).

DESALINATION

In certain parts of the world, notably in hot, arid regions bordered by the sea, the most economical way to obtain fresh water is to remove the dissolved salts from sea water. The process of desalination is also used to purify brackish water (i.e., water with a relatively low salt content), which is common in much of the southwestern part of the United States. A variety of methods have been tested for removing all or nearly all of the ions (Table 11.7) from salt water; we will discuss a few of the more promising of these.

DISTILLATION.   The oldest method of desalination, distillation, is the one that is used most widely today. It accounts for about 95% of the total amount of fresh water recovered from the oceans. The price of desalinated

TABLE 11.7   Concentrations (Molalities) of Principal Ions in Sea Water

CATIONS		ANIONS	
Na^+	0.457	Cl^-	0.535
Mg^{2+}	0.0556	SO_4^{2-}	0.0277
Ca^{2+}	0.0100	HCO_3^-	0.0023
K^+	0.0097	Br^-	0.00081
Sr^{2+}	0.0001	CO_3^{2-}	0.00027

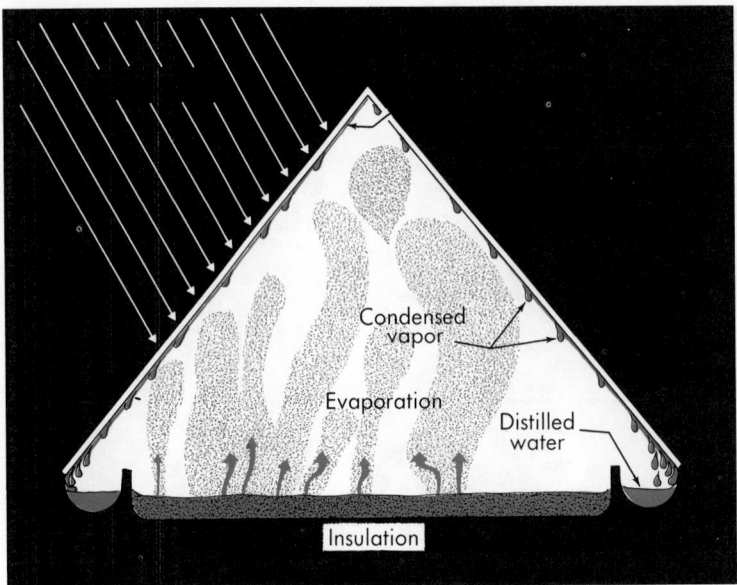

Figure 11.15 Solar still for purification of salt water.

water has decreased from $4 to 25¢ per 1000 gallons, largely because of improved distillation techniques worked out by engineers.

In most distillation units, oil or natural gas is burned to supply the heat required to vaporize the water. A cheaper source of heat would be solar energy (Chapter 4), which is available in unlimited quantities in the arid regions where desalination is most vital. Solar stills of the type shown in Figure 11.15 have been built and tested on a pilot-plant basis. One such still, installed on the Greek island of Patmos, supplies one half of the fresh water requirement of the 2000 people living on the island.

FREEZING. When salt water is frozen, the product is virtually pure ice. This suggests that freezing might be an effective way of obtaining fresh water from the sea. In practice, freezing as a desalination process appears to have two distinct disadvantages. In the first place, it is a relatively slow process; more important, the energy requirement is relatively high. Work has to be done to operate the refrigeration unit that cools salt water down to its freezing point ($-1.8°C$).

REVERSE OSMOSIS. We saw in Chapter 10 that when a solution is separated by a semipermeable membrane from a sample of pure water, there is a tendency for the water to move through the membrane to the solution side by the process of osmosis. Clearly, ordinary osmosis could hardly be used to desalinate water; the movement is in the wrong direction.

You will recall, however, that osmosis can be prevented by applying, on the solution side, a pressure just equal to the osmotic pressure of the solution. If the applied pressure *exceeds* the osmotic pressure, water moves out of the solution to the pure water side of the membrane. This process, called reverse osmosis, accomplishes the objective of desalination—the extraction of pure water from salt water.

A major problem in reverse osmosis is to construct a membrane strong enough to withstand the high pressures required and, at the same time, be

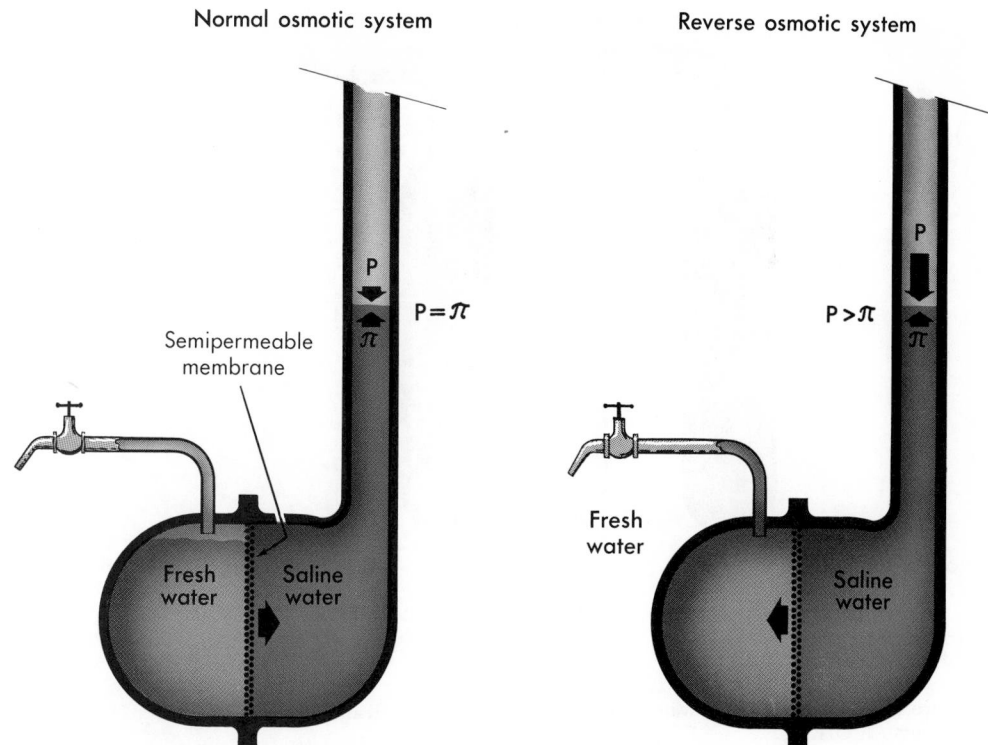

Figure 11.16 Purification of water by reverse osmosis. Sufficient pressure is applied to the solution to drive water through the membrane against the force of osmosis.

impermeable to the ions in sea water. Membranes which meet these requirements have recently been prepared; they consist of a relatively thin film of cellulose acetate. Pilot plants using these membranes are now desalinating 20,000 to 40,000 gallons of sea water per day.

ION EXCHANGE. Water of very high purity can be obtained by the ion exchange process shown schematically in Figure 11.17. The column at the left contains a synthetic cation exchange resin similar to that discussed earlier except that the exchangeable cation is H^+ rather than Na^+. The second column is packed with an anion exchange resin in which replaceable OH^- ions are embedded in a cationic network. Referring to the two resins as HR (cation exchanger) and R'OH (anion exchanger), we can write the following equations to describe what happens when water containing a dissolved salt MX is passed successively through the two columns

1st column: $\quad\quad\quad\quad M^+(aq) + HR(s) \rightarrow MR(s) + H^+(aq)$

2nd column: $\quad\quad\quad\quad X^-(aq) + R'OH(s) \rightarrow R'X(s) + OH^-(aq)$

The H^+ and OH^- ions produced react with each other to give water

$$H^+(aq) + OH^-(aq) \rightarrow H_2O(l)$$

We see from these equations that there are no ions in solution at the end of the process, which is referred to as *deionization*. In principle at least, any salt present in the water will be removed; M^+ can be any cation and X^- any anion. To regenerate the first column, it can be flushed with a

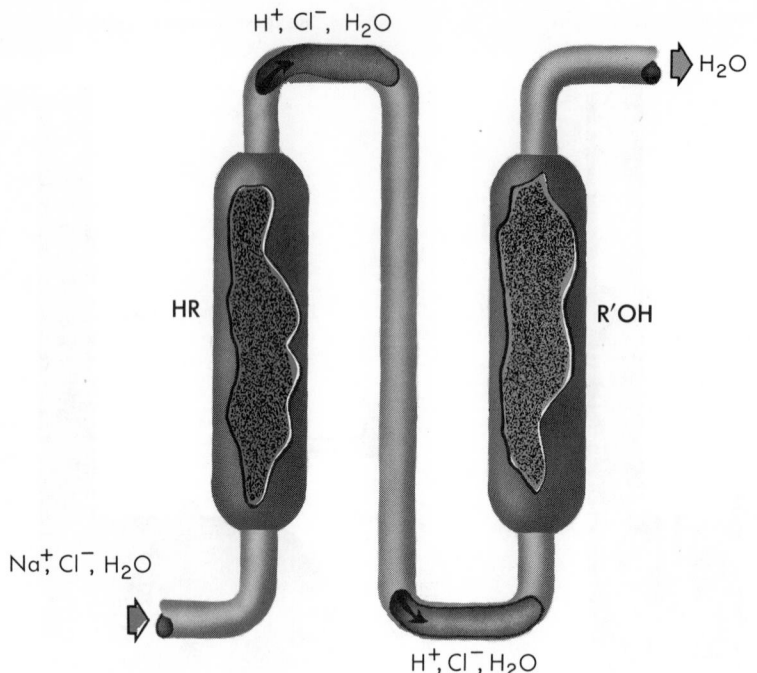

Figure 11.17   Purification of water using a cation exchanger HR and an anion exchanger R'OH.

strong solution of hydrochloric acid, which acts as a source of H⁺ ions. The second column is returned to its original state by flushing with sodium hydroxide.

The major drawback to the ion-exchange method is the cost of the chemicals used to regenerate the resins. They are expensive enough to rule out on an economic basis the use of this method for desalination except for brackish waters in which the concentration of salt is 300 parts per million or less (the concentration in sea water is 35,000 ppm). On the other hand, this method is widely used to purify water in the laboratory. Water which has been passed through ion-exchange columns has an electrical conductivity an order of magnitude lower than that of ordinary distilled water.

Distilled water, however, has the advantage that it doesn't contain any bacteria

PROBLEMS

11.1   Recently, there was considerable controversy about the existence of a new form of liquid water, called "anomalous" water. It was reported to have a vapor pressure much lower than that of ordinary water, a freezing point considerably below 0°C, a high viscosity and a density of 1.4 g/ml. Speculate upon what life would be like if all water were "anomalous."

11.17   One form of ice, stable at high pressures, has a density of 1.34 g/ml and melts at 32°C. Speculate as to what our environment would be like if this were the stable form of ice at atmospheric pressure.

11.2 Explain, in your own words, why water, unlike other hydrogen-bonded liquids, expands upon freezing.

11.3 Summarize the arguments for and against the idea that there are microcrystals of ice in liquid water.

11.4 If the parameter i in Equations 11.2 to 11.4 were equal to the number of moles of ions per mole of electrolyte, what would be the value of i for KNO_3? $CaSO_4$? $Al_2(SO_4)_3$? Would you expect the observed values of i to be larger or smaller than those you have calculated? For which salt would you expect the deviation to be the greatest?

11.5 Explain why DDT and other halogenated insecticides are concentrated in the food chain.

11.6 Suggest at least one method of solving the "phosphate" problem in detergents which was not mentioned in the text.

11.7 The total surface area of the oceans is 3.61×10^8 km². Estimate the distance by which the average level of the oceans would be raised if 50% of the ice and snow in the world melted (cf. Table 11.1).

11.8 In 1934 Bernal and Fowler suggested that liquid water might be a mixture of ordinary ice (d = 0.91 g/ml) and a high pressure form of ice (d = 1.08 g/ml). If this were true, what would have to be the fraction of ordinary ice in liquid water at 0°C?

11.9 The freezing point of a 0.010 M solution of H_2SO_4 is −0.052°C. Decide, by calculation, whether H_2SO_4 at this concentration ionizes predominantly as

$$H_2SO_4(aq) \rightarrow H^+(aq) + HSO_4^-(aq)$$

or

$$H_2SO_4(aq) \rightarrow 2\ H^+(aq) + SO_4^{2-}(aq)$$

11.10 The concentration of dissolved oxygen in a sample of waste water decreases from 6.8×10^{-3} g/liter to 5.2×10^{-3} g/liter after five days. Calculate the BOD of the sample in mg/liter.

11.11 Twenty kilograms of acetaldehyde, CH_3CHO, is discharged into a water supply which has a total mass of 1.5×10^6 liters. By how much does this increase the BOD (mg/liter) of the water? Assume the following reaction occurs:

$$CH_3CHO(aq) + 2\ O_2(aq) \rightarrow 2\ CO_2(aq) + 2\ H_2O$$

11.18 Why does water have a higher boiling point than either HF or NH_3, both of which have comparable molecular weights and are hydrogen-bonded?

11.19 Describe in your own words the "flickering cluster" model for liquid water.

11.20 Explain in terms of the Debye-Hückel theory why the parameter i

    a. is smaller for $MgCl_2$ than for NH_4Cl at a given concentration.
    b. decreases with increasing concentration.

11.21 Discuss the advantages and disadvantages of substituting parathion for DDT.

11.22 Discuss the relative merits of using "high-phosphate" as against "low-phosphate" detergents in a rural home where waste water goes into a septic tank.

11.23 Every day the earth receives about 1.8×10^{18} kcal of energy from the sun; about 30% of this energy is used to evaporate water. How many kg of water are evaporated each day?

11.24 Pople (Proceedings of the Royal Society of London, A 205: 163, 1951) suggested a model for liquid water in which the hydrogen bonds in ice are bent so that the O—H—O angle is no longer linear. Suggest how the peculiar behavior of water between −10 and +10°C could be explained on the basis of this model.

11.25 The freezing point of a 3.0×10^{-3} M solution of HNO_2 is −0.0061°C. What fraction of the nitrous acid is ionized?

$$HNO_2(aq) \rightarrow H^+(aq) + NO_2^-(aq)$$

11.26 The BOD of the water in a certain river is 5.3 mg/liter. The solubility of oxygen in pure water saturated with air is 8.9×10^{-3} g/liter. What is the concentration of O_2 in this water when equilibrium is reached with the BOD?

11.27 A certain electroplating company discharges CN^- into a river. An analyst finds that every time a batch of cyanide is released, the BOD of the river water increases by 4.0 mg/liter. Assuming the reaction

$$CN^-(aq) + \tfrac{3}{2}\ O_2(g) + H^+(aq) \rightarrow$$
$$CO_2(aq) + \tfrac{1}{2}\ N_2(g) + \tfrac{1}{2}\ H_2O$$

what is the concentration of CN^- in the river?

11.12 If the concentration of mercury in a certain brand of canned tuna fish is 0.62 part per million, how many grams of mercury are there in a one-ounce can? If all the mercury is present as $Hg(CH_3)_2$, how much of this compound is there in the can?

11.13 How many grams of carbon would be required to react with one cubic meter of water in which the concentration of chlorine is 0.65 ppm?

11.14 A water supply contains the following ions in the indicated concentrations. Determine the number of moles of Na_2CO_3 and $Ca(OH)_2$ that should be added to soften one liter of this water.

    a. 1.6×10^{-4} M Ca^{2+}, 3.2×10^{-4} M HCO_3^-
    b. 1.6×10^{-4} M Ca^{2+}, 3.2×10^{-4} M Cl^-
    c. 1.6×10^{-4} M Ca^{2+}, 1.2×10^{-4} M HCO_3^-, 2.0×10^{-4} M Cl^-

11.15 Write balanced equations for

    a. the removal of Zn^{2+} from water using the zeolite NaZ.
    b. the removal of dissolved Na_2SO_4 by passing water through a deionizer containing the resins HR and R'OH.

11.16 Sea water can be considered to be a 3.5% by weight solution of NaCl. Calculate the pressure required to extract pure water from sea water by reverse osmosis. (Assume $i = 1.9$, $T = 25°C$.)

11.28 The concentration of Cd^{2+} in a certain water supply is 0.0060 part per million. What is the molarity of Cd^{2+}?

11.29 A column containing 100 g of activated charcoal is used to remove chlorine at a concentration of 1.0 part per milllion in water. Referring to Equation 11.11, how many liters of water can be "dechlorinated" by the column?

11.30 A municipal water supply contains the following ions in the indicated concentrations. Calculate the number of grams of $Ca(OH)_2$ and Na_2CO_3 that should be added to soften 10 liters of this water.

    a. 6.0×10^{-4} M Ca^{2+}, 6.0×10^{-4} M HCO_3^-, 3.0×10^{-4} M SO_4^{2-}
    b. 5.0×10^{-4} M Ca^{2+}, 1.0×10^{-3} M HCO_3^-
    c. 4.0×10^{-4} M Ca^{2+}, 2.0×10^{-4} M SO_4^{2-}

11.31 A principal objective of water softening is the removal of Ca^{2+}; yet, in the lime-soda process, more Ca^{2+} ions are added in the form of $Ca(OH)_2$. Explain why this is done.

11.32 Considering sea water to be a 3.5% by weight solution of NaCl, calculate

    a. the temperature at which salt water starts to freeze (take $i = 1.8$).
    b. the freezing point of the solution that remains after 60% of the water has been removed as ice (take $i = 1.7$).

*11.33 Suggest several experiments which might be carried out to determine the effect of a particular salt on the water structure, i.e., how it changes the ratio of "flickering clusters" to free molecules.

*11.34 Each day a certain power plant takes in 2.0×10^6 liters of water at 25°C from a river and uses it to condense 50,000 kg of steam.

    a. What is the temperature of the water when it is discharged?
    b. If the river into which this water is discharged has a flow of 4.0×10^7 liter/day, estimate the increase in temperature to be expected downstream.

*11.35 Many of the properties of sea water have been determined using synthetic samples, made by adding pure salts to water. Using the data in Table 11.7, write out a set of directions telling a technician how to prepare such a sample, using one kg of water and the appropriate pure salts.

12 SPONTANEITY OF REACTION; ΔG AND ΔS

In our discussion of thermochemistry in Chapter 4 we concentrated upon reactions carried out at constant pressure and temperature in such a way that no useful work is done. Under these conditions, which are the ones that we encounter most frequently in the world around us, the heat flow is equal to the difference in enthalpy, ΔH, between products and reactants.

In this chapter we return to a discussion of reactions carried out at constant pressure and temperature. This time we shall be interested in one crucial question: how can we predict in advance whether a given reaction would occur under these conditions, given sufficient time? Putting it another way, how can we decide whether a reaction, at a certain temperature and pressure, will be spontaneous?

We shall find that a knowledge of the sign and magnitude of the enthalpy change, ΔH, will help us to answer this basic question. However, in order to establish a completely general criterion of spontaneity, it will be necessary to introduce two new functions; these are the free energy change, ΔG, and the entropy change, ΔS.

12.1 CRITERION OF SPONTANEITY. MAXIMUM USEFUL WORK

We know from experience that certain chemical and physical changes are spontaneous in the sense that they occur "by themselves" without any exertion on our part. An ice cube melts when we take it out of the refrigerator. Iron exposed to moist air rusts. A mixture of methane and oxygen burns when we set a match to it. Certain other processes are clearly nonspontaneous. No one has ever observed ice cubes forming in a glass of water at 25°C. A hammer left out on the lawn all winter does not lose its coating of rust when spring comes. The products of combustion of natural gas do not recombine to form methane and oxygen. In summary, at room temperature and atmospheric pressure, the following processes are spontaneous, whereas the reverse processes are nonspontaneous:

$$H_2O(s) \rightarrow H_2O(l) \tag{12.1}$$

$$2\ Fe(s) + \tfrac{3}{2}\ O_2(g) + 3\ H_2O(l) \rightarrow 2\ Fe(OH)_3(s) \tag{12.2}$$

$$CH_4(g) + 2\ O_2(g) \rightarrow CO_2(g) + 2\ H_2O(l) \tag{12.3}$$

It is not always easy to tell whether a reaction is spontaneous; appearances can be deceiving. A mixture of hydrogen and oxygen can be maintained for long periods of time without any apparent reaction. This might lead us to conclude mistakenly that the reaction

$$H_2(g) + \tfrac{1}{2}\ O_2(g) \rightarrow H_2O(l) \tag{12.4}$$

Some spontaneous reactions must be initiated, or occur very slowly in the absence of a catalyst

is nonspontaneous. However, if we bring a lighted match or a platinum wire up to the mixture, reaction 12.4 occurs, often with explosive violence. This tells us that the formation of liquid water from the elements is a spontaneous process, since it is *capable* of proceeding by itself once it is initiated.

It would be very convenient to be able to predict in advance whether a given reaction is potentially spontaneous. Consider, for example, the reaction

$$CO(g) + NO(g) \rightarrow CO_2(g) + \tfrac{1}{2}\ N_2(g) \tag{12.5}$$

If this reaction could be shown to be spontaneous at, let us say 25°C and 1 atm pressure, it might offer an ideal way to remove two of the most serious air pollutants, carbon monoxide and nitric oxide. Certainly it would be worth looking for a suitable catalyst or other means of initiating the reaction. If, on the other hand, it can be shown by calculation that it is impossible to make this reaction occur at any reasonable temperature and pressure, we may as well forget about it and concentrate upon other processes for removing these species from air.

A hundred years ago many chemists felt that they had a general criterion for predicting reaction spontaneity. The prevailing idea, espoused by P. M. Berthelot in Paris and Julius Thomsen in Copenhagen, was that all spontaneous reactions are exothermic. If this were true, all we would have to do to predict reaction spontaneity would be to calculate the enthalpy change, ΔH, and look at its sign. If ΔH turned out to be negative, we could assume that the reaction must be spontaneous; if ΔH were positive, the reaction could not occur by itself.

It turns out that almost all exothermic *chemical reactions* are spontaneous *at 25°C and 1 atm*. For reactions 12.2 to 12.4, ΔH is −189 kcal, −213 kcal, and −68.3 kcal, respectively; each of these reactions is spontaneous at room temperature and atmospheric pressure. On the other hand, this simple rule fails for many familiar phase changes. An example is the melting of ice (Equation 12.1) which takes place at 25°C and 1 atm even though ΔH is a positive quantity, +1.4 kcal. In another case we find that potassium nitrate dissolves when added to water at room temperature even though the solution process

$$KNO_3(s) \rightarrow K^+(aq) + NO_3^-(aq); \quad \Delta H = +8.4\ kcal \tag{12.6}$$

is endothermic.

There is still another basic objection to using the sign of ΔH as a completely general criterion for spontaneity. We often find that endothermic reactions which are nonspontaneous at room temperature become spontaneous when the temperature is raised. Consider, for example, the reaction by which carbon dioxide and quicklime are prepared from limestone

$$CaCO_3(s) \rightarrow CaO(s) + CO_2(g); \quad \Delta H = +42.5 \text{ kcal} \qquad (12.7)$$

The spontaneity of a reaction may depend on both temperature and pressure

At 25°C and 1 atm, this reaction is nonspontaneous, as shown by the existence of the white cliffs of Dover over eons of time. However, if the temperature is raised to about 850°C, the limestone decomposes to give off carbon dioxide at 1 atm pressure. If we operate at lower pressures, calcium carbonate decomposes even more readily; at 0.1 mm Hg, Reaction 12.7 becomes spontaneous at 500°C. We see then that it is possible to make this endothermic reaction spontaneous by increasing the temperature or reducing the pressure, even though ΔH remains virtually constant at +42.5 kcal, nearly independent of both temperature and pressure. Clearly, even though the sign of ΔH is a fairly reliable indicator of reaction spontaneity at room temperature and atmospheric pressure, it cannot be the general criterion that we are looking for.

Late in the 19th century J. Willard Gibbs, professor of mathematical physics at Yale, showed that the proper criterion for the spontaneity of a chemical reaction is its capacity to produce useful work. He proved that **if, at constant temperature and pressure, a reaction can, in principle or in practice, be harnessed to perform useful work, that reaction is spontaneous. If work has to be supplied from the surroundings to make the reaction occur, it cannot be spontaneous.**

To illustrate the meaning of Gibbs' criterion of spontaneity, we will take as our first example a simple physical change, the vaporization of water:

$$H_2O(l, 101°C, 1 \text{ atm}) \rightarrow H_2O(g, 101°C, 1 \text{ atm}) \qquad (12.8)$$

We know from experience that this process is spontaneous; water heated above 100°C in a beaker or saucepan boils to produce steam at atmospheric pressure. Thomas Newcomen demonstrated, when he designed a practical steam engine over 200 years ago, that this process can perform useful work. Notice that when we boil water in an open container in the laboratory or in the kitchen, no useful work is produced; the force of the steam escaping into the atmosphere is not harnessed in any way. The fact that it is *possible* to make this process do work in a machine such as a steam engine proves its spontaneity.

Let us now consider a chemical reaction, the combustion of methane

$$CH_4(g) + 2\ O_2(g) \rightarrow CO_2(g) + 2\ H_2O(l)$$

As pointed out earlier, this reaction occurs spontaneously when we light a Bunsen burner or a gas range. Under these conditions, no useful work is done; all that is observed is the evolution of heat. However, there are at least two ways in which we can persuade this reaction to do work for us. One is to carry out the reaction in an internal combustion engine, thereby generating mechanical energy. Natural gas is now being used as a fuel to drive fleets of trucks operated by private companies and government

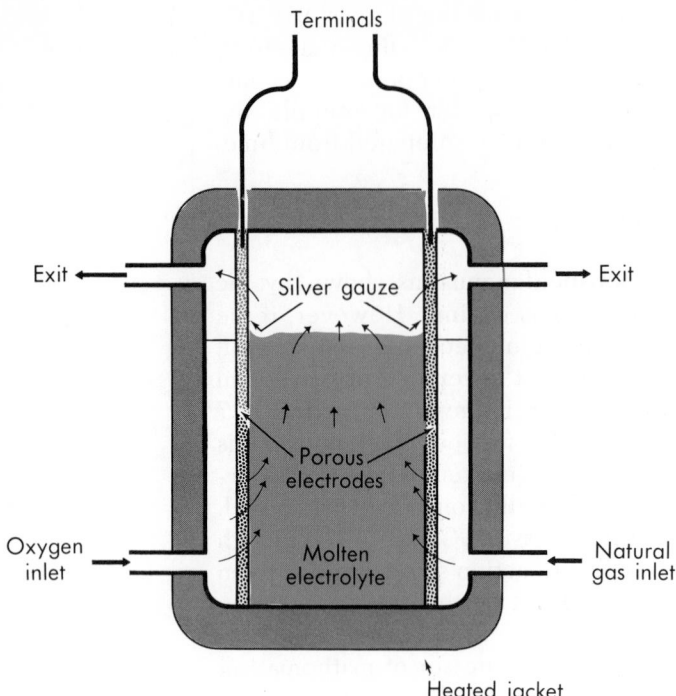

Figure 12.1 Generation of electrical energy in a fuel cell using the reaction between methane and oxygen.

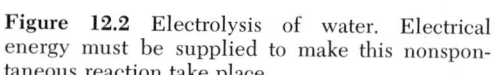

Figure 12.2 Electrolysis of water. Electrical energy must be supplied to make this nonspontaneous reaction take place.

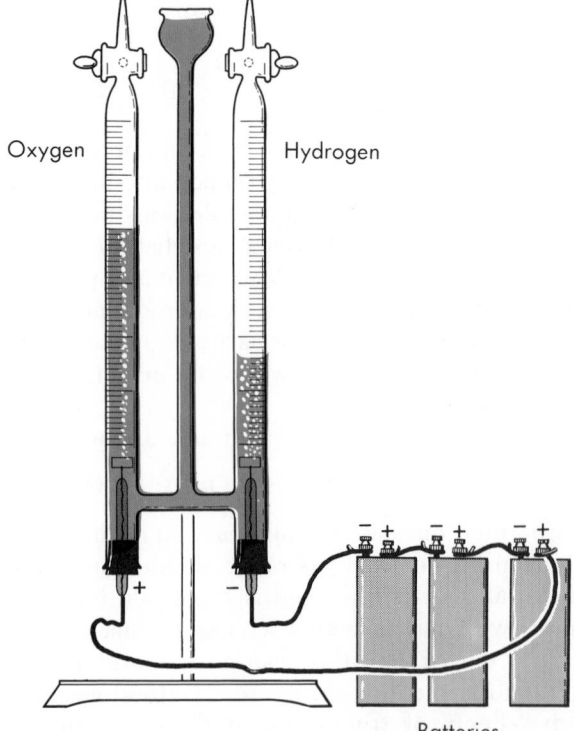

agencies in California. Alternatively the reaction of methane with oxygen can be carried out in a fuel cell, a device that generates electrical energy (Fig. 12.1). Fuel cells utilizing this or other spontaneous reactions are now being used.

In contrast to the two reactions just discussed, consider the decomposition of water to the elements.

$$H_2O(l) \rightarrow H_2(g) + \tfrac{1}{2} O_2(g) \tag{12.9}$$

This reaction can be made to take place in the electrolysis cell shown in Figure 12.2. However, in order for this to happen, electrical energy must be supplied from an external source, which may be a fuel cell, a lead storage battery, or a series of dry cells. Since work has to be done to make the reaction go, we conclude that it is nonspontaneous. Notice that "nonspontaneous" does not imply "impossible"; we often find that we can make nonspontaneous reactions occur, but it always costs us work to do so.

MAXIMUM USEFUL WORK

The amount of useful work that we can get out of a reaction depends to some extent upon the efficiency of the machine used to harness the reaction. When natural gas burns in an internal combustion engine, we seldom get more than 25 to 50 kcal of useful work per mole of methane. In a fuel cell we can do somewhat better, perhaps obtaining as much as 160 kcal of work per mole of methane. We find, however, that there is an upper limit to the amount of work that can be obtained from Reaction 12.3. No matter how ingenious or how economical we are, we can never get more than 195.6 kcal of useful work per mole of methane consumed at 25°C and 1 atm. We may get no work at all, we may get 100 kcal or 190 kcal or even, conceivably, 195 kcal, but no scheme that we can devise will enable us to get 200 kcal of useful work out of this reaction.

The behavior just described is characteristic of all reactions. For every reaction taking place at constant temperature and pressure, there is a certain maximum amount of useful work (W_m) that can be obtained from a given amount of reactant. As you might expect, this maximum work is produced when the reaction is carried out slowly against a resisting force only slightly less than the force driving the reaction.

Values of W_m are listed below for a few of the reactions discussed up to this point.

$$CH_4(g) + 2\ O_2(g) \rightarrow CO_2(g) + 2\ H_2O(l); \quad W_m = +195.6\ \text{kcal} \tag{12.3}$$

$$CO(g) + NO(g) \rightarrow CO_2(g) + \tfrac{1}{2}\ N_2(g); \quad W_m = +82.2\ \text{kcal} \tag{12.5}$$

$$H_2O(l) \rightarrow H_2(g) + \tfrac{1}{2}\ O_2(g); \quad W_m = -56.7\ \text{kcal} \tag{12.9}$$

We interpret the value of W_m in Reaction 12.5 to mean that we can obtain up to 82.2 kcal of useful work from the reaction of one mole of carbon monoxide with nitric oxide. For Reaction 12.9 the negative sign of W_m tells us that we have to supply at least 56.7 kcal of work to decompose a mole of water.

Previously we related the spontaneity of a reaction to its capacity to do useful work. From a slightly different point of view we can relate re-

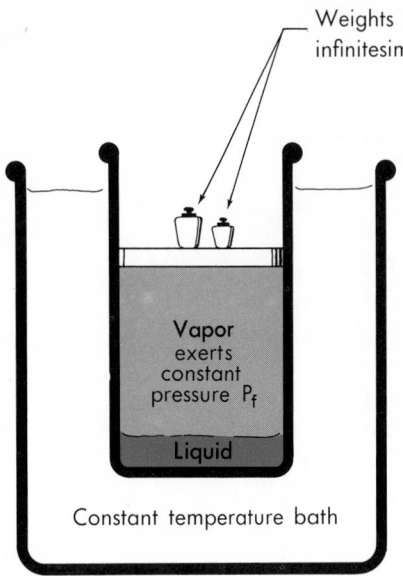

Weights exert a pressure
infinitesimally less than P_f.

Vapor
exerts
constant
pressure P_f

Liquid

Constant temperature bath

Figure 12.3 Reversible vaporization of a liquid. The maximum amount of work is obtained when the driving force is only infinitesimally greater than the opposing force.

action spontaneity to the sign of W_m. A reaction carried out at constant temperature and pressure is spontaneous if W_m is a positive quantity and nonspontaneous if it is negative. That is

if $W_m > 0$; reaction spontaneous; useful work can be produced.
if $W_m < 0$; reaction nonspontaneous; work must be consumed to make the reaction occur.

If, by chance, W_m is zero, the reaction system is at equilibrium and will undergo no change.

12.2 FREE ENERGY CHANGE, ΔG

You will recall that the heat flow in a reaction carried out at constant pressure and temperature can be related to the enthalpies of products and reactants. If, for example, the enthalpy of the products is 10 kcal less than that of the reactants ($\Delta H = -10$ kcal), the reaction is exothermic and 10 kcal of heat is produced in the surroundings. If, on the other hand, the enthalpy of the products is 10 kcal greater than that of the reactants ($\Delta H = +10$ kcal), the reaction is endothermic and 10 kcal of heat must be supplied to make the reaction go.

In a similar manner the capacity of a reaction, carried out at constant pressure and temperature, to produce useful work can be interpreted in terms of a fundamental property of products and reactants known as their **free energy** and given the symbol **G**. If, for example, the free energy of the products is 10 kcal less than that of the reactants ($\Delta G = -10$ kcal), as much as 10 kcal of work can be produced in the surroundings ($W_m = 10$ kcal), and the reaction is spontaneous. If, on the other hand, the free energy of the products is 10 kcal greater than that of the reactants ($\Delta G = +10$ kcal), then at least 10 kcal of work must be supplied to make the nonspontaneous reaction go ($W_m = -10$ kcal).

The free energy of a substance reflects its capacity to do useful work in a chemical reaction

From this discussion it should be clear that, in general,

$$\Delta G = -W_m \tag{12.10}$$

and, hence, that

if ΔG is $-$, reaction is spontaneous at constant P,T.
if ΔG is $+$, reaction is nonspontaneous; reverse reaction is spontaneous.
if ΔG is 0, reaction is at equilibrium and will not proceed in either direction.

The argument which we have gone through to correlate reaction spontaneity with ΔG is an abbreviated version of that presented by Gibbs nearly a century ago. His English is more elegant than ours and his logic a great deal tighter, but Gibbs concluded as we do that it is the sign of ΔG which is the general criterion for predicting the spontaneity of a reaction taking place at a given temperature and pressure.

FREE ENERGY CHANGE AT 1 atm, $\Delta G^{1\,atm}$

The free energy change, ΔG, unlike the enthalpy change, ΔH, can vary considerably with both temperature and pressure. Indeed, this must be the case if the sign of ΔG is to be the criterion of reaction spontaneity. We have seen that there is a shift in the direction in which the reaction

$$CaCO_3(s) \rightarrow CaO(s) + CO_2(g)$$

proceeds as we change the temperature and pressure. At 25°C and 1 atm this reaction is nonspontaneous ($\Delta G > 0$). It can be made spontaneous ($\Delta G < 0$) by raising the temperature or lowering the pressure. In Table 12.1 we have tabulated values of ΔG at various temperatures and pressures. Notice that it changes from $+31.1$ kcal at 25°C and 1 atm to -5.9 kcal at 1000°C and 1 atm and to -41.5 kcal at 1000°C and 10^{-6} atm (0.00076 mm Hg). In contrast, ΔH for this reaction remains virtually constant at $+42.5$ kcal over the entire range of temperatures and pressures.

We shall discuss later in this chapter the variation of ΔG with temperature. We shall not, however, consider further its variation with pressure. In all our calculations we shall work with the **free energy change at one atmosphere pressure**, for which we shall write the symbol $\Delta G^{1\,atm}$. You should keep in mind, however, that ΔG and, hence, the direction of spontaneity of reaction might be quite different if we were to allow the pressure

Knowing the $\Delta G^{1\,atm}$ for a reaction allows us to determine its spontaneity at 1 atm pressure

TABLE 12.1   Variation of ΔG with Temperature and Pressure
for the Reaction $CaCO_3(s) \rightarrow CaO(s) + CO_2(g)$

T(°C)	P = 1 atm	P = 10^{-2} atm	P = 10^{-4} atm	P = 10^{-6} atm
25	+31.1	+28.4	+25.6	+22.9
200	+24.5	+20.2	+15.8	+11.5
400	+16.9	+10.7	+4.5	−1.7
600	+9.3	+1.2	−6.9	−15.0
800	+1.7	−8.3	−18.2	−28.2
1000	−5.9	−17.8	−29.6	−41.5

to vary. From Table 12.1 we see that at 600°C, $\Delta G^{1\,atm} = +9.3$ kcal, but ΔG at 10^{-6} atm $= -15.0$ kcal.

FREE ENERGIES OF FORMATION

We saw in Chapter 4 how changes in enthalpy, ΔH, can be calculated from heats of formation, ΔH_f, of products and reactants. It is possible to carry out analogous calculations for free energy changes, ΔG, in chemical reactions if the free energies of formation, ΔG_f, of products and reactants are available. The free energy of formation of a pure substance is taken to be the free energy change when one mole of the substance is formed from the elements at 1 atm pressure; it is assigned the symbol $\Delta G_f^{1\,atm}$. In Table 12.2 we have listed values of $\Delta G_f^{1\,atm}$ for a variety of compounds at 25°C.

From examination of Table 12.2 you will note that $\Delta G_f^{1\,atm}$ is negative for most compounds at 25°C. This reflects the fact that most compounds are stable with respect to decomposition into their elements at 25°C and 1 atm pressure. In general, any compound with a negative free energy of formation can be formed directly from the elementary substances. For example, in the reaction

$$\tfrac{1}{2}\,N_2(g) + \tfrac{3}{2}\,H_2(g) \rightarrow NH_3(g), \ \Delta G^{1\,atm} = \Delta G_f^{1\,atm}\,NH_3 = -4.0 \text{ kcal} \quad (12.11)$$

The free energy change $\Delta G^{1\,atm}$ in the reaction, which by definition is equal

TABLE 12.2   Free Energies of Formation (kcal/mole) at 25°C, 1 atm

AgBr(s)	−22.9	CO(g)	−32.8	H_2O(g)	−54.6	NH_4Cl(s)	−48.7
AgCl(s)	−26.2	CO_2(g)	−94.3	H_2O(l)	−56.7	NO(g)	+20.7
AgI(s)	−15.9	C_2H_2(g)	+50.0	H_2S(g)	−7.9	NO_2(g)	+12.4
Ag_2O(s)	−2.6	C_2H_4(g)	+16.3	HgO(s)	−14.0	NiO(s)	−51.7
Ag_2S(s)	−9.6	C_2H_6(g)	−7.9	HgS(s)	−11.7	$PbBr_2$(s)	−62.1
Al_2O_3(s)	−376.8	C_3H_8(g)	−5.6	KBr(s)	−90.6	$PbCl_2$(s)	−75.0
$BaCl_2$(s)	−193.8	CoO(s)	−51.0	KCl(s)	−97.6	PbO(s)	−45.1
$BaCO_3$(s)	−272.2	Cr_2O_3(s)	−250.2	$KClO_3$(s)	−69.3	PbO_2(s)	−52.3
BaO(s)	−126.3	CuO(s)	−30.4	KF(s)	−127.4	Pb_3O_4(s)	−147.6
$BaSO_4$(s)	−323.4	Cu_2O(s)	−35.0	$MgCl_2$(s)	−141.6	PCl_3(g)	−68.4
$CaCl_2$(s)	−179.3	CuS(s)	−11.7	$MgCO_3$(s)	−246	PCl_5(g)	−77.6
$CaCO_3$(s)	−269.8	$CuSO_4$(s)	−158.2	MgO(s)	−136.1	SiO_2(s)	−192.4
CaO(s)	−144.4	Fe_2O_3(s)	−177.1	$Mg(OH)_2$(s)	−199.3	$SnCl_4$(l)	−113.3
$Ca(OH)_2$(s)	−214.3	Fe_3O_4(s)	−242.4	$MgSO_4$(s)	−280.5	SnO(s)	−61.5
$CaSO_4$(s)	−315.6	HBr(g)	−12.7	MnO(s)	−86.8	SnO_2(s)	−124.2
CCl_4(l)	−16.4	HCl(g)	−22.8	MnO_2(s)	−111.4	SO_2(g)	−71.8
CH_4(g)	−12.1	HF(g)	−64.7	NaCl(s)	−91.8	SO_3(g)	−88.5
$CHCl_3$(l)	−17.1	HI(g)	0.3	NaF(s)	−129.3	ZnO(s)	−76.1
CH_3OH(l)	−38.7	HNO_3(l)	−19.1	NH_3(g)	−4.0	ZnS(s)	−47.4

to the molar free energy of formation of ammonia, $\Delta G_f^{1\,atm}\ NH_3(g)$, is equal to −4.0 kcal. By our thermodynamic criteria, the reaction by which ammonia is formed from the elements is spontaneous. (In practice, Reaction 12.11 occurs very slowly, so that at 25°C and 1 atm a mixture of N_2 and H_2 may exist for many years without producing a significant amount of NH_3. Many spontaneous reactions occur very slowly, particularly at low temperatures.)

Free energy changes at 1 atm can be calculated from $\Delta G_f^{1\,atm}$ data in a manner completely analogous to that used to find enthalpy changes from ΔH_f data. The relation we use is:

$$\Delta G^{1\,atm} = \Sigma\ \Delta G_f^{1\,atm}\ \text{products} - \Sigma\ \Delta G_f^{1\,atm}\ \text{reactants} \qquad (12.12)$$

Example 12.1 Find the free energy change $\Delta G^{1\,atm}$ at 25°C for the following reactions:

  a. $Al_2O_3(s) + 3\ H_2(g) \rightarrow 2\ Al(s) + 3\ H_2O(l)$
  b. $4\ NH_3(g) + 5\ O_2(g) \rightarrow 4\ NO(g) + 6\ H_2O(l)$

Solution

  a. $\Delta G^{1\,atm} = \Sigma\ \Delta G_f^{1\,atm}\ \text{products} - \Sigma\ \Delta G_f^{1\,atm}\ \text{reactants}$
     $= 2\ \Delta G_f^{1\,atm}\ Al(s) + 3\ \Delta G_f^{1\,atm}\ H_2O(l)$
     $-(\Delta G_f^{1\,atm}\ Al_2O_3(s) + 3\ \Delta G_f^{1\,atm}\ H_2(g))$

  The free energies of formation of the elementary substances are, by the definition of $\Delta G_f^{1\,atm}$, zero. Therefore,

$$\Delta G^{1\,atm} = 3(-56.7)\ \text{kcal} - (-376.8)\ \text{kcal} = +206.7\ \text{kcal}$$

  This reaction does not occur at 25°C and 1 atm.

  b. $\Delta G^{1\,atm} = 4\ \Delta G_f^{1\,atm}\ NO(g) + 6\ \Delta G_f^{1\,atm}\ H_2O(l) - 4\ \Delta G_f^{1\,atm}\ NH_3(g)$
     $= 4(20.7)\ \text{kcal} + 6(-56.7)\ \text{kcal} - 4(-4.0)\ \text{kcal} = -241.4\ \text{kcal}$

  The reaction is spontaneous at 25°C and 1 atm.

12.3 ENTROPY CHANGE, ΔS

Let us consider once again the reaction

$$CaCO_3(s) \rightarrow CaO(s) + CO_2(g) \qquad (12.7)$$

In Figure 12.4 we have plotted the free energy change at 1 atm pressure as a function of temperature, using data taken from Table 12.1. In the same plot we show ΔH, which remains nearly constant at +42.5 kcal over the entire temperature range.

There are two features of this plot which are particularly interesting: (1) ΔG appears to be a linear function of temperature; (2) if we extrapolate the straight-line plot to 0°K, it appears that, at that temperature, ΔG would become equal to ΔH.

These observations are summarized by the following equation

$$\Delta G = \Delta H - aT \qquad (12.13)$$

where a is a constant whose value can be determined from the slope of the straight line.

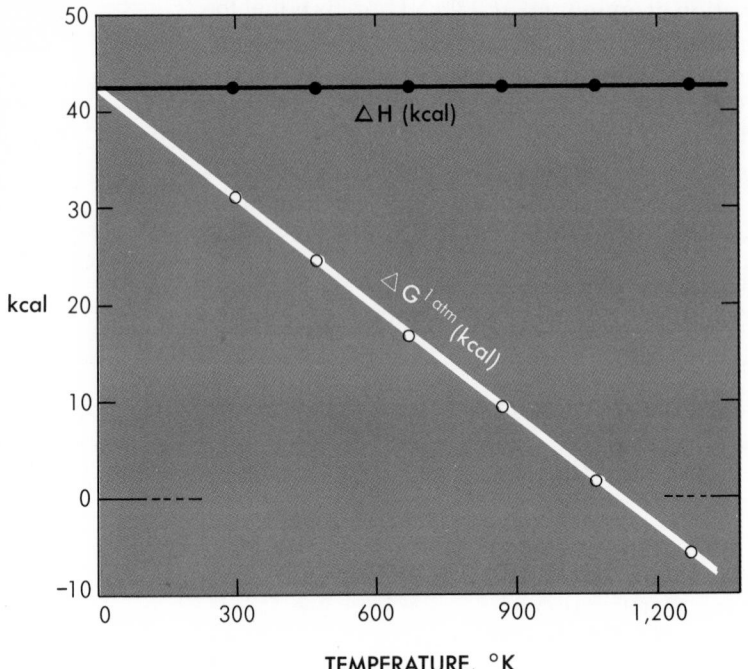

Figure 12.4 ΔH and $\Delta G^{1\,atm}$ as a function of temperature for the reaction $CaCO_3(s) \rightarrow CaO(s) + CO_2(g)$.

Is this reaction spontaneous at 1000°C and 1 atm?

Example 12.2 Using Figure 12.4, calculate "a" for Reaction 12.7.

Solution. Solving Equation 12.13 for "a," we obtain

$$a = \frac{\Delta H - \Delta G}{T}$$

We can apply this equation at any point; let us, for convenience, choose 1000°K, where $\Delta H = +42.5$ kcal and ΔG appears to be about +4.5 kcal

$$a = \frac{(42.5 - 4.5)\,\text{kcal}}{1000°K} = 0.0380\,\frac{\text{kcal}}{°K} = 38.0\,\frac{\text{cal}}{°K}$$

The quantity we have just calculated has more than a trivial importance since, through Equation 12.13, it relates the enthalpy change, ΔH, associated with the heat flow in a reaction to the free energy change, ΔG, which tells us how much work the reaction can produce. The quantity "a," like ΔH and ΔG, is equal to the change in a fundamental property characteristic of the reaction. This property is called the **entropy** and is given the symbol S. Associated with every reaction, there is a change in entropy, ΔS, equal to the entropy of the products minus the entropy of the reactants. In Reaction 12.7, ΔS is equal to the sum of the entropies of one mole of calcium oxide and one mole of carbon dioxide minus the entropy of one mole of calcium carbonate. That is:

$$\Delta S = S\ CaO(s) + S\ CO_2(g) - S\ CaCO_3(s) = 38.0\ \text{cal}/°K$$

The analysis that we have just applied to Reaction 12.7 can be generalized to describe the relationship between ΔG and ΔH in any reaction. The general equation is written in the form

$$\Delta G = \Delta H - T\Delta S \qquad (12.14)$$

where T is the absolute temperature in °K and ΔS is the difference in entropy between products and reactants, i.e.,

$$\Delta S = \Sigma S_{products} - \Sigma S_{reactants}$$

We shall have a great deal more to say later in this chapter about the usefulness of Equation 12.14. In order to appreciate its significance, we must have a clear physical picture of what is meant by the entropy of a pure substance.

NATURE OF ENTROPY

The entropy of a substance, like its enthalpy, free energy, or volume, is one of its characteristic properties. Qualitatively, entropy is a measure of *disorder* or *randomness*. Substances which are highly disordered have high entropies; low entropy is associated with strongly ordered substances. "Order," in the sense we use it here, describes the extent to which the particles of a substance are confined to a given region of space. In a crystal, where the atoms, molecules, or ions are fixed in position, the entropy is relatively low. When a solid melts, any particle is free to move through the entire liquid volume and the entropy of the substance increases. Upon vaporization the particles acquire still greater freedom to move about and there is a large increase in entropy.

Considering the rather nebulous way in which we have described entropy, you may be surprised to learn that we can determine absolute values for the molar entropies of pure substances. Actually, it is possible to define entropy very precisely in a thermodynamic sense and measure it by calorimetric means. The details of how this is done are beyond the level of this text, but the results are readily interpreted. In Table 12.3 we have listed the molar entropies of some common solids, liquids, and gases.

At 0°K all pure crystalline substances are completely ordered and have zero entropy. (This is the Third Law of Thermodynamics)

Our simple interpretation of entropy in terms of disorder is well borne out by the data in Table 12.3. We see that solids as a group have lower entropies than liquids, which in turn have lower entropies than gases. Among substances in the same physical state, it is clear that molar entropy goes up with the number of atoms per formula unit; complex molecules have higher entropies than simple ones because there are more atoms to move about.

TABLE 12.3   Molar Entropies (cal/°K) at 1 atm, 25°C

SOLIDS		LIQUIDS		GASES	
C (diamond)	0.6	H_2O	16.7	He	30.1
Si	4.5	Hg	18.5	Ar	37.0
Al	6.8	CH_3OH	30.3	O_2	49.0
Na	12.2	Br_2	36.4	Cl_2	53.3
Pb	15.5	C_2H_5OH	38.4	NO_2	57.7
NaCl	17.3	$CHCl_3$	48.5	Br_2	58.6
RbBr	25.6	CCl_4	51.3	N_2O_4	72.7

Among solids, those that are very hard like diamond have lower entropies than those that are soft like sodium or lead.

As we might expect, the entropy of a substance increases as the temperature rises. The increased kinetic energy weakens the forces holding the particles together and gives them greater freedom of motion. For simple molecules, the effect is relatively small. The molar entropy of oxygen at 1 atm, $S^{1\,atm}$, increases by less than 4 per cent when the temperature is raised from 25 to 100°C. An increase in pressure decreases the entropy of a substance because it confines the molecules to a smaller volume. For liquids and solids the effect is very small, but for gases it is appreciable. Increasing the pressure on a gas from 1 to 10 atm decreases its entropy by about 4.6 cal/°K per mole.

ΔS FOR PHASE CHANGES

When a substance passes from a more ordered to a less ordered state by melting or vaporizing, its entropy increases. We can calculate the entropy change of vaporization at 1 atm pressure by realizing that $\Delta G^{1\,atm}_{vap}$ must be zero at the normal boiling point, since the liquid and vapor are in equilibrium at this temperature and pressure. Setting ΔG equal to zero in Equation 12.14 we obtain

$$\Delta S^{1\,atm}_{vap} = \frac{\Delta H_{vap}}{T_b} \qquad (12.15)$$

where T_b is the normal boiling point in °K. Similarly,

$$\Delta S^{1\,atm}_{fus} = \frac{\Delta H_{fus}}{T_f} \qquad (12.16)$$

where T_f is the melting point at 1 atm pressure.

You may recall from Chapter 9 that the quantity $\Delta H_{vap}/T_b$ is about 21 cal/°K for most liquids. This means that the ΔS should be about 21 cal/°K for vaporization at 1 atm. Referring to Table 12.3 we note that the entropy of vaporization of bromine is rather close to the predicted value

$$\Delta S^{1\,atm}_{vap} = \Delta S^{1\,atm}\ Br_2(g) - \Delta S^{1\,atm}\ Br_2(l)$$
$$= (58.6 - 36.4)\ cal/°K = 22.2\ cal/°K$$

Example 12.3 Using Equations 12.15 and 12.16, calculate $\Delta S^{1\,atm}$ for the vaporization of water and the melting of ice.

Solution. The heats of vaporization and fusion of H_2O are about 540 and 80 cal/g, or 9720 and 1440 cal/mole. For water, $T_b = 373$°K and $T_f = 273$°K.

$$\Delta S^{1\,atm}_{vap} = \frac{9720\ cal}{373°K} = 26\ cal/°K$$

$$\Delta S^{1\,atm}_{fus} = \frac{1440\ cal}{273°K} = 5.3\ cal/°K$$

The entropy of vaporization of water is significantly higher than the predicted value of 21 cal/°K because of the unusually strong intermolecular bonds in liquid water.

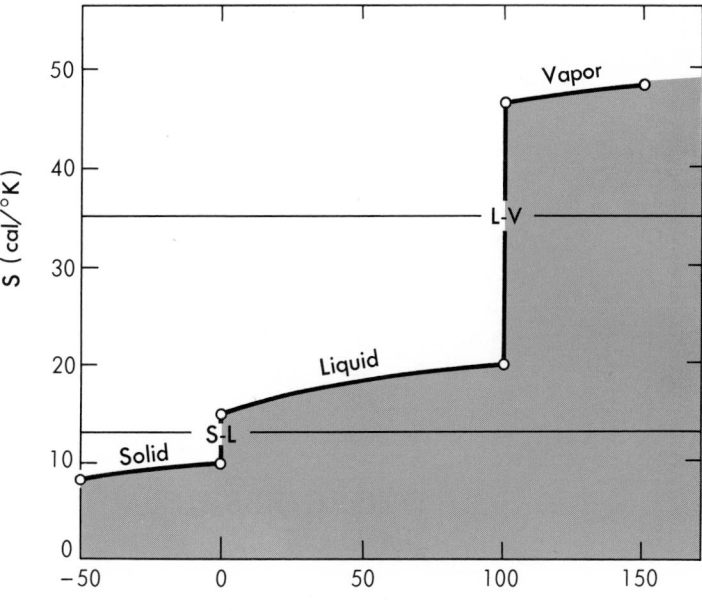

Figure 12.5 Variation with temperature of $S^{1\,atm}$ for one mole of water. Note large increases at 0°C (fusion) and 100°C (vaporization).

ΔS FOR CHEMICAL REACTIONS

A knowledge of the factors that lead to either high or low entropy can help us to predict the sign of ΔS for a chemical reaction. Almost invariably *a reaction which results in an increase in the number of moles of gas is accompanied by an increase in entropy; if the number of moles of gas decreases, ΔS will be a negative quantity.* This rule is illustrated by reactions 1 to 3 of Table 12.4. In Reaction 4, where there is the same number of moles of gas on both sides of the equation, ΔS is relatively small.

The molar entropies of gases are much larger than those of solids or liquids

For reactions in which no gases are involved (reactions 5 to 8, Table 12.4), we frequently, but not always, find that an increase in the total number of moles leads to an increase in entropy. As Table 12.4 implies, we can be confident that this rule will apply only if Δn_{tot} is large (Reaction 5) or if a liquid is produced at the expense of a solid (Reactions 5 and 6).

The entropy change for a reaction, at least above 25°C, is nearly independent of temperature. This is true despite the fact that the entropy of an

TABLE 12.4   Values of $\Delta S^{1\,atm}$ in cal/°K

REACTION	$\Delta S^{1\,atm}$	Δn_{gas}	Δn_{tot}
1. $CaCO_3(s) \rightarrow CaO(s) + CO_2(g)$	+38.4	+1	
2. $N_2O_4(g) \rightarrow 2\ NO_2(g)$	+42.2	+1	
3. $CH_4(g) + 2\ O_2(g) \rightarrow CO_2(g) + 2\ H_2O(l)$	−58.0	−2	
4. $2\ HCl(g) \rightarrow H_2(g) + Cl_2(g)$	−4.7	0	
5. $CuSO_4 \cdot 5\ H_2O(s) \rightarrow CuSO_4(s) + 5\ H_2O(l)$	+37.6		+5
6. $PbBr_2(s) \rightarrow Pb(s) + Br_2(l)$	+13.3		+1
7. $PbI_2(s) \rightarrow Pb(s) + I_2(s)$	+1.1		+1
8. $2\ AgI(s) \rightarrow 2\ Ag(s) + I_2(s)$	−6.2		+1

individual substance invariably increases with temperature. In most cases the increase in entropy of the products just about balances that of the reactants and ΔS does not change appreciably with temperature. Much the same argument can be used, incidentally, to justify neglecting the temperature dependence of ΔH, the enthalpy change of a reaction.[*]

On the other hand, ΔS for a reaction may vary appreciably with pressure if, as is often the case, there is a change in the number of moles of gas. We can calculate (cf. Problem 12.29) that ΔS for the decomposition of calcium carbonate, which is about +38 cal/°K at 1 atm, changes to +47 cal/°K at a pressure of 10^{-2} atm and +56 cal/°K at 10^{-4} atm. For this reason among others, we shall work with the **entropy change at one atmosphere pressure, ΔS$^{1\ atm}$**, in all our calculations.

12.4 THE GIBBS-HELMHOLTZ EQUATION

Having examined the concept of entropy and shown how changes in entropy can be predicted or calculated, we are now in a position to investigate Equation 12.14 in some depth. This equation, which is by all odds one of the most fundamental mathematical relationships in all of chemistry, is called the Gibbs-Helmholtz equation, in honor of the two men most responsible for its development.

$$\Delta G = \Delta H - T\Delta S \qquad (12.14)$$

This is one of the few truly profound relations in all of science

Like most fundamental equations, it has a very simple form. In words, it tells us that the driving force for a chemical reaction (ΔG) depends upon two quantities: the energy change due to making and breaking of bonds (ΔH), and the product of the change in randomness (ΔS) times the absolute temperature. The two factors which tend to make ΔG negative and, hence, lead to a spontaneous reaction are:

1. A NEGATIVE VALUE OF ΔH. Exothermic reactions (ΔH < 0) will tend to be spontaneous inasmuch as they contribute to a negative value of ΔG. On the molecular level this means that there will be a tendency to form "strong" bonds at the expense of "weak" ones.

2. A POSITIVE VALUE OF ΔS. If the entropy change is positive (ΔS > 0), the term −TΔS will make a negative contribution to ΔG. Consequently there will be a tendency for reactions to be spontaneous if the products are less ordered than the reactants.

If we think about the matter carefully, we realize that there is an inherent tendency, common to inanimate substances (and probably to animate objects like students and professors), to become disorganized. Water in a pond slowly evaporates, a beaker shatters when we drop it on a stone lab bench, rivers become polluted, and wilderness trails get cluttered with

---

[*] So far as Equation 12.14 is concerned, there is an added justification for ignoring the temperature dependence of ΔH and ΔS. These two quantities always change in the same direction as temperature changes (e.g., if ΔH becomes more positive, so also does ΔS). This means that the two effects tend to cancel each other and the true value of ΔG at any temperature is about the same as the one we calculate by taking ΔH and ΔS to be constant, independent of temperature.

litter. All these processes involve an increase in entropy and are spontaneous, as anyone who tries to reverse them soon discovers.

In many physical processes the entropy increase is the major driving force. An example is the formation of a solution. When nitrogen diffuses into oxygen or benzene dissolves in toluene, the enthalpy change is virtually zero, but ΔS is a positive quantity because the solution is less ordered than the pure substances from which it is formed. Another physical process driven by the entropy factor is osmosis. The movement of water through a semipermeable membrane from a more dilute to a more concentrated solution equalizes the two concentrations and, hence, leads to a state of maximum entropy.

In a few chemical reactions ΔS is nearly zero and the enthalpy change is the only important component of the driving force. Most of the reactions in this category are ones which involve no change in the number of moles of gas. An example is the synthesis of hydrogen fluoride from the elements,

$$\tfrac{1}{2} H_2(g) + \tfrac{1}{2} F_2(g) \rightarrow HF(g) \tag{12.17}$$

for which $\Delta H = -64.2$ kcal and $\Delta S^{1\,atm}$ is 0.0016 kcal/°K. The free energy change, $\Delta G^{1\,atm}$, is -64.7 kcal at 25°C, virtually identical with ΔH.

The more common case is one in which both ΔH and $T\Delta S$ make significant contributions to ΔG; both must be considered to decide upon the direction of spontaneity. In particular, their signs and magnitudes will determine what effect, if any, an increase in temperature at constant pressure will have upon reaction spontaneity.

EFFECT OF TEMPERATURE ON REACTION SPONTANEITY

When the temperature of a reaction system is increased at constant pressure the direction in which the reaction proceeds spontaneously may or may not change, depending upon the sign of ΔH and ΔS. The four possible situations, deduced from the Gibbs-Helmholtz equation, are summarized in Table 12.5.

Notice that if ΔH and ΔS have opposite signs (I and II), it is impossible by a change in temperature alone to reverse the direction in which the reaction proceeds spontaneously. The two terms ΔH and $T\Delta S$ reinforce one another, and ΔG has the same sign at all temperatures. Reactions of this

TABLE 12.5 Effect of Temperature on Reaction Spontaneity at a Given Pressure

	ΔH	**ΔS**	**ΔG = ΔH − TΔS**	**REMARKS**
I	−	+	always −	Spontaneous at all T; reverse reaction always nonspontaneous
II	+	−	always +	Nonspontaneous at all T; reverse reaction occurs
III	+	+	+ at low T − at high T	Nonspontaneous at low T; becomes spontaneous as T is raised
IV	−	−	− at low T + at high T	Spontaneous at low T; at high T, reverse reaction becomes spontaneous

type are relatively uncommon. One important example is the thermal decomposition of carbon monoxide

$$CO(g) \rightarrow C(s) + \tfrac{1}{2} O_2(g) \tag{12.18}$$

for which ΔH is +26.4 kcal and $\Delta S^{1 \text{ atm}}$ is -0.0215 kcal/°K. Substituting into the Gibbs-Helmholtz equation, we have

$$\Delta G^{1 \text{ atm}} = \Delta H - T\Delta S^{1 \text{ atm}}$$
$$= 26.4 \text{ kcal} + 0.0215 \text{ kcal/°K}$$

Clearly, $\Delta G^{1 \text{ atm}}$ is positive at all temperatures; the reaction cannot take place spontaneously at one atmosphere pressure regardless of temperature. In practical terms this means that it would be futile to hope that the carbon monoxide in automobile exhausts might be eliminated by thermal decomposition to the elements.

Could we hope to get rid of CO by burning it in O_2? $\Delta H = -$, $\Delta S = -$

It is more common to find that ΔH and ΔS have the same sign (types III and IV). When this happens, the enthalpy and entropy factors oppose each other, ΔG changes sign as the temperature increases, and the direction in which the reaction proceeds spontaneously reverses. At low temperatures ΔH predominates and the exothermic reaction occurs; as the temperature rises the quantity $T\Delta S$ eventually predominates and the reaction which leads to an increase in entropy occurs. In most cases 25°C is a "low" temperature, at least when the pressure is maintained at 1 atm; this explains the empirical rule mentioned earlier that exothermic reactions are usually spontaneous at room temperature and atmospheric pressure.

CALCULATION OF ΔG AT TEMPERATURES OTHER THAN 298°K

We saw in Section 12.2 that $\Delta G^{1 \text{ atm}}$ could be calculated directly at 298°K from a table of free energies of formation. Provided ΔH is also known, $\Delta G^{1 \text{ atm}}$ can be obtained at any temperature by making use of the Gibbs-Helmholtz equation. To do this we solve Equation 12.14 to obtain $\Delta S^{1 \text{ atm}}$.

This is the relation to use for finding $\Delta S^{1 \text{ atm}}$

$$\Delta S^{1 \text{ atm}} = \frac{\Delta H - \Delta G^{1 \text{ atm}} \text{ at } 298°K}{298} \tag{12.19}$$

Then we apply the equation directly to obtain $\Delta G^{1 \text{ atm}}$ at the desired temperature, T:

$$\Delta G^{1 \text{ atm}} \text{ at } T = \Delta H - T\Delta S^{1 \text{ atm}}$$

Example 12.4   For the vaporization of water at 1 atm pressure

$$H_2O(l, 1 \text{ atm}) \rightarrow H_2O(g, 1 \text{ atm})$$

obtain, using data in Tables 4.1 and 12.2,

　　a. ΔH b. $\Delta G^{1 \text{ atm}}$ at 298°K   c. $\Delta S^{1 \text{ atm}}$
　　d. a general equation relating $\Delta G^{1 \text{ atm}}$ to T.

Solution

    a. From Table 4.1:

$$\Delta H = \Delta H_f \; H_2O(g) - \Delta H_f \; H_2O(l)$$
$$= -57.8 \text{ kcal} - (-68.3 \text{ kcal}) = +10.5 \text{ kcal}$$

    b. From Table 12.2:

$$\Delta G^{1 \text{ atm}} \text{ at } 298°K = \Delta G_f^{1 \text{ atm}} \; H_2O(g) - \Delta G_f^{1 \text{ atm}} \; H_2O(l)$$
$$= -54.6 \text{ kcal} - (-56.7 \text{ kcal}) = +2.1 \text{ kcal}$$

    c. Solving the Gibbs-Helmholtz equation for ΔS:

$$\Delta S^{1 \text{ atm}} = \frac{\Delta H - \Delta G^{1 \text{ atm}} \text{ at } 298°K}{298°K} = \frac{+10.5 \text{ kcal} - 2.1 \text{ kcal}}{298°K} = 0.028 \frac{\text{kcal}}{°K}$$

    d. $\Delta G^{1 \text{ atm}} = \Delta H - T\Delta S^{1 \text{ atm}} = +10.5 \text{ kcal} - 0.028 \frac{\text{kcal}}{°K} \times T$

In Table 12.6 we list values of $\Delta G^{1 \text{ atm}}$ calculated from the general equation given in Example 12.4 at 25° temperature intervals between 0°C (273°K) and 125°C (398°K). For convenience, values of ΔH and $T\Delta S^{1 \text{ atm}}$ are also included. These data are plotted in Figure 12.6 below.

We see that only at temperatures above 100°C, where the TΔS term predominates, can liquid water boil spontaneously at 1 atm pressure. At temperatures below 100°C the ΔH term prevails, $\Delta G^{1 \text{ atm}}$ is positive, and water in an open beaker refuses to boil. Instead, the reverse reaction,

Figure 12.6 ΔH and TΔS for the vaporization of water at 1 atm. At 373°K (100°C), ΔH = TΔS, ΔG = 0; liquid and vapor are in equilibrium at 1 atm.

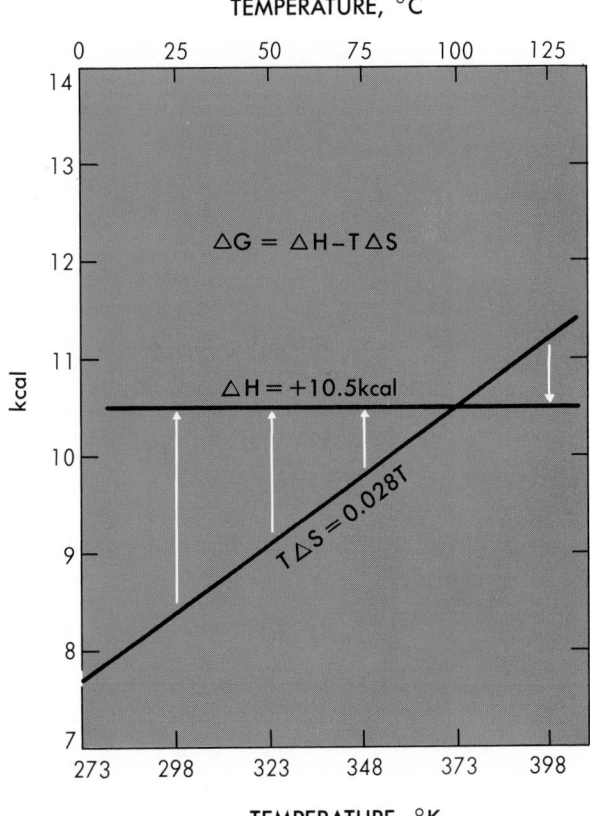

TABLE 12.6 $H_2O(l, 1 \text{ atm}) \rightarrow H_2O(g, 1 \text{ atm})$

T(°C)	T(°K)	ΔH(kcal)	TΔS(kcal)	ΔG(kcal)
0	273	+10.5	+7.7	+2.8
25	298	+10.5	+8.4	+2.1
50	323	+10.5	+9.1	+1.4
75	348	+10.5	+9.8	+0.7
100	373	+10.5	+10.5	0.0
125	398	+10.5	+11.2	−0.7

which amounts to the condensation of supercooled steam, occurs. At precisely 100°C, ΔH and TΔS become equal, ΔG is zero, and the system is at equilibrium, balanced on a knife edge; there is no tendency for reaction to occur in either direction.

$$<100°C; \Delta H > T\Delta S; \Delta G > 0; H_2O(l, 1 \text{ atm}) \leftarrow H_2O(g, 1 \text{ atm})$$
$$100°C; \Delta H = T\Delta S; \Delta G = 0; H_2O(l, 1 \text{ atm}) \rightleftarrows H_2O(g, 1 \text{ atm})$$
$$>100°C; \Delta H < T\Delta S; \Delta G < 0; H_2O(l, 1 \text{ atm}) \rightarrow H_2O(g, 1 \text{ atm})$$

CALCULATION OF TEMPERATURE AT WHICH $\Delta G^{1 \text{ atm}} = 0$

The analysis that we have just carried out suggests a simple way to obtain an important quantity: the temperature at which a reaction is at equilibrium at one atmosphere pressure. All that is required is to determine the temperature at which $\Delta G^{1 \text{ atm}}$ becomes zero. To illustrate, let us consider the reaction

$$CaO(s) + SO_3(g) \rightarrow CaSO_4(s) \tag{12.20}$$

for which $\Delta H = -96.0$ kcal and $\Delta G^{1 \text{ atm}} = -82.7$ kcal at 298°K. This reaction has been suggested as a method of removing sulfur trioxide from furnace gases, thereby cutting down on atmospheric pollution. Clearly the reaction is spontaneous at room temperature, but we need to know whether it will remain spontaneous at the high temperatures reached in coal furnaces. Perhaps the most general approach to this problem is to calculate the temperature at which the reaction is at equilibrium at 1 atm pressure.

Example 12.5   Calculate the temperature at which the reaction

$$CaO(s) + SO_3(g) \rightarrow CaSO_4(s)$$

is at equilibrium at one atmosphere pressure.

Solution.   We first derive a general expression for $\Delta G^{1 \text{ atm}}$ as a function of temperature. We proceed as in Example 12.4 (c) and (d).

$$\Delta S^{1 \text{ atm}} = \frac{\Delta H - \Delta G^{1 \text{ atm}} \text{ at } 298°K}{298°K}$$
$$= \frac{-96.0 \text{ kcal} - (-82.7 \text{ kcal})}{298°K} = -0.0446 \text{ kcal/°K}$$

Hence:
$$\Delta G^{1\,atm} = \Delta H - T\Delta S^{1\,atm}$$
$$= -96.0 \text{ kcal} + 0.0446 \frac{\text{kcal}}{^\circ K} \times T$$

We now solve for the temperature at which $\Delta G^{1\,atm}$ becomes zero.

$$0 = -96.0 + 0.0446 \text{ T}$$
$$T = 96.0/0.0446 = 2150^\circ K$$

We conclude from Example 12.5 that Reaction 12.20 is spontaneous at any temperature below 2150°K (1880°C). Since furnace temperatures are ordinarily well below 1000°C, the addition of quicklime, calcium oxide, to coal furnaces would appear to be a practical way of tying up sulfur trioxide and preventing its escape into the air. Indeed, this process is now being used and offers one of the more promising methods of solving this particular problem in the area of air pollution.

THERMODYNAMIC ANALYSIS OF THE CONVERSION OF Fe_2O_3 TO IRON

We conclude our discussion of the Gibbs-Helmholtz equation by showing how it can be applied to a very practical problem, the conversion of hematite iron ore, Fe_2O_3, to metallic iron. Even though methods of reducing iron ore were worked out many years ago by an empirical approach, it is interesting to see what light thermodynamic analysis can shed on the problem.

We can think of several different ways in which ferric oxide might be converted to iron. Conceivably, heat alone might be sufficient; that is, the reaction

$$Fe_2O_3(s) \rightarrow 2 \text{ Fe}(s) + \tfrac{3}{2} O_2(g) \tag{12.21}$$

might be spontaneous at a reasonable temperature. (After all, Priestley produced mercury from mercuric oxide by thermal decomposition.) Perhaps, however, it would be better to use a reducing agent which has a strong affinity for oxygen; carbon (i.e., charcoal), hydrogen gas, and aluminum metal are three which come to mind.

This kind of analysis can be applied to any chemical reactions for which thermodynamic data are available

$$Fe_2O_3(s) + \tfrac{3}{2} C(s) \rightarrow 2 \text{ Fe}(s) + \tfrac{3}{2} CO_2(g) \tag{12.22}$$

$$Fe_2O_3(s) + 3 H_2(g) \rightarrow 2 \text{ Fe}(s) + 3 H_2O(g) \tag{12.23}$$

$$Fe_2O_3(s) + 2 \text{ Al}(s) \rightarrow 2 \text{ Fe}(s) + Al_2O_3(s) \tag{12.24}$$

The values of ΔH and $\Delta S^{1\,atm}$ for these four reactions are listed in Table 12.7.

TABLE 12.7 Conversion of $Fe_2O_3 \rightarrow Fe$

REACTION	ΔH	$\Delta S^{1\,atm}$
$Fe_2O_3(s) \rightarrow 2 \text{ Fe}(s) + \tfrac{3}{2} O_2(g)$	+196.5 kcal	+65.1 cal/°K
$Fe_2O_3(s) + \tfrac{3}{2} C(s) \rightarrow 2 \text{ Fe}(s) + \tfrac{3}{2} CO_2(g)$	+55.4	+66.0
$Fe_2O_3(s) + 3 H_2(g) \rightarrow 2 \text{ Fe}(s) + 3 H_2O(g)$	+23.1	+32.7
$Fe_2O_3(s) + 2 \text{ Al}(s) \rightarrow 2 \text{ Fe}(s) + Al_2O_3(s)$	−202.6	−9.8

Before proceeding further it may be well to reflect upon the magnitudes of ΔH and $\Delta S^{1\,atm}$ for these reactions. Clearly the principal advantage of using a reducing agent is that it lowers the value of ΔH by forming an oxide (CO_2, H_2O, or Al_2O_3) which has a negative heat of formation. We see from the equations that the decrease in ΔH is accounted for by the heat of formation of 3/2 mole CO_2 (−141.1 kcal), 3 moles of H_2O (−173.4 kcal), or 1 mole of Al_2O_3 (−399.1 kcal). The effect is greatest with aluminum oxide because, for a given amount of oxygen, it has the largest negative value of ΔH_f.

From an entropy standpoint, the direct thermal decomposition is attractive since it is accompanied by a large increase in entropy, +65.1 kcal. The entropy change for carbon reduction is nearly the same, as one might expect, since both reactions result in the same increase in the number of moles of gas, 3/2. For reduction with hydrogen, where the number of moles of gas does not change, ΔS is less favorable; its positive value reflects the greater complexity of the H_2O molecule as compared to H_2. Finally, in the reduction with aluminum, where only solids are involved, ΔS is small and, as it happens, negative.

Using the Gibbs-Helmholtz equation, we can calculate ΔG as a function of temperature, with the results shown in Table 12.8.

TABLE 12.8 $\Delta G^{1\,atm}$ for Conversion of $Fe_2O_3(s)$ to 2 Fe(s)

	$\Delta G^{1\,atm}$ (kcal)		
	300°K	1000°K	2000°K
Heat alone	+177.1	+131.4	+66.3
Reduction by C	+35.6	−10.6	−76.6
Reduction by H_2	+13.3	−9.6	−42.3
Reduction by Al	−199.8	−192.8	−183.0

Before the development of this approach, rational decisions on the feasibility and control of industrial processes were very hard to make

Obviously we can forget about producing iron by thermal decomposition of Fe_2O_3; this reaction is nonspontaneous at 2000°K, well above any temperature that we could hope to achieve in a blast furnace. On the other hand, reduction by carbon or hydrogen is clearly feasible, since temperatures below 1000°K are sufficient. We are encouraged to calculate the temperatures at which these two reactions become spontaneous, using the data in Table 12.7.

Reduction by C:
$$T = \frac{\Delta H}{\Delta S} = \frac{55.4 \text{ kcal}}{0.0660 \text{ kcal/°K}} \approx 840°K \approx 570°C$$

Reduction by H_2:
$$T = \frac{\Delta H}{\Delta S} = \frac{23.1 \text{ kcal}}{0.0327 \text{ kcal/°K}} \approx 710°K \approx 440°C$$

Both these temperatures are well within the realm of possibility for commercial furnaces. In the case of aluminum it appears that we do not even have to raise the temperature; the reaction is spontaneous ($\Delta G^{1\,atm} = -199.8$ kcal) at room temperature.

Experience bears out the conclusions that we have reached. Ferric oxide can indeed be converted to iron by carbon, hydrogen, or aluminum. In the case of aluminum the reaction occurs spontaneously at room temperature; a mixture of powdered ferric oxide and aluminum produces a vigorous exothermic reaction when set off by a magnesium fuse. Carbon is used commercially to reduce iron ore, primarily because it is the cheapest of the suitable reducing agents. Hydrogen is used to produce small amounts of chemically pure iron by reaction 12.23.

A HISTORICAL PERSPECTIVE

JOSIAH WILLARD GIBBS (1839–1903)

A century ago chemistry was primarily an empirical science. The outstanding chemists of that era were experimentalists who isolated and characterized new substances. The principles of chemistry were descriptive or correlative in nature, as illustrated by the atomic theory of Dalton and the Periodic Table of Mendeleev. Two theoreticians working in the latter half of the 19th century changed the very nature of chemistry by deriving the mathematical laws that govern the behavior of matter undergoing physical or chemical change. One of these was James Clerk Maxwell, whose contributions to kinetic theory were discussed in Chapter 5. The other was J. Willard Gibbs, professor of mathematical physics at Yale from 1871 until his death in 1903.

In 1876 Gibbs published in the Transactions of the Connecticut Academy of Sciences the first portion of a remarkable paper entitled, "On the Equilibrium of Heterogeneous Substances." When the paper was completed in 1878 (it was 323 pages long), the foundation had been established for the science of chemical thermodynamics. Here for the first time appeared the concepts of maximum work and free energy that we have used in this chapter to discuss the spontaneity of chemical reactions. Included as well were the basic principles of chemical equilibrium to which we shall devote Chapter 13. Sections of the paper went on to apply the laws of thermodynamics to develop principles of phase equilibria (Chapter 9), laws of dilute solutions (Chapter 10), the nature of adsorption at surfaces (Chapter 14), and the mathematical relationships governing energy changes in voltaic cells (Chapter 21).

If Gibbs had never published another paper, this single contribution would have placed him among the greatest theoreticians in the history of science. Generations of experimental scientists have established their reputations by demonstrating in the laboratory the validity of the relationships that Gibbs derived at his desk. Many of these relationships were rediscovered by others; an example is the Gibbs-Helmholtz equation developed in 1882 by Helmholtz, who was completely unaware of Gibbs' work.

In the 25 years that remained to him, Gibbs made substantial contributions in chemistry, physics, astronomy, and mathematics. Among these were two papers published in 1881 and 1884 that established the discipline known today as vector analysis. His last work, published in 1901, was a book entitled "Elementary Principles in Statistical Mechanics." Here

Gibbs used the statistical principles that govern the behavior of systems to develop thermodynamic equations that he had derived from an entirely different point of view at the beginning of his career. Here, too, we find the "randomness" interpretation of entropy that has received so much attention in the social as well as the natural sciences.

J. Willard Gibbs is often cited as an example of the "prophet without honor in his own country." His colleagues at New Haven and elsewhere in the United States seem not to have realized the significance of his work until late in his life. During his first 10 years as a professor at Yale he received no salary. In 1920 when he was first proposed for the Hall of Fame of distinguished Americans at New York University, he received nine votes out of a possible 100. Not until 1950 was he elected to that body. Even today the name of J. Willard Gibbs is generally unknown among educated Americans outside of those interested in the natural sciences.

Admittedly, Gibbs was himself largely responsible for the fact that for many years his work did not attract the attention it deserved. He made little effort to publicize it; the *Transactions of the Connecticut Academy of Sciences* was hardly the leading scientific journal of its day. Gibbs was one of those rare individuals who seem to have no inner need for recognition by contemporaries. His satisfaction came from solving a problem in his mind; having done so, he was ready to pass on to other problems without being concerned whether people understood what he had done. His papers are not easy to read; he seldom cites examples to illustrate his abstract reasoning. Frequently the implications of the laws that he derives are left for the reader to grasp on his own. One of his colleagues at Yale confessed many years later that none of the members of the Connecticut Academy of Sciences understood his paper on thermodynamics; as he put it, "We knew Gibbs and took his contributions on faith."

Gibbs achieved recognition in Europe long before his work was generally appreciated in this country. Maxwell, the pre-eminent theoretician of his time, somehow came across a copy of Gibbs' paper on thermodynamics, saw its significance, and referred to it repeatedly in his own publications. Wilhelm Ostwald, who said of Gibbs, "To physical chemistry, he gave form and content for a hundred years," translated the paper into German in 1892. Seven years later, Le Châtelier translated it into French.

Muriel Rukeyser, in her fascinating biography of Gibbs, tells an anecdote that reveals a great deal about the man and the scientist. In one of Gibbs' early papers there was a discussion of the phase equilibria between ice, liquid water, and water vapor. In Miss Rukeyser's words, "Again, Willard Gibbs had stopped at the bare idea and left undone that step which might have bridged the gap between himself and his audience. Maxwell added the last personal impression which must have touched and delighted Gibbs more than any other gift." The eminent English theoretician made a plaster model showing graphically the thermodynamic relationships involved and sent it to New Haven. Gibbs took the model to class but never referred to it in his lectures. One day, a student asked where it had come from.

"A friend sent it to me," said Gibbs with his own punishing modesty.

"Who is the friend?" the boy asked, knowing very well who it was. But all that Gibbs would say was, "A friend in England."

PROBLEMS

12.1 Which of the following processes are spontaneous? nonspontaneous? neither?

    a. the expansion of a gas into a vacuum.
    b. the freezing of water at 0°C and 1 atm.
    c. climbing Mount Washington.
    d. the decomposition of hydrogen peroxide into the elements at 25°C and 1 atm. (ΔG_f $H_2O_2 = -27.2$ kcal)

12.2 Cite examples of endothermic, spontaneous processes other than those described in this chapter.

12.3 Describe ways of obtaining useful work from the following spontaneous reactions:

    a. the combustion of coal.
    b. the combustion of sugar.
    c. $Zn(s) + Cu^{2+}(aq) \rightarrow Zn^{2+}(aq) + Cu(s)$

12.4 A fuel cell utilizes the reaction

$$H_2(g) + \tfrac{1}{2} O_2(g) \rightarrow H_2O(l)$$

What is W_m for this process at 25°C and 1 atm? If the fuel cell actually produces 40.0 kcal of useful work per mole of water formed, what is its efficiency?

12.5 Discuss the variation with temperature and pressure of the three quantities, ΔH, ΔS, and ΔG. Why do we write $\Delta G^{1\ atm}$ and $\Delta S^{1\ atm}$ but not $\Delta H^{1\ atm}$?

12.6 Calculate the free energy change at 25°C and 1 atm for the reactions

    a. $NO(g) + \tfrac{1}{2} O_2(g) \rightarrow NO_2(g)$
    b. $Fe_2O_3(s) + 3 H_2(g) \rightarrow$ $2 Fe(s) + 3 H_2O(g)$
    c. $CHCl_3(l) + 3 HCl(g) \rightarrow$ $CH_4(g) + 3 Cl_2(g)$

Which of these reaction(s) are spontaneous at 25°C and 1 atm? Of those which are nonspontaneous, which one(s) might be expected to become spontaneous at high temperature?

12.15 Which of the following processes are spontaneous? nonspontaneous? neither?

    a. the detonation of an explosive.
    b. the removal of detergents from a water supply.
    c. the formation of a solution of sodium hydroxide in water.
    d. the decomposition of acetylene, C_2H_2, into the elements at 25°C, 1 atm.

12.16 Can you think of processes which occur spontaneously despite a decrease in entropy?

12.17 How might the following processes, all of which are nonspontaneous, be accomplished?

    a. $6 CO_2(g) + 6 H_2O(l) \rightarrow C_6H_{12}O_6(s) + 6 O_2(g)$
    b. sugar solution $\rightarrow$ sugar + water, at 25°C.
    c. $NaCl(l) \rightarrow Na(l) + \tfrac{1}{2} Cl_2(g)$

12.18 How much work would have to be supplied to make the reaction

$$N_2(g) + O_2(g) \rightarrow 2 NO(g)$$

take place at 25°C and 1 atm? If the sign of W_m for this reaction were +, what would be the effect on our environment?

12.19 Consider the reaction described in Table 12.1. At 25°C and 10^{-6} atm, what is the value of ΔH? ΔS? Explain why ΔS is a positive quantity. which increases as the pressure is reduced.

12.20 For the reaction

$$N_2H_4(l) + 2 O_2(g) \rightarrow N_2(g) + 2 H_2O(l)$$

ΔG at 25°C and 1 atm is -149.0 kcal.

    a. What is the free energy of formation of N_2H_4?
    b. Explain why this reaction evolves less heat when water vapor is formed rather than water liquid.
    c. Why is $\Delta G^{1\ atm}$ at 25°C less negative when $H_2O(g)$ is formed rather than $H_2O(l)$?

12.7 Given the following data for the reaction

$$Cu_2O(s) + \tfrac{1}{2} O_2(g) \rightarrow 2\ CuO(s)$$

T(°K)	0	300
$\Delta G^{1\ atm}$ (kcal)	−34.4	−25.8

a. calculate ΔH and $\Delta S^{1\ atm}$.
b. explain why $\Delta S^{1\ atm}$ has a negative sign.
c. If copper is exposed to air at room temperature, it becomes coated with a black oxide. What is the formula of this oxide?
d. When a copper wire is heated in a Meker burner, it forms a red oxide. Explain.

12.8 Estimate the sign of ΔS for each of the following processes:

a. the condensation of water vapor at 1 atm pressure.
b. the typing of a term paper by a monkey.
c. the conversion of an egg to an omelet.
d. $2\ CO(g) + O_2(g) \rightarrow 2\ CO_2(g)$
e. $ZnO(s) + H_2(g) \rightarrow Zn(s) + H_2O(l)$
f. $C_{15}H_{30}(s) + Br_2(l) \rightarrow C_{15}H_{30}Br_2(s)$

12.9 Calculate ΔS for the fusion of one mole of benzene which freezes at 5°C with a heat of fusion of 2.55 kcal/mole.

12.10 Consider the following reactions, each at 1 atm pressure.

	ΔH	$\Delta S^{1\ atm}$
$\tfrac{1}{2} H_2(g) + \tfrac{1}{2} I_2(s) \rightarrow HI(g)$	+6.2 kcal	+19.8 cal/°K
$2\ NO_2(g) \rightarrow N_2O_4(g)$	−13.9	−42.2
$H_2(g) + S(s) \rightarrow H_2S(g)$	−4.8	+10.4

Which of these reactions are

a. spontaneous at all temperatures?
b. spontaneous at low temperatures, nonspontaneous at high temperatures?
c. spontaneous at high temperatures, nonspontaneous at low temperatures?

12.11 For the reaction $C(s) + CO_2(g) \rightarrow 2\ CO(g)$ calculate

a. ΔH
b. $\Delta G^{1\ atm}$ at 298°K
c. $\Delta S^{1\ atm}$
d. $\Delta G^{1\ atm}$ at 500°C

12.21 The reaction that occurs in a lead storage battery is

$$PbO_2(s) + Pb(s) + 2\ H_2SO_4(aq) \rightarrow 2\ PbSO_4(s) + 2\ H_2O(aq)$$

A student measures the voltage of this cell as a function of temperature and obtains the following values for $\Delta G^{1\ atm}$:

T(°K)	298	323	348
$\Delta G^{1\ atm}$ (kcal)	−84.2	−83.3	−82.3

a. Plot $\Delta G^{1\ atm}$ as a function of temperature.
b. Estimate $\Delta S^{1\ atm}$ and ΔH for this reaction.

12.22 Estimate the sign of ΔS for each of the following processes

a. the formation of polyethylene, a polymer of empirical formula $(CH_2)_n$, where n is a very large number, from ethylene, C_2H_4.
b. the translation of Homer's Iliad into English.
c. the reaction which occurs when pyrites, FeS_2, is burned in air to give sulfur dioxide and Fe_2O_3.
d. $Co^{3+}(aq) + 6\ NH_3(aq) \rightarrow Co(NH_3)_6^{3+}(aq)$

12.23 Calculate $\Delta S^{1\ atm}$ for the vaporization of benzene (normal bp = 80°C, $\Delta H_{vap} = 7.30$ kcal/mole). Compare this value of ΔS with that calculated for water in Example 12.3 and comment on the relative magnitudes of the two quantities.

12.24 Classify each of the following statements as always true, usually true, always false or usually false.

a. An endothermic reaction is spontaneous at 25°C, 1 atm.
b. A reaction for which ΔH and ΔS are both negative is spontaneous at all temperatures.
c. An exothermic reaction is spontaneous at 0°K.
d. A reaction which results in an increase in the number of moles leads to an increase in entropy.

12.25 For the reaction

$$MgCO_3(s) \rightarrow MgO(s) + CO_2(g)$$

obtain a general expression for $\Delta G^{1\ atm}$ as a function of temperature and use it to calculate ΔG at 1 atm and 1000°C.

12.12 Estimate the temperature at which the reaction

$$N_2(g) + 3\ H_2(g) \rightarrow 2\ NH_3(g)$$

will be at equilibrium at one atmosphere pressure.

12.13 Nitric acid can be prepared by the reaction

$$3\ NO_2(g) + H_2O \rightarrow 2\ HNO_3(l) + NO(g)$$

Is this reaction at 1 atm pressure spontaneous at 25°C? at higher temperatures? at lower temperatures?

12.14 Consider the reduction of Cr_2O_3 to chromium metal. Carry out a thermodynamic analysis similar to that given in the text for Fe_2O_3 to decide upon the feasibility of reduction of Cr_2O_3 by heat, C, H_2 and Al (i.e., prepare tables similar to 12.7 and 12.8).

12.26 Hydrated plaster of Paris, $CaSO_4\cdot 2\ H_2O$, crumbles when heated strongly because of the reaction

$$CaSO_4\cdot 2\ H_2O(s) \rightarrow CaSO_4(s) + 2\ H_2O(g)$$

Taking the heat of formation and standard free energy of formation at 298°K of $CaSO_4\cdot 2\ H_2O$ to be -483.1 and -429.2 kcal, respectively, estimate the temperature at which this reaction becomes spontaneous.

12.27 It has been suggested that when a metal oxide is heated with carbon, the actual reducing agent is CO rather than C. Calculate $\Delta G^{1\ atm}$ for the reduction of one mole of ZnO by CO and C at 25°C; 1000°C. Comment on the meaning of your calculations.

12.28 Barium oxide, BaO, might be used rather than calcium oxide to remove SO_3 from furnace gases. Carry out an analysis similar to that used in Example 12.5 to decide whether this process would be feasible and, if so, whether BaO would have any advantage over CaO.

*12.29 Consider the data in Table 12.1

a. Show that at any given temperature, the variation of ΔG with pressure is given by the expression

$$\Delta G^P = \Delta G^{1\ atm} + a \log P$$

i.e., that ΔG is a linear function of log P.
b. Evaluate a at 25°C.
c. Calculate ΔS at pressures of 1 atm, 10^{-2} atm, and 10^{-4} atm at 25°C.
d. Show that $\Delta S^P = \Delta S^{1\ atm} - \dfrac{a}{T} \log P$.

*12.30 Consider the reaction

$$6\ CO_2(g) + 6\ H_2O(l) \rightarrow C_6H_{12}O_6(s) + 6\ O_2(g)$$

For glucose, $\Delta H_f = -311.2$ kcal and $\Delta G_f^{1\ atm}$ at 25°C $= -215.8$ kcal.

a. Calculate $\Delta G^{1\ atm}$ for this reaction at 25°C.
b. The reaction referred to represents a net result of the process of photosynthesis. Do you conclude that photosynthesis is a spontaneous process?
c. Explain how photosynthesis takes place in the world around us, in view of your answer to (a).

*12.31 A major difficulty with the methane-oxygen fuel cell mentioned in this chapter is that it must be operated at high temperatures, about 600°C. Calculate the maximum amount of work that can be obtained from the combusion of methane at 600°C and 1 atm and compare to that at 25°C. (Note that the physical state of water will differ in the two cases.) Can you think of other reasons why a high temperature would be undesirable from a practical point of view?

*12.32 For the reactions

$$NiO(s) + H_2(g) \rightarrow Ni(s) + H_2O(g)$$
$$ZnO(s) + H_2(g) \rightarrow Zn(s) + H_2O(g)$$
$$\tfrac{1}{2} SnO_2(s) + H_2(g) \rightarrow \tfrac{1}{2} Sn(s) + H_2O(g)$$
$$\tfrac{1}{3} Fe_2O_3(s) + H_2(g) \rightarrow \tfrac{2}{3} Fe(s) + H_2O(g)$$

a. Calculate ΔH and $\Delta S^{1\,atm}$.
b. Can you suggest why $\Delta S^{1\,atm}$ is nearly the same for all these reactions?
c. Can you suggest why the temperature required to reduce a metal oxide by hydrogen increases steadily as the heat of formation of the oxide, per GAW of oxygen, becomes more negative?
d. Predict the relative ease of reduction by hydrogen of Al_2O_3, Fe_3O_4, HgO, and Cu_2O.
e. How could you test your prediction in (d) in the laboratory?

13 CHEMICAL EQUILIBRIUM IN GASEOUS SYSTEMS

In Chapter 12 we showed how one can use the sign of the free energy change, ΔG, to make predictions concerning reaction spontaneity. The calculations we made were designed to answer one very important question: in which direction will a reaction proceed spontaneously *when reactants and products are at one atmosphere pressure?* As important as this question is, it raises other, more general questions. In particular, we would like to be able to predict the direction and extent to which a reaction will proceed at a given temperature when reactants and products are at *any* given pressure or concentration.

It is possible to answer these questions by a rigorous thermodynamic development which considers the variation of ΔG with pressure or concentration. Instead of using that approach, we shall follow a more empirical route, studying the behavior of chemical systems as they approach equilibrium at constant temperature. Once this point is reached, the driving force for further change disappears ($\Delta G = 0$) and there is no tendency for reaction to take place spontaneously in either direction.

To illustrate what chemical equilibrium implies, consider what happens when a sample of calcium carbonate is introduced into a closed container at 850°C. Some of the calcium carbonate decomposes by way of the reaction

$$CaCO_3(s) \rightarrow CaO(s) + CO_2(g)$$

As the concentration of carbon dioxide builds up in the gas phase, some of the CO_2 molecules react with calcium oxide

$$CaO(s) + CO_2(g) \rightarrow CaCO_3(s)$$

Eventually the rates of these two competing reactions become equal: the concentration of carbon dioxide does not change with time, and we have arrived at equilibrium. Experiment shows that at 850°C (1123°K), the position of this equilibrium is such that the pressure of CO_2 is one atmosphere; its concentration, calculated from the Ideal Gas Law, now is

$$\frac{n}{V} = \frac{P}{RT} = \frac{(1.00 \text{ atm})}{\left(0.0821 \dfrac{\text{lit atm}}{\text{mole °K}}\right)(1123°K)} = 0.0108 \text{ mole/lit}$$

The chemical equilibrium

$$CaCO_3(s) \rightleftharpoons CaO(s) + CO_2(g)$$

is a particularly simple one in that its position can be described by citing a single concentration, that of carbon dioxide. In that sense it resembles the physical equilibrium

At any given temperature the vapor pressure of water has a constant value

$$H_2O(l) \rightleftharpoons H_2O(g)$$

whose position at any temperature can be described by giving the concentration or pressure of water vapor.

Unfortunately most systems at equilibrium cannot be described as simply as this. To illustrate, consider the system

$$N_2O_4(g) \rightleftharpoons 2 NO_2(g)$$

Here the equilibrium concentration of nitrogen dioxide, NO_2, will not have a fixed value at a particular temperature. Instead, it may take on any of an infinite number of values, depending upon the concentration of the other gaseous species present, dinitrogen tetroxide, N_2O_4 (Table 13.1).

TABLE 13.1   Equilibrium Measurements in the N_2O_4-NO_2 System at 120°C

EXPERI- MENT	VOLUME (liters)	INITIAL No. OF MOLES		EQUILIBRIUM No. OF MOLES		EQUILIBRIUM CONCENTRATION	
		N_2O_4	NO_2	N_2O_4	NO_2	N_2O_4	NO_2
1	4.0	1.00	0.00	0.40	1.20	0.10	0.30
2	4.0	0.00	2.00	0.40	1.20	0.10	0.30
3	4.0	1.00	2.00	1.04	1.92	0.26	0.48
4	10.0	1.00	0.00	0.25	1.50	0.025	0.15

We always find that in a system at equilibrium involving two or more species in solution, the concentrations of these species are interrelated. Throughout the remainder of this chapter, we shall be concerned with the nature of this relationship as embodied in the concept of the equilibrium constant, K_c. Here we shall deal mainly with equilibria in gaseous systems. In later chapters we shall apply the principles governing equilibrium to aqueous solutions.

13.1   THE N_2O_4-NO_2 EQUILIBRIUM SYSTEM. CONCEPT OF K_c

To illustrate the principles of chemical equilibrium, let us consider the gas phase reaction at 120°C:

$$N_2O_4(g) \rightleftharpoons 2\ NO_2(g) \qquad (13.1)$$

To study this equilibrium experimentally, we might start by admitting 1.00 mole of N_2O_4 into an evacuated, 4-liter container at 120°C. As time passes we find that a reddish brown color develops. The color is associated with the NO_2 molecule, an odd-electron species; N_2O_4 is colorless. Eventually the intensity of the color becomes constant, implying that we have reached equilibrium. At this point, we withdraw a sample of the gaseous mixture for analysis and find that there are 1.20 moles of NO_2 present. We note from Equation 13.1 that 1 mole of N_2O_4 is required to form 2 moles of NO_2. It follows that the number of moles of N_2O_4 that must have decomposed to form 1.20 moles of NO_2 is

$$1.20\ \text{mole}\ NO_2 \times \frac{1\ \text{mole}\ N_2O_4}{2\ \text{moles}\ NO_2} = 0.60\ \text{mole}\ N_2O_4$$

Since we started with 1.00 mole of N_2O_4, we must have left:

$$(1.00 - 0.60)\ \text{mole}\ N_2O_4 = 0.40\ \text{mole}\ N_2O_4$$

The total mass of sample in the container remains fixed

Knowing the numbers of moles of NO_2 and N_2O_4 at equilibrium and the volume of the container (4.0 liters), we can readily calculate the equilibrium concentrations.

$$[NO_2] = 1.20\ \text{moles}/4.0\ \text{liter} = 0.30\ \text{mole/liter}$$
$$[N_2O_4] = 0.40\ \text{mole}/4.0\ \text{liter} = 0.10\ \text{mole/liter}$$

The square brackets, here and elsewhere throughout the remainder of this text, are used to represent equilibrium concentrations in moles/liter.

The data from this experiment are recorded in a convenient form in the first horizontal row of Table 13.1. In successive rows of this table are data from three other experiments, carried out under different conditions. In Experiment 2, we start with 2.00 moles of NO_2 rather than 1.00 mole of N_2O_4; in Experiment 3, we start with a mixture of NO_2 and N_2O_4, while in Experiment 4 we change the volume of the container from 4 to 10 liters. There are several features of these data that are worth commenting upon:

   1. In each case, the change in the number of moles of NO_2 as we pro-

Sample for analysis

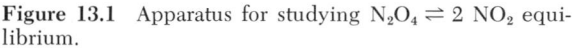
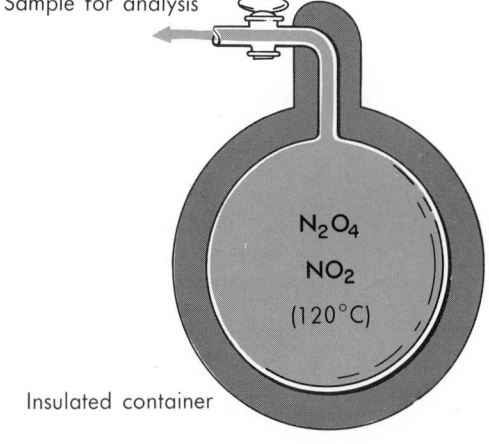

Figure 13.1 Apparatus for studying $N_2O_4 \rightleftharpoons 2\ NO_2$ equilibrium.

N_2O_4

NO_2

(120°C)

Insulated container

ceed to equilibrium is exactly twice that for N_2O_4. In Experiments 2 and 3 the number of moles of NO_2 decreases by 0.80 and 0.08, respectively, while the number of moles of N_2O_4 increases by 0.40 and 0.04, respectively. In Experiment 4 the number of moles of NO_2 increases by 1.50, while that of N_2O_4 decreases by 0.75. This 2:1 relationship is, of course, required by the coefficients of the balanced equation.

2. The final concentrations in Experiment 2 are the same as those in Experiment 1. This is hardly surprising in view of the fact that the volume is constant and the overall composition is the same: 2 gram atomic weights of nitrogen and 4 gram atomic weights of oxygen. The only difference between the two experiments is that in the first case we started with pure N_2O_4 while in the second we started with an equivalent amount of pure NO_2. We approached equilibrium from opposite directions and, logically enough, reached the same final state. Experiments 1 and 2 taken together illustrate the general principle that the final equilibrium state of a system of fixed overall composition and volume is independent of the nature of the species initially present.

3. In Experiment 4, where the volume is increased from 4 to 10 liters, more NO_2 is present at equilibrium (1.50 moles) than in Experiment 1 (1.20 moles). In other words, the increase in volume has shifted the position of the equilibrium to favor the formation of NO_2. We will have more to say about the significance of this observation in Section 13.4.

Looking at the data in Table 13.1, we might wonder whether there could be any quantitative relationship between the equilibrium concentrations of NO_2 and N_2O_4 that would be valid for all four experiments. Amazingly enough, there is. In particular, the quotient: $[NO_2]^2/[N_2O_4]$ is the same, about 0.90, for all four experiments.

Expt. 1, 2:
$$\frac{[NO_2]^2}{[N_2O_4]} = \frac{(0.30)^2}{0.10} = \frac{0.090}{0.10} = 0.90$$

Expt. 3:
$$\frac{[NO_2]^2}{[N_2O_4]} = \frac{(0.48)^2}{0.26} = \frac{0.23}{0.26} = 0.89$$

Expt. 4:
$$\frac{[NO_2]^2}{[N_2O_4]} = \frac{(0.15)^2}{(0.025)} = \frac{0.0225}{0.025} = 0.90$$

Experimentally this simple relationship is found to hold for any equilibrium system containing NO_2 and N_2O_4 at 120°C. Regardless of the amounts of the two gases we start with or the volume of the container, we eventually arrive at an equilibrium whose position is described by the condition that

$$\frac{[NO_2]^2}{[N_2O_4]} = 0.90 \text{ (at } 120°C)$$

Further experiments with many different systems containing NO_2 and N_2O_4 at various temperatures leads to the general conclusion:

At any given temperature, the quantity

$$\frac{[NO_2]^2}{[N_2O_4]}$$

is a constant, independent of the amounts of N_2O_4 and NO_2 that we start *with, the volume of the container, or the total pressure.* This constant is referred to as the **equilibrium constant,** K_c, for the reaction

$$N_2O_4(g) \rightleftharpoons 2\ NO_2(g)$$

At 120°C the numerical value of K_c is 0.90; at 150°C it is a considerably larger number, 3.2.

13.2 THE GENERAL FORM OF K_c

We have seen that for the reaction

$$N_2O_4(g) \rightleftharpoons 2\ NO_2(g),\ K_c = \frac{[NO_2]^2}{[N_2O_4]}$$

At this stage you might well ask how you can predict, for a given reaction, what the mathematical form of the expression for K_c will be. The answer to this question may be a little more obvious when it is found that for the reaction

$$2\ HI(g) \rightleftharpoons H_2(g) + I_2(g)$$

the ratio

$$\frac{[H_2] \times [I_2]}{[HI]^2}$$

has a constant value at a given temperature. That is

$$K_c = \frac{[H_2] \times [I_2]}{[HI]^2}$$

and that for the reaction: $N_2(g) + 3\ H_2(g) \rightleftharpoons 2\ NH_3(g)$

$$K_c = \frac{[NH_3]^2}{[N_2] \times [H_2]^3}$$

The constant, K_c, puts a condition on the concentrations of species present in an equilibrium system

These examples imply the Law of Chemical Equilibrium, which tells us that for the general gas phase reaction

$$aA(g) + bB(g) \rightleftharpoons cC(g) + dD(g)$$

where A, B, C, and D represent different substances and a, b, c, and d are their coefficients in the balanced equation,

$$K_c = \frac{[C]^c \times [D]^d}{[A]^a \times [B]^b} \tag{13.2}$$

It is worth pointing out that the condition on the equilibrium concentrations of products and reactants in a given reaction system can be expressed in a variety of ways. Consider, for example, the NO_2-N_2O_4 system. We have chosen to express the equilibrium condition at 120° by writing

$$\frac{[NO_2]^2}{[N_2O_4]} = 0.90$$

It would be equally valid to invert this expression and say that

$$\frac{[N_2O_4]}{[NO_2]^2} = \frac{1}{0.90} = 1.1$$

Alternatively, we could take the square root of both sides and write

$$\frac{[NO_2]}{[N_2O_4]^{1/2}} = (0.90)^{1/2} = 0.95$$

Each of the three equations we have written, and many others that we could write, is a valid way to express the condition governing the equilibrium concentrations of NO_2 and N_2O_4 at 120°C. Clearly, it would be ambiguous at the very least to say that for this system at 120°C "the equilibrium constant is 0.90." The equilibrium constant takes on meaning only when it is associated with a chemical equation. Thus, we have

$$N_2O_4(g) \rightleftharpoons 2\ NO_2(g) \quad K_c = \frac{[NO_2]^2}{[N_2O_4]} = 0.90$$

All these expressions are equivalent in fixing the equilibrium condition

or

$$2\ NO_2(g) \rightleftharpoons N_2O_4(g) \quad K_c' = \frac{[N_2O_4]}{[NO_2]^2} = 1.1$$

or

$$\tfrac{1}{2}\ N_2O_4(g) \rightleftharpoons NO_2(g) \quad K_c'' = \frac{[NO_2]}{[N_2O_4]^{1/2}} = 0.95$$

Equilibria involving gases often include pure solids or liquids as reactants or products. Examples include

$$CaCO_3(s) \rightleftharpoons CaO(s) + CO_2(g) \tag{13.3}$$

$$SnO_2(s) + 2\ CO(g) \rightleftharpoons Sn(s) + 2\ CO_2(g) \tag{13.4}$$

$$CO_2(g) + H_2(g) \rightleftharpoons CO(g) + H_2O(l) \tag{13.5}$$

For such reactions we find that the position of the equilibrium is independent of the amount of solid or liquid present and that the concentrations of these species need not and ordinarily do not appear in the expression for K_c. For the above reactions, we would write

13.3: $$K_c = [CO_2]$$

13.4: $$K_c = \frac{[CO_2]^2}{[CO]^2}$$

13.5: $$K_c = \frac{[CO]}{[CO_2] \times [H_2]}$$

omitting, in each case, the terms for solids or liquids.

To understand why it is possible to make this simplification, consider Reaction 13.5 taking place, let us say, at 50°C. The equation tells us that liquid water is present in the equilibrium system. We found in Chapter 9 that at a given temperature the pressure of water vapor in equilibrium with the liquid has a fixed value, independent of the presence of other gases. At 50°C this pressure is 92.5 mm Hg, equivalent (according to the Ideal Gas Law) to a concentration of 0.00460 mole/liter. This will be the concentration of water vapor at 50°C regardless of what the concentrations of the other species may be. At constant temperature nothing that we can do to the equilibrium system can change the concentration of water vapor provided some liquid water is present. Hence, the general expression for the equilibrium constant

$$K = \frac{[CO] \times [H_2O]}{[CO_2] \times [H_2]}$$

can be simplified to read

$$K = \frac{[CO] \times 0.00460}{[CO_2] \times [H_2]} \quad \text{or} \quad \frac{K}{0.00460} = \frac{[CO]}{[CO_2] \times [H_2]}$$

It is convenient to define K_c to be the ratio of the two numbers K/0.00460.

$$K_c = \frac{K}{0.00460} = \frac{[CO]}{[CO_2] \times [H_2]}$$

This practice is ordinarily followed in dealing with equilibria in which pure liquids or solids are present. The gas-phase concentration of such a substance, which must be a constant at a particular temperature, is incorporated into K_c and, hence, does not appear explicitly.

> Although pure solids and liquids do not appear in the K_c expression, they *must be present* in the equilibrium system

13.3 APPLICATIONS OF K_c

A knowledge of the equilibrium constant for a particular reaction at a given temperature is valuable to the chemist who wishes to carry out the reaction in the laboratory or to predict whether it will occur in nature at that temperature. In particular, the equilibrium constant can be used to decide:

1. The direction in which a chemical system will move to reach equilibrium.

2. The extent to which a reaction will occur.

3. The effect that a change in conditions will have on a chemical system at equilibrium.

We shall now consider, in turn, each of these applications of K_c. (The third application is discussed in Section 13.4.)

PREDICTION OF THE DIRECTION OF REACTION

We have seen that for the general gas phase reaction

$$aA(g) + bB(g) \rightleftharpoons cC(g) + dD(g)$$

the equilibrium concentrations of products and reactants must satisfy the condition

$$\frac{[C]^c \times [D]^d}{[A]^a \times [B]^b} = K_c$$

where K_c is a number characteristic of the reaction at a particular temperature. When we carry out a reaction, the original concentration quotient

$$\frac{(\text{orig. conc. C})^c \times (\text{orig. conc. D})^d}{(\text{orig. conc. A})^a \times (\text{orig. conc. B})^b}$$

will seldom be equal numerically to K_c. If it is not, reaction will occur in one direction or the other so as to bring the concentrations of products and reactants to the ratio required at equilibrium. We can distinguish two possibilities:

1. If
$$\frac{(\text{orig. conc. C})^c \times (\text{orig. conc. D})^d}{(\text{orig. conc. A})^a \times (\text{orig. conc. B})^b} < K_c$$

the reaction will proceed from left to right, i.e.,

$$aA(g) + bB(g) \rightarrow cC(g) + dD(g)$$

In this way, the concentrations of products increase and those of reactants decrease. As this happens, the concentration quotient increases until it becomes equal to K_c, at which point reaction ceases.

2. If
$$\frac{(\text{orig. conc. C})^c \times (\text{orig. conc. D})^d}{(\text{orig. conc. A})^a \times (\text{orig. conc. B})^b} > K_c$$

we conclude that the concentrations of products are "too high" and those of the reactants "too low" to meet the equilibrium condition. Reaction must proceed from right to left, i.e.:

$$cC(g) + dD(g) \rightarrow aA(g) + bB(g)$$

to reduce the concentration quotient to its equilibrium value.

Example 13.1   Among the products that come out of the exhaust system of an automobile are carbon dioxide and the extremely toxic gas, carbon monoxide. In the presence of oxygen, the following equilibrium is established:

$$CO_2(g) \rightarrow CO(g) + \tfrac{1}{2} O_2(g)$$

At 500°C, K_c for this reaction is about 4×10^{-16}. Assume that in the gas coming out of the exhaust the concentration of CO_2 is 10^{-3} mole/liter and that of CO is 10^{-5} mole/liter. Predict the direction in which the reaction will occur to reach equilibrium when
   a. there is no oxygen present originally.
   b. the original concentration of oxygen is 10^{-2} mole/liter (its normal concentration in air).

Solution
   a. If there is *no* oxygen present, the original concentration quotient

$$\frac{(\text{orig. CO}) \times (\text{orig. conc. O}_2)^{1/2}}{(\text{orig. conc. CO}_2)}$$

   must be zero. Clearly, $0 < K_c$, so reaction must proceed from left to right (i.e., some CO_2 must decompose) to establish equilibrium. Note that this happens despite the fact that K_c for this reaction is a very small number. Actually, the amount of CO_2 that decomposes is minute; the concentration of CO at equilibrium is only slightly greater than it was to start with. From a practical standpoint, we conclude that, in the absence of oxygen, lethal amounts of CO can be present in the exhaust gases.
   b. Here, we need to compare the original concentration quotient to K_c.

$$\frac{(\text{orig. conc. CO}) \times (\text{orig. conc. O}_2)^{1/2}}{(\text{orig. conc. CO}_2)} = \frac{(10^{-5}) \times (10^{-2})^{1/2}}{10^{-3}}$$

$$= \frac{(10^{-5}) \times (10^{-1})}{10^{-3}} = 1 \times 10^{-3}$$

The original concentration quotient, 1×10^{-3}, is considerably larger than K_c, 4×10^{-16}. This means that the reaction will proceed from right to left, converting CO to CO_2. In this way the concentrations of CO and O_2 decrease and that of CO_2 increases until the concentration quotient drops to 4×10^{-16}. As a matter of fact, virtually all the carbon monoxide will be consumed when equilibrium is reached. We conclude that to reduce the concentration of CO in exhaust gases, we should make sure that there is plenty of oxygen around for it to react with. In practical terms, this amounts to using an excess of air in burning the fuel.

PREDICTION OF EXTENT OF REACTION

Knowing the equilibrium constant K_c for a particular reaction and the original concentrations of reactants, it is possible to calculate the extent to which a reaction occurs. The basic approach that we will use involves a three-step process. The first step requires that we express the equilibrium concentrations of all species in terms of a single unknown, x. These concentrations are then substituted into the expression for K_c to give an equation which can be solved for x. Having found the numerical value of x, we can then readily obtain the amounts and/or concentrations of all the species in the equilibrium mixture. This approach is illustrated in Example 13.2.

Example 13.2   At 2000°C $K_c = 0.10$ for the reaction

$$N_2(g) + O_2(g) \rightleftharpoons 2\ NO(g)$$

a. If we start with one mole of N_2 and one mole of O_2 in a 10-liter container, at 2000°C, how much NO will be present at equilibrium?
b. If we start with 4.0 moles of N_2 and 1.0 mole of O_2 in a 10-liter container at 2000°C, what will be the concentration of NO at equilibrium?

Solution

a. We must first express the equilibrium concentrations of N_2, O_2, and NO in terms of a single unknown, x. Let us choose x to be the number of moles of N_2 that react. The balanced equation tells us that an equal number of moles, x, of oxygen must be consumed in the reaction. Furthermore, since two moles of NO are formed for every mole of N_2 that reacts, we must form 2x moles of NO. Summarizing our reasoning in the form of a table:

No. of Moles	Original	Change	Present at Equilibrium
N_2	1	$-x$	$1 - x$
O_2	1	$-x$	$1 - x$
NO	0	$+2x$	$2x$

This table *must* be constructed properly. Note that the numbers in the Change column are the *coefficients* in the equation for the reaction. Their *signs* reflect whether the species is used up $(-)$ or produced $(+)$

Expressions for the equilibrium concentrations are readily set up, knowing that the volume of the container is 10 liters:

$$[NO] = \frac{2x}{10}; \quad [N_2] = \frac{1-x}{10}; \quad [O_2] = \frac{1-x}{10}$$

We are now ready to substitute these quantities into the expression for K_c and, ultimately, solve for x.

$$K_c = 0.10 = \frac{[NO]^2}{[N_2] \times [O_2]} = \frac{(2x/10)^2}{\left(\dfrac{1-x}{10}\right)\left(\dfrac{1-x}{10}\right)}$$

Simplifying: $0.10 = \dfrac{(2x)^2}{(1-x)^2}$

There is a very easy way to solve this equation for x; take the square root of both sides of the equation

$$\frac{2x}{1-x} = (0.10)^{1/2} = 0.31$$

or: $2x = 0.31 - 0.31x; \quad 2.31x = 0.31; \quad x = 0.13$

To complete the problem, we note that we were asked to calculate the number of moles of NO at equilibrium. The table reminds us that this quantity is 2x.

$$\text{no. of moles NO} = 2x = 0.26$$

b. Here the first step is entirely analogous to (a).

No. of Moles	Original	Change	Present at Equilibrium
N_2	4.0	$-x$	$4.0 - x$
O_2	1.0	$-x$	$1.0 - x$
NO	0.00	$+2x$	$2x$

Since we are dealing with a 10-liter container:

$$[NO] = \frac{2x}{10}; \quad [N_2] = \frac{4.0 - x}{10}; \quad [O_2] = \frac{1.0 - x}{10}$$

As before, we substitute into the expression for K_c and obtain, after simplification,

$$0.10 = \frac{(2x/10)^2}{\left(\dfrac{4.0-x}{10}\right)\left(\dfrac{1.0-x}{10}\right)} = \frac{(2x)^2}{(4.0-x)(1.0-x)}$$

This time we cannot solve for x by extracting the square root of both sides of the equation, since the denominator of the right-hand side is not a perfect square. Instead, we have to use the general method of solving a quadratic equation which involves rearranging to the form

$$ax^2 + bx + c = 0$$

and applying the quadratic formula

$$x = \frac{-b \pm \sqrt{b^2 - 4\,ac}}{2a}$$

Following this procedure, we obtain

$(2x)^2 = (4.0 - x)(1.0 - x)0.10$
$\quad 4x^2 = (4 - 5x + x^2)0.10 = 0.40 - 0.50x + 0.10x^2$
$3.9x^2 + 0.50x - 0.40 = 0$: i.e., $a = 3.9$, $b = 0.50$, $c = -0.40$

$$x = \frac{-0.50 \pm \sqrt{(0.50)^2 + 4(3.9)(0.40)}}{2(3.9)}$$

$$= \frac{-0.50 \pm \sqrt{6.49}}{7.8} = \frac{-0.50 \pm 2.55}{7.8} = 0.26 \text{ or } -0.39$$

Of the two answers, only 0.26 is plausible. A value of -0.39 requires a negative number of moles of NO at equilibrium, which is physically impossible.

Finally, we use the value of x just calculated to obtain the equilibrium concentration of nitric oxide

$$NO = \frac{2x}{10} = \frac{0.52}{10} = 0.052 \text{ mole/liter}$$

The algebra in these calculations is less formidable than it may appear to be. Try to understand why the calculation steps are made, as well as what is done

The calculations that we have just gone through suggest three general comments:

1. There are many different ways in which we may choose our unknown in a problem of this type. It is immaterial what quantity we take provided we are consistent in relating equilibrium concentrations to it. In Example 13.2a we might have taken the unknown, y, to be the number of moles of NO formed, in which case the numbers of moles of N_2, O_2, NO at equilibrium would have been $1 - y/2$, $1 - y/2$ and y, respectively. Had we made this choice, our equations would have looked slightly different, but the final answer would have been the same.

2. Some equilibrium calculations are more difficult than others. Frequently we have to solve quadratic or higher order equations to find the equilibrium state of a system. Unfortunately systems in nature are seldom simple. For example, the system in Example 13.2b is more realistic than that in Example 13.2a; the mole ratio of N_2 to O_2 in air is about $4:1$ rather than $1:1$.

3. At 2000°C appreciable quantities of nitric oxide are formed from nitrogen and oxygen. Fortunately in combustion processes involving air we work at lower temperatures where K_c is a smaller number ($K_c = 10^{-30}$ at 25°C), and, hence, form much smaller amounts of NO.

13.4 EFFECT OF CHANGES IN CONDITIONS UPON THE POSITION OF AN EQUILIBRIUM

Once a system has attained chemical equilibrium, it is possible to change the position of that equilibrium in various ways. We shall consider three ways in which an equilibrium may be disturbed:

1. Adding a reactant or product.
2. Changing the volume of the system.
3. Changing the temperature.

Each of these effects will be considered separately, the assumption being that the other two factors remain unchanged.

CHANGE IN THE NUMBER OF MOLES OF REACTANTS OR PRODUCTS

Using the expression for K_c, we can readily deduce the direction in which an equilibrium system will shift it we add or remove a substance involved in the reaction. Consider the general gaseous system

$$aA(g) + bB(g) \rightleftharpoons cC(g) + dD(g)$$

$$K_c = \frac{[C]^c \times [D]^d}{[A]^a \times [B]^b}$$

Let us suppose that after equilibrium has been established, we add a finite amount of substance A, thereby temporarily increasing its concentration. The system is no longer at equilibrium; we have reduced the concentration quotient to a value less than that which must apply at equilibrium.

Addition of A

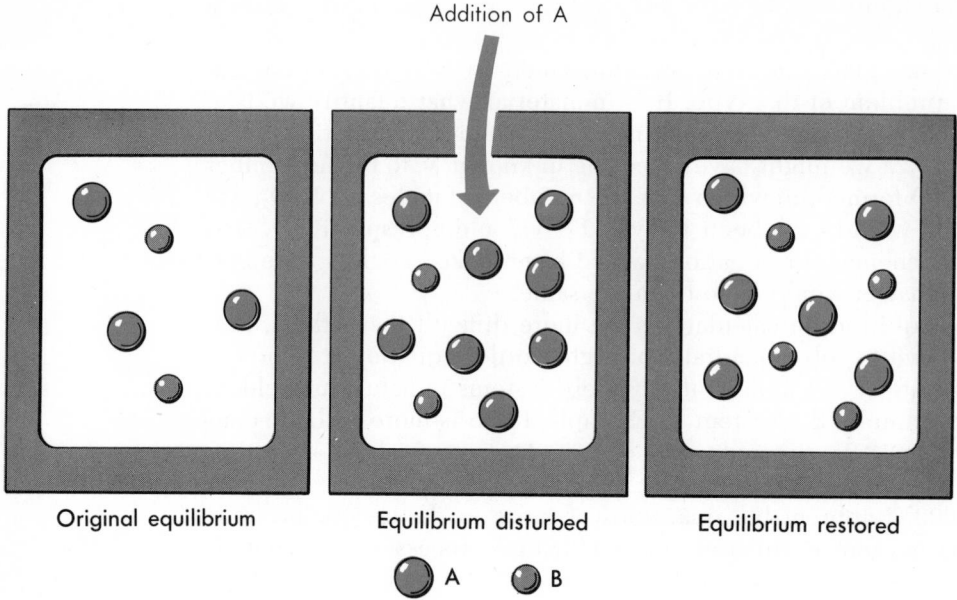

Original equilibrium   Equilibrium disturbed   Equilibrium restored

$\bigcirc$ A $\bigcirc$ B

Figure 13.2   Effect of adding reactant to the equilibrium system: $A(g) \rightleftharpoons B(g)$. Part of the added A reacts to form B.

$$\frac{(\text{conc. C})^c \times (\text{conc. D})^d}{(\text{conc. A})^a \times (\text{conc. B})^b} < K_c$$

To restore equilibrium, part of the A added must react

$$aA(g) + bB(g) \rightarrow cC(g) + dD(g)$$

How do the new concentrations of A, B, C and D compare with their values before the system was disturbed?

As reaction occurs, the concentrations of A and B decrease from the values they had immediately after the equilibrium was disturbed; those of C and D increase by equivalent amounts. The concentration quotient increases until it becomes equal to K_c, whereupon equilibrium is re-established and no further reaction occurs.

By the same line of reasoning, the addition of a product (C or D) to the equilibrium mixture would make the concentration quotient larger than K_c.

$$\frac{(\text{conc. C})^c \times (\text{conc. D})^d}{(\text{conc. A})^a \times (\text{conc. B})^b} > K_c$$

and the reverse reaction

$$cC(g) + dD(g) \rightarrow aA(g) + bB(g)$$

would occur to restore equilibrium. In the general case, we conclude that: **The addition of a species (reactant or product) to a system at equilibrium disturbs that system by temporarily changing the concentration quotient from its equilibrium value, K_c. To restore equilibrium, the system reacts in such a direction as to consume part of the added species.**

Example 13.3 It is believed that one of the ways in which sulfur dioxide in the air is converted to sulfur trioxide and eventually to droplets of sulfuric acid is by means of the reaction

$$SO_2(g) + NO_2(g) \rightarrow SO_3(g) + NO(g)$$

In a certain mixture at 700°C, the equilibrium concentration of SO_3, NO, SO_2 and NO_2 are 1.0×10^{-4}, 1.0×10^{-4}, 4.0×10^{-4} and 1.0×10^{-4} mole/liter, respectively.

    a. Determine K_c.

    b. Determine the equilibrium concentrations of each species if enough NO_2 is added to raise its concentration temporarily to 4.0×10^{-4} mole/liter.

Solution

  a. $K_c = \dfrac{[SO_3] \times [NO]}{[SO_2] \times [NO_2]} = \dfrac{(1.0 \times 10^{-4})(1.0 \times 10^{-4})}{(4.0 \times 10^{-4})(1.0 \times 10^{-4})} = \dfrac{1.0 \times 10^{-8}}{4.0 \times 10^{-8}} = 0.25$

  b. Here it is convenient to take our "original" concentrations to be those that exist immediately after the NO_2 is added (see table below).

        In order to re-establish equilibrium, the reaction must proceed in the forward direction, i.e., $SO_2(g) + NO_2(g) \rightarrow SO_3(g) + NO(g)$. We will let x be the increase in concentration of SO_3 that results. Noting that all the coefficients in the equation are 1, we have

Concentrations	Original	Change	Final
SO_2	4.0×10^{-4}	$-x$	$4.0 \times 10^{-4} - x$
NO_2	4.0×10^{-4}	$-x$	$4.0 \times 10^{-4} - x$
SO_3	1.0×10^{-4}	$+x$	$1.0 \times 10^{-4} + x$
NO	1.0×10^{-4}	$+x$	$1.0 \times 10^{-4} + x$

Note that the concentration of SO_3 will increase by the same amount that the concentration of SO_2 decreases

Our unknown x must have a value such that it satisfies the equation

$$\frac{(1.0 \times 10^{-4} + x)(1.0 \times 10^{-4} + x)}{(4.0 \times 10^{-4} - x)(4.0 \times 10^{-4} - x)} = K_c = 0.25$$

To solve for x, we take the square root of both sides of this equation and obtain

$$\frac{1.0 \times 10^{-4} + x}{4.0 \times 10^{-4} - x} = (0.25)^{1/2} = 0.50$$

$1.0 \times 10^{-4} + x = 2.0 \times 10^{-4} - 0.50x$; $1.50x = 1.0 \times 10^{-4}$; $x = 0.7 \times 10^{-4}$

Hence $[SO_3] = [NO] = 1.0 \times 10^{-4} + x = 1.7 \times 10^{-4}$ mole/lit

             $[SO_2] = [NO_2] = 4.0 \times 10^{-4} - x = 3.3 \times 10^{-4}$ mole/lit

Notice that only *part* of the added NO_2 reacted; its final concentration is intermediate between its value before the equilibrium was disturbed (1.0×10^{-4}) and immediately afterward (4.0×10^{-4}).

CHANGES IN VOLUME

It is possible to deduce from the equilibrium constant expression what will happen to a system at equilibrium when the volume is changed. To illustrate the principle involved, let us return to the equilibrium discussed earlier

$$N_2O_4(g) \rightleftharpoons 2\, NO_2(g)$$

$$K_c = \frac{[NO_2]^2}{[N_2O_4]}$$

Since we are interested in the effect of a change in volume upon the position of this equilibrium, we shall introduce the volume explicitly into the expression for K_c. Noting that the concentration of any species in moles/liter is equal to the number of moles, n, divided by the volume, V, we have

$$[NO_2] = \frac{n\ NO_2}{V}; \quad [N_2O_4] = \frac{n\ N_2O_4}{V}$$

$$K_c = \frac{(n\ NO_2/V)^2}{(n\ N_2O_4/V)}$$

Simplifying, we obtain:

$$K_c \times V = \frac{(n\ NO_2)^2}{n\ N_2O_4}$$

Now suppose that in an equilibrium mixture of NO_2 and N_2O_4 we increase the volume, perhaps from one to two liters. This increases the product $K_c \times V$. The ratio

$$\frac{(n\ NO_2)^2}{n\ N_2O_4}$$

must also increase to restore equilibrium. The only way that this can happen is for the reaction

$$N_2O_4(g) \rightarrow 2\ NO_2(g)$$

to take place, increasing the number of moles of NO_2 and decreasing the number of moles of N_2O_4.

Following the same reasoning, if we were to decrease the volume available to the system at equilibrium, the ratio $(n\ NO_2)^2/n\ N_2O_4$ would have to decrease to compensate for this change. To accomplish this, NO_2 would be consumed via the reaction

$$2\ NO_2(g) \rightarrow N_2O_4(g)$$

In Table 13.2 we list the numbers of moles of NO_2 and N_2O_4 present at equilibrium when one starts with one mole of N_2O_4 at 120°C ($K_c = 0.90$) in containers ranging in volume from 1 to 1000 liters. Notice that the mole fraction of NO_2 at equilibrium increases steadily from 0.54 in a 1 l container to more than 0.99 in a 1000 liter container.

TABLE 13.2   Effect of Change in Volume on the Position of the Equilibrium at 120°C:

$N_2O_4(g) \rightleftarrows 2\ NO_2(g)$; n N_2O_4 (orig.) = 1.00; n NO_2 orig. = 0.00

	1	10	100	1000
V (lit)	1	10	100	1000
n N_2O_4 (equil.)	0.63	0.25	0.04	0.0044
n NO_2 (equil.)	0.74	1.50	1.92	1.99
Mole fraction NO_2	0.54	0.86	0.98	0.99+

TABLE 13.3   Effect of a Change in Volume
upon the Position of Gaseous Equilibria

SYSTEM	V IN-CREASES	V DE-CREASES
1. $N_2O_4(g) \rightleftharpoons 2\ NO_2(g)$	$\rightarrow$	$\leftarrow$
2. $SO_2(g) + \frac{1}{2}\ O_2(g) \rightleftharpoons SO_3(g)$	$\leftarrow$	$\rightarrow$
3. $N_2(g) + 3\ H_2(g) \rightleftharpoons 2\ NH_3(g)$	$\leftarrow$	$\rightarrow$
4. $C(s) + H_2O(g) \rightleftharpoons CO(g) + H_2(g)$	$\rightarrow$	$\leftarrow$
5. $N_2(g) + O_2(g) \rightleftharpoons 2\ NO(g)$	0	0

The analysis we have carried out for the NO_2-N_2O_4 system can be applied to any equilibrium system.

When the position of an equilibrium is disturbed by an increase in the volume, reaction takes place in a direction so as to increase the number of moles of gas (e.g., $N_2O_4 \rightarrow 2\ NO_2$); **when the volume is decreased, the reaction which decreases the number of moles of gas** (e.g., $2\ NO_2 \rightarrow N_2O_4$) **takes place.**

The application of this principle to various reactions involving gases is shown in Table 13.3. Notice particularly that it is the change in the number of moles of *gas* which determines which way the equilibrium shifts when the volume is increased (system 4), and that when there is no change in the number of moles of gas (system 5), a change in volume has no effect upon the position of the equilibrium. (You may recall that in Example 13.2, where we studied the N_2-O_2-NO system, the volumes cancelled out of the expression for K_c.)

In System 4, C(s) does not appear in K_c, and so does not influence the shift in equilibrium

Changes in volume of gaseous systems at equilibrium ordinarily result in changes in pressure. For example, if we cut the volume of the N_2O_4-NO_2 system in half, we momentarily increase the pressure by a factor of

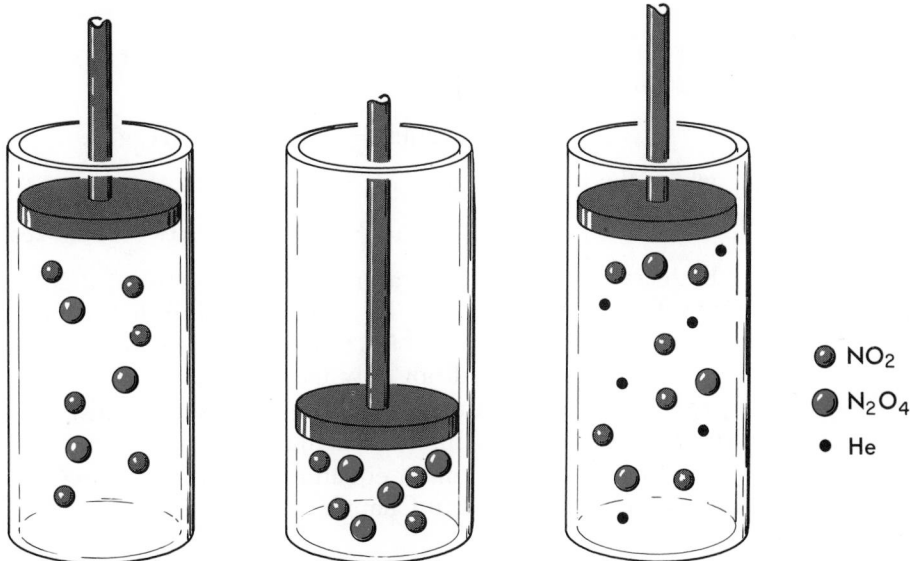

NO₂ : NO_2
N₂O₄ : N_2O_4
He

Figure 13.3   Effect of increase in pressure on $N_2O_4(g) \rightleftharpoons 2\ NO_2(g)$ system. When the pressure is increased by reducing the volume (center cylinder), some NO_2 is converted to N_2O_4. Addition of an inert gas (cylinder at right) has no effect on the concentrations of N_2O_4 and NO_2.

two. Instead of saying that the shift in the position of the equilibrium comes about because of a change in volume, we might ascribe the shift to the pressure change that accompanies the volume decrease. Specifically, we could say that an *increase in pressure shifts the position of the equilibrium in such a way as to decrease the number of moles of gas* $(2 NO_2 \rightarrow N_2O_4)$.

In discussing the effect of pressure upon the position of an equilibrium, we must be very careful to specify how the pressure change is achieved. There are many different ways in which we might change the pressure of the N_2O_4-NO_2 system without changing the volume. We might, for example, add an inert gas such as helium, which would increase the total number of moles of gas, and hence the total pressure, without affecting the volume. If we do this, there is no change in the position of the equilibrium, assuming ideal gas behavior. The concentrations of N_2O_4 and NO_2 remain constant, K_c is still satisfied, and no reaction occurs.

CHANGES IN TEMPERATURE

So far our attention has been directed toward changes in equilibrium systems at constant temperature, where the equilibrium constant K_c is a fixed number, independent of volume, pressure, or the numbers of moles of reactants or products. If the temperature changes, the magnitude of K_c changes as well. As is always the case, an increase in temperature favors the endothermic process. This means that *if the forward reaction is endothermic, an increase in temperature will result in an increase in* K_c. An example is the system

$$N_2(g) + O_2(g) \rightleftharpoons 2 NO(g)$$

where the forward reaction is endothermic

$$N_2(g) + O_2(g) \rightarrow 2 NO(g); \quad \Delta H = +43.2 \text{ kcal}$$

Large temperature changes can cause enormous changes in the value of K_c

For this reaction, K_c increases from about 10^{-30} at room temperature to 0.1 at 2000°C. In contrast, for the system

$$N_2(g) + 3 H_2(g) \rightleftharpoons 2 NH_3(g)$$

where the forward reaction is exothermic

$$N_2(g) + 3 H_2(g) \rightarrow 2 NH_3(g); \quad \Delta H = -22.0 \text{ kcal}$$

K_c decreases as the temperature rises, falling from 5×10^8 at room temperature to about 0.5 at 400°C. In general, we find that *if the forward reaction is exothermic, an increase in temperature leads to a decrease in* K_c.

We are interested not only in the direction in which the equilibrium constant changes as the temperature increases, but also the extent to which it changes. It is possible to calculate the magnitude of the temperature effect by making use of an equation analogous to the Clausius-Clapeyron equation * introduced in Chapter 9.

---

* Strictly speaking, this equation would apply exactly only if ΔH were independent of temperature and, for *gaseous* equilibria, if the expression for K involved partial pressures rather than concentrations in moles/liter. For these reasons among others, we cannot expect values of K_c calculated from this equation to be accurate to more than one significant figure.

$$\log_{10} \frac{K_2}{K_1} = \frac{\Delta H}{(2.30)(1.99)} \left[\frac{T_2 - T_1}{T_2 T_1} \right] \qquad (13.6)$$

where K_2 and K_1 are the equilibrium constants at temperatures T_2 and T_1, respectively (°K), and ΔH is the enthalpy change in *calories* for the forward reaction.

Example 13.4   In the Haber Process by which elementary nitrogen is converted to ammonia

$$N_2(g) + 3 \ H_2(g) \rightleftharpoons 2 \ NH_3(g)$$

ΔH is -22.0 kcal. Taking the equilibrium constant to be 5×10^8 at room temperature (25°C), calculate its value at 400°C.

Solution.   We shall take T_2 to be the higher temperature.

$$T_2 = (273 + 400)°K = 673°K; \ \ T_1 = (273 + 25)°K = 298°K$$

Substituting these values and that of ΔH in Equation 13.6, we obtain

$$\log_{10} \frac{K_2}{K_1} = \frac{-22{,}000}{(2.30)(1.99)} \left[\frac{673 - 298}{673 \times 298} \right] = -9.0$$

To find the ratio K_2/K_1, we note that the number whose logarithm is -9.0 is the fraction 1×10^{-9}.

$$\frac{K_2}{K_1} = 1 \times 10^{-9}; \ \ K_2 = K_1 \ (1 \times 10^{-9})$$
$$= 5 \times 10^8 \times 1 \times 10^{-9} = 0.5$$

Noting that the equilibrium constant for this reaction decreases by a factor of 10^9 in going from room temperature to 400°C, we infer that to obtain a high yield of ammonia by the Haber Process, it would be desirable to use as low a temperature as possible.

Le CHATELIER'S PRINCIPLE

We have discussed changes in equilibrium systems, both qualitatively and quantitatively, in terms of the expression for the equilibrium constant. Alternatively, we can deduce qualitatively how the position of an equilibrium will shift with a change in conditions by applying Le Chatelier's Principle, which states that:

If a system in equilibrium is altered in any way, the system will shift so as to minimize the effect of the change.

To illustrate the use of this principle, let us apply it to the three changes in conditions that we have considered in this section.

1. CHANGE IN NUMBER OF MOLES OF REACTANT OR PRODUCT.   If to the equilibrium system

$$SO_2(g) + NO_2(g) \rightleftharpoons SO_3(g) + NO(g)$$

we add NO_2, the Principle tells us that the position of the equilibrium will shift to consume some of the added reactant. In other words, the forward reaction will occur to restore equilibrium.

2. CHANGE IN VOLUME.   If we compress the equilibrium system

$$N_2O_4(g) \rightleftharpoons 2 \ NO_2(g)$$

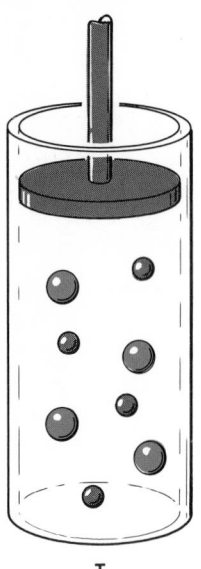
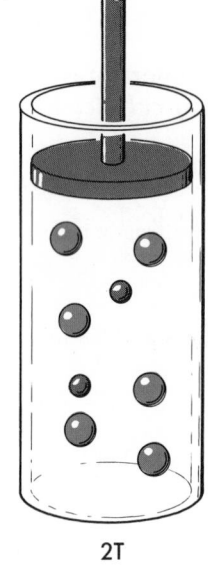

T   2T

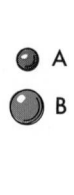

A

B

Figure 13.4 Effect of increase in temperature on A(g) $\rightleftharpoons$ B(g); $\Delta H > 0$. Part of A is converted to B.

Since the reaction as written absorbs heat the equilibrium will shift to the right to counteract an increase in temperature

Le Chatelier's Principle implies that the position of the equilibrium will shift to minimize the increase in pressure. This will be achieved if some of the NO_2 dimerizes to N_2O_4, thereby decreasing the total number of moles of gas. In other words, the reverse reaction will occur to restore equilibrium.

3. CHANGE IN TEMPERATURE. If we raise the temperature of the system

$$N_2(g) + 3\ H_2(g) \rightleftharpoons 2\ NH_3(g);\ \ \Delta H = -22.0\ kcal$$

the Principle predicts that a reaction will occur which absorbs heat, thereby minimizing the effect of the temperature increase. Since it is the reverse reaction which is endothermic, that reaction will be favored at high temperatures; some of the ammonia will decompose to the elements upon heating.

You will recall that we came to precisely these conclusions by arguments based on the form of K_c and the equation for its variation with temperature. Moreover, we were able to calculate the extent as well as the direction of the changes involved.

13.5 RELATION BETWEEN THE FREE ENERGY CHANGE AND THE EQUILIBRIUM CONSTANT

In Chapter 12, we pointed out that the free energy change, ΔG, is the fundamental criterion of spontaneity. Reactions occur spontaneously at a given temperature and pressure if ΔG is a negative quantity. In this chapter we have discussed the direction and the extent to which chemical reactions take place spontaneously in terms of the equilibrium constant, K_c. As you might expect, these two quantities are intimately related. It can be shown by a thermodynamic argument which will not be presented here that

$$\Delta G^{1M} = -RT \ln K_c$$
$$= -(2.30)(1.99) \, T \, \log_{10} K_c \quad (\Delta G^{1M} \text{ in calories}) \qquad (13.7)$$

where ΔG^{1M} can be taken to be *the free energy change for the reaction where all species in solution are at a concentration of 1 mole/liter*. In other words for the reaction

$$aA(g) + bB(g) \rightarrow cC(g) + dD(g)$$

ΔG^{1M} is the free energy change when a moles of A and b moles of B, both at a concentration of 1M, react to form c moles of C and d moles of D at that same concentration.

Looking at Equation 13.7, we see that if ΔG^{1M} for a reaction is a negative quantity, $\log K_c$ must be positive and, hence, K_c must be greater than one. A moment's reflection should convince you that this is reasonable. A negative value of ΔG^{1M} means that the reaction proceeds spontaneously in the forward direction when all species are at unit concentration, i.e., when

$$\frac{(\text{conc. C})^c \times (\text{conc. D})^d}{(\text{conc. A})^a \times (\text{conc. B})^b} = \frac{(1)^c \times (1)^d}{(1)^a \times (1)^b} = 1$$

But, from the argument advanced in Section 13.3, this can be the case only if $K_c > 1$, i.e., if it is greater than the original concentration quotient.

In a similar manner we can rationalize the observation that a positive value of ΔG^{1M} corresponds to a reaction with an equilibrium constant less than one ($\log_{10} K_c < 0$). If ΔG^{1M} is positive, the reverse reaction occurs spontaneously at unit concentrations. This is consistent with a value of $K_c < 1$; to achieve equilibrium, the concentration of products must decrease and those of reactants increase, thereby reducing the concentration quotient from unity to the equilibrium value set by K_c.

If ΔG^{1M} happens to be zero for a reaction at a particular temperature, $\log K_c = 0$ and $K_c = 1$. Under these conditions the reaction will be at equilibrium at unit concentrations and will show no tendency to occur spontaneously in either direction. Summarizing,

if $\Delta G^{1M} < 0$; $K_c > 1$; forward reaction spontaneous at unit concentrations
$\quad \Delta G^{1M} > 0$; $K_c < 1$; reverse reaction spontaneous at unit concentrations
$\quad \Delta G^{1M} = 0$; $K_c = 1$; reaction at equilibrium at unit concentrations

Equation 13.7 also enables us to give a physical interpretation to the magnitude of the free energy change. If ΔG^{1M} is a large negative number, K_c will be much greater than one and the forward reaction, under ordinary conditions, will go virtually to completion. Conversely, if ΔG^{1M} is a large positive number, K_c will be a small fraction and the reverse reaction will go nearly to completion.

Example 13.5 Calculate K_c for reactions which, at 25°C, have ΔG^{1M} values of
    a. −20.0 kcal   b. +20.0 kcal

Solution. Solving Equation 13.7 for $\log K_c$, we have

$$\log_{10} K_c = \frac{-\Delta G^{1M}}{(2.30)(1.99)T}$$

At 25°C, T = 298, and $\log_{10} K_c = \dfrac{-\Delta G^{1M}}{(2.30)(1.99)(298)} = \dfrac{-\Delta G^{1M}}{1360}$

a. $\qquad \log_{10} K_c = -\dfrac{(-20,000)}{1,360} = 14.7$

Taking antilogs, $\log_{10} K_c = 0.7 + 14.0$; $K_c = 5 \times 10^{14}$

b. $\qquad \log_{10} K_c = -\dfrac{20,000}{1,360} = -14.7$

Taking antilogs, $\log_{10} K_c = 0.3 - 15.0$; $K_c = 2 \times 10^{-15}$

From Example 13.5 we see that values of −20 kcal and +20 kcal for ΔG^{1M} lead to equilibrium constants that are very large (5×10^{14}) and very small (2×10^{-15}), respectively. Table 13.4 lists K_c calculated from Equation 13.7 for a spectrum of ΔG^{1M} values at 25°C ranging from +50 to −50 kcal. From the table it is clear that unless ΔG^{1M} at 25°C falls in a relatively narrow range between about +10 and −10 kcal, K_c will be either an extremely small or an extremely large number.

To complete our discussion of the relationship between the standard free energy change and the equilibrium constant, we should point out that ΔG^{1M} will ordinarily differ somewhat from the quantity $\Delta G^{1\,atm}$ discussed extensively in Chapter 12. Remember that ΔG^{1M} was defined as the free energy change when reactants and products are at a *concentration of 1 mole/liter*, while $\Delta G^{1\,atm}$ is the free energy change when reactants and products are at a *pressure of 1 atmosphere*. These two sets of conditions are not identical; one can show, using the Ideal Gas Law, that a concentration of 1 mole/liter corresponds, at 25°C, to a pressure of about 24 atm. By considering the way in which ΔG varies with pressure, it is possible to show (cf. Problem 13.37) that the relationship between the two quantities at 298°K is

$$\Delta G^{1M} = \Delta G^{1\,atm} + 1900\Delta n_{gas} \qquad (13.8)$$
$$\text{(valid at 298°K; } \Delta G^{1M} \text{ and } \Delta G^{1\,atm} \text{ in cal)}$$

where Δn_{gas} is the change in the number of moles of gas in the reaction. This equation implies and the data in Table 13.5 confirm that the difference between ΔG^{1M} and $\Delta G^{1\,atm}$ is quite small unless the reaction produces a large change in the number of moles of gas. If there is no change in the number of moles of gas, $\Delta G^{1\,atm}$ is exactly equal to ΔG^{1M}.

TABLE 13.4   Magnitude of K_c for Various Values of ΔG^{1M} at 298°K

ΔG^{1M} (kcal)	K_c	ΔG^{1M} (kcal)	K_c
+50	2×10^{-37}	−50	5×10^{36}
+20	2×10^{-15}	−20	5×10^{14}
+10	4.5×10^{-8}	−10	2.2×10^{7}
+5	0.00021	−5	4700
+2	0.033	−2	30
+1	0.18	−1	5.5
0	1.0	0	1.0

TABLE 13.5   Relation between ΔG^{1M} and $\Delta G^{1\,atm}$ at 298°K *

REACTION	Δn_{gas}	$\Delta G^{1\,atm}$	ΔG^{1M}
$N_2(g) + O_2(g) \rightleftharpoons 2\ NO(g)$	0	+41,400 cal	+41,400 cal
$C(s) + O_2(g) \rightleftharpoons CO_2(g)$	0	−94,300	−94,300
$CaCO_3(s) \rightleftharpoons CaO(s) + CO_2(g)$	+1	+31,100	+33,000
$N_2O_4(g) \rightleftharpoons 2\ NO_2(g)$	+1	+1,300	+3,200
$N_2(g) + 3\ H_2(g) \rightleftharpoons 2\ NH_3(g)$	−2	−8,000	−11,800

* $\Delta G^{1\,atm}$ calculated from data in Table 12.2, Chapter 12; ΔG^{1M} calculated from Equation 13.8.

PROBLEMS

13.1 Formulate expressions for the equilibrium constant, K_c, for each of the following systems:

   a. $2\ NO(g) + O_2(g) \rightleftharpoons 2\ NO_2(g)$.
   b. $H_2(g) + S(s) \rightleftharpoons H_2S(g)$.
   c. $2\ C_2H_6(g) + 7\ O_2(g) \rightleftharpoons 4\ CO_2(g) + 6\ H_2O(l)$.
   d. The decomposition of 1 mole of $NOCl(g)$ to the elements.

13.2 The equilibrium constant for the reaction

$$2\ HI(g) \rightleftharpoons H_2(g) + I_2(g)$$

is 0.016 at 520°C. At this same temperature, what is K_c for

   a. $H_2(g) + I_2(g) \rightleftharpoons 2\ HI(g)$?
   b. $HI(g) \rightarrow \frac{1}{2} H_2(g) + \frac{1}{2} I_2(g)$?

13.3 Consider the equilibrium $N_2(g) + 3\ H_2(g) \rightleftharpoons 2\ NH_3(g)$. At 220°C a certain equilibrium mixture in a 3.0-liter container consists of 0.60 mole of NH_3, 0.30 mole of H_2 and 0.45 mole of N_2. Calculate K_c.

13.4 One mole of N_2O_4 is placed in a 5.0-liter container at 100°C. Part of it dissociates to form NO_2; at equilibrium 1.00 mole of NO_2 is present. Calculate K_c for the reaction.

$$N_2O_4(g) \rightleftharpoons 2\ NO_2(g)$$

13.5 For the reaction given in Problem 13.4, $K_c = 0.90$ at 120°C. Predict the direction in which each of the following systems will move to achieve equilibrium at 120°C.

   a. Conc. $N_2O_4 = 0.20$ mole/liter; no NO_2 present
   b. Conc. $N_2O_4 = 0.20$ mole/liter; conc. $NO_2 = 0.20$ mole/liter
   c. Conc. $N_2O_4 = 0.20$ mole/liter; conc. $NO_2 = 0.50$ mole/liter

13.18 Formulate expressions for the equilibrium constant K_c for the following

   a. $2\ H_2(g) + O_2(g) \rightleftharpoons 2\ H_2O(g)$.
   b. $H_2(g) + \frac{1}{2} O_2(g) \rightleftharpoons H_2O(g)$.
   c. $H_2(g) + \frac{1}{2} O_2(g) \rightleftharpoons H_2O(l)$.
   d. the decomposition of 1 mole of $N_2H_4(l)$ to the elements.

13.19 The equilibrium constant for the reaction

$$I_2(g) + Cl_2(g) \rightleftharpoons 2\ ICl(g)$$

is 2.0×10^5 at 25°C. At this temperature, calculate K_c for

   a. $2\ ICl(g) \rightleftharpoons I_2(g) + Cl_2(g)$.
   b. $\frac{1}{2} I_2(g) + \frac{1}{2} Cl_2(g) \rightleftharpoons ICl(g)$.

13.20 Consider the equilibrium

$$2\ NO(g) + O_2(g) \rightleftharpoons 2\ NO_2(g)$$

At 460°C, an equilibrium concentration of O_2 of 0.014 mole/liter converts half of the NO in an air sample to NO_2. Calculate K_c at 460°C.

13.21 For the reaction in Problem 13.20, it is found that when 0.40 mole of NO and 0.60 mole of O_2 are added to a 3.0-liter container at 620°C, 0.16 mole of NO_2 is formed at equilibrium. Calculate K_c for the reaction at 620°C.

13.22 For the system $N_2(g) + 3\ H_2(g) \rightleftharpoons 2\ NH_3(g)$, K_c is 0.50 at 400°C. Predict the direction in which each of the following systems will move to achieve equilibrium at 400°C:

   a. conc. $NH_3 = 1.0 \times 10^{-6}$ mole/lit, conc. $H_2 = 1.2$ mole/liter, no N_2.
   b. a mixture containing 1.0 mole/lit of H_2, N_2 and NH_3.
   c. a mixture of 2.0×10^{-3} mole of NH_3, 1.0×10^{-2} mole N_2 and 1.0×10^{-1} mole of H_2 in a ten-liter container.

13.6 For the reaction

$$2 SO_2(g) + O_2(g) \rightleftharpoons 2 SO_3(g)$$

$K_c = 440$ at 700°C. What must be the equilibrium concentration of O_2 in a mixture in which $[SO_3] = 0.50$ mole/liter; $[SO_2] = 0.050$ mole/liter?

13.7 For the reaction

$$2 HI(g) \rightleftharpoons H_2(g) + I_2(g)$$

$K_c = 0.016$ at 520°C. Calculate the concentrations of all species at equilibrium in a 6.0-liter container starting with

    a. 0.40 mole of HI.
    b. 0.20 mole of H_2, 0.20 mole of I_2.
    c. 0.40 mole of H_2, 0.40 mole of I_2.

13.8 Consider the reaction

$$CaCO_3(s) \rightarrow CaO(s) + CO_2(g)$$

for which $K_c = 6.0 \times 10^{-3}$ at 740°C. Starting with 0.100 mole of $CaCO_3$ in a 10.0-liter container, calculate, at equilibrium

    a. the concentration of CO_2.
    b. the number of moles of CO_2.
    c. the fraction of the calcium carbonate converted to CaO.
    d. the pressure of CO_2, using the Ideal Gas Law.

13.9 The concentrations of CO and CO_2 in a sample taken from an automobile exhaust are 3.0×10^{-4} and 3.0×10^{-3} mole/liter, respectively. If this exhaust were passed through an afterburner at 1600°C in which the concentration of O_2 is maintained at a constant value of 4.0×10^{-4} mole/liter, what would be the equilibrium concentration of CO? At 1600°C, K_c for the reaction

$$CO(g) + \tfrac{1}{2} O_2(g) \rightarrow CO_2(g)$$

is about 1×10^4.

13.10 For the reaction

$$N_2(g) + O_2(g) \rightarrow 2 NO(g)$$

$K_c = 1 \times 10^{-30}$ at 25°C and 0.10 at 2000°C. Starting with 0.040 mole/liter of N_2 and 0.010 mole/liter of O_2, calculate the equilibrium concentration of NO

    a. at 2000°C.
    b. at 25°C. (Note that K_c is very small at this temperature. The equilibrium concentrations of N_2 and O_2 will be virtually equal to their original concentrations.)

13.23 For the reaction given in Problem 13.22, what must be the equilibrium concentration of H_2 in a mixture in which the concentrations of N_2 and NH_3 are 1.0×10^{-2} mole/liter?

13.24 For the reaction

$$SO_2(g) + NO_2(g) \rightleftharpoons SO_3(g) + NO(g)$$

K_c is 9.0 at 700°C. Calculate the concentrations of all species at equilibrium in a 500-ml flask if one starts with

    a. 2.0×10^{-3} moles of SO_2 and NO_2.
    b. 2.0×10^{-3} moles of SO_3 and NO.
    c. 2.0×10^{-3} moles of SO_2, NO_2, SO_3 and NO.

13.25 Small amounts of very pure iron ore are produced by the reaction

$$Fe_2O_3(s) + 3 H_2(g) \rightleftharpoons 2 Fe(s) + 3 H_2O(g)$$

for which $K_c = 0.064$ at 340°C.

    a. Calculate K_c for the reaction $\tfrac{1}{3} Fe_2O_3(s) + H_2(g) \rightleftharpoons \tfrac{2}{3} Fe(s) + H_2O(g)$.
    b. The reaction is carried out so that p $H_2 = 1.00$ atm. Using the Ideal Gas Law, calculate $[H_2]$.
    c. From the results of (a) and (b), calculate $[H_2O]$.

13.26 A certain gaseous mixture contains 2.0×10^{-1} mole of PCl_5 and 4.0×10^{-1} mole of PCl_3 in a 100-ml container at 240°C. It is desired to convert 90% of the PCl_3 present to PCl_5 by taking advantage of the reaction

$$PCl_3(g) + Cl_2(g) \rightleftharpoons PCl_5(g)$$

for which K_c is 20 at 240°C. To accomplish this:

    a. what must be the equilibrium concentration of Cl_2?
    b. what is the *total* number of moles of Cl_2 that must be added?

13.27 When chlorine gas is heated, it decomposes according to the reaction

$$Cl_2(g) \rightleftharpoons 2 Cl(g)$$

K_c for this reaction is 1.2×10^{-6} at 1000°C and 3.6×10^{-2} at 2000°C. Starting with a concentration of Cl_2 of 0.010 mole/liter, what will be the equilibrium concentration of atomic chlorine at

    a. 2000°C?
    b. 1000°C? (Note that K_c at 1000°C is so small that $[Cl_2]$ will be very nearly equal to its original concentration.)

13.11   At 70°C, K_c for the reaction

$$N_2O_4(g) \rightleftharpoons 2\ NO_2(g)$$

is 0.090. Starting with 1.00 mole of N_2O_4 in a two-liter container at this temperature, determine

    a. the equilibrium concentrations of NO_2 and N_2O_4.
    b. the ratio of the total pressure at equilibrium to the original pressure.

13.12   Consider the equilibrium:

$$2\ NH_3(g) \rightleftharpoons N_2(g) + 3\ H_2(g); \quad \Delta H = +22.0\ kcal$$

A mixture of these three substances is allowed to reach equilibrium in a 10.0-liter container at 150°C. Predict the direction in which the system will move to re-establish equilibrium if

    a. one mole of H_2 is added.
    b. the total pressure of the mixture is increased by adding nitrogen.
    c. the total pressure of the mixture is increased by adding argon.
    d. the temperature is increased to 300°C.

13.13   Water gas, a commercial fuel, is made by the reaction of hot coke with steam: $C(s) + H_2O(g) \rightarrow CO(g) + H_2(g)$. When equilibrium is established at 800°C in a 50-liter vessel, the concentrations of CO, H_2, and H_2O are 4.0×10^{-2}, 4.0×10^{-2} and 1.0×10^{-2} mole/liter, respectively.

    a. Calculate K_c for the reaction.
    b. What will be the final equilibrium concentrations of CO and H_2 if one mole of steam is added to the mixture?

13.14   For the reaction

$$PCl_5(g) \rightleftharpoons PCl_3(g) + Cl_2(g)$$

K_c is 0.050 at 240°C. Starting with 0.20 mole of PCl_5 at this temperature, what will be the number of moles of Cl_2 at equilibrium in a

    a. 1-liter container?
    b. 10-liter container?

13.15   Consider the Haber Process

$$N_2(g) + 3\ H_2(g) \rightleftharpoons 2\ NH_3(g)$$

for which $K_c = 3.8 \times 10^8$ at 27°C. Estimate K_c for this reaction at 127°C (the heat of formation of ammonia is −11.0 kcal/mole).

13.16   For two different reactions, (1) and (2), $\Delta G_2^{1M} - \Delta G_1^{1M} = +5.0$ kcal at 298°K. Calculate the ratio, K_{c2}/K_{c1} of the two equilibrium constants.

13.28   For the reaction in Problem 13.27 at 2000°C if we start with 1.0×10^{-2} mole/liter of both Cl and Cl_2:

    a. which way will the system move to establish equilibrium?
    b. what will be the equilibrium concentration of atomic chlorine?

13.29   Consider the decomposition of hydrogen peroxide to the elements

$$H_2O_2(l) \rightleftharpoons H_2(g) + O_2(g); \quad \Delta H = +44.8\ kcal$$

Equilibrium is established in a 100-ml flask at room temperature. Predict the direction in which the reaction will go if

    a. a small amount of H_2O_2 is added (assume constant volume).
    b. 50 ml of an inert solid is added to the flask.
    c. hydrogen gas is added to the flask.
    d. the temperature is raised to 500°C.

13.30   For the reaction

$$CO_2(g) + H_2(g) \rightleftharpoons CO(g) + H_2O(g)$$

equilibrium is established at 1400°C in a 5.0-liter vessel in which 2.0×10^{-3} mole of CO, 1.0×10^{-3} mole of H_2O, 5.0×10^{-4} mole of CO_2 and 1.6×10^{-3} mole of H_2 is present.

    a. Calculate K_c.
    b. If 1.0×10^{-3} mole of H_2O is added, how much of it will have decomposed when equilibrium is re-established?

13.31   For the system

$$N_2O_4(g) \rightleftharpoons 2\ NO_2(g)$$

$K_c = 0.90$ at 120°C. Starting with 1.00 mole of N_2O_4 at 120°C, calculate the number of moles of NO_2 at equilibrium in a

    a. one-liter container.
    b. ten-liter container.

Compare your answers with those given in Table 13.2.

13.32   Consider the reaction

$$ZnO(s) + H_2(g) \rightleftharpoons Zn(s) + H_2O(g)$$

for which K_c is about 2×10^{-16} at 25°C. Using heat of formation data, calculate the temperature at which $K_c = 1$.

13.33   Using Equation 13.7, calculate ΔG^{1M} values at 298°K for reactions in which K_c is 100, 1.0 and 0.010, respectively.

13.17   From the data in Table 13.5, calculate K_c for the reaction at 25°C

$$N_2O_4(g) \rightleftharpoons 2\ NO_2(g)$$

13.34   For the reaction

$$2\ NO(g) + O_2(g) \rightarrow 2\ NO_2(g)$$

a. calculate $\Delta G^{1\ atm}$ at 298°K (Table 12.2).
b. calculate ΔG^{1M}, using Equation 13.8.
c. calculate K_c at 298°K.

*13.35   Consider the reactions

$$N_2(g) + O_2(g) \rightleftharpoons 2\ NO(g);\ \ K_1$$

$$2\ NO(g) + O_2(g) \rightleftharpoons 2\ NO_2(g);\ \ K_2$$

Show that for the reaction

$$N_2(g) + 2\ O_2(g) \rightarrow 2\ NO_2(g);\ \ K_3$$

$K_3 = K_2 \times K_1$. This problem illustrates an important principle; the equilibrium constant for a reaction that can be considered to be the sum of two other reactions is the product of the K_c's of the two reactions.

*13.36   For the reaction

$$N_2(g) + 3\ H_2(g) \rightleftharpoons 2\ NH_3(g)$$

at 300°C, $K_c = 10$. Starting with 0.40 mole of N_2 and 1.20 mole of H_2 in a 5.0-liter container, calculate the concentration of NH_3 at equilibrium.

*13.37   The variation with pressures of the free energy change for a reaction is given by the equation

$$\Delta G\ (at\ P_2) = \Delta G\ (at\ P_1) + \Delta n_{gas}\ RT\ \ln \frac{P_2}{P_1}$$

where P_2 and P_1 are two different pressures. Show that, at 298°K, this expression leads to Equation 13.8. (Hint: Take P_1 to be 1 atm and P_2 to be the pressure exerted at 298°K by a gas at a concentration of 1 mole/liter.)

*13.38   The equilibrium constant that we have worked with in this chapter, K_c, is not the only one that can be used to describe gaseous equilibria. Another constant that is often used is K_p, in which concentrations in moles per liter are replaced by partial pressures in atomspheres. For example, for the reaction

$$aA(g) + bB(g) \rightleftharpoons cC(g) + dD(g)$$

$$K_p = \frac{(pC)^c(pD)^d}{(pA)^a(pB)^b}$$

where pA, pB, pC and pD are partial pressures. Show, using the Ideal Gas Law, that $K_p = K_c\ (RT)^{\Delta n_{gas}}$.

*13.39   It has been suggested that the generation of nitric oxide in the upper atmosphere by the combustion of fuel in a supersonic transport might destroy the layer of ozone which protects us from ultraviolet light. The reaction is

$$NO(g) + O_3(g) \rightarrow NO_2(g) + O_2(g)$$

a. Given that the free energy of formation of ozone is +39.1 kcal/mole at 298°K and 1 atm and using other data from Table 12.2, Chapter 12, calculate $\Delta G^{1\ atm}$, ΔG^{1M}, and K_c for this reaction at 25°C.
b. Assuming that the concentrations of NO, O_3, and O_2 in the upper atmosphere before reaction occurs are 2×10^{-9}, 1×10^{-9}, and 2×10^{-3} mole/liter, respectively, calculate the equilibrium concentration of O_3. (Assume there is no NO_2 present originally.)
c. According to your calculations, what percentage of the O_3 would be destroyed? (Note that you are assuming that the rate of reaction is great enough to quickly establish equilibrium, which may not be the case.)

14 RATES OF REACTION

In the past two chapters we have shown how, from a knowledge of the free energy change or the equilibrium constant, it is possible to predict the direction and extent of a reaction. Problems of this type fall in the general area of *chemical thermodynamics*. An understanding of thermodynamic principles is important to the chemist because it guides him in choosing reactant concentrations and temperatures at which a reaction is feasible.

We often find, however, that reactions which are spontaneous in principle occur so slowly as to be of little practical value. Consider, for example, the reaction

$$CO(g) + NO(g) \rightarrow CO_2(g) + \tfrac{1}{2} N_2(g)$$

We can calculate that ΔG for this reaction, at 25°C and 1 atm, is −82.2 kcal, indicating that the reaction is spontaneous under these conditions. Indeed, K_c at 25°C is so large ($\sim 10^{60}$) that these two toxic gases, even at the low concentrations found in automobile exhausts, should combine almost completely to form carbon dioxide and elementary nitrogen. Unfortunately, however, this reaction takes place so slowly under ordinary conditions that it does not offer a practical method of removing carbon monoxide and nitric oxide from polluted air.

Many of the most familiar substances in our environment are unstable from a thermodynamic viewpoint. The fossil fuels—coal, petroleum, and natural gas—should, according to thermodynamic calculations, be converted to carbon dioxide and water upon exposure to air. The same is true of the organic compounds that make up the living cells of our bodies. Life persists because reactions that are in principle spontaneous occur at an infinitesimal rate under the conditions of temperature and pressure that prevail on the earth's surface.

It's somewhat sobering to realize that people are thermodynamically unstable

We conclude that there is no direct correlation between the rate of a reaction and the thermodynamic driving force as expressed by the free energy change or the equilibrium constant. In order to predict how rapidly a reaction will occur, we must become familiar with a different set of principles which fall in the area of *chemical kinetics*. We shall develop and discuss certain of these principles in this chapter, where our primary concern will be with reactions involving gases. In later chapters we shall have more to say about reaction rates in water solution.

14.1 MEANING OF REACTION RATE

In order to discuss reaction rate in a meaningful way, we must know precisely what is meant by the term. The rate of reaction is a positive quantity that tells us how the concentration of a reactant or product changes with time. To see what this statement means, consider the reaction between carbon monoxide and nitrogen dioxide

$$CO(g) + NO_2(g) \rightarrow CO_2(g) + NO(g) \tag{14.1}$$

The rate of this reaction can be taken to be the change in concentration per unit time of one of the products, let us say, carbon dioxide

$$\text{rate} = \frac{\text{change in conc. } CO_2}{\text{time interval}} = \frac{\Delta \text{ conc. } CO_2}{\Delta t} \tag{14.1a}$$

Alternatively, the rate could be expressed in terms of the disappearance of a reactant

$$\text{rate} = \frac{-\Delta \text{ conc. } CO}{\Delta t} \tag{14.1b}$$

Notice that, because of the 1:1 stoichiometry of Equation 14.1, these two expressions for the rate of reaction are equivalent. For every mole of CO_2 formed, one mole of CO disappears. If, in one second, the concentration of CO_2 were to increase by 0.020 mole/liter (Δ conc. CO_2 = +0.020 mole/lit), that of CO would have to decrease by the same amount (Δ conc. CO = −0.020 mole/lit). The rate, calculated from either 14.1a or 14.1b, would be +0.020 mole/lit sec.

We note from the rate expressions that the units of reaction rate will be those of concentration divided by time. We shall consistently express concentrations in moles/liter. Time, on the other hand, may be given in seconds, minutes, hours, days, or years. Consequently the units of reaction rate may be

mole/lit sec, mole/lit min, . . .

Let us consider how we might experimentally determine the rate of the reaction between CO and NO_2, as expressed by Equation 14.1b. Clearly we need to know how the concentration of CO changes with time. To find out, we might start by introducing 0.10 mole of CO and 0.10 mole of NO_2 into a 1-liter container at 400°. Every 10 seconds we withdraw a small sample of gas from the reaction mixture, cool it quickly to stop the reaction, and analyze for CO, perhaps by vapor phase chromatography (Chapter 1). The results of such an experiment are listed in Table 14.1 and shown graphically in Figure 14.1.

TABLE 14.1   Rate of Reaction $CO(g) + NO_2(g) \rightarrow CO_2(g) + NO(g)$

(Data taken at 400°C with initial concentrations of CO = NO_2 = 0.10 mole/liter)

Conc. CO_2	0.100	0.067	0.050	0.040	0.033	...	0.017	...	0.002
Time (sec)	0	10	20	30	40	...	100	...	1000

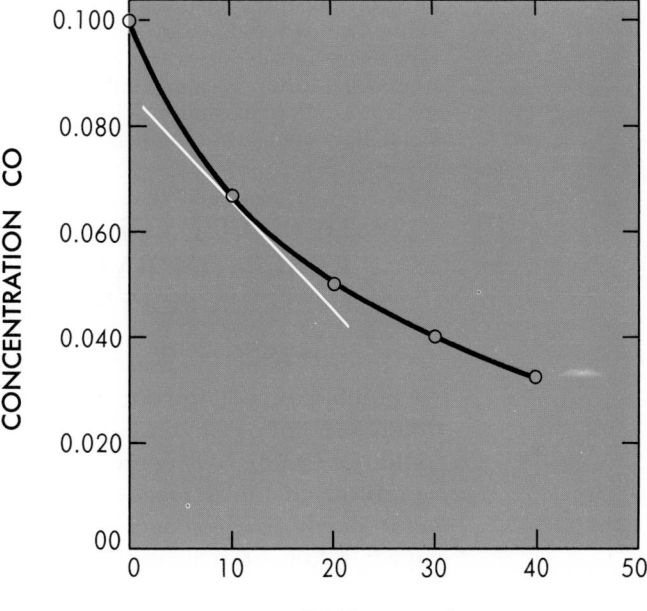

Figure 14.1 Variation of concentration of CO with time in the reaction: $CO(g) + NO_2(g) \rightarrow CO_2(g) + NO(g)$. The rate at any point can be determined by drawing the tangent to the curve.

You will notice, from both the table and the figure, that the concentration of carbon monoxide drops rapidly in the initial stages of the reaction. As the reaction proceeds, the concentration of CO decreases more and more slowly. Clearly the rate of reaction is itself a function of time. Over the first 10-second period, the average rate is

$$\frac{-\Delta \text{ conc. CO}}{\Delta t} = \frac{-(0.067 - 0.100)}{10 - 0} = \frac{0.033}{10} = 0.0033 \ \frac{\text{mole}}{\text{liter}}$$

In the period between 10 and 20 seconds, the average rate is considerably smaller

$$\frac{-\Delta \text{ conc. CO}}{\Delta t} = \frac{-(0.050 - 0.067)}{10} = \frac{0.017}{10} = 0.0017 \ \frac{\text{mole}}{\text{lit sec}}$$

If we require the rate of reaction at a particular instant, let us say at $t = 10$ sec, we might guess that it would be halfway between these two rates, i.e.,

$$\frac{0.0033 + 0.0017}{2} = 0.0025 \ \frac{\text{mole}}{\text{lit sec}}$$

A better way to estimate the rate at this point would be to draw a tangent to the curve of Figure 14.1 at $t = 10$ sec and find its slope. If we do this carefully, we find an instantaneous rate of 0.0022 mole/lit sec.

$\text{rate} = -\text{slope}$
$\quad = -\left(\dfrac{-.044}{20}\right)$
$\quad = 0.0022$

An even better way to obtain the rate of reaction at a given time uses the principles of calculus. The instantaneous rate of reaction, in derivative notation, is given by

$$\text{rate} = \frac{-d \ (\text{conc. CO})}{dt}$$

This relationship suggests that to find the rate at t = 10 sec we could fit the data in Table 14.1 to an algebraic equation relating conc. CO to time, obtain a general expression for the derivative, and evaluate it at t = 10 sec. We shall not pursue this approach further because experience indicates that manipulations of this type are not familiar to students in general chemistry, even those who have had an elementary course in calculus.

14.2 DEPENDENCE OF REACTION RATE UPON CONCENTRATION

In discussing the reaction of carbon monoxide with nitrogen dioxide, we pointed out that the rate of reaction decreases with time. From a slightly different point of view, we could say that the rate decreases as the concentration of CO or NO_2 decreases. This observation is generally valid for a variety of chemical reactions. We ordinarily find that reactions proceed more slowly as the concentrations of reactants decrease. Increasing the concentration of reactants ordinarily increases the reaction rate. These observations can be interpreted quite simply in terms of the collision theory of reaction rates (Section 14.5). The higher the concentration of reactant molecules, the more frequently they will collide and be converted to products.

One way to study the effect of concentration upon reaction rate is to obtain the initial rate (i.e., the rate at t = 0) as a function of the concentration of a particular reactant, holding the concentrations of all other reactants constant. We might, for example, conduct a series of experiments in which we measure the initial rate of the CO—NO_2 reaction at different concentrations of CO, holding the concentration of NO_2 constant. Data for three such series are presented in Table 14.2.

Looking at the vertical columns in Table 14.2, we observe that the rate is directly proportional to the concentration of CO. If, for example, the concentration of CO is doubled (e.g., from 0.10 to 0.20 mole/lit), the rate also doubles (from 0.005 to 0.010 mole/lit sec in series 1, from 0.010 to 0.020 mole/lit sec in series 2, and so forth). In a similar way we can deduce the effect of NO_2 concentration upon rate by examining the horizontal rows of data in Table 14.2. Notice, for example, that when the concentration of CO is held constant at 0.10 mole/lit (first horizontal row) the rate increases

TABLE 14.2 Initial Rates of Reaction (mole/lit sec)
$CO(g) + NO_2(g) \rightarrow CO_2(g) + NO(g)$ at 400°C

SERIES 1			SERIES 2			SERIES 3		
conc. CO	conc. NO_2	rate	conc. CO	conc. NO_2	rate	conc. CO	conc. NO_2	rate
0.10	0.10	0.005	0.10	0.20	0.010	0.10	0.30	0.015
0.20	0.10	0.010	0.20	0.20	0.020	0.20	0.30	0.030
0.30	0.10	0.015	0.30	0.20	0.030	0.30	0.30	0.045
0.40	0.10	0.020	0.40	0.20	0.040	0.40	0.30	0.060

in direct proportion to the concentration of NO_2. We conclude that the rate of this reaction is directly proportional to the concentrations of both CO and NO_2 or that

$$\text{rate} = k \text{ (conc. CO)(conc. NO}_2) \tag{14.2}$$

All the rates in Table 14.2 can be calculated with this equation

The constant of proportionality k in Equation 14.2 is referred to as the **rate constant** for the reaction. For a particular reaction, k is a function only of temperature; it is independent of the concentrations of reactants. It can, of course, be calculated from the observed rate at known reactant concentrations. Substituting in Equation 14.2 at a point where the concentrations of CO and NO_2 are both 0.10 mole/liter, we have

$$0.005 \frac{\text{mole}}{\text{lit sec}} = k \left(0.10 \frac{\text{mole}}{\text{lit}}\right)\left(0.10 \frac{\text{mole}}{\text{lit}}\right)$$

Solving:

$$k = \frac{0.005 \text{ mole/lit sec}}{(0.10 \text{ mole/lit})^2} = 0.5 \frac{\text{lit}}{\text{mole sec}}$$

Reaction rates at 400°C can now be calculated from the expression

$$\text{rate} = 0.5 \text{ (conc. CO)(conc. NO}_2)$$

Rate expressions have been established experimentally for a large number of reactions. In general, for the reaction

$$aA(g) + bB(g) \rightarrow \text{products}$$

the rate expression takes the form

$$\text{rate} = k \text{ (conc. A)}^m\text{(conc. B)}^n \tag{14.3}$$

The powers to which the concentrations of reactants are raised in the rate expression describe the **order** of the reaction. If m in Equation 14.3 is 1, we say that the reaction is "first order with respect to A." If n = 2, the reaction is said to be "second order in B." The overall order of the reaction is the sum of the exponents m and n. A "first order reaction" is one in which m + n = 1, a "second order reaction" one in which m + n = 2, and so on.

Example 14.1 The initial rate of decomposition of acetaldehyde

$$CH_3CHO(g) \rightarrow CH_4(g) + CO(g)$$

was measured at a series of different concentrations with the following results

conc. CH_3CHO (mole/lit)	0.10	0.20	0.30	0.40
rate (mole/lit sec)	0.020	0.081	0.182	0.318

Using these data, determine
     a. the order of the reaction with respect to CH_3CHO.

     b. the rate constant.

     c. the rate of reaction at a conc. of CH_3CHO of 0.15 mole/lit.

Solution
     a. Perhaps the simplest approach here is to write the general rate expression at two different concentrations

$$\text{rate}_2 = k \text{ (conc.}_2)^m$$
$$\text{rate}_1 = k \text{ (conc.}_1)^m$$

Taking the ratio of the rates we obtain: $\dfrac{\text{rate}_2}{\text{rate}_1} = \left(\dfrac{\text{conc.}_2}{\text{conc.}_1}\right)^m$

Choosing our two concentrations to be 0.20 and 0.10, our equation becomes

$$\frac{0.081}{0.020} = \left(\frac{0.20}{0.10}\right)^m$$

This is certainly the easiest way to find the order of a reaction

Simplifying: $4.0 = 2^m$

Clearly m is two; the same result would be obtained within experimental error if we had chosen any other pair of points.

b. The rate expression must be

$$\text{rate} = k \, (\text{conc. } CH_3CHO)^2; \quad k = \frac{\text{rate}}{(\text{conc. } CH_3CHO)^2}$$

Substituting values at one particular concentration, let us say at a conc. of 0.30 mole/liter:

$$k = \frac{0.182 \text{ mole/lit sec}}{(0.30 \text{ mole/liter})^2} = \frac{0.182}{0.090} \frac{\text{lit}}{\text{mole sec}} = 2.0 \text{ lit/mole sec}$$

A more elegant approach to (a) and (b) would involve taking the logarithm of both sides of the rate expression to obtain

$$\log \text{rate} = \log k + m \, (\log \text{conc. } CH_3CHO)$$

One could then plot log rate vs log conc. CH_3CHO and draw the best straight line through the points. The slope of the straight line would give m directly, while k could be calculated from the intercept. This approach seems unnecessary here but might be advisable if the rate expression were a more complicated one.

c. Having found the rate expression

$$\text{rate} = 2.0 \frac{\text{lit}}{\text{mole sec}} \; (\text{conc. } CH_3O)^2$$

we need only substitute 0.15 for the concentration of CH_3CHO and solve to obtain a rate of 0.045 mole/lit sec.

Example 14.1 illustrates the fact that the *order of a reaction must be determined experimentally and cannot, in general, be deduced from the coefficients of the balanced equation.* We found that the decomposition of acetaldehyde was second order even though the coefficient of CH_3CHO in the balanced equation was 1. Other examples of this general principle include the rate expressions for the reactions

$$2 \, N_2O_5(g) \rightarrow 4 \, NO_2(g) + O_2(g) \quad \text{rate} = k \, (\text{conc. } N_2O_5)^1$$

$$4 \, HBr(g) + O_2(g) \rightarrow 2 \, Br_2(g) + 2 \, H_2O(g) \quad \text{rate} = k \, (\text{conc. } HBr)^1 (\text{conc. } O_2)^1$$

$$CHCl_3(g) + Cl_2(g) \rightarrow CCl_4(g) + HCl(g) \quad \text{rate} = k \, (\text{conc. } CHCl_3)^1 (\text{conc. } Cl_2)^{1/2}$$

Frequently it is found that the exponents m and n in Equation 14.3 are integers (0, 1, or 2); in many reactions, however, they are simple or complex fractions. We shall discuss in some detail only one particular type of rate expression, that in which a single species decomposes via a first order reaction. Such reactions are relatively common in the gas phase and in aqueous solution. Perhaps the most important type of first order reaction is radioactive decay, discussed in Chapter 22.

FIRST ORDER REACTIONS

An example of a first order gas phase reaction is the thermal decomposition of dinitrogen pentoxide, N_2O_5:

$$2 N_2O_5(g) \rightarrow 4 NO_2(g) + O_2(g) \qquad (14.4)$$

The rate constant for this reaction has the value 0.35 min^{-1} at 67°C.

$$\text{rate} = 0.35 \text{ min}^{-1} \text{ (conc. } N_2O_5\text{)} \qquad (14.4a)$$

Equation 14.4a enables us to calculate the reaction rate at any desired concentration. It does not, however, provide a simple relationship which would be even more valuable to us, that between concentration and time. Looking at the data in Table 14.3 or the curve at the left of Figure 14.2, it is not obvious that there is any simple equation relating concentration of N_2O_5 to time. However, if we plot the logarithm of the N_2O_5 concentration versus t, as at the right of Figure 14.2, it is clear that there is a linear relationship between these two quantities which is ordinarily expressed by the equation

$$\log_{10}X = \log_{10}X_0 - \frac{kt}{2.30}$$

This equation expresses, admittedly indirectly, how the concentration, X, varies with time

or, in a somewhat more compact form, as

$$\log_{10} \frac{X_0}{X} = \frac{kt}{2.30} \qquad (14.5)$$

Here, X_0 and X represent reactant concentrations at zero time and t, respectively; k is the first order rate constant (e.g., 0.35 min^{-1} for the decomposition of N_2O_5 at 67°C).

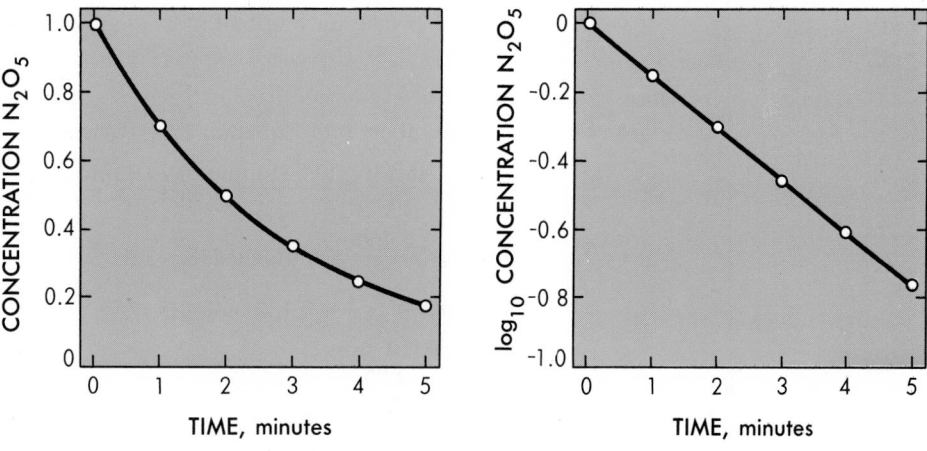

Figure 14.2 Variation of concentration of N_2O_5 with time in the reaction: $2 N_2O_5(g) \rightarrow 4 NO_2(g) + O_2(g)$. For this or any other first order reaction, a plot of log (conc.) vs time is linear.

TABLE 14.3 Concentration-Time Data for the Reaction
$2 N_2O_5(g) \rightarrow 4 NO_2(g) + O_2(g)$ at 67°C

t (min)	0	1	2	3	4	5
conc. N_2O_5	1.000	0.705	0.497	0.349	0.246	0.173
log conc. N_2O_5	0.000	−0.152	−0.304	−0.457	−0.609	−0.762

Equation 14.5 is readily derived from the principles of calculus. For a first order reaction we have

$$\text{rate} = \frac{-dX}{dt} = k\,X$$

Rearranging,

$$\frac{dX}{X} = -k\,dt$$

Integrating from $t = 0$ to t,

$$\int_{X_0}^{X} \frac{dX}{X} = -k \int_0^t dt; \quad \ln \frac{X}{X_0} = -kt$$

where "ln X/X_0" represents the natural logarithm (base e) of the ratio X/X_0.

Or,

$$\ln \frac{X_0}{X} = kt, \text{ and finally } \log_{10} \frac{X_0}{X} = \frac{kt}{2.30}$$

For a first order reaction, Equation 14.5 can be used to calculate the concentration of a reactant remaining after a given time (Example 14.2a), the time required for reactant concentration to drop to a certain level (Example 14.2b), or the time required for a given fraction of a sample to react (Example 14.2c).

Example 14.2 The decomposition of hydrogen peroxide to water and oxygen

$$2\ H_2O_2(l) \rightarrow 2\ H_2O(l) + O_2(g)$$

is a first order reaction with a rate constant of 0.0410 min^{-1}.

a. If we start with a 0.200 M solution of H_2O_2, what will be its concentration after 10.0 minutes?

b. How long will it take for the concentration of H_2O_2 to drop from 0.50 M to 0.10 M?

c. How long will it take for one half of a sample of H_2O_2 to decompose?

Solution

a. Substituting numbers into Equation 14.5, we obtain

$$\log_{10} \frac{0.200}{X} = \frac{(0.0410\ \text{min}^{-1})(10.0\ \text{min})}{2.30} = 0.178$$

Taking antilogs, $\dfrac{0.200}{X} = $ antilog $0.178 = 1.51$

Solving, $X = 0.200/1.51 = 0.132$ mole/liter

b. $\log_{10} \dfrac{0.50}{0.10} = \dfrac{(0.0410\ \text{min}^{-1})t}{2.30}$

Simplifying: $\log_{10} 5.0 = 0.018\ \text{min}^{-1}\,t$

$$t = \left(\frac{\log_{10} 5.0}{0.018}\right) \text{min} = \frac{0.699}{0.018} \text{min} = 39 \text{ minutes}$$

c. When half the sample has decomposed, $X = \frac{1}{2} X_0$ or $X_0 = 2\,X$

$$\log_{10} \frac{2\,X}{X} = \log_{10} 2 = 0.301 = \frac{kt}{2.30}$$

Solving for t:

$$t = \frac{2.30 \times 0.301}{k} = \frac{0.693}{k} = \frac{0.693}{0.0410} \text{min} = 16.9 \text{ minutes}$$

The analysis of Example 14.2c shows that *the time required for a given fraction of a reactant to decompose via a first order reaction is independent of concentration.* Specifically, the time required for one half the sample to decompose, often referred to as the **half-life** of the reaction, is

$$t_{1/2} = \frac{0.693}{k} \qquad (14.6)$$

Note that the half-life, $t_{1/2}$, is inversely proportional to the rate constant, k. A "fast" reaction, for which k is large, will have a short half-life; a "slow" reaction (small value of k) will be characterized by a comparatively long half-life.

For a first order reaction, if $t_{1/2} = 20$ sec:

At	Fraction unreacted
$t = 0$	1.0
$t = 20$	$1/2$
$t = 40$	$1/4$
$t = 60$	$1/8$
$t = 80$	$1/16$

REACTION OF OTHER INTEGRAL ORDERS

Among gas phase reactions, second order processes are perhaps more common than those of first order. Zero order reactions, in which the rate is independent of reactant concentration, are less common, being confined largely to reactions taking place at solid surfaces. One such reaction is the thermal decomposition of hydrogen iodide on a gold surface:

$$2\ HI(g) \xrightarrow{\ Au\ } H_2(g) + I_2(g); \quad rate = k$$

which is found to occur at a constant rate independent of the concentration of HI. Third order reactions are quite rare: one important example, involved in smog formation, is the reaction between nitric oxide and oxygen

$$2\ NO(g) + O_2(g) \rightarrow 2\ NO_2(g); \quad rate = k\ (conc.\ NO)^2(conc.\ O_2)$$

Rather than consider each of these reaction types in detail, we list in Table 14.4 some of the characteristics of the more common types, i.e., zero, first, or second order reactions. All the relations apply to the decomposition of a single reactant

$$A \rightarrow products$$

In the last column of Table 14.4, we list the functions (X, $\log_{10}X$ and $1/X$) which, when plotted against t, give a straight line for zero, first, and second order reactions, respectively. These relations offer a simple method of deciding upon the order of a reaction which has been studied in the laboratory.

TABLE 14.4 Characteristics of Zero, First, and Second Order Reactions

$$A \rightarrow products$$
(X, X_0 = conc. A at t and t = 0, respectively)

ORDER	RATE EXPRESSION	CONC.-TIME RELATION	$t_{1/2}$	LINEAR PLOT
0	rate = k	$X_0 - X = kt$	$X_0/2k$	X vs t
1	rate = kX	$\log_{10} \frac{X_0}{X} = \frac{kt}{2.30}$	$0.693/k$	$\log_{10}X$ vs t
2	rate = kX^2	$\frac{1}{X} - \frac{1}{X_0} = kt$	$1/kX_0$	$\frac{1}{X}$ vs t

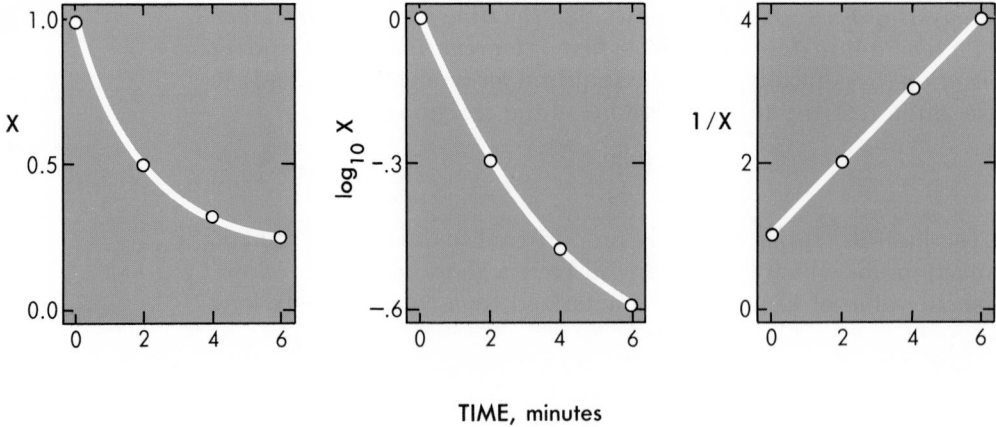

Figure 14.3 Decomposition of HI. Since the only linear plot is that of 1/X vs t, the reaction must be second order.

Example 14.3 The following data were obtained for the gas phase decomposition of hydrogen iodide.

t (hr)	0	2	4	6
conc. HI	1.00	0.50	0.33	0.25

Is this reaction zero, first, or second order in HI?

Solution. It will be useful to prepare a table in which we list X, logX and 1/X as a function of time.

t	X	$\log_{10}X$	1/X
0	1.00	0.00	1.0
2	0.50	−0.30	2.0
4	0.33	−0.48	3.0
6	0.25	−0.60	4.0

If it is not obvious from the table above that the only linear plot will be that of 1/X vs t, that point should be clear from Figure 14.3. We conclude that we are dealing with a second order reaction.

14.3 DEPENDENCE OF REACTION RATE UPON TEMPERATURE

The rates of most chemical reactions increase as the temperature rises. A housewife in a hurry to prepare dinner employs this principle when she uses a pressure cooker to cook eggs, apples, or a pot roast (hopefully, not all at the same time). By storing the leftovers in a refrigerator, she slows down the chemical reactions responsible for food spoilage. As a general and very approximate rule, it is often stated that an increase in temperature of 10°C doubles the reaction rate. If this rule holds, foods should cook twice as fast in a pressure cooker at 110°C as in an open saucepan and deteriorate four times as rapidly at room temperature (25°C) as they do in a refrigerator at 5°C.

TABLE 14.5   Temperature Dependence of the Rate Constant for the
Reaction $CO(g) + NO_2(g) \rightarrow CO_2(g) + NO(g)$

TEMPERATURE (°K)	k (MOLE/LIT)$^{-1}$ SEC^{-1}
600	0.028
650	0.22
700	1.3
750	6.0
800	23

Qualitatively, we can explain the effect of temperature upon reaction
rate by recalling from Chapter 5 that raising the temperature greatly increases the fraction of molecules having high kinetic energies. These are
the molecules that are most likely to react when they collide (Section 14.5).
To derive a quantitative relationship between reaction rate and temperature, let us consider the reaction between CO and NO_2 discussed earlier.
If we determine the rate constant of this reaction over a series of temperatures between 600 and 800°K, the data in Table 14.5 are obtained.

As usual, we search for a linear relationship between some function of
the rate constant, k, and temperature. It turns out that a plot of $\log_{10}k$ vs $1/T$
is a straight line (Fig. 14.4). This tells us that

$$\log_{10}k = A - \frac{B}{T} \tag{14.7}$$

where A and B are constants for a particular reaction that can be determined
from the slope and intercept of the plot. The general validity of Equation

Straight line
graphs are often
helpful, as here, in
showing relation-
ships between ex-
perimental quanti-
ties

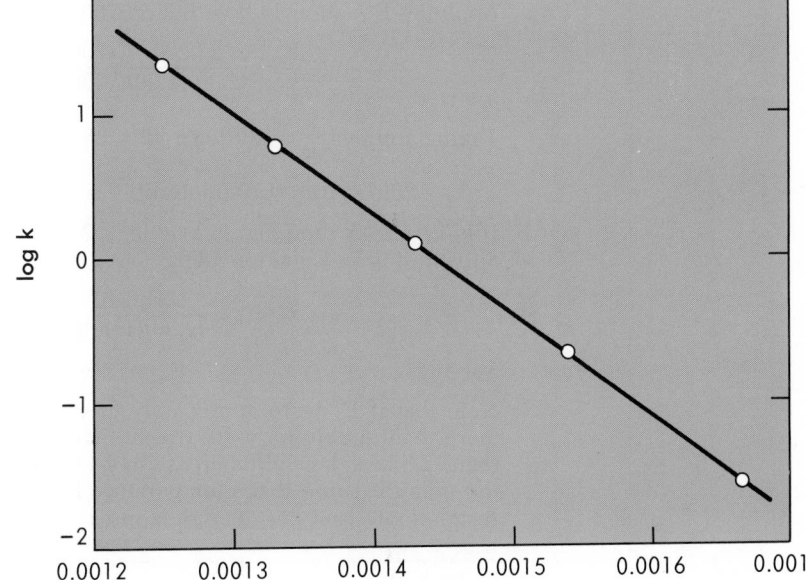

Figure 14.4   Dependence of rate constant, k, on temperature. Plot of log k vs
$1/T$ is linear.

14.7 was first demonstrated by the Swedish physical chemist Svandte Arrhenius in 1887.

The constant B in Equation 14.7 is directly related to an important parameter known as the activation energy, E_a, for the reaction. It turns out that $B = E_a/2.30\ R$, where R is the gas law constant, 1.99 cal/°K. Making this substitution in Equation 14.7, we obtain

$$\log_{10}k = A - \frac{E_a}{(2.30)(1.99)T} \tag{14.8}$$

The basis for this relation lies in the collision theory of reaction rates (Section 14.5)

Equation 14.8 can be used directly to obtain the activation energy for a reaction from a plot of $\log_{10}k$ vs $1/T$ (slope $= -E_a/(2.30)(1.99)$). Alternatively, it can be manipulated (cf. Equations 9.3 and 9.4, Chapter 9) to give the following relationship between the rate constants, k_2 and k_1, at two different temperatures T_2 and T_1:

$$\log_{10}\frac{k_2}{k_1} = \frac{E_a}{(2.30)(1.99)}\left(\frac{T_2 - T_1}{T_2 T_1}\right) \tag{14.9}$$

where E_a, the activation energy, is expressed in calories. The usefulness of this relation is illustrated in Example 14.4.

Example 14.4

a. The activation energy for the reaction $2\ NO_2(g) \rightarrow 2\ NO(g) + O_2(g)$ is 27,200 cal. At 600°K, $k = 0.75$ (mole/liter)$^{-1}$ sec^{-1}. Calculate k at 700°K.

b. What must be the value of E_a if the rate constant for a reaction is to double when the temperature increases from 27 to 37°C?

Solution

a. Applying Equation 14.9 with $T_2 = 700°K$, $T_1 = 600°K$.

$$\log_{10}\frac{k_2}{k_1} = \frac{27,200}{(2.30)(1.99)}\left(\frac{700 - 600}{700 \times 600}\right) = 1.42$$

Taking antilogs, $\frac{k_2}{k_1} = $ antilog $1.42 = 26$

$$k_2 = 26k_1 = 26 \times 0.75\ (\text{mole/lit})^{-1}\ \text{sec}^{-1} = 20\ (\text{mole/lit})^{-1}\ \text{sec}^{-1}$$

b. If $k_2/k_1 = 2.00$, then $\log_{10}k_2/k_1 = \log_{10}2 = 0.301$
Substituting in Equation 14.9:

$$0.301 = \frac{E_a(310 - 300)}{(2.30)(1.99)(310)(300)}$$

Solving, $\quad\quad\quad\quad\quad\quad E_a = 12,800$ cal

Note that if E_a were appreciably greater than 12.8 kcal, k would more than double for a 10° rise in temperature; if E_a were smaller than 12.8 kcal, k would increase by less than a factor of two. Clearly the empirical rule that a temperature increase of 10°C doubles the reaction rate is at best a crude approximation.

The activation energy associated with a reaction can be given a simple physical interpretation. In order for a reaction to occur between stable molecules, a certain amount of energy must be absorbed to weaken the bonds holding the reactant molecules together. The quantity E_a represents the energy required to bring the reactants to the point where they can rearrange to form products. As we might expect, very fast reactions are characterized by small activation energies. Looking at Equation 14.8, we see that if E_a is small, the constant A, which is a positive number, will predominate and k will be relatively large.

Expanding upon this physical picture, we can say that stable molecules, before being converted to products, must pass through an unstable, high energy, intermediate species. This transient, highly reactive species is referred to as an *activated complex*. Its exact nature is difficult to determine; in the reaction between CO and NO_2 it might be a "pseudomolecule" made up of CO and NO_2 molecules in close contact. We might visualize the path of reaction somewhat as follows

$$O\equiv C + O-N \diagdown_O \quad \rightarrow \quad O\equiv C\cdots O\cdots N \diagdown_O \quad \rightarrow \quad O\equiv C\equiv O + N\equiv O$$

$$\text{reactants} \qquad\qquad \text{activated complex} \qquad\qquad \text{products}$$

The dotted lines represent "partial bonds" in the activated complex; the N—O bond in the NO_2 molecule has been partially broken and a new bond between carbon and oxygen atoms has started to form.

According to this picture, we interpret the activation energy to be the difference in energy between the activated complex and the reactant molecules. This concept is shown schematically in Figure 14.5 for the CO—NO_2 reactions. Notice that the activated complex has an energy 16 kcal greater than that of the reactants, CO and NO_2, and 70 kcal greater than that of the products. In other words, the activation energies for the forward reaction, E_a, and the reverse reaction, E_a' are, respectively,

$$E_a = 16 \text{ kcal: } E_a' = 70 \text{ kcal}$$

Activation energy is the barrier which prevents some thermodynamically spontaneous reactions from occurring rapidly

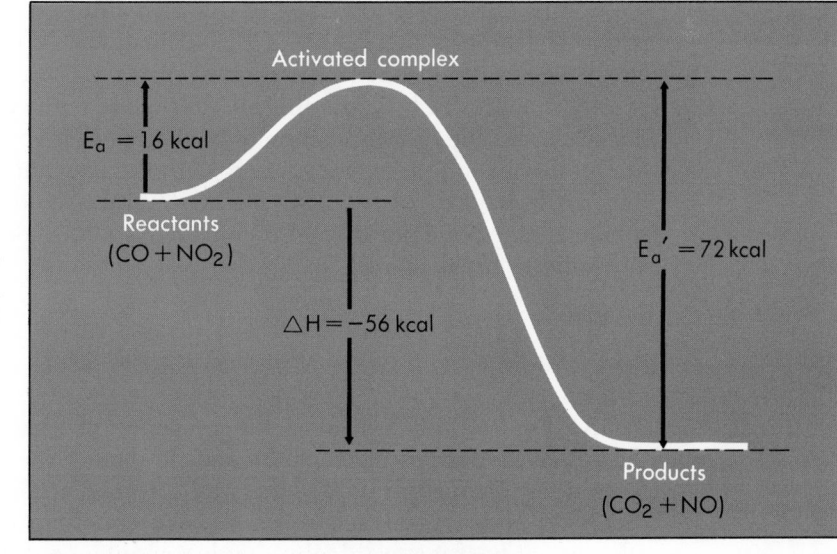

PATH OF REACTION

Figure 14.5 Concept of activation energy. Notice that $E_a - E_a' = \Delta H$.

The difference between these two quantities represents the energy (or enthalpy) difference between products and reactants

$$CO(g) + NO_2(g) \rightarrow CO_2(g) + NO(g): \quad \Delta H = -54 \text{ kcal}$$

In general, for any reaction

$$\Delta H = E_a - E_a' \tag{14.10}$$

For an exothermic reaction ($\Delta H < 0$), the activation energy for the reverse reaction, E_a', must be greater than that for the forward reaction, E_a. If, on the other hand, the products have a higher enthalpy than the reactants ($\Delta H > 0$), $E_a > E_a'$.

14.4 CATALYSIS

It has long been known that certain substances called *catalysts* can increase the rate of a reaction without being consumed by it. A familiar example of a reaction which is extremely susceptible to catalysis is the decomposition of hydrogen peroxide:

$$H_2O_2(aq) \rightarrow H_2O(l) + \tfrac{1}{2} O_2(g) \tag{14.11}$$

This occurs rather slowly under ordinary conditions (recall Example 14.2) but takes place almost instantaneously if a pinch of manganese dioxide, MnO_2, is added. All the MnO_2 can be recovered when the reaction is over, indicating that it is a true catalyst. Another reaction of this type is the decomposition of nitrous oxide

$$N_2O(g) \rightarrow N_2(g) + \tfrac{1}{2} O_2(g) \tag{14.12}$$

which can be accelerated by bringing the N_2O into contact with a metal such as gold.

The effectiveness of a catalyst in increasing the rate of a reaction is explained in terms of a lowering of the activation energy (Fig. 14.6). Compare, for example, the values of E_a for Reaction 14.12: 60 kcal for the uncatalyzed reaction vs 29 kcal on a gold surface. The reduction in activation energy is achieved by providing an alternative pathway of lower energy for the reaction. The decomposition of N_2O on gold involves chemical adsorption of the gas on the metal surface, with the formation of a weak bond between the oxygen of the N_2O molecule and a gold atom. This, in turn, weakens the bond joining nitrogen to oxygen, making it easier for the molecule to break apart.

$$\tfrac{1}{2} O_2 + Au$$
$$\uparrow$$
$$N\equiv N\!-\!O + Au \rightarrow N\equiv N\cdots O\cdots Au \rightarrow N\equiv N + O\cdots Au$$

Catalysts can only speed up the attainment of chemical equilibrium; they cannot change the position of the equilibrium

As Figure 14.6 indicates, the presence of a catalyst does not affect the relative energies of reactants and products. The overall enthalpy and free energy changes as well as the equilibrium constant are the same for the catalyzed and uncatalyzed reactions. We conclude that a catalyst can be effective only if we are dealing with a spontaneous reaction. It would be futile to search for a catalyst to bring about a nonspontaneous reaction; ΔG and K would still be unfavorable and the reaction could not occur.

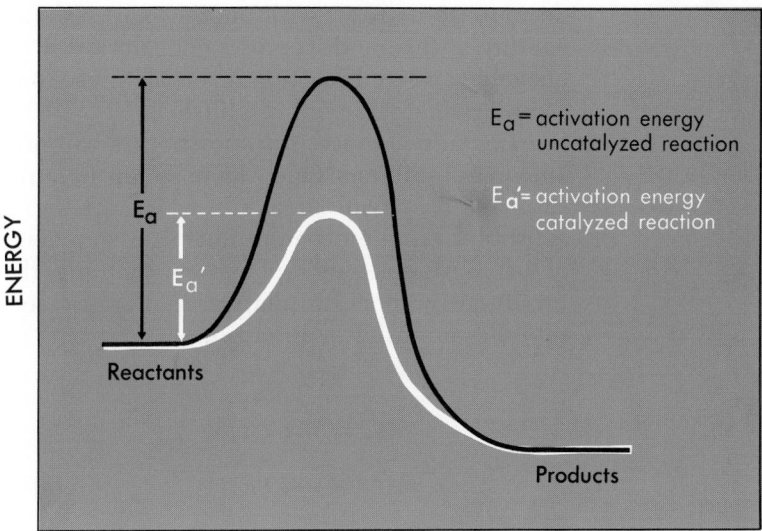

PATH OF REACTION

Figure 14.6 Lowering of activation energy by adding a catalyst.

Many reactions which take place slowly, if at all, under ordinary laboratory conditions occur readily in the body in the presence of catalysts called enzymes, which are proteins with high molecular weights. An example is the combustion of sugar

$$C_{12}H_{22}O_{11}(s) + 12 \ O_2(g) \rightarrow 12 \ CO_2(g) + 11 \ H_2O(l)$$

which is difficult to bring about directly, as, for example, by heating a sample of sugar in a test tube. In the body, sugar is metabolized at 37°C (98.6°F) in a series of biochemical reactions which give carbon dioxide and water as end products. Each step in the sequence is catalyzed by a particular enzyme specifically adapted for that purpose.

The effectiveness of an enzyme in catalyzing biochemical reactions can be interpreted crudely in terms of the "lock and key" analogy shown in Figure 14.7. The reactant known in biochemical jargon as the "substrate" fits into a specific site on the enzyme surface, where it is held in position by intermolecular forces. The substrate-enzyme complex can then either

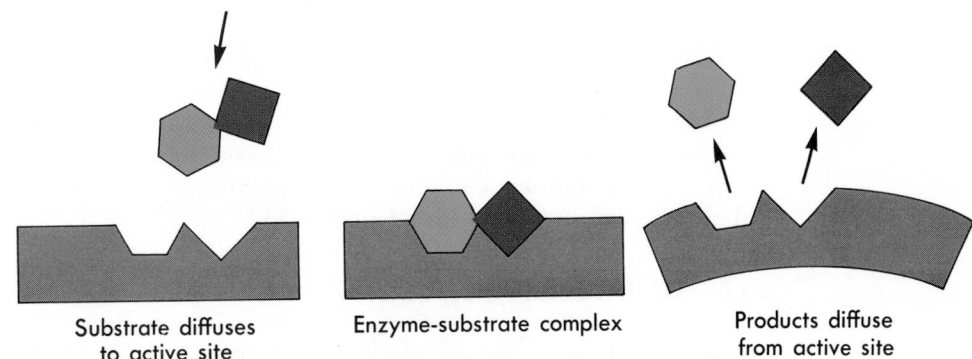

Substrate diffuses to active site

Enzyme-substrate complex

Products diffuse from active site

Figure 14.7 "Lock and key" relation between enzyme and substrate, used to rationalize enzyme catalysis.

decompose or react with another species such as a water molecule. The validity of this model gains some support from the fact that the kinetics of enzyme-catalyzed reaction resemble those found in surface catalysis where a similar model is well established.

Enzyme activity is diminished in the presence of certain substances known as inhibitors. One way in which an inhibitor can operate is to occupy sites on an enzyme molecule which are supposed to be reserved for the substrate. Frequently inhibitors have geometries closely resembling those of the substrates they replace. For example, the metabolism of citric acid is inhibited by its close relative, fluorocitric acid, which can presumably fit into the same slot in an enzyme.

$$
\begin{array}{cc}
\text{H} & \text{F} \\
| & | \\
\text{H—C—COOH} & \text{H—C—COOH} \\
| & | \\
\text{HO—C—COOH} & \text{HO—C—COOH} \\
| & | \\
\text{H—C—COOH} & \text{H—C—COOH} \\
| & | \\
\text{H} & \text{H} \\
\text{citric acid} & \text{fluorocitric acid}
\end{array}
$$

In another case we find that the insecticide parathion inhibits the hydrolysis of the acetylcholine ion, which is produced at the end of a nerve cell when an impulse passes through it on its way to or from the brain.

$$
\begin{array}{cc}
\underset{\text{parathion}}{C_2H_5\!-\!O\!-\!\overset{\displaystyle S}{\underset{\displaystyle O-C_2H_5}{P}}\!-\!O\!-\!\bigcirc\!-\!NO_2} &
\underset{\text{acetylcholine ion}}{H_3C\!-\!\overset{\displaystyle CH_3}{\underset{\displaystyle CH_3}{N}}\!-\!CH_2\!-\!CH_2\!-\!O\!-\!\overset{\displaystyle O}{C}\!-\!CH_3}
\end{array}
$$

The presence of parathion leads to a buildup of acetylcholine ions at nerve cell junctions. This blocks the transmission of further nerve impulses, leading to unconsciousness and, eventually, death. Presumably the parathion operates by occupying the active site on the enzyme molecule responsible for breaking down the acetylcholine ion.

Until quite recently we knew almost nothing about the molecular structure of enzymes; the "lock and key" model was little more than a convenient way to rationalize the kinetics of enzyme-controlled reactions. Within the past few years it has been possible to establish by x-ray crystallography the structures and configurations of some of the simpler enzymes. Even more remarkably, the structures of some enzyme-substrate complexes have been determined. Research in this area led in 1969 to the first laboratory synthesis of an enzyme, ribonuclease, for which Drs. Stein and Moore of Rockefeller University and Anfinsen of NIH won the 1972 Nobel prize in chemistry. (See Chapter 23.)

14.5 REACTION MECHANISMS

As we have seen, kinetic studies can furnish us with practical information as to how reaction rates change with concentration, temperature, and the presence of a catalyst. An equally important objective of chemical kinetics is to help us unravel the *mechanism* of a reaction. From the rate equation for a reaction we can frequently extract information as to the detailed path followed when reactants are converted to products. Since the mechanisms of homogeneous gas phase reactions such as

$$CO(g) + NO_2(g) \rightarrow CO_2(g) + NO(g)$$

differ considerably from those of surface catalyzed reactions, e.g.,

$$N_2O(g) \xrightarrow{\text{Au}} N_2(g) + \tfrac{1}{2} O_2(g)$$

we shall discuss these two topics separately.

The mechanism of a reaction is the path, a sequence of steps, by which the reaction occurs

HOMOGENEOUS GAS PHASE REACTIONS

COLLISION THEORY. It seems reasonable to suppose that in order for two gas molecules to react with each other, they must first collide. Indeed, for very simple bimolecular reactions of the type

$$A + B \rightarrow \text{products} \tag{14.13}$$

we might assume, somewhat naively, that every collision would result in reaction. If this were the case, the rate of reaction expressed in the proper units would be equal to the number of collisions per second, Z

$$\text{rate} = Z$$

The collision number, Z, can be calculated from the kinetic theory of gases. It turns out to be a very large number; typically at room temperature and atmospheric pressure, Z is of the order of 10^{27} molecules per second. From the magnitude of this number we see that if every collision resulted in reaction, gas phase reactions would occur essentially instantaneously. In practice, of course, they do not.

There are at least two reasons why, for a bimolecular process such as

$$CO(g) + NO_2(g) \rightarrow CO_2(g) + NO(g),$$

the rate of reaction is ordinarily much smaller than the collision number, Z. In the first place, in order for reaction to occur, the colliding molecules must have a total kinetic energy at least equal to the activation energy, E_a (16 kcal/mole). Collisions between low energy molecules will be ineffective since they simply cannot provide sufficient energy to push the reaction over the energy barrier. Moreover, the two reactant molecules must be favorably oriented with respect to one another if they are to react upon collision. If, when CO and NO_2 molecules collide, the carbon and nitrogen atoms come in contact with one another (Fig. 14.8), it is unlikely that the necessary transfer of an oxygen atom will take place.

Taking account of these two factors, we can modify our rate expression to read

$$\text{rate} = p \times Z \times f \tag{14.14}$$

where p is the probability that the colliding molecules will be favorably oriented for reaction and f is the fraction of molecules that possess sufficient energy to react upon collision. Since both p and f are less than unity, we

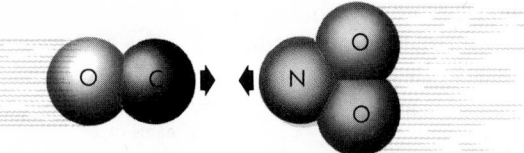

"Ineffective" collision

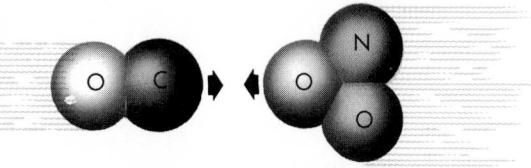

"Effective" collison

Figure 14.8 Collision between CO and NO_2 molecules is effective only if they are oriented properly when they collide.

can expect the rate of reaction to be significantly less than the collision number Z.

It is difficult to compare experimentally measured reaction rates with those predicted by Equation 14.14 because the factor p can seldom be estimated accurately. We can, however, use this equation to rationalize the temperature dependence of reaction rate. You may recall from our discussion of the Maxwell distribution of molecular energies in Chapter 5 that the fraction of molecules having energies equal to or greater than a certain energy, for example, E_a, is

$$f = e^{-E_a/RT} \tag{14.15}$$

E_a is usually large as compared to RT, making f very small

where e is the base of natural logarithms, R is the gas constant, and T the absolute temperature. Making this substitution in Equation 14.14, we obtain

$$rate = p \times Z \times e^{-E_a/RT}$$

Taking the natural logarithm of both sides of this equation

$$\ln rate = \ln (p \times Z) - \frac{E_a}{RT}$$

Converting to base 10 logs and substituting for R its value in calories per degree, 1.99

$$\log_{10} rate = \log_{10}(p \times Z) - \frac{E_a}{(2.30)(1.99)T} \tag{14.16}$$

Since the rate constant k is equal to the rate at unit concentrations, we see that this equation becomes identical with Equation 14.8 if we take $A = \log_{10}(p \times Z)$. We conclude that the collision theory satisfactorily explains the dependence of reaction rate upon temperature.

MULTISTEP MECHANISMS. The simple picture that we have presented suffers from one serious deficiency. If a reaction of the type

$$A + B \rightarrow products$$

is a one-step process involving a simple collision between molecules of A and B, we might expect all gas phase reactions to show second order kinetics. Clearly the rate at which molecules of A and B collide will increase in direct proportion to their concentration. Doubling the concentration of A or B should double the number of collisions and, hence, the rate of reaction; in general, we would expect to find that

By the simple collision theory, reactions would all be second order

$$\text{rate} = k \, (\text{conc. A}) \times (\text{conc. B})$$

In practice a great many gas phase reactions do not show second order kinetics. Many reactions have other integral orders (0, 1, 3) or fractional orders (e.g., 1/2, 3/2). In still other cases we find the rate expression to be a complicated one, involving perhaps the concentrations of products as well as reactants.

These observations reflect the important fact that *very few reactions take place in a single bimolecular step.* Instead, they pass through a series of two or more steps before yielding the final products. These steps may involve successive bimolecular collisions or, perhaps, the dissociation of an unstable species. We might, for example, have a two-step mechanism of the type

$$A + B \rightarrow C \rightarrow \text{products}$$

The first step involves a simple collison between molecules A and B to give an intermediate species C. This intermediate then decomposes to products in a second step.

Rate expressions for reactions which occur by multistep mechanisms can be quite complicated, particularly if all the steps occur at about the same rate. Frequently, however, we find that one step in the mechanism is much slower than any of the others. If this is the case, the rate of that step determines the overall reaction rate in much the same way that one unusually slow runner in a relay race can, in effect, determine the time posted by his entire team.

If one of the steps is slow enough to be rate-determining, the rate expression often turns out to be relatively simple. To illustrate how this can happen, consider the thermal decomposition of ozone, for which the overall reaction is

$$2 \, O_3(g) \rightarrow 3 \, O_2(g)$$

Experimentally it is found that the rate expression for this reaction is

$$\text{rate} = \frac{k \, (\text{conc. } O_3)^2}{(\text{conc. } O_2)} \qquad (14.17)$$

The fact that the rate is inversely proportional to the concentration of the product, O_2, tells us that the mechanism cannot involve only a bimolecular collision between two ozone molecules. However, the rate expression is consistent with the following two-step mechanism:

1. A *rapid* equilibrium established between O_3 molecules, O_2 molecules, and oxygen atoms:

Equilibria of this sort occur in many reaction mechanisms

$$O_3 \rightleftharpoons O_2 + O \quad \text{fast}$$

2. A *slow* step in which an oxygen atom collides with an ozone molecule:

$$O + O_3 \rightarrow 2\ O_2 \quad \text{slow}$$

The sum of these two steps gives us the overall stoichiometry of the reaction; i.e., $2\ O_3(g) \rightarrow 3\ O_2(g)$. To show that this mechanism is consistent with the observed rate law, we first write the expression for the slow step, (2). Since this is a simple collision process, we have

$$\text{rate} = k_2\ (\text{conc. } O_3) \times (\text{conc. } O)$$

But the concentration of oxygen atoms is governed by the position of the equilibrium in (1); that is:

$$K_1 = \frac{(\text{conc. } O_2) \times (\text{conc. } O)}{(\text{conc. } O_3)}, \quad \text{or} \quad (\text{conc. } O) = \frac{K_1\ (\text{conc. } O_3)}{(\text{conc. } O_2)}$$

Substituting this expression for conc. O in the rate expression:

O$_2$, the product, inhibits the reaction

$$\text{rate} = k_2 K_1\ \frac{(\text{conc. } O_3)^2}{(\text{conc. } O_2)}$$

This equation is, of course, equivalent to 14.17 if we take "k" to be the product of the rate constant, k_2, and the equilibrium constant, K_1.

Many gas phase reactions are initiated by the formation, at very low concentrations, of an extremely reactive species which sets off a series of reactions leading to the formation of products. Such processes are referred to as **chain reactions**; typically they occur very rapidly after a short induction period to allow for the formation of the reactive species. An important example of a chain reaction is the formation of hydrogen chloride from the elements. A mixture of hydrogen and chlorine stored at room temperature in the dark shows no evidence of reaction over long periods of time. However, if the mixture is exposed to ultraviolet light or heated to 200°C, a vigorous reaction occurs. The first step in this reaction, referred to as **chain initiation**, involves the reversible dissociation of a chlorine molecule into atoms:

$$Cl_2 \rightleftharpoons 2\ Cl \tag{14.18a}$$

The chlorine atoms formed are extremely reactive toward hydrogen molecules:

$$Cl + H_2 \rightarrow HCl + H \tag{14.18b}$$

This reaction forms another highly reactive species, a hydrogen atom, which attacks a chlorine molecule:

$$H + Cl_2 \rightarrow HCl + Cl \tag{14.18c}$$

For every Cl atom formed in the initiation step, 10^6 HCl molecules may be produced

In this way, the chlorine atoms are regenerated; the **chain propagation**, represented by Equations 14.18b and 14.18c, occurs over and over again until the H_2 and Cl_2 are almost completely converted to HCl.

The hydrogen and chlorine atoms, which act as **chain carriers**, can be consumed by reaction with each other:

$$H + Cl \rightarrow HCl; \quad Cl + Cl \rightarrow Cl_2; \quad H + H \rightarrow H_2 \tag{14.18d}$$

These processes represent **chain termination**, since they break the chain mechanism.

As you can well imagine, kinetic studies of chain reactions are not easy to carry out in the laboratory. Frequently a chemist trying to establish the order of a chain reaction finds that it is over before he has had time to make any measurements. In extreme cases the only result of his efforts may be shattered glassware and broken or badly bent apparatus. The H_2-Cl_2 reaction is particularly notorious in this respect. Sometimes, however, it is possible to choose conditions under which the reaction occurs slowly enough to be studied by conventional methods. Alternatively, one can use special techniques worked out to follow the kinetics of very fast reactions. One chain reaction for which the rate expression has been established is that between chlorine and chloroform (Example 14.5).

Example 14.5 The mechanism for the Cl_2-$CHCl_3$ reaction is believed to be similar to that for the Cl_2-H_2 reaction. In particular, the first two steps are:

    a. $Cl_2 \rightleftharpoons 2Cl$ fast

    b. $Cl + CHCl_3 \rightarrow CCl_4 + H$ slow, rate-determining

Show that this mechanism is consistent with the observed rate expression:

$$\text{rate} = k \, (\text{conc. } CHCl_3) \times (\text{conc. } Cl_2)^{1/2}$$

Is this a chain reaction?

Solution. Since (b) involves a simple bimolecular collision, we have

$$\text{rate} = k_b \, (\text{conc. } CHCl_3)(\text{conc. } Cl)$$

But, from equilibrium considerations applied to (a)

$$\frac{(\text{conc. } Cl)^2}{(\text{conc. } Cl_2)} = K_a$$

Solving for (conc. Cl); $(\text{conc. } Cl)^2 = K_a \, (\text{conc. } Cl_2)$; conc. $Cl = K_a^{1/2} \, (\text{conc. } Cl_2)^{1/2}$

Substituting in the rate expression:

$$\begin{aligned}
\text{rate} &= k_b \, (\text{conc. } CHCl_3)(\text{conc. } Cl) \\
&= k_b K_a^{1/2} \, (\text{conc. } CHCl_3)(\text{conc. } Cl_2)^{1/2}
\end{aligned}$$

which is equivalent to the experimental rate law with $k = k_b K_a^{1/2}$

The type of analysis illustrated in Example 14.5 can always be used to test a postulated mechanism against the experimentally observed rate equation. We first write out the rate expression for the slow step in the mechanism. Such an equation will often involve an unstable intermediate (e.g., Cl). To obtain the final rate expression, the concentration of the intermediate must be related to those of stable species participating in the reaction (e.g., Cl_2). Hopefully, our final equation will agree with experiment. If it doesn't, the mechanism must be wrong; we proceed to modify it and try again.

It is important to point out one limitation of this type of analysis. *The fact that a particular mechanism agrees with the observed rate expression does not prove that it is correct.* There may well be another plausible mechanism that leads to the same rate law. A classic example is the reaction

$$H_2(g) + I_2(g) \rightarrow 2 \, HI(g) \tag{14.19}$$

$$\text{rate} = k \, (\text{conc. } H_2)(\text{conc. } I_2)$$

which was cited for many years as an example of a simple bimolecular reaction. Experimental evidence now suggests a more complex two-step mechanism

$$\text{(a)} \qquad I_2 \rightleftharpoons 2\,I \quad \text{fast} \tag{14.19a}$$

$$\text{(b)} \quad H_2 + 2\,I \rightarrow 2\,HI \quad \text{slow} \tag{14.19b}$$

This mechanism also leads to second order kinetics since

$$\text{rate} = k_b\,(\text{conc. } H_2)(\text{conc. } I)^2$$

But $$(\text{conc. } I)^2 = K_a\,(\text{conc. } I_2)$$

So $$\text{rate} = k_b K_a\,(\text{conc. } H_2)(\text{conc. } I_2)$$

SURFACE REACTIONS

When a reaction occurs at a solid surface rather than in the gas phase, the activation energy and the rate expression ordinarily change. This implies a difference in mechanism in the two types of reactions. Mechanisms of surface reactions can be quite complicated; we shall consider only the simplest possible process (Fig. 14.9) in which a molecule adsorbed on the surface of a solid catalyst decomposes to give two or more product molecules. An example of such a reaction is the decomposition of N_2O on gold.

$$N_2O(g) \xrightarrow{\text{Au}} N_2(g) + \tfrac{1}{2}\,O_2(g)$$

The rate of the decomposition reaction must be directly proportional

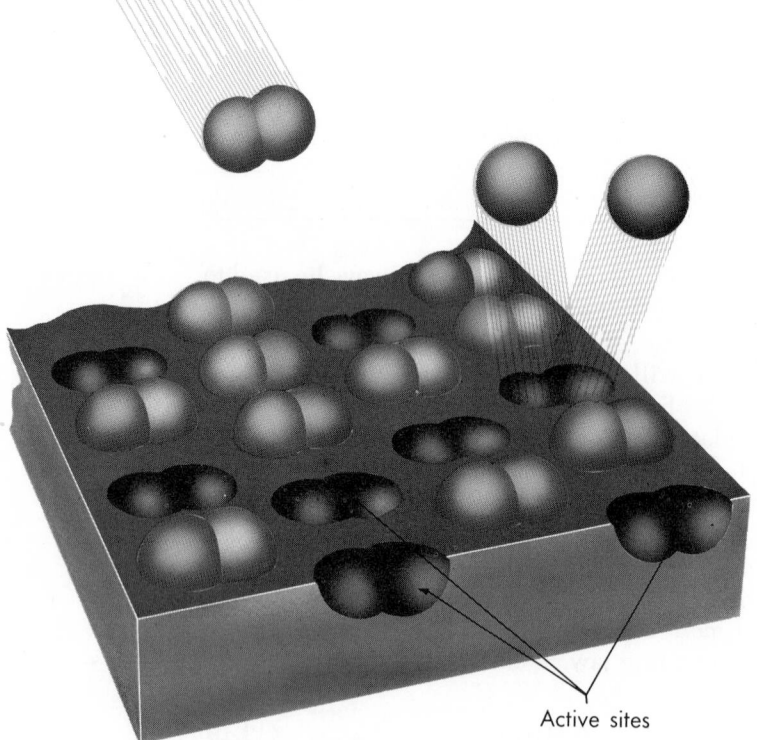

Active sites

Figure 14.9 Mechanism of a simple decomposition reaction occurring at a solid surface.

to the concentration of adsorbed molecules or, from a slightly different point of view, to the fraction, θ, of the surface covered by these molecules.

$$\text{rate} = k\,\theta$$

In order to deduce the order of the reaction with respect to the gaseous reactant (e.g., N_2O), we must relate θ to concentration in the gas phase. One way to do this is to study experimentally the variation of the amount of adsorption with gas concentration. In the simplest case such studies give the relation

$$\theta = \frac{b\,(\text{conc. R})}{1 + b\,(\text{conc. R})}$$

where R represents the gaseous reactant and b is a proportionality constant which depends upon the nature of the gas and the solid surface. Physically we interpret this equation to mean that:

1. At low concentrations, where $b\,(\text{conc. R}) \ll 1$, the fraction of the surface covered by adsorbed molecules is directly proportional to their concentration in the gas phase (i.e., $\theta = b \times \text{conc. R}$).

2. At high concentrations, where $(b \times \text{conc. R}) \gg 1$, the surface becomes saturated with adsorbed gas. Further increases in concentration have no effect on the amount of gas adsorbed; all the active sites on the surface are occupied and $\theta = 1$.

Substituting for θ in the rate expression, we obtain

$$\text{rate} = \frac{k\,b\,(\text{conc. R})}{1 + b\,(\text{conc. R})} \tag{14.20}$$

This equation implies that the order of a surface reaction will change with an increase in concentration of reactant. At low concentrations, the

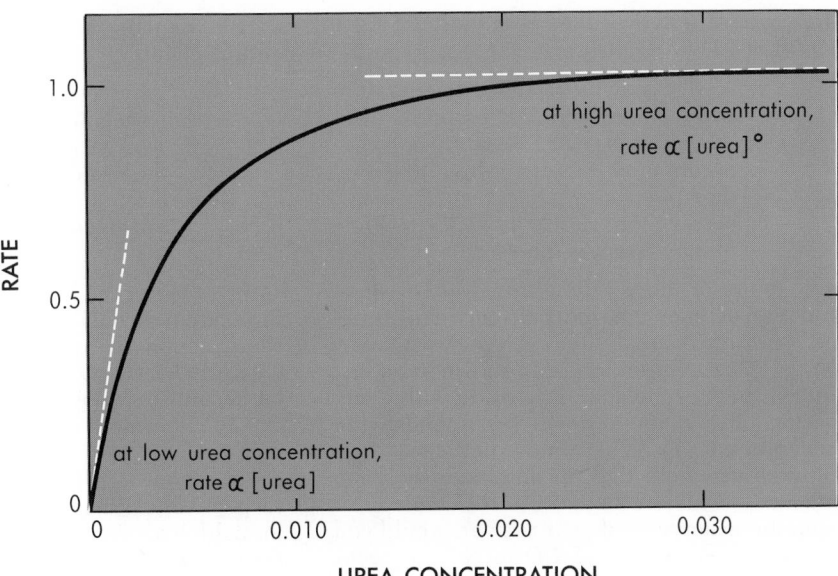

Once the enzyme surface is saturated, its effect on reaction rate cannot be increased by increasing substrate concentration

Figure 14.10 Rate of hydrolysis of urea as a function of urea concentration. At low concentration, rate is first order; it approaches zero order at high concentrations.

second term in the denominator is negligible compared to 1 and the expression becomes

$$\text{rate of reaction} = k \, b \, (\text{conc. R})$$

But, at high concentrations, the second term predominates and we have

$$\text{rate of reaction} = k \, b \, (\text{conc. R})/b \, (\text{conc. R}) = k$$

In other words, Equation 14.20 predicts a first order process at low reactant concentrations, shifting to a zero order reaction at high concentrations.

This kinetic behavior is typical of many reactions at solid surfaces. In the decomposition of N_2O on Au, the reaction is first order at low concentrations of N_2O; at high concentrations the rate approaches a constant value. Equation 14.20 can also be applied to enzyme-catalyzed reactions where "R" is now the substrate and the "surface" is that of the enzyme molecule. Biochemists refer to this equation, usually written in a slightly different form, as the Michaelis-Menten equation. Figure 14.10 shows its application to the hydrolysis of urea catalyzed by the enzyme urease.

$$H_2N\!-\!\underset{\underset{O}{\|}}{C}\!-\!NH_2(aq) + H_2O \xrightarrow{\text{urease}} 2 \, NH_3(aq) + CO_2(g) \qquad (14.21)$$

The fact that the kinetics of this and many other enzymatic reactions strongly resembles that of solid-catalyzed gas reactions emphasizes once again the similarity in mechanism of these two processes.

PROBLEMS

14.1 Explain why the reaction: $CO(g) + NO_2(g) \rightarrow CO_2(g) + NO(g)$

    a. occurs slowly at room temperature and atmospheric pressure, even though ΔG is -53.2 kcal.

    b. occurs more rapidly at high temperatures.

    c. occurs more slowly when the volume of the system is increased.

14.2 One of the reactions involved in smog formation is known to be that between ozone and nitric oxide: $O_3(g) + NO(g) \rightarrow O_2(g) + NO_2(g)$. This reaction has been shown to be first order in both ozone and nitric oxide with a rate constant of 1.2×10^7 liter/mole sec. Calculate the concentration of NO_2 formed per second in polluted air where the concentrations of ozone and nitric oxide are both 1.0×10^{-7} mole/liter. From the magnitude of your answer, would you expect the conversion of NO to NO_2 to occur rapidly or slowly?

14.16 Explain, in terms of concepts introduced in this chapter,

    a. why every collision between gas molecules does not lead to reaction.

    b. why a catalyst cannot be used to make a nonspontaneous reaction occur.

    c. how an inhibitor slows down an enzyme-catalyzed reaction.

14.17 One of the major eye irritants in smog is formaldehyde, CH_2O, which may be formed by the reaction of ozone with ethylene: $O_3(g) + C_2H_4(g) \rightarrow 2 \, CH_2O(g) + O(g)$, which is known to be second order with a rate constant of about 2×10^3 liter/mole sec. The concentrations of O_3 and C_2H_4 in heavily polluted air are estimated to be about 1×10^{-7} and 1×10^{-8} mole/liter, respectively. What is the rate of production of formaldehyde in mole/lit sec? How long will it take to build up a formaldehyde concentration of 1×10^{-8} mole/liter, the threshold above which eye irritation becomes noticeable?

14.3 For a reaction of the type: A → products, the rate is found to be 0.012 mole/liter sec when conc. A = 0.10 mole/liter. What will be the rate when conc. A = 0.30 mole/liter if the reaction is

    a. zero order in A?
    b. first order in A?
    c. second order in A?

14.4 For the reaction: A + B → products, the following data are obtained

conc. A	conc. B	initial rate
0.10 M	0.20 M	0.030 mole/lit sec
0.20	0.20	0.059 mole/lit sec
0.20	0.30	0.060 mole/lit sec
0.30	0.30	0.090 mole/lit sec
0.30	0.50	0.089 mole/lit sec

Estimate the order with respect to both A and B.

14.5 For the decomposition of N_2O_5 at 45°C, the following data were obtained

t (sec)	0	200	400	600	800
conc. N_2O_5 (mole/lit)	2.50	2.22	1.96	1.73	1.53

    a. Show from a plot of log conc. vs t that this reaction is first order.
    b. From your plot, evaluate the first order rate constant.
    c. Using the equation: rate = k (conc. N_2O_5), calculate the rate at t = 400 sec and compare to the average rate over the interval 0–800 sec.

14.6 The first order rate constant for the decomposition of C_2H_5Cl at 700°K is 2.50×10^{-3} min^{-1}. Starting with a concentration of 0.0100 mole/liter, calculate

    a. the concentration of C_2H_5Cl after ten minutes; after one day.
    b. the time required for the concentration of C_2H_5Cl to drop to 1/2 of its original value.

14.7 In the first order decomposition of acetone at 600°C it is found that 40% of a sample has decomposed after 60 seconds. How long will it take for 80% of the acetone to decompose?

14.8 The half-life of a certain reaction is 50 minutes when the initial concentration is 1.2 mole/liter. Using the information in Table 14.4, predict the half-life when the initial concentration is 3.0 mole/liter if the reaction is

    a. zero order.
    b. first order.
    c. second order.

14.18 The rate of a certain reaction: A → products is 1.0×10^{-3} mole/liter sec when conc. A = 0.020 M. What is the rate constant, k, if the reaction is

    a. zero order?
    b. first order?
    c. second order?

14.19 For the reaction A + B → products, the following data are obtained

conc. A	conc. B	initial rate
1.00 M	1.00 M	2.50×10^{-3} mole/lit sec
2.00 M	2.00 M	5.00×10^{-3}
2.00 M	4.00 M	7.05×10^{-3}

Estimate the order of the reaction with respect to both A and B.

14.20 For a certain first order reaction, the following data are available

t (min)	0	10	20	30
conc. (mole/liter)	.0500	.0397	.0315	.0250

    a. Use a plot of concentration versus time to estimate the concentration at 50 minutes.
    b. Use a plot of log (conc.) vs time to estimate the concentration at 50 minutes.
    c. Which of your answers (a) or (b) would you expect to be the more accurate? Explain.

14.21 For the first order decomposition of nitromethane at 300°C, the rate constant is 0.23 sec^{-1}. Starting with a concentration of 0.40 mole/liter, calculate

    a. the concentration after 10 seconds; after one minute.
    b. the time required for the concentration to drop to 0.10 mole/liter.

14.22 For the first order decomposition of N_2O_5 at a certain temperature it is found that the concentration is 0.0250 mole/liter after 10 minutes and 0.0100 mole/liter after 20 minutes. Calculate the rate constant, k, and the original concentration of N_2O_5.

14.23 Given the following data for a certain reaction

t (sec)	0	10	20	30
conc. (mole/lit)	0.64	0.52	0.40	0.28

determine whether the reaction is zero, first, or second order.

14.9 A housewife finds that it takes 30 minutes to boil potatoes at 100°C in an open saucepan and only 12 minutes to boil them in a pressure cooker at 110°C. Assuming that the rate constant is inversely proportional to cooking time, estimate the activation energy for cooking potatoes, which involves the conversion of cellulose into starch.

14.10 For the decomposition of N_2O_5, the activation energy is 24.7 kcal. At 27°C the rate constant for the reaction is 4.0×10^{-5} sec^{-1}. Calculate the rate constant at 37°C.

14.11 The activation energy for the reaction: $H_2(g) + Cl_2(g) \rightarrow 2\ HCl(g)$ is about 37 kcal. Using data in Table 4.1, Chapter 4, calculate the activation energy for the reverse reaction and prepare a diagram similar to Figure 14.4 for the reaction.

14.12 Suppose you were put in charge of a research project designed to determine how the acetylcholine ion "fits into" the structure of the enzyme that promotes its hydrolysis. Suggest some simple experiments that you might carry out with different inhibitors to solve this problem.

14.13 The rate law for the reaction: $2\ NO(g) + O_2(g) \rightarrow 2\ NO_2(g)$ is known to be: rate = k (conc. $NO)^2 \times$ (conc. O_2). Which of the following mechanisms is consistent with this rate law?

  a. $2\ NO + O_2 \rightarrow 2\ NO_2$
  b. $2\ NO \rightleftharpoons N_2O_2$ fast
   $N_2O_2 + O_2 \rightarrow 2\ NO_2$ slow
  c. $2\ NO \rightleftharpoons N_2 + O_2$ fast
   $N_2 + 2\ O_2 \rightarrow 2\ NO_2$ slow

14.14 Under certain conditions, the rate law for the reaction $H_2(g) + Br_2(g) \rightarrow 2\ HBr(g)$ is

  rate = k (conc. H_2)(conc. $Br_2)^{1/2}$

Which of the following mechanisms is consistent with this law?

  a. $H_2 + Br_2 \rightarrow 2\ HBr$
  b. $H_2 \rightleftharpoons 2\ H$ fast
   $H + Br_2 \rightarrow HBr + Br$ slow
   $Br + H_2 \rightarrow HBr(g) + H$ fast
  c. $Br_2 \rightleftharpoons 2\ Br$ fast
   $Br + H_2 \rightarrow HBr + H$ slow
   $H + Br_2(g) \rightarrow HBr + Br$ fast

14.15 For the enzymatic hydrolysis of urea, the rate is found to approach 1.0×10^{-3} mole/liter sec at very high concentrations: at very low concentrations the rate is 1.0×10^{-4} sec^{-1} (conc. urea). Calculate k and b in Equation 14.20 and estimate the rate when the concentration of urea is 0.40 M.

14.24 A bottle of milk stored at 30°C sours in 36 hours. Stored in a refrigerator at 5°C it has not turned sour after a week. Assuming the rate constant to be inversely related to the souring time, estimate a minimum value for the activation energy of the chemical reaction involved in the souring of milk.

14.25 The activation energy of a certain enzyme-catalyzed reaction going on in the body is 10.0 kcal. By what factor is the rate of this reaction increased when you have a fever of 104.0°F, assuming a normal temperature to be 98.6°F?

14.26 Taking the activation energy for the decomposition of N_2O to the elements to be 60 kcal and its heat of formation to be +19 kcal, construct an energy diagram similar to Figure 14.4.

14.27 It is frequently found that the activation energy of an enzyme-catalyzed reaction is a function of temperature, becoming negative at high temperature. Can you suggest an explanation for this phenomenon?

14.28 The reaction: $CO(g) + Cl_2(g) \rightleftharpoons COCl_2(g)$ has the rate law: rate = k (conc. CO) (conc. $Cl_2)^{3/2}$. Show that this rate law is consistent with the mechanism

  $Cl_2 \rightleftharpoons 2\ Cl$ fast
  $Cl + CO \rightleftharpoons COCl$ fast
  $Cl_2 + COCl \rightarrow COCl_2 + Cl$ slow

14.29 At low temperatures, the rate law for the reaction $CO(g) + NO_2(g) \rightarrow CO_2(g) + NO(g)$ is: rate = k (conc. $NO_2)^2$. Which of the following mechanisms is consistent with this rate law?

  a. $CO + NO_2 \rightarrow CO_2 + NO$
  b. $2\ NO_2 \rightleftharpoons N_2O_4$ fast
   $N_2O_4 + 2\ CO \rightarrow 2\ CO_2 + 2\ NO$ slow
  c. $2\ NO_2 \rightarrow NO_3 + NO$ slow
   $NO_3 + CO \rightarrow NO_2 + CO_2$ fast
  d. $2\ NO_2 \rightarrow N_2 + 2\ O_2$ slow
   $2\ CO + O_2 \rightarrow 2\ CO_2$ fast
   $N_2 + O_2 \rightarrow 2\ NO$ fast

14.30 For a certain enzyme-catalyzed reaction the rate is measured as a function of substrate concentration with the following results:

conc. S (mole/liter)	0.10	0.20	
rate (mole/lit min)	0.025	0.043	
conc. S (mole/liter)	0.50	1.00	2.00
rate (mole/lit min)	0.075	0.100	0.120

From a plot of these data, determine k and b in Equation 14.20.

*14.31 The ozone present in polluted air is formed by the two-step reaction

$$NO_2(g) \xrightarrow{k_1} NO(g) + O(g) \quad \text{(1st order)}$$

$$O(g) + O_2(g) \xrightarrow{k_2} O_3(g) \quad \text{(2nd order)}$$

It is known that $k_1 = 6.0 \times 10^{-3}$ sec^{-1} and $k_2 = 1.0 \times 10^6$ lit/mole sec. The concentrations of NO_2 and O_2 in polluted air are approximately 4.0×10^{-9} and 1.0×10^{-2} moles/lit, respectively. It is generally assumed that the concentration of atomic oxygen reaches a "steady state"; i.e., it is consumed by the second reaction at the same rate that it is produced by the first reaction.

  a. Calculate the steady state concentration of $O(g)$ and the rate of formation of O_3 in polluted air.
  b. Using the Ideal Gas Law, calculate the partial pressure of $O(g)$ in polluted air at 25°C and its mole fraction, X.
  c. The concentration of pollutants in air is often expressed in "parts per hundred million" (pphm), where 1 pphm $= 10^8$ X. What is the concentration of $O(g)$ in pphm? the concentration of NO_2?

*14.32 Suggest how one might distinguish experimentally between the two possible mechanisms suggested on p. 392 for the H_2-I_2 reaction. Note that both mechanisms lead to the same rate expression. The second mechanism is discussed in an article by J. H. Sullivan in the Journal of Chemical Physics, Vol. 46, p. 73, 1967.

*14.33 In a first order reaction, let us suppose that a quantity, X, of reactant is added at regular intervals of time, Δt. At first the amount of reactant in the system builds up; eventually, however, it levels off at a "saturation value" given by the expression:

$$\frac{X}{1 - 10^{-a}}, \text{ where } a = 0.30 \frac{\Delta t}{t_{1/2}}$$

This analysis applies to the intake of mercury into the body, where one takes in a certain amount each day. The half-life for elimination of mercury appears to be about 70 days.

Suppose that a person eats 2 oz of fish containing 0.5 part per million of mercury each day. Using the above equation, calculate the number of ounces of Hg in his body at "saturation," and compare to the value at which symptoms of mercury poisoning appear, about 5×10^{-4} oz. How many ounces of fish would he need to eat per day to reach the toxic limit?

*14.34 The decomposition of N_2O_5 is believed to occur by the following mechanism:

$$2 N_2O_5 \rightleftharpoons N_2O_5{}^* + N_2O_5 \quad \text{equilibrium constant} = K_1$$

$$N_2O_5{}^* \rightleftharpoons NO_2 + NO_3 \quad \text{equilibrium constant} = K_2$$

$$NO_2 + NO_3 \xrightarrow{k_1} NO_2 + O_2 + NO$$

$$NO + NO_3 \xrightarrow{k_2} 2 NO_3$$

  where $N_2O_5{}^*$ represents an activated molecule
Show that the rate of formation of O_2 is directly proportional to the concentration of N_2O_5.

15 THE ATMOSPHERE

Life on this planet depends upon the presence of the relatively thin layer of air that surrounds it. Even though the atmosphere accounts for only about 0.0001% of the total mass of the earth, it is the reservoir from which we draw oxygen essential to metabolism, carbon dioxide for photosynthesis, and nitrogen, whose compounds are essential to plant growth. Our climate is governed by the movement of water vapor from the earth's surface into the atmosphere and back again.

Even trace constituents of the atmosphere can have beneficial or detrimental effects on the delicate balance of life. Small amounts of ozone at an elevation of about 30 km absorb most of the harmful ultraviolet radiation of the sun. On the other side of the ledger, as little as 0.2 part per million of ozone near the earth's surface promotes the photochemical reactions responsible for smog formation.

In this chapter we shall consider some of the physical and chemical properties of the components of the atmosphere. With that background, we shall discuss the general problem of air pollution, how it comes about, and how it can be controlled.

15.1 THE COMPOSITION OF THE ATMOSPHERE

In Table 15.1 are given the mole fractions of the various components of the atmosphere. Two species are omitted: water vapor, whose mole

TABLE 15.1 Composition of Clean, Dry Air at Sea Level

COMPONENT	MOLE FRACTION	COMPONENT	MOLE FRACTION	COMPONENT	MOLE FRACTION
N_2	0.7808	Ne	1.82×10^{-5}	SO_2	$<1 \times 10^{-6}$
O_2	0.2095	He	5.24×10^{-6}	O_3	$<1 \times 10^{-7}$
Ar	0.00934	CH_4	$2 \ \ \times 10^{-6}$	NO_2	$<2 \times 10^{-8}$
CO_2	0.00314	Kr	1.14×10^{-6}	I_2	$<1 \times 10^{-8}$
		H_2	$5 \ \ \times 10^{-7}$	NH_3	$<1 \times 10^{-8}$
		N_2O	$5 \ \ \times 10^{-7}$	CO	$<1 \times 10^{-8}$
		Xe	$8.7 \ \times 10^{-8}$	NO	$<1 \times 10^{-8}$

fraction may vary from 0.02 in the tropics to 0.0005 in polar regions, and suspended particles (e.g., dust, smoke), which vary greatly both in concentration and in chemical composition.

The composition of air, indeed that of any gaseous mixture, can be expressed in a number of ways. Several of the more common concentration units used for solutions of gases are related to mole fraction in Table 15.2. The units "parts per million" (ppm) and "parts per billion" (ppb) are commonly used for minor constituents.

TABLE 15.2 Concentration Units for Gases as Related to Mole Fraction (X)

Volume % = mole % = 100 X

Weight % = volume % $\times \dfrac{MW}{aver.\ MW}$ *

Partial pressure (atm) = X (total pressure)

Moles/liter = $\dfrac{partial\ pressure}{(0.0821)T}$

Parts per million (ppm) = 10^6 X

Parts per billion (ppb) = 10^9 X

* The average molecular weight of dry air is 29.0.

Example 15.1 Using the information in Tables 15.1 and 15.2, calculate
a. the volume %, weight %, partial pressure at 1 atm total pressure, and concentration in moles/liter at 25°C of N_2 in clean, dry air.
b. the concentration of krypton in ppm and ppb.

Solution
a. Volume % = 100(0.7808) = 78.08

Weight % = 78.08 $\times \dfrac{28.01}{29.0}$ = 75.4   (note that O_2 and Ar raise the average molecular weight of air above that of N_2)

P_{N_2} = 0.7808 × 1 atm = 0.7808 atm

moles/liter = $\dfrac{0.7808}{(0.0821)(298)}$ = 0.0320

b. ppm = $10^6(1.14 \times 10^{-6})$ = 1.14
ppb = $10^9(1.14 \times 10^{-6})$ = 1140

The two major components of air, nitrogen and oxygen, are obtained by the fractional disillation (Chapter 1) of liquid air. One method of liquefying air is shown in Figure 15.1. Notice that cooling is accomplished by two quite different methods. In part, compressed air is cooled by passing it through a heat exchanger around which a cold fluid (e.g., water, ammonia, or liquid air) is circulated. Further cooling is achieved by allowing the compressed air to expand suddenly. When the gas molecules move farther apart, heat must be absorbed to overcome the attractive forces between these molecules. The absorption of heat lowers the kinetic energy and, hence, the temperature of the gas.

7.5×10^6 tons of air were liquefied in the U.S. by this process in 1971

When liquid air is allowed to warm up slowly, nitrogen (bp = 77°K) evaporates first, leaving a residue which is mostly liquid oxygen (bp = 90°K). Of the noble gases, helium (bp = 4°K) and neon (bp = 27°K) pass off in the nitrogen fraction. Most of the argon (bp = 87°K) and all the heavier noble gases are combined with the liquid oxygen, from which they can be

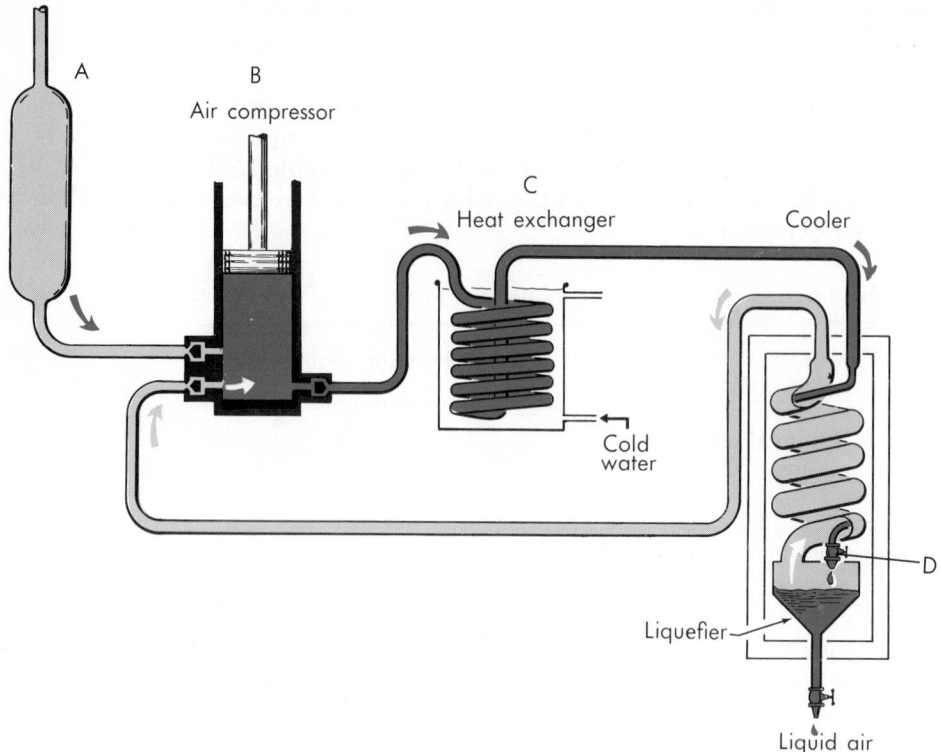

Figure 15.1 Liquefaction of air. Carbon dioxide and water are removed by passing through a chamber (A) packed with NaOH. The air is then compressed (B) and cooled by passing through a heat exchanger (C). Further cooling takes place upon expansion at D, a throttle valve. After passing through the cycle several times, the air begins to condense at (D).

removed and purified by repeated fractionation. Neon is extracted from the nitrogen fraction in a similar way. Helium is obtained from natural gas, where its concentration is much higher than in air.

15.2 NITROGEN

We shall begin our discussion of the properties of the principal components of the atmosphere by considering the most abundant species, elementary nitrogen.

CHEMICAL PROPERTIES

For all practical purposes N_2 at 25°C is an inert gas

At room temperature and atmospheric pressure, elementary nitrogen fails to react with any other element. Its inertness can be attributed to the strength of the triple bond holding the N_2 molecule together

$$:N\equiv N:(g) \rightarrow 2 \; :\dot{N}\cdot(g); \quad \Delta H = +225 \text{ kcal}$$

This factor tends to make it difficult to form binary compounds of nitrogen from the standpoint of reaction rate (high activation energy) and/or reaction spontaneity (unfavorable ΔH_f, ΔG_f).

At high temperatures, nitrogen reacts with a few very reactive metals such as lithium and magnesium.

$$6 \text{ Li(s)} + N_2(g) \rightarrow 2 \text{ Li}_3N(s) \qquad (15.1)$$

$$3 \text{ Mg(s)} + N_2(g) \rightarrow Mg_3N_2(s) \qquad (15.2)$$

These compounds contain the nitride ion, $(:\ddot{N}:)^{3-}$. Nitrogen also reacts with boron and aluminum at high temperatures to form compounds with empirical formulas BN and AlN, but here the bonding is predominantly covalent rather than ionic. Boron nitride and aluminum nitride have macromolecular structures similar to those of graphite and diamond.

Nitrogen reacts directly with only two nonmetals, oxygen and hydrogen.

$$N_2(g) + O_2(g) \rightarrow 2 \text{ NO(g)} \qquad (15.3)$$

$$N_2(g) + 3 \text{ H}_2(g) \rightarrow 2 \text{ NH}_3(g) \qquad (15.4)$$

The reaction with oxygen is spontaneous only at temperatures above 2500°C; the product, nitric oxide, readily decomposes to the elements upon cooling. Ammonia, NH_3, is considerably more stable than nitric oxide. However, as we shall see, it can be formed at a reasonable rate only at relatively high temperatures in the presence of special catalysts.

The uses of elementary nitrogen reflect its chemical inertness at ordinary temperatures. Liquid nitrogen is used as a coolant in preference to liquid air because the gas given off on evaporation does not support combustion. Certain foods, including instant coffee, peanuts, and potato chips, are packed under nitrogen to prevent oxidation and loss of flavor. Nitrogen gas under pressure is used to force flammable liquid fuels and their oxidizers into rocket motors.

NITROGEN FIXATION

Combined nitrogen in the form of protein is essential to both plant and animal life. The key step in the nitrogen cycle in nature, shown in Figure 15.2, is the fixation of elementary nitrogen, i.e., its conversion to useful compounds such as ammonia or nitric acid. Small amounts of elementary nitrogen are converted to nitric oxide by lightning discharges; subsequent reaction with oxygen and water forms nitric acid, HNO_3.

A more important natural process of nitrogen fixation involves the action of certain bacteria found in the roots of leguminous plants such as peas, beans, clover, and alfalfa. These bacteria contain enzymes that catalyze the conversion of elementary nitrogen to ammonia. Although the mechanism of this process is not thoroughly understood, it appears that two different enzymes are involved. Both contain iron atoms; one of the enzymes contains molybdenum as well. It is generally believed that molecular nitrogen forms a weak complex with the metal atoms of the enzymes; this is then converted chemically to a more stable ammonia complex.

Faced with an expanding population and inadequate food supplies, man has attempted for centuries to speed up the nitrogen cycle in nature by adding nitrogen compounds directly to the soil. Prior to this century the only way to do this was to add "organic nitrogen" (i.e., manure). In

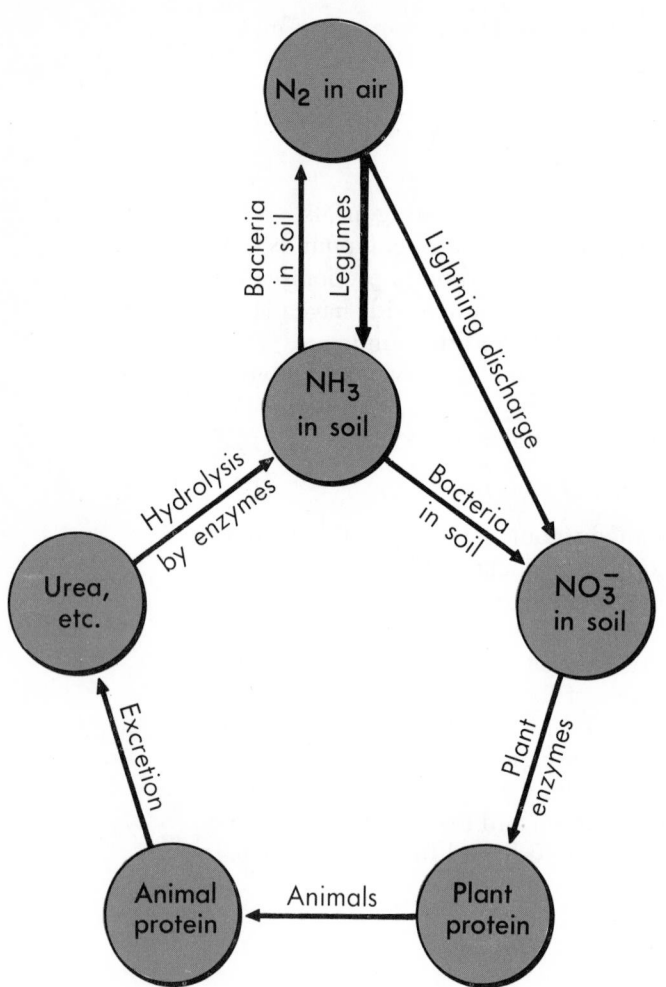

Figure 15.2 Nitrogen cycle in nature. Nitrogen enters the soil either as NO_3^- (from NO formed by lightning) or as NH_3 (fixation by bacteria in roots of legumes). Plants convert NO_3^- to proteins, which are eaten by animals. Metabolism releases nitrogen in the form of urea and other waste products. These products are eventually converted to N_2 by soil bacteria.

1908 Fritz Haber in Germany convinced himself from preliminary experiments and calculations that it would be feasible to fix atmospheric nitrogen by reacting it with hydrogen to form ammonia.

$$N_2(g) + 3\ H_2(g) \rightleftharpoons 2\ NH_3(g); \quad \Delta H = -22.0 \text{ kcal} \qquad (15.4)$$

His research was supported by German industrialists who were interested in using the ammonia to produce nitric acid, the starting material for such explosives as nitroglycerine and trinitrotoluene, TNT. In 1913 the first large-scale ammonia plant went into production. Throughout World War I the Haber process produced sufficient quantities of ammonia to make Germany independent of foreign supplies of nitrates, which were cut off by the British blockade. Virtually all the ammonia was converted to nitrate explosives rather than fertilizers. The resultant decrease in agricultural production led to widespread hunger, which contributed to the collapse of the German army in 1918 and was a major factor in the widespread unrest after the war.

The Haber process is now the main source of fixed nitrogen in the world. Its feasibility depends upon choosing experimental conditions

TABLE 15.3 Effect of Temperature and Pressure on the Yield of
Ammonia in the Haber Process ($[H_2] = 3 [N_2]$)

MOLE % NH_3 IN EQUILIBRIUM MIXTURE

°C	K_c	10 atm	50 atm	100 atm	300 atm	1000 atm	
200	650	51	74	82	90	98	Note the large
300	9.5	15	39	52	71	93	effect of pressure
400	0.5	4	15	25	47	80	on the yield
500	0.08	1	6	11	26	57	of NH_3
600	0.014	0.5	2	5	14	13	

under which nitrogen will react rapidly with hydrogen to give a reasonably
high yield of ammonia. At room temperature and atmospheric pressure, the
position of the equilibrium strongly favors the formation of ammonia
($K_c = 4 \times 10^9$) but the rate of reaction is virtually zero. Equilibrium can be
reached more rapidly by increasing the temperature. However, since Reaction 15.4 is exothermic, high temperatures reduce the yield of ammonia
(Table 15.3). High pressures, on the other hand, have a favorable influence
on both the rate of the reaction and the position of the equilibrium. An
increase in pressure makes it possible for the system to attain equilibrium
more rapidly because it increases the concentrations of the gases involved.
It also increases the relative amount of ammonia present at equilibrium,
since the formation of NH_3 results in a decrease in the number of moles of
gas.

Much of Haber's research was devoted to finding a catalyst that would
enable Reaction 15.4 to be carried out at a reasonable rate without going to
excessively high temperatures. The first catalyst he came up with, a
mixture of the oxides of osmium and uranium, was prohibitively expensive.
In a modern ammonia synthesis plant the catalyst used is a specially prepared mixture of iron, potassium oxide, and aluminum oxide. The synthesis
is carried out at a temperature of 400 to 450°C and pressures of 200 to 600
atm. Ammonia (bp $= -33$°C) is condensed out as a liquid from the gaseous
mixture; unreacted hydrogen and nitrogen are recycled to raise the yield
of ammonia.

Much of the ammonia produced these days by the Haber process is
used as a fertilizer, either directly or in the form of ammonium salts such
as $(NH_4)_2SO_4$ or $(NH_4)_3PO_4$. A portion of the ammonia is converted to nitric
acid by the Ostwald process, which involves the following series of reactions:

$$4\ NH_3(g) + 5\ O_2(g) \rightarrow 4\ NO(g) + 6\ H_2O(g) \quad \text{(1000°C, Pt catalyst)} \quad (15.5)$$

$$2\ NO(g) + O_2(g) \rightarrow 2\ NO_2(g) \quad \text{(NO allowed to cool in air)} \quad (15.6)$$

$$3\ NO_2(g) + H_2O(l) \rightarrow 2\ HNO_3(l) + NO(g) \quad \text{(NO_2 passed into water)} \quad (15.7)$$

15.3 OXYGEN

Elementary oxygen is much more reactive than nitrogen, reflecting
both its higher electronegativity (3.0 vs 2.5 for nitrogen) and the fact that
the bond in the O_2 molecule is considerably weaker than that in N_2. Among

the elements, only the noble gases, the halogens, and a few inactive metals (Pd, Au, Pt) fail to react directly with oxygen.

Pure O_2 is much more reactive than air

Liquid oxygen (LOX) is finding increased use as an oxidizer for liquid fuels. Oxygen gas under pressure replaces air where high temperature flames are required, as in blowing Pyrex glass or welding with an acetylene torch. Enormous amounts of oxygen are used in the manufacture of iron, partially replacing air in the blast furnace reaction and increasing plant capacity by more than enough to justify the added expense. Oxygen is also employed in steel production processes to burn out impurities such as manganese and silicon.

REACTION WITH METALS

When oxygen reacts with a metal, the usual product is an ionic oxide containing the oxide ion, O^{2-}

$$4 \text{ Li(s)} + O_2(g) \rightarrow 2 \text{ Li}_2O(s) \tag{15.8}$$

$$2 \text{ Mg(s)} + O_2(g) \rightarrow 2 \text{ MgO(s)} \tag{15.9}$$

These reactions occur relatively slowly at ordinary temperatures, since they require that the covalent bond in the O_2 molecule be broken. Certain of the 1A and 2A metals react with oxygen at ordinary temperatures to form compounds containing the peroxide ion, O_2^{2-}

$$2 \text{ Na(s)} + O_2(s) \rightarrow \text{Na}_2O_2(s) \tag{15.10}$$

$$\text{Ba(s)} + O_2(s) \rightarrow \text{BaO}_2(s) \tag{15.11}$$

These peroxides and superoxides react violently with water

or the superoxide ion, O_2^-

$$K(s) + O_2(s) \rightarrow \text{KO}_2(s) \tag{15.12}$$

Since the O—O bond remains unbroken, Reactions 15.10 to 15.12 occur more rapidly than 15.8 and 15.9.

Some of the reactions of these three oxygen-containing ions are given, along with their electronic structures, in Table 15.4. Concentrated solutions of hydrogen peroxide can be prepared by adding sodium peroxide to water

$$\text{Na}_2O_2(s) + 2 \text{ H}_2O(l) \rightarrow \text{H}_2O_2(aq) + 2 \text{ NaOH(s)} \tag{15.13}$$

TABLE 15.4 Properties of O^{2-}, O_2^{2-}, O_2^- Ions

Ion	Electronic Structure	Reaction with Water	Thermal Decomposition
Oxide	$(:\ddot{O}:)^{2-}$	$O^{2-} + H_2O \rightarrow 2 \text{ OH}^-$	stable
Peroxide	$(:\ddot{O}—\ddot{O}:)^{2-}$	$O_2^{2-} + 2 \text{ H}_2O \rightarrow 2 \text{ OH}^- + H_2O_2$	$O_2^{2-} \rightarrow O^{2-} + \frac{1}{2} O_2$
Superoxide	$(:\ddot{O}—\ddot{O}:)^-$	$2 O_2^- + 2 \text{ H}_2O \rightarrow 2 \text{ OH}^- + H_2O_2 + O_2$	$2 O_2^- \rightarrow O_2^{2-} + O_2$

TABLE 15.5 Oxides of the Metals of the First Transition Series

Cation	Sc	Ti	V	Cr	Mn	Fe	Co	Ni	Cu	Zn
		TiO_2	V_2O_5	CrO_3	Mn_2O_7					
			VO_2		MnO_2					
M^{3+}	Sc_2O_3	Ti_2O_3	V_2O_3	Cr_2O_3	Mn_2O_3	Fe_2O_3	Co_2O_3			
$2\,M^{3+},\,M^{2+}$					Mn_3O_4	Fe_3O_4	Co_3O_4			
M^{2+}		TiO	VO	CrO	MnO	FeO	CoO	NiO	CuO	ZnO
M^+									Cu_2O	

or by adding barium peroxide to sulfuric acid

$$BaO_2(s) + H_2SO_4(aq) \rightarrow H_2O_2(aq) + BaSO_4(s) \qquad (15.14)$$

Many of the transition and post-transition metals form more than one stable oxide (Table 15.5). In certain of these oxides the bonding is predominantly covalent rather than ionic. Titanium dioxide, TiO_2, and manganese dioxide, MnO_2, have macromolecular structures in which there are no discrete +4 cations. The highest oxide of manganese, Mn_2O_7, is a reddish liquid that freezes below $-20°C$, indicating a molecular structure.

Among the ionic oxides of the transition metals, charges of +3 or +2 for the metal ions are most common. Manganese, iron, and cobalt form oxides of empirical formula, M_3O_4, in which four oxide ions, $4(-2) = -8$, are balanced by two +3 ions and one +2 ion, $2(+3) + 2 = +8$.

When a metal forms more than one ionic oxide, it is invariably the higher oxide (i.e., the one containing the greatest percentage of oxygen) that is formed at low temperatures. For example, powdered copper, exposed to air at room temperature, is very slowly converted to black copper(II) oxide

$$2\,Cu(s) + O_2(g) \rightarrow 2\,CuO(s) \qquad (15.15)$$

This oxide can be converted to copper(I) oxide, Cu_2O, by heating to about 900°C

$$2\,CuO(s) \rightarrow Cu_2O(s) + \tfrac{1}{2}\,O_2(g) \qquad (15.16)$$

This behavior is readily understood in terms of the thermodynamic principles introduced in Chapter 12. For Reaction 15.16, $\Delta G^{1\,atm}$ at 298°K = +25.5 kcal, so CuO does not convert to Cu_2O under these conditions. However, since $\Delta S^{1\,atm}$ for this reaction is also positive (+28.8 cal/°K), $\Delta G^{1\,atm}$ becomes smaller as the temperature is increased. At about 900°C (cf. Problem 15.9), $\Delta G^{1\,atm}$ goes to zero; above this temperature, Cu_2O is the stable oxide.

REACTION WITH NONMETALS

Table 15.6 lists the oxides of the nonmetals that have negative free energies of formation and, hence, in principle at least, can be made by direct reaction with oxygen. Notice that none of the oxides of nitrogen (N_2O_5, N_2O_4, NO_2, N_2O_3, NO and N_2O) meet this criterion, which implies

TABLE 15.6   Stable Oxides of the Nonmetals ($\Delta G_f^{1 \text{ atm}}$ negative at 298°K)

3A	4A	5A	6A	7A
$B_2O_3(s)$	$CO_2(g)$ $CO(g)$			
	$SiO_2(s)$	$P_4O_{10}(s)$ $P_4O_6(s)$	$SO_3(g)$ $SO_2(g)$	
	$GeO_2(s)$	$As_2O_5(s)$ $As_4O_6(s)$	$SeO_3(s)$ $SeO_2(s)$	
		$Sb_2O_5(s)$ $Sb_4O_6(s)$	$TeO_3(s)$ $TeO_2(s)$	$I_2O_5(s)$

that they are potentially unstable with respect to decomposition to the elements. The fact that a compound such as NO_2 persists in our environment can be attributed to kinetic factors. The activation energy for the decomposition of NO_2 to the elements is relatively high, about 30 kcal, which means that the rate of decomposition is small at ordinary temperatures.

Among nonmetals forming more than one oxide, it is normally the higher oxide that is formed when the element burns at ordinary temperatures in an excess of oxygen. For example, phosphorus and carbon form P_4O_{10} and CO_2 when ignited in air.

$$P_4(s) + 5\,O_2(g) \rightarrow P_4O_{10}(s) \quad \text{low T, excess } O_2 \tag{15.17}$$

$$C(s) + O_2(g) \rightarrow CO_2(g) \quad \text{low T, excess } O_2 \tag{15.18}$$

Formation of the lower oxide (P_4O_6, CO) is favored by high temperatures or a limited supply of oxygen; carbon monoxide is formed under precisely these conditions when a fuel-rich mixture burns in an automobile engine.

The reaction of sulfur with oxygen is a particularly important one, since it is the first step in the preparation of sulfuric acid. When sulfur is heated in air, it catches fire, forming a choking gas made up principally of sulfur dioxide.

$$S(l) + O_2(g) \rightarrow SO_2(g) \tag{15.19}$$

Up to 2% of sulfur trioxide is formed simultaneously. To prepare this compound in good yield, sulfur dioxide is reacted further with oxygen.

$$SO_2(g) + \tfrac{1}{2}\,O_2(g) \rightleftharpoons SO_3(g); \quad \Delta H = -23.5 \text{ kcal} \tag{15.20}$$

The rate of this reaction and the equilibrium yield of sulfur trioxide depend

TABLE 15.7   Effect of Temperature on the Equilibrium
$$SO_2(g) + \tfrac{1}{2}\,O_2(g) \rightleftharpoons SO_3(g)$$

T(°C)	400	500	600	700	800
K_c	2300	400	70	20	7
Mole % SO_3°	96	88	66	40	20

° Under the condition that $[O_2] = \tfrac{1}{2}\,[SO_2]$, and $p_{tot} = 1$ atm.

upon temperature and pressure. The principles involved here are the same as those discussed previously in connection with the Haber process. At low temperatures, the equilibrium constant for the formation of sulfur trioxide is large, but equilibrium is reached very slowly. As the temperature is raised, the rate increases, but since the reaction is exothermic, the yield of sulfur trioxide drops off. High pressures would tend to increase both the yield of sulfur trioxide and the rate of reaction.

In the so-called contact process for the manufacture of sulfuric acid, the equilibrium represented by Equation 15.20 is reached rapidly by passing sulfur dioxide and oxygen, at atmospheric pressure, over a solid catalyst at a temperature of about 650°C. The equilibrium mixture is then recycled at a lower temperature, 400 to 500°C, to increase the yield of sulfur trioxide. The two catalysts which have proved most effective are vanadium pentoxide, V_2O_5, and finely divided platinum. The sulfur trioxide produced can be converted to sulfuric acid by reaction with water.

$$SO_3(g) + H_2O(l) \rightarrow H_2SO_4(l) \qquad (15.21)$$

This reaction is ordinarily carried out in sulfuric acid solution; if sulfur trioxide is added directly to water, sulfuric acid is formed as a fog of tiny particles which are difficult to recover. Sulfuric acid is produced in larger amounts than any other industrial chemical. In 1970 about 30 million tons of H_2SO_4 were manufactured in this country. You can see why sulfuric acid is called a "heavy chemical."

What do you think all that acid is used for?

15.4 THE NOBLE GASES

All the noble gases (Periodic Group 8A) except radon are found in the atmosphere. Argon is by far the most abundant, accounting for more than 99.7% of the total noble gas content of air. Small amounts of helium and radon are found in association with radioactive minerals (Chapter 22). Natural gas, from which helium is extracted commercially, sometimes contains as much as six mole % of that gas.

Helium, because of its low density (about 1/7 that of air) and chemical inertness, is used in dirigibles and in synthetic atmospheres to make breathing easier for patients suffering from certain types of lung disorders. Smaller quantities are used as a carrier gas in gas chromatography and as a liquid coolant to achieve very low temperatures (bp = −269°C). Argon and, more recently, krypton are used to provide an inert atmosphere in light bulbs and thereby retard decomposition of the tungsten filament. In illuminated "neon signs" a high voltage is passed through a glass tube containing neon at a very low pressure. The red glow emitted under these conditions corresponds to an intense line in the neon spectrum at 6400 Å.

The first person to isolate one of the noble gases was the English scientist Sir Henry Cavendish. He subjected a sample of atmospheric nitrogen to repeated electrical discharges in the presence of oxygen, thereby forming oxides of nitrogen which were dissolved out in water. Cavendish reported that a bubble of gas, about 1% of the original volume, remained undissolved. Nothing came of his observation for over a hundred years. In 1892 Lord Rayleigh, Professor of Physics at the Cavendish Labora-

TABLE 15.8   Properties of Binary Compounds of the Noble Gases

	ΔH_f	* $\Delta G_f^{1\,atm}$ at 298°K	MOLECULAR GEOMETRY
XeF_2	−26 kcal	−18 kcal	linear
XeF_4	−52	−33	square planar
XeF_6	−70	−41	distorted octahedron
XeO_3	+100	−	−
KrF_2	+14	−	linear

° Calculated by extrapolation from higher temperatures.

tory in Cambridge, found that nitrogen prepared by removing oxygen from air had a density about 0.5% greater than that of nitrogen prepared synthetically by the decomposition of ammonium nitrite, NH_4NO_2. A Scotsman, Sir William Ramsey, separated the heavier impurity from atmospheric nitrogen; from a study of its emission spectrum, he identified it as a new element which he called argon, from the Greek word for "lazy." Over the next five years, Rayleigh and Ramsey, working together, isolated the remaining members of the noble gas family. Helium had previously been detected in the solar spectrum in 1868, but was unknown on earth.

Up until about 10 years ago the noble gases were more frequently referred to as "inert gases," since they were believed to be completely unreactive toward other elements and compounds. In 1962 Neil Bartlett, a 29 year old chemist doing research on platinum-fluorine compounds at the University of British Columbia, isolated a reddish solid which he showed to be $O_2^+(PtF_6)^-$. Realizing that the ionization energy of xenon (270 kcal/mole) is virtually identical with that of molecular oxygen (265 kcal/mole), he prepared by an analogous method a substance which he postulated to be $Xe^+(PtF_6)^-$. The success of Bartlett's experiments opened up a new era in noble gas chemistry. Within a few months a group at the Argonne National Laboratories near Chicago prepared the first stable binary compound of a noble gas, XeF_4, by direct reaction of the elements under pressure at 400°C. Xenon tetrafluoride is a white molecular solid with a melting point of 140°C.

Since 1962 several different compounds of xenon with fluorine and oxygen have been reported. The binary compounds of xenon whose existence is well established are listed in Table 15.8. The three fluorides of xenon are thermodynamically stable, as indicated by their negative free energies of formation. On the other hand, the oxide, XeO_3, is unstable and decomposes violently on the slightest provocation. The lighter noble gases are much less reactive than xenon: only one binary compound of krypton, KrF_2, has been isolated. There have been attempts in many laboratories to prepare fluorides or oxides of argon, but all results so far have been negative. Radon would presumably be more reactive than krypton, but the fact that the most stable isotope of this element has a half-life of only 3.8 days has discouraged investigation of its chemistry.

XeF_6 is not a regular octahedron. Can you suggest why?

The geometries of the xenon fluorides appear to agree with the predictions of electron pair repulsion (Problem 7.45, Chapter 7, and Problem 15.44).

15.5 CARBON DIOXIDE

The percentage of carbon dioxide in the air is too small for it to be extracted profitably. Instead, it is obtained commercially by the combustion of coke or other carbon-containing fuels,

$$C(s) + O_2(g) \rightarrow CO_2(g) \qquad (15.22)$$

by heating limestone or other carbonate minerals,

$$CaCO_3(s) \rightarrow CaO(s) + CO_2(g) \qquad (15.23)$$

or as a by-product of fermentation processes, e.g.,

$$C_6H_{12}O_6(aq) \rightarrow 2\ C_2H_5OH(aq) + 2\ CO_2(g) \qquad (15.24)$$

We are familiar with the solid form of carbon dioxide (Dry Ice) and its use as a refrigerant. A common type of fire extinguisher contains liquid CO_2 at its equilibrium vapor pressure, about 59 atm at 20°C. When the cylinder valve is opened, the liquid boils. The amount of heat absorbed in this process is great enough to lower the temperature of the escaping gas to the point (−78°C) where it forms a finely divided "snow" of solid carbon dioxide. Since pure CO_2 is a nonconductor of electricity, extinguishers of this type are particularly useful in dealing with electrical fires. They cannot be used to put out fires of burning magnesium or aluminum, as Londoners found out to their dismay during the fire-bombings in World War II, since these metals react exothermically with the carbon dioxide.

How could you put out such a fire?

$$2\ Mg(s) + CO_2(g) \rightarrow 2\ MgO(s) + C(s);\ \ \Delta H = -193.5\ kcal \quad (15.25)$$

The average concentration of carbon dioxide in the atmosphere has increased by about 10% in the past century. Increased consumption of fossil fuels is primarily responsible for this trend. Extensive land clearing, which reduces the amount of CO_2 consumed in photosynthesis, is also a factor. It is estimated that by the year 2050, unless steps are taken to restore the CO_2 balance, its concentration may be twice what it is today.

There has been considerable speculation that increasing concentrations of CO_2 may have a significant effect on our climate. Carbon dioxide molecules absorb infrared radiation given off by the earth, thereby acting as an insulating blanket to prevent heat from escaping into outer space. From 1850 to 1950, worldwide temperatures rose by 0.5°C; to what extent this increase can be correlated with CO_2 concentration is debatable. Interestingly enough, temperatures appear to have dropped slightly since 1950, despite a rising carbon dioxide level. This cooling trend may be due to increased amounts of suspended particles, formed by fuel combustion and industrial processes. These particles reflect solar radiation back into space, thus offsetting the "greenhouse effect" of CO_2.

15.6 WATER VAPOR

The concentration of water vapor in the air is often expressed in terms of the *relative humidity,* which is defined as

$$R.H. = \frac{P_{H_2O}}{P^\circ_{H_2O}} \times 100\% \qquad (15.26)$$

where P_{H_2O} is the partial pressure of water vapor in the air and $P°_{H_2O}$ is the equilibrium vapor pressure of water at the same temperature. On a day when the temperature is 25°C ($P°_{H_2O} = 23.8$ mm Hg) and the partial pressure of water vapor is 20.0 mm Hg, the relative humidity is

$$\frac{20.0}{23.8} \times 100\% = 84.0\%$$

Since the equilibrium vapor pressure of water increases with rising temperature, Equation 15.26 implies that the relative humidity will drop when cold air is warmed up (Example 15.2).

Example 15.2 On a cold winter day, when the outside temperature is 32°F (0°C), the vapor pressure of water in the air is 3.0 mm Hg. Calculate the relative humidity of the outside air and the relative humidity of the same air when it is brought into a house and warmed to 68°F (20°C).

Solution. The equilibrium vapor pressure of water is 4.6 mm Hg at 0°C and 19.8 mm Hg at 20°C. Consequently

$$\text{R.H. outside} = \frac{3.0}{4.6} \times 100\% = 65\%$$

$$\text{R.H. inside} = \frac{3.0}{19.8} \times 100\% = 15\%$$

Conversely, if a warm air mass is suddenly cooled, the relative humidity increases, perhaps to 100% or more. When this happens, liquid water condenses; this is precisely the way in which clouds are formed in the atmosphere. Clouds consist of many billions of tiny droplets of liquid water, on the average about 0.1 mm in diameter. Droplets of this size are too small to fall under the influence of gravity. The growth of small water droplets at temperatures above 0°C is ordinarily a very slow process. The rate of growth is increased in the presence of dust particles, which act as nuclei upon which small droplets can condense. This explains why volcanic eruptions are often followed by rainstorms.

More frequently ice crystals formed in the colder upper regions of clouds act as nuclei for precipitation. In principle, ice crystals should form at 0°C; in practice, because of supercooling, they seldom develop unless the temperature drops to at least −15°C. To stimulate the formation of ice crystals, clouds are sometimes seeded with Dry Ice. The sublimation of solid carbon dioxide absorbs enough heat from the cloud to reduce the temperature below that required for ice crystal formation. Another substance that is frequently used is finely divided silver iodide, which has a crystal structure similar to that of ice. The presence of silver iodide tends to prevent supercooling and, hence, allows ice crystals to form at temperatures close to 0°C.

Cloud seeding is about as far as we have gone with weather control

Silver iodide has also been used with some success in seeding hurricanes. The objective here is to add so much silver iodide that an enormous number of tiny ice crystals form. These crystals are too small to bring about precipitation, but the heat evolved in their formation tends to dissipate the storm clouds at the eye of the hurricane. One difficulty is that hurricane seeding may simply shift the storm off course without weakening it ap-

Figure 15.3 Cloud seeding with dry ice.

preciably. For this reason, seeding experiments are limited to storms that are far removed from populated areas.

Another problem in weather modification is that of dispersing fogs, which are a major navigational problem at many airports. Cold fogs (i.e., fogs in which the temperature is below 0°C) are relatively easy to disperse by conventional cloud-seeding techniques. Warm fogs pose a more serious problem. One approach involves spraying small crystals of sodium chloride into the fog. Since the vapor pressure of a saturated solution of sodium chloride is less than that of pure water, drops of salt solution act as nuclei for condensation of water to precipitate the fog.

15.7 THE UPPER ATMOSPHERE

Twenty years ago very little was known about species in the atmosphere above 30 km, the upper limit to which balloons can ascend. The development of rockets and satellites made it possible to explore the upper atmosphere and obtain information of the type shown in Figure 15.4. Small-scale mass spectrometers and spectrophotometers carried on these vehicles are used to detect species such as O_3, O and O_2^+, which are unstable in the laboratory. There is, however, a great deal yet to be learned about the properties and composition of the upper atmosphere. The region between 50 and 85 km is particularly difficult to study. It is too high to be reached by balloons and most aircraft and too low for earth satellites. What little information we have about this area comes from rockets which pass through it in about 20 seconds.

The regions of the atmosphere listed in Figure 15.4 are defined in terms of their temperature profile. In the *troposphere*, which extends up to

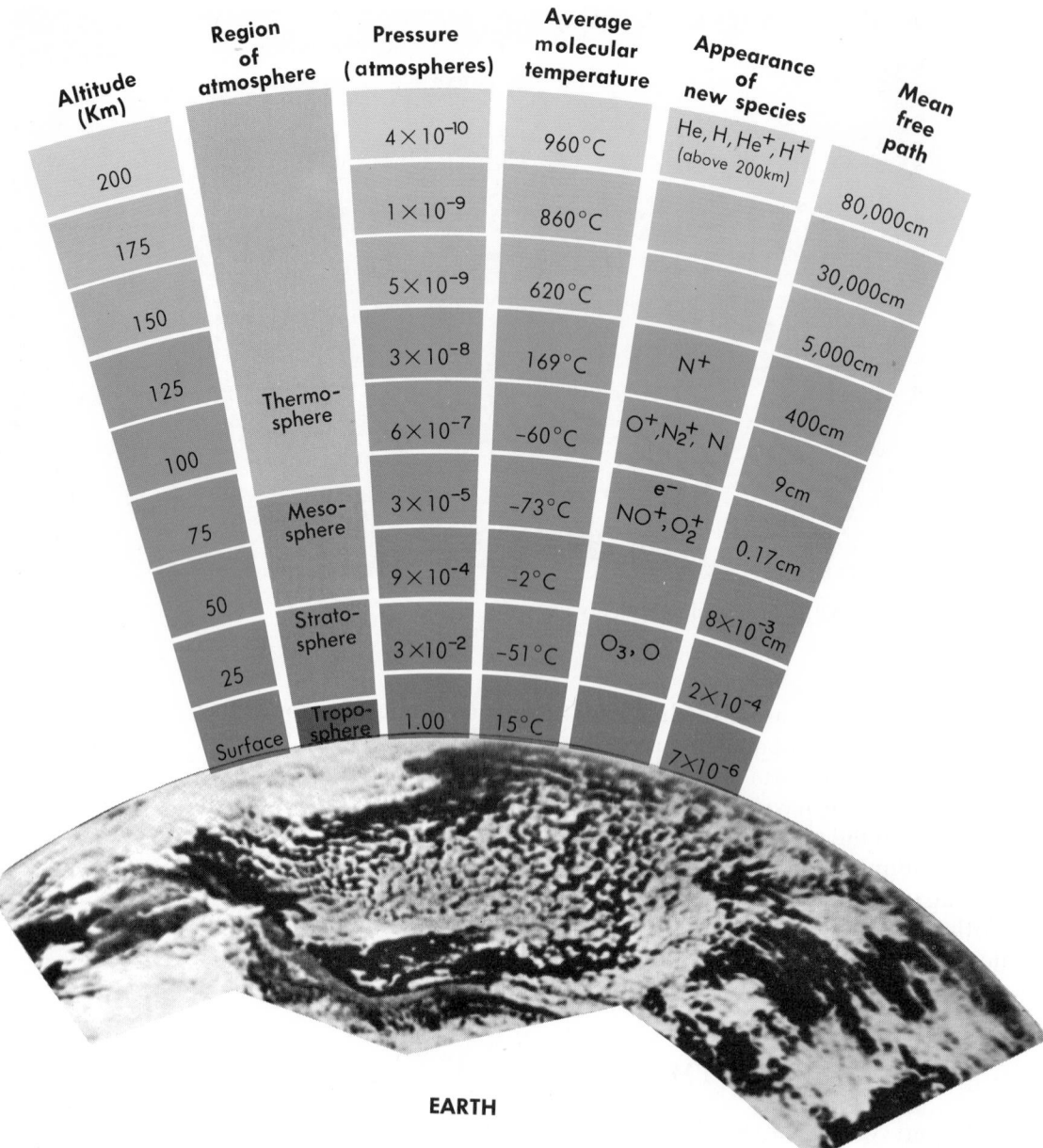

Figure 15.4 Properties of the atmosphere. Adapted from "Chemistry and the Environment: The Atmosphere," Special Report, Chemical and Engineering News, 1967.

11 km, the temperature decreases linearly with distance from the main source of heat, the earth's surface.

$$T = T_0 - 6.4 \, h \qquad (15.27)$$

(T = temperature at height h in km, T_0 = surface temperature)

Taking an average value of 15°C for T_0, we calculate a temperature of about −55°C at the upper limit of the troposphere.

In the *stratosphere* the temperature rises to a maximum of about 0°C

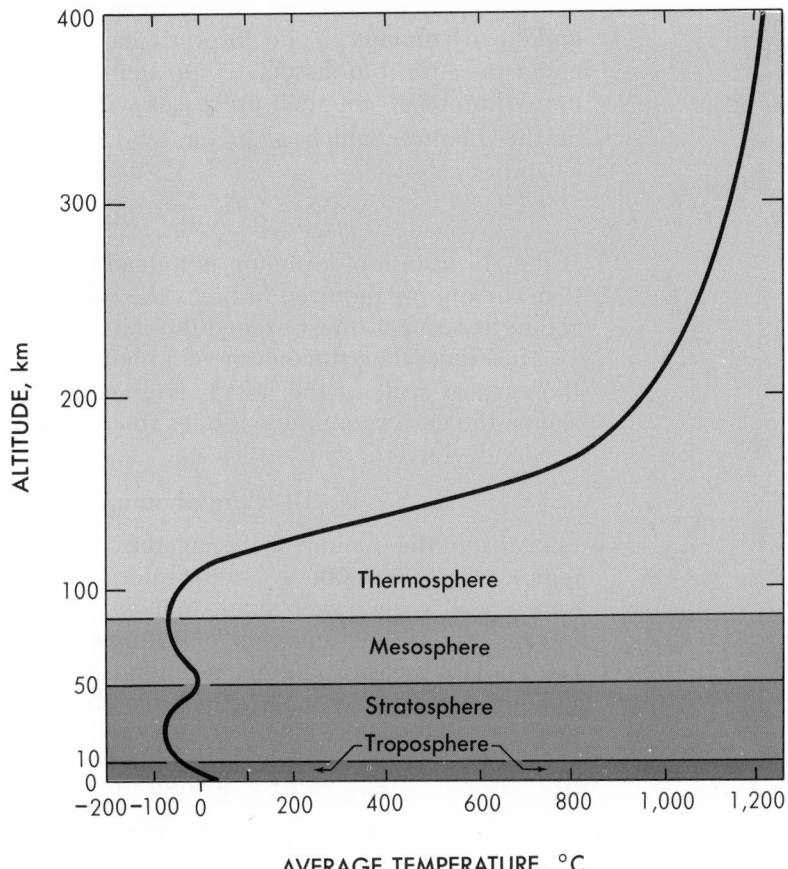

Figure 15.5 Variation of temperature with altitude in the atmosphere.

at an altitude of 50 km. due largely to the absorption of ultraviolet radiation by ozone. Above the ozone layer the temperature drops to a minimum of about −90°C at the upper limit of the *mesosphere.*

Figure 15.5 shows a steadily increasing temperature in the lower portion of the *thermosphere,* approaching a constant value of about 1200°C at 400 km. However, we should emphasize that "temperature" in this region has lost much of its physical meaning. For example, 1200°C is simply a calculated temperature based on the very high speeds acquired by such species as NO^+, O_2^+, and O^+ when they absorb solar radiation. Since the concentrations of these species are very low, of the order of 10^{-13} mole/liter, the total kinetic energy of the particles in a given volume of the thermosphere is only a tiny fraction of that at the surface of the earth. A thermometer stuck through the wall of a satellite passing through this region would actually read, in the shade, a temperature very close to 0°K. It would lose energy by radiation to space much more rapidly than it could acquire it from an occasional molecule or ion that passed by.

The presence of thermodynamically unstable species such as O_3, O, and O_2^+ at high altitudes is explained in terms of two characteristics of the upper atmosphere.

1. The Availability of High Energy Solar Radiation in the Ultraviolet and Far Ultraviolet. The absorption of a photon of light energy by a molecule can cause it to dissociate into atoms ($O_2 \rightarrow 2\ O$) or ionize ($O_2 \rightarrow O_2^+ + e^-$). Whether or not such processes will occur depends upon the energy of the photon, which is given by Einstein's equation (Equation 6.2, Chapter 6):

$$E = hc/\lambda \qquad\qquad (15.28)$$

If the absorption of a photon is to lead to dissociation, E must be greater than the energy required to break the bond within the molecule. For ionization to occur, E must exceed the ionization energy.

In Figure 15.6 the energy of a photon is plotted as a function of λ. On the vertical scale at the left, E is given in ergs/photon. At the right are shown the corresponding energies in kcal/mole, calculated using the conversion factor

$$1 \times 10^{-12}\ \text{erg/photon} = 14.4\ \text{kcal/mole} \qquad (15.29)$$

Most of the photons that reach the surface of the earth are in the visible region ($\lambda = 8000$–4000 Å), with energies (36–72 kcal/mole) too low to ionize or dissociate such stable molecules as O_2 or N_2. At elevations above 30 km, significant amounts of ultraviolet radiation ($\lambda = 4000$–2000 Å) are available, corresponding to energies in the range 72–144 kcal/mole. Photons with these wavelengths bring about the formation of ozone by the two-step process

$$O_2(g) \rightarrow 2\ O(g); \quad \Delta H = +119\ \text{kcal}$$

$$O_2(g) + O(g) \rightarrow O_3(g)$$

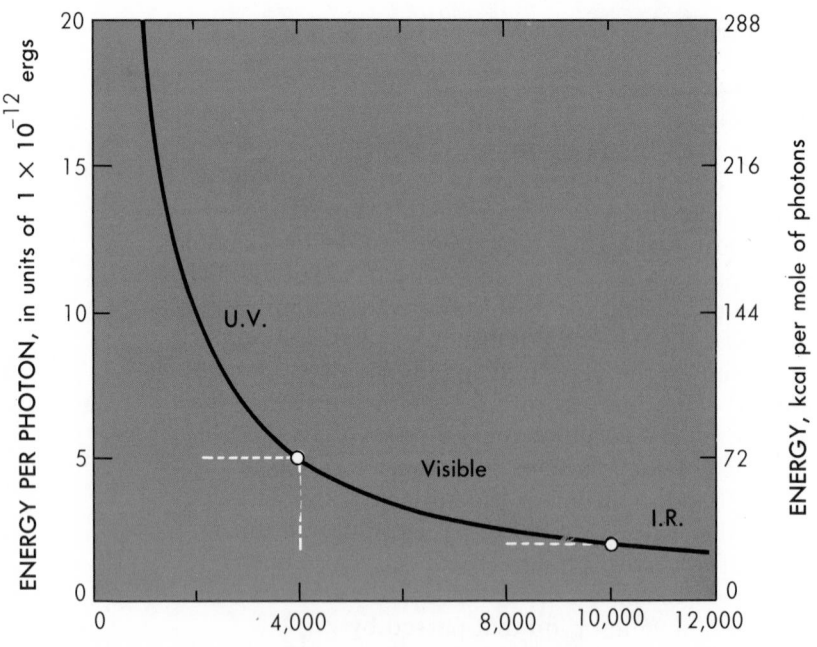

The energies of photons increase tremendously as their wavelengths become short

Figure 15.6 Relation between the energy of a photon and its wavelength: $E = \dfrac{hc}{\lambda}$.

At still higher altitudes, where the air is thinner, photons in the far UV have sufficient energies to ionize O_2 or even N_2:

$$O_2(g) \rightarrow O_2^+(g) + e^-: \quad \Delta H = +265 \text{ kcal}$$

$$N_2(g) \rightarrow N_2^+(g) + e^-: \quad \Delta H = +361 \text{ kcal}$$

Example 15.3 Using Equations 15.28 and 15.29, find the longest wavelength of a photon that could bring about the reactions

  a. $O_2(g) + h\nu \rightarrow 2\ O(g)$

  b. $O_2(g) + h\nu \rightarrow O_2^+(g) + e^-$

  c. $N_2(g) + h\nu \rightarrow N_2^+(g) + e^-$

Solution

  a. Noting that $E = 119$ kcal/mole (p. 414), we convert to ergs/molecule using Equation 15.29.

$$119\ \frac{\text{kcal}}{\text{mole}} \times \frac{10^{-12}\ \text{erg/molecule}}{14.4\ \text{kcal/mole}} = 8.26 \times 10^{-12}\ \text{erg/molecule}$$

Substituting in Equation 15.28 and solving for λ:

$$\lambda = \frac{hc}{E} = \frac{(6.62 \times 10^{-27}\ \text{erg sec}) \times (3.00 \times 10^{10}\ \text{cm/sec})}{8.26 \times 10^{-12}\ \text{erg}}$$
$$= 2.40 \times 10^{-5}\ \text{cm} = 2400\ \text{Å}; \quad (1\ \text{Å} = 10^{-8}\ \text{cm})$$

  b. and c. In precisely the same way ($E = 265$ and 361 kcal/mole, respectively)

$$\lambda = 1080\ \text{Å}; \quad \lambda = 791\ \text{Å}$$

which correspond to the far UV. Actually the photons absorbed will be at lower wavelengths (higher energies) than those calculated. The excess energy is carried off by the products (O_2^+, N_2^+) as kinetic energy, accounting for the "high temperature" in the upper portions of the thermosphere.

---

2. The Large Distance Between Particles. One of the parameters shown in Figure 15.4 is the mean free path of a gas molecule calculated from the kinetic theory of gases. (The mean free path represents the average distance that a molecule travels before it collides with another.) At the earth's surface, where the concentration of molecules is relatively high, the mean free path is small, about 7×10^{-6} cm. Molecules collide frequently, reactions occur relatively rapidly, and the concentrations of different species such as $O_2(g)$ and $O(g)$ are determined by equilibrium considerations. At high altitudes the air is much thinner and the situation quite different. A mean free path of 80,000 cm (about one-half mile) at an altitude of 200 km means that collisions between particles are extremely infrequent. Reactions occur very slowly, so that unstable species are frequently present at concentrations far greater than we would predict from equilibrium calculations.

Chemical equilibria can exist only where molecular collisions are very numerous

To illustrate the importance of this factor, consider the dissociation of O_2

$$O_2(g) \rightleftarrows 2\ O(g)$$

We have seen that the forward reaction is brought about by the absorption of UV radiation in the upper atmosphere. We might wonder, however, why the oxygen atoms do not immediately recombine, since ΔG^{1M} for the forward reaction at 300°K is +112 kcal, corresponding to an equilibrium constant of 10^{-82}

$$\log_{10}K_c = \frac{-112{,}000}{(2.30)(1.99)(300)} = -82: \; K_c = \frac{[O]^2}{[O_2]} = 10^{-82}$$

Near the surface of the earth, equilibrium is established rapidly and the fraction of O_2 dissociated is negligibly small. At an altitude of 120 km, however, the concentration of oxygen atoms is about equal to that of O_2, some 10^{36} times greater than that calculated from the expression for K_c (cf. Problem 15.35). Once an oxygen atom is formed in this region of the atmosphere, it may be a long time before it finds another one to collide with to re-form an O_2 molecule. Moreover, even when two oxygen atoms do meet, they are usually moving so rapidly that collision will not result in reaction. Instead, they bounce off each other and continue on their merry way through space.

15.8 AIR POLLUTION

Since man discovered fire, he has polluted the atmosphere with noxious gases and soot. One can imagine our distant ancestors huddled around a soggy wood fire grumbling about the smoke. Not until the 14th century, when coal began to be used as a fuel, did the problem become one of public concern. During the reign of Edward II (1307–1327), the English Parliament passed a law restricting the use of coal in London. Apparently they were serious about it; a coal merchant was tortured and hanged for creating a "pestilential odor." This does not seem to have solved the problem; London and other industrial cities were still being blackened by coal smoke 600 years later. Increased fuel consumption by industry, concentration of population in urban areas, and the advent of motor vehicles have, over the years, made the problem worse. Today the principal cause of pollution in our atmosphere is the gasoline engine.

Any substance whose addition to the atmosphere produces a measurable effect on man or his environment can be classified as a pollutant. A

TABLE 15.9   Major Air Pollutants

SPECIES	AVERAGE CONCENTRATION (ppm) IN URBAN ATMOSPHERE [*]
1. Oxides of sulfur (SO_2, SO_3)	0.06
2. Oxides of nitrogen (NO, NO_2)	0.04
3. Carbon monoxide (CO)	8
4. Hydrocarbons (paraffins, olefins, aromatics)	3
5. Suspended particles (dust, smoke, etc.)	variable

[*] Average for six cities for the year 1965. Peak concentrations are much higher than these.

host of materials fit this broad definition. Radioactive species produced by nuclear fallout qualify, as do toxic gases released accidentally into the atmosphere. In this chapter we shall limit our attention to the five major types of pollutants listed in Table 15.9.

SOURCES OF POLLUTANTS

In Table 15.10 are listed the principal sources of the five major air pollutants. Notice that automobiles produce virtually all the CO, most of the hydrocarbons and nearly half of the oxides of nitrogen.

Suspended particles in the atmosphere vary in size from those found in cigarette smoke, which have diameters as small as 0.01 μ ($\mu = 1$ micron $= 10^{-4}$ cm) to the large dust particles produced by a cement kiln, many of which have diameters greater than 100 μ. A major contributor to this type of pollutant is the combustion of coal, where much of the ash and some unburned carbon mixes with stack gases. Fuel oil is cleaner burning; natural gas is even better in this respect. A properly adjusted burner using natural gas produces virtually no smoke or soot.

Sulfur oxides are also produced by the combustion of solid fuels. Most of the coal burned in power plants in this country contains from 1 to 3% sulfur, largely in the form of minerals such as pyrites, FeS_2. Upon combustion the sulfur is converted to sulfur dioxide

$$4 \ FeS_2(s) + 11 \ O_2(g) \rightarrow 2 \ Fe_2O_3(s) + 8 \ SO_2(g)$$

Most hydrocarbon fuels contain very little sulfur. An exception is the residual oil left after the distillation of petroleum. This tarry material, which is used in many industrial and municipal heating plants, contains from 1 to 2% sulfur. The sulfur, present largely as organic compounds, is converted to sulfur dioxide upon combustion.

Metallurgical processes are another major source of sulfur oxides. Many of the heavy metals occur in nature as sulfide ores (Cu_2S, ZnS, HgS, PbS). These ores are "roasted" in air to convert them to the oxide or the free metal. In either case, SO_2 is a byproduct:

$$2 \ PbS(s) + 3 \ O_2(g) \rightarrow 2 \ PbO(s) + 2 \ SO_2(g) \tag{15.30}$$

$$Cu_2S(s) + O_2(g) \rightarrow 2 \ Cu(s) + SO_2(g) \tag{15.31}$$

Steel mills consume tremendous amounts of coal and in some cities are the main source of air pollution

TABLE 15.10   Sources of Air Pollutants in the United States (1965)
(Millions of Tons per Year)

	SUSPENDED PARTICLES	SO$_2$, SO$_3$	NO, NO$_2$	HYDR.	CO	TOTAL	% OF TOTAL
Automobiles	1	1	6	12	66	86	60
Industry	6	9	2	4	2	23	17
Electric plants	3	12	3	1	1	20	14
Space heating	1	3	1	1	2	8	6
Refuse disposal	1	1	1	1	1	5	3
	12	26	13	19	72	142	

From equilibrium considerations we might expect sulfur dioxide in the air to be converted to sulfur trioxide.

$$SO_2(g) + \tfrac{1}{2} O_2(g) \rightarrow SO_3(g); \quad K_c = 6 \times 10^{12} \text{ at } 25°C$$

However, as pointed out previously, this reaction takes place very slowly in the absence of a catalyst. Ordinarily, less than 1% of the SO_2 in polluted air is converted to SO_3. Suspended solids in the atmosphere, such as those found in coal smoke, can raise this fraction to 5% or more by catalyzing this reaction. The further reaction of sulfur trioxide with water droplets (Reaction 15.21) gives a mist of sulfuric acid.

Small amounts of **oxides of nitrogen,** primarily nitric oxide, are formed from the elements when fuels are burned at high temperatures. In air, nitric oxide, NO, is slowly converted to nitrogen dioxide, NO_2, by oxygen

$$2\ NO(g) + O_2(g) \rightarrow 2\ NO_2(g); \quad k = 7 \times 10^{-3} \text{ lit}^2/\text{mole}^2 \text{ sec at } 25°C \quad (15.32)$$

more rapidly by ozone

$$NO(g) + O_3(g) \rightarrow NO_2(g) + O_2(g); \quad k = 1.2 \times 10^7 \text{ lit/mole sec at } 25°C \quad (15.33)$$

Most of the NO_2 produced reacts ultimately with water to form nitric acid (Reaction 15.7).

Hydrocarbons enter the atmosphere directly as vented gases from petroleum refineries or by evaporation from the fuel tanks of automobiles. A more important source is automobile exhaust, which contains significant quantities of unburned hydrocarbons. A high percentage of the hydrocarbons in the exhaust are olefins produced by thermal decomposition ("cracking") of paraffins in the gasoline range (C_5 to C_{10}). The following reaction is typical.

$$\underset{\text{octane}}{C_8H_{18}(l)} \rightarrow \underset{\text{methane}}{CH_4(g)} + \underset{\text{ethylene}}{2\ C_2H_4(g)} + \underset{\text{propylene}}{C_3H_6(g)} \quad (15.34)$$

Carbon monoxide is produced when carbonaceous fuels undergo incomplete combustion. The principal culprit is again the automobile, where reactions such as

$$C_8H_{18}(l) + 11\ O_2(g) \rightarrow 5\ CO_2(g) + 3\ CO(g) + 9\ H_2O(l) \quad (15.35)$$

are typical, particularly when the fuel mixture is too "rich," i.e., contains too much fuel and too little air. In principle the carbon monoxide should be converted to CO_2 in the atmosphere

$$CO(g) + \tfrac{1}{2} O_2(g) \rightarrow CO_2(g): \quad K_c = 3 \times 10^{45} \text{ at } 25°C \quad (15.36)$$

but this reaction, like the conversion of SO_2 to SO_3, is ordinarily quite slow.

Many of the pollutants that enter the atmosphere undergo subsequent reactions which yield even more noxious species. An important example is the reaction sequence involved in the formation of *photochemical smog,* which made its reputation originally in Los Angeles but has now become infamous in cities as widely separated as Honolulu, Denver, and Washington, D.C. Smog formation is initiated by the photochemical decomposition of nitrogen dioxide

$$NO_2(g) + h\nu \rightarrow NO(g) + O(g) \quad (15.37)$$

This nonspontaneous reaction ($\Delta G^{1\,atm}$ at 25°C = +63.3 kcal) is brought about by the absorption of a photon of visible light (λ = 4000 Å). The oxygen atoms produced can react with NO by the reverse of Reaction 15.37, but most of them react with the more abundant species O_2

$$O(g) + O_2(g) \rightarrow O_3(g) \tag{15.38}$$

to produce ozone, a major component of photochemical smog. Both of these very reactive species, oxygen atoms and ozone molecules, attack hydrocarbons, primarily of the olefin type. A variety of products are eventually formed, including alcohols, ketones, and aldehydes. Olefinic hydrocarbons can also react with either NO or NO_2 to produce organic nitrogen derivatives such as

$$H_3C\!-\!O\!-\!\underset{\underset{O}{\|}}{N}\!-\!O \quad \text{and} \quad H_3C\!-\!\underset{\underset{O}{\|}}{C}\!-\!O\!-\!O\!-\!\underset{\underset{O}{\|}}{N}\!-\!O$$

methyl nitrate   peroxyacetyl nitrate (PAN)

both of which are powerful lachrymators.

EFFECTS OF AIR POLLUTION

Anyone who lives in a major city (and many who do not) is familiar with those characteristics of air pollution which directly affect our senses of sight, smell, and taste. Reduced visibility (Fig. 15.7), caused by absorption and scattering of light by suspended particles, is perhaps the most

In a sense, smog is a luxury that a city cannot afford

Figure 15.7 New York City smog.

obvious effect. Residents of Manhattan have long since ceased to be surprised when the upper floors of the Empire State Building disappear from sight. Haze, smoke, and smog have seriously curtailed astronomical observations at the world famous Mount Wilson Observatory in California, which is located outside the city limits of Los Angeles at an elevation of 5700 feet.

Eye irritation is a familiar feature of photochemical smog. Several different lachrymators have been identified, among them peroxyacetyl nitrate (PAN) and two aldehydes

$$
\begin{array}{cc}
\overset{\displaystyle H}{\underset{}{\mid}} & \overset{\displaystyle H\ \ H}{\underset{}{\mid\ \ \mid}} \\
H\!-\!C\!=\!O & H_2C\!=\!C\!-\!C\!=\!O \\
\text{formaldehyde} & \text{acrolein}
\end{array}
$$

These and other organic compounds can also cause the acrid odor and burning taste associated with smog and smoke. Other offenders include the oxides and acids of sulfur (SO_2, SO_3, H_2SO_4) and, to a lesser extent, those of nitrogen (NO_2, HNO_3).

We now know that air pollution plays a major role in chronic lung and heart disorders. Studies have shown that the condition of patients suffering from asthma, bronchitis, and emphysema improves when the level of air pollution drops and deteriorates when it worsens. In every major disaster caused by air pollution, the highest mortality rate has been among people with lung ailments. This was true at Donora, Pennsylvania, where 20 died in 1948, and in the "black fog" which was responsible for 4000 deaths in London in 1952.

Lung damage by air pollutants can come about in several different ways. Certain gases, including sulfur dioxide, ozone, and nitrogen dioxide, attack lung tissue directly; silica dust, SiO_2, has a similar effect. The disease called silicosis, characterized by the progressive deterioration of lung tissue, was responsible for the fact that very few granite cutters of a generation ago lived past age 50.

Carbon monoxide acts in quite a different way by reducing the ability of the blood to transport oxygen throughout the body. It does this by forming a complex with the hemoglobin of the blood which is more stable than that formed by oxygen (Chapter 19).

Although N_2 is similar to CO in many of its properties, it does not interact with hemoglobin

$$CO(g) + Hem \cdot O_2(aq) \rightleftharpoons O_2(g) + Hem \cdot CO(aq): \quad K_c = 210 \quad (15.39)$$

$$(\text{Hem} = \text{hemoglobin})$$

A person suffering from acute or chronic carbon monoxide poisoning has to breathe more air to deliver a given amount of oxygen to body cells. If as much as 10 to 20% of hemoglobin is tied up by carbon monoxide, the resulting strain on the heart and lungs can be fatal.

Air pollutants are as injurious to vegetation as they are to human beings. Smog in Southern California is responsible for an estimated loss of $10,000,000 a year in agricultural products. Crops ranging from lettuce to lima beans to oranges are damaged by certain components of smog, notably ozone and PAN. Shrubs and trees are also susceptible to attack by air pollutants. Pine trees are particularly sensitive to sulfur dioxide, which

Cleaning a building like this requires sandblasting

Figure 15.8 Building being cleaned shows accumulation of dirt due to air pollution (U.S. Public Health Service).

causes the needles to become brown and brittle. Forests of white pine along the Cumberland Plateau in Tennessee have been killed by SO_2 fumes emitted from nearby industries and the atomic energy center at Oak Ridge.

The adverse effect of air pollutants on building materials is evident when one looks at the soot-covered buildings and statuary in cities around the world. Less well known but equally serious is the deterioration of paper caused by sulfur dioxide in the air. Curiously enough, books and documents printed before 1750 are almost immune to attack. At about that time modern methods of paper making were introduced; they leave traces of metal oxides which catalyze the conversion of SO_2 to SO_3 and, hence, to sulfuric acid.

REDUCING AIR POLLUTION

AUTOMOBILE EMISSIONS. There are two basically different approaches to reducing air pollution caused by automobiles. One is to replace the gasoline-burning internal combustion engine by a "cleaner" source of energy. An obvious possibility is to use electrical energy, producing no noxious gases whatsoever. Many milk delivery trucks in England run on batteries, quietly and cleanly. A method that appears to be more popular in this country, at least at the present, is to substitute natural or bottled gas for gasoline (Chapter 4). The combustion of gaseous hydrocarbons produces virtually no carbon monoxide or olefins.

Hydrogen gas is thought by some to be the motor fuel of the future

A less drastic approach is to modify the combustion system of the automobile to reduce the amount of pollutants that come out of the exhaust. Up to the present time most of the effort in this area has been devoted to reducing emissions of carbon monoxide and hydrocarbons, since these are the pollutants whose maximum levels are set by federal standards. Car manufacturers have met these standards by relatively minor changes whose objective is to insure complete combustion of the fuel. This is accomplished by increasing the ratio of air to fuel, modifying the design of the combustion chamber to obtain better mixing, and retarding the spark that ignites the fuel-air mixture. These changes have not been enthusiastically received by many drivers because they tend to reduce gas mileage and make starting more difficult, particularly in cold weather.

It is clear that more extensive modifications will be necessary to meet the stringent standards which are to go into effect in 1975. The so-called "Clean Air Act" of 1970 sponsored by Senator Edmund Muskie requires that emissions of CO and hydrocarbons be reduced by 90% over a five-year period. Two possible ways of achieving this are shown in Figure 15.9. One uses a "catalytic muffler" in which these pollutants are made to react with oxygen on the surface of a solid catalyst. The major problem here is to find catalysts which remain effective for long periods (up to 50,000 miles of driving). Those now available are readily "poisoned" by adsorption of foreign substances, notably lead compounds derived from $Pb(C_2H_5)_4$ added to gasoline to raise its octane number. Another approach incorporates a "thermal reactor" into the exhaust system. Here, carbon monoxide and unburned hydrocarbons are ignited at a high temperature producing, hopefully, nothing but carbon dioxide and water vapor.

The Muskie bill further requires that emissions of oxides of nitrogen be reduced by 90% in 1975 car models. Current methods of controlling exhaust emissions tend, if anything, to produce more NO, since its formation from the elements is promoted by using a lean fuel mixture and a high combustion temperature. One way to reduce the concentration of NO is to recycle up to 30% of the exhaust gases through the engine, thereby reducing the temperature. Another approach is to find a catalyst which could be introduced into the exhaust system to promote the decomposition of NO, perhaps by the reaction

$$2\ CO(g) + 2\ NO(g) \rightarrow 2\ CO_2(g) + N_2(g)$$

This reaction is spontaneous ($\Delta G^{1\ atm}$ at 25°C $= -82.2$ kcal) but comparatively slow at the "parts per million" level at which CO and NO are formed in the automobile engine.

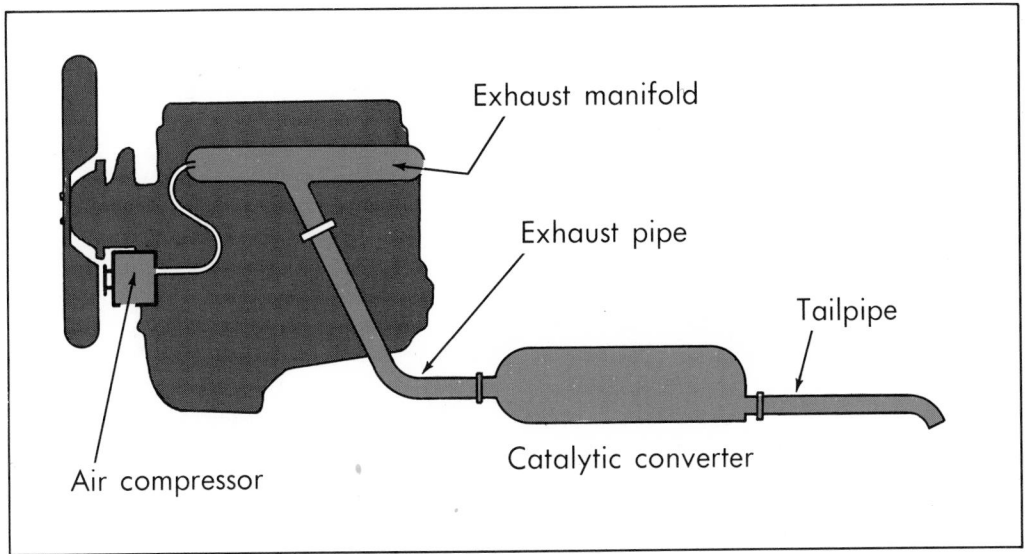

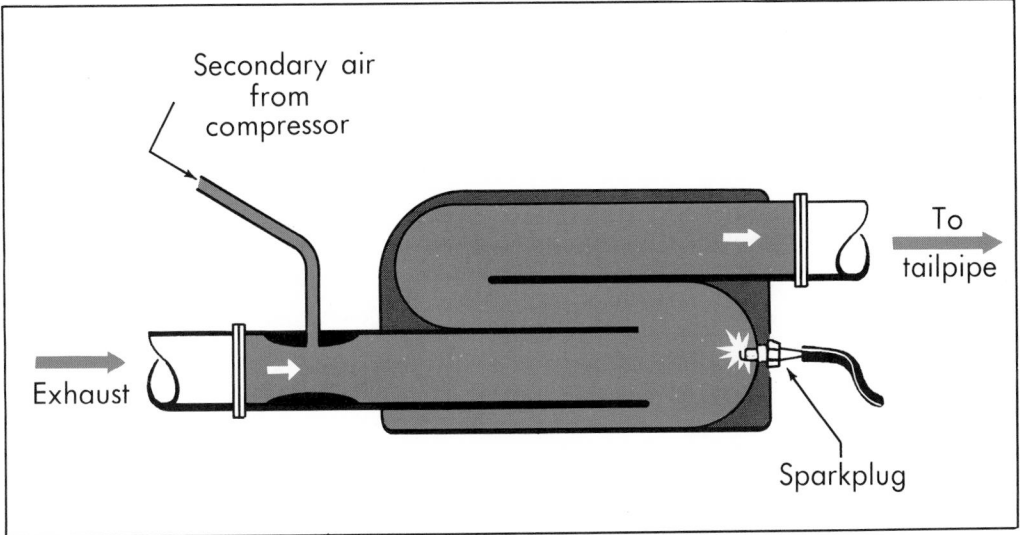

Figure 15.9 Methods of removing CO and hydrocarbons from automobile exhaust. (a) = catalytic muffler. (b) = thermal reactor.

INDUSTRIAL AND POWER PLANT EMISSIONS. Sulfur dioxide and suspended particles are not of serious concern in automobile exhaust emissions, but they are the major pollutants produced by combustion of fuels in electric power plants and many industries. An obvious way to prevent SO_2 emission is to remove sulfur compounds from the fuel before it is burned. In the case of coal, where the sulfur is present largely as the mineral pyrites, FeS_2, the sulfur content can be reduced by "washing" with a concentrated solution of calcium chloride (d = 1.35 g/ml). The coal (d = 1.2 g/ml) floats in this solution, but pyrites (d = 4.9 g/ml) sink to the bottom. With finely divided coal, the process of flotation is used to remove sulfur compounds (Fig. 15.10).

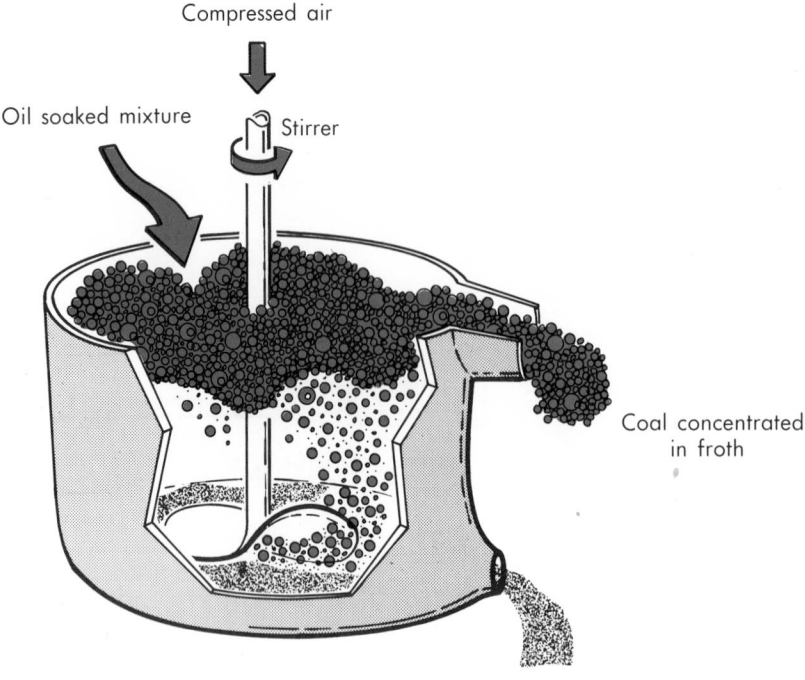

Figure 15.10 Flotation process of removing pyrites and other mineral impurities from finely divided coal.

Another way to reduce sulfur dioxide emissions is to remove it from stack gases by reaction with calcium or magnesium oxide.

$$MgO(s) + SO_2(g) + \tfrac{1}{2} O_2(g) \rightarrow MgSO_4(s) \qquad (15.40)$$

The oxides may be employed as solids or as aqueous suspensions.

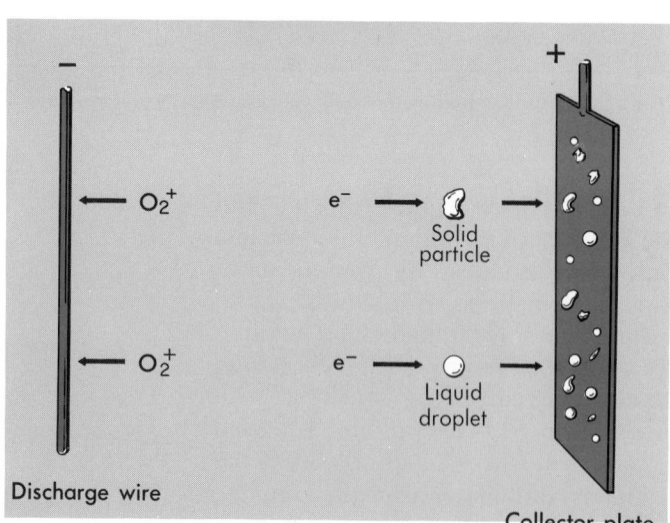

Figure 15.11 Electrolytic precipitation, used to remove suspended particles from stack gases.

This kind of device is now used in many industrial smokestacks

Many different devices are used to remove smoke particles and dust from stack gases. One of the most effective is the electrostatic precipitator shown schematically in Figure 15.11. The voltage used is high enough to ionize gas molecules such as oxygen.

$$O_2(g) \rightarrow O_2^+(g) + e^-$$

Cations move to the discharge wire where they are neutralized. Electrons move in the opposite direction, collide with solid or liquid particles, and transfer their charge to them. These particles, now carrying a negative charge, drift to the collector plate where they accumulate and are removed at regular intervals.

PROBLEMS

15.1 Explain why air cools when it expands suddenly. Does carbon dioxide behave similarly? Would an ideal gas cool on expansion?

15.2 Using Tables 15.1 and 15.2, calculate

   a. the volume and weight % of O_2 in the atmosphere.
   b. the concentration of He in mole/liter at 25°C; in ppm; in ppb.

15.3 Explain why

   a. liquid nitrogen is safer to use than liquid air.
   b. higher flame temperatures are obtained when oxygen is substituted for air.
   c. helium is substituted for nitrogen in synthetic atmospheres for breathing.
   d. carbon dioxide rather than nitrogen is used in bottling soft drinks.

15.4 Discuss the effect of changes in temperature and applied pressure on the position of the following equilibria and the rate at which they are attained

   a. $N_2(g) + O_2(g) \rightleftarrows 2\ NO(g)$
   b. $N_2(g) + 3\ H_2(g) \rightleftarrows 2\ NH_3(g)$

15.5 Which of the following species would you expect to be predominantly ionic and which predominantly covalent?

   a. Na_2O_2 c. Cu_2O
   b. CrO_3 d. Fe_3O_4

15.21 Which of the following substances are obtained commercially from air?

   a. N_2 b. He   c. CO_2
   d. O_2 e. Ar

15.22 Using data from Table 15.9, calculate the average concentration of CO in urban air in

   a. ppb
   b. volume %
   c. mole/liter (25°C)

15.23 Would it be feasible to

   a. substitute argon for nitrogen in packaging instant coffee?
   b. substitute oxygen gas for liquid oxygen as a component of rocket propellants?
   c. make a balloon rise by filling it with pure nitrogen? pure argon?
   d. use compressed nitrogen to extinguish burning magnesium?

15.24 Discuss how one might increase the rate and/or the yield of product in the following reactions

   a. $SO_2(g) + \frac{1}{2} O_2(g) \rightarrow SO_3(g)$
   b. $Xe(g) + 2\ F_2(g) \rightarrow XeF_4(s)$

15.25 Which of the following species would you expect to be predominantly ionic?

   a. Li_3N c. SO_2
   b. AlN   d. Ca_3N_2

15.6   Describe how you would prepare

  a. Mg_3N_2 from Mg
  b. H_2O_2 from H_2O
  c. Cu_2O from Cu
  d. XeF_4 from Xe

15.7   Write balanced equations to explain what happens when

  a. sodium is heated with nitrogen.
  b. sodium is exposed to oxygen.
  c. potassium superoxide is treated with water.

15.8   Show the electronic structures of

  a. O_2^{2-} c. OH^-
  b. O_2^+ d. OH
                    e. XeF_2

15.9   Using data on heats of formation in Table 4.1, Chapter 4 and free energies of formation, Table 12.2, Chapter 12, show that the temperature at which the reaction $2\,CuO(s) \rightarrow Cu_2O(s) + \frac{1}{2}\,O(g)$ becomes spontaneous is approximately 900°C.

15.10   Consider the data in Table 15.3.

  a. From the table, determine the number of moles of NH_3, H_2, and N_2 in an equilibrium mixture containing one mole of gas at 300°C and 50 atm. Note that $[H_2] = 3\,[N_2]$.
  b. Using the Ideal Gas Law, calculate the volume occupied by one mole of gas under these conditions.
  c. Combining your answers to (a) and (b), calculate the equilibrium concentrations in mole/liter of NH_3, H_2, and N_2.
  d. From your answer to (c), calculate K_c at 300°C and compare to the value given in the table.

15.11   Suggest

  a. factors other than the increase in CO_2 concentration which might be responsible for an increase in worldwide temperatures.
  b. substances other than Dry Ice or silver iodide which might stimulate precipitation in the atmosphere.
  c. how an increase in CO_2 concentration in the air might lead to an increase in the concentration of water vapor.

15.12   The relative humidity on a hot summer day (30°C) is 50%. What will be the relative humidity when this air is brought into a house maintained at 23°C, assuming no condensation occurs? (The saturated vapor pressures of water are 31.8 mm Hg at 30°C and 21.1 mm Hg at 23°C.)

15.26   How would you accomplish the following conversions?

  a. $CO \rightarrow CO_2$ c. $N_2 \rightarrow HNO_3$
  b. $S \rightarrow H_2SO_4$ d. $O_2 \rightarrow O_3$

15.27   Write balanced equations for the reactions that occur when

  a. a high energy photon collides with an NO molecule.
  b. ammonia burns in air.
  c. sodium peroxide is heated.

15.28   Write electronic structures for

  a. N^{3-}
  b. O_2^-
  c. XeF_6

15.29   Using data from Tables 4.1 and 12.2, calculate the maximum temperature at which SO_2 can be removed from stack gases by adding MgO (Reaction 15.40).

15.30   Consider the data in Table 15.7.

  a. Given that the mole % of SO_3 at 700°C is 40 and that $[O_2] = \frac{1}{2}\,[SO_2]$, calculate the mole % of all three gases in the equilibrium mixture at this temperature.
  b. Using the Ideal Gas Law, find the number of moles of gas in a one-liter container at one atm and 700°C.
  c. Combining your results in (a) and (b), find the equilibrum concentrations of all three gases in a one-liter container at one atm and 700°C.
  d. From your answers to (c), find K_c at 700°C and compare to the value given in the table.

15.31   Explain why

  a. variations in the concentration of CO_2 or O_3 in the atmosphere could profoundly affect our environment.
  b. the mole fraction of water in the air is higher in the tropics than it is in the arctic.
  c. the concentration of suspended particles in the air can affect our weather.

15.32   Using a table of water vapor pressures (Appendix 1), calculate the temperature to which warm air at a temperature of 80°F and a relative humidity of 60% must be cooled before condensation can occur.

15.13 Explain what is meant by a "temperature of 1200°C" in the thermosphere.

15.14 Estimate the altitude at which Ar^+ ions will be found in the atmosphere. (Ionization energy $Ar = 363$ kcal/mole.) See Figure 15.4.

15.15 Explain why the concentration of nitrogen atoms in the upper atmosphere is higher than one would predict from the equilibrium constant for the reaction: $N_2(g) \rightarrow 2\ N(g)$.

15.16 Calculate the maximum wavelength of a photon that can bring about the reaction: $N_2(g) \rightarrow 2\ N(g)$; $\Delta H = +225$ kcal.

15.17 Explain why

a. most of the CO but very little of the SO_2 in the atmosphere is produced by automobiles.
b. residual oil contains more sulfur compounds than natural gas.
c. SO_2 and CO are found in polluted air, even though they are thermodynamically unstable with respect to SO_3 and CO_2.

15.18 Describe and explain several different ways of reducing emissions of CO and unburned hydrocarbons from an automobile exhaust.

15.19 Discuss methods of removing sulfur compounds from coal; petroleum; natural gas.

15.20 Taking the concentrations of NO, O_3, and O_2 in polluted air to be $4 \times 10^{-8}, 2 \times 10^{-8}$, and 1.0×10^{-2} mole/liter, respectively, calculate and compare the rates of Reactions 15.32 and 15.33 under these conditions.

15.33 Using Equation 15.27, estimate the temperature outside a jetliner traveling at an altitude of 16,000 feet.

15.34 Explain why the species O, O_2^+, and O^+ are found at successively higher altitudes.

15.35 At an altitude of 120 km the concentration of O_2 is about 1×10^{-10} mole/liter. Calculate the equilibrium concentration of oxygen atoms, taking K_c for the reaction $O_2(g) \rightarrow 2\ O(g)$ to be 10^{-82}, and compare to the observed concentration, about the same as that of O_2.

15.36 For the first step in the formation of photochemical smog: $NO_2(g) \rightarrow NO(g) + O(g)$, $\Delta H = 72.7$ kcal. Can this reaction be brought about by visible light?

15.37 Explain why

a. the percentage of olefins in the hydrocarbons from automobile exhaust is much higher than that in gasoline itself.
b. oxides of nitrogen are necessary for smog formation.
c. methods of reducing the CO content of automobile exhaust usually reduce the concentration of hydrocarbons.

15.38 Explain how it is possible to reduce NO emissions by recirculating some of the exhaust gases through the engine of an automobile.

15.39 Discuss several alternative methods for reducing air pollution by automobiles. What are some of the advantages and disadvantages of using $Pb(C_2H_5)_4$ in gasoline?

15.40 Taking k for the reaction: $2\ CO(g) + 2\ NO(g) \rightarrow 2\ CO_2(g) + N_2(g)$ to be 3×10^{-27} lit²/mole² sec, calculate the rate of this reaction when the concentrations of CO and NO are 4×10^{-6} and 4×10^{-9} mole/liter, respectively. (The reaction is first order in CO and second order in NO.)

*15.41 Given that K_c for Reaction 15.39 is 210, calculate the fraction of hemoglobin which will be in the CO·Hem form in an individual who is breathing polluted air in which the concentrations of CO and O_2 are 4×10^{-6} and 0.0085 mole/liter, respectively.

*15.42 Suppose that a molecule of O_2 absorbs a photon at a wavelength of 2200 Å and subsequently decomposes to 2 oxygen atoms. Taking ΔH for this reaction to be $+119$ kcal, how much "extra energy" in kcal will the oxygen atoms have? If all this energy is absorbed in raising the kinetic energy of O(g) and, hence, its temperature, what will the final "apparent" temperature be? (The heat capacity of O(g) is 3 cal/mole °K.)

*15.43   A certain type of coal contains 2.0% by weight of sulfur.

   a. What weight of sulfur dioxide will be produced by burning 100 grams of this coal?
   b. What volume of SO_2 at 25°C and one atm will be produced?
   c. If 1000 liters of air are used to burn the coal, what will be the concentration of SO_2 in the stack gas in ppm?

*15.44   How many pairs of electrons are there around the xenon atom in XeF_2? Using the electron pair repulsion principle, predict the orientation of these electron pairs. On this basis, suggest possible values for the bond angle in XeF_2. Can you explain why the observed angle is 180°?

16 PRECIPITATION REACTIONS

Most of the reactions considered to this point have been ones that take place in the gas phase or at the solid-gas interface. As important as such reactions are, reactions taking place in aqueous solution are perhaps of even greater significance. In the oceans, on rain-soaked fields and mountains, and in living organisms occur a multitude of chemical reactions that influence our world and our lives in important ways.

In this and succeeding chapters we shall look at the characteristics of certain types of reactions taking place in water solution. Most of these reactions involve dissolved ions as reactants or products. In this chapter we shall consider what is perhaps the simplest type of reaction that can occur between dissolved ions, that of precipitation. When certain pairs of solutions of such substances as nickel chloride, $NiCl_2$, and sodium hydroxide, $NaOH$, are mixed, ions combine to form an insoluble solid. We shall develop principles to predict under what conditions such reactions will occur, what their products will be, and how they can be represented by chemical equations.

16.1 NET IONIC EQUATIONS

If 0.1 M solutions of nickel chloride, $NiCl_2$, and sodium hydroxide, $NaOH$, are mixed, a green, gelatinous precipitate forms. In principle, the green precipitate might be either sodium chloride, $NaCl$, resulting from the interaction of Na^+ ions of the $NaOH$ solution with Cl^- ions of the $NiCl_2$ solution, or nickel hydroxide, $Ni(OH)_2$, formed when Ni^{2+} ions of the $NiCl_2$ solution come in contact with OH^- ions of the $NaOH$ solution. Experience enables us to make a choice between these two possibilities: we know that sodium chloride, ordinary table salt, is neither green nor insoluble. By elimination, we deduce that nickel hydroxide, $Ni(OH)_2$, must be the product of the reaction.

Having deduced the nature of the reaction, it is now possible to write an equation for it. The product is solid nickel hydroxide, $Ni(OH)_2$; the reactants are Ni^{2+} and OH^- in aqueous solution. Consequently the balanced equation is

$$Ni^{2+}(aq) + 2\ OH^-(aq) \rightarrow Ni(OH)_2(s) \qquad (16.1)$$

429

This equation, representing a reaction between ions in water solution, is an example of a **net ionic equation.** Such equations, referred to briefly in Chapter 3, will be encountered throughout the remainder of this text. We shall use them to describe many different types of reactions. In writing net ionic equations, the same conventions are followed as in writing any chemical equation. Only those species which actually take part in the reaction are included in the equation. In Equation 16.1 neither the Na^+ nor the Cl^- ions are included, since they do not take part in the reaction. By the same token, an equation for the combustion of carbon in air does not include the N_2 molecules present in the mixture, since they do not enter into the reaction.

To illustrate further how net ionic equations can be deduced from experimental observations, let us consider what happens when dilute aqueous solutions of aluminum sulfate, $Al_2(SO_4)_3$, and barium bromide, $BaBr_2$, are mixed. A white granular precipitate forms. To decide upon its composition, we note that the ions originally present are

$$Al_2(SO_4)_3 \text{ solution:} \quad Al^{3+}, SO_4^{2-}$$
$$BaBr_2 \text{ solution:} \quad Ba^{2+}, Br^-$$

Logically, we could obtain either barium sulfate, $BaSO_4$, or aluminum bromide, $AlBr_3$, as a precipitate. To choose between these possibilities, we could obtain bottles of the two solids from the storeroom and test their solubilities. Alternatively, the identity of the precipitate could be decided by testing other pairs of solutions. Suppose, for example, solutions of $Al_2(SO_4)_3$ and NaBr are mixed. If this is done, it is found that no precipitate forms, indicating that aluminum bromide is soluble in water. On the other hand, mixing solutions of $Ba(NO_3)_2$ and $Al_2(SO_4)_3$ gives a precipitate. Indeed, whenever two solutions, one containing Ba^{2+} ions and the other containing SO_4^{2-} ions, are mixed, a white precipitate is formed. We deduce that the process occurring when solutions of $Al_2(SO_4)_3$ and $BaBr_2$ are mixed must be the reaction of Ba^{2+} ions with SO_4^{2-} ions to form solid barium sulfate, $BaSO_4$.

$$Ba^{2+}(aq) + SO_4^{2-}(aq) \rightarrow BaSO_4(s) \qquad (16.2)$$

Precipitation reactions usually occur essentially instantaneously: $E_a \cong O$

Sometimes, two precipitation reactions occur when two solutions are mixed. Suppose, for example, a solution of barium hydroxide, $Ba(OH)_2$, is added to a solution of nickel sulfate, $NiSO_4$. It has been pointed out that both of the possible products, $Ni(OH)_2$ and $BaSO_4$, are insoluble in water. Hence, we would predict that both of these compounds should precipitate. Experiment confirms this deduction; if we look at the precipitate under a microscope, it is possible to distinguish white crystals of $BaSO_4$ from green particles of $Ni(OH)_2$. In representing this double precipitation reaction, we write two net ionic equations, since two entirely different reactions are taking place:

$$Ba^{2+}(aq) + SO_4^{2-}(aq) \rightarrow BaSO_4(s)$$

and
$$Ni^{2+}(aq) + 2 OH^-(aq) \rightarrow Ni(OH)_2(s)$$

Net ionic equations such as those written above can be given a quantitative meaning by following the principles outlined in Chapter 3. Example 16.1 illustrates how this is done.

Example 16.1 When 300 ml of 0.10 M Na_2SO_4 is added to 200 ml of 0.20 M $BaCl_2$, a white precipitate forms.

    a. Write a net ionic equation for the reaction.

    b. Calculate the number of moles of precipitate formed.

    c. Determine the number of moles of each ion remaining in solution after precipitation.

Solution

    a. The ions present are Na^+, SO_4^{2-}, Ba^{2+}, and Cl^-. In principle there are two possible precipitates: NaCl and $BaSO_4$. Knowing that NaCl is soluble and $BaSO_4$ insoluble, we deduce that the equation must be

$$Ba^{2+}(aq) + SO_4^{2-}(aq) \rightarrow BaSO_4(s)$$

    b. It will be helpful to start by calculating the number of moles of each ion present in solution before precipitation takes place. Noting that we have 300 ml (0.300 liter) of solution containing 0.10 mole of Na_2SO_4 per liter and that 1 mole $Na_2SO_4 \simeq 2$ moles Na^+, we calculate the number of moles of Na^+ to be

$$0.300 \text{ liter} \times \frac{0.10 \text{ mole } Na_2SO_4}{\text{liter}} \times \frac{2 \text{ moles } Na^+}{1 \text{ mole } Na_2SO_4} = 0.060 \text{ mole } Na^+$$

Applying the same reasoning, we obtain

$$0.300 \text{ liter} \times \frac{0.10 \text{ mole } Na_2SO_4}{\text{liter}} \times \frac{1 \text{ mole } SO_4^{2-}}{1 \text{ mole } Na_2SO_4} = 0.030 \text{ mole } SO_4^{2-}$$

$$0.200 \text{ liter} \times \frac{0.20 \text{ mole } BaCl_2}{\text{liter}} \times \frac{1 \text{ mole } Ba^{2+}}{1 \text{ mole } BaCl_2} = 0.040 \text{ mole } Ba^{2+}$$

$$0.200 \text{ liter} \times \frac{0.20 \text{ mole } BaCl_2}{\text{liter}} \times \frac{2 \text{ moles } Cl^-}{1 \text{ mole } BaCl_2} = 0.080 \text{ mole } Cl^-$$

In systems such as this the reaction can be assumed to go to completion

We note from Equation 16.2 that *one* mole of Ba^{2+} reacts with *one* mole of SO_4^{2-} to form *one* mole of $BaSO_4$. The limiting factor is the number of moles of SO_4^{2-} available. Assuming that all the SO_4^{2-} reacts:

$$0.030 \text{ mole } SO_4^{2-} + 0.030 \text{ mole } Ba^{2+} \rightarrow 0.030 \text{ mole } BaSO_4$$

    c. Subtracting the number of moles of ions reacting from that present originally, we obtain the number of moles remaining

$$Na^+: 0.060 \text{ mole} - 0.000 \text{ mole} = 0.060 \text{ mole}$$
$$SO_4^{2-}: 0.030 \text{ mole} - 0.030 \text{ mole} = 0.000 \text{ mole}$$
$$Ba^{2+}: 0.040 \text{ mole} - 0.030 \text{ mole} = 0.010 \text{ mole}$$
$$Cl^-: 0.080 \text{ mole} - 0.000 \text{ mole} = 0.080 \text{ mole}$$

Notice that all the Na^+ and Cl^- ions remain in solution. Part of the Ba^{2+} and virtually all the SO_4^{2-} react. Actually, as we shall see in Section 16.3, equilibrium considerations require that a trace of SO_4^{2-} be left in solution, but the number of moles of SO_4^{2-} is much less than 0.001

16.2 SOLUBILITIES OF IONIC COMPOUNDS

Solubility data obtained from an experimental study of a limited number of precipitation reactions can be used to predict results of a great many other reactions. For example, having established the fact that nickel hydroxide is insoluble in water, one can predict that mixing solutions of $Ni(NO_3)_2$ and $NaOH$, $NiSO_4$ and KOH, or $NiBr_2$ and $Ca(OH)_2$ will result in the formation of a precipitate of $Ni(OH)_2$, according to Equation 16.1. Again, if it is established through experiment that sodium chloride and potassium nitrate are both soluble in water, it follows that no precipitation reaction will occur when solutions of $NaNO_3$ and KCl are mixed. It might appear that all we need to predict the results of possible precipitation reactions is a table of solubilities in which every ionic solid is neatly classified as soluble or insoluble. Unfortunately there are difficulties associated with any attempt to set up such a simple classification scheme.

Ionic solids do not fall neatly into the categories "soluble" and "insoluble" with a sharp dividing line between them. Instead, they cover an enormous range of solubility. One of the most soluble salts known is lithium chlorate, $LiClO_3$, which dissolves to the extent of 35 moles/lit at room temperature. We can safely predict that this compound will not be the product of a precipitation reaction. At the other extreme is mercuric sulfide, HgS; one can calculate that a saturated solution should contain only about 10^{-26} mole/lit of Hg^{2+} and S^{2-} ions. Quite clearly, we can expect to get a precipitate of mercuric sulfide whenever water solutions containing Hg^{2+} and S^{2-} ions are mixed, even if the solutions are extremely dilute. Lead chloride, $PbCl_2$, is an example of a compound of intermediate solubility. When equal volumes of 0.1 M $Pb(NO_3)_2$ and 0.1 M $NaCl$ are mixed, the solubility of $PbCl_2$ is exceeded and it precipitates. If, on the other hand, the solutions mixed are somewhat more dilute, say 0.04 M, no precipitate is formed. Compounds such as lead chloride are difficult to classify as soluble or insoluble; it is perhaps begging the question to classify them as slightly soluble.

Bearing these qualifications in mind, we can classify the more common ionic solids on the basis of their solubility behavior according to the rules outlined in Table 16.1. The use of solubility rules to predict the results of precipitation reactions is illustrated by Example 16.2.

TABLE 16.1   Solubility Rules

NO_3^-	All nitrates are soluble.
Cl^-	All chlorides are soluble except $AgCl$, Hg_2Cl_2 and $PbCl_2$.*
SO_4^{2-}	All sulfates are soluble except $CaSO_4$,* $SrSO_4$, $BaSO_4$, Hg_2SO_4, $HgSO_4$, $PbSO_4$, and Ag_2SO_4.
CO_3^{2-}	All carbonates are insoluble except those of the 1A elements and NH_4^+.
OH^-	All hydroxides are insoluble except those of the 1A elements, $Sr(OH)_2$ and $Ba(OH)_2$. ($Ca(OH)_2$ is slightly soluble.)
S^{2-}	All sulfides except those of the 1A and 2A elements and NH_4^+ are insoluble.

It would be helpful to you to memorize these rules

*Insoluble compounds are those which precipitate upon mixing equal volumes of solutions 0.1 M in the corresponding ions. Compounds which fail to precipitate at concentrations slightly below 0.1 M are starred.

Example 16.2 Write balanced equations for the reactions, if any, that occur when equal volumes of 0.1 M solutions of the following compounds are mixed:

   a. $AgNO_3$ and Na_2CO_3 b. Na_2SO_4 and KOH   c. $CuSO_4$ and $Ba(OH)_2$

Solution. In each case we first write down the formulas of the two possible precipitates and then decide, on the basis of the solubility rules, whether one or both of these compounds will precipitate.

   a. Possible precipitates: Ag_2CO_3, $NaNO_3$. Table 16.1 indicates that Ag_2CO_3 is insoluble (Ag is not a 1A element!), whereas $NaNO_3$ is soluble (all nitrates are soluble). Consequently, we have

$$2\ Ag^+(aq) + CO_3^{2-}(aq) \rightarrow Ag_2CO_3(s)$$

   b. Possible precipitates: NaOH, K_2SO_4. From the solubility rules, it is clear that both of these compounds are soluble. No precipitation reaction occurs.

   c. Possible precipitates: $Cu(OH)_2$, $BaSO_4$. Of these compounds, $BaSO_4$ is specifically listed as insoluble; it can be deduced that $Cu(OH)_2$ is also insoluble. Two precipitation reactions occur simultaneously:

$$Ba^{2+}(aq) + SO_4^{2-}(aq) \rightarrow BaSO_4(s)$$

and

$$Cu^{2+}(aq) + 2\ OH^-(aq) \rightarrow Cu(OH)_2(s)$$

Unfortunately there is no simple theory which enables us to explain the solubility rules listed in Table 16.1. The difficulties involved in predicting relative solubilities of ionic compounds are inherent in the nature of the solution process, in which solvent water molecules are intimately involved. Normally when a solid dissolves in a liquid, we expect both the enthalpy change, ΔH, and the entropy change, ΔS, to be positive, as pointed out in Chapters 10 and 12. When an ionic solute dissolves in water, this may not be the case. The formation of hydrated ions (Fig. 16.1) evolves

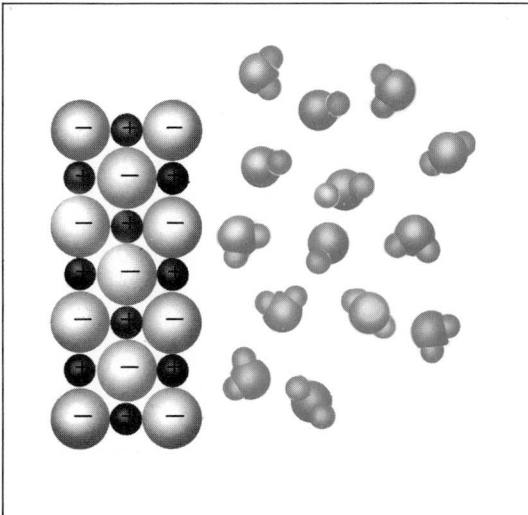
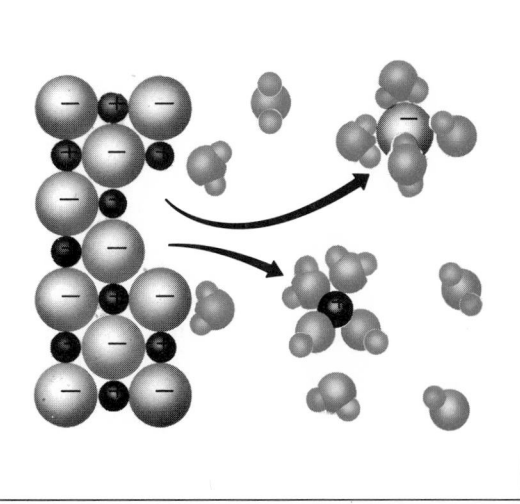

Figure 16.1 Dissolving of ionic solute in water. The solution process involves an exchange of neighbors; polar water molecules take the place of ions of opposite charge.

heat, which tends to make ΔH negative, and orients water molecules in fixed positions around ions, which may lead to a decrease in entropy. These complications, among others, make it difficult to predict from first principles the relative solubilities of different ionic compounds in water. The fact that we have to depend upon empirical rules to predict solubilities is another illustration of our basic lack of knowledge concerning the structure of electrolyte solutions and, indeed, of water itself.

In most cases we find that the solubilities of ionic compounds are inversely related to the charge densities (charge to size ratio) of their ions. Compounds containing ions of low charge density (e.g., large ions of +1 or −1 charge) tend to be more soluble than those containing highly charged small ions. We find, for example, that virtually all the compounds of K^+ (charge = +1, r = 1.33 Å) are soluble in water, while many Ca^{2+} (charge = +2, r = 0.99 Å) salts such as CaF_2, $CaCO_3$, and $CaSO_4$ are relatively insoluble. Again virtually all the salts containing NO_3^-, a large ion of low charge, are soluble. In contrast, many compounds containing anions of higher charge (e.g., CO_3^{2-}) or smaller size (e.g., OH^-) are insoluble in water.

Salts in which the cation and anion differ greatly in size often have abnormally high solubilities. An example is $MgSO_4$, which dissolves in water to the extent of about two moles per liter at 20°C, despite the high charges of its ions (+2, −2). The magnesium ion (radius = 0.65 Å) is so much smaller than the sulfate ion (effective radius ≈ 1.8 Å) that there is anion-anion contact in a crystal of anhydrous $MgSO_4$. The repulsion between adjacent ions of like charge weakens the crystal lattice of magnesium sulfate and makes it more soluble than one would expect on the basis of charge density considerations. In barium sulfate, where the ions are more nearly equal in size (radius Ba^{2+} = 1.35 Å), this effect disappears and the solubility drops to a very low value, about 3×10^{-5} mole/liter.

16.3 SOLUBILITY EQUILIBRIA

QUALITATIVE ASPECTS: COMMON ION EFFECT

When silver acetate is dissolved in pure water to form a saturated solution, an equilibrium is established between the solid and its ions in solution.

$$AgC_2H_3O_2(s) \rightleftharpoons Ag^+(aq) + C_2H_3O_2^-(aq) \qquad (16.3)$$

The concentrations of Ag^+ and $C_2H_3O_2^-$ ions under these conditions are equal to each other and have a fixed value, independent of the amount of silver acetate or water used in preparing the saturated solution. At 20°C the equilibrium concentrations of Ag^+ and $C_2H_3O_2^-$ in a solution prepared by dissolving silver acetate in pure water are about 0.045 mole/lit.

The solubility in water of $AgC_2H_3O_2$ is 0.045 mole/lit

It is possible to change the relative concentrations of Ag^+ and $C_2H_3O_2^-$ ions in equilibrium with solid silver acetate in various ways. In particular, we might accomplish this by adding to the solution:

1. *An electrolyte containing a common ion, Ag^+ or $C_2H_3O_2^-$.* Suppose, for example, that we add a concentrated solution of a more soluble silver

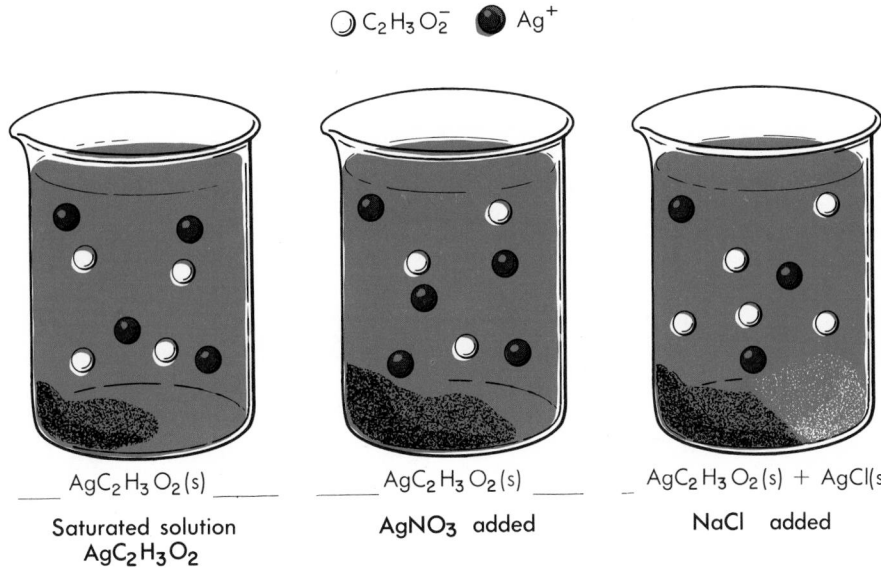

Figure 16.2 Effect of added ions on the equilibrium: $AgC_2H_3O_2(s) \rightleftharpoons Ag^+(aq) + C_2H_3O_2^-(aq)$. Addition of solution of $AgNO_3$ precipitates $AgC_2H_3O_2$, lowering concentration of $C_2H_3O_2^-$. Addition of solution of NaCl precipitates AgCl, lowers concentration of Ag^+.

salt, such as silver nitrate. Le Chatelier's Principle tells us that the system will respond in such a way as to counteract the increase in the concentration of Ag^+; some of the added Ag^+ ions combine with the $C_2H_3O_2^-$ ions in solution to precipitate silver acetate. This behavior is referred to as the **common ion** effect; addition of another silver salt (which has the Ag^+ ion in common with $AgC_2H_3O_2$) to a saturated solution of silver acetate, disturbs the solubility equilibrium so as to bring about precipitation. When equilibrium is re-established, the concentration of $C_2H_3O_2^-$ ion is reduced from its original value while that of Ag^+ ion is increased (Fig. 16.2).

2. *An electrolyte which forms a precipitate with one of the ions.* If one adds sodium chloride to a saturated solution of silver acetate, the very insoluble salt, silver chloride, precipitates. As Ag^+ ions are removed from solution as AgCl, more silver acetate dissolves to restore equilibrium between Ag^+ and $C_2H_3O_2^-$ ions. The net effect is to increase the concentration of $C_2H_3O_2^-$ ions in solution and decrease the concentration of Ag^+ ions (Fig. 16.2).

QUANTITATIVE TREATMENT: SOLUBILITY PRODUCT

As the above discussion implies, there is an inverse relationship between the equilibrium concentrations of ions in contact with the electrolyte from which they are derived. Any process which shifts the equilibrium so as to increase the concentration of one of these ions decreases the concentration of the other. The quantitative relationship between these concentrations can be introduced by considering a slightly soluble 1:1 electrolyte, MX, in equilibrium with its ions in aqueous solution

$$MX(s) \rightleftarrows M^+(aq) + X^-(aq) \qquad (16.4)$$

For this reaction the equilibrium constant expression takes the form

$$K_{sp} = [M^+] \times [X^-] \qquad (16.5)$$

where the square brackets, as always, are used to denote equilibrium concentrations, in moles/liter, of the ions M^+ and X^- in the solution. The concentration of solid MX does not appear in the equilibrium expression. (Recall the discussion of gas-solid equilibrium in Section 13.2, Chapter 13.)

Some solid, however, must be present if equilibrium is to exist

The quantity K_{sp} appearing in Equation 16.5 is a particular type of equilibrium constant, referred to as the **solubility product.** For any given electrolyte K_{sp} has a fixed value, independent of the concentrations of the individual ions in equilibrium with the solid.

The concept of the solubility product is readily extended to electrolytes of any valence type. For an electrolyte MX_2,

$$MX_2(s) \rightleftarrows M^{2+}(aq) + 2 X^-(aq) \qquad (16.6)$$

we can write

$$K_{sp} = [M^{2+}] \times [X^-]^2$$

Thus, we have, for example,

$$K_{sp} \text{ of } PbCl_2 = [Pb^{2+}] \times [Cl^-]^2 = 1.7 \times 10^{-5}$$

Again, for a compound such as arsenic sulfide, As_2S_3,

$$As_2S_3(s) \rightleftarrows 2 As^{3+}(aq) + 3 S^{2-}(aq) \qquad (16.7)$$

$$K_{sp} = [As^{3+}]^2 \times [S^{2-}]^3 = 5 \times 10^{-27}$$

In general, the solubility product principle may be stated as follows: **In any water solution in equilibrium with a slightly soluble ionic compound, the product of the ion concentrations, each raised to a power equal to the coefficient of the ion in the solubility equation, has a constant value.**

In principle, at least, the numerical value of the solubility product can be calculated from the measured solubility, or vice-versa (Example 16.3). In practice, values of K_{sp}, particularly for electrolytes of very low solubility, are ordinarily calculated by less direct methods, one of which is mentioned in Chapter 21 in connection with voltage measurements on electrical cells.

Example 16.3

a. The measured solubility of $AgC_3H_3O_2$ at 20°C is 0.045 mole/liter. Calculate K_{sp} for this salt.

b. The solubility product of the mineral fluorite, CaF_2, is known to be 2×10^{-10}. Estimate the solubility, in mole/liter, of this mineral.

One must be careful to distinguish between solubility and solubility product

Solution

a. From Equation 16.3, it is evident that for every mole of $AgC_2H_3O_2$ that dissolves, a mole of Ag^+ and a mole of $C_2H_3O_2^-$ enter the solution. Hence, when 0.045 mole of solid dissolves per liter of solution, the equilibrium concentrations of Ag^+ and $C_2H_3O_2^-$ must both be 0.045 M.

$$K_{sp} = [Ag^+] \times [C_2H_3O_2^-] = (0.045) \times (0.045) = 2.0 \times 10^{-3}$$

b. The solubility product expression is

$$[Ca^{2+}] \times [F^-]^2 = 2 \times 10^{-10}$$

From the equation: $CaF_2(s) \rightarrow Ca^{2+}(aq) + 2\ F^-(aq)$

we see that for every mole of CaF_2 that dissolves, one mole of Ca^{2+} and two moles of F^- enter the solution. Consequently, if we let

$$S = \text{solubility } CaF_2 \text{ (mole/liter)}$$

then: $\qquad [Ca^{2+}] = S, \text{ and } [F^-] = 2\ S$

Substituting in the expression for K_{sp}

$$(S)(2\ S)^2 = 2 \times 10^{-10}; \quad 4\ S^3 = 2 \times 10^{-10}; \quad S^3 = 0.5 \times 10^{-10} = 50 \times 10^{-12}$$

Solving, $\qquad S = (50)^{1/3} \times 10^{-4} = 3.7 \times 10^{-4} \text{ mole/liter.}$

We shall use solubility products to determine: (1) whether or not a precipitate will form when two solutions are mixed (Example 16.4); (2) the extent to which a precipitation reaction occurs, e.g., the fraction of an ion left in solution after the reaction is over (Example 16.5).

Example 16.4 Hard water containing Ca^{2+} ions frequently precipitates calcium sulfate, $CaSO_4$. When the concentration of Ca^{2+} is 0.01 M, will a precipitate of $CaSO_4$ form if

    a. enough SO_4^{2-} is added to make its concentration 0.001 M?

    b. 400 ml of 0.02 M Na_2SO_4 is added to 800 ml of the hard water?

Solution

    a. To decide whether or not a precipitate forms, we must compare the actual concentration product to that required for equilibrium ($K_{sp} = 3 \times 10^{-5}$)

$$\text{conc. } Ca^{2+} \times \text{conc. } SO_4^{2-} = (0.01) \times (0.001) = 1 \times 10^{-5}$$

Since the concentration product is smaller than K_{sp}, we conclude that there are not enough ions in the solution to establish equilibrium. In other words, no $CaSO_4$ forms.

If the conc. product is less than K_{sp}, we do not get a precipitate

    b. This problem is analogous to (a), except for one important factor. We must calculate the concentrations of Ca^{2+} and SO_4^{2-} *after* the two solutions are mixed, since the volume available to the ions has increased. To obtain the concentration of Ca^{2+}, we note that the 800 ml of solution 0.01 M in Ca^{2+} has, in effect, been diluted to 1200 ml by the addition of the Na_2SO_4 solution. Consequently the concentration of Ca^{2+} has decreased by a factor of 800/1200:

$$\text{conc. } Ca^{2+} = 0.01 \times \frac{800}{1200} = 0.007 \text{ M}$$

Similarly: $\qquad \text{conc. } SO_4^{2-} = 0.02 \times \frac{400}{1200} = 0.007 \text{ M}$

Consequently: $(\text{conc. } Ca^{2+}) \times (\text{conc. } SO_4^{2-}) = (7 \times 10^{-3})(7 \times 10^{-3})$
$$= 49 \times 10^{-6} = 5 \times 10^{-5}$$

If the conc. product exceeds K_{sp}, a precipitate will form and equilibrium will be established

Clearly the concentration product, 5×10^{-5}, is greater than K_{sp}, 3×10^{-5}. Solid $CaSO_4$ will precipitate until the concentrations of Ca^{2+} and SO_4^{2-} drop to the point at which their product becomes equal to 3×10^{-5}.

TABLE   16.2   Solubility Products

Acetate	$AgC_2H_3O_2$	2×10^{-3}	Iodides	AgI	1×10^{-16}
				PbI_2	1×10^{-8}
Bromides	$AgBr$	1×10^{-13}			
	$PbBr_2$	5×10^{-6}	Sulfates	$BaSO_4$	1.5×10^{-9}
				$CaSO_4$	3×10^{-5}
Carbonates	$BaCO_3$	1×10^{-9}		$PbSO_4$	1×10^{-8}
	$CaCO_3$	5×10^{-9}			
	$MgCO_3$	2×10^{-8}	Sulfides	Ag_2S	1×10^{-49}
	$PbCO_3$	1×10^{-13}		CdS	1×10^{-26}
				CoS	1×10^{-21}
Chlorides	$AgCl$	1.6×10^{-10}		CuS	1×10^{-25}
	Hg_2Cl_2	1×10^{-18}		FeS	2×10^{-17}
	$PbCl_2$	1.7×10^{-5}		HgS	1×10^{-52}
				MnS	1×10^{-13}
Chromates	Ag_2CrO_4	1×10^{-12}		NiS	1×10^{-22}
	$BaCrO_4$	2×10^{-10}		PbS	1×10^{-27}
	$PbCrO_4$	2×10^{-14}		ZnS	1×10^{-23}
Fluorides	BaF_2	2×10^{-6}			
	CaF_2	2×10^{-10}			
	PbF_2	4×10^{-8}			
Hydroxides	$Al(OH)_3$	5×10^{-33}			
	$Cr(OH)_3$	1×10^{-30}			
	$Fe(OH)_2$	1×10^{-15}			
	$Fe(OH)_3$	5×10^{-38}			
	$Mg(OH)_2$	1×10^{-11}			
	$Zn(OH)_2$	5×10^{-17}			

Example 16.5   The concentration of Mg^{2+} in sea water is 5×10^{-2} M. It is extracted from sea water by adding OH^- to precipitate $Mg(OH)_2$ ($K_{sp} = 1 \times 10^{-11}$). Enough OH^- is added to make its final concentration 1×10^{-3} M.

    a. What is the equilibrium concentration of Mg^{2+} at this point?

    b. What percentage of the Mg^{2+} originally present remains in the sea water?

Solution

    a. From the equation:
$$Mg(OH)_2(s) \rightleftarrows Mg^{2+}(aq) + 2\ OH^-(aq)$$
$$K_{sp} = [Mg^{2+}] \times [OH^-]^2 = 1 \times 10^{-11}$$

Since $[OH^-] = 1 \times 10^{-3}$ M, we have:

$$[Mg^{2+}] \times (1 \times 10^{-3})^2 = 1 \times 10^{-11}$$

$$[Mg^{2+}] = \frac{1 \times 10^{-11}}{1 \times 10^{-6}} = 1 \times 10^{-5}$$

    b. Originally, the concentration of Mg^{2+} was 5×10^{-2} M; it is now 1×10^{-5} M

$$\%Mg^{2+} \text{ remaining} = \frac{1 \times 10^{-5}}{5 \times 10^{-2}} \times 100 = 0.02\%$$

In other words 99.98% of the Mg^{2+} has been precipitated out of the sea water as $Mg(OH)_2$.

LIMITATIONS ON K_{sp}: ACTIVITY COEFFICIENTS OF IONS

The solubility product principle, in the form we have stated it, is strictly applicable only in very dilute aqueous solutions. We find, for example, if we study the solubility of AgCl in potassium nitrate solutions of different concentrations (Table 16.3) that the product

$$[Ag^+] \times [Cl^-]$$

which has the limiting value 1.6×10^{-10} in pure water, is about 50% greater in 0.04 M KNO_3.

The failure of the ion concentration product to behave as a true constant can be attributed to the interionic attractions discussed in Chapter 11. The effect of these forces can be taken into account by writing the expression for K_{sp} of AgCl in the form

$$\gamma_{Ag^+} \times \gamma_{Cl^-} \times [Ag^+] \times [Cl^-] = 1.6 \times 10^{-10} \qquad (16.8)$$

The quantities γ_{Ag^+} and γ_{Cl^-} are known as **activity coefficients.** In very dilute solution, where the forces between ions play a minor role, the activity coefficients approach unity and we obtain the simple expression

$$[Ag^+] \times [Cl^-] = 1.6 \times 10^{-10}$$

As the total concentration of ions is increased by adding KNO_3, interionic forces become more significant and the activity coefficients deviate from unity. In moderately dilute solutions, γ_{Ag^+} and γ_{Cl^-} are less than one. We see from Equation 16.8 that this requires that the product $[Ag^+] \times [Cl^-]$ become greater than 1.6×10^{-10} (Table 16.3).

Unless otherwise specified, you may assume in working problems that all activity coefficients equal unity

We can use the Debye-Hückel theory, discussed in Chapter 11, to predict the way in which the activity coefficients of Ag^+ and Cl^- ions will vary with the concentration of KNO_3 solution. For this particular system, the relation is

$$\log (\gamma_{Ag^+} \times \gamma_{Cl^-}) = -1.0 \ M^{1/2}$$

where M is the molar concentration of KNO_3. Similar, but more complex, relationships are predicted by the Debye-Hückel theory for systems involving electrolytes of higher valence types (e.g., 2:1, 2:2; see Problem 16.39).

It is interesting to apply this equation to 0.010 M KNO_3 solution. Here, we have

$$\log (\gamma_{Ag^+} \times \gamma_{Cl^-}) = -1.0 \ (0.010)^{1/2} = -0.10 = 0.90 - 1$$
$$\gamma_{Ag^+} \times \gamma_{Cl^-} = 8.0 \times 10^{-1} = 0.80$$

TABLE 16.3   Solubility of AgCl in KNO_3 Solutions

CONCENTRATION KNO_3	SOLUBILITY AgCl	$[Ag^+] \times [Cl^-]$
0	1.28×10^{-5}	1.64×10^{-10}
0.0010	1.33×10^{-5}	1.77×10^{-10}
0.0050	1.38×10^{-5}	1.90×10^{-10}
0.010	1.43×10^{-5}	2.04×10^{-10}
0.020	1.49×10^{-5}	2.22×10^{-10}
0.040	1.55×10^{-5}	2.40×10^{-10}

Substituting in Equation 16.8,

$$(0.80)\,[Ag^+] \times [Cl^-] = 1.6 \times 10^{-10}$$
$$[Ag^+] \times [Cl^-] = 2.0 \times 10^{-10}$$

which is remarkably close to the observed value (Table 16.3). At higher concentrations the deviation between theory and experiment increases. We find, as we did in discussing colligative properties in Chapter 11, that the simple Debye-Hückel theory gives accurate results only in quite dilute solution.

16.4 PRECIPITATION REACTIONS IN ANALYTICAL CHEMISTRY

QUANTITATIVE ANALYSIS

Precipitation reactions are a useful tool to the analyst who wants to know the percentage of a particular element or compound in a mixture. In the branch of analytical chemistry known as **gravimetric analysis,** a weighed sample is first brought into solution by adding a suitable solvent. A reagent is then added which precipitates only the species being analyzed for. To complete the analysis, the precipitate is separated and weighed. The calculations involved in this type of analysis are illustrated in Example 16.6.

Example 16.6 A sample of insecticide weighing 1.200 g is analyzed for arsenic by dissolving it in acid and adding hydrogen sulfide in excess. When this is done, 0.260 g of arsenic sulfide, As_2S_3, is obtained. Calculate the percentage of arsenic in the insecticide.

Solution. From the formula As_2S_3, we see that

$$2 \text{ moles As} \simeq 1 \text{ mole As}_2S_3$$
$$149.8 \text{ g As} \simeq 245.8 \text{ g As}_2S_3$$

The weight of arsenic in 0.260 g of As_2S_3 must then be

$$0.260 \text{ g As}_2S_3 \times \frac{149.8 \text{ g As}}{245.8 \text{ g As}_2S_3} = 0.158 \text{ g As}$$

Knowing the weight of As and the total weight of the sample, we can readily calculate the per cent of As in the sample.

$$\%\text{As} = \frac{\text{weight As}}{\text{weight sample}} \times 100 = \frac{0.158 \text{ g}}{1.200 \text{ g}} \times 100 = 13.2\% \text{ As}$$

Another way to determine the percentage of a substance in a mixture is to measure the volume of a reagent of known concentration required to react exactly with it. Analyses based on volume measurements of this type fall in the general area of **volumetric analysis.** They may involve precipitation (Example 16.7) or other types of reactions (e.g., acid-base, Chapter 18; complex ion formation, Chapter 19; or oxidation-reduction, Chapter 21).

Example 16.7 The percentage of Cl^- ion in a household bleach is found by measuring the volume of 0.184 M $AgNO_3$ required to react with it. It is found that 15.0 ml of $AgNO_3$ is required to react with a sample weighing 0.800 g. Calculate the per cent of Cl^- in the bleach.

Solution. Knowing the volume and concentration of the $AgNO_3$ solution, we can readily calculate the number of moles of $AgNO_3$ added

$$\text{no. moles } AgNO_3 = 0.0150 \text{ liter} \times 0.184 \frac{\text{mole}}{\text{lit}} = 0.00276 \text{ mole}$$

But this must be the number of moles of Ag^+ (1 mole $Ag^+ \simeq$ 1 mole $AgNO_3$) and also the number of moles of Cl^- (1 mole Ag^+ reacts with 1 mole of Cl^-)

$$\text{no. moles } Cl^- = 0.00276 \text{ mole}$$

Having calculated the number of moles of Cl^-, we next obtain the number of grams of Cl^- and finally its percentage

$$\text{no. g } Cl^- = 0.00276 \text{ mole } Cl^- \times \frac{35.45 \text{ g } Cl^-}{1 \text{ mole } Cl^-} = 0.0978 \text{ g } Cl^-$$

$$\% Cl^- = \frac{\text{g } Cl^-}{\text{g sample}} \times 100 = \frac{0.0978 \text{ g}}{0.800 \text{ g}} \times \ 00 = 12.2\% \ Cl^-$$

In this method for the determination of chloride, as in all volumetric analyses, it is essential to know the exact point at which reaction is complete. In principle, we could do this by noting the point at which the pre-

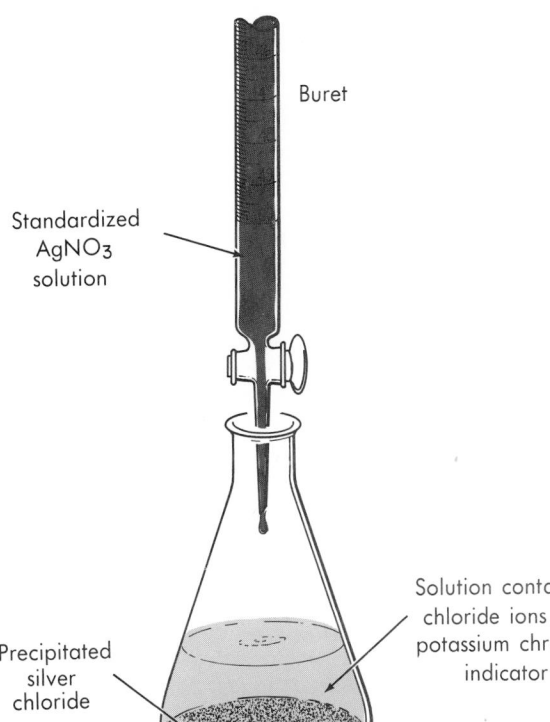

Buret

Standardized
$AgNO_3$
solution

Figure 16.3 Chloride titration (Mohr).

In this titration Ag_2CrO_4 (s, red) forms at the endpoint, when essentially all the Cl^- ion present has been precipitated as AgCl (s, white)

Solution containing
chloride ions and
potassium chromate
indicator

Precipitated
silver
chloride

cipitate of silver chloride stops forming. In practice, it is more convenient to add a substance known as an *indicator*, which changes color when the equivalence point is reached, that is, when chemically equivalent quantities of precipitating agent and sample are present. A few drops of potassium chromate solution added to a chloride sample serves as a suitable indicator for titration with silver ions. At the equivalence point, when essentially all the chloride ion has been removed as silver chloride, a dark red precipitate of silver chromate, Ag_2CrO_4, forms. Since this compound requires a somewhat higher concentration of Ag^+ to precipitate than does silver chloride, it does not form as long as an appreciable amount of chloride ion remains in solution.

The experimental setup for the quantitative determination of Cl^- ion by this method, referred to as a Mohr titration, is shown in Figure 16.3.

QUALITATIVE ANALYSIS

The qualitative detection of ions in a mixture is commonly accomplished by a scheme of analysis in which precipitation reactions play a major role. The general procedure in such a scheme is illustrated in Example 16.8.

Example 16.8 Develop a scheme based on the information given in Table 16.1 to separate and identify the ions Ag^+, Cu^{2+}, and Ca^{2+} in a water solution.

Solution. We look first for a reagent to precipitate one ion, leaving the others in solution. Clearly, Cl^- qualifies, since it precipitates Ag^+ but not Cu^{2+} or Ca^{2+}. The first step in the scheme might then be the addition of Cl^- in the form of a solution of hydrochloric acid. Formation of a precipitate would indicate Ag^+; if no precipitate forms, Ag^+ must be absent.

We next choose a reagent to distinguish between Cu^{2+} and Ca^{2+}. We see from Table 16.1 that S^{2-} is a suitable choice. Addition of a solution of sodium sulfide would precipitate Cu^{2+} if it is present, leaving Ca^{2+} in solution.

At this point, we need only test for Ca^{2+}. This could be done by adding a solution of Na_2CO_3, which will precipitate Ca^{2+} as $CaCO_3$. Failure to obtain a precipitate at this point would show Ca^{2+} to be absent. In summary

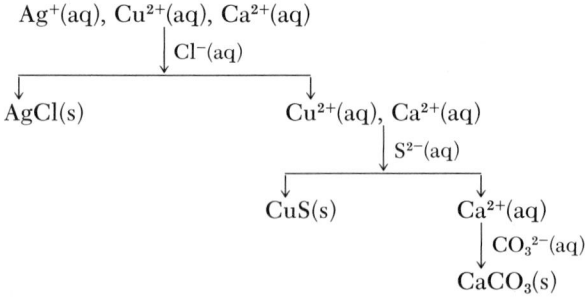

The scheme developed in Example 16.8 illustrates two important points concerning qualitative analysis.

1. The order in which reagents are added is crucial. It would have been useless to have added CO_3^{2-} to the original solution since it gives a precipitate with all three cations.

2. All reactions must be essentially quantitative. Any Ag^+ left after the first step would interfere with the tests for Cu^{2+} and Ca^{2+}, since both Ag_2S and Ag_2CO_3 are insoluble.

Several different analytical schemes have been devised to separate and identify the 25 or more ions commonly included in a laboratory course in qualitative analysis. Needless to say, these schemes are a great deal more complex than that outlined in Example 16.8. Many other types of reactions in addition to precipitation are used. We shall consider later how acid-base reactions (Chapter 18), complex-ion formation reactions (Chapter 19), and oxidation-reduction reactions (Chapter 21) are utilized in qualitative analysis. At this stage, it may be useful to indicate the scope of qualitative analysis by giving a brief outline of one of the simpler schemes of cation analysis.

The cations are first separated into five major groups on the basis of the solubilities of their compounds.

GROUP 1 (Ag^+, Pb^{2+}, Hg_2^{2+}). Separated as insoluble chlorides by addition of dilute HCl. All other cations remain in solution, since their chlorides are soluble.

GROUP 2 (Cu^{2+}, Bi^{3+}, Cd^{2+}, Hg^{2+}, As^{3+}, Sb^{3+}, Sn^{4+}). All these ions form extremely insoluble sulfides (e.g., $K_{sp}CuS = 1 \times 10^{-25}$). The sulfides are precipitated by adding H_2S in acidic solution, in which the concentration of S^{2-} is very small.

GROUP 3 (Al^{3+}, Cr^{3+}, Fe^{3+}, Zn^{2+}, Ni^{2+}, Co^{2+}, Mn^{2+}). These sulfides are somewhat more soluble than those of Group 2 (e.g., $K_{sp}MnS = 1 \times 10^{-13}$). The ions are precipitated by adding H_2S in basic solution, where the above equilibrium is shifted to the right to give a relatively high concentration of S^{2-}. Under these conditions, Al^{3+} and Cr^{3+} precipitate as the hydroxides, the other ions as sulfides.

GROUP 4 (Mg^{2+}, Ca^{2+}, Sr^{2+}, Ba^{2+}). May be precipitated as carbonates.

GROUP 5 (Na^+, K^+, NH_4^+). Special reagents are ordinarily used to identify these ions. The violet color imparted to a flame by K^+ is a useful test for that ion.

In qualitative analysis the Groups are removed from a solution in order; first Group 1, then Group 2, then Group 3, and so on

Once the cations have been separated into groups, the problem becomes one of further subdividing the groups. We shall not attempt at this point to describe the chemistry involved in these separations. A single example may, however, be instructive. Lead can be separated from the other group 1 ions by boiling the chloride precipitate with water, filtering the hot mixture, and testing the hot filtrate with CrO_4^{2-}. The formation of a yellow precipitate of lead chromate, $PbCrO_4$, which is much less soluble than $PbCl_2$, indicates the presence of lead.

$$PbCl_2(s) \rightarrow Pb^{2+}(aq) + 2\ Cl^-(aq)$$

$$Pb^{2+}(aq) + CrO_4^{2-}(aq) \rightarrow PbCrO_4(s)$$

$$\overline{PbCl_2(s) + CrO_4^{2-}(aq) \rightarrow 2\ Cl^-(aq) + PbCrO_4(s)} \qquad (16.9)$$

16.5 PRECIPITATION REACTIONS IN INORGANIC PREPARATIONS

Precipitation reactions are a convenient source of insoluble ionic compounds. Barium sulfate can be prepared by adding a slight excess of sulfuric acid to a solution of a barium salt

$$Ba^{2+}(aq) + SO_4^{2-}(aq) \rightarrow BaSO_4(s) \tag{16.10}$$

The precipitate is filtered, washed to remove foreign ions, and dried at 100°C to give a product of high purity. A similar technique is used to prepare the insoluble silver halides (AgCl, AgBr, AgI), used in photography, from silver nitrate. As noted earlier, an important industrial process, the extraction of magnesium from sea water, depends upon a precipitation reaction. Calcium hydroxide is added to precipitate magnesium hydroxide, thereby separating Mg^{2+} from the more abundant Na^+ ions.

$$Mg^{2+}(aq) + 2 OH^-(aq) \rightarrow Mg(OH)_2(s) \tag{16.11}$$

In addition to their use in preparing insoluble inorganic compounds, precipitation reactions offer a means of converting one soluble ionic compound to another. Suppose, for example, we wish to convert Na_2CO_3 to NaOH. One way to do this is to add a solution of $Ca(OH)_2$. A precipitate of $CaCO_3$ forms, leaving Na^+ and OH^- ions in solution. Evaporation of this solution gives solid sodium hydroxide.

Precipitation:

$$2 Na^+(aq) + CO_3^{2-}(aq) + Ca^{2+}(aq) + 2 OH^-(aq) \rightarrow$$
$$CaCO_3(s) + 2 Na^+(aq) + 2 OH^-(aq)$$

Evaporation:

$$2 Na^+(aq) + 2 OH^-(aq) \rightarrow 2 NaOH(s) \tag{16.12}$$

About one fifth of the 5 million tons of sodium hydroxide prepared annually in the United States is made by this two-step sequence.

The method just described can be applied to a wide variety of synthetic problems in inorganic chemistry. Metal chlorides, bromides, or iodides can be converted to the corresponding nitrates by adding $AgNO_3$, filtering off the insoluble silver halide, and evaporating the solution remaining. As an example, consider the conversion of CsCl to $CsNO_3$.

Precipitation:

One would not use excess $AgNO_3$. Why?

$$Cs^+(aq) + Cl^-(aq) + Ag^+(aq) + NO_3^-(aq) \rightarrow AgCl(s) + Cs^+(aq) + NO_3^-(aq)$$

Evaporation:

$$Cs^+(aq) + NO_3^-(aq) \rightarrow CsNO_3(s) \tag{16.13}$$

The main drawback here is that $AgNO_3$ is expensive, so that applications of this sort are limited to cases where small amounts of otherwise unavailable substances are needed.

PROBLEMS

16.1 Write net ionic equations for the reactions, if any, that occur when 0.1 M solutions of the following compounds are mixed.

a. NaOH and $CuSO_4$
b. KNO_3 and $Ba(OH)_2$
c. BaS and $ZnSO_4$
d. $(NH_4)_2CO_3$ and $ZnCl_2$
e. ZnS and $NaNO_3$
f. Na_2SO_4 and $Pb(NO_3)_2$

16.2 Suggest an explanation for the fact that

a. NaCl is more soluble than CuS.
b. LiI is more soluble than CsI.
c. AgCl is more soluble in KNO_3 solution than in pure water.
d. $PbSO_4$ is less soluble in Na_2SO_4 solution than in pure water.

16.3 Describe, in your own words,

a. what is meant by the common ion effect.
b. why K_2CrO_4 is used as an indicator in titrating Cl^- with Ag^+.
c. how it is possible to convert one soluble salt to another by means of a precipitation reaction.

16.4 Classify the following statements as always true, sometimes true, or always false.

a. A precipitate of $PbCl_2$ will form when the product (conc. Pb^{2+}) × (conc. Cl^-)^2 exceeds K_{sp}.
b. In a solution in equilibrium with $Ni(OH)_2(s)$, $[OH^-] = 2 [Ni^{2+}]$.

16.5 300 ml of 0.20 M $Pb(NO_3)_2$ solution is mixed with 500 ml of 0.10 M Na_2SO_4.

a. Calculate the number of moles of each ion present before reaction.
b. Write a net ionic equation for the reaction that takes place.
c. Calculate the number of moles of each ion present after reaction.

16.6 What volume of 0.10 M $AgNO_3$ is required to react exactly with

a. 100 ml of 0.10 M Na_2S?
b. 0.200 g of NaCl?
c. 0.200 mole of K_2CrO_4?

16.7 Given the following solubilities, calculate the corresponding value of K_{sp}

a. $CaSO_4$: 5×10^{-3} mole/lit
b. PbF_2: 0.53 g/liter

16.18 Write net ionic equations for the reactions, if any, that occur when 0.1 M solutions of the following compounds are mixed.

a. $AgNO_3$ and K_2S
b. NaOH and K_2SO_4
c. $Al(NO_3)_3$ and KOH
d. CuS and NaCl
e. $NiSO_4$ and $Sr(OH)_2$
f. NaCl and $Pb(NO_3)_2$

16.19 Suggest an explanation for the fact that

a. $MgSO_4$ is more soluble than $BaSO_4$.
b. $Ni(OH)_2$ is less soluble than KOH.
c. the addition of NaCl to a saturated solution of Ag_2CrO_4 causes a precipitate to form.

16.20 Describe in your own words

a. precisely what is meant by the statement that $K_{sp}AgCl = 1.6 \times 10^{-10}$.
b. how Group 1 is separated from the other groups in the qualitative analysis of cations.
c. why AgI is less soluble in KI solution than in pure water.

16.21 Classify each of the following statements as always true, always false, or sometimes true and sometimes false.

a. In a solution formed by adding $BaSO_4$ to pure water, $[Ba^{2+}] = [SO_4^{2-}]$.
b. Of two slightly soluble electrolytes, the one with the larger value of K_{sp} is the more soluble in water.

16.22 200 ml of 0.30 M $NiCl_2$ is added to 600 ml of 0.10 M KOH. Calculate

a. the number of moles of each ion present before reaction.
b. the number of grams of $Ni(OH)_2(s)$ formed.
c. the number of moles of each ion present after reaction.

16.23 What volume of 0.200 M $Pb(NO_3)_2$ is required to react exactly with

a. 100 ml of 0.10 M NaCl?
b. 0.200 g of Na_2SO_4?
c. 0.150 mole of Na_2S?

16.24 Given the following solubilities, calculate the corresponding values of K_{sp}

a. PbI_2: 1.4×10^{-3} mole/liter
b. $MgCO_3$: 1.2×10^{-2} g/liter

16.8 From the values of K_{sp} given in Table 16.2 calculate the solubility of

    a. $PbCl_2$ (mole/liter)
    b. HgS (g/liter)

16.9 Decide whether or not precipitates will form under the following conditions:

    a. Enough $SO_4{}^{2-}$ is added to a solution 0.02 M in Pb^{2+} to make the concentration of $SO_4{}^{2-}$ 1×10^{-5} M.
    b. Hard water in which the concentration of Ca^{2+} is 1×10^{-3} M is fluoridated to make the concentration of F^- 5×10^{-5} M.
    c. Equal volumes of solutions 0.04 M in Pb^{2+} and Cl^- are mixed.

16.10 The concentration of Pb^{2+} in a certain water supply is 1×10^{-6} M. If it is desired to remove 90% of the Pb^{2+} by adding $CrO_4{}^{2-}$, what must be $[CrO_4{}^{2-}]$?

16.11 If Ag^+ is slowly added to a solution 0.050 M in Na_2CrO_4,

    a. At what concentration of Ag^+ does a precipitate start to form?
    b. When $[Ag^+] = 1 \times 10^{-4}$ M, what percentage of the $CrO_4{}^{2-}$ is left in solution?
    c. What concentration of Ag^+ will be required to precipitate 99% of the $CrO_4{}^{2-}$?

16.12 Hydrochloric acid is slowly added to a solution 0.10 M in Pb^{2+} and 0.01 M in Ag^+.

    a. Which precipitate forms first, $PbCl_2$ or $AgCl$?
    b. What is the concentration of the first ion precipitated when the second precipitate starts to form?

16.13 An analyst determines the percentage of F^- in a type of rat poison by precipitating it as CaF_2. He finds that a sample weighing 0.100 g gives 0.028 g of CaF_2. What is the percentage of F^- present?

16.14 A student determines the per cent of Cl^- in a mixture by weighing out 0.500 g, dissolving in 20 cc of water, and titrating with 24.0 ml of 0.120 M $AgNO_3$ solution. What is the per cent of Cl^- in the mixture?

16.15 Develop a scheme, based on Table 16.1, for separating and identifying the ions in a mixture containing

    a. Ag^+, Ni^{2+}, Ca^{2+}
    b. Cl^-, $SO_4{}^{2-}$, $CO_3{}^{2-}$

16.25 From the values of K_{sp} in Table 16.2, calculate the solubility of

    a. Ag_2CrO_4 in mole/liter
    b. $PbCO_3$ in g/liter

16.26 Decide whether or not precipitates will form under the following conditions:

    a. Enough Ag^+ is added to a solution 1×10^{-3} M in $CrO_4{}^{2-}$ to make the concentration of Ag^+ 2×10^{-3} M.
    b. Well water which is 1×10^{-4} M in Fe^{3+} is adjusted to a OH^- concentration of 1×10^{-7} M.
    c. 20 ml of 1.0×10^{-2} M $Pb(NO_3)_2$ is added to 40 ml of 1.0×10^{-2} M HBr.

16.27 The concentration of Mg^{2+} in sea water is 5×10^{-2} M. What must be $[OH^-]$ necessary to remove 60% of the Mg^{2+}?

16.28 500 ml of 0.0020 M $BaCl_2$ solution is treated with an equal volume of 0.0030 M Na_2SO_4.

    a. How many moles of $SO_4{}^{2-}$ remain in solution after precipitation is complete?
    b. What is the final $[SO_4{}^{2-}]$?
    c. What is the final $[Ba^{2+}]$?

16.29 A solution is 0.020 M in Cl^- and 0.010 M in $CrO_4{}^{2-}$. If Ag^+ ions are slowly added, What precipitate forms first? What is the concentration of the first ion when the second ion starts to precipitate? On the basis of your answers to these questions, suggest why $CrO_4{}^{2-}$ ion might serve as a suitable indicator for the titration of Cl^- with Ag^+.

16.30 The arsenic content of a sample can be determined by precipitation as As_2S_3. A sample weighing 1.000 g gives 0.0581 g of As_2S_3. What is the percentage of arsenic in the sample?

16.31 A student finds that 22.0 ml of 0.200 M H_2SO_4 is required to completely precipitate a certain sample weighing 2.000 g. If the sample contains Ba^{2+}, what is the % of Ba^{2+}?

16.32 Develop a scheme, based on Table 16.1, to separate and identify

    a. Cu^{2+}, Ba^{2+}, Pb^{2+}
    b. S^{2-}, Cl^-, OH^-

16.16 Explain why, in separating Group 2 cations in qualitative analysis, the solution is made acidic.

16.17 Describe, in some detail, how the following conversions might be accomplished.

   a. $Ni(NO_3)_2 \rightarrow NiS$
   b. $ZnCl_2 \rightarrow ZnSO_4$
   c. $AgNO_3 \rightarrow KNO_3$
   d. $H_2SO_4 \rightarrow HCl$

16.33 A certain qualitative analysis unknown contains ions from each of the five groups. A student, instead of adding HCl in the first step, adds H_2S in acidic solution. What problems, if any, will this cause?

16.34 Describe, in some detail, how the following conversions might be accomplished:

   a. $Na_2SO_4 \rightarrow NaCl$
   b. $KOH \rightarrow KNO_3$
   c. $K_2S \rightarrow KCl$
   d. $Pb(NO_3)_2 \rightarrow Cu(NO_3)_2$

*16.35 The white pigment in certain paints contains Pb^{2+} ions, which react with H_2S to form black, insoluble PbS. If the concentration of H_2S in air is 10 parts per million, how many liters of air at 25°C and one atmosphere must come in contact with one gram of Pb^{2+} to convert half of it to PbS? (Consult Table 15.1, Chapter 15, for the appropriate conversion factors.)

*16.36 The inside of a teakettle is coated with 10.0 g of $CaCO_3$. With how many liters of pure water must the teakettle be washed to remove the $CaCO_3$?

*16.37 The concentrations of various cations in sea water in moles/liter are:

ion	Na^+	Mg^{2+}	Ca^{2+}	Al^{3+}	Fe^{3+}
M	0.46	0.050	0.01	4×10^{-7}	2×10^{-7}

   a. At what conc. of OH^- does $Mg(OH)_2$ start to precipitate?
   b. At this concentration, will any of the other ions precipitate?
   c. If enough OH^- is added to precipitate 50% of the Mg^{2+}, what percentage of each of the other ions will be precipitated?
   d. Under the conditions in (c), what weight of precipitate will be obtained from a gallon of sea water?

*16.38 Using the Debye-Hückel theory, calculate the value of the product $[Ag^+] \times [Cl^-]$ in each of the KNO_3 solutions listed in Table 16.3 and compare to the observed values.

*16.39 The general Debye-Hückel expression for the activity coefficient product of a 1:1 electrolyte in a salt solution of molarity M is

$$\log \gamma_+ \gamma_- = -1.0 \; I^{1/2}$$

where $I = \frac{1}{2} (M_c Z_c^2 + M_a Z_a^2)$

(M_c, M_a = molarities of cation and anion in salt solution; Z_c, Z_a = charges of cation and anion in salt solution)

   a. Show that this equation reduces to $-1.0 \; M^{1/2}$ when the salt is KNO_3.
   b. Show that when the salt is Na_2SO_4: $\log \gamma_+ \gamma_- = -1.7 \; M^{1/2}$.
   c. Calculate the solubility of AgCl in 0.010 M $Ca(NO_3)_2$ solution.

*16.40 An alloy of copper and silver is dissolved in HNO_3 and treated with H_2S to precipitate both CuS and Ag_2S. It is found that a 0.500 g sample gives 0.730 g of sulfide. What is the percentage of Ag in the mixture?

17 ACIDS AND BASES

In the world of the chemist the most important solvent, as you have perhaps already recognized, is water. We have studied in the previous chapter the properties of one class of reactions, namely, those involving precipitations, as they occur in water solution. Water is so common a solvent for all sorts of reactions that we often forget that in many reactions its role is as important as that of the solutes. In one group of reactions in particular, water has an especially significant part to play; these reactions involve those species known as acids and bases.

17.1 THE DISSOCIATION OF WATER; NATURE OF ACIDS AND BASES

One might realistically say that acids and bases exist because of one of the key properties of water, namely, its dissociation to ions according to the equation:

$$H_2O(l) \rightleftarrows H^+(aq) + OH^-(aq) \qquad (17.1)$$

In any system containing liquid water the above equilibrium between water and its ions, H^+ and OH^-, is, as we shall see, an exceedingly important one. The reaction to equilibrium is very rapid in both directions. It is known, for example, that a given proton spends less than 1×10^{-3} seconds attached to any given water molecule; the protons participate vigorously in what is called a hydrogen exchange reaction.

Although the equilibrium reaction is very rapid, it does not go very far to the right, so that at any moment only a relatively few H_2O molecules are dissociated. Following the general rules of the law of equilibrium, we would formulate the equilibrium expression for Reaction 17.1 as

$$K_c = \frac{[H^+] \times [OH^-]}{[H_2O]}$$

In aqueous solution the concentration of water is typically very large, about 55M, and remains essentially constant for dilute solutions. Rather than carry its concentration in calculations, we invariably include it in K_c; under such conditions the equilibrium expression takes the form

$$K_w = [H^+] \times [OH^-] = 1.0 \times 10^{-14} \qquad (17.2)$$

The dissociation constant of water is given the symbol K_w, and is, as you can see, very small, being about 1.0×10^{-14} at 25°C.

By Equation 17.2 any water solution at 25°C will contain H^+ and OH^- ions at such concentrations that their product will be 1.0×10^{-14}. If one is dealing with pure water, it is easy to calculate the concentrations of these two ions, since by Equation 17.1 they will form in equal amounts:

in pure H_2O $[H^+] = [OH^-]$ $[H^+] \times [OH^-] = [H^+]^2 = 1.0 \times 10^{-14}$

$$[H^+] = 1.0 \times 10^{-7} \text{ M} = [OH^-]$$

A solution in which $[H^+]$ equals $[OH^-]$ is said to be *neutral*. Such solutions turn out to be relatively rare, but in principle at least, pure water is neutral.

Ordinarily the concentrations of hydrogen and hydroxide ions in a solution are not equal. A water solution in which the hydrogen ion concentration is larger than the hydroxide ion concentration is said to be *acidic*. On the other hand, a water solution in which the hydroxide ion concentration is larger than that of hydrogen ion is *basic*. As the concentration of one ion goes up, that of the other must go down proportionally so that Equation 17.2 will be satisfied. This means that any solution in which $[H^+]$ is greater than 1.0×10^{-7} M is acidic; a solution in which $[OH^-]$ is greater than 1.0×10^{-7} M is basic.

Acidic solutions are formed by adding to water any one of a class of substances called *acids*. Another class of substances, called *bases*, tend to make their water solution basic. With the use of these two kinds of substances it is possible to prepare solutions in which either $[H^+]$ or $[OH^-]$ is quite high, even larger than 10 M. In all these solutions, however, Equation 17.2 will be obeyed, so that it is a relatively easy matter to calculate the concentration of one of these ions if the other is known. In Table 17.1 we have indicated some possible combinations of concentrations of these two ions. Solutions 1 through 4 would be classified as acidic; solutions 5 through 8 are progressively more basic.

TABLE 17.1

SOLUTION	No. 1	No. 2	No. 3	No. 4	No. 5	No. 6	No. 7	No. 8
$[H^+]$	1	10^{-2}	10^{-4}	10^{-6}	10^{-8}	10^{-10}	10^{-12}	10^{-14}
$[OH^-]$	10^{-14}	10^{-12}	10^{-10}	10^{-8}	10^{-6}	10^{-4}	10^{-2}	1
pH	0	2	4	6	8	10	12	14

17.2 pH

We have seen that it is possible to describe quantitatively the acidity or basicity of a water solution by specifying the concentration of H^+ ion. In 1909 Sörensen proposed an alternative method of accomplishing this purpose, making use of a term known as pH, defined as follows:

$$pH \equiv -\log_{10}[H^+] = \log_{10} 1/[H^+] \qquad (17.3)$$

A great many important solutions have a pH between 1 and 14, corresponding to a variation in hydrogen ion concentration of 10^{-1} to 10^{-14} M. For such
solutions it is perhaps more convenient to express acidity in terms of pH
rather than H^+ ion concentration, thereby avoiding the use of either small
fractions or negative exponents.

It follows from the definition that *the lower the pH, the more acidic a
solution is;* conversely, the higher the pH, the more basic is the solution.
Just as it is possible to differentiate between neutral, basic, and acidic
solutions on the basis of the concentrations of H^+ or OH^- ions, so one can
make this distinction in terms of pH:

neutral solution: $[H^+] = 10^{-7}$ M, pH $= 7.0$
acidic solution: $[H^+] > 10^{-7}$ M, pH < 7.0
basic solution: $[H^+] < 10^{-7}$ M, pH > 7.0

Example 17.1 illustrates the calculation of the pH of water solutions and
the interrelationships between pH, $[H^+]$, and $[OH^-]$.

The pH, $[H^+]$, or $[OH^-]$ of a water solution may be determined experimentally in a number of different ways, one of which involves the use
of acid-base indicators, which undergo a color change over a rather narrow
pH range. Litmus, which is red in acidic and blue in basic solution, was one
of the first indicators and is still widely used. By the judicious use of one or
more of the indicators listed in Table 17.3, it is possible to bracket quite
accurately the pH or $[H^+]$ of a solution. For example, if one finds that a
solution gives a red (basic) color with phenolphthalein but a yellow (acidic)
color with alizarine yellow, its pH must be approximately 10.

Example 17.1   Calculate
    a. The pH of a solution 0.020 M in H^+.
    b. The pH of a solution 2.5×10^{-3} M in OH^-.
    c. The $[H^+]$ in sea water.

Solution.   The conversion of $[H^+]$ to pH and vice versa involves a simple use
of logarithms.

    a. $[H^+] = 0.020$ M $= 2.0 \times 10^{-2}$ M

$$pH = -\log_{10}[H^+] = -\log_{10}(2.0 \times 10^{-2}) = -(0.3 - 2.0)$$
$$= 1.7$$

        or alternatively: $pH = \log \dfrac{1}{[H^+]} = \log \dfrac{1}{2 \times 10^{-2}}$

$$= \log 50 = \log (5 \times 10) = .7 + 1 = 1.7$$

    b. Before calculating the pH, we need to know $[H^+]$. We can obtain
       this by making use of the relation $[H^+] \times [OH^-] = 1.0 \times 10^{-14}$.

$$[H^+] = \frac{1.0 \times 10^{-14}}{[OH^-]} = \frac{1.0 \times 10^{-14}}{2.5 \times 10^{-3}} = 0.40 \times 10^{-11} = 4.0 \times 10^{-12}$$

$$pH = -\log_{10}(4.0 \times 10^{-12}) = -(0.6 - 12.0) = 11.4$$

    c. From Table 17.2 we note that the pH of sea water is 8.3.

$$-\log_{10}[H^+] = 8.3$$

$$\log_{10}[H^+] = -8.3 = 0.7 - 9.0$$

    Taking antilogs, $[H^+] = 5 \times 10^{-9}$ M

TABLE 17.2 pH of Common Materials

Vinegar	3.0	Human urine	5–8
Apples	3.2	Sea water	8.3
Sauerkraut	3.5	0.1 M NaHCO₃	8.4
Tomatoes	4.2	Satd Mg(OH)₂	10.5
Carrots	5	0.1 M NH₃	11.2
Saliva	6.5–7.5	0.1 M Na₂CO₃	11.7
Milk	6.6	Satd Ca(OH)₂	12.4

A universal indicator made by combining several acid-base indicators may be used to determine, within about one unit, the pH of any water solution. This mixture of indicators shows an entire spectrum of colors ranging from deep red in strongly acidic solution to deep blue in strongly basic solution. A similar principle is used to prepare pH paper, widely used to test the acidity of biological fluids. Strips of paper impregnated with a mixture of indicators can be designed to give gradations of color over a wide or narrow pH range.

The most accurate way to measure pH is with a pH meter (Chap. 21). Such a device can determine pH to ±0.01 units

In a more qualitative way, the acidity or basicity of a solution can be established by some very simple tests. Dilute acidic solutions are characteristically sour; lemon juice, vinegar, and rhubarb get at least part of their taste from the acids they contain. It wouldn't be advisable to test-taste solutions which are strongly acidic, since there are various undesirable side effects, to say the least. Such solutions, however, will react with zinc or magnesium metal, evolving hydrogen gas. They will also react readily with carbonates, in which case bubbles of carbon dioxide are given off.

Basic solutions typically feel slippery. A solution of lye, sodium hydroxide, feels very slippery, but this is because it is dissolving a surface layer of skin. Probably a better test for bases is to add a small amount of a solution of a magnesium salt such as $MgCl_2$. The formation of a white precipitate of magnesium hydroxide indicates the solution is basic.

17.3 STRONG ACIDS AND BASES

In order for a solute to make a solution acidic it must somehow release H^+ ion to the solution. The simplest kind of substance which can do this is

TABLE 17.3 Typical Acid-Base Indicators

NAME	pH INTERVAL	ACID COLOR	BASE COLOR
Methyl violet	0.0–1.6	yellow	violet
Methyl yellow	2.9–4.0	red	yellow
Methyl orange	3.1–4.4	red	yellow
Methyl red	4.8–6.2	red	yellow
Bromthymol blue	6.0–8.0	yellow	blue
Thymol blue	8.0–9.6	yellow	blue
Phenolphthalein	8.2–10.0	colorless	pink
Alizarine yellow	10.1–12.0	yellow	red

one like HCl, which on being dissolved in water undergoes the following reaction:

For this reaction we can assume that K_c is infinite

$$HCl(aq) \rightarrow H^+(aq) + Cl^-(aq) \qquad (17.4)$$

Strong acids, of which HCl is a classic example, ionize essentially completely in water in a reaction like 17.4, producing hydrogen ion and an anion. In a solution of a strong acid there are assumed to be no acid molecules, but only hydrogen ions and the anions resulting from the dissociation. In a 0.5 M HCl solution, made by dissolving a half mole of HCl(g) per liter of solution, $[H^+]$ is 0.5 M, $[Cl^-]$ is 0.5 M, and [HCl] is virtually zero. In solution, then, HCl behaves as a strong electrolyte. It conducts the electric current very well and has the colligative properties of an ionized substance.

There are only a few strong acids; the more important ones are listed in Table 17.4. Since all these acids are very soluble in water, it is possible to use them to prepare solutions in which $[H^+]$ is very high. Concentrated HCl is about 12 M and concentrated HNO_3 is about 16 M. Sulfuric acid is furnished commercially as essentially pure H_2SO_4; that liquid contains almost no free water, is a very strong dehydrating agent and a good oxidizing agent, as well as having acid properties. HCl, HNO_3, and H_2SO_4 are all very important industrial chemicals; each has its own characteristic properties, but they share in common the capacity to furnish high concentrations of hydrogen ion in water solutions. The "workhorse" acids of the chemistry laboratory are 6 M HCl, 6 M HNO_3, and 3 M H_2SO_4, all referred to as "dilute" acids.

Analogous to the strong acids are the **strong bases,** of which NaOH is the most common example. Sodium hydroxide dissolves very readily in water to give a solution containing Na^+ and OH^- ions.

$$NaOH(s) \rightarrow Na^+(aq) + OH^-(aq) \qquad (17.5)$$

In solution NaOH is effectively 100 per cent ionized, as are all the strong bases. In 0.8 M NaOH solution, made by dissolving 0.8 moles of NaOH per liter of solution, $[Na^+]$ is 0.8 M, $[OH^-]$ is 0.8 M, and [NaOH], the concentration of NaOH molecules, is just about zero.

There are not many strong bases, and most of them are listed in Table 17.4. Of these bases, only NaOH and KOH are ordinarily used in the chemistry laboratory. All compounds of lithium, rubidium, and cesium are expensive; barium hydroxide is used for some special purposes but has limited solubility, whereas NaOH and KOH are very soluble. Metal oxides and hydroxides for the most part tend to form basic solutions, but, except for those listed in Table 17.4, they are all relatively insoluble. $Mg(OH)_2$ and $Ca(OH)_2$ are slightly soluble and do behave as strong bases.

TABLE 17.4   The Strong Acids and Bases

Hydrochloric acid	HCl	Sodium hydroxide	NaOH
Nitric acid	HNO_3	Potassium hydroxide	KOH
Sulfuric acid	H_2SO_4	Barium hydroxide	$Ba(OH)_2$
Perchloric acid	$HClO_4$	Lithium hydroxide	LiOH
Hydrobromic acid	HBr	Rubidium hydroxide	RbOH
Hydriodic acid	HI	Cesium hydroxide	CsOH

Using stock solutions of 1 M HCl and 1 M NaOH it is possible to prepare, by dilution, acidic and basic solutions of nearly any strength.

Example 17.2   How would you proceed to prepare, from 1 M HCl and 1 M NaOH stock solutions, one liter of a solution in which [H$^+$] is

a. 4×10^{-3} M?   b. 5×10^{-13} M?

Solution

a. Since HCl is completely ionized, the solution to be made will have to be 4×10^{-3} M HCl. One liter of such a solution will contain 4×10^{-3} moles HCl. We need to find the volume of 1 M HCl which will contain that number of moles; we use the relation:

$$\text{no. moles HCl} = M \times V; \; V = \frac{4 \times 10^{-3} \text{ moles}}{1 \text{ mole/lit}} = 4 \times 10^{-3} \text{ lit} = 4 \text{ ml}$$

We measure out 4 ml 1 M HCl and dilute to a volume of one liter, mixing the final solution well.

b. We cannot prepare this solution by dilution of 1 M HCl with water, since the best we can do by that approach is to get essentially pure water, which has a [H$^+$] of 10^{-7} M. Since the solution is basic, we make it from the stock 1 M NaOH.

> This may come as a surprise, until you think about it for a while

We first need to find [OH$^-$], using the expression for K$_w$:

$$[H^+] \times [OH^-] = K_w = 1.0 \times 10^{-14}$$

$$[OH^-] = \frac{1.0 \times 10^{-14}}{5 \times 10^{-13}} = 0.2 \times 10^{-1} \text{ M} = 2 \times 10^{-2} \text{ M}$$

To make one liter of this solution, we need 2×10^{-2} moles NaOH. By the procedure used in part a, we find that we would require 20 ml 1 M NaOH. We measure out that volume of the basic stock solution and dilute to one liter with water, not forgetting to mix well.

When a solution of an acid is mixed with that of a base, a chemical reaction invariably occurs. The net ionic equation for the reaction between strong acids and bases is

$$H^+(aq) + OH^-(aq) \rightleftarrows H_2O \quad K = 1/K_w = 1.0 \times 10^{14} \qquad (17.6)$$

This reaction is the reverse of the dissociation of water and has an enormous equilibrium constant. Reaction 17.6 is a very important one and the next chapter will be devoted to exploring some of its many applications.

17.4   WEAK ACIDS

Most acids and bases are not strong, i.e., completely ionized in water, but fall in the category called weak. A *weak acid* or a *weak base* is characterized by the fact that in solution its reaction to form H$^+$ or OH$^-$ ion proceeds only to a small extent.

A typical example of a weak acid is hydrofluoric acid, made by dissolving hydrogen fluoride, HF, in water. Rather than dissociating completely, like HCl, HF in solution ionizes only partially:

$$HF(aq) \rightleftarrows H^+(aq) + F^-(aq)$$

In the final solution the concentrations of H^+ and F^- ions are low as compared with the concentration of undissociated HF molecules. For example, in 0.1 M HF solution, only about 8 per cent of the HF exists in ionized form. This property of weak acids makes their behavior considerably different from that of strong acids.

SPECIES WHICH BEHAVE AS WEAK ACIDS

There are many substances which in water solution behave as weak acids. These include a small multitude of organic acids, of which acetic acid, the acid component of vinegar, $HC_2H_3O_2$, is the most common example. All the organic acids contain at least one acidic hydrogen atom and dissociate in a manner similar to acetic acid:

$$HC_2H_3O_2(aq) \rightleftarrows H^+(aq) + C_2H_3O_2^-(aq)$$

Inorganic weak acids are also fairly common and include many molecular species, such as carbonic acid, H_2CO_3, sulfurous acid, H_2SO_3, hydrogen sulfide, H_2S, and hydrocyanic acid, HCN, as well as HF. Many of these acids are solutions of a gas in water; in some cases the gas is an oxide. Carbonic acid and sulfurous acid are simply aqueous solutions of CO_2 and SO_2, respectively. In other weak acids the solute gas does not contain oxygen, as with H_2S, HCN, and HF. Some weak acids contain two or more ionizable hydrogen atoms and dissociate in two or more steps, the first of which always occurs much more easily than the second or third. Carbonic acid is an acid of this sort:

Tap water is often acidic due to dissolved CO_2 from the air

$$H_2CO_3(aq) \rightleftarrows H^+(aq) + HCO_3^-(aq) \qquad \text{first ionization}$$

$$HCO_3^-(aq) \rightleftarrows H^+(aq) + CO_3^{2-}(aq) \qquad \text{second ionization}$$

It is possible to prepare salts like $NaHSO_3$ or NaH_2PO_4, which contain anions with acidic hydrogens. These anions ionize by a reaction like that of the HCO_3^- ion.

Some inorganic cations exhibit weak acid properties in solution. The most important of these is the ammonium ion, NH_4^+:

$$NH_4^+(aq) \rightleftarrows H^+(aq) + NH_3(aq)$$

Surprisingly enough, even some metallic cations, particularly Zn^{2+} and Al^{3+}, are also weak acids. In solution these cations are hydrated, and the attraction of the positive charge of the central ion for electrons weakens the bonds in the hydrated water molecules sufficiently to liberate hydrogen ion. The reactions can be represented by the equations:

$$Al(H_2O)_6^{3+} \rightleftarrows H^+ + Al(H_2O)_5(OH)^{2+}$$

$$Zn(H_2O)_4^{2+} \rightleftarrows H^+ + Zn(H_2O)_3(OH)^+$$

Experimentally one can distinguish between strong and weak acids by measuring:

1. The conductivities of their water solutions. Strong acids behave as strong electrolytes; solutions of weak molecular acids such as hydrofluoric acid or acetic acid are relatively poor electrical conductors.

2. The colligative properties of their water solutions. The freezing point of a 0.10 M solution of HCl is almost exactly the same as that of 0.10 M NaCl. In contrast, the freezing point of a 0.10 M solution of acetic acid is comparable to that of 0.10 M solutions of nonelectrolytes such as sugar or urea.

3. The pH of their water solutions. A 0.10 M solution of perchloric acid has a pH of 1, indicating complete dissociation. The pH of a 0.10 M solution of HF is about 2.1, indicating that relatively few of the HF molecules are ionized.

4. The rate of reaction with metals such as zinc or magnesium (Fig. 17.1).

The equilibria between species in solutions of weak acids can be handled quantitatively in a manner analogous to that developed previously in connection with gases (Chapter 13) and slightly soluble salts (Chapter 16). In particular, it is possible to derive an expression for the equilibrium constant, or the ionization constant, for the dissociation of a weak acid in water. We shall now consider how this equilibrium constant, symbol K_a, is expressed, how it is determined experimentally, and how it is used in practical calculations.

EXPRESSION FOR K_a

Consider the dissociation of the weak acid HX:

$$HX(aq) \rightleftarrows H^+(aq) + X^-(aq) \qquad (17.7)$$

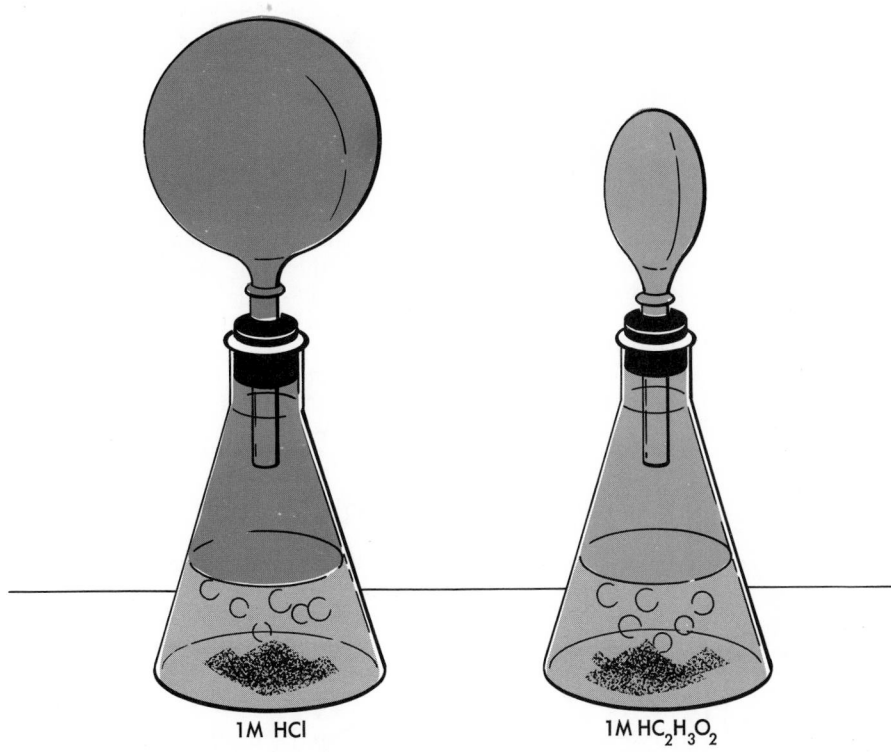

Figure 17.1   Reaction of zinc with acids. H_2 is generated more rapidly with HCl.

Formulating the expression for the equilibrium constant, K_a, for this reaction according to the rules we have used previously, we obtain

$$K_a = \frac{[H^+] \times [X^-]}{[HX]} \qquad (17.8)$$

The equilibrium constant, K_a, is called the *acid constant* or the *ionization constant* of the weak acid HX.

To illustrate the form taken by K_a for various weak acids, consider the three species HF, NH_4^+, and H_2CO_3:

$$HF(aq) \rightleftarrows H^+(aq) + F^-(aq) \qquad K_a = \frac{[H^+] \times [F^-]}{[HF]}$$

Note that the concentration of undissociated acid appears in the expression for K_a

$$NH_4^+(aq) \rightleftarrows H^+(aq) + NH_3(aq) \qquad K_a = \frac{[H^+] \times [NH_3]}{[NH_4^+]}$$

$$H_2CO_3(aq) \rightleftarrows H^+(aq) + HCO_3^-(aq) \qquad K_1 = \frac{[H^+] \times [HCO_3^-]}{[H_2CO_3]}$$

$$HCO_3^-(aq) \rightleftarrows H^+(aq) + CO_3^{2-}(aq) \qquad K_2 = \frac{[H^+] \times [CO_3^{2-}]}{[HCO_3^-]}$$

EXPERIMENTAL DETERMINATION OF K_a

It is possible to measure the ionization constants of weak acids by several different methods. Two of these are illustrated in the following examples.

Example 17.3 Some of the common organic acids were originally detected by their odors and were given names which indicated their source. A case in point is caproic acid, found in the skin secretions of goats (L. caper, goat). Caproic acid, $CH_3(CH_2)_4COOH$, is similar to acetic acid but has a longer carbon chain; only the last hydrogen in the formula is acidic, so we can call the acid HCap.

The concentration of hydrogen ion in a 0.030 M HCap solution was measured and found to be 6.5×10^{-4} M. Find K_a for caproic acid.

Solution. The dissociation of caproic acid would occur as follows:

$$HCap(aq) \rightleftarrows H^+(aq) + Cap^-(aq) \qquad K_a = \frac{[H^+] \times [Cap^-]}{[HCap]}$$

To find the numerical value of K_a, we must know the equilibrium concentrations of H^+, Cap^-, and HCap. That of H^+ has been determined to be 6.5×10^{-4} M. From the equation for the dissociation of HCap, it is evident that one mole of caproate ion, Cap^-, is formed for every mole of H^+ ion. The only effective source of H^+ ion is the HCap molecule (the dissociation of water produces a negligible amount of H^+), so we can say that in the solution

$$[H^+] = [Cap^-] = 6.5 \times 10^{-4} \text{ M}$$

To obtain the equilibrium concentration of HCap, we note that for every mole of H^+ formed, a mole of HCap must dissociate. Basing our calculation on a liter of solution, which would contain 0.030 moles of caproic acid originally, we can see that [HCap] at equilibrium can be found by subtracting $[H^+]$ from the original concentration of HCap:

$$[HCap] = 3.0 \times 10^{-2} \text{ M} - 6.5 \times 10^{-4} \text{ M}$$
$$= 3.0 \times 10^{-2} \text{ M} - 0.065 \times 10^{-2} \text{ M}$$
$$= 2.9 \times 10^{-2} \text{ M}$$

Substituting:

$$K_a = \frac{[H^+] \times [Cap^-]}{[HCap]} = \frac{(6.5 \times 10^{-4}) \times (6.5 \times 10^{-4})}{2.9 \times 10^{-2}} = 1.5 \times 10^{-5}$$

The procedure illustrated by Example 17.3 can be applied to any weak acid. Two quantities must be measured—the original concentration of the weak acid and the equilibrium concentration of H^+.

A more general method for determining the ionization constant of a weak acid, in which one need not know its concentration or even its chemical identity, is outlined in Example 17.4.

Example 17.4 A student is instructed to determine the ionization constant of an acid by the following procedure: A sample of the weak acid, HA, is first dissolved in water to give 50 ml of solution. This solution is then split into two equal portions of 25 ml each. One of these portions is neutralized with sodium hydroxide and then added to the unneutralized portion. Show how the ionization constant of the acid can be calculated from the measured $[H^+]$ of the final solution.

Solution. By definition,

$$K_a = \frac{[H^+] \times [A^-]}{[HA]}$$

By the design of the experiment, $[A^-] = [HA]$, since exactly half the HA molecules originally present were converted to A^- ions by neutralization. Substituting in the expression for K_a,

$$\frac{[A^-]}{[HA]} = 1; \quad K_a = [H^+]$$

In the neutralization, OH^- ions react with HA, forming A^- and H_2O

Consequently, the ionization constant of the acid is equal to the $[H^+]$ in the final solution.

(This method requires that K_a be small, say 1×10^{-4} or smaller, if [HA] and $[A^-]$ are to be equal after mixing.)

The numerical value of the ionization constant of an acid gives a qualitative measure of its strength. The stronger the acid, the larger will be the value of K_a. From Table 17.5 (p. 462) we find that acetic acid, $HC_2H_3O_2(K_a = 1.8 \times 10^{-5})$, is weaker than hydrofluoric acid ($K_a = 7.0 \times 10^{-4}$) but stronger than boric acid ($K_a = 5.8 \times 10^{-10}$). In equimolar solutions of two different acids the concentration of H^+ and the fraction of the weak acid dissociated will be greater for the acid of larger ionization constant.

USE OF K_a IN CALCULATIONS

Examples 17.5 and 17.6 illustrate how ionization constants can be used in calculations involving solutions of weak acids.

Example 17.5 Calculate the $[H^+]$ in a 1.0 M solution of $HC_2H_3O_2$.

Solution. From Table 17.5 we have

$$HC_2H_3O_2(aq) \rightleftarrows H^+(aq) + C_2H_3O_2^-(aq)$$

$$K_a = \frac{[H^+] \times [C_2H_3O_2^-]}{[HC_2H_3O_2]} = 1.8 \times 10^{-5}$$

Let x represent the unknown equilibrium concentration of H^+. It is evident from the equation for the dissociation of acetic acid that a mole of acetate ions is produced for every mole of H^+ formed. Neglecting the small amount of H^+ produced from the water, the concentration of $C_2H_3O_2^-$ must also be x. The formation of x moles of H^+ consumes an equal number of moles of acetic acid; the equilibrium concentration of $HC_2H_3O_2$ must then be $1.0 - x$. Substituting,

$$\frac{(x)(x)}{1.0 - x} = 1.8 \times 10^{-5}$$

This equation could be rearranged to the form $x^2 + bx + c = 0$ and solved for x, using the quadratic formula. Such a procedure is tedious and, in this case, unnecessary. Since $HC_2H_3O_2$ is a weak acid, only slightly dissociated in water, the equilibrium concentration of $HC_2H_3O_2$, $1.0 - x$, must be very nearly equal to its original concentration before dissociation, namely, 1.0 M. Making this approximation, we obtain

$$\frac{x^2}{1.0} = 1.8 \times 10^{-5}; \quad x^2 = 1.8 \times 10^{-5} = 18 \times 10^{-6}$$

Solving for x by extracting the square root of both sides, we have

$$x = [H^+] = 4.2 \times 10^{-3}$$

The fact that the concentration of H^+, 0.0042, is so much less than the original concentration of $HC_2H_3O_2$ justifies the approximation made earlier, i.e., $1.0 - x = 1.0$. In general, the expression for K_a is rarely valid to better than ±5 per cent. Consequently, in the expression

$1.0 - 0.0042 =$ 1.0 to 2 significant figures

$$K_a = \frac{x^2}{a - x}$$

in which $x = [H^+]$ and a = concentration of weak acid prior to dissociation, one is justified in setting $(a - x)$ equal to a, provided this approximation does not introduce an error of more than about 5 per cent. In practice it is ordinarily simplest to make the approximation, calculate x, and compare to a. If the value of x thus obtained is less than 5 per cent of a (in this problem, x is 0.42 per cent of a), the approximation is valid. If x is greater than about 5 per cent of a, one can go back to the original equation and solve by means of the quadratic formula or, alternatively, by the method of successive approximations (Example 17.6).

Example 17.6 Lactic acid, $CH_3CHOHCOOH$, gets its name from sour milk, from which it was first isolated in 1780 (L. *lactis*, milk). K_a for HLac is 8.4×10^{-4}. Find the $[H^+]$ in a sample of sour milk containing 0.100 M HLac as its acid component.

Solution. Lactic acid dissociates according to the equation:

$$HLac(aq) \rightleftarrows H^+(aq) + Lac^-(aq) \quad K_a = \frac{[H^+] \times [Lac^-]}{[HLac]} = 8.4 \times 10^{-4}$$

Letting $[H^+] = x$, $[Lac^-]$ will also equal x, and we obtain

$$\frac{(x)(x)}{0.100 - x} = 8.4 \times 10^{-4}$$

If we make the same approximation as before, i.e., $0.10 - x \cong 0.10$,

$$x^2 = 8.4 \times 10^{-5} = 84 \times 10^{-6} \quad x = 9.2 \times 10^{-3} \cong [H^+]$$

We note that in this case the calculated concentration of H^+ is more than 5 per cent of the original concentration of undissociated acid;

$$\frac{9.2 \times 10^{-3}}{0.10} = 9.2 \times 10^{-2} = 9.2\%$$

To refine our calculation, we can make a second approximation, more nearly valid than the first. We use the value of $[H^+]$ just calculated to find a better approximation to the true value of $[HLac]$.

If $$[H^+] \cong 9.2 \times 10^{-3}$$

then $$HLac \cong 0.100 - 9.2 \times 10^{-3}$$
$$= 0.100 - 0.0092 \cong 0.091 \text{ M}$$

Substituting this value for $[HLac]$ into the expression for K_a, we have

$$\frac{x^2}{0.091} = 8.4 \times 10^{-4}$$

$$x^2 = 7.6 \times 10^{-5} = 76 \times 10^{-6} \quad x = 8.7 \times 10^{-3} \text{ M} = [H^+]$$

This value of $[H^+]$ is closer to the true $[H^+]$ in the solution, since 0.091 M is a better approximation to the equilibrium concentration of HLac than was 0.100 M. If we are still not satisfied, we can attempt a further improvement, using the value of $[H^+]$ we just calculated to obtain a still better value for $[HLac]$. If we do this, we find that $[HLac]$ keeps the value 0.091 M; that is, if

$$[H^+] = 8.7 \times 10^{-3} \text{ M}$$

then $$HLac = 0.100 - 0.0087 = 0.091 \text{ M}$$

This means that if we were to solve again for $[H^+]$ we would get the same answer. In other words, "we have gone about as far as we can go."

This method of *successive approximations* is a very useful one for working problems of this type involving equilibria of all sorts. Usually it is found that the first approximation, (i.e., $a - x = a$), is excellent. Occasionally it is desirable to make a further refinement by the method illustrated here.

This method avoids some of the math errors that occur easily when the quadratic formula is used

The extent of dissociation of a weak acid depends on two factors, the acid constant K_a of the acid and the molarity of the acid in the solution. We have seen that a relatively large value of K_a will make for greater dissociation. It is also true that per cent dissociation will increase as the acid becomes more dilute. The actual degree of dissociation can be calculated by the methods in Examples 17.5 and 17.6. In Figure 17.2 we have plotted the per cent dissociation of lactic acid as a function of its molarity in solution. At high concentrations the per cent dissociation is low, and the approximation in Example 17.5 is valid.

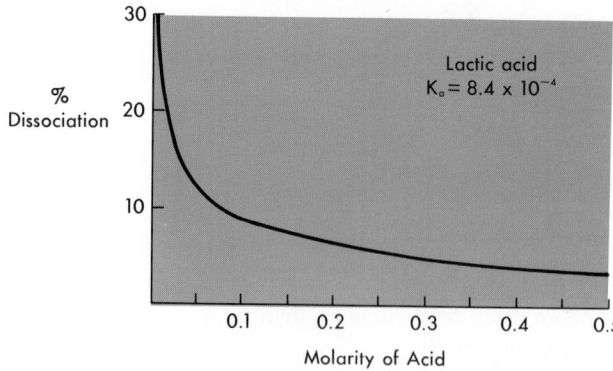

Figure 17.2 Degree of dissociation of a weak acid as a function of concentration.

17.5 WEAK BASES

There is a rather large family of substances called *weak bases,* paralleling the weak acids, which react in solution to create appreciable concentrations of hydroxide ion. The behavior of weak bases is analogous to that of the weak acids. A weak base on being added to water reacts in such a way that the solution becomes basic, but the concentration of hydroxide ion is much lower than it would be if the reaction to form OH⁻ ion went to completion.

As with the weak acids, there are a great many weak bases. In fact, it turns out that *for every weak acid, there is a corresponding weak base.* That base is the dissociation product of the acid, formed at the same time as the hydrogen ion. For the weak acid HX, where

$$HX(aq) \rightleftarrows H^+(aq) + X^-(aq) \quad (K_a \text{ is typically small})$$

the X⁻ ion in solution will *behave as a weak base.* The X⁻ ion is called the *conjugate base* of the acid HX.

To understand why the anion X⁻ is a weak base, we must consider the situation in a solution in which a salt containing X⁻ ion is added directly to *water.* The X⁻ ion will combine, if it can, with H⁺ from whatever source it can find, since the reaction to form HX has a strong tendency to proceed. The only obvious source is the water molecule, with which the following reaction can, and does, occur:

$$X^-(aq) + H_2O \rightleftarrows HX(aq) + OH^-(aq) \tag{17.9}$$

We would not expect Reaction 17.9 to go very far to the right, since water is very reluctant to release hydrogen ions. However, Reaction 17.9 does go to some extent; its K value is small, but since some OH⁻ ion is produced, the solution becomes basic.

In a very real sense the equilibrium position of Reaction 17.9 reflects the relative ease with which H_2O and HX give up a proton. As a matter of fact, we can readily show that the equilibrium constant for that reaction is a very simple function of K_w and K_a for HX. In a solution of X⁻ in water there are three important equilibria:

$$\text{I} \quad H_2O \rightleftarrows H^+(aq) + OH^-(aq) \qquad K_I = K_w = [H^+] \times [OH^-] \tag{17.1}$$

> What is the conjugate base of hydrocyanic acid, HCN?

II $X^-(aq) + H^+(aq) \rightleftarrows HX(aq)$ $K_{II} = 1/K_a = \dfrac{HX}{[H^+] \times [X^-]}$ (17.10)

III $X^-(aq) + H_2O \rightleftarrows HX(aq) + OH^-(aq)$ $K_{III} = K_b = \dfrac{[HX] \times [OH^-]}{[X^-]}$ (17.11)

(In formulating K_{III}, we incorporated, as before, $[H_2O]$ into the equilibrium constant.)

Reaction III, the reaction in which we are interested is, as you can see, the sum of Reactions I and II. We find that this puts a condition on K_{III}, relating it to K_I and K_{II}:

$$K_I = [H^+] \times [OH^-]; \; K_{II} = \frac{[HX]}{[H^+] \times [X^-]}$$

$$K_I \times K_{II} = [\cancel{H^+}] \times [OH^-] \times \frac{[HX]}{[\cancel{H^+}] \times [X^-]} = \frac{[HX] \times [OH^-]}{[X^-]}$$

But, we note from Equation 17.11 that: $K_{III} = \dfrac{[HX] \times [OH^-]}{[X^-]}$

Therefore, it must be true that

$$K_{III} = K_I \times K_{II} \qquad (17.12)$$

This analysis can be applied to any equilibrium system; indeed, we have derived here a general rule which can be stated in words as follows:

If, in a system involving multiple equilibria, one reaction can be expressed as the sum of other reactions, the equilibrium constant for that reaction will be equal to the product of the equilibrium constants for the reactions which are added. Mathematically,

if   Reaction III = Reaction I + Reaction II

Then $K_{III} = K_I \times K_{II}$

This relationship, which we shall call the *Multiple Equilibria Rule*, is a very useful one, and we shall apply it in several instances later in this text.

Referring back now to the solution of X^-, and applying the Multiple Equilibria Rule, we find that, since K_{III} equals K_1 times K_{II},

$$K_b = (K_w) \times (1/K_a) \quad \text{or} \quad K_a K_b = K_w \qquad (17.13)$$

The equilibrium constant K_b for Reaction 17.9:

$$X^-(aq) + H_2O \rightleftarrows HX(aq) + OH^-(aq)$$

is called the *base dissociation constant* for the X^- ion. If we know K_a for the weak acid, HX, we can always find K_b for its conjugate base X^- by using Equation 17.13.

Example 17.7   Lactic acid, HLac, is known to have a dissociation constant K_a equal to 8.4×10^{-4}. Find K_b for the lactate ion, Lac^-.

Solution.   Substituting into Equation 17.13,

$$K_b = K_w/K_a = \frac{1.0 \times 10^{-14}}{8.4 \times 10^{-4}} = 0.12 \times 10^{-10} = 1.2 \times 10^{-11}$$

TABLE 17.5   Dissociation Constants of Weak Acids and Bases

	ACID		K_a	BASE		K_b
Acetic acid	$HC_2H_3O_2$		1.8×10^{-5}	$C_2H_3O_2^-$		5.6×10^{-10}
Ammonium ion	NH_4^+		5.6×10^{-10}	NH_3		1.8×10^{-5}
Benzoic acid	$HC_7H_5O_2$		6.6×10^{-5}	$C_7H_5O_2^-$		1.5×10^{-10}
Boric acid	H_3BO_3		5.8×10^{-10}	$H_2BO_3^-$		1.7×10^{-5}
Carbonic acid	H_2CO_3		4.2×10^{-7}	HCO_3^-		2.4×10^{-8}
	HCO_3^-		4.8×10^{-11}	CO_3^{2-}		2.1×10^{-4}
Formic acid	$HCHO_2$		2.1×10^{-4}	CHO_2^-		4.8×10^{-11}
Hydrocyanic acid	HCN		4.0×10^{-10}	CN^-		2.5×10^{-5}
Hydrofluoric acid	HF		7.0×10^{-4}	F^-		1.4×10^{-11}
Hydrogen sulfide	H_2S		1×10^{-7}	HS^-		1×10^{-7}
	HS^-		1×10^{-15}	S^{2-}		1×10^{1}
Hypochlorous acid	$HClO$		3.2×10^{-8}	ClO^-		3.1×10^{-7}
Nitrous acid	HNO_2		4.5×10^{-4}	NO_2^-		2.2×10^{-11}
Phosphoric acid	H_3PO_4		7.5×10^{-3}	$H_2PO_4^-$		1.3×10^{-12}
	$H_2PO_4^-$		6.2×10^{-8}	HPO_4^{2-}		1.6×10^{-7}
	HPO_4^{2-}		1.7×10^{-12}	PO_4^{3-}		5.9×10^{-3}
Propionic acid	$HC_3H_5O_2$		1.4×10^{-5}	$C_3H_5O_2^-$		7.1×10^{-10}
Hydrogen sulfate	HSO_4^-		1.2×10^{-2}	SO_4^{2-}		8.3×10^{-13}
Sulfurous acid	H_2SO_3		1.7×10^{-2}	HSO_3^-		5.9×10^{-13}
	HSO_3^-		5.6×10^{-8}	SO_3^{2-}		1.8×10^{-7}

For the acids:
$$HX(aq) \rightleftarrows$$
$$H^+(aq) + X^-(aq)$$
$$K_a = \frac{[H^+][X^-]}{[HX]}$$

For the bases:
$$X^-(aq) + H_2O \rightleftarrows$$
$$HX(aq) + OH^-(aq)$$
$$K_b = \frac{[HX][OH^-]}{[X^-]}$$

In Table 17.5 we have listed values of K_a and K_b for some common weak acids and their conjugate bases.

Because of the fact that K_b is finite for the conjugate bases of weak acids, solutions containing the sodium salts of weak acids will be basic. Sodium lactate solution, containing the lactate ion, Lac^-, the conjugate base of lactic acid, HLac, is basic. Sodium carbonate solution, Na_2CO_3, containing a salt of carbonic acid, H_2CO_3, is also basic. The strength of a weak base depends on its value of K_b in the same way as the strength of a weak acid depends on K_a. A relatively large K_b value implies a relatively strong base. Cyanide ion, CN^-, for which K_b equals 2.5×10^{-5}, is a much stronger base than acetate ion, $C_2H_3O_2^-$, whose K_b equals 5.6×10^{-10}.

Since K_a and K_b for an acid and its conjugate base must have a product equal to K_w, it is true that the conjugate base of a moderately strong acid, with a large value of K_a, will be relatively weak since its K_b value will have to be small. The cyanide ion is a relatively strong base because hydrocyanic acid, HCN, is such a weak acid ($K_a = 4.0 \times 10^{-10}$). Lactate ion will behave as a very weak base because lactic acid is relatively strong ($K_a = 8.4 \times 10^{-4}$).

The strong acids, whose K_a values are very large, essentially infinite for our purposes, will have conjugate bases which are so weak as to be neutral, $K_b \cong 0$. This explains why Cl^-, Br^-, I^-, NO_3^-, and ClO_4^- ions are neutral ions in solution. Similarly, since K_b for NaOH and KOH is very large, K_a values for Na^+ and K^+ are effectively zero, making those ions also neutral in their solutions. In Table 17.6 we have classified the commonly encountered ions according to their acidic or basic properties in water solutions. In that table we have included some ions from acids which can furnish more than one hydrogen ion; such species, like HSO_4^- and HCO_3^-, can either ionize further, tending to make an acid solution, or

TABLE 17.6 Acid-Base Properties of Ions in Aqueous Solution

	NEUTRAL	ACIDIC	BASIC	
Anions	Cl^-	HSO_4^-	OH^-	HS^-
	Br^-	$H_2PO_4^-$	CO_3^{2-}	CN^-
	I^-		HCO_3^-	F^-
	NO_3^-		PO_4^{3-}	$C_2H_3O_2^-$
	ClO_4^-		HPO_4^-	
	SO_4^{2-}		S^{2-}	
Cations	Na^+	H^+	none	
	K^+	Al^{3+}		
	Li^+	NH_4^+		
	Ca^{2+}	Zn^{2+}		
	Mg^{2+}	Cu^{2+}		
	Ba^{2+}	other transition metal ions		

react as conjugate bases. Since H_2SO_4 is a strong acid, K_b for HSO_4^- is about zero, so the only route for HSO_4^- is to ionize to form an acidic solution. The HCO_3^- ion has finite values for both K_a and K_b (4.8×10^{-11} and 2.4×10^{-8}, respectively); since K_b is a little larger than K_a, a solution of $NaHCO_3$ is slightly basic.

Example 17.8 Classify each of the following 0.1 M solutions as acidic, neutral, or basic:

$$NaNO_3, \quad KCN, \quad Na_2CO_3$$
$$NH_4Br, \quad AlCl_3, \quad NH_4F$$

Solution.

$NaNO_3$	neutral cation, neutral anion	neutral solution
KCN	neutral cation, basic anion	basic solution
Na_2CO_3	neutral cation, basic anion	basic solution
NH_4Br	acidic cation, neutral anion	acidic solution
$AlCl_3$	acidic cation, neutral anion	acidic solution
NH_4F	acidic cation, basic anion	(comparison necessary)

K_a for $NH_4^+ = 5.6 \times 10^{-10}$; K_b for $F^- = 1.4 \times 10^{-11}$

NH_4^+ is a stronger acid than F^- is a base; $K_a > K_b$

The solution would be very slightly acidic.

Table 17.6 allows one to predict the acidic or basic properties of most of the common salt solutions

---

Most sodium salts of weak acids are such weak bases that they are not very effective as laboratory sources of hydroxide ion. An important exception is sodium carbonate, Na_2CO_3, which is a reasonably strong base and is also very soluble. (Sodium carbonate is sometimes called washing soda, because in the days when soap was used for laundry purposes sodium carbonate was frequently added to the wash water to improve its solvent properties. Washing soda is not a serious pollutant, since carbonates occur naturally in surface waters; the difficulty with carbonates as cleaning agents is that they form insoluble $CaCO_3$ with hard water, and this gums up the wash pretty badly.)

The most commonly encountered weak base is ammonia, NH_3. Ammonia is made by dissolving NH_3 gas in water; it has a reasonably large base dissociation constant, $K_b = 1.8 \times 10^{-5}$, and is very soluble, so it makes a useful base in the chemistry lab. The stock solutions are 6 M NH_3 and 17 M NH_3. Ammonia has some other desirable properties; it is relatively volatile, so that it can be driven, if necessary, from solution by boiling; it also forms stable complexes with many metallic cations, which makes it an important reagent in qualitative analysis. (See Chapter 19.)

CALCULATIONS INVOLVING K_b

Calculations of the equilibrium properties of solutions of weak bases are made in a manner very similar to that used with weak acids.

Example 17.9 Find $[OH^-]$, $[H^+]$, and the concentration of unreacted base in

   a. 0.50 M NH_3 b. 0.50 M $NaC_2H_3O_2$

Solution

 a. NH_3 in solution acts as a weak base:

$$NH_3(aq) + H_2O \rightleftarrows NH_4^+(aq) + OH^-(aq)$$

$$K_b = \frac{[NH_4^+] \times [OH^-]}{[NH_3]} = 1.8 \times 10^{-5}$$

Since the reaction produces NH_4^+ and OH^- ions in equal amounts, we can say that, at equilibrium,

$$[NH_4^+] = [OH^-] = x$$

We can see that K_b is not very large, so only a small percentage of the NH_3 would be expected to react, which would leave its concentration essentially unchanged, at 0.50 M. Substituting into the expression for K_b, we obtain

$$K_b = \frac{x^2}{0.50} = 1.8 \times 10^{-5}; \quad x^2 = 0.90 \times 10^{-5} = 9.0 \times 10^{-6}$$

$$x = 3.0 \times 10^{-3} = [OH^-] = [NH_4^+]$$

Since $[OH^-]/[NH_3]$ is small, $3 \times 10^{-3}/0.50$, about 0.006, very little of the NH_3 reacts, and the approximation is a good one; $[NH_3]$ remains at essentially 0.50 M.

 A tiny bit of water dissociates to furnish H^+ ion to satisfy the condition:

$$[H^+] \times [OH^-] = 1.0 \times 10^{-14} = [H^+] \times 3.0 \times 10^{-3}$$

$$[H^+] = \frac{1.0 \times 10^{-14}}{3.0 \times 10^{-3}} = \frac{10 \times 10^{-15}}{3.0 \times 10^{-3}} = 3.3 \times 10^{-12} \text{ M}$$

 b. In $NaC_2H_3O_2$ solution, the Na^+ ion will be neutral and the basic character of the solution will be due to the $C_2H_3O_2^-$ ion:

$$C_2H_3O_2^-(aq) + H_2O \rightleftarrows HC_2H_3O_2(aq) + OH^-(aq)$$

$$K_b = \frac{[HC_2H_3O_2] \times [OH^-]}{[C_2H_3O_2^-]} = 5.6 \times 10^{-10}$$

As before, K_b is very small; $[HC_2H_3O_2] = [OH^-] = x$; $[C_2H_3O_2^-] \cong$ 0.50. Substituting,

$$K_b = \frac{x^2}{0.50} = 5.6 \times 10^{-10} \quad x^2 = 2.8 \times 10^{-10}$$

$$x = 1.7 \times 10^{-5} = [OH^-] = [HC_2H_3O_2]$$

The solution is less basic than NH_3 of similar molarity, since the acetate ion is a weaker base (smaller value of K_b) than is ammonia. For the H^+:

$$[H^+] = K_w/[OH^-] = 5.9 \times 10^{-10} \text{ M}$$

17.6 FACTORS INFLUENCING STRENGTHS OF ACIDS AND BASES

We have seen in the previous sections that the degrees of dissociation of acids and bases vary widely, from essentially 100 per cent to nearly zero. In this section we shall discuss some of the reasons for these differences, starting with an examination of what it is that makes one substance an acid when another, which superficially resembles it, behaves as a base.

Nearly every acid and all the strong bases can be represented in a schematic way by the formula

$$-\overset{|}{\underset{|}{M}}-O-H$$

where M is a central atom, metallic or nonmetallic, to which may be attached, in addition to one or more OH groups, other atoms such as oxygen or larger molecular fragments. In Table 17.7 we have drawn the structures of several substances that can be described by the general formula. Most of the central elements M are from the third period in the Periodic Table, which furnishes several well-known acids and bases.

Among the substances listed in the table, all the acids lose protons on ionization, while the two bases lose the hydroxide ion. Since all the species contain an M—O—H group, clearly the factor which makes one substance an acid and another a base is the relative weakness of either the O—H bond or the M—O bond. Among the compounds listed, only NaOH and $Mg(OH)_2$ behave as bases. The bond between the metallic cation and the hydroxide ion is primarily ionic and much more likely to be broken in solution than is the covalent bond holding the hydroxide ion together. In NaOH essentially complete ionization occurs. The large charge and small size of the Mg^{2+} ion are sufficient to make $Mg(OH)_2$ rather insoluble (about 10 mg/lit), so even its saturated solution is not very basic. The solubility is even lower for $Al(OH)_3$, where the high charge density of the Al^{3+} ion holds the OH^- groups very tightly (aluminum hydroxide has, it turns out, both weakly acidic and basic properties). The strong bases, then, are limited to those few hydroxides in which the charge density and electronegativity of the cation are low, as in Na^+, and K^+, and to some extent the 2A cations. Since many of the other metallic cations have high charge to size ratios, their hydroxides are insoluble and their basic character limited. It is also true that many metals have sufficiently high electronegativities to make the M—O bond in the hydroxides largely covalent.

In the strong bases the M—OH bond is essentially ionic

TABLE 17.7 Structures of Some Representative Acids and Bases

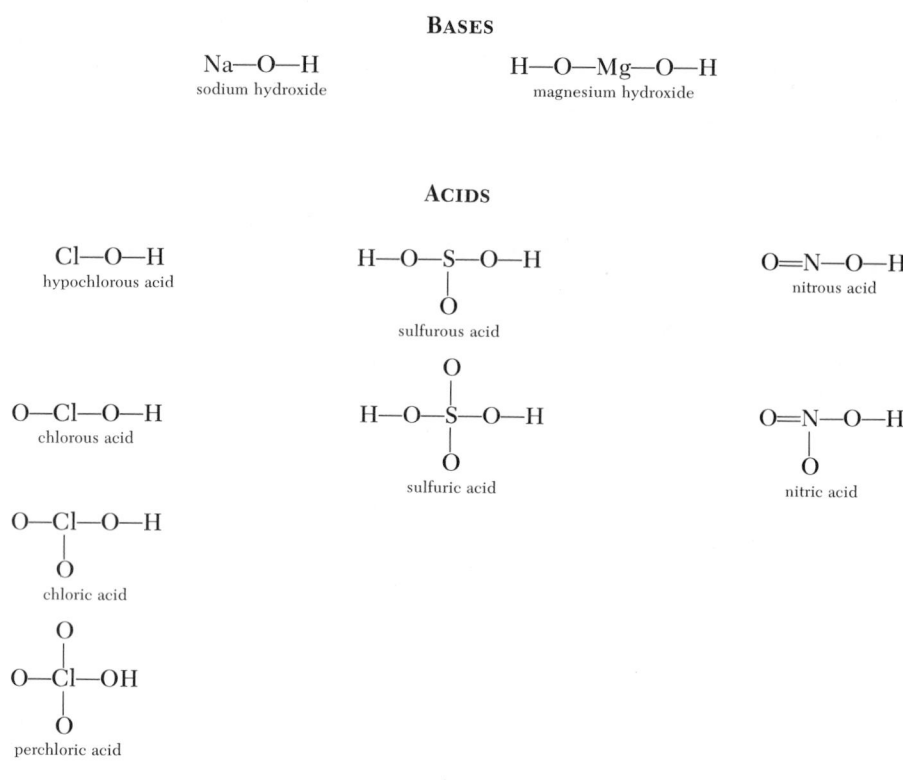

BASES

Na—O—H
sodium hydroxide

H—O—Mg—O—H
magnesium hydroxide

ACIDS

Cl—O—H
hypochlorous acid

O—Cl—O—H
chlorous acid

O—Cl—O—H
|
O
chloric acid

O
‖
O—Cl—OH
|
O
perchloric acid

H—O—S—O—H
|
O
sulfurous acid

O
‖
H—O—S—O—H
|
O
sulfuric acid

O=N—O—H
nitrous acid

O=N—O—H
|
O
nitric acid

As one selects M from groups 4A, 5A, 6A, and 7A in the Periodic Table, its electronegativity gradually increases with group number, making for more covalent character in the M—O bond. The increased electronegativity of the M atom causes a shift in the electrons in the M—O bond toward M; this tends to some extent to pull electrons in the O—H bond toward the O atom and weakens the O—H bond. If M is a nonmetal, the electron shift is large enough to make M—O—H behave as an acid. Hypochlorous acid, HClO, formally very similar in its formula to NaOH, behaves as a weak acid.

As more oxygen atoms are added to the central atom, they also tend to draw electrons from the central atom, which weakens the O—H bond still more and makes the resultant acid stronger. The effect is particularly dramatic for the oxyacids formed by chlorine, where the acid strength varies in the order $HClO < HClO_2 < HClO_3 < HClO_4$, but is observed with sulfur and nitrogen as well. (See Table 17.8.) HNO_3 and H_2SO_4 are strong acids, while the analogous acids with fewer oxygen atoms are weak. It appears that for an acid to be strong, it must have at least *two* oxygen atoms bonded to the central atom but not to hydrogen.

All the strong oxyacids follow this rule

TABLE 17.8   Strength of Oxyacids

Variation with Number of Unprotonated Oxygen Atoms

Acid	K_a	Acid	K_a	Acid	K_a
$HClO_4$	10^7	H_2SO_4	very large	HNO_3	very large
$HClO_3$	large	H_2SO_3	1.7×10^{-2}	HNO_2	4.5×10^{-4}
$HClO_2$	1×10^{-2}			$H_2N_2O_2$	9×10^{-8}
$HClO$	3.2×10^{-8}				

Variation with Electronegativity

Acid	K_a	Acid	K_a	Acid	K_a
$HClO$	3.2×10^{-8}	H_2SO_3	1.7×10^{-2}	HNO_3	very large
$HBrO$	2×10^{-9}	H_2SeO_3	3×10^{-3}	H_3PO_4	7.5×10^{-3}
HIO	5×10^{-13}	H_2TeO_3	6×10^{-6}	H_3AsO_4	5×10^{-3}

From what we have said we can formulate a general rule for relative strength of the acid —M—O—H in terms of the electron attracting power of the central atom M. **The acid strength increases with increasing electronegativity of the central atom and is enhanced if other electronegative atoms, such as oxygen, are also attached to the central atom.**

The rule we have just formulated can be tested very nicely. For example, it predicts that within a group in the Periodic Table the strengths of acids with the same general formula would be largest for central atoms with the highest electronegativities. This is borne out by the data in Table 17.8, where we see that HClO is a stronger acid than HBrO(EN Cl = 3.0, EN Br = 2.8), and H_2SO_3 is a stronger acid than H_2SeO_3(EN S = 2.5, EN Se = 2.4).

In organic molecules containing the —C̤—O—H group, the C—O and O—H bonds are both covalent and strong, so that in the absence of highly electronegative atoms attached to the carbon atom, the molecules behave as exceedingly weak acids; methanol, CH_3OH, is essentially neutral as far as acid-base properties go. All the organic acids contain the group O=C̤—O—H, in which the double-bonded oxygen atom is strongly electron-attracting; this results ultimately in sufficient weakening of the O—H bond to make the molecule behave as an acid. In trichloracetic acid, Cl_3C—COOH, the chlorine atoms are so electron-attracting as to make the acid quite strong ($K_a = 2 \times 10^{-1}$).

Many of the common acids and bases can be considered to be derived from oxides. As we noted earlier, carbonic acid, H_2CO_3, is really a water solution of CO_2; solutions of sulfuric acid, H_2SO_4, and sodium hydroxide, NaOH, can be made by dissolving SO_3 and Na_2O, respectively, in water. In the early days chemists spoke more often than they do now of acidic and basic oxides in the above sense. The "dehydrated" species, the oxide, is sometimes called an anhydride, especially in the case of the oxyacids. In Table 17.9 we have listed some of the more important oxyacids, the oxyanions, and the associated acid anhydrides.

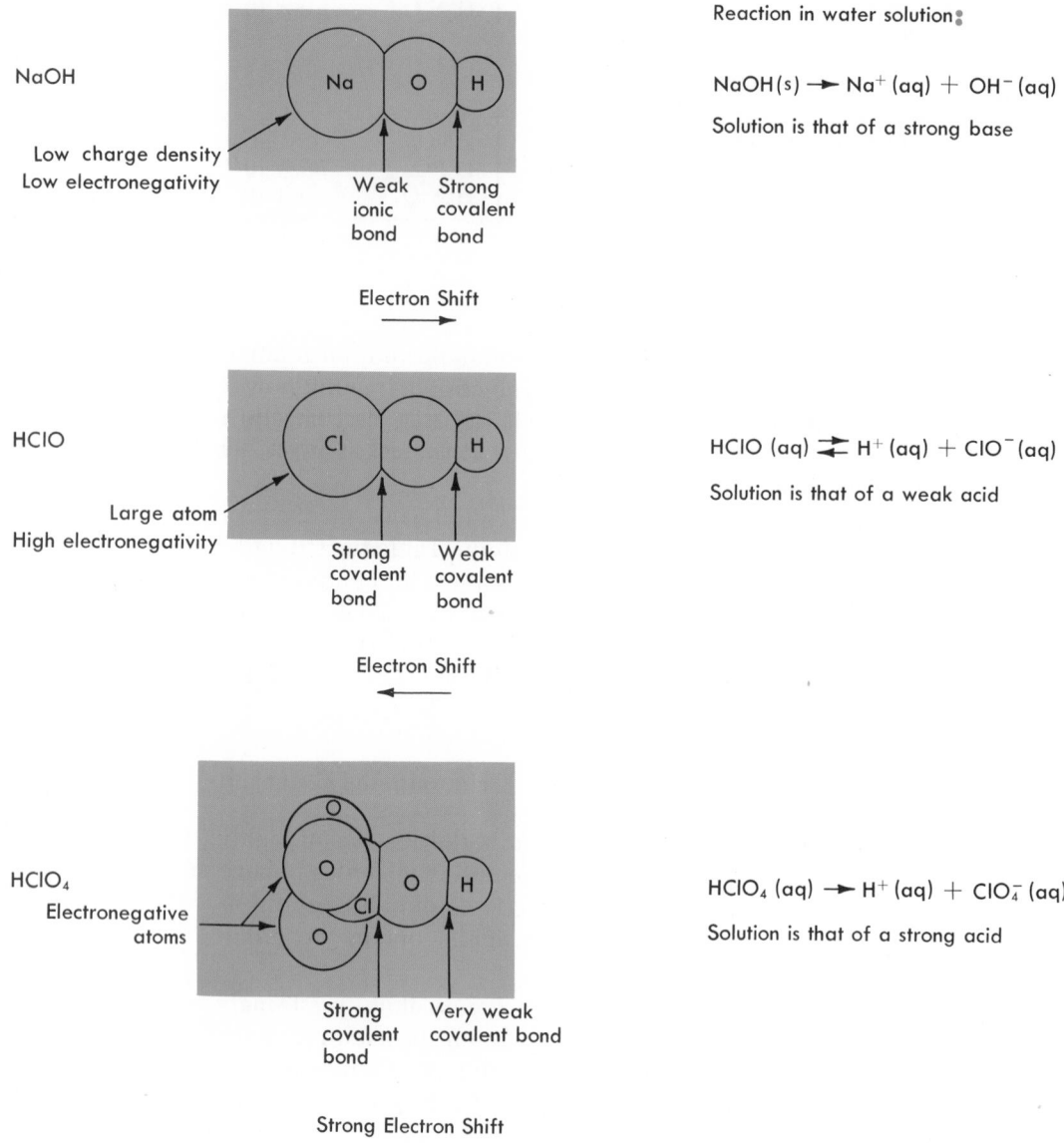

NaOH

Low charge density
Low electronegativity

Weak ionic bond
Strong covalent bond

Electron Shift

HClO

Large atom
High electronegativity

Strong covalent bond
Weak covalent bond

Electron Shift

$HClO_4$
Electronegative atoms

Strong covalent bond
Very weak covalent bond

Strong Electron Shift

Reaction in water solution:

$$NaOH(s) \longrightarrow Na^+ (aq) + OH^- (aq)$$

Solution is that of a strong base

$$HClO (aq) \rightleftharpoons H^+ (aq) + ClO^- (aq)$$

Solution is that of a weak acid

$$HClO_4 (aq) \longrightarrow H^+ (aq) + ClO_4^- (aq)$$

Solution is that of a strong acid

Figure 17.3 Factors influencing acid-base strength.

TABLE 17.9   Acid Anhydrides, Oxyacids, and Oxyanions

Acid Anhydride	Oxyacid	Oxyanion
Cl_2O_7	$HClO_4$	ClO_4^-
Cl_2O	$(HClO)$ *	ClO^-
SO_3	H_2SO_4	HSO_4^-
SO_2	(H_2SO_3)	HSO_3^-
N_2O_5	HNO_3	NO_3^-
P_2O_5	H_3PO_4	$H_2PO_4^-$
CO_2	(H_2CO_3)	HCO_3^-

* The oxyacids enclosed in parentheses cannot be isolated from water solution as pure compounds. Indeed, in the case of carbonic acid, H_2CO_3, the evidence for its existence is fragmentary at best

17.7   GENERAL THEORIES OF ACIDS AND BASES

Chemical theorists have always been interested in the fundamental properties of acids and bases. Acids and bases have been known since the beginnings of chemistry, and the fact that their behavior is so closely linked to water, the solvent, tended to confuse matters. So far in this chapter we have considered an acid to be a substance which in water solution produces an excess of H^+ ion. A base was similarly defined to be a substance which, directly or indirectly, forms excess OH^- ion in solution. This approach to acids and bases is indeed a very practical one and was first proposed by Arrhenius in 1884.

The Arrhenius concept defines acids and bases in terms of the species they produce on addition to water. This approach tends to minimize to some extent the role that the solvent plays in acid-base systems, and in so doing sacrifices some rather useful insights into the relationships between acids and bases. An alternate, more general picture, suggested independently by Brönsted in Denmark and Lowry in England in 1923, shows the relationships more clearly.

BRÖNSTED-LOWRY CONCEPT

According to the Brönsted-Lowry idea, an acid-base reaction is one in which there is a *proton transfer* from one species to another. The species which gives up, or *donates*, the proton is referred to as an acid; the molecule or ion which *accepts* the proton is a *base*.

To illustrate the application of the Brönsted-Lowry concept, let us re-examine as an example of an acid-base reaction the ionization of HF in water solution:

$$HF(aq) \rightleftarrows H^+(aq) + F^-(aq) \quad K_a = 7.0 \times 10^{-4} \qquad (17.7)$$

We immediately run into a snag, since although HF appears to be the proton donor, there isn't any obvious acceptor. According to Brönsted and Lowry, in such a reaction a proton acceptor must be present. In this case the acceptor is the water molecule itself. Recognizing that, we can write the reaction for the ionization in the following way:

$$HF(aq) + H_2O \rightleftarrows H_3O^+(aq) + F^-(aq) \qquad (17.7A)$$
$$\underset{\text{acid}}{} \quad \underset{\text{base}}{}$$

This equation satisfies the Brönsted-Lowry picture; it tells us that in the reaction the acid, HF, the proton donor, reacts with the base, water, the proton acceptor, to form a hydrated proton and a fluoride ion.

One advantage in writing the reaction in this way is that it shows more clearly than Equation 17.7 why the ionization occurs at all. Certainly the breaking of the bond in HF takes energy; in fact, the bond-breaking reaction is so endothermic that if it were the only change that occurred, K_a would be virtually zero. We would, however, expect that the H^+ ion produced in solution would react strongly with water to form a hydrated species, and Equation 17.7A emphasizes that hydration. We have written the formula of the hydrated proton as H_3O^+, although it is very likely that there are several hydrates, such as $H_5O_2^+$ or $H_9O_4^+$, present in the acidic solution.

Another feature of the Brönsted-Lowry approach becomes clear when we write Reaction 17.7A in the reverse direction:

$$H_3O^+(aq) + F^-(aq) \rightleftarrows HF(aq) + H_2O \qquad (17.7A')$$
$$\underset{\text{acid}}{} \quad \underset{\text{base}}{} \quad \underset{\text{acid}}{} \quad \underset{\text{base}}{}$$

In this reaction the fluoride ion is clearly a base, since it accepts a proton from the acid, H_3O^+. Using the Arrhenius approach we showed that F^- ion was a base by a rather different argument; here its role as a base is clear.

It is possible to look at all acid-base reactions in terms of the Brönsted-Lowry idea. In every such reaction an acid reacts with a base to form another acid and another base which are conjugate to the original ones. We might write several other examples, taken from earlier portions of this chapter:

$$HC_2H_3O_2(aq) + H_2O \rightleftarrows H_3O^+(aq) + C_2H_3O_2^-(aq)$$
$$\underset{\text{acid}}{} \qquad \underset{\text{base}}{} \qquad \underset{\text{acid}}{} \qquad \underset{\text{base}}{}$$

$$NH_4^+(aq) + H_2O \rightleftarrows H_3O^+(aq) + NH_3(aq)$$
$$\underset{\text{acid}}{} \qquad \underset{\text{base}}{} \qquad \underset{\text{acid}}{} \qquad \underset{\text{base}}{}$$

$$F^-(aq) + H_2O \rightleftarrows HF(aq) + OH^-(aq)$$
$$\underset{\text{base}}{} \qquad \underset{\text{acid}}{} \qquad \underset{\text{acid}}{} \qquad \underset{\text{base}}{}$$

$$NH_3(aq) + H_2O \rightleftarrows NH_4^+(aq) + OH^-(aq)$$
$$\underset{\text{base}}{} \qquad \underset{\text{acid}}{} \qquad \underset{\text{acid}}{} \qquad \underset{\text{base}}{}$$

What is the conjugate acid of H_2O? the conjugate base of H_2O?

On examining the above reactions, one notes that water, according to the Brönsted-Lowry approach, can act as both an acid and a base. In both roles it is very weak. There are some other species which can act as both acid and base; these are all derived from substances having at least two available acid protons. The HCO_3^- ion is an example; in reaction with sodium hydroxide solution it is an acid, whereas with hydrofluoric acid it behaves as a base:

$$HCO_3^-(aq) + OH^-(aq) \rightleftarrows H_2O + CO_3^{2-}(aq)$$
$$\underset{\text{acid}}{} \qquad \underset{\text{base}}{} \qquad \underset{\text{acid}}{} \qquad \underset{\text{base}}{}$$

$$HCO_3^-(aq) + HF(aq) \rightleftharpoons H_2CO_3(aq) + F^-(aq)$$
$$\underset{\text{base}}{} \qquad \underset{\text{acid}}{} \qquad \underset{\text{acid}}{} \qquad \underset{\text{base}}{}$$

An important advantage of the Brönsted-Lowry concept of acids and bases is that it can be applied to solutions in which the solvent is not water. In liquid ammonia, for example, sodium hydride ionizes and participates in the following acid-base reaction:

$$\underset{\text{acid}}{NH_3} + \underset{\text{base}}{H^-} \rightarrow \underset{\text{acid}}{H_2} + \underset{\text{base}}{NH_2^-}$$

Here the ammonia molecule acts as an acid, donating a proton to the base, H^-.

In Table 17.10 we have listed a group of Brönsted-Lowry acids and bases, along with their dissociation constants. Included are the strong acids and some strong bases, along with the weak acids listed in Table 17.5. The dissociation constants, where they can be measured, are as in Table 17.5. Acid strength decreases as you go down the left column; the strong acids are arranged in order of strength as determined in a solvent like H_2O_2, which is a weaker base than water. In water the strong acids are completely ionized, but in H_2O_2 they are not and their relative dissociation constants can be estimated. The bases increase in strength as you go down the right column; the anions of the strong acids, at the top of the column, are neutral species, since their K_b values are effectively zero.

TABLE 17.10 Brönsted-Lowry Acids and Bases

K_a	ACID	BASE	K_b
very large	$HClO_4$	ClO_4^-	very small
very large	HI	I^-	very small
very large	HBr	Br^-	very small
very large	HCl	Cl^-	very small
very large	H_2SO_4	HSO_4^-	very small
very large	HNO_3	NO_3^-	very small
very large	$H_3O_2^+$	H_2O_2	very small
very large	H_3O^+	H_2O	very small
1.7×10^{-2}	H_2SO_3	HSO_3^-	5.9×10^{-13}
1.2×10^{-2}	HSO_4^-	SO_4^{2-}	8.3×10^{-13}
7.5×10^{-3}	H_3PO_4	$H_2PO_4^-$	1.3×10^{-12}
7.0×10^{-4}	HF	F^-	1.4×10^{-11}
4.5×10^{-4}	HNO_2	NO_2^-	2.2×10^{-11}
1.8×10^{-5}	$HC_2H_3O_2$	$C_2H_3O_2^-$	5.6×10^{-10}
4.2×10^{-7}	H_2CO_3	HCO_3^-	2.4×10^{-8}
1×10^{-7}	H_2S	HS^-	1×10^{-7}
5.6×10^{-8}	HSO_3^-	SO_3^{2-}	1.8×10^{-7}
3.2×10^{-8}	$HOCl$	OCl^-	3.1×10^{-7}
5.6×10^{-10}	NH_4^+	NH_3	1.8×10^{-5}
4.0×10^{-10}	HCN	CN^-	2.5×10^{-5}
4.8×10^{-11}	HCO_3^-	CO_3^{2-}	2.1×10^{-4}
2.6×10^{-12}	H_2O_2	HO_2^-	3.8×10^{-3}
4.4×10^{-13}	HPO_4^{2-}	PO_4^{3-}	2.3×10^{-2}
1.0×10^{-14}	H_2O	OH^-	1
very small	CH_3OH	CH_3O^-	very large
very small	NH_3	NH_2^-	very large
very small	H_2	H^-	very large

Decreasing acid strength ↓   Decreasing base strength ↑

THE LEWIS CONCEPT

We have seen that the Brönsted-Lowry picture represents a considerable extension of the Arrhenius concept of acids and bases. However, the Brönsted-Lowry picture is restricted in one important respect: it can be applied only to reactions involving a proton transfer. In particular, in order to act as a Brönsted-Lowry acid, a species must contain an ionizable hydrogen atom.

The Lewis acid-base concept, first proposed by the American physical chemist G. N. Lewis in 1923, removes this restriction. The Lewis concept considers an *acid* to be a species that can *accept* an *electron pair; a base* is a substance that can *donate* an *electron pair*. According to the Lewis concept, any reaction which leads to the formation of a coordinate covalent bond is an acid-base reaction.

From a structural point of view, the Lewis concept of a base does not differ in any essential way from the Brönsted concept. In order for a species to accept a proton and thereby act as a Brönsted base it must possess an unshared pair of electrons. Consider, for example, the NH_3 molecule, the H_2O molecule, and the F^- ion, all of which can act as Brönsted bases:

$$H-\overset{\cdot\cdot}{N}-H, \quad H-\overset{\cdot\cdot}{\underset{\cdot\cdot}{O}}-H, \quad (:\overset{\cdot\cdot}{\underset{\cdot\cdot}{F}}:)^-$$
$$\vert$$
$$H$$

Each of these species contains an unshared pair of electrons that is utilized in accepting a proton to form the NH_4^+ ion, the H_3O^+ ion, or the HF molecule.

$$\left[\begin{array}{c} H \\ H:\overset{\cdot\cdot}{N}:H \\ \overset{\cdot\cdot}{H} \end{array} \right]^+ \quad \left[H:\overset{\cdot\cdot}{\underset{\cdot\cdot}{O}}:H \atop \overset{}{H} \right]^+ \quad H:\overset{\cdot\cdot}{\underset{\cdot\cdot}{F}}:$$

Clearly, NH_3, H_2O, and F^- can also act as Lewis bases since they possess an unshared electron pair which can be donated to an acid. We see then that the Lewis concept does not significantly change the number of species which can act as bases.

On the other hand, the Lewis concept greatly increases the number of species which can be considered to be acids. The substance which accepts an electron pair and therefore acts as a Lewis acid can be a proton:

In each case the Lewis base furnishes the pair of electrons which form the bond with H^+

$$H^+(aq) + H_2O \rightarrow H_3O^+(aq)$$
$$\text{acid} \qquad \text{base}$$

$$H^+(aq) + NH_3(aq) \rightarrow NH_4^+(aq)$$
$$\text{acid} \qquad \text{base}$$

It can equally well be a cation, such as Zn^{2+}, which is capable of forming a coordinate covalent bond with a Lewis base:

$$Zn^{2+}(aq) + 4\ H_2O \rightarrow Zn(H_2O)_4^{2+}(aq)$$
$$\text{acid} \qquad \text{base}$$

$$Zn^{2+}(aq) + 4\ NH_3 \rightarrow Zn(NH_3)_4^{2+}(aq)$$
$$\text{acid} \qquad \text{base}$$

Another important class of Lewis acids comprises molecules containing an incomplete octet of electrons. A classic example is boron trifluoride, BF_3, which reacts readily with ammonia, accepting a pair of electrons:

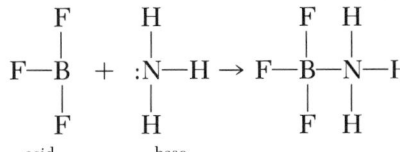

acid   base

Lewis acids are often used as catalysts in organic chemistry

In most cases when a chemist speaks of an acid or base he uses the terms in the Arrhenius or the Brönsted-Lowry sense. The term "Lewis acid" is more restricted and is generally taken to refer to a species like BF_3, which is a powerful electron pair acceptor.

PROBLEMS

17.1   Distinguish between

     a. an acid and a base.
     b. a weak base and a strong base.
     c. a Brönsted acid and a Lewis acid.
     d. an acid and its anhydride.
     e. K_a and K_b.
     f. pH and $[H^+]$.

17.2   Find the pH of solutions with the following $[H^+]$:

     a. 1×10^{-4} M   c. 3×10^{-6} M
     b. 1 M   d. 5×10^{-10} M

17.3   Calculate $[H^+]$ and $[OH^-]$ in solutions having the following pH:

     a. 6.0   c. -0.50
     b. 3.60   d. 9.80

17.4   Find $[H^+]$, $[OH^-]$, and the pH of the following solutions:

     a. 0.25 M HNO_3
     b. 0.20 M KOH
     c. a solution made by diluting 50 ml 0.2 M HNO_3 to 300 ml.
     d. a solution made by dissolving 10 g NaOH in water to make 200 ml of solution.

17.5   Give two examples of

     a. an acid that is made by dissolving an oxide in water.
     b. a weak acid that does not contain oxygen.
     c. an acid that is weaker than $HC_2H_3O_2$.
     d. a base that is stronger than H_2O.
     e. a salt which dissolves in water to form a basic solution.
     f. a salt which dissolves in water to yield an acidic solution.

17.21   Define briefly but clearly each of the following:

     a. an acid.   e. a Lewis acid.
     b. a weak base.   f. a strong acid.
     c. a conjugate base.   g. an acid anhydride.
     d. a Brönsted base.   h. a basic water solution.

17.22   Find the pH of solutions with the following $[H^+]$:

     a. 1×10^{-2} M
     b. 2 M
     c. 4×10^{-8} M
     d. 6.2×10^{-3} M

17.23   Find $[H^+]$ and $[OH^-]$ in solutions having the following pH; indicate for each solution whether it is acidic or basic:

     a. 2.0   c. -1.00
     b. 6.5   d. 13.20

17.24   Find $[H^+]$, $[OH^-]$, and the pH of the following solutions:

     a. 2 M NaOH
     b. 0.3 M HBr
     c. a solution made by dissolving 10 g HNO_3 in two liters of water.
     d. a solution made by diluting 25 ml 6.0 M NaOH to 400 ml.

17.25   Give two examples of

     a. a Lewis acid.
     b. a strong acid.
     c. a substance with two acid hydrogen atoms.
     d. an acid that is stronger than H_2O.
     e. a salt which dissolves in water to yield a neutral solution.
     f. an oxide that would dissolve in water to form a basic solution.

17.6   In each pair choose the species that would be the stronger acid:

  a. NH_4^+ or NH_3
  b. $HClO$ or HIO
  c. $HClO_3$ or $HClO_2$
  d. CH_3COOH or $CH_2ClCOOH$
  e. H_3PO_4 or H_3AsO_4
  f. H_2O or H_2S

17.7   Write balanced net ionic equations for the reaction, if any, of each of the following ions in water solution; indicate whether the final solution would be acidic or basic:

  a. NH_4^+ d. K^+
  b. HSO_4^- e. $Cu(H_2O)_4^{2+}$
  c. CO_3^{2-} f. CN^-

17.8   Six solutions are prepared, each containing a half mole of one of the substances listed dissolved in one liter of water; for each solution state whether it would be acidic, basic, or neutral, and write a net ionic equation to explain why:

  a. $LiOH$ d. Na_2CO_3
  b. NH_4NO_3 e. $KHSO_4$
  c. Na_2S f. KBr

17.9   Butyric acid, $CH_3CH_2CH_2COOH$, is one of the decomposition products in rancid butter and has a well-deserved malodorous reputation. In a well-vented hood a student measured the pH of 500 ml of a solution containing 0.20 mole/ butyric acid and found it to be 2.60. Find the dissociation constant K_a for butyric acid.

17.10   Having isolated a sample of an organic acid which he believed to contain only one ionizable hydrogen atom, a chemist dissolved 0.50 g of the sample in 10 ml water. He divided the solution into two equal volumes and titrated one of them with NaOH until it was neutral. He added the neutral solution to the other portion of acid and measured the pH of the mixture, finding it to be 6.2. Estimate K_a of the unknown acid.

17.11   Find the $[H^+]$, $[OH^-]$, the pH, and the per cent dissociation of the acid in each of the following solutions:

  a. 0.30 M HI
  b. 0.30 M NH_4I
  c. 0.30 M HCN
  d. 0.30 M $HC_2H_3O_2$

17.12   Calculate the $[OH^-]$, $[H^+]$, the pH, and the concentrations of other solute species in

  a. 0.45 M KOH
  b. 0.45 M NH_3
  c. 0.45 M KCN
  d. 0.45 M sodium benzoate

17.26   In each pair select the species that would be the stronger acid:

  a. HBr or H_2S
  b. H_2SO_3 or H_2SO_4
  c. CCl_3COOH or CBr_3COOH
  d. NH_4^+ or NH_3
  e. $HBrO$ or $HBrO_3$

17.27   Write balanced net ionic equations for the reactions, if any, of each of the following ions in water solution; indicate whether as a result of the reaction the solution would be acidic or basic:

  a. Li^+ d. $Zn(H_2O)_4^{2+}$
  b. S^{2-} e. Br^-
  c. HCO_3^- f. $C_2H_3O_2^-$

17.28   Ten grams of each of the following substances is dissolved in separate 500-ml volumes of water. In each case indicate whether the solution would be acidic, basic, or neutral, and write a net ionic equation to explain your answer:

  a. KCl d. Na_3PO_4
  b. $NaC_2H_3O_2$ e. $KHCO_3$
  c. NH_4I f. KF

17.29   If you get bitten by ants or wander into nettles, the burning sensation comes from injection of a tiny amount of formic acid, $HCOOH$, the simplest organic acid. Sketch the molecular structure, using Table 17.7 as a guide. If 4.0 g HCOOH are dissolved in a liter of water, the pH of the resulting solution is 2.40. Estimate the dissociation constant K_a.

17.30   Lactic acid contains only one acidic hydrogen atom. If 0.05 mole of lactic acid is dissolved in 100 ml water, it is found by conductivity measurements that 4.0 per cent of the acid molecules dissociate. Find K_a for lactic acid from this information.

17.31   Find the $[H^+]$, $[OH^-]$, the pH, and the per cent dissociation of the acid in each of the following solutions:

  a. 0.50 M $NaNO_3$
  b. 0.50 M $HClO$
  c. 0.50 M $KHSO_4$
  d. 0.50 M NH_4SO_3

17.32   Calculate the $[OH^-]$, $[H^+]$, the pH, and the concentration of the other solute species in

  a. 0.25 M RbOH
  b. 0.25 M sodium formate, $NaCHO_2$
  c. 0.25 M NH_3
  d. 0.25 M $NaNO_2$

17.13 Find the value of K_b for the conjugate base of each of the following acids:

    a. gallic acid, present in tea, $K_a = 3.9 \times 10^{-5}$
    b. mandelic acid, extracted first from bitter almonds (Ger. *mandel*, almond), $K_a = 1.4 \times 10^{-4}$

17.14 How would the experimental properties of 0.5 M HCl and 0.5 M HF differ? Give three simple experiments you might perform, each of which would tell you whether you had one or the other.

17.15 Given a substance with the molecular formula MO_2H_2, what factors would determine whether the compound would act as an acid or a base?

17.16 For each of the following reactions, indicate the Brönsted acids and bases:

    a. $HSO_4^-(aq) + H_2O$
        $\rightleftarrows H_3O^+(aq) + SO_4^{2-}(aq)$
    b. $CN^-(aq) + H_3O^+(aq)$
        $\rightleftarrows HCN(aq) + H_2O$
    c. $NH_3(aq) + H_2O$
        $\rightleftarrows NH_4^+(aq) + OH^-(aq)$
    d. $HSO_3^-(aq) + HCO_3^-(aq)$
        $\rightleftarrows H_2CO_3(aq) + SO_3^{2-}(aq)$

17.17 Find K for each of the reactions in Problem 17.16 using Table 17.5 and, if necessary, the Multiple Equilibria Rule; predict whether the equilibrium state of the system would tend to favor products or reactants.

17.18 Which of the following species can act as Brönsted acids? Brönsted bases? Lewis acids? Lewis bases?

    a. NH_4^+ c. Cu^{2+}
    b. H_2O d. CN^-

17.19 Arrange the following 0.1 M solutions in order of increasing pH:

    NaBr KOH HF HNO_3
  NH_3 $NaC_2H_3O_2$ $HC_2H_3O_2$ NH_4Cl

17.20 It is desired to prepare two solutions, each with a volume of 5 liters and each with a pH of 2.0. How many moles of each solute would it take if one of the solutions is to be made with HNO_3 and the other with HLac, lactic acid ($K_a = 8.4 \times 10^{-4}$)?

17.33 Find the value of K_a for the conjugate acid of each of the following organic bases:

    a. dimethylamine, used to make the insecticide Sevin, $K_b = 5.2 \times 10^{-4}$
    b. aniline, an important dye intermediate, $K_b = 3.8 \times 10^{-10}$

17.34 How would the experimental properties of 0.5 M KOH and 0.5 M NH_3 differ? Cite two simple experiments you might perform, each of which would tell you unequivocally which solution you had.

17.35 In solution, KOH behaves as a strong base, but BrOH, with much the same formula, is a weak acid. How do we explain the different properties of these substances?

17.36 For each of the following reactions, indicate the Brönsted acids and bases:

    a. $H_3O^+(aq) + NO_2^-(aq)$
        $\rightleftarrows H_2O + HNO_2(aq)$
    b. $H_2PO_4^-(aq) + H_2O$
        $\rightleftarrows H_3PO_4(aq) + OH^-(aq)$
    c. $C_3H_5O_2^-(aq) + H_2O$
        $\rightleftarrows HC_3H_5O_2(aq) + OH^-(aq)$
    d. $H_2CO_3(aq) + SO_4^{2-}(aq)$
        $\rightleftarrows HCO_3^-(aq) + HSO_4^-(aq)$

17.37 Find K for each of the reactions in Problem 17.36 using Table 17.5 and, if necessary, the Multiple Equilibria Rule. Predict whether the reactions would tend to go to the right under ordinary conditions.

17.38 Which of the following species can act as Brönsted acids? Brönsted bases? Lewis acids? Lewis bases?

    a. H_3O^+ c. CO_3^{2-}
    b. HS^- d. SO_2

17.39 Arrange the following 0.1 M solutions in order of increasing $[H^+]$:

    NH_3 KCl KF H_2SO_4
  $KHSO_4$ HNO_2 HI NH_4NO_3

17.40 A student is asked to make two solutions, each with a volume of 2 liters and each with a pH of 11.0. How many moles of each solute would it take if one of the solutions is to be made with KOH and the other with NH_3?

*17.41 Show that for an acid H_2X, which ionizes in two steps,

$$\frac{[H^+]^2 \times [X^{2-}]}{[H_2X]} = K_1K_2$$

where K_1 and K_2 are the first and second ionization constants, respectively. Hint: Write the total dissociation as the sum of the two steps and apply the Multiple Equilibria Rule.

*17.42 In an acid like H_2S or H_2CO_3 the second ionization constant is always much smaller than the first. As a result of this, the $[H^+]$ is almost completely established by the first ionization alone, since that reaction is dominant. Using this approach find the $[H^+]$ and the pH of 0.1 M H_2S and 0.1 M H_2CO_3.

*17.43 It is found that 0.1 M solutions of the three sodium salts, NaX, NaY, and NaZ, have pH's of 7.0, 8.0, and 9.0, respectively. Arrange the acids HX, HY, and HZ in order of increasing strength. Where you can, find the ionization constants of the acids.

*17.44 Many organic acids contain two ionizable hydrogen atoms. One of these is ascorbic acid, vitamin C, whose formula we may write as H_2A. For this acid $K_1 = 8 \times 10^{-5}$ and $K_2 = 1.6 \times 10^{-12}$. In a 0.1 M H_2A solution find $[H_2A]$, $[HA^-]$, $[A^{2-}]$, $[H^+]$, and the pH. Use the approximation described in Problem 17.42.

*17.45 Among the important biochemical substances are the amino acids, of which glycine is an example:

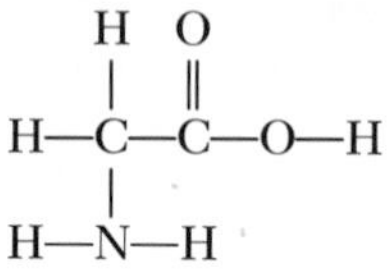

$$
\begin{array}{ccc}
 & H & O \\
 & | & \| \\
H- & C-C & -O-H \\
 & | & \\
 & H-N-H &
\end{array}
$$

In each amino acid there is a weak acid group, —COOH, and a weak base group, —NH$_2$. Into what species would glycine tend to convert in very acid solution? What species would it form in basic solution? Estimate the pH of a 0.1 M solution of glycine if K_a for —COOH and K_b for —NH$_2$ are about equal.

18 ACID-BASE REACTIONS

In the previous chapter we considered many chemical changes that could be called, at least in the Brönsted-Lowry sense, reactions between acids and bases. In most of those changes, however, the extent of reaction was small and involved the dissociation to equilibrium of a weak acid or base. In general, when a chemist speaks of an acid-base reaction, he means the sort of change which occurs when acidic and basic solutions are mixed. Such reactions typically have very large equilibrium constants and so proceed far to the right.

18.1 TYPES OF ACID-BASE REACTIONS

REACTIONS OF STRONG ACIDS AND BASES

We mentioned earlier the reaction that takes place when a solution of a strong acid like HCl, HNO_3, or H_2SO_4 is mixed with that of a strong base like $NaOH$ or KOH. If, say, nitric acid is mixed with sodium hydroxide, the H^+ ions in the completely ionized acid solution will react with the OH^- ions in the ionized base:

$$H^+(aq) + OH^-(aq) \rightleftarrows H_2O \quad K = 1/K_w = 1.0 \times 10^{14} \qquad (18.1)$$

Explain why $K = \dfrac{1}{K_w}$

This reaction is often called a *neutralization* reaction, since if just enough base is added to react with all the acid, the solution becomes neutral. If, for example, exactly 200 ml 0.500 M HNO_3 are mixed with exactly 200 ml 0.500 M $NaOH$, the H^+ ion from the acid will be just neutralized by the OH^- ion from the base, and the final solution will contain only the nitrate ions from the acid and sodium ions from the base.

Since K for Reaction 18.1 is enormous, the reaction between H^+ and OH^- ions is, for all practical purposes, quantitatively complete. Of course, it is not necessary that equal amounts of acid and base be mixed for a reaction to occur. In many cases an excess of one or the other will be used,

477

and the final solution will contain a salt and the reagent present in excess. Mixing 200 ml 0.500 M HNO_3 with 300 ml 0.500 M NaOH would result in a solution equivalent to one made from $NaNO_3$ and NaOH, since the NaOH is in excess. Under such circumstances the solution would be basic and, in fact, it is easy to show that in that solution $[OH^-]$ will be 0.1 M.

Example 18.1   Find $[OH^-]$ in a solution made by mixing 200 ml 0.500 M HNO_3 with 300 ml 0.500 M NaOH.

Solution.   When the solutions are mixed the reaction which occurs is

$$H^+(aq) + OH^-(aq) \rightleftarrows H_2O$$

In the acid:

$$\text{no. moles } H^+ = M_{HNO_3}V_{HNO_3} = 0.500 \; \frac{\text{moles } H^+}{\text{lit}} \times 0.200 \text{ lit} = 0.100 \text{ moles}$$

In the base:

$$\text{no. moles } OH^- = M_{NaOH}V_{NaOH} = 0.500 \text{ moles } OH^-/\text{lit} \times 0.300 \text{ lit} = 0.150 \text{ moles}$$

Assuming complete reaction, 0.100 moles H^+ in the acid will be neutralized by 0.100 moles of OH^- from the base, forming 0.100 moles H_2O and leaving an excess of 0.050 moles OH^- in the 500 ml of mixed solution.

$$[OH^-] = \frac{\text{no. moles } OH^-}{\text{no. lit solution}} = \frac{0.050}{0.500} = 0.10 \text{ M}$$

REACTIONS OF WEAK ACIDS WITH STRONG BASES

When a solution of a weak acid is mixed with that of a strong base, an acid-base reaction again occurs. Indeed, H^+ ions from the dissociated part of the weak acid react with OH^- ions from the base. This depletes the hydrogen ion concentration and more weak acid dissociates to restore equilibrium. The net change which occurs is the disappearance of HX, the weak acid, by reaction with OH^- ion to form water and the X^- ion:

$$HX(aq) + OH^-(aq) \rightleftarrows H_2O + X^-(aq) \tag{18.2}$$

This reaction also typically goes far to the right. We can find its equilibrium constant by considering that the reaction proceeds much as we described it, in two steps; first, the ionization of the acid and, second, the reaction of hydrogen and hydroxide ions:

$$\begin{array}{lll} \text{I} & HX(aq) \rightleftarrows H^+(aq) + X^-(aq) & K_I = K_a \\ \text{II} & H^+(aq) \; + OH^-(aq) \rightleftarrows H_2O & K_{II} = 1/K_w \end{array} \tag{18.1}$$

$$\begin{array}{llll} \text{III} & HX(aq) \; + OH^-(aq) \rightleftarrows H_2O + X^-(aq) & K_{III} = K \end{array} \tag{18.2}$$

We see that the overall reaction, III, is the sum of Reactions I and II, so we can apply the Multiple Equilibria Rule and say that K_{III} will be equal to the product of K_I and K_{II}. Therefore, the equilibrium constant, K, for the reaction of a weak acid and a strong base will be given by the equation:

$$K = \frac{K_a}{K_w}, \quad \text{since } K_{III} = K_I \times K_{II} \tag{18.3}$$

$$k_a \times \frac{1}{k_w}$$

Since K_a is ordinarily much, much larger than K_w, K itself is usually very large, so Reaction 18.2 tends to go quantitatively.

If equivalent amounts of HX and OH^- are mixed, the final solution will be that of a salt, as before, but in this case it will be that of a salt of a weak acid and will not be neutral. If, for example, we mix 20 ml 1.0 M $HC_2H_3O_2$, acetic acid, with 20 ml 1.0 M NaOH, the final solution will contain the salt, sodium acetate. Since acetate ion behaves as a weak base, the solution will be basic. In general, at the equivalence point of the reaction between any weak acid and a strong base, the solution will be basic.

Example 18.2

   a. Calculate K for the reaction which occurs when a solution of $HC_2H_3O_2$ is mixed with a solution of NaOH.
   b. Find the concentrations of OH^- ion and molecular $HC_2H_3O_2$ at equilibrium in a solution made by mixing exactly 20 ml 1.00 M $HC_2H_3O_2$ with 20 ml 1.00 M NaOH.

Solution

   a. The reaction is: $HC_2H_3O_2(aq) + OH^-(aq) \rightarrow H_2O + C_2H_3O_2^-(aq)$

   For $HC_2H_3O_2$, $K_a = 1.8 \times 10^{-5}$

   By Equation 18.3: $K = \dfrac{K_a}{K_w} = \dfrac{1.8 \times 10^{-5}}{1.0 \times 10^{-14}} = 1.8 \times 10^9$

   b. In 20 ml 1.00 M $HC_2H_3O_2$ there are 0.020 moles $HC_2H_3O_2$.

In 20 ml 1.00 M NaOH there are 0.020 moles OH^-, and 0.020 moles Na^+. Again assuming the reaction proceeds to completion, 0.020 moles $HC_2H_3O_2$ react with 0.020 moles OH^- to form:

$$0.020 \text{ moles } H_2O \text{ and } 0.020 \text{ moles } C_2H_3O_2^-$$

The final solution contains 0.020 moles Na^+ and 0.020 moles $C_2H_3O_2^-$ in 0.040 liters. It is, therefore, completely equivalent to a 0.50 M $NaC_2H_3O_2$ solution (i.e., 0.020 mole/0.040 liter = 0.50 M).

The $[OH^-]$ in 0.50 M $NaC_2H_3O_2$ was determined in the previous chapter, Example 17.8, and was found to be 1.7×10^{-5} M. This is the result of the basic properties of the $C_2H_3O_2^-$ ion:

$$C_2H_3O_2^-(aq) + H_2O \rightleftarrows HC_2H_3O_2(aq) + OH^-(aq)$$

Since one mole of $HC_2H_3O_2$ is formed per mole of OH^-, $[HC_2H_3O_2] = [OH^-] = 1.7 \times 10^{-5}$ M. $[HC_2H_3O_2]$ is clearly extremely small as compared to its initial value of 0.1 M, so the reaction goes essentially to completion.

Note that we consider this equilibrium after the main acid-base reaction has occurred

---

With weak acids containing more than one acidic hydrogen, reaction with OH^- ions will occur stepwise, neutralizing the most easily ionized hydrogen atoms first. An important example is the reaction of carbonic acid, H_2CO_3, with a solution of strong base such as NaOH:

$$H_2CO_3(aq) + OH^-(aq) \rightleftarrows H_2O + HCO_3^-(aq) \qquad (18.4)$$

$$HCO_3^-(aq) + OH^-(aq) \rightleftarrows H_2O + CO_3^{2-}(aq) \qquad (18.5)$$

The ultimate product depends on the relative quantities of acid and base which are used. The first reaction goes nearly to completion before the second one starts. If one mole of OH^- ions is added per mole of H_2CO_3, the product is the HCO_3^- ion. Addition of another mole of OH^- removes the second proton to give the carbonate ion, CO_3^{2-}. Reaction 18.5 would occur directly if a solution of a strong base were added to a solution made by dissolving sodium hydrogen carbonate, $NaHCO_3$, in water.

REACTIONS OF STRONG ACIDS AND WEAK BASES

If a solution of a weak base, such as NH_3, is mixed with a solution of a strong acid, such as HCl, the net overall reaction would be:

$$NH_3(aq) + H^+(aq) \rightleftarrows NH_4^+(aq) \tag{18.6}$$

If we write this reaction as occurring in two steps, we perhaps come closer to describing the path by which it occurs:

The net ionic equation is III, not II

I $\quad NH_3(aq) + H_2O \rightleftarrows NH_4^+(aq) + OH^-(aq) \quad K_I = K_b = 1.8 \times 10^{-5}$

II $\quad H^+(aq) + OH^-(aq) \rightleftarrows H_2O \qquad\qquad\quad K_{II} = 1/K_w = 1.0 \times 10^{14}$

III $\quad NH_3(aq) + H^+ \rightleftarrows NH_4^+ \qquad\qquad\qquad\quad K_{III} = K$

Again we see that Reaction 18.6 is the sum of Reactions I and II, so by the Multiple Equilibria Rule, its equilibrium constant K is given by the equation:

$$K = \frac{K_b}{K_w} \tag{18.7}$$

For the reaction of NH_3 with a strong acid,

$$K = \frac{1.8 \times 10^{-5}}{1.0 \times 10^{-14}} = 1.8 \times 10^9$$

In any acid-base reaction of this sort, K_b of the weak base is usually much larger than K_w, so the equilibrium constant K is, as with the $NH_3 + H^+$ reaction, very large.

When equivalent amounts of a weak base and strong acid are mixed, the resulting solution will be, as you may have guessed by now, that of a salt. When ammonia and hydrochloric acid react, a solution of ammonium chloride, NH_4Cl, is produced; at the equivalence point such a solution will be slightly acidic. (Why?)

If an anion is capable of acquiring two protons, the reaction with a strong acid occurs in two steps:

$$H^+(aq) + CO_3^{2-}(aq) \rightleftarrows HCO_3^-(aq) \tag{18.8}$$

$$HCO_3^-(aq) + H^+(aq) \rightleftarrows H_2CO_3(aq) \rightleftarrows CO_2(g) + H_2O \tag{18.9}$$

Addition of excess acid to a solution containing CO_3^{2-} ion may lead to the evolution of CO_2 gas if the initial carbonate ion concentration is high enough. Carbon dioxide may also be formed if sodium hydrogen carbonate is added to an acidic solution. This is, of course, what happens in the stomach when bicarbonate of soda, $NaHCO_3$, is taken to relieve acid indigestion.

You may have noticed there is a symmetry among our analyses of the various sorts of acid-base reactions. This is not accidental, and you will

TABLE 18.1 Characteristics of Acid-Base Reactions

TYPE	EXAMPLE	K	SPECIES PRESENT AT EQUIVALENCE POINT
Strong acid + strong base HCl-NaOH	$H^+(aq) + OH^-(aq) \rightleftarrows H_2O$	1.0×10^{14}	Na^+, Cl^- neutral solution
Weak acid + strong base $HC_2H_3O_2$-NaOH	$HC_2H_3O_2(aq) + OH^-(aq) \rightleftarrows$ $H_2O + C_2H_3O_2^-(aq)$	1.8×10^9	$Na^+, C_2H_3O_2^-$ basic solution
Weak base + strong acid NH_3-HCl	$NH_3(aq) + H^+(aq) \rightleftarrows NH_4^+(aq)$	1.8×10^9	NH_4^+, Cl^- acidic solution

find that if you can see the pattern of similarity of the approach to each kind of reaction it will be a lot easier to work with acid-base reactions in general.

REACTIONS OF STRONG ACIDS WITH SOLIDS

Unlike the previous acid-base reactions, all of which occur between solutions, there is an important class of reactions in which the source of the base is a solid. Many of the compounds of the metals contain anions which are basic; a majority of these compounds are insoluble in water. These include the hydroxides, the carbonates, phosphates, and sulfides of most of the metals except those in Group IA. These substances dissolve in water to only a small extent to achieve equilibrium with their ions in solution as discussed in Chapter 16. Barium carbonate is a typical example:

$$BaCO_3(s) \rightleftarrows Ba^{2+}(aq) + CO_3^{2-}(aq)$$

If to the equilibrium mixture of $BaCO_3$ and its ions a strong acid is added, the balance is greatly disturbed, since H^+ ions will react as in Equations 18.8 and 18.9, converting CO_3^{2-} ion to HCO_3^- ion and carbonic acid. The overall reaction between $BaCO_3$ and acid can be represented as the result of the series of reactions we have just mentioned:

I Dissolving of $BaCO_3$ to attain equilibrium with its ions:
$BaCO_3(s) \rightleftarrows Ba^{2+}(aq) + CO_3^{2-}(aq)$ $\qquad K_I = K_{sp}BaCO_3$

II Reaction of CO_3^{2-} ion with acid:
$CO_3^{2-}(aq) + H^+(aq) \rightleftarrows HCO_3^-(aq)$ $\qquad K_{II} = 1/K_2$ of H_2CO_3

III Reaction of HCO_3^- ion with acid:
$HCO_3^-(aq) + H^+(aq) \rightleftarrows H_2CO_3(aq)$ $\qquad K_{III} = 1/K_1$ of H_2CO_3

IV Overall reaction of $BaCO_3$ with acid:
$BaCO_3(s) + 2 H^+(aq) \rightleftarrows Ba^{2+}(aq) + H_2CO_3(aq)$ $\quad K = K_{IV}$

(18.10)

An approach like this may help you in deriving correct net ionic equations

The analysis of this rather complex system is immensely simplified once we recognize that Reaction IV is the sum of Reactions I, II, and III. We can therefore apply the Multiple Equilibria Rule and find K for the overall reaction:

$$K_{IV} = K_I \times K_{II} \times K_{III} = K$$

$$K = \frac{K_{sp}}{K_1 K_2} \qquad (18.11)$$

Knowing K_{sp} for $BaCO_3$ and the two acid constants for carbonic acid, we can find the magnitude of the equilibrium constant K for Reaction 18.10:

$$K_{sp} = 1 \times 10^{-9}, \quad K_1 = 4.2 \times 10^{-7}, \quad K_2 = 4.8 \times 10^{-11}$$

$$K = \frac{K_{sp}}{K_1 K_2} = \frac{1 \times 10^{-9}}{(4.2 \times 10^{-7})(4.8 \times 10^{-11})} = 5 \times 10^7$$

So, for the reaction of $BaCO_3$ with acid,

$$BaCO_3(s) + 2\ H^+(aq) \rightleftharpoons Ba^{2+}(aq) + H_2CO_3(aq)$$

$$K = \frac{[Ba^{2+}] \times [H_2CO_3]}{[H^+]^2} = 5 \times 10^7$$

Just by looking at the magnitude of K you would judge from its large size that probably $BaCO_3$ would tend to dissolve in acid. That conclusion is indeed correct; $BaCO_3$, and carbonates in general, all dissolve in acid solutions, since K for the solution reaction of a carbonate is always large. If we examine phosphates and hydroxides and sulfides by the same kind of procedure we used to get Equation 18.11, we find that for those substances as well, K is usually large. The only exceptions are some of the very insoluble sulfides; these will be discussed in Section 18.5. However, we find that, in general:

Insoluble inorganic compounds containing basic anions tend to be soluble in solutions of strong acids.

Example 18.3 What will be the solubility of $BaCO_3$ in moles per liter in a solution whose pH is maintained at 3.0, assuming $[H_2CO_3] = 0.03$ M?

Solution. The reaction which occurs is:

$$BaCO_3(s) + 2\ H^+(aq) \rightleftharpoons Ba^{2+}(aq) + H_2CO_3(aq)$$

For every mole of $BaCO_3$ which dissolves, one mole of Ba^{2+} is formed. Hence,

$$\text{solubility } BaCO_3 = [Ba^{2+}]$$

To calculate $[Ba^{2+}]$, we rearrange the expression previously given for K:

$$K = \frac{[Ba^{2+}] \times [H_2CO_3]}{[H^+]^2} = 5 \times 10^7 \quad [Ba^{2+}] = 5 \times 10^7 \times \frac{[H^+]^2}{[H_2CO_3]}$$

At pH = 3.0, $[H^+] = 1 \times 10^{-3}$ M.

Substituting, we obtain $\quad [Ba^{2+}] = \dfrac{5 \times 10^7 (1 \times 10^{-3})^2}{3 \times 10^{-2}} = 2 \times 10^3$

The fact that this number is so large means that for all practical purposes $BaCO_3$ is completely soluble in a solution maintained at pH = 3, say, by addition of acid as the $BaCO_3$ dissolves. Indeed, we could readily show that even at a pH of 5 (a very weakly acidic solution), as much as 0.2 moles of $BaCO_3$ would dissolve per liter. It is certainly very reasonable to say that $BaCO_3$ is "soluble in acid."

There are, as you may know, some insoluble salts which contain anions which are essentially neutral. Among the most common of these are AgCl, BaSO$_4$, and PbSO$_4$. If you attempt to dissolve these substances in acid, you find that for the most part nothing happens; they remain insoluble. The reason for this becomes apparent if you carry out the development we did for BaCO$_3$, using, for example, PbSO$_4$. Equation 18.11 again results, but K$_1$K$_2$ for H$_2$SO$_4$ is very, very large, since sulfuric acid is a strong acid. This makes K for the reaction of PbSO$_4$ with acid just about zero, which implies what we observe: the insolubility of lead sulfate in acids.

Would oxides tend to dissolve in acidic solutions?

18.2 ACID-BASE TITRATIONS

Since the equilibrium constants for many acid-base reactions are very large, these reactions are often applied in quantitative analysis. In particular, they are used in titrations to determine concentrations of acids and bases in solution and to aid in the analysis of mixtures. The general procedure is similar to that used with titrations involving precipitation reactions. One measures with a buret the volume of a standardized acidic or basic solution which is required to just react with a measured volume of unknown base or acid or with a known mass of a base or acid. The problem of knowing when stoichiometrically equivalent amounts of acid and base have been mixed is handled in general by the use of suitable indicators or a pH meter. Procedures for carrying out acid-base titrations have been developed to the point where they are among the most precise methods of chemical analysis. Examples 18.4 and 18.5 indicate the calculations which are involved.

Example 18.4   It is found that 22.3 ml of 0.240 M NaOH is required to react with a 50.0 ml sample of vinegar, a solution of acetic acid in water. Calculate the concentration of acetic acid in the vinegar.

Solution.   From the equation for the reaction of OH$^-$ ions with acetic acid

$$HC_2H_3O_2(aq) + OH^-(aq) \rightarrow C_2H_3O_2^-(aq) + H_2O$$

it is evident that the reactants are consumed in a 1:1 mole ratio. That is,

$$\text{no. moles OH}^- = \text{no. moles HC}_2\text{H}_3\text{O}_2$$

But

$$\text{no. moles of OH}^- = (\text{M NaOH})(\text{volume NaOH})$$

$$\text{no. moles HC}_2\text{H}_3\text{O}_2 = (\text{M HC}_2\text{H}_3\text{O}_2)(\text{volume HC}_2\text{H}_3\text{O}_2 \text{ solution})$$

Hence

$$(\text{M NaOH})(\text{volume NaOH}) = (\text{M HC}_2\text{H}_3\text{O}_2)(\text{volume HC}_2\text{H}_3\text{O}_2 \text{ solution})$$

$$0.240 \frac{\text{mole}}{\text{lit}} (0.0223 \text{ lit}) = (\text{M HC}_2\text{H}_3\text{O}_2)(0.0500 \text{ lit})$$

Solving, $M \ HC_2H_3O_2 = 0.240 \frac{\text{mole}}{\text{lit}} \times \frac{0.0223}{0.0500} = 0.107 \text{ M}$

Example 18.5 A chemist synthesizes a substance which he believes is barbituric acid, MW = 128.1, a precursor for many sleeping tablets. Barbituric acid has one acidic hydrogen, $K_a = 9.8 \times 10^{-5}$. To help with the identification, he titrates a 0.5000 g crystalline sample with 0.1000 M NaOH and finds that at the equivalence point he has added 39.10 ml of the base. Is the sample likely to be barbituric acid?

Solution. Since the acid has one acidic hydrogen atom, one mole of the acid will react with one mole of OH^-. Therefore, at the equivalence point,

$$\text{no. moles acid} = \text{no. moles } OH^- \text{ added}$$

$$\frac{\text{no. grams acid}}{\text{GMW}} = M_{NaOH} \times V_{NaOH} = \frac{0.100 \text{ moles}}{1 \text{ lit}} \times 0.03910 \text{ lit}$$
$$= 3.910 \times 10^{-3} \text{ moles}$$

$$\text{GMW} = \frac{\text{no. grams acid}}{\text{no. moles } OH^-} = \frac{0.5000 \text{ g}}{3.910 \times 10^{-3} \text{ moles}} = 127.9 \text{ g/mole}$$

Since the molecular weight of the sample is very nearly that reported for barbituric acid, the material may well be barbituric acid. Further evidence, such as melting point and infrared spectrum, would have to be obtained to confirm the identification.

NORMALITY: GRAM EQUIVALENT WEIGHTS OF ACIDS AND BASES

The concentrations of solutions used in titrations are sometimes expressed in terms of their normalities rather than their molarities. For any solute A its normality in solution is defined as:

$$\text{Normality of A} = \frac{\text{no. GEW of A}}{\text{no. lit solution}}; \quad N_A = \frac{\text{no. GEW of A}}{V_A} \qquad (18.12)$$

One gram equivalent weight, GEW, or one equivalent of A is that mass of A which, in a reaction between A and B, will react with one GEW of the species B. This approach makes for very simple calculations of titration results, once the normality of A or B has been established. For the reaction mixture at the equivalence point,

$$\text{no. GEW A added} = \text{no. GEW B added}$$

Using Equation 18.12: $\qquad N_A \times V_A = N_B \times V_B \qquad (18.13)$

Knowing the normality N_A and the volume of the solution of A required to react exactly with a measured volume of the solution of B, the normality of B is very easily found. Equation 18.13 can be applied to any titration reaction, including acid-base reactions.

The only difficulty in using normalities of solutions is to establish a simple rational definition for the amount of substance in one gram equivalent weight. For acids and bases the GEW is defined as follows:

1 GEW acid = weight of acid which reacts with one mole of OH^-
1 GEW base = weight of base which reacts with one mole of H^+

By this definition,

$$1 \text{ GEW HCl} = 36.5 \text{ g HCl} = 1 \text{ mole HCl}$$
$$1 \text{ GEW NaOH} = 40 \text{ g NaOH} = 1 \text{ mole NaOH}$$
$$1 \text{ GEW H}_2\text{SO}_4 = 49 \text{ g H}_2\text{SO}_4 = \tfrac{1}{2} \text{ mole H}_2\text{SO}_4$$
$$1 \text{ GEW Ca(OH)}_2 = 37 \text{ g Ca(OH)}_2 = \tfrac{1}{2} \text{ mole Ca(OH)}_2$$

This means that the normality and molarity of a solution are related; for example,

$$1.0 \text{ N HCl} = 1.0 \text{ M HCl} \qquad 0.5 \text{ N NaOH} = 0.5 \text{ M NaOH}$$
$$1.0 \text{ N H}_2\text{SO}_4 = 0.5 \text{ M H}_2\text{SO}_4 \quad 0.1 \text{ N Ca(OH)}_2 = 0.05 \text{ M Ca(OH)}_2$$

In some cases the relationship between normality and molarity is not quite so apparent. It is sometimes even the case that a given reagent solution of a certain molarity may have several possible normalities, depending on the reaction in which the reagent participates. If, for instance, we react phosphoric acid, H_3PO_4, with a base, we can stop the titration when we have removed one, or two, or three protons:

I $\quad H_3PO_4(aq) + OH^-(aq) \rightleftarrows H_2PO_4^-(aq) + H_2O \qquad N_{H_3PO_4} = M_{H_3PO_4}$

II $\quad H_3PO_4(aq) + 2\ OH^-(aq) \rightleftarrows HPO_4^{2-}(aq) + 2\ H_2O \quad N_{H_3PO_4} = M_{H_3PO_4} \times 2$

III $\quad H_3PO_4(aq) + 3\ OH^-(aq) \rightleftarrows PO_4^{3-}(aq) + 3\ H_2O \quad N_{H_3PO_4} = M_{H_3PO_4} \times 3$

A liter of 1 M H_3PO_4 would, in Reaction I, react with one mole OH^- ion and, hence, would contain one equivalent, and would be 1 N; in Reaction II, a liter of the *same solution* reacts with *two* moles of OH^- ion and so would be 2 N H_3PO_4; in Reaction III, the same solution would be 3 N H_3PO_4.

The fact that the normality of a given reagent *depends on the reaction in which it participates* is a serious deficiency of the whole idea of normality. In recent years some chemists have suggested that concentrations never be expressed in normalities because of the ambiguities we have just noted. It is certainly true that molarities are easier to understand, are clearly defined, and are the preferred concentration units. However in cases where one is working with a particular reaction under routine conditions it may be worthwhile, because of the simplicity of Equation 18.13, to use normalities.

ACID-BASE INDICATORS

The end point of an acid-base titration is established by a change in color in an acid-base indicator which is added, before the reaction starts, to the solution being titrated. An acid-base indicator is typically a very highly colored organic dye. The indicator itself is a weak acid, in which the undissociated acid molecule has a color which is different from that of its conjugate base. Bromthymol blue, whose formula we will write simply as HIn, is yellow when present in solution as HIn, and blue when present as In^-. Like any weak acid, bromthymol blue in solution undergoes reaction to equilibrium:

$$\underset{\text{yellow}}{HIn(aq)} \rightleftarrows H^+(aq) + \underset{\text{blue}}{In^-(aq)} \qquad K_a = \frac{[H^+] \times [In^-]}{[HIn]} = 1 \times 10^{-7} \quad (18.14)$$

Let us consider what color bromthymol blue will have in solutions of different pH. Depending on the $[H^+]$, the ratio of $[In^-]/[HIn]$ will take on whatever value is required to satisfy Equation 18.14. Solving that equation for the ratio, we get

$$\frac{[In^-]}{[HIn]} = \frac{1 \times 10^{-7}}{[H^+]} \tag{18.14'}$$

Clearly if $[H^+]$ in the solution is 1×10^{-7} M, the ratio of $[In^-]/[HIn]$ will be unity, and the undissociated acid and its conjugated base will be present in equal concentrations; the color of the solution will be green, resulting from the mixture of yellow and blue species.

If in the solution in which we have bromthymol blue, $[H^+] = 1 \times 10^{-8}$ M, pH 8, then the ratio $[In^-]/[HIn]$ will have to equal 10, to satisfy Equation 18.14; the indicator will be about 90 per cent in the form of In^- and will appear essentially blue. Similarly, if the pH of the solution is 6, $[H^+]$ equals 1×10^{-6} M, and the ratio $[In^-]/[HIn]$ will be 0.1; the color will be yellow. At pH values of 8 or greater, bromthymol blue will be blue, and whenever the pH is 6 or less the solution will look yellow. Bromthymol blue will change color gradually from yellow to blue as the pH goes from 6 to 8. Like most indicators, it has a range of about 2 pH units over which it changes color. We say that bromthymol blue is an indicator with an **end point** at pH 7, since at that pH the change in color is most easily detected.

What is the color of phenol-phthalein at pH 11? pH 7? The end point observed with any particular indicator will depend on the magnitude of its acid constant K_a. The pH at the end point will equal pK_a of the indicator where $pK_a = -\log_{10}K_a$. Methyl red ($K_a = 1 \times 10^{-5}$) changes color from red to yellow at pH 5; phenolphthalein ($K_a = 1 \times 10^{-9}$) turns from colorless to pink at pH 9.

In carrying out an acid-base titration, we try to select an indicator which changes color at or very near the pH of the equivalence point of the reaction; i.e., the point at which equal quantities of acid and base have been added. A successful titration, one with good quantitative results, is possible only if the **end point** as established by the indicator actually occurs at the **equivalence point** in the titration. Since the pH at the equivalence point in an acid-base reaction depends on the relative strengths of the acid and base involved, one cannot always use the same indicator for different kinds of acid-base titrations.

1. TITRATION OF A STRONG ACID WITH A STRONG BASE. If a solution of a strong base like NaOH is slowly added to a solution of a strong acid like HCl, the pH of the solution will be initially low and will gradually rise as the acid is neutralized (Fig. 18.1). In the vicinity of the equivalence point the pH changes more rapidly with added base. At the equivalence point itself the change in pH is extremely rapid, going up by at least 6 pH units on addition of one drop of base. The actual equivalence point is at pH 7, since at that point the solution contains only a neutral salt, such as NaCl. Bromthymol blue would serve well as an indicator of the end point, but since the pH of the solution at the end point is so sensitive to added base, one can equally well use phenolphthalein (end point at pH 9) or methyl red (end point at pH 5) in the titration. The pH of the solution at a few points during the titration is calculated in the following example.

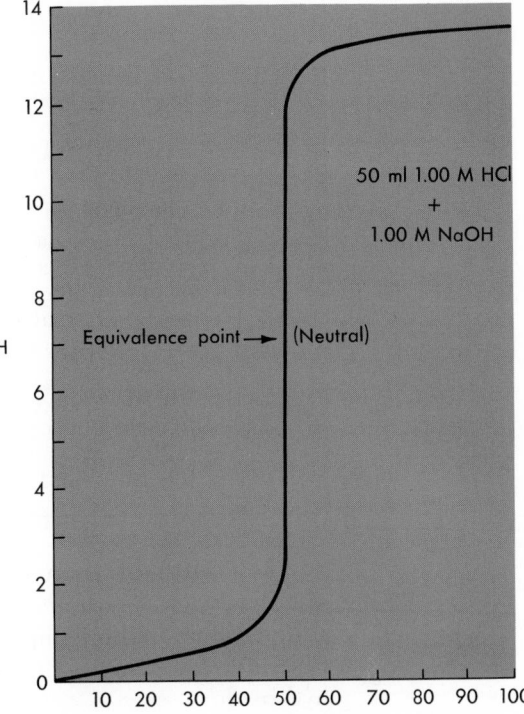

Figure 18.1 Titration of a strong acid with a strong base.

At the equivalence point the pH of the solution changes very rapidly as base is added

50 ml 1.00 M HCl
+
1.00 M NaOH

Equivalence point ⟶ (Neutral)

ml NaOH added

Example 18.6 50.00 ml of 1.000 M HCl is titrated with 1.000 M NaOH. Find the pH of the solution after the following volumes of 1.000 M NaOH have been added:

 a. 40.00 ml b. 49.99 ml c. 50.00 ml

Solution. During the titration the following reaction occurs:

$$H^+(aq) + OH^-(aq) \rightleftarrows H_2O \quad K = 1.0 \times 10^{14}$$

a. The number of moles of H^+ available for reaction is fixed by the volume and molarity of the HCl:

$$\text{no. moles } H^+ = M_{H^+} \times V_{H^+} = 1.000 \text{ moles/lit} \times 0.05000 \text{ lit}$$
$$= 0.05000 \text{ moles}$$

As the basic solution is added, the OH^- ions react quantitatively with the H^+ ions already present:

$$\text{no. moles } OH^- \text{ added} = M_{OH^-} \times V_{OH^-}$$
$$= 1.000 \text{ moles/lit} \times 0.04000 \text{ lit}$$
$$= 0.04000 \text{ moles}$$

Reaction occurs, using up 0.04000 moles OH^- and 0.04000 moles H^+, leaving 0.01000 moles H^+ in about 90 ml (i.e., 50 ml + 40 ml) of solution:

$$[H^+] = \frac{\text{no. moles } H^+ \text{ remaining}}{\text{no. lit solution}} = \frac{0.01}{0.09} = 0.11 \text{ M}; \quad pH = 0.95$$

b. The problem is essentially the same, except more NaOH has been added:

no. moles OH$^-$ added = 1.000 moles/lit $\times$ 0.04999 lit
$$= 0.4999 \text{ moles}$$

The reaction consumes 0.04999 moles OH$^-$ and 0.04999 moles H$^+$, leaving 0.00001 moles H$^+$ in essentially 100 ml of solution:

$$[H^+] = \frac{\text{no. moles H}^+ \text{ remaining}}{\text{no. lit solution}} = \frac{0.00001}{0.100} = 0.0001 \text{ M}; \quad pH = 4.0$$

c. The 50.00 ml 1 M NaOH completely neutralize the 50.00 ml 1 M HCl, leaving a solution of 0.05 moles NaCl in 100 ml, a 0.50 M NaCl solution. Since Na$^+$ and Cl$^-$ are both neutral species, the pH is that of pure water, namely 7. This is the equivalence point of the titration. Note that 0.01 ml of base is all that is required at the equivalence point to move the pH of the solution from 4 to 7. This volume is much less than that of 1 drop of reagent. If one more drop of NaOH were added to the neutral solution, OH$^-$ ion would be present in excess and the pH would go up to over 10. Any indicator with an end point between pH 5 and 9 would be satisfactory for this titration. Phenolphthalein is often used.

2. TITRATION OF A WEAK ACID WITH A STRONG BASE.

If a solution of a weak acid is titrated with a strong base, the change of pH as base is added differs in some important ways from that observed with a strong acid. Consider, for example, the titration of acetic acid with a solution of sodium hydroxide:

$$HC_2H_3O_2(aq) + OH^-(aq) \rightleftarrows H_2O + C_2H_3O_2^-(aq) \qquad (18.15)$$

Since the acid is weak, the pH at the beginning of the titration will be

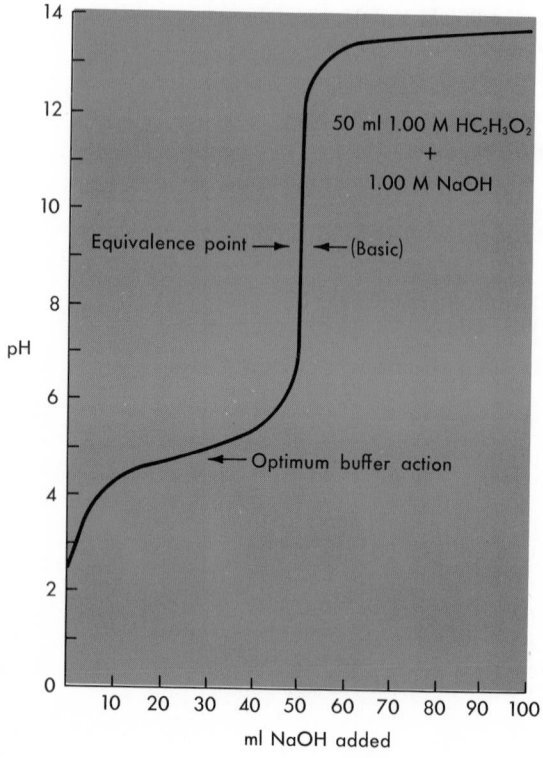

Figure 18.2 Titration of a weak acid with a strong base.

The steep portion of this titration curve is considerably shorter than when a strong acid is titrated

typically higher by a few units than with a strong acid of the same molarity (Fig. 18.2). The pH rises more rapidly at the early stages of the titration than it did before and goes through a somewhat smaller change of pH in the immediate vicinity of the equivalence point. At the equivalence point the solution contains only sodium acetate, which we have seen previously is basic. These factors make the choice of indicator more crucial. In the case of the titration of acetic acid, the equivalence point occurs at about pH 9, which is the pH of dilute $NaC_2H_3O_2$ solution. Phenolphthalein (end point at pH 9) would be an excellent indicator for the titration. If one attempted to use methyl red (end point at pH 5), it would change color much too early when the titration was only about 65 per cent complete (Fig. 18.2), hardly a good time to stop adding base.

With a weak acid having a very small ionization constant, the beginning portion of the titration curve rises still higher than it does with acetic acid. This has the effect of making the detection of the equivalence point more difficult, since the sharp change of pH at the equivalence point, which makes possible a precise end point, may be lost. If one titrates a solution of HCN ($K_a = 4 \times 10^{-10}$) with NaOH he observes a dependence of pH on volume of base added like that in Figure 18.3; the equivalence point is actually at pH 11.5, but an indicator with an end point at that pH would not show a rapid color change at the equivalence point since the pH is changing so gradually. It is actually impossible to precisely determine the molarity of an HCN solution by using an acid-base indicator in a direct titration.

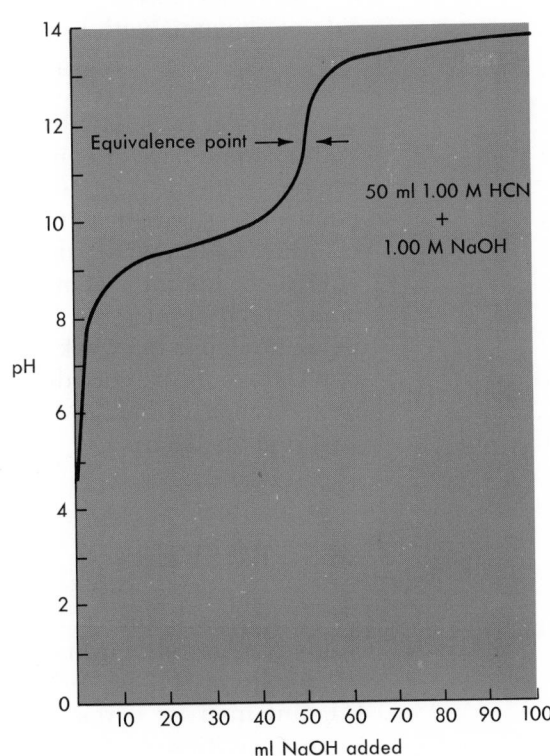

Figure 18.3 Titration of a very weak acid with a strong base.

If the titration curve is to be steep at the equivalence point, it is necessary that K for the acid-base reaction be large, $>10^7$

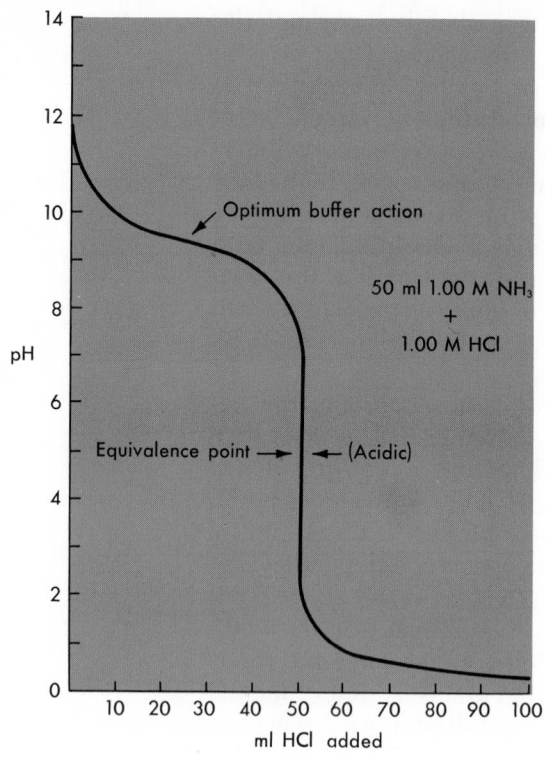

Figure 18.4 Titration of a weak base with a strong acid.

Why is phenolphthalein not a suitable indicator for this titration?

3. **TITRATION OF A WEAK BASE WITH A STRONG ACID.** In the titration of ammonia with hydrochloric acid, the acid-base reaction is

$$NH_3(aq) + H^+(aq) \rightleftarrows NH_4^+(aq) \tag{18.6}$$

The solution is initially basic, but with a lower pH than one would have with a strong base of equal molarity (Fig. 18.4). The pH drops during the titration, with the steepest drop occurring as we go through the equivalence point. The solution at the equivalence point contains only ammonium chloride, and is acidic due to the presence of NH_4^+ ion. The pH at the equivalence point is about 5, so methyl red (end point at pH 5), or methyl orange (end point at pH 4) would be good end point indicators. If the base being titrated is too weak, the pH changes too slowly near the equivalence point. It would be very difficult to titrate the acetate ion ($K_b = 5.6 \times 10^{-10}$) in a solution of $NaC_2H_3O_2$, with hydrochloric acid, since the end point would not be sharp.

18.3 BUFFERS

It is a well-known fact that the pH of blood and many other body fluids is relatively insensitive to the addition of acid or base. If one adds 0.01 moles of HCl or NaOH to 1 liter of blood, its pH changes by less than 0.1 unit from its normal value of 7.4. In contrast, the addition of these same quantities of acid or base to pure water changes its pH by about 5 units (from 7 to 2 with 0.01 mole of HCl and from 7 to 12 with 0.01 mole of NaOH).

A solution whose pH changes on addition of acid or base by much less than would pure water is said to be *buffered.* For buffer action a solution ordinarily contains two species, one capable of reacting with H^+ ions and the other with OH^- ions; in addition, the species in the buffer must not react with each other. Usually a buffer consists of a mixture of an acid and its conjugate base. One of the simplest buffers can be prepared by adding acetic acid to a solution of sodium acetate. If a strong base is added to the resulting mixture, the OH^- ions will react with acetic acid molecules:

$$OH^-(aq) + HC_2H_3O_2(aq) \rightleftarrows H_2O + C_2H_3O_2^-(aq)$$

As we have seen, this reaction is nearly quantitative, so the added OH^- ions are almost completely removed.

Addition of a strong acid to the buffer results in reaction of the H^+ ions of the acid with acetate ions of the sodium acetate:

$$H^+(aq) + C_2H_3O_2^-(aq) \rightleftarrows HC_2H_3O_2(aq)$$

In either case the added OH^- or H^+ ions are consumed and are unable to produce the drastic change in pH that occurs when a strong acid or base is added to water or an unbuffered solution.

The pH of a buffer depends on the value of K_a of the weak acid it contains and the relative concentrations of that acid and its conjugate base. Addition of strong acid or base will change the relative concentrations and so will cause a small pH change, the magnitude of which can be calculated as illustrated in Example 18.7.

Example 18.7 A liter of buffered solution contains 0.10 mole $HC_2H_3O_2$ and 0.10 mole $C_2H_3O_2^-$.
  a. Calculate the pH of the buffer.
  b. Calculate the pH after addition of 0.01 mole H^+ ion.
  c. Calculate the pH after addition of 0.01 mole OH^- ion.

Solution
  a. In the buffer the pH is established by the equilibrium between acetic acid and acetate ion:

$$K_a = \frac{[H^+] \times [C_2H_3O_2^-]}{[HC_2H_3O_2]} = 1.8 \times 10^{-5} \quad [H^+] = 1.8 \times 10^{-5} \times \frac{[HC_2H_3O_2]}{[C_2H_3O_2^-]}$$

In the buffer $[C_2H_3O_2^-]$ and $[HC_2H_3O_2]$ are both equal to 0.1 M, so

$$[H^+] = K_a = 1.8 \times 10^{-5} \text{ M} \quad pH = 4.74$$

This is the optimum pH of the buffer, where concentrations of the weak acid and conjugate base are equal.
  b. When 0.01 mole of H^+ is added, this will react with 0.01 mole of $C_2H_3O_2^-$ to form 0.01 mole $HC_2H_3O_2$. This means that in the final solution there are 0.09 moles of $C_2H_3O_2^-$ ion and 0.11 moles of $HC_2H_3O_2$, since we started with 0.10 mole of each. Therefore

$$[H^+] = 1.8 \times 10^{-5} \times \frac{[HC_2H_3O_2]}{[C_2H_3O_2^-]} = 1.8 \times 10^{-5} \times \frac{0.11}{0.09} = 2.2 \times 10^{-5}$$
$$pH = 4.66$$

The addition of 0.01 mole of H^+ to the buffer lowers the pH by 0.08 unit.

Note that we let the reaction of H^+ and $C_2H_3O_2^-$ proceed to completion and *then* find $[H^+]$ from the amounts of $HC_2H_3O_2$ and $C_2H_3O_2^-$ in the *reacted* mixture

c. In this case, addition of 0.01 mole OH⁻ results in its reaction with 0.01 mole of $HC_2H_3O_2$, producing 0.01 mole $C_2H_3O_2^-$. In the final solution there will be 0.09 mole $HC_2H_3O_2$ and 0.11 mole $C_2H_3O_2^-$.

$$[H^+] = 1.8 \times 10^{-5} \times \frac{[HC_2H_3O_2]}{[C_2H_3O_2^-]} = 1.8 \times 10^{-5} \times \frac{0.09}{0.11} = 1.5 \times 10^{-5}$$
$$pH = 4.82$$

Again we find that the pH changes by less than 0.1 unit.

The buffer we have considered in our example is approximately the same solution as is obtained during the titration of $HC_2H_3O_2$ with NaOH in Figure 18.2, at the stage when 25 ml of NaOH have been added. At that point the acid is half neutralized, and $[HC_2H_3O_2]$ equals $[C_2H_3O_2^-]$. You will note that at that point in the titration the pH is changing relatively slowly with added base; a buffer takes advantage of the fact that a mixture containing roughly equal amounts of weak acid and its conjugate base has a relatively stable pH, at least as compared to the unbuffered system, which in Figure 18.2 corresponds to the point where 50 ml NaOH have been added. There are two properties of buffers that are nicely shown by Figure 18.2:

1. The capacity of a buffer to absorb H⁺ or OH⁻ ions is limited. If enough strong acid or base is added, all the conjugate base or weak acid in the buffer will be used up and the strong acid or base will then take over the system. Relating our example to Figure 18.2, the addition of 0.01 mole H⁺ or OH⁻ is equivalent to removing or adding about 2.5 ml of the NaOH solution. If we should add ten times that amount of NaOH we would exhaust the buffer and would go to a pH of about 9. Similarly, addition of HCl in amount sufficient to react with all the acetate ion would lower the pH to about 2.8.

What would be the optimum pH of an NH_4^+-NH_3 buffer?

2. The pH range of a buffer is limited, usually to about one unit above or one unit below the optimum value, where the weak acid and its conjugate base have equal concentrations. Referring again to Figure 18.2, you can see that in order to change the acetate-acetic acid buffer from its optimum pH of about 4.74 to about 5.7 we would have to add roughly 22 ml of the NaOH solution. Such a solution would still be rated as a buffer but would contain very little $HC_2H_3O_2$, since most of it has been neutralized. This means that the capacity of that buffer to absorb OH⁻ ion is very limited. Similarly, if one wished to lower the pH of the buffer by one unit, he would make the buffer by adding roughly 2.5 ml of NaOH to 25 ml of $HC_2H_3O_2$ of the same molarity. This buffer would absorb OH⁻ ion very well, but would have very low capacity for H⁺ ion. For these reasons one tries to find buffer systems whose optimum pH is at about the value that is required.

There are many common buffers, some of which are used by our bodies in maintaining the pH of its many fluids. In the latter class are the H_2CO_3-HCO_3^- buffer and a phosphate buffer, $H_2PO_4^-$-HPO_4^{2-}. In the blood the pH is maintained for the most part by a rapid-acting H_2CO_3-HCO_3^- buffer. This buffer has a pH of about 6.4 if the concentrations of H_2CO_3 and HCO_3^- are equal. The actual pH of blood is kept close to 7.4 by the fact that the ratio $[H_2CO_3]/[HCO_3^-]$ is kept at about 0.05; this gives a calculated pH of about 7.7, but this is lowered a bit by other buffers, particularly the phosphate buffer, which has a normal pH of 7.2. The capacity of the H_2CO_3-HCO_3^- buffer in the blood is greatest for acid, as is desirable since acid, particularly lactic acid, is produced by exercise and must be removed quickly if the pH in the body cells is not to drop so low as to kill the cells. Carbon dioxide itself is also produced in metabolism and must be continuously absorbed by the blood and vented through the lungs. There are several crucial equilibria related to the mechanics of transporting oxygen to the tissues and CO_2 to the lungs, and these depend in a remarkable way on the proper blood pH. Without a simple buffer available, pH control in the blood would be impossible.

How would the H_2CO_3-HCO_3^- buffer in the blood be affected by added CO_2 produced during exercise?

Buffers have many applications in analytical and physical chemistry. Electrophoresis, used to separate and purify proteins, depends on very accurate pH control and is always carried out in a buffer. Many reactions involving organic and inorganic systems are studied in buffers, since their rates depend markedly on pH. In the previous chapter we cited a method for determination of the ionization constant of an acid which involves its partial neutralization with base. Since the resulting solution is buffered, its pH can be determined quite accurately even in the presence of acidic or basic impurities.

18.4 APPLICATION OF ACID-BASE REACTIONS IN INORGANIC SYNTHESES

Acid-base reactions are used very frequently in the procedures by which inorganic compounds are prepared. They lend themselves to the preparation of a variety of salts; in addition, acid-base reactions can be applied in the laboratory to prepare small amounts of certain volatile acids and bases.

PREPARATION OF SALTS

One of the most common industrial methods for the preparation of metallic salts takes advantage of the reaction between strong acids and solids containing basic anions. Let us assume, for example, that we wish to make some pure crystalline $CuCl_2$ and that we have available a solution containing some Cu^{2+} ion in the presence of various unidentified anions.

Our procedure is to precipitate either the hydroxide or carbonate of copper by adding a solution of NaOH or Na_2CO_3:

$$Cu^{2+}(aq) + 2\ OH^-(aq) \rightleftarrows Cu(OH)_2(s)$$
$$Cu^{2+}(aq) + CO_3^{2-}(aq) \rightleftarrows CuCO_3(s)$$

The solid product can be separated from the rest of the solution and washed with distilled water to remove any soluble contaminants, leaving essentially pure $Cu(OH)_2$ or $CuCO_3$. To the solid we add hydrochloric acid, slowly, with stirring, to dissolve the solid:

$$Cu(OH)_2(s) + 2\ H^+(aq) \rightleftarrows Cu^{2}(aq) + 2\ H_2O \qquad (18.16)$$

$$CuCO_3(s) + 2\ H^+(aq) \rightleftarrows Cu^{2+}(aq) + CO_2(g) + H_2O \qquad (18.17)$$

As soon as all the solid has dissolved, we stop adding acid and should have, at that point, a solution containing only Cu^{2+} ions, Cl^- ions from the acid, and possibly some dissolved CO_2. We boil the solution down and recover pure crystalline copper(II) chloride as a dihydrate:

$$Cu^{2+}(aq) + 2\ Cl^-(aq) + 2\ H_2O \rightarrow CuCl_2 \cdot 2\ H_2O(s) \qquad (18.18)$$

By essentially the same approach one can prepare salts of all the metallic cations which form insoluble hydroxides or carbonates. The anions of choice are introduced by simply selecting the proper acid—nitric acid for nitrates, sulfuric for sulfates, and so on.

PREPARATION OF VOLATILE ACIDS OR BASES

A convenient way to prepare small quantities of ammonia is to heat an ammonium salt or a solution containing NH_4^+ ions with a strong base like sodium hydroxide:

$$NH_4^+(aq) + OH^-(aq) \rightarrow NH_3(g) + H_2O \qquad (18.19)$$

Anhydrous NH_3 could be made by passing the evolved gas over a suitable drying agent.

Several volatile weak acids are easily prepared by treating one of their salts or solutions thereof with a strong nonvolatile acid such as sulfuric acid. Carbon dioxide, sulfur dioxide, and hydrogen sulfide are frequently made in the laboratory by this kind of reaction:

$$CO_3^{2-}(aq) + 2\ H^+(aq) \rightleftarrows H_2CO_3(aq) \rightleftarrows CO_2(g) + H_2O \qquad (18.20)$$

$$SO_3^{2-}(aq) + 2\ H^+(aq) \rightleftarrows H_2SO_3(aq) \rightleftarrows SO_2(g) + H_2O(aq) \qquad (18.21)$$

$$FeS(s) + 2\ H^+(aq) \rightleftarrows H_2S(g) + Fe^{2+}(aq) \qquad (18.22)$$

The usefulness of Reactions 18.20 to 18.22 depends on the acid formed being volatile, not on its being weak. Indeed, one can prepare volatile

TABLE 18.2   Volatility of Acids from Water Solution

ACID	BOILING POINT (°C)	
Hydrogen sulfide	100	H_2S escapes on warming
Carbonic acid	100	CO_2 escapes on warming
Sulfurous acid	100	SO_2 escapes on warming
Hydrochloric acid	110	constant boiling mixture (20% HCl)
Hydrofluoric acid	120	constant boiling mixture (35% HF)
Nitric acid	121	constant boiling mixture (68% HNO_3)
Hydrobromic acid	126	constant boiling mixture (47% HBr)
Hydriodic acid	127	constant boiling mixture (57% HI)
Phosphoric acid	213	$2\ H_3PO_4 \rightarrow H_4P_2O_7 + H_2O$
Sulfuric acid	330	$H_2SO_4 \rightarrow SO_3 + H_2O$

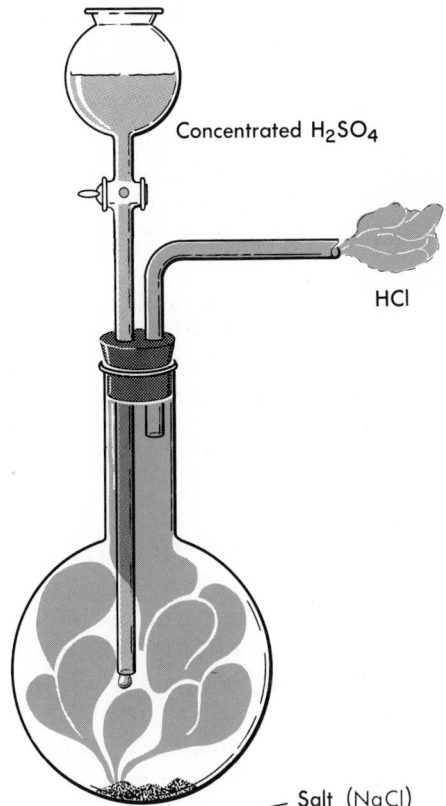

Figure 18.5 Preparation of HCl from NaCl.

Phosphoric acid, H_3PO_4, is used instead of H_2SO_4 to prepare HBr(g) or HI(g). Can you suggest why?

strong acids by reactions entirely analogous to these. For example, Hydrogen chloride can be formed by allowing sulfuric acid to drop on sodium chloride (Fig. 18.5):

$$NaCl(s) + H_2SO_4(l) \rightarrow NaHSO_4(s) + HCl(g) \qquad (18.23)$$

18.5 APPLICATIONS OF ACID-BASE REACTIONS IN QUALITATIVE ANALYSIS

Acid-base reactions are widely used in qualitative analysis for either of two different purposes. An acid-base reaction may be used to test for a specific ion by converting that ion into a volatile, easily detected product. Alternatively, such a reaction is often applied to separate one cation from another, where advantage is taken of the ability of acidic solutions to dissolve certain water-insoluble salts while leaving others unaffected.

TESTS FOR SPECIFIC IONS

Some of the reactions in the previous section work very well for the identification of certain ions commonly found in solution. Reaction 18.19

is used for detection of NH_4^+ ion. The NH_3 formed is detected either by its odor or by its effect on moist litmus; of all the gases commonly found in the analytical laboratory, only NH_3 is basic to litmus.

Reactions 18.20, 18.21, and 18.22 are used in anion analysis to detect carbonates or hydrogen carbonates, sulfites, and sulfides. Carbon dioxide has no odor but is easily detected by passing it into a solution of barium hydroxide, producing a white precipitate of barium carbonate:

$$Ba^{2+}(aq) + 2\ OH^-(aq) + CO_2(g) \rightarrow BaCO_3(s) + H_2O \qquad (18.24)$$

Sulfur dioxide has the characteristic odor of burning sulfur and also will decolorize $KMnO_4$ solution. Hydrogen sulfide is hard to miss, since its odor is bad and characteristic; it will also darken a piece of filter paper moistened with lead nitrate solution:

$$Pb^{2+}(aq) + H_2S(g) \rightarrow Pb(s) + 2\ H^+(aq) \qquad (18.25)$$

SEPARATION OF IONS

If the precipitate completely dissolves in acid, what can you conclude about the sample?

The use of an acid-base reaction to separate one ion from another is illustrated by a procedure commonly used in anion analysis to distinguish between CO_3^{2-} and SO_4^{2-}. These ions are first precipitated as the barium salts. The mixed precipitate of $BaCO_3$ and $BaSO_4$ is then treated with hydrochloric acid. The carbonate is brought into solution by Reaction 18.10, while the sulfate is unaffected. (Why is the sulfate immune to acid?)

$$CO_3^{2-},\ SO_4^{2-}$$
$$\downarrow Ba^{2+}$$
$$BaCO_3\ BaSO_4(s)$$
$$\downarrow H^+$$
$$Ba^{2+},\ CO_2(g) \qquad\qquad BaSO_4(s)$$

A more subtle application of this principle is involved in the reaction of hydrochloric acid with a mixed precipitate of CuS and CoS. The cobalt sulfide dissolves in acid while copper sulfide does not. This separation is more commonly carried out in reverse; that is, by adding hydrogen sulfide in acidic solution to a mixture of Co^{2+} and Cu^{2+} ions, CuS is precipitated while Co^{2+} remains in solution:

$$Co^{2+},\ Cu^{2+}$$
$$\downarrow H_2S,\ H^+(.3\ M)$$
$$Co^{2+} \qquad\qquad CuS(s)$$

As noted in Chapter 16, ions that form water-insoluble sulfides can be separated into two groups in this manner. The ions which form extremely insoluble sulfides (Group 2 = Cu^{2+}, Bi^{3+}, As^{3+}, Sb^{3+}, Sn^{2+}, Sn^{4+}, Cd^{2+}) are precipitated by hydrogen sulfide in acidic solution, while those sulfides which are somewhat more soluble (Co^{2+}, Fe^{2+}, Fe^{3+}, Ni^{2+}, Zn^{2+}, Mn^{2+} in Group 3) are precipitated only after the solution is made basic.

The general principles governing reactions of this sort were discussed in the section on reactions of strong acids with basic solids. Equations 18.10 and 18.11 can be easily rederived and applied to the situation in which a sulfide MS dissolves in acid; for the reaction:

$$MS(s) + 2\ H^+(aq) \rightleftarrows H_2S(aq) + M^{2+}(aq) \qquad (18.26)$$

$$K = \frac{[H_2S] \times [M^{2+}]}{[H^+]^2} = \frac{K_{sp}}{K_1 K_2} = \frac{K_{sp}}{1 \times 10^{-22}} \qquad (18.27)$$

where K_{sp} is the solubility product of the sulfide and K_1 and K_2 are the first and second dissociation constants of H_2S.

Equation 18.27 must be satisfied by a system containing a solid sulfide MS in equilibrium with M^{2+}, H_2S, and H^+. In a solution saturated with H_2S, $[H_2S]$ is about 0.1 M. The hydrogen ion concentration is readily controlled with strong acid. The solubility of the sulfide in the medium will be equal to $[M^{2+}]$. Several factors will affect that solubility, as implied by Equation 18.27:

1. *Effect of the solubility product of the sulfide.* The equilibrium constant for the overall reaction varies directly with the solubility product, so that, as we might expect, the more soluble the sulfide is in water the more soluble it will be in acid. Cobalt sulfide, $CoS(K_{sp} = 1 \times 10^{-21})$, is more soluble in water than copper sulfide, $CuS(K_{sp} = 1 \times 10^{-25})$, so we would expect it to be more soluble in acid.

2. *Effect of acid concentration.* We see from Equation 18.27 that the concentration of metal ion in solution varies directly with the square of the concentration of H^+. That is:

$$[M^{2+}] = \frac{K \times [H^+]^2}{[H_2S]}$$

Therefore, as $[H^+]$ goes up, $[M^{2+}]$ must also go up if equilibrium is to be maintained. This means that the solubility of a given sulfide will increase markedly as the solution is made more acidic.

The quantitative treatment of these ideas is illustrated in Example 18.8.

Example 18.8
  a. Find the solubilities in moles/lit of CoS and CuS in a solution saturated with H_2S in which $[H^+]$ is 0.3 M.
  b. At what concentration of H^+ would 0.1 M Co^{2+} begin to precipitate from a solution saturated with H_2S?

Solution
  a. Let us deal first with CuS, finding K for its reaction with acid:

$$CuS(s) + 2\ H^+(aq) \rightleftarrows Cu^{2+}(aq) + H_2S(aq) \quad K = \frac{K_{sp}}{K_1 K_2} = \frac{1 \times 10^{-25}}{(1 \times 10^{-7})(1 \times 10^{-15})}$$

$$K = 1 \times 10^{-3} = \frac{[Cu^{2+}] \times [H_2S]}{[H^+]^2}$$

Since K is small, we wouldn't expect much sulfide to dissolve:

$$[Cu^{2+}] = \frac{1 \times 10^{-3} \times [H^+]^2}{[H_2S]} = \frac{1 \times 10^{-3} \times (0.3)^2}{0.1} = 1 \times 10^{-3}\ M$$

This means that about 1×10^{-3} moles of CuS would dissolve in a liter of 0.3 M HCl saturated with H_2S. Or, conversely, if acid and H_2S were added to, for example, 0.1 M $CuCl_2$ solution, CuS would form until $[Cu^{2+}]$ fell to a very small value indeed. It turns out, experimentally, that CuS is essentially insoluble in nonoxidizing acids and will precipitate readily from acidic solution containing any H_2S. With cobalt sulfide the situation is somewhat different:

$$CoS(s) + 2\ H^+(aq) \rightleftarrows Co^{2+}(aq) + H_2S(aq) \quad K = \frac{K_{sp}}{K_1K_2} = \frac{1 \times 10^{-21}}{1 \times 10^{-22}}$$

$$K = 10 = \frac{[Co^{2+}][H_2S]}{[H^+]^2}$$

Here K is much larger, so the solubility is higher than for CuS:

$$[Co^{2+}] = \frac{10\ [H^+]^2}{[H_2S]} = \frac{10 \times (0.3)^2}{0.1} = 9\ M$$

In principle anyway, 9 moles of CoS would dissolve in a liter of acid solution maintained at 0.3 M in H^+ ion, so CoS is soluble. More practically, in order to precipitate CoS from a solution in which $[H^+]$ was 0.3 M and $[H_2S]$ was 0.1 M, $[Co^{2+}]$ would have to be very high; 0.1 M Co^{2+} would not precipitate.

We have seen, then, that in hydrogen sulfide solution kept at $[H^+]$ of about 0.3 M, 0.1 M Cu^{2+} will precipitate quantitatively and Co^{2+} will remain in solution. These are the conditions under which the Group II cations like Cu^{2+}, all of which have very insoluble sulfides, are separated from those in Group III like Co^{2+}.

b. This problem is best interpreted as asking at what $[H^+]$ will CoS be in equilibrium with Co^{2+} at a concentration of 0.1 M. Solving Equation 18.27 with K = 10, $[Co^{2+}]$ = 0.1 M and $[H_2S]$ = 0.1 M, we have:

$$[H^+]^2 = \frac{Co^{2+} \times [H_2S]}{K} = \frac{(0.1)(0.1)}{10} = 1 \times 10^{-3} = 10 \times 10^{-4}$$

$$[H^+] = 0.03\ M$$

At concentrations of H^+ ion less than about 0.03 M we would expect that CoS would precipitate from 0.1 M Co^{2+} solution. Group III is actually brought down at a $[H^+]$ of about 10^{-9}, where CoS precipitates quantitatively.

This analysis shows why Group 2 can be separated from Group 3 by addition of H_2S under moderately acidic conditions

18.6 AN INDUSTRIAL APPLICATION OF ACID-BASE REACTIONS: THE SOLVAY PROCESS

A commercially important process in which acid-base reactions play an important part is the so-called Solvay Process for the manufacture of sodium hydrogen carbonate, $NaHCO_3$, and sodium carbonate, Na_2CO_3. The principal economic advantage of the Solvay Process is that it uses as raw materials sodium chloride and limestone, both of which are cheap and available from naturally occurring deposits.

PREPARATION OF NaHCO₃

The first product formed in the Solvay Process is sodium hydrogen carbonate, $NaHCO_3$. To prepare this compound, advantage is taken of its comparatively low solubility at temperatures near the freezing point of water. Carbon dioxide is bubbled through a concentrated brine solution saturated with ammonia and maintained at a temperature of approximately 0°C. Under these conditions, finely divided crystals of sodium hydrogen carbonate precipitate. The equation for the overall reaction may be written:

$$CO_2(g) + H_2O + NH_3(g) + Na^+(aq) + Cl^-(aq) \rightarrow$$
$$NaHCO_3(s) + NH_4^+(aq) + Cl^-(aq) \quad (18.28)$$

The sodium hydrogen carbonate is filtered from the solution of ammonium chloride. It is possible to prepare $NaHCO_3$ in this manner because its solubility at 0°C (0.82 mole/lit) is considerably less than that of any of the other possible products: NH_4Cl (5.5 moles/lit), $NaCl$ (6.1 moles/lit), or NH_4HCO_3 (1.5 moles/lit).

A reaction such as 18.28, involving as it does a large number of species, can best be understood if we break it down into a series of relatively simple steps. In this case, three such steps may be considered:

1. Bubbling carbon dioxide through water establishes the equilibrium

$$CO_2(g) + H_2O \rightleftarrows H_2CO_3(aq) \rightleftarrows H^+(aq) + HCO_3^-(aq) \quad (18.28a)$$

The concentration of HCO_3^- ion produced by this reaction (0.00018 M) is far below that required to precipitate $NaHCO_3$ (about 0.10 M).

2. To increase the concentration of HCO_3^- ions in solution, it is necessary to add a reagent that will shift the equilibrium in 18.28a to the right. This may be accomplished by adding ammonia, which removes H^+ ions by converting them to NH_4^+ ions:

$$NH_3(g) + H^+(aq) \rightarrow NH_4^+(aq) \quad (18.28b)$$

Adding, $\qquad CO_2(g) + NH_3(g) + H_2O \rightarrow NH_4^+(aq) + HCO_3^-(aq)$

3. Under these conditions the HCO_3^- ions are present at a sufficiently high concentration to be precipitated by the sodium ions of the sodium chloride solution:

$$Na^+(aq) + Cl^-(aq) + HCO_3^-(aq) \rightarrow NaHCO_3(s) + Cl^-(aq) \quad (18.28c)$$

Adding all three equations, we get Equation 18.28

$$CO_2(g) + H_2O + NH_3(g) + Na^+(aq) + Cl^-(aq) \rightarrow$$
$$NaHCO_3(s) + NH_4^+(aq) + Cl^-(aq) \quad (18.28)$$

PREPARATION OF Na₂CO₃ FROM NaHCO₃

The greater part of the sodium hydrogen carbonate prepared by the Solvay Process is converted to the more widely used salt, sodium carbonate. This conversion is accomplished by heating $NaHCO_3$ to about 300°C; at this temperature carbon dioxide and water vapor are given off and a white residue of sodium carbonate remains:

$$2 \text{ NaHCO}_3(s) \rightarrow \text{Na}_2\text{CO}_3(s) + \text{CO}_2(g) + \text{H}_2\text{O}(g) \qquad (18.29)$$

The carbon dioxide formed is recycled to prepare more NaHCO_3 by Reaction 18.28.

PREPARATION OF CO₂ AND RECOVERY OF NH₃

For every mole of NaHCO_3 or Na_2CO produced by the Solvay Process, one mole of CO_2 is consumed. The carbon dioxide is produced by heating limestone:

$$\text{CaCO}_3(s) \rightarrow \text{CaO}(s) + \text{CO}_2(g) \qquad (18.30)$$

The economics of the Solvay Process, like those of all industrial operations, depend upon the effective use of all the products. In particular, it is important that the calcium oxide produced by Reaction 18.30 be utilized. Furthermore, the solution of ammonium chloride remaining after the precipitation of sodium hydrogen carbonate is far too valuable to be discarded. Indeed, if the ammonia consumed in Reaction 18.28 were not recovered, the Solvay Process would be economically impossible, since the ammonia costs more than the sodium hydrogen carbonate is worth. Fortunately these two problems—the utilization of the CaO produced in Reaction 18.30 and the recovery of the NH_3 used in 18.28—can be solved simultaneously. It was pointed out earlier that ammonia can be formed from ammonium salts by heating with a base:

$$2 \text{ NH}_4^+(\text{aq}) + 2 \text{ OH}^-(\text{aq}) \rightarrow 2 \text{ NH}_3(g) + 2 \text{ H}_2\text{O}$$

Could CaO be used instead of NH_3 to bring about Reaction 18.28?

Calcium oxide, upon addition to water, gives a moderately basic solution:

$$\text{CaO}(s) + \text{H}_2\text{O} \rightarrow \text{Ca}^{2+}(\text{aq}) + 2 \text{ OH}^-(\text{aq})$$

Consequently, if the solution of NH_4Cl remaining after the precipitation of NaHCO_3 is heated with CaO, the reaction

$$\text{CaO}(s) + 2 \text{ NH}_4^+(\text{aq}) + 2 \text{ Cl}^-(\text{aq}) \rightarrow$$
$$\text{Ca}^{2+}(\text{aq}) + 2 \text{ Cl}^-(\text{aq}) + 2 \text{ NH}_3(g) + \text{H}_2\text{O} \quad (18.31)$$

occurs, liberating ammonia and leaving as a final by-product a solution of calcium chloride.

SUMMARY OF THE SOLVAY PROCESS

The raw materials used in the Solvay Process are sodium chloride, water, and limestone. The first product formed is sodium hydrogen carbonate, NaHCO_3, which precipitates at 0°C from a sodium chloride solution saturated with CO_2 (produced by the thermal decomposition of CaCO_3) and NH_3 (recovered by heating the solution remaining with CaO). The NaHCO_3 may be sold for use in medicine, as a baking powder, and so on, or it may be converted to Na_2CO_3 by heating. More than six million tons of

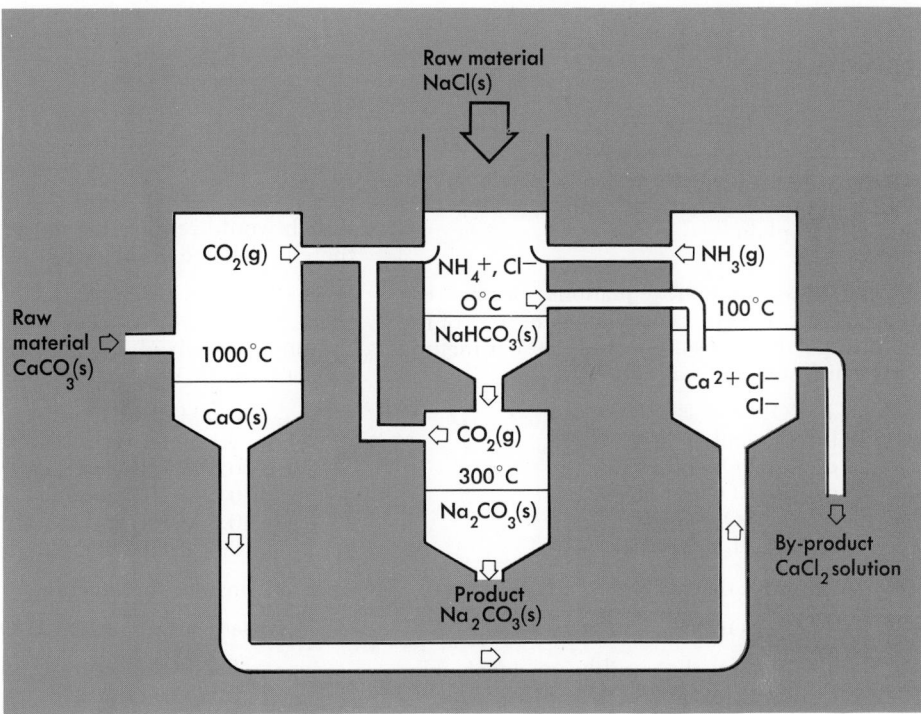

Figure 18.6 The Solvay Process.

sodium carbonate are used annually in this country in the manufacture of glass, paper, soap, and other chemicals. Calcium chloride, formed as a by-product, is used as a drying agent and in ice removal from highways.

Sodium carbonate, sometimes called washing soda or soda ash, has had an interesting industrial history. In the 18th century the French government, dependent on the ashes of wood and plants as a source of Na_2CO_3, offered a very substantial prize for a process to make the material from natural sources. Leblanc developed such a process in 1791, using common salt, limestone, coal, and sulfuric acid. In the process salt was heated with sulfuric acid, producing salt cake, Na_2SO_4, and HCl. The Na_2SO_4 was then heated to about 900°C with a mixture of coal and limestone, forming a solid mixture of Na_2CO_3, CaS, and various impurities called "black ash." The sodium carbonate was leached from the solid with water and recrystallized.

Unfortunately before Leblanc could capitalize on his process, the French Revolution broke out; Leblanc's plant was destroyed and he was a suicide. However, his process was implemented in England and was the main source of the world's Na_2CO_3 during much of the nineteenth century. Following its development in 1865, the Solvay Process gradually superseded the Leblanc method and became the source of most of the Na_2CO_3 produced during this century. In the past decade some localized deposits of nearly pure trona, $Na_2CO_3 \cdot NaHCO_3 \cdot 2\ H_2O$, discovered in 1938 near Green River, Wyoming, have been very extensively mined and currently furnish about half the sodium carbonate needs of this country.

It is usually cheaper to mine an industrially important substance than to produce it by a chemical process

PROBLEMS

18.1 When solutions of NaOH and HCl are mixed, we write the equation for the reaction in a form which differs from that we write to describe the reaction between solutions of NaOH and HF. Explain.

18.2 Write net ionic equations for the reactions, if any, which occur when solutions of the following substances are mixed. Do not include ions which serve only as "observers":

    a. KOH and HBr
    b. NH_3 and $HClO_4$
    c. NH_4NO_3 and HCl
    d. $NaHCO_3$ and HNO_3
    e. $NaHSO_3$ and KOH
    f. $NaC_2H_3O_2$ and HI

18.3 Consider the reaction:

$$HF(aq) + OH^-(aq) \rightleftarrows H_2O + F^-(aq)$$

    a. Calculate the equilibrium constant for this reaction.
    b. Find the concentration at equilibrium of HF when 0.50 moles NaOH are added to 1.0 liter of 0.50 M HF. What is $[H^+]$ in the solution?

18.4 Calculate equilibrium constants for the following reactions using Table 17.5, p. 462, and Table 16.2, p. 438.

a. $H^+(aq) + NO_2^-(aq) \rightleftarrows HNO_2(aq)$
b. $HCHO_2(aq) + OH^-(aq) \rightleftarrows H_2O + CHO_2^-(aq)$
c. $MgCO_3(s) + 2\ H^+(aq) \rightleftarrows Mg^{2+}(aq) + H_2CO_3(aq)$

18.5 What volume of 0.25 M NaOH would be required to react completely with 46.3 ml 0.40 M H_2SO_4?

18.6 Two hundred and fifty ml 0.10 M NaOH are added to 250 ml 0.25 M HNO_3. Find the number of moles of each ion and its concentration in the solution after the reaction is complete.

18.7 How many moles of $BaCO_3$ would you expect to dissolve in one liter 0.1 M HNO_3?

18.8 One can find the gram equivalent weight (GEW) of an acid by titration against a standardized NaOH solution. A 1.30 g crystalline sample of citric acid, found in oranges and lemons, was found to require 38.5 ml 0.177 M NaOH for complete neutralization. Find the GEW of citric acid. In a freezing-point depression experiment the molecular weight of citric acid was found to be about 190 g. How many acid hydrogens are there in a citric acid molecule? What is the GMW of citric acid?

18.20 You have a liter of each of the following:

Solution	0.1 M HNO_3	0.1 M $HC_2H_3O_2$	0.1 M HCN
pH	1	2.9	5.2

Which would require the most base to titrate to the equivalence point? Why?

18.21 Write net ionic equations for the reactions, if any, which occur when the following solutions are mixed. Do not include ions which do not participate:

    a. 0.1 M NaOH and 0.1 M $HCHO_2$
    b. 0.1 M KOH and 0.2 M NH_4Br
    c. 0.3 M $NaC_2H_3O_2$ and 0.1 M NH_3
    d. 0.5 M HCN and 0.4 M KOH
    e. 0.2 M HF and 0.3 M KCN
    f. 0.5 M HCl and 0.03 M $Ba(OH)_2$

18.22 Consider the reaction:

$$HNO_2(aq) + OH^-(aq) \rightleftarrows H_2O + NO_2^-(aq)$$

    a. Find the equilibrium constant for this reaction.
    b. Find $[HNO_2]$ at equilibrium if 100 ml 0.3 M HNO_2 are mixed with 100 ml 0.3 M NaOH. What is the pH of the final solution?

18.23 Calculate equilibrium constants for the following reactions:

a. $NH_4^+(aq) + OH^-(aq) \rightleftarrows NH_3(aq) + H_2O$
b. $H^+(aq) + C_2H_3O_2^-(aq) \rightleftarrows HC_2H_3O_2(aq)$
c. $CaCO_3(s) + 2\ H^+(aq) \rightleftarrows Ca^{2+}(aq) + H_2CO_3(aq)$
d. $MnS(s) + 2\ H^+(aq) \rightleftarrows H_2S(aq) + Mn^{2+}(aq)$

18.24 In a titration of an NaOH solution with 0.1007 M HBr, it required 35.48 ml of the NaOH to neutralize 25.00 ml of the HBr. Find the molarity of the NaOH.

18.25 A mixture containing 200 ml 0.2 M $HC_2H_3O_2$ and 300 ml 0.1 M NaOH is prepared. Find $[HC_2H_3O_2]$, $[C_2H_3O_2^-]$, and the pH of the solution after the system has reached equilibrium. Is this solution a buffer?

18.26 Calculate the solubility, in moles/lit of $BaCO_3$ when $[H^+] = 10^{-4}$, $[H_2CO_3] = 0.03$.

18.27 NaOH solutions are most often standardized by titration to a phenolphthalein end point against weighed samples of potassium hydrogen phthalate, KHPh, an organic crystalline acid with one acidic hydrogen. This material has a very high molar weight, 204.22 g/mole, and can be obtained in very pure form. A 2.319 g sample of KHPh required 21.66 ml of NaOH before the end point in the titration was reached. Find the molarity of the NaOH solution.

18.9 Cresol red, one of the common acid-base indicators, has a K_a equal to 1×10^{-8}. Its acid form is yellow, while its conjugate base is red. What color would the indicator have in a solution of pH 6? 7? 8? 9? 14? Would cresol red be a good choice for an equivalence point detector in an NaOH-HNO$_3$ titration?

18.10 In each of the following reactions the underlined reagent is 0.10 M. What is its normality?

a. $\underline{HNO_2(aq)} + OH^-(aq) \rightleftarrows H_2O + NO_2^-(aq)$
b. $\underline{H_2SO_3(aq)} + OH^-(aq) \rightleftarrows H_2O + HSO_3^-(aq)$
c. $\underline{H_3PO_4(aq)} + 2\ OH^-(aq) \rightleftarrows 2\ H_2O + HPO_4^{2-}(aq)$
d. $\underline{H_2CO_3(aq)} + 2\ OH^-(aq) \rightleftarrows 2\ H_2O + CO_3^{2-}(aq)$

18.11 Fifty ml 0.1000 M HCl is titrated with 0.1000 M NaOH. Calculate the pH of the solution after the following volumes of NaOH have been added:

a. 00.00 ml	e. 50.01 ml
b. 25.00 ml	f. 50.10 ml
c. 49.90 ml	g. 75.00 ml
d. 49.99 ml	h. 100.00 ml

From your data, construct a titration curve similar to that in Figure 18.1.

18.12 Given three acid-base indicators — methyl red (end point at pH 5), bromthymol blue (end point at pH 7), and phenolphthalein (end point at pH 9) — which would you select for each of the following titrations? Why?

a. HNO$_3$ with NaOH
b. NaCN with HCl
c. NH$_3$ with HNO$_3$
d. HC$_2$H$_3$O$_2$ with KOH

18.13 A buffer solution is made by adding 0.40 moles formic acid, HCHO$_2$, and 0.20 moles sodium hydroxide to one liter of water.

a. Find the pH of the buffer.
b. What is the pH after 0.02 mole of HCl is added?
c. What is the pH after 0.02 mole of NaOH is added?
d. What is the pH after 0.20 mole of HCl is added?

18.14 Blood is buffered mainly by the H$_2$CO$_3$-HCO$_3^-$ system, in which the concentration ratio of [HCO$_3^-$] to [H$_2$CO$_3$] is about 20:1.

a. What would the pH of blood be if this were the only buffer?
b. What would the pH become if the relative amount of H$_2$CO$_3$ present were doubled, so that the concentration ratio was 10:1.
c. Is the carbonic acid-hydrogen carbonate buffer in the blood more effective against addition of acid than of base? Why?

18.28 Congo Red, a well-known acid-base indicator, has a K_a value of 1×10^{-4}. Its acid form is blue, its basic form is red. What color would Congo Red be in a solution of pH 1? pH 3? pH 4? pH 5? pH 10? What is the pH at the end point of Congo Red? For which titration would Congo Red be most satisfactory: NaOH and HNO$_3$; NH$_4$Cl and NaOH; or NH$_3$ and HBr?

18.29 In each of the following reactions the underlined reagent is 0.4 M. What is its normality?

a. $\underline{H_2CO_3(aq)} + OH^-(aq) \rightleftarrows HCO_3^-(aq) + H_2O$
b. $\underline{H_2Ox(aq)} + 2\ OH^-(aq) \rightleftarrows Ox^{2-}(aq) + 2\ H_2O$
c. $\underline{H_3PO_4(aq)} + 3\ OH^-(aq) \rightleftarrows PO_4^{3-}(aq) + 3\ H_2O$
d. $\underline{HCN(aq)} + OH^-(aq) \rightleftarrows CN^-(aq) + H_2O$

18.30 Fifty ml 0.2000 M HC$_2$H$_3$O$_2$ is titrated with 0.2000 M NaOH. Calculate the pH of the solution after the following volumes of NaOH have been added:

a. 00.00 ml	e. 50.01 ml
b. 25.00 ml	f. 50.10 ml
c. 49.90 ml	g. 75.00 ml
d. 49.99 ml	h. 100.00 ml

Which of the above solutions would be classified as buffers?

18.31 At the equivalence point in the following titrations, will the solution be acidic, basic or neutral?

a. HBr with NH$_3$
b. NaC$_2$H$_3$O$_2$ with HCl
c. HNO$_3$ with KOH
d. NaHCO$_3$ with HClO$_4$

Which titration would be easiest to accomplish to a precise end point? most difficult?

18.32 A buffer solution is made by adding 0.30 moles NH$_3$ and 0.30 moles NH$_4$Cl to a liter of water.

a. Find the pH of the buffer.
b. Find the pH after addition of 0.05 moles HCl.
c. Find the pH after addition of 0.05 moles NH$_3$.
d. Find the pH after addition of 0.30 moles HCl.

18.33 One of the functions of the H$_2$CO$_3$-HCO$_3^-$ buffer in the blood is to quickly remove lactic acid, HLac, from tissues after it is formed during active exercise. K_aHLac $= 8.4 \times 10^{-4}$. Find K for the reaction:

$$HLac(aq) + HCO_3^-(aq) \rightleftarrows H_2CO_3(aq) + Lac^-(aq)$$

In normal blood [H$_2$CO$_3$] $= 0.0014$ M and [HCO$_3^-$] $= 0.027$ M. Find the pH of blood, assuming the only buffer is H$_2$CO$_3$-HCO$_3^-$. Find the pH after addition of 0.006 moles HLac per liter.

18.15 Describe in some detail how you might prepare

    a. $CaBr_2$ from $Ca(NO_3)_2$
    b. $Ni(OH)_2$ from $NiCO_3$
    c. NH_4Cl from NH_3
    d. KBr from KOH
    e. Na_2CO_3 from $NaOH$
    f. $CuSO_4$ from $CuCl_2$

18.16 How could you use acid-base reactions to separate the following ions:

    a. Cl^- and CO_3^{2-}
    b. SO_3^{2-} and SO_4^{2-}
    c. NH_4^+ and Ca^{2+}
    d. S^{2-} and NO_3^-

18.17 How could you distinguish between the following, all of which are white solids:

    a. K_2SO_4 and K_2CO_3
    b. NH_4Cl and $NaCl$
    c. KHS and $KHCO_3$
    d. $KHSO_4$ and K_2SO_4
    e. $BaCl_2$ and $NaNO_3$
    f. KCN and KCl

18.18 In the Solvay process,

    a. what is the fate of the CO_2?
    b. what is the role of the CaO?
    c. why is ammonia used?

18.19 $CdS(K_{sp} = 1 \times 10^{-26})$ is the one of the more soluble of the Group II sulfides.

    a. Find K for the reaction of CdS with acid.
    b. Find the solubility of cadmium sulfide in a solution that is 0.1 M in H_2S and in which $[H^+]$ is 0.3 M.
    c. What fraction of the Cd^{2+} in a solution in which its molarity is 0.1 will precipitate under the conditions in b?

18.34 Describe in some detail how you might prepare

    a. $MnSO_4$ from $MnCl_2$
    b. NH_3 from NH_4Br
    c. $AgNO_3$ from Ag_2O
    d. $CoCO_3$ from $Co(NO_3)_2$
    e. HCl from $NaCl$
    f. KI from KOH

18.35 How could you use acid-base or precipitation reactions to resolve each of the following ions from a mixture (obtain four fractions, each containing one ion):

$$SO_4^{2-} \quad CO_3^{2-} \quad Cl^- \quad NH_4^+$$

18.36 Write equations for the reactions, if any, which occur when

    a. carbon dioxide is bubbled through a solution of sodium hydroxide.
    b. baking soda, $NaHCO_3$, is added to a very acid stomach.
    c. a solution of NH_4Cl is made strongly basic and heated.
    d. H_2S gas is bubbled through an acidified solution of $CuCl_2$.

18.37 The overall reaction in the Solvay Process for the preparation of Na_2CO_3 is:

$$2\ Na^+(aq) + CaCO_3(s) \rightarrow Na_2CO_3(s) + Ca^{2+}(aq).$$

Write balanced equations for each step of the process and add them to give this equation as the net result.

18.38 Using Table 16.2 for sulfide solubility products, predict the order in which Co^{2+}, Fe^{2+}, Mn^{2+}, Ni^{2+} and Zn^{2+} would precipitate from a solution saturated with H_2S if all the cations were present at 0.1 M and the pH of the solution were very slowly increased from 0 to 9. At pH 7 what fraction, if any, of the Fe^{2+} present would be in the form of FeS?

*18.39 State whether a precipitate of MnS will form when enough OH^- ion is added to a solution which is 0.1 M in Mn^{2+} and 0.1 M in H_2S to make the pH equal to 7. $K_{sp}MnS = 1 \times 10^{-13}$.

*18.40 Ammonium chloride solutions are slightly acidic, so are better solvents than water for insoluble substances such as $Mg(OH)_2$. Find the solubility of $Mg(OH)_2$ in one liter of 2.0 M NH_4Cl and compare with the solubility in water. Hint: Find K for the reaction:

$$Mg(OH)_2(s) + 2\ NH_4^+(aq) \rightleftarrows Mg^{2+}(aq) + 2\ NH_3(aq) + 2\ H_2O(l)$$

*18.41 The amino acid glycine, like all amino acids, has both a basic and an acidic group in its molecule, NH_2—CH_2—$COOH$. In acid solution, glycine takes the form NH_3—CH_2—$COOH^+$, while in a base it exists as NH_2—CH_2—COO^-. The intermediate form is uncharged, but is not thought to be the molecule, but rather the zwitterion, NH_3—CH_2—COO.

$$NH_3\text{—}CH_2\text{—}COOH^+(aq) \rightleftarrows H^+(aq) + NH_3\text{—}CH_2\text{—}COO(aq) \quad K_1 = 4.5 \times 10^{-3}$$
$$NH_3\text{—}CH_2\text{—}COO(aq) \rightleftarrows H^+(aq) + NH_2\text{—}CH_2\text{—}COO^-(aq) \quad K_2 = 1.7 \times 10^{-10}$$

a. Which hydrogen in the completely protonated species comes off more easily? Does this seem reasonable in view of the properties of acetic acid and ammonium ion?

b. Calculate the pH at which the concentrations of $NH_3\text{—}CH_2\text{—}COOH^+$ and $NH_2\text{—}CH_2\text{—}COO^-$ are equal. (Hint: multiply K_1 by K_2). This pH is called the isoelectric point; at that pH glycine would not migrate under electrolysis.

c. Could glycine solutions serve as buffers? At what pH would they be most useful?

19 COMPLEX IONS; COORDINATION COMPOUNDS

If white, anhydrous copper sulfate is exposed to ammonia gas, a deep blue crystalline product is formed. Analysis reveals that this product contains four moles of ammonia for every mole of copper sulfate. The positive ion in this compound consists of a Cu^{2+} ion bonded to four ammonia molecules, i.e., $Cu(NH_3)_4^{2+}$. The reaction between the Cu^{2+} ion and the NH_3 molecules is represented in electron dot notation as

$$Cu^{2+} + 4 \ NH_3 \rightarrow Cu(NH_3)_4^{2+} \tag{19.1}$$
colorless · deep blue

The nitrogen atom of each ammonia molecule contributes a pair of unshared electrons to form a coordinate covalent bond with the Cu^{2+} ion. In this reaction ammonia is acting as a Lewis base; Cu^{2+} is a Lewis acid. In this sense, Reaction 19.1 resembles that between a proton and an ammonia molecule to form an ammonium ion.

Lewis acid · Lewis base

The $Cu(NH_3)_4^{2+}$ ion is commonly referred to as a complex ion. In the broadest sense, a complex ion may be considered to be a charged aggregate consisting of more than one atom. Strictly speaking, such oxyanions as nitrate (NO_3^-) and sulfate (SO_4^{2-}) can be classified as complex ions. However, we shall use the term in a more restricted sense to refer to a **charged aggregate in which a metal atom is joined by coordinate covalent bonds to neutral molecule(s) and/or negative ions.** Species such as $Cu(H_2O)_4^{2+}$ and $Zn(H_2O)_3(OH)^+$, encountered in previous chapters, are properly considered to be complex ions.

The metal atom in a complex ion (Cu, Zn, . . .) is referred to as the **central atom.** The molecules (NH_3, H_2O, . . .) or anions (OH^-, Cl^-, . . .) attached to the central atom are known as coordinating groups or **ligands.** In order to act as a ligand, a molecule or anion must ordinarily have an unshared electron pair which can be donated to the central atom to form a coordinate covalent bond. The number of bonds (e.g., 2, 4, 6) formed by the central atom is called its **coordination number.** The nomenclature of complex ions and their salts is discussed in Appendix 3.

[$Cu(NH_3)_4$]SO_4 is called tetramminecopper(II) sulfate

19.1 CHARGES OF COMPLEX IONS. NEUTRAL COMPLEXES

The Pt^{2+} ion forms a series of five different complexes in which the ligands are ammonia molecules and/or chloride ions. The formulas of these complexes, along with those of typical salts in which they are present, are listed in Table 19.1 (the square brackets are used to enclose the complex).

The structures of these five compounds can be deduced from experimental observations on the behavior of their aqueous solution. In particular, we find:

1. When a direct electric current is passed through a water solution of compound 1 or 2, the platinum-containing species migrates toward the negative electrode. This suggests that the platinum in these species is part of a complex *cation.* In compounds 4 and 5, it is found that the concentration of platinum increases around the positive electrode, implying that it is present as a complex *anion.*

2. Solutions of compounds 1, 2, 4, and 5 conduct an electric current. With compounds 1 and 5, the conductivities per mole of platinum are relatively high, comparable to the molar conductivities of simple 2:1 or 1:2 electrolytes such as $CaCl_2$ or K_2SO_4. The molar conductivities of compounds 2 and 4 are about equal to those of simple 1:1 salts such as KCl or KNO_3. Compound 3 is quite insoluble in water and appears to be a nonelectrolyte.

TABLE 19.1 Complexes Formed by Pt^{2+} with NH_3 and Cl^-

SALT	COMPLEX	CHARGE OF COMPLEX	ANALOGOUS SIMPLE SALT
1. [$Pt(NH_3)_4$]Cl_2	$Pt(NH_3)_4^{2+}$	+2	$CaCl_2$
2. [$Pt(NH_3)_3Cl$]Cl	$Pt(NH_3)_3Cl^+$	+1	KCl
3. [$Pt(NH_3)_2Cl_2$]	$Pt(NH_3)_2Cl_2$	0	—
4. K[$Pt(NH_3)Cl_3$]	$Pt(NH_3)Cl_3^-$	−1	KNO_3
5. K_2[$PtCl_4$]	$PtCl_4^{2-}$	−2	K_2SO_4

3. When $AgNO_3$ in the cold is added to solutions of 5, 4, and 3, no precipitate is formed, indicating that all the Cl^- ions in these complexes are tightly bound within the complex. With compound 2, addition of $AgNO_3$ precipitates 1 mole of AgCl per mole of Pt; with compound 1, two moles of AgCl are formed per mole of platinum.

Once the formula of a complex has been established, we can readily obtain its charge in a formal way by taking *the algebraic sum of the charge of the metal ion and those of the ligands.* Thus, for the complexes of Pt^{2+} listed in Table 19.1, realizing that NH_3 is a neutral molecule and that the Cl^- ion has a charge of -1, we obtain:

1. $+2 + 4(0) = +2$
2. $+2 + 3(0) + 1(-1) = +1$
3. $+2 + 2(0) + 2(-1) = 0$
4. $+2 + 1(0) + 3(-1) = -1$
5. $+2 + 4(-1) = -2$

Species such as $[Pt(NH_3)_2Cl_2]$ or $[Zn(H_2O)_2(OH)_2]$ in which the charge of the metal ion (+2) is exactly balanced by the total charge of the ligands (-2) are called neutral complexes. Compounds of this type are typically insoluble in water. We find, for example, that when enough OH^- ions are added to a solution of a Zn^{2+} salt to give an appreciable concentration of the neutral complex $Zn(H_2O)_2(OH)_2$, a precipitate of "hydrated zinc hydroxide" forms. Many of the transition metal hydroxides precipitate out of aqueous solution, not as simple unhydrated species (e.g., $Zn(OH)_2$) but as hydrated compounds with compositions close to that of a neutral complex.

The industrial method of refining nickel (the Mond process) involves forming a neutral complex between a nickel atom and four carbon monoxide molecules. When carbon monoxide at 60°C is passed over nickel containing iron and cobalt impurities, the following reaction occurs

This is one of the few volatile complexes

$$Ni(s) + 4\ CO(g) \rightarrow Ni(CO)_4(g) \qquad (19.2)$$

Under these conditions, none of the impurities react. The so-called nickel carbonyl can be separated from excess CO by cooling; at 43°C it condenses to a colorless liquid. To recover the nickel, the neutral complex is heated to 200°C to reverse Reaction 19.2.

19.2 COMPOSITION OF COMPLEX IONS. GENERAL PRINCIPLES

CENTRAL METAL ATOM

In general, we find a correlation between the charge density (charge to size ratio) of a metal ion and its ability to form complex ions. Cations of high charge and small size, such as the +2 and +3 ions of the metals in groups 6B to 2B of the Periodic Table, form a large number of stable complex ions. The poorest complex formers are large metal ions of +1 charge, specifically such alkali metal cations as Na^+ and K^+.

Among the transition metals which form more than one cation, it is usually the ion of higher charge which is the better complex former. We find, for example, that the complexes formed by Cr^{3+}, Fe^{3+}, and Co^{3+} are both more numerous and more stable than those of the +2 ions of these metals, Cr^{2+}, Fe^{2+}, and Co^{2+}.

In solution the cations of transition metals usually exist as complex ions

The correlation between charge density and complexing ability implies that electrostatic forces between a metal ion and the unpaired electrons of a ligand are a major factor in complex ion formation. We shall have more to say on this topic when we discuss bonding in complex ions in Section 19.4. It is well to point out, however, that factors other than charge density must be considered in assessing the complexing ability of a metal ion. The Ag^+ ion ($r = 1.26$ Å) forms many more stable complexes than either Na^+ ($r = 0.95$ Å) or Ca^{2+} ($r = 0.99$ Å), both of which have higher charge densities. If you have ever developed your own photographic film, you have probably made one of the more stable complexes of silver, the $Ag(S_2O_3)_2^{3-}$ ion. This ion is formed when the developed film is "fixed" by washing with sodium thiosulfate, $Na_2S_2O_3$, to remove unreacted silver halide.

LIGANDS. CHELATING AGENTS

In principle, any molecule or anion possessing an unshared pair of electrons can donate them to a metal to form a complex ion. In practice, the atom within the ligand which furnishes these electrons is ordinarily derived from one of the more electronegative elements (C, N, O, S, F, Cl, Br, I). Hundreds of different ligands containing one or more of these atoms are known. Among the ligands most frequently encountered in general chemistry are the NH_3 and H_2O molecules and the OH^-, Cl^-, and CN^- ions (Table 19.2).

The relative abilities of different ligands to coordinate with metal ions depend upon a great many factors. One of the most important of these is the basicity of the ligand. It is perhaps not too surprising to find that molecules or ions which have a strong attraction for a proton are among the better coordinating agents. The species NH_3, OH^-, and CN^-, all of which are strong bases in the Brönsted-Lowry or Lewis sense, form stable complexes with a wide variety of transition metal ions. The ClO_4^- ion, which shows no tendency to acquire a proton in water solution, is a notoriously poor coordinating agent; the NO_3^- and HSO_4^- ions, both derived from strong acids, form relatively few stable complexes. It should be pointed out, however, that the Cl^- ion, which does not act as a base in water, forms stable complexes with many transition metal ions (cf. Table 19.2).

TABLE 19.2 A Few Complex Ions Involving Simple Ligands

H_2O	NH_3	OH^-	Cl^-	CN^-
$Cu(H_2O)_4^{2+}$	$Ag(NH_3)_2^+$	$Zn(OH)_4^{2-}$	$AgCl_2^-$	$Ag(CN)_2^-$
$Cr(H_2O)_6^{3+}$	$Cu(NH_3)_4^{2+}$	$Al(OH)_6^{3-}$	$CuCl_4^{2-}$	$Ni(CN)_4^{2-}$
$Co(H_2O)_6^{3+}$	$Zn(NH_3)_4^{2+}$	$Cr(OH)_6^{3-}$	$AlCl_4^-$	$Zn(CN)_4^{2-}$
$Ni(NH_3)_6^{2+}$	$Co(NH_3)_6^{3+}$		$PtCl_4^{2-}$	$Fe(CN)_6^{3-}$
$Zn(H_2O)_6^{2+}$	$Ni(NH_3)_6^{2+}$		$PtCl_6^{2-}$	$Fe(CN)_6^{4-}$

One of the first ligands to be studied extensively was the ethylenediamine molecule:

$$\begin{array}{cccc} & H & H & \\ & | & | & \\ H-\ddot{N}-C-C-\ddot{N}-H \\ & | & | & | & | \\ & H & H & H & H \end{array}$$

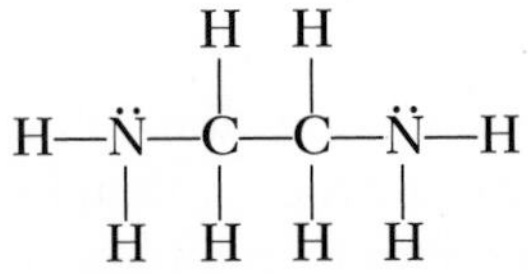

This molecule, containing two nitrogen atoms, each with an unshared pair of electrons, forms two coordinate covalent bonds with metal atoms. These bonds are extremely stable, as shown by the fact that the addition of ethylenediamine to a solution containing the $Cu(NH_3)_4^{2+}$ ion results in the displacement of the four NH_3 molecules by two $H_2N-CH_2-CH_2-NH_2$ molecules.

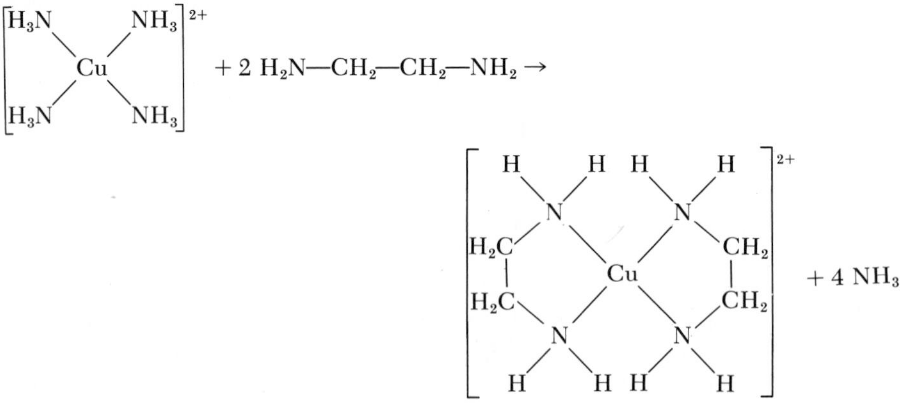

In writing a chemical equation to represent this reaction, we frequently abbreviate the ethylenediamine molecule as "en"

$$Cu(NH_3)_4^{2+}(aq) + 2 \text{ en}(aq) \rightarrow Cu(en)_2^{2+}(aq) + 4 \text{ } NH_3(aq) \quad (19.3)$$

The fact that ethylenediamine is a powerful coordinating agent is due in part to the high stability of the five-membered rings that it forms with metal ions. Methylenediamine, $H_2N-CH_2-NH_2$, is not nearly as good a complexing ligand. Another factor which tends to make reactions such as 19.3 spontaneous is the increase in the number of solute molecules (2 en → 4 NH_3). As we saw in Chapter 12, a reaction in which the number of molecules increases usually results in an increase in entropy. A positive value of ΔS tends to make ΔG negative ($\Delta G = \Delta H - T\Delta S$), corresponding to a spontaneous reaction.

A great many molecules and anions in addition to ethylenediamine form more than one bond with a metal ion. Ligands which act in this way are known as chelating agents: the complexes formed are called **chelates** from the Greek word meaning a crab's claw. Two anions which can act as chelating agents are the oxalate ion, $C_2O_4^{2-}$ and the carbonate ion, CO_3^{2-}. (The numbers (1) and (2) in the diagram on p. 511 indicate the atoms involved in chelate formation.)

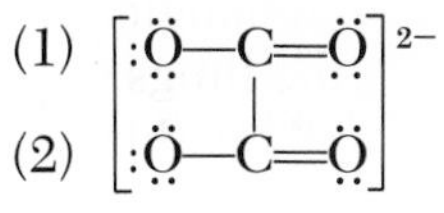

$$(1)\ \left[\ddot{\underset{..}{O}}\!-\!C\!=\!\ddot{\underset{..}{O}}\right]^{2-} \qquad (1)\ \left[\begin{array}{c} \ddot{\underset{..}{O}} \\ \diagdown \\ \end{array}\right.$$

$$(2)\ \left[\ddot{\underset{..}{O}}\!-\!C\!=\!\ddot{\underset{..}{O}}\right] \qquad \qquad C\!=\!\ddot{\underset{..}{O}}$$

$$(2)\ \left.\ddot{\underset{..}{O}} \diagup \right]^{2-}$$

$$C_2O_4{}^{2-} \qquad\qquad\qquad CO_3{}^{2-}$$

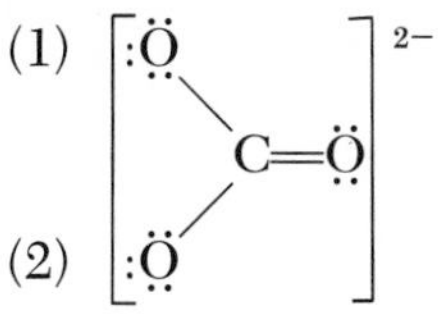

Chelating agents are abundant in nature and play an important role in many processes essential to plant and animal life. Certain species of soybeans are known to synthesize and secrete organic chelating agents that extract iron from insoluble compounds in the soil, thereby making it available for metabolic processes in the plant. Mosses and lichens growing on boulders use a similar process to obtain the metal ions they need for growth.

Many important natural products are chelates in which a central metal atom is bonded into a large organic molecule. In chlorophyll, the green coloring matter of plants, the central atom is magnesium; in the essential vitamin B_{12}, it is cobalt. The structure of heme, the pigment responsible for the red color of blood, is shown in Figure 19.1. We see that there is an Fe^{2+} ion at the center surrounded by four nitrogen atoms at the corners of a square. A fifth coordination position around the iron is occupied by a high molecular weight organic molecule (globin) which, in combination with heme, gives the proteinaceous material that we refer to as hemoglobin. The composition of the globin molecule is extremely important. A relatively minor variation in a tiny subsection of the molecule is responsible for the disease called sickle cell anemia, in which misshapen red blood cells clog capillaries, causing blood clots and depriving tissues of oxygen.

In hemoglobin, heme lies in a crevice on the surface of the large globin group

The sixth coordination position appears to be occupied by a water molecule which can be replaced reversibly by oxygen to give a derivative known as oxyhemoglobin.

$$\text{Hemoglobin} + O_2 \rightleftharpoons \text{Oxyhemoglobin} + H_2O$$

The position of this equilibrium is sensitive to the pressure of oxygen. In the lungs, where the blood is saturated with air (partial pressure $O_2 = 0.2$

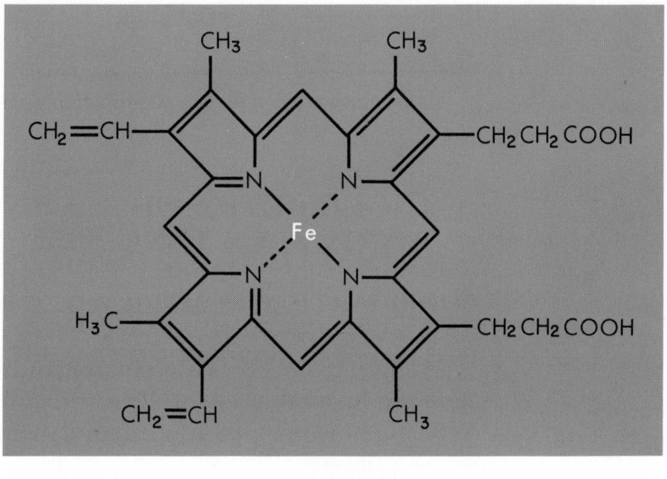

Figure 19.1 Structure of heme. Fe^{2+} ion is at the center of an octahedron, surrounded by four nitrogen atoms, a globin molecule and a water molecule.

atm), the hemoglobin is almost completely converted to the oxidized form. In the tissues serviced by arterial blood, the partial pressure of oxygen drops and the oxyhemoglobin breaks down to release elementary oxygen essential for the combustion of food. By this reversible process, hemoglobin acts as an oxygen carrier, absorbing oxygen in the lungs and liberating it to the tissues.

Unfortunately, hemoglobin forms a complex with carbon monoxide that is considerably more stable than oxyhemoglobin. The carbon monoxide complex is formed preferentially in the lungs even at CO concentrations as low as one part per thousand. When this happens, the flow of oxygen to the tissues is cut off, resulting eventually in muscular paralysis and death.

COORDINATION NUMBER

Coordination numbers are almost always even

As may be seen from Table 19.3, the most common coordination number shown by metals in complex ions is 6. A coordination number of 4 is somewhat less often found, while a coordination number of 2 is restricted largely to the +1 ions of the 1B elements. Odd coordination numbers (1, 3, 5) are relatively rare.

Certain metal ions show only one coordination number regardless of the ligands involved in the complex ion. For example, Pt^{2+} invariably forms four bonds, giving complex ions such as $Pt(NH_3)_4^{2+}$ and $PtCl_4^{2-}$. Similarly, Cr^{3+} and Co^{3+} always show a coordination number of 6. Other metal ions show variable coordination numbers, depending upon the nature of the ligand. The Ni^{2+} ion, for example, has two different coordination numbers, 4 and 6, shown respectively in the complex ions $Ni(CN)_4^{2-}$ and $Ni(H_2O)_6^{2+}$. Aluminum (Al^{3+}) and zinc (Zn^{2+}) behave similarly.

TABLE 19.3   Coordination Number and Geometry of Complex Ions

COORDINATION NO.	GEOMETRY	EXAMPLES
2	linear	Cu^+, **Ag^+**, Au^+
4	square planar	Cu^{2+}, Ni^{2+}, Pt^{2+}, Pd^{2+}
4	tetrahedral	Al^{3+}, Ni^{2+}, Co^{2+}, Zn^{2+}, Cd^{2+}
6	octahedral	Al^{3+}, **Cr^{3+}**, **Fe^{2+}**, **Fe^{3+}**, Co^{2+}
		Co^{3+}, Ni^{2+}, Cu^{2+}, Zn^{2+}, Cd^{2+}, **Pt^{4+}**

Ions which show only one coordination number are in heavy type.

19.3   GEOMETRY OF COMPLEX IONS

COORDINATION NUMBER = 2

Complex ions in which two ligands are coordinated to a central metal atom are invariably linear. The structures of the $Ag(NH_3)_2^+$, $Ag(CN)_2^-$ and $Au(CN)_2^-$ ions may be represented as follows:

$(H_3N—Ag—NH_3)^+$, $(N\equiv C—Ag—C\equiv N)^-$, $(N\equiv C—Au—C\equiv N)^-$

COORDINATION NUMBER = 4

For a complex in which the metal ion forms four bonds, two different geometries are possible. The four coordinating groups may be located at the corners of a square to give what is known, redundantly, as a *square planar complex*. In other cases, the four bonds are directed toward the corners of a regular tetrahedron (*tetrahedral complex*). The complexes of Pt^{2+} are of the square planar type, whereas the four-coordinated complexes of Zn^{2+} are tetrahedral (Fig. 19.2).

Certain square complexes can exist in two different forms with quite different properties. Consider, for example, the complex $[Pt(NH_3)_2Cl_2]$. Two forms of this compound, differing in absorption spectrum, water solubility, and chemical reactivity, have been prepared. One of these, made by reacting ammonia with the $PtCl_4^{2-}$ ion, has a structure in which the two ammonia molecules are located at adjacent corners of a square. In the other form, prepared by reacting the $Pt(NH_3)_4^{2+}$ ion with hydrochloric acid, the two ammonia molecules are located at opposite corners of the square:

Isolation of isomers such as these is possible only if the complexes do not exchange ligands rapidly with the surrounding solution

$$
\begin{array}{ccc}
H_3N & \diagdown\diagup & Cl \\
 & Pt & \\
H_3N & \diagup\diagdown & Cl \\
 & \text{Cis} &
\end{array}
\qquad
\begin{array}{ccc}
H_3N & \diagdown\diagup & Cl \\
 & Pt & \\
Cl & \diagup\diagdown & NH_3 \\
 & \text{Trans} &
\end{array}
$$

The two forms of $Pt(NH_3)_2Cl_2$ are called **geometrical isomers.** From a structural standpoint, they differ only in the spatial arrangement of the groups coordinated about the central atom. The form in which like groups are as close together as possible is called the **cis** isomer; the form in which like groups are far apart is referred to as the **trans** isomer. Geometrical isomerism can occur with any square planar complex of general formula Ma_2b_2 or Ma_2bc, in which M refers to the central atom and a, b, c represent ligands.

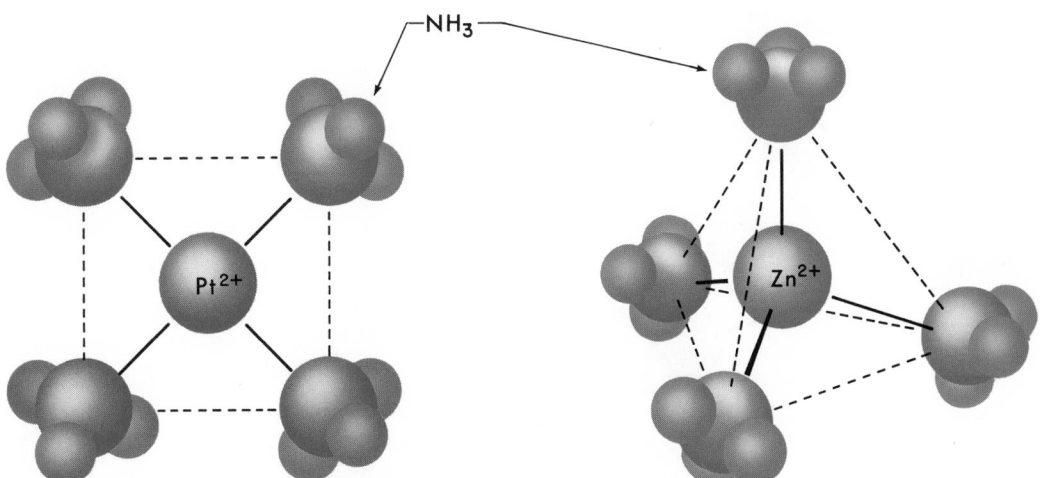

Figure 19.2   Structure of $Pt(NH_3)_4^{2+}$ (square planar) and $Zn(NH_3)_4^{2+}$ (tetrahedral).

The assignment of a *cis* or *trans* configuration to a particular isomer of a complex ion is by no means a simple experimental problem. Physical methods are perhaps most reliable for this purpose. X-ray diffraction studies are definitive but difficult to apply. A careful comparison of the ultraviolet absorption spectra of the two isomers in water solution can yield useful information, particularly if the spectra of several similar complexes of known configuration are available. Historically, chemical methods were widely used to assign cis and trans configurations. When the *cis* isomer of $[Pt(NH_3)_2Cl_2]$ is reacted with oxalate ions in solution, two Cl^- ions are displaced by an $C_2O_4^{2-}$ ion to form a chelate.

$$
\begin{array}{c}
H_3N \qquad Cl \\
\diagdown \diagup \\
Pt \\
\diagup \diagdown \\
H_3N \qquad Cl
\end{array}
\;+\;
\begin{bmatrix} O-C=O \\ | \\ O-C=O \end{bmatrix}^{2-}
\;\rightarrow\;
\begin{array}{c}
H_3N \qquad O-C=O \\
\diagdown \diagup \qquad | \\
Pt \\
\diagup \diagdown \qquad | \\
H_3N \qquad O-C=O
\end{array}
\;+\; 2\,Cl^- \qquad (19.4)
$$

Chelation cannot occur with the *trans* isomer since the oxalate ion cannot be sufficiently distorted to attach itself to platinum at two points trans to each other. Unfortunately this approach leads to misleading results with many complex ions where changes of configuration can occur in water solution.

COORDINATION NUMBER = 6

The six groups surrounding the metal ion in such complexes as $Fe(CN)_6^{3-}$ and $Co(NH_3)_6^{3+}$ are located at the corners of a regular octahedron, a figure with six corners and eight faces, all of which are equilateral triangles (Fig. 19.3). The metal ion is located at the center of the octahedron.

The spatial distribution of ligands in octahedral complexes is often shown by skeleton structures such as those in Figure 19.4. The drawing at

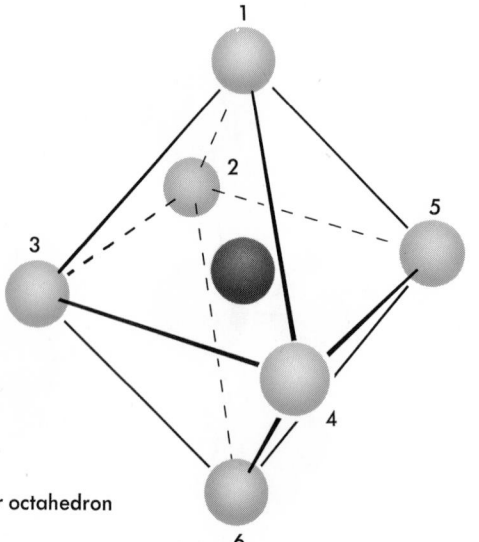

Regular octahedron

Figure 19.3 Regular octahedron. Note 6 apices with atom at center.

In this structure all the ligands occupy equivalent sites

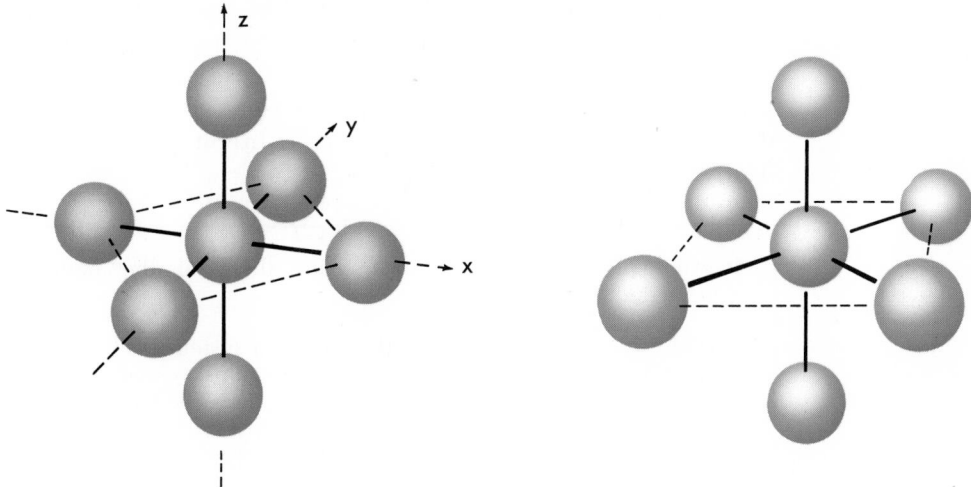

Figure 19.4   Two different views of an octahedron.

the left emphasizes that the six ligands are located along three axes at right angles to each other (x, y, and z axes), *equidistant from the metal atom at the center.* The skeleton structure at the right of Figure 19.4 is easier to draw and serves to emphasize another characteristic of an octahedral complex; it can be visualized as a derivative of a square planar complex in which the two additional ligands are located along a line perpendicular to the square at its center.

Geometrical isomerism can occur in octahedral as well as square complexes. Notice from Figure 19.3 that for any given position of a ligand in an octahedral complex, there are four equivalent positions equidistant from the first, and one at a greater distance. If, for example, we choose position 1 as a point of reference, groups located at 2, 3, 4, and 5 will be equidistant from it, while a group at position 6 will be farther away. We may refer to positions 1 and 2, 1 and 3, 1 and 4, or 1 and 5 as being **cis** to each other while positions 1 and 6 are **trans.** Consequently, an ion such as $Co(NH_3)_4Cl_2{}^+$ can

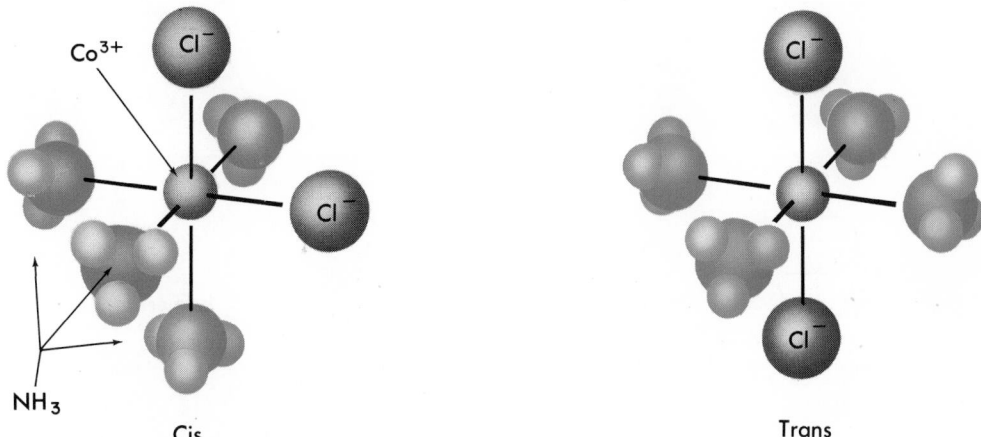

Figure 19.5   Isomers of $Co(NH_3)_4Cl_2{}^+$. Note that Cl^- ions are closer together in the cis than in the trans isomer.

exist in two isomeric forms, one in which the two chloride ions are in a **cis** relationship to each other and another in which they are in a **trans** configuration (Fig. 19.5). These two isomers, like those of $[Pt(NH_3)_2Cl_2]$, have different physical and chemical properties. The most striking difference is their color; trans-$[Co(NH_3)_4Cl_2]Cl$ is a bright green solid, while cis-$[Co(NH_3)_4Cl_2]Cl$ is violet.

Example 19.1 How many isomers are there of the neutral complex $[Co(NH_3)_3Cl_3]$?

Solution. It is best to approach this problem systematically. We might start by putting two NH_3 molecules in trans positions, perhaps at the "top" and "bottom" of the octahedron (Fig. 19.6a). We then ask ourselves: In how many spatially different positions can we place the third NH_3 molecule? A moment's reflection should convince you that there is really only one choice. All four of the remaining positions are equivalent in that they are cis to the two groups that we have already located. Choosing one of these positions arbitrarily, we get our first isomer (Fig. 19.6b).

To see if there are other isomers we start again, this time locating two NH_3 molecules cis to each other (Fig. 19.6c). If we were to place the third NH_3 molecule at one of the other corners of the square, we would simply reproduce the first isomer. (Remember that the symmetry of a regular octahedron requires that the distance across the diagonal of the square be the same as that from "top" to "bottom.") We are left with two equivalent positions. Placing the third NH_3 molecule arbitrarily at the "top" we arrive at a second isomer (Fig. 19.6d), distinctly different from the first because all three NH_3 molecules are cis to one another.

We have now exhausted in a logical manner all the possibilities for geometrical isomerism, finding two isomers. There are no others.

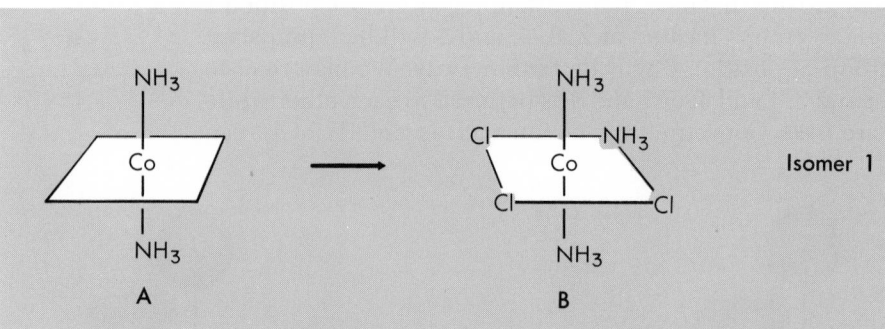

Isomer 1

Isomer 2

It is easy to overestimate the number of isomeric forms of complexes

Figure 19.6 Isomers of $Co(NH_3)_3Cl_3$. (Example 19.1).

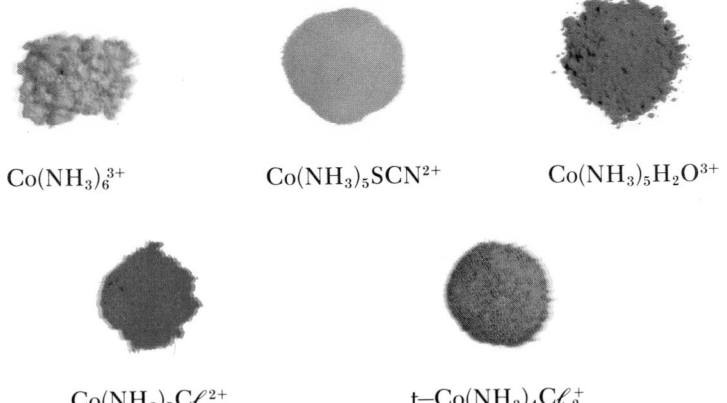

Co(NH₃)₆³⁺   Co(NH₃)₅SCN²⁺   Co(NH₃)₅H₂O³⁺

Co(NH₃)₅Cℓ²⁺   t—Co(NH₃)₄Cℓ₂⁺

Spectrochemical series in complex ions of Co^{3+} (see Table 19–4, p. 522). As NH_3 molecules are replaced by ligands of lower field strength ($NH_3 > SCN^- > H_2O > Cℓ^-$), the absorption shifts to higher wavelengths with a corresponding change in color.

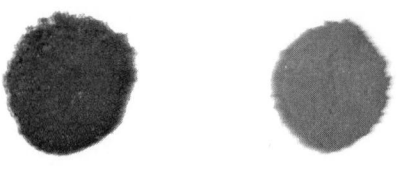

Cis—Co(en)₂Cℓ₂⁺   trans—Co(en)₂Cℓ₂⁺

Geometrical isomers often differ in color

Liquid oxygen suspended between the poles of an electromagnet. (Courtesy of S. Ruven Smith, Chemistry Department, University of Connecticut.) Both the paramagnetism and the blue color of oxygen are due to the unpaired electrons in the O_2 molecule.

19.4 ELECTRONIC STRUCTURE OF COMPLEX IONS

The *atomic orbital* or *valence bond* approach presented in Chapter 7 can be extended to describe the electronic structures and rationalize the geometries of complex ions. Alternatively, one can use the molecular orbital approach discussed briefly in Chapter 7. Still another model, based on what is known as *crystal-field theory,* has been developed to explain certain of the physical and chemical properties of coordination compounds which are difficult to understand in terms of the valence-bond model. Here we shall describe only the valence-bond and crystal-field treatments.

VALENCE-BOND (ATOMIC ORBITAL) MODEL

In this model it is assumed that *electron pairs donated by the ligands enter hybrid orbitals associated with the central metal ion.* We shall illustrate this concept by applying it to complexes formed by certain of the metal ions in the first transition series. Before proceeding further, you may wish to review the electronic structures of these ions, discussed on p. 164, and the concept of hybridization, discussed in Section 7.6, Chapter 7.

COORDINATION NUMBER = 2: sp HYBRIDIZATION. Complex ions in which the central metal atom has a coordination number of 2 can be assigned electronic structures analogous to that postulated for the BeF_2 molecule (p. 182). The two pairs of bonding electrons are assumed to occupy two hybrid orbitals formed by combining an s and a p orbital. These sp orbitals should be oriented at angles of 180° to each other, in agreement with the experimental observation that complex ions of this type have a linear structure.

As an illustration of the formation of sp hybrid bonds, consider the electronic structure of the complex ions of Cu^+ in which the coordination number is 2.

	no. of e⁻	3d	4s	4p
$_{29}Cu^+$ *	28	[↑↓][↑↓][↑↓][↑↓][↑↓]	[ ]	[ ][ ][ ]
$Cu(NH_3)_2{}^+$	32	[↑↓][↑↓][↑↓][↑↓][↑↓]	<u>[↑↓]</u>	<u>[↑↓]</u>[ ][ ]

(In this structure and those that follow, the 18 electrons of the argon shell are not shown; heavy horizontal lines are drawn above and below the hybrid bonding orbitals.)

COORDINATION NUMBER = 4: sp³ HYBRIDIZATION. Complexes in which zinc or aluminum has a coordination number of 4 show the tetrahedral geometry characteristic of such compounds as CH_4 and CCl_4. In the $Zn(NH_3)_4{}^{2+}$ ion, as in the CH_4 molecule, the four pairs of bonding electrons can be accommodated in four equivalent sp³ hybrid orbitals.

---

* Recall that when a transition metal ion is formed, the outer s electrons (e.g., 4s) are removed first. Consequently we can always derive the electronic structure of the bare metal ion by simply adding the appropriate number of electrons (in this case $28 - 18 = 10$) to the d orbitals.

	no. of e⁻	3d	4s	4p
₃₀Zn²⁺	28	[↑↓][↑↓][↑↓][↑↓][↑↓]	[ ]	[ ][ ][ ]
Zn(NH₃)₄²⁺	36	[↑↓][↑↓][↑↓][↑↓][↑↓]	[↑↓]	[↑↓][↑↓][↑↓]

COORDINATION NUMBER = 4: dsp² HYBRIDIZATION. The square planar complexes formed, for example, by Ni²⁺ differ in electronic structure as well as in geometry from tetrahedral complexes. In the valence-bond model, the four orbitals occupied by bonding electrons in a square planar complex are described as *dsp²* hybrids, formed by combining a d, an s, and two p orbitals. Looking at the structure of the Ni²⁺ ion

In some, but not all, complexes the central atom attains a noble gas structure

	no. of e⁻	3d	4s	4p
₂₈Ni²⁺	26	[↑↓][↑↓][↑↓][↑][↑]	[ ]	[ ][ ][ ]

it appears that the two odd electrons in the 3d atomic orbitals have to be paired to free a d orbital for dsp² hybridization.

	no. of e⁻	3d	4s	4p
Ni²⁺ (valence state)	26	[↑↓][↑↓][↑↓][↑↓][ ]	[ ]	[ ][ ][ ]
Ni(CN)₄²⁻	34	[↑↓][↑↓][↑↓][↑↓][↑↓]	[↑↓]	[↑↓][↑↓][ ]

COORDINATION NUMBER = 6: d²sp³ HYBRIDIZATION. In the octahedral complex Fe(CN)₆⁴⁻, the six pairs of electrons contributed by the ligands can be located in six hybrid d²sp³ orbitals, formed by combining two 3d, one 4s and three 4p orbitals.

	no. of e⁻	3d	4s	4p
₂₆Fe²⁺	24	[↑↓][↑][↑][↑][↑]	[ ]	[ ][ ][ ]
Fe²⁺ (valence state)	24	[↑↓][↑↓][↑↓][ ][ ]	[ ]	[ ][ ][ ]
Fe(CN)₆⁴⁻	36	[↑↓][↑↓][↑↓][↑↓][↑↓]	[↑↓]	[↑↓][↑↓][↑↓]

The fact that compounds such as K₄Fe(CN)₆ are diamagnetic tends to confirm a structure such as this in which there are no unpaired electrons.

Example 19.2 Using the valence-bond approach, write electronic structures for the octahedral complexes formed by Co³⁺ and Cr³⁺.

Solution. We start by counting the number of electrons in Co³⁺. Since there are 27 electrons in an atom of cobalt (at. no. = 27), we deduce that there are 24 electrons in Co³⁺. At this point, we can save ourselves a lot of time if we recall that this is precisely the number of electrons in the Fe²⁺ ion! Consequently we would expect octahedral complexes of Co³⁺ to have the same electronic structure as that written above for Fe(CN)₆⁴⁻.

In Cr³⁺ we have (at. no. Cr = 24) a total of 24 − 3 = 21 electrons to work with. Subtracting the 18e⁻ of Ar leaves three 3d electrons in Cr³⁺, which we would expect to find in three separate orbitals according to Hund's rule.

This leaves two empty 3d orbitals to hybridize with a 4s and three 3p orbitals, giving a total of six d^2sp^3 hybrid orbital, each of which will be filled by a pair of bonding electrons in an octahedral complex such as $Cr(NH_3)_6^{3+}$.

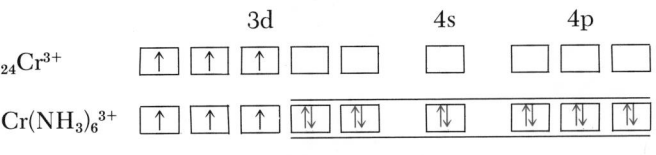

From the analyses that we have gone through, we might expect all of the octahedral complex ions formed by Fe^{2+} and Co^{3+} to be diamagnetic, since they contain no unpaired electrons. However, certain complexes in this category are paramagnetic. In particular, it is known that the $Fe(H_2O)_6^{2+}$ ion has four unpaired electrons, the same number as are found in the bare Fe^{2+} ion. One way to rationalize this in terms of valence-bond theory is to postulate that the d orbitals involved in bonding are those in the 4th principal energy level rather than the third.

Complexes of this type in valence-bond notation are referred to as *outer complexes* to distinguish them from *inner complexes* such as $Fe(CN)_6^{4-}$ where the d orbitals utilized in bonding are "inner orbitals" (e.g., 3d as opposed to 4d).

CRYSTAL-FIELD MODEL

Although the valence bond approach has proved extremely useful in explaining and correlating the geometries, electronic structures, and many of the properties of complex ions, it is deficient in certain important respects. For example, it cannot explain the wide variety of brilliant colors characteristic of so many coordination compounds. Again, although the valence bond model can rationalize the existence of two different kinds of octahedral complexes of Fe^{2+}, it cannot explain why the CN^- ion forms one type of complex and the H_2O molecule another.

It has long been recognized that many of the properties of complex ions can best be explained in terms of an electrostatic rather than a covalent model of bonding between metal cation and ligand. The fact that the ability of a metal ion to form complexes seems to be directly related to its charge density implies that it forms ionic or ion-dipole bonds, with anions or molecules acting as ligands. In many cases the strength of the metal-ligand bonds follows the same trend and lends itself to the same explanation.

The crystal-field model starts with the assumption that the attractive forces holding a complex ion together are primarily electrostatic rather than covalent. It then considers the modifications in the electronic structure of the metal ion due to electrostatic interactions between the ligands and the d electrons of the metal. To illustrate how these interactions arise, let us consider a specific example, the formation of the $Fe(CN)_6^{4-}$ ion.

$$Fe^{2+}(aq) + 6\ CN^-(aq) \rightarrow Fe(CN)_6^{4-}(aq) \qquad (19.5)$$

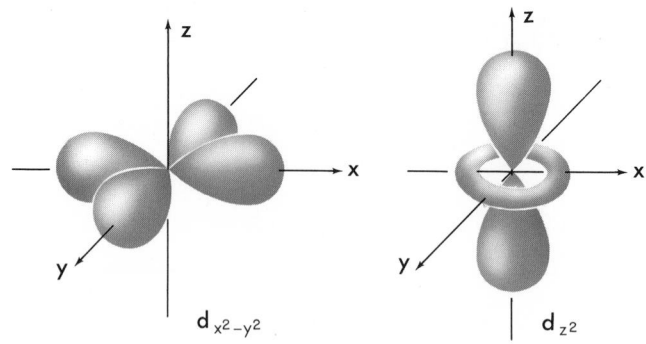

A pair of electrons would have the same energy in any of these orbitals

Figure 19.7 Spatial orientation of d orbitals. Note that $d_{x^2-y^2}$ and d_{z^2} orbitals are oriented toward ligands approaching corners of octahedron.

In the bare Fe^{2+} ion, all the 3d orbitals have the same energy. However, when six CN^- ions approach the Fe^{2+} ion to form an octahedral complex, geometrical considerations suggest that these orbitals should be split into two groups of different energies. To understand why this is the case, let us examine the orientation of the electron density clouds associated with the five d orbitals (Fig. 19.7).

Imagine now six CN^- ions approaching these orbitals, two along the Z

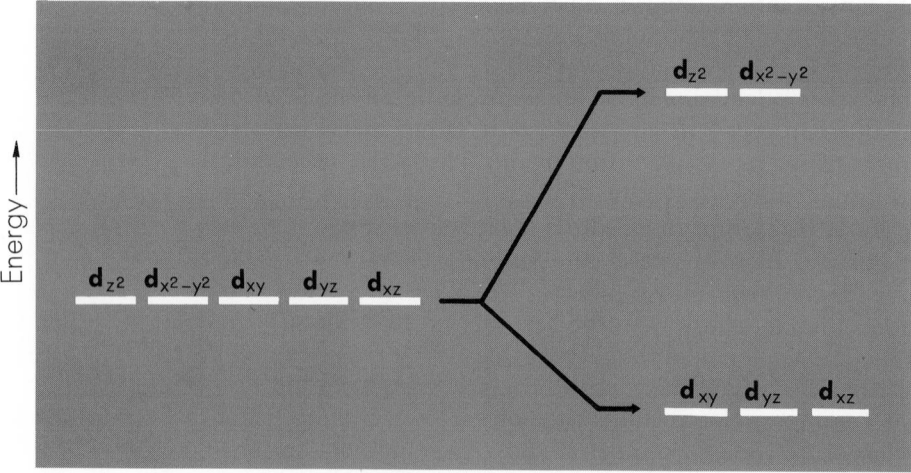

Figure 19.8 Splitting of d orbitals in octahedral field. Two of the orbitals ($d_{x^2-y^2}$, d_{z^2}) are raised in energy, while three (d_{xy}, d_{yz}, d_{xz}) are lowered.

axis (north and south), two along the X axis (east and west), and two along the Y axis. When they move in close to either of the two orbitals labeled "d_{z^2}" or "$d_{x^2-y^2}$", a strong electrostatic repulsion is set up because the electron density in these orbitals is concentrated along the z, x, and y axes. In contrast, the electrons in the other three orbitals are less strongly repelled since their densities are concentrated between the axes rather than along them. In other words, in the "octahedral field" created by the approach of CN^- ions, the five 3d orbitals of Fe^{2+}, or indeed of any transition metal ion, are split into two groups. One group, comprising the d_{xy}, d_{yz}, and d_{xz} orbitals, is lower in energy than the other group of two orbitals, d_{z^2} and $d_{x^2-y^2}$. Schematically, we have the situation shown in Figure 19.8.

As a result of the splitting of d orbitals, it seems reasonable to postulate a rearrangement of the electronic structures of Fe^{2+} when it forms a complex with CN^-. The six 3d electrons of the Fe^{2+} will tend to pair in the three lower energy levels, in "violation" of Hund's rule.

$$Fe^{2+}(in\ Fe(CN)_6{}^{4-})$$

The extent to which the energies of the d orbitals are modified by complex ion formation depends upon how strongly the ligands interact with the electrons in these orbitals. The H_2O molecule, which is a weaker base than the CN^- ion, repels electrons to a lesser extent and, hence, gives a smaller energy separation. As a result, the electron distribution in the Fe^{2+} ion is retained in $Fe(H_2O)_6{}^{2+}$. In other words, water molecules do not interact strongly enough with the d electrons of Fe^{2+} to overcome their tendency to remain unpaired insofar as possible.

In crystal field theory the electrons on the ligands are not shared with the central atom

$$Fe^{2+}(in\ Fe(H_2O)_6{}^{2+})$$

Looking at the structures we have written for the Fe^{2+} ion in the $Fe(CN)_6{}^{4-}$ and $Fe(H_2O)_6{}^{2+}$ complexes, we see that the crystal field model explains why the first complex is diamagnetic and the second paramagnetic with four unpaired electrons. To be sure, valence-bond theory also explains these observations in terms of inner and outer complexes. The advantage of crystal-field theory is that it suggests why a strong base such as CN^- should form a *"low spin"* complex while a weak base such as water may form a *"high spin"* complex with the maximum number of unpaired electrons.

Example 19.3 Following the crystal-field model, derive the electronic structure of the Co^{2+} ion in high spin and low spin octahedral complexes.

Solution. We first determine the number of d electrons. Since the atomic number of Co is 27, we have $27 - 2 = 25$ electrons in Co^{2+}. Of these, 18 are located in the argon core, leaving 7 d electrons. In the high spin complex, Hund's rule is followed. This leaves 3 of the electrons unpaired, putting 4 of them in two of the lower d orbitals. In low spin complexes, the three lower levels are completely filled, leaving only one unpaired electron

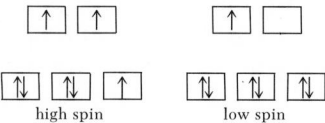

high spin   low spin

Both types of complexes are known. The $Co(H_2O)_6^{2+}$ ion is of the high spin type with three unpaired electrons; $Co(CN)_6^{4-}$ has only one unpaired electron.

Crystal-field theory also explains why most complexes formed from metal ions with unfilled d orbitals are colored. The splitting of the d orbitals creates a situation in which electrons can move into higher energy levels by absorbing visible light. The situation is particularly simple in $_{22}Ti^{3+}$, where there is only one d electron available. Consider, for example, the $Ti(H_2O)_6^{3+}$ ion, which has an intense purple color. Here, the splitting between d levels amounts to 3.9×10^{-12} erg. Using Einstein's equation, we can calculate the wavelength of light which must be absorbed by an electron to raise it from a lower to a higher d orbital.

Any colored species must have at least one energy level about $(4 \pm 1) \times 10^{-12}$ ergs above the ground state

$$\lambda = \frac{hc}{E} = \frac{(6.62 \times 10^{-27} \text{ erg sec})(3.00 \times 10^{10} \text{ cm/sec})}{(3.9 \times 10^{-12} \text{ erg})}$$
$$= 5.1 \times 10^{-5} \text{ cm} = 5100 \text{ Å}$$

The absorption at 5100 Å is in the green region. The purple (red-blue) color that we see when we look through a solution of $Ti(H_2O)_6^{3+}$ is what is left when the green component is subtracted from the visible spectrum.

The extent to which a ligand splits the d orbitals of a metal ion will determine the wavelength of light absorbed when an electron moves from a lower to a higher level. This means that the absorption and, hence, the color of a series of complexes of a metal ion will vary, depending upon the nature of the ligand. This effect is illustrated in Table 19.4 for certain complexes of Co^{3+}. We see that when we substitute for NH_3, ligands such as SCN^-, H_2O, or Cl^-, which produce smaller d orbital splittings, the light absorbed shifts to longer wavelengths (lower energy). On the basis of these and other observations, it is possible to arrange various ligands in order of decreasing tendency to split the d orbitals of a metal ion. An abbreviated version of such a *spectrochemical* series is:

$$CN^- > NO_2^- > en > NH_3 > SCN^- > H_2O > F^- > Cl^-$$

TABLE 19.4   Colors of Complex Ions of Co^{3+}

COMPLEX	COLOR OBSERVED	COLOR ABSORBED	APPROXIMATE WAVELENGTH (Å) ABSORBED
$Co(NH_3)_6^{3+}$	yellow	violet	4300
$Co(NH_3)_5SCN^{2+}$	orange	blue	4700
$Co(NH_3)_5H_2O^{3+}$	red	blue-green	5000
$Co(NH_3)_5Cl^{2+}$	purple	yellow-green	5300
trans $Co(NH_3)_4Cl_2^+$	green	red	6800

Lest it be supposed that the crystal field theory can explain all the properties of complex ions, we should point out at least one of its deficiencies. If one thinks of the bonding in complex ions as being primarily electrostatic, it is hard to explain why certain molecules that have very small dipole moments, such as CO, can be effective coordinating agents. In order to explain this, it is necessary to modify the crystal field theory to take into account covalent as well as ionic bonding. A more sophisticated version of the electrostatic approach, which we shall not attempt to describe here, known as **ligand field theory,** has been developed with this in mind.

19.5 RATE OF COMPLEX ION FORMATION

When an inorganic chemist wishes to prepare a particular complex ion, he most frequently uses a substitution reaction in which one ligand is replaced by another. This process is often carried out in aqueous solution. For example, to prepare the $Cu(NH_3)_4{}^{2+}$ ions, we might add aqueous ammonia to a solution of copper sulfate or copper nitrate containing the $Cu(H_2O)_4{}^{2+}$ cation.

$$Cu(H_2O)_4{}^{2+}(aq) + 4\ NH_3(aq) \rightarrow Cu(NH_3)_4{}^{2+}(aq) + 4\ H_2O \qquad (19.6)$$
light blue   deep blue

As ammonia is added, the light blue color of the $Cu(H_2O)_4{}^{2+}$ ion is replaced by the deeper blue, almost purple color of $Cu(NH_3)_4{}^{2+}$.

Complex ions differ greatly in the rate at which they undergo substitution reactions. Consider, for example, what happens when nitric acid is added to two different solutions, one containing the $Cu(NH_3)_4{}^{2+}$ ion, the other the analogous Co^{3+} complex, $Co(NH_3)_6{}^{3+}$. In the first case, substitution takes place instantaneously to regenerate $Cu(H_2O)_4{}^{2+}$.

$$Cu(NH_3)_4{}^{2+}(aq) + 4\ H^+(aq) + 4\ H_2O \rightarrow Cu(H_2O)_4{}^{2+}(aq) + 4\ NH_4{}^+(aq)$$
purple   blue

$$(19.7)$$

In contrast, days or even weeks may go by before we detect any color change or other evidence for reaction in the solution containing the Co^{3+} complex. This is true even though the reaction

$$Co(NH_3)_6{}^{3+}(aq) + 6\ H^+(aq) + 6\ H_2O \rightarrow Co(H_2O)_6{}^{3+}(aq) + 6\ NH_4{}^+(aq)$$
$$(19.8)$$

is thermodynamically spontaneous; the equilibrium constant for Reaction 19.8 is of the order of 10^{20}.

Complexes such as those of Cu^{2+} which exchange ligands virtually instantaneously are described as **labile.** Complexes such as $Co(NH_3)_6{}^{3+}$ which undergo substitution reactions at a rate which is measurable by ordinary techniques are referred to as nonlabile or **inert.** There is no sharp dividing line between these two categories: inert complexes can have half-lives ranging from a few minutes to several years.

Among the relatively few cations which consistently form inert complexes are Cr^{3+}, Co^{3+}, Pt^{2+}, and Pt^{4+}. It is no coincidence that complexes of these ions were among the first to be studied in the laboratory and that much of our knowledge about the structures and properties of complex

It is not possible to use the electronic structure of a complex to predict whether it will be inert

ions are based upon their behavior in aqueous solution. The great advantage of working with inert complexes is that they retain their identity in solution at least long enough for their chemical and physical properties to be studied. Once formed, they are relatively easy to separate from solution as pure compounds. Labile complexes such as $Cu(NH_3)_4^{2+}$ are much more difficult to work with; when added to water solution they immediately react to give an equilibrium mixture of many different species such as $Cu(H_2O)_4^{2+}$, $Cu(NH_3)_2(H_2O)_2^{2+}$, ... You can imagine how difficult it would be to obtain the information given in Table 19.1 or to establish the existence of isomers such as those shown in Figure 19.5 if the metal ion involved were Cu^{2+} rather than Pt^{2+} or Co^{3+}.

19.6 COMPLEX ION EQUILIBRIA

In the preceding section we compared the (kinetic) labilities of different complex ions. It is equally important to be able to compare their (thermodynamic) stabilities. One way to do this is to formulate and measure the equilibrium constant for the reaction that occurs when a compound containing a particular complex ion is added to water. In doing this, it is customary to treat the reaction as if it were a simple dissociation, ignoring the water molecules involved. Thus, for the equilibrium which is set up when a compound containing $Ag(NH_3)_2^+$ ions is added to water, we write

$$Ag(NH_3)_2^+(aq) \rightleftharpoons Ag^+(aq) + 2\ NH_3(aq); \quad K_c = \frac{[Ag^+] \times [NH_3]^2}{[Ag(NH_3)_2^+]} \quad (19.9)$$

even though a more appropriate equation would be

$$Ag(NH_3)_2^+(aq) + 2\ H_2O \rightarrow Ag(H_2O)_2^{2+}(aq) + 2\ NH_3(aq)$$

The fact that the equilibrium constant for Reaction 19.9 is a very small number, 4×10^{-8}, means that $Ag(NH_3)_2^+$ dissociates to only a small extent when added to water. Looking at it another way, even in quite dilute solutions of ammonia, the concentration of Ag^+ will be low relative to that of $Ag(NH_3)_2^+$ (Example 19.4).

Example 19.4 Calculate the ratio of the concentration of Ag^+ to that of $Ag(NH_3)_2^+$ in a solution prepared by adding a small amount of $AgNO_3$ to a large volume of 0.02 M NH_3.

Solution. From the way the problem is worded, it seems legitimate to take the equilibrium concentration of NH_3 to be 0.02 M (i.e., neglect the small amount of NH_3 consumed in forming the complex). Substituting in the expression for K_c (Equation 19.9):

$$K_c = \frac{[Ag^+] \times (0.02)^2}{[Ag(NH_3)_2^+]} = 4 \times 10^{-8}$$

Solving: $\quad \dfrac{[Ag^+]}{[Ag(NH_3)_2^+]} = \dfrac{4 \times 10^{-8}}{4 \times 10^{-4}} = 1 \times 10^{-4}$

In other words, even in this very dilute solution, about 99.99% of the silver is complexed by ammonia.

TABLE 19.5 Dissociation Constants of Complex Ions

MA$_2$		MA$_4$		MA$_6$		
AgCl$_2^-$	1×10^{-6}	CdCl$_4^{2-}$	4×10^{-3}	Cr(OH)$_6^{3-}$	1×10^{-38}	The small values
Ag(NH$_3$)$_2^+$	4×10^{-8}	Cu(NH$_3$)$_4^{2+}$	2×10^{-13}	Co(NH$_3$)$_6^{3+}$	1×10^{-35}	of many of these
Ag(SCN)$_2^-$	1×10^{-10}	Cu(CN)$_4^{2-}$	1×10^{-25}	Co(CN)$_6^{3-}$	1×10^{-64}	constants reflect
Ag(S$_2$O$_3$)$_2^{3-}$	1×10^{-13}	Ni(CN)$_4^{2-}$	1×10^{-14}	Fe(CN)$_6^{3-}$	1×10^{-31}	the high stability
Ag(CN)$_2^-$	1×10^{-21}	PdCl$_4^{2-}$	1×10^{-13}			of the associated
CuCl$_2^-$	3×10^{-6}	Zn(NH$_3$)$_4^{2+}$	3×10^{-10}			complex ions
Cu(NH$_3$)$_2^+$	1×10^{-7}	Zn(OH)$_4^{2-}$	3×10^{-16}			
		Zn(CN)$_4^{2-}$	1×10^{-17}			

The relative stabilities of different complex ions of silver can be estimated from a knowledge of their dissociation constants. For example, when we find that the equilibrium constant for the reaction

$$\text{Ag(CN)}_2^-(aq) \rightleftharpoons \text{Ag}^+(aq) + 2\text{ CN}^-(aq) \quad K_c = \frac{[\text{Ag}^+] \times [\text{CN}^-]^2}{[\text{Ag(CN)}_2^-]} \quad (19.10)$$

is only about 1×10^{-21}, we deduce that the Ag(CN)$_2^-$ complex is even more stable than Ag(NH$_3$)$_2^+$. We would expect that the addition of CN$^-$ ions to a solution containing Ag(NH$_3$)$_2^+$ would convert most of these ions to the more stable Ag(CN)$_2^-$ complex. This is shown experimentally by the fact that AgI dissolves more readily in a solution of NaCN than in aqueous ammonia.

19.7 COMPLEX IONS IN ANALYTICAL CHEMISTRY

QUALITATIVE ANALYSIS

Reactions involving the formation of complex ions are widely used in qualitative analysis for either of two purposes. The ability of a metal ion to form a colored complex or a precipitate with a particular complexing agent may be used as a specific test for that ion. Alternatively, two ions may be separated from one another by adding a complexing agent that forms a complex with only one of them. Not infrequently these two purposes are achieved simultaneously by adding the proper complexing agent at a particular stage in an analysis.

TESTS FOR SPECIFIC IONS. An extremely sensitive test for the Cu^{2+} ion in water solution involves its ability to form a deep blue complex with ammonia. The color of the Cu(NH$_3$)$_4^{2+}$ ion is much more intense than that of the light blue Cu(H$_2$O)$_4^{2+}$ ion; it can be detected at concentrations of Cu^{2+} as low as 10^{-4} M. Certain other ions interfere with this test; nickel, for example, also forms a deep blue complex ion with ammonia.

The Fe^{3+} ion is readily detected by adding a solution of potassium thiocyanate, KSCN, to give the blood red color characteristic of the FeSCN^{2+} complex ion (more exactly, Fe(H$_2$O)$_5$SCN^{2+}). A precipitate known as Prussian blue is also given by Fe^{3+} upon addition of a solution of potassium hexacyanoferrate, K$_4$Fe(CN)$_6$. Despite the fact that Prussian blue was the first coordination compound to be prepared (in the first decade of the 18th century), its exact composition has never been established. A plausible equation for its formation is

$$Fe^{3+}(aq) + K^+(aq) + Fe(CN)_6^{4-}(aq) \rightarrow KFe[Fe(CN)_6](s) \quad (19.11)$$

Chelating agents, because of their ability to form extremely stable complexes, are widely used to test for specific cations. Dimethyl glyoxime

$$\begin{array}{c} H_3C-C-C-CH_3 \\ \parallel \quad \parallel \\ HO-\underset{\cdot\cdot}{N} \quad \underset{\cdot\cdot}{N}-OH \end{array}$$

is one such compound. It uses the unshared electron pairs on the two nitrogen atoms to form chelates with many metal ions. The formation of a red, insoluble complex of dimethyl glyoxime with Ni^{2+} is a very sensitive test for that ion.

SEPARATION OF IONS. Cations are often separated from one another by taking advantage of differences in their tendencies to form complex ions with a particular ligand. To illustrate the method, consider the two ions Fe^{3+} and Al^{3+}, which are ordinarily precipitated in the same group in cation analysis. If sodium hydroxide is added to a solution containing these ions, they precipitate as $Fe(OH)_3$ and $Al(OH)_3$. However, as more sodium hydroxide is added, the aluminum hydroxide dissolves to form a complex ion

$$Al(OH)_3(s) + 3\ OH^-(aq) \rightarrow Al(OH)_6^{3-}(aq) \quad (19.12)$$

Iron(III) hydroxide fails to dissolve and is thus separated from Al^{3+}.

Many transition metal ions, like Al^{3+}, form sufficiently stable complex ions with OH^- to go into solution in concentrated sodium hydroxide. Among the hydroxides which dissolve in a strong base are $Zn(OH)_2$, $Cr(OH)_3$ and, to a lesser extent, $Cu(OH)_2$. Compounds such as these, which are capable of reacting with OH^- ions as well as H^+ ions are said to be **amphoteric.**

Another complexing agent which is frequently used to separate metal ions is the ammonia molecule, NH_3. In the analysis of the group 1 cations, advantage is taken of the stability of the $Ag(NH_3)_2^+$ complex to separate silver from mercury. Treatment of a precipitate containing $AgCl$ with dilute ammonia leads to the reaction

$$AgCl(s) + 2\ NH_3(aq) \rightarrow Ag(NH_3)_2^+(aq) + Cl^-(aq)$$

bringing the silver into solution in the form of the complex ion. To confirm the presence of Ag^+, one can add nitric acid to the solution. The hydrogen ions from the acid destroy the complex by converting NH_3 molecules to NH_4^+ ions.

$$Ag(NH_3)_2^+(aq) + Cl^-(aq) + 2\ H^+(aq) \rightarrow AgCl(s) + 2\ NH_4^+(aq)$$

The ability of silver salts to form a complex ion with ammonia may be used in a more subtle way to separate the three anions Cl^-, Br^-, and I^- from each other. Addition of silver nitrate to a solution containing these three ions gives a mixed precipitate of $AgCl$, $AgBr$, and AgI. Silver chloride dissolves in dilute ammonia to give the $Ag(NH_3)_2^+$ complex. Silver bromide goes into solution only in concentrated ammonia, while silver iodide is insoluble even at very high ammonia concentrations. Consequently, the addition of silver nitrate, followed first by dilute and then concentrated ammonia, serves to separate Cl^-, Br^-, and I^- ions from each other.

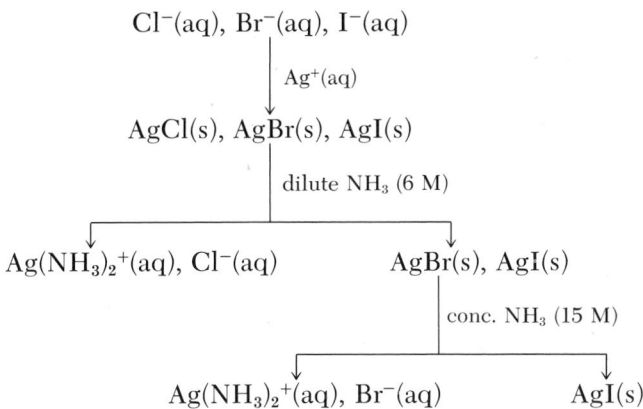

To understand the principles underlying the separation of metal ions by complex formation, we must consider the equilibria involved. In the reaction of ammonia with a silver halide, AgX, we have

$$AgX(s) \rightleftharpoons Ag^+(aq) + X^-(aq); \quad K_1 = K_{sp}AgX \tag{19.13a}$$

$$Ag^+(aq) + 2\ NH_3(aq) \rightleftharpoons Ag(NH_3)_2^+(aq); \quad K_2 = \frac{1}{K_cAg(NH_3)_2^+} = 2.5 \times 10^7 \tag{19.13b}$$

$$AgX(s) + 2\ NH_3(aq) \rightleftharpoons Ag(NH_3)_2^+(aq) + X^-(aq); \quad K_3 = K_1K_2 = 2.5 \times 10^7 \times K_{sp}AgX \tag{19.13}$$

Looking at these relations, we can deduce that the solubility of a silver halide in water, and indeed the solubility of any solid in a complexing agent, will depend upon:

1. *The solubility of the solid in water* (19.13a). The greater the solubility in water (i.e., the larger the value of K_{sp}), the greater will be the value of K_3 and the solubility in a complexing agent. For the three silver halides, AgCl, AgBr, and AgI with ammonia, we have

AgCl: $K_{sp} = 1.6 \times 10^{-10}$; $K_3 = (2.5 \times 10^7)(1.6 \times 10^{-10}) = 4.0 \times 10^{-3}$

AgBr: $K_{sp} = 1 \times 10^{-13}$; $K_3 = (2.5 \times 10^7)(1 \times 10^{-13}) = 2.5 \times 10^{-6}$

AgI: $K_{sp} = 1 \times 10^{-16}$; $K_3 = (2.5 \times 10^7)(1 \times 10^{-16}) = 2.5 \times 10^{-9}$

We see that K_3 for Reaction 19.13 becomes successively smaller as we move from AgCl to AgI. This agrees with the experimental observation that, of these three compounds, AgCl is most soluble in ammonia and AgI least soluble.

2. *The concentration of complexing agent* (19.13b). The greater the concentration of complexing agent, the greater will be the tendency of the solid to dissolve to form a complex ion. As pointed out earlier, silver bromide is soluble in concentrated ammonia but insoluble in dilute ammonia. An increase in the concentration of ammonia from 6 M to 15 M exerts a sufficient influence on the equilibrium in Reaction 19.13b to bring a significant amount of silver bromide into solution.

3. *The stability of the complex ion formed.* In many cases a solid

which does not dissolve in one complexing agent can be brought into solution by using a reagent which forms a more stable complex. Looking at Reaction 19.13 we see that K_3 for the overall reaction is inversely related to the dissociation constant of the complex ion. Silver iodide, which is insoluble in ammonia ($K_c Ag(NH_3)_2{}^+ = 4 \times 10^{-8}$), dissolves readily in potassium cyanide solution to form the $Ag(CN)_2{}^-$ complex ($K_c Ag(CN)_2{}^- = 1 \times 10^{-21}$) or in sodium thiosulfate to form the $Ag(S_2O_3)_2{}^{3-}$ complex ($K_c Ag(S_2O_3)_2{}^{3-} = 1 \times 10^{-13}$).

QUANTITATIVE ANALYSIS

The reaction of a metal ion with a complexing agent bears at least a superficial resemblance to the reaction of an H^+ ion with an OH^- ion. Comparing the two equations

$$Cu^{2+}(aq) + 4\ NH_3(aq) \rightarrow Cu(NH_3)_4{}^{2+}(aq)$$

$$H^+(aq) + OH^-(aq) \rightarrow H_2O$$

we note that in both cases a positive ion is converted to an extremely stable, covalently bonded species. It might seem, then, that we could determine the concentration of a metal ion by titrating with a complexing agent in much the same way that an acid is titrated with a base.

In practice, it is seldom possible to use ordinary complexing agents to analyze quantitatively for metal ions in solution. The difficulty is that, as previously noted, the formation of a metal complex is a stepwise process. If we add ammonia to a solution of a copper salt, the hydrated Cu^{2+} ion is not converted directly to the $Cu(NH_3)_4{}^{2+}$ ion. Instead, intermediate species containing one, two, or three ammonia molecules are formed. As the concentration of ammonia increases, the various equilibria gradually shift to lower the Cu^{2+} ion concentration. There is no sharp change in "free" Cu^{2+} ion concentration analogous to the abrupt change in H^+ ion concentration that one observes at the equivalence point of an acid-base titration. The end point, instead of being sharp and precise, is drawn out and diffuse.

Analytical chemists have developed a series of reagents that react with metal ions to give extremely stable 1:1 complexes and, hence, are suitable for metal-ion titrations. These substances are chelating agents; the best known is the sodium salt of ethylenediaminetetracetic acid, commonly called **EDTA**. The anion of this salt has the following structure:

EDTA is a hexadentate ligand

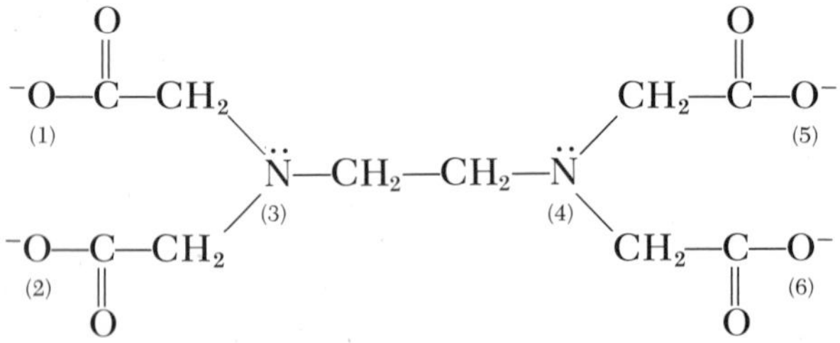

A single EDTA anion can attach itself to a metal ion through as many as six different atoms (numbered 1 to 6 in the foregoing structural formula), filling all its coordination requirements. Difficulties inherent in stepwise complex formation are thereby avoided; EDTA titrations yield a sharp, easily observed end point.

EDTA is among the most effective complexing agents known; it forms stable 1:1 chelates with a wide variety of metal ions. One of its earliest applications was in the determination of the alkaline-earth metals calcium and magnesium found in hard water. Several hundred papers, appearing over the past 20 years, have described the use of EDTA titrations in the determination of over 60 different elements.

To illustrate how an EDTA titration may be carried out, let us consider the use of this chelating agent in the determination of Fe^{3+}. The reaction that occurs may be represented as:

$$Fe^{3+}(aq) + EDTA^{4-}(aq) \rightarrow Fe(EDTA)^{-}(aq) \qquad (19.14)$$

The SCN^- ion, which forms a blood-red complex with Fe^{3+}, can be used as an indicator. As EDTA is added, the thiocyanate complex of iron(III) is converted to the more stable EDTA complex. At the equivalence point, the SCN^- ions attached to iron are quantitatively displaced and the color changes from deep red to yellow. Knowing the concentration of the EDTA solution and the volume which must be added to reach the end point, we can readily calculate the amount of Fe^{3+} present (Problem 19.20).

When EDTA forms the complex with Fe^{3+}, any and all attached SCN^- ions are displaced

The uses of EDTA are not confined to analytical chemistry. It is an effective antidote for heavy metal poisoning. Children who have ingested toxic amounts of lead compounds, which were at one time a major component of the pigment in ordinary house paint, are treated by an intramuscular injection of a solution of EDTA. This brings Pb^{2+} into solution as a complex ion and, hence, leads to its elimination from the body. EDTA has also been used to remove radioactive isotopes of metallic elements, notably plutonium, from body tissues. One of the most dangerous of these isotopes, strontium-90, can be eliminated by treatment with a derivative of EDTA known as "BAETA," which is unusual in that it forms a more stable chelate with strontium than with calcium.

AN HISTORICAL PERSPECTIVE

ALFRED WERNER AND SOPHUS MADS JORGENSEN

The basic ideas concerning the structure and geometry of complex ions presented in Sections 19.1 to 19.3 were developed by one of the most gifted individuals in the history of inorganic chemistry, Alfred Werner. His theory of coordination chemistry was published in 1893 when Werner was 26 years old, holding the equivalent of an associate professorship at the University of Zurich. In his paper Werner made the revolutionary suggestion that metal ions such as Co^{3+} could show two different kinds of valences. For the compound $Co(NH_3)_6Cl_3$, Werner postulated a central Co^{3+} ion joined by "primary valences" (ionic bonds) to 3 Cl^- ions and by "secondary valences" (coordinate covalent bonds) to 6 NH_3 molecules. Moreover, he made the inspired guess that the 6 secondary valences were directed toward the corners of a regular octahedron. In Table 19.6 we list the familiar Werner structures for the series of compounds $Co(NH_3)_xCl_3$, where $x = 6, 5, 4,$ or 3. All these compounds were known at the time, and many of their properties had been established. In particular, it was known from conductivity studies and by precipitation with $AgNO_3$ that the first three members of the series yielded 3, 2, and 1 moles of Cl^-, respectively, when dissolved in water. This evidence was, of course, in complete agreement with Werner's theory.

Werner's structures, which seem so obvious to us today, aroused little enthusiasm among his contemporaries. The opposition was led by Sophus Mads Jorgensen, a 56 year old professor of chemistry at the University of Copenhagen. Jorgensen was convinced that Co^{3+} could form no more than three bonds. To rationalize the existence of "addition compounds" of $CoCl_3$ containing 6, 5, 4, or 3 NH_3 molecules, he invoked the chain structures shown in Table 19.6, where NH_3 molecules are linked together much like CH_2 groups in hydrocarbons. The differing extents of ionization of these compounds in water were explained by assuming that only those chlorine atoms bonded to NH_3 groups could ionize. Chlorines attached directly to cobalt were supposed to be held so tightly that they could not ionize in water.

From the data in Table 19.6 it would appear that the controversy between Werner and Jorgensen could have been settled quite simply by studying the behavior in water solution of the compound $Co(NH_3)_3Cl_3$.

TABLE 19.6   Structure and Properties of the Compounds $Co(NH_3)_xCl_3$

			MOLES Cl^- per MOLE COMPOUND		
	STRUCTURE				
x	Werner	Jorgensen	Werner	Jorgensen	Observed
6	$[Co(NH_3)_6]^{3+}$, 3 Cl^-	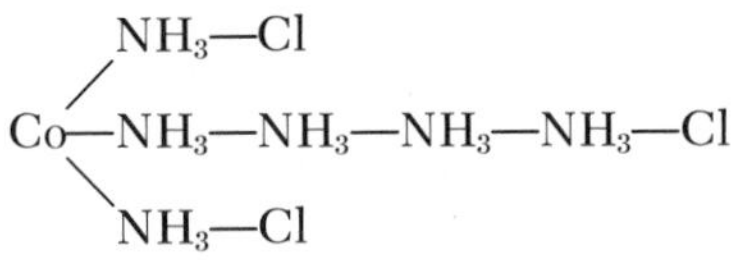 $Co\diagup^{NH_3-Cl} -NH_3-NH_3-NH_3-NH_3-Cl \diagdown_{NH_3-Cl}$	3	3	3
5	$[Co(NH_3)_5Cl]^{2+}$, 2 Cl^-	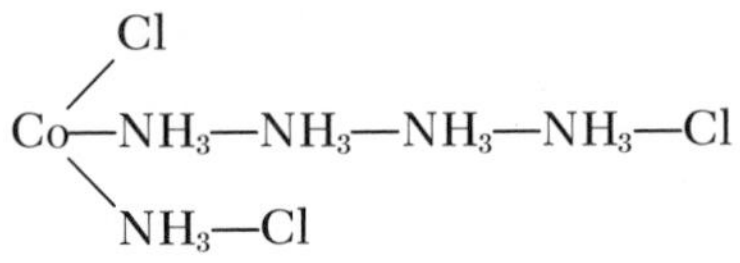 $Co\diagup^{Cl} -NH_3-NH_3-NH_3-NH_3-Cl \diagdown_{NH_3-Cl}$	2	2	2
4	$[Co(NH_3)_4Cl_2]^+$, Cl^-	$Co\diagup^{Cl} -NH_3-NH_3-NH_3-NH_3-Cl \diagdown_{Cl}$	1	1	1
3	$[Co(NH_3)_3Cl_3]^0$	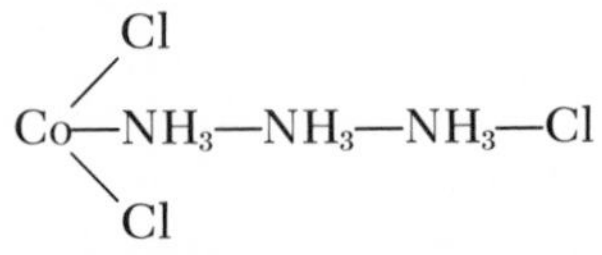 $Co\diagup^{Cl} -NH_3-NH_3-NH_3-Cl \diagdown_{Cl}$	0	1	?

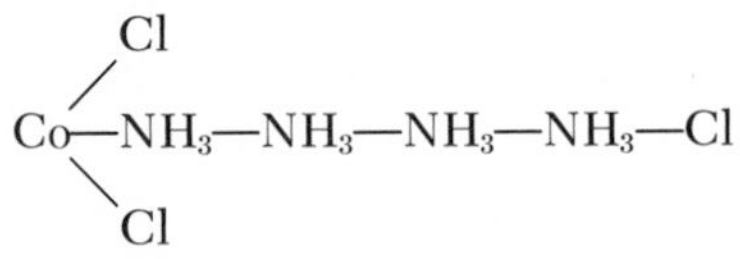

Werner's structure required that this species be a nonelectrolyte with no ionizable chlorine. In contrast, the chain structure of Jorgensen implied one ionizable chlorine, i.e., a 1:1 electrolyte similar to NaCl. Unfortunately the evidence was ambiguous; at 25°C, a water solution of $Co(NH_3)_3Cl_3$ has a conductivity intermediate between that of a nonelectrolyte and a 1:1 salt.

Werner's assumption of octahedral coordination around the Co^{3+} ion offered further opportunities for testing his ideas against those of Jorgensen. If Werner were correct, there should be two isomeric forms (cis and trans) of the compound $Co(NH_3)_4Cl_3$. At the time, only one compound of this formula was known. All of Werner's early attempts to prepare a second isomer failed, thereby weakening his position.

As the years passed, the weight of evidence began to shift toward Werner's structures. Studies at 0°C gave a very low value for the conductivity of $Co(NH_3)_3Cl_3$, which tended to support Werner's contention that the anomalous conductivity at 25°C was due to the reaction

$$Co(NH_3)_3Cl_3(s) + H_2O \rightarrow [Co(NH_3)_3(H_2O)Cl_2]^+(aq) + Cl^-(aq)$$

Moreover, Werner showed that the compound $Co(NH_3)_3(NO_2)_3$, which is entirely analogous to $Co(NH_3)_3Cl_3$, behaves as a true nonelectrolyte even at 25°C.

In 1907 Werner, after years of effort, finally succeeded in preparing a second isomer of the compound $Co(NH_3)_4Cl_3$. Jorgensen graciously accepted this new evidence as conclusive proof of Werner's structures, and the chain theory of coordination chemistry faded away. Six years later, in 1913, Alfred Werner received the Nobel prize in chemistry.

PROBLEMS

19.1 What are the charges of complexes of Co^{3+} in which the ligands are

a. 3 NH_3 molecules, 3 Cl^- ions?
b. 3 NH_3 molecules, 3 H_2O molecules?
c. 3 en molecules?
d. 1 NH_3 molecule, 5 NO_2^- ions?

19.2 For each compound at the left, choose the compound at the right which would approach it most closely in molar conductivity in water.

a. $[Co(NH_3)_4Cl_2]Cl$ CH_3OH
b. $[Co(en)_3]Cl_3$ $NaCl$
c. $K[Co\ en\ (NO_2)_4]$ $Mg(NO_3)_2$
d. $[Co(NH_3)_3Cl_3]$ $Al(NO_3)_3$
e. $[Co(NH_3)_5Cl](NO_3)_2$ $Al_2(SO_4)_3$
f. $Na_3[Co(NO_2)_6]$ $MgSO_4$
g. $[Co(NH_3)_5Cl]SO_4$

19.21 What must be the charge of the central ion in each of the following

a. $Fe(CN)_6^{4-}$
b. $PdCl_4^{2-}$
c. $Pt(NH_3)_4Cl_2$
d. $Co(H_2O)_2Cl_4^{2-}$

19.22 Arrange the following compounds in order of increasing molar conductivity, assuming that the complexes are inert in water solution

a. $[Cr(NH_3)_3Cl_3]$
b. $[Cr\ en_3]Br_3$
c. $K_2[Cr(NH_3)(NO_2)_5]$
d. $[Cr\ en_2\ Cl_2]Cl$

Discuss how the conductivities of these solutions would change if one or more of the ligands were replaced by water molecules.

19.3 The addition of $AgNO_3$ in the cold to a compound of Co^{3+} precipitates only ionic halide (Cl, Br, or I). If the solution is heated to boiling, halide inside as well as outside the coordination sphere is precipitated. A one-gram sample of $[Co(en)_2Cl_2]I$ is dissolved in water and titrated with 0.100 M $AgNO_3$. If the titration is carried out near 0°C, how many ml of $AgNO_3$ will be required? If the solution is then heated to boiling, how many more ml of $AgNO_3$ will be required?

19.4 A chemist is interested in studying the properties in solution of the complex $[Cr(NH_3)_5SO_4]^+$. The salt of this ion that he happens to have available is $[Cr(NH_3)_5SO_4]Cl$. He decides to convert it to the perchlorate salt $[Cr(NH_3)_4SO_4]ClO_4$. Can you suggest why it might be desirable to make this conversion? How might the conversion be accomplished (cf. Ch. 16)?

19.5 Which of the following would you expect to be effective chelating agents

    a. $H_3C\!-\!OH$
    b. $HO\!-\!OH$
    c. $H_3C\!-\!\overset{\displaystyle H}{N}\!-\!\overset{\displaystyle H}{N}\!-\!CH_3$
    d. PH_3
    e. $H_3C\!-\!\underset{\displaystyle \underset{O}{\|}}{C}\!-\!\overset{\displaystyle H}{C}\!=\!\overset{\displaystyle H}{C}\!-\!O^-$

19.6 It has been stated that the coordination number of a metal ion is usually twice its charge. What evidence can you cite in support of or in opposition to this generalization?

19.7 Every time you breathe, you take in about 400 ml of air (21% O_2 by volume) at approximately 20°C and one atmosphere pressure. If all the oxygen in this air entered the bloodstream, how many molecules of hemoglobin would be involved?

19.8 Sketch the geometry of

    a. $Fe(CN)_6^{4-}$
    b. cis-$[Pd\ en\ Br_2]$
    c. $Pt(NH_3)_5Cl^{3+}$
    d. trans-$[Cr(H_2O)_4Cl_2]^+$
    e. $Zn(OH)_4^{2-}$
    f. $Cu(NH_3)_2^+$

19.9 Draw structures for all the isomers of

    a. $Co(NH_3)_4ClBr^+$
    b. $Cr(NH_3)_3(NO_2)_3$
    c. $Pt\ en_2\ Cl_2^{2+}$
    d. $Co(NH_3)_3Cl_2Br$

19.23 A certain compound has the empirical formula $CoN_4H_{12}ClBrI$. When a 1.000 g sample is dissolved in water at 0°C and treated with $AgNO_3$, a precipitate weighing 0.509 g forms. Identify the complex ion present. If the solution is heated with $AgNO_3$, how many more grams of precipitate will form? (See Problem 19.3.)

19.24 Explain in your own words why

    a. CN^- is a better coordinating agent than NO_3^-.
    b. SO_4^{2-} is a better coordinating agent than HSO_4^-.
    c. chelates complexes are unusually stable.
    d. carbon monoxide is extremely toxic.

19.25 Suggest an experimental method by which you might be able to

    a. distinguish between the compounds $H_3C\!-\!\overset{\displaystyle H}{N}\!-\!\overset{\displaystyle H}{N}\!-\!CH_3$ and $H_2N\!-\!CH_2\!-\!CH_2\!-\!NH_2$, both of which have the same empirical and molecular formulas.
    b. determine whether the CO_3^{2-} ion is acting as a chelating agent or a simple (monodentate) ligand.

19.26 Can you explain why a metal ion that forms square planar complexes often forms octahedral complexes as well?

19.27 What is the percentage by weight of iron in heme (Figure 19.1)?

19.28 Indicate the geometry of

    a. $Co(NO_2)_6^{3-}$
    b. $Pt(NH_3)_3Br^+$
    c. cis-$Co(en)_2Cl_2^+$
    d. $Ag(CN)_2^-$
    e. $Cr\ en\ Cl_4^-$
    f. $Zn(H_2O)_3OH^+$

19.29 How many geometrical isomers are there for complexes with the following general formulas (M = central metal atom; A, B, C = ligands)

    a. MA_4BC
    b. MA_3B_3
    c. $MA_2B_2C_2$
    d. MA_3B_2C
    e. MA_2BC (square planar)
    f. MA_2BC (tetrahedral)

19.10 Using the valence-bond model draw orbital diagrams to indicate the electronic structure around the central ion in:

a. $Co(en)_3^{3+}$
b. $Fe(CN)_6^{3-}$
c. $AlCl_4^-$
d. $Ni(NH_3)_6^{2+}$

Assume inner complexes wherever possible.

19.11 It is usually found that outer complexes of a particular ion are less stable than inner complexes of the same ion. Explain, first in terms of the valence-bond model and then of crystal-field theory.

19.12 Using crystal-field theory, draw electronic structures for the high spin and low spin forms of the octahedral complexes listed in Problem 19.10.

19.13 Based on the data in Table 19.4, what do you predict the color of the $Co(NH_3)_5NO_2^{2+}$ ion to be?

19.14 Drying agents which are used to remove water vapor from air are often coated with $CoCl_2$, which is blue when anhydrous and red or pink when hydrated. Explain this color change in terms of crystal-field splitting.

19.15 Devise a lecture demonstration to illustrate the lability of the $Ni(NH_3)_6^{2+}$ ion as contrasted to the inertness of the $Cr(NH_3)_6^{3+}$ ion.

19.16 Using the data in Table 19.5, calculate the ratio of the concentration of $Cu(NH_3)_4^{2+}$ to free Cu^{2+} ion when $[NH_3] = 0.10$ M.

19.17 Complete and balance the following equations

a. $Cr^{3+}(aq) + NH_3(aq) \rightarrow$
b. $AgI(s) + NH_3(aq) \rightarrow$
c. $Ni(NH_3)_6^{2+}(aq) + en(aq) \rightarrow$
d. $Ca^{2+}(aq) + EDTA^{4-}(aq) \rightarrow$
e. $Cr(OH)_3(s) + OH^-(aq) \rightarrow$

19.18 Consider the reaction

$$AgI(s) + 2\ S_2O_3^{2-}(aq) \rightarrow Ag(S_2O_3)_2^{3-}(aq) + I^-(aq)$$

Using the solubility product for silver iodide and the dissociation constant for $Ag(S_2O_3)_2^{3-}$, calculate K for this reaction.

19.19 Suggest a suitable reagent to bring each of the following compounds into solution. (Consider acid-base reactions as well as complex ion formation.)

a. $CaCO_3$
b. $Zn(OH)_2$
c. AgI
d. $Al(OH)_3$

19.30 Using the valence-bond model, give the orbital diagram and number of unpaired electrons in complexes of the first two ions listed in Table 19.3 opposite linear, square planar, tetrahedral, and octahedral complexes.

19.31 All the complexes of Cr^{3+} have three unpaired electrons. Explain this observation, first in terms of the valence-bond model and then of crystal-field theory.

19.32 Using the crystal-field model, predict which of the ions listed opposite octahedral complexes in Table 19.3 could form both high and low spin complexes.

19.33 Based on the spectrochemical series given on p. 522, how many unpaired electrons would you expect to find in the FeF_6^{4-} complex?

19.34 The complex ions of Cu^{2+} with NH_3, H_2O, and Cl^- are colored violet, blue, and green, respectively. Rationalize this trend in colors in terms of crystal-field splitting.

19.35 Consider the reaction

$$Co(NH_3)_5Cl^{2+}(aq) + H_2O \rightarrow$$
$$Co(NH_3)_5H_2O^{3+}(aq) + Cl^-(aq)$$

Suggest at least three different ways in which you might follow experimentally the rate of this reaction.

19.36 Calculate the concentration of NH_3 required to convert 50% of Zn^{2+} to $Zn(NH_3)_4^{2+}$.

19.37 Write balanced equations for the reactions, if any, that occur when

a. dilute ammonia is added to AgI.
b. a solution of sodium hydroxide is added to a solution of $Al(NO_3)_3$ (two steps).
c. silver bromide is dissolved in ammonia and then reprecipitated with HNO_3 (two reactions).

19.38 Calculate K for the reaction

$$CuS(s) + 4\ NH_3(aq) \rightarrow Cu(NH_3)_4^{2+}(aq) + S^{2-}(aq)$$

Do you think it would be feasible to try to dissolve CuS by adding ammonia?

19.39 How would you accomplish the following conversions

a. $AgCl(s) \rightarrow Ag(S_2O_3)_2^{3-}(aq)$
b. $AgCl(s) \rightarrow AgI(s)$
c. $Zn(NH_3)_4^{2+}(aq) \rightarrow Zn(H_2O)_4^{2+}(aq)$
d. $Al(H_2O)_6^{3+}(aq) \rightarrow Al(H_2O)_3(OH)_3(s) \rightarrow$
$Al(OH)_6^{3-}(aq)$

19.20 A sample of iron ore weighing 0.150 g is brought into solution and titrated with EDTA. 15.0 ml of 0.110 M EDTA is required. What is the percentage of Fe^{3+} in the sample?

19.40 How many ml of 0.10 M EDTA should be injected into the soil to bring the iron in one gram of Fe_2O_3 into solution?

*19.41 If we represent by Δ the energy difference between the higher and lower d orbitals in an octahedral crystal field, it will turn out that:

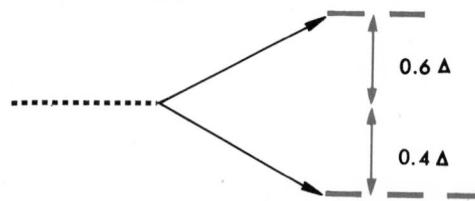

0.6 Δ

0.4 Δ

    a. How does the total energy of the d orbitals after splitting compare to that before splitting?
    b. Express, in terms of Δ, the energy gained as a result of crystal-field splitting in ions with 1, 2, 3, 8, 9, and 10 d electrons.

*19.42 If we represent by E the energy absorbed when an electron is paired in violation of Hund's rule,

    a. How much energy must be absorbed to pair electrons when we go from the high spin to the low spin complex of Co^{3+}?
    b. Referring to Problem 19.41, how much energy is evolved as a result of crystal-field splitting in going from the high spin to the low spin complex of Co^{3+}?
    c. In a particular complex of Co^{3+}, $E = 0.4\Delta$. Would you expect this to be a high spin or low spin complex?

*19.43 A solution containing Ni^{2+} and Al^{3+} ions is treated with aqueous ammonia. A bluish precipitate forms at first; as more ammonia is added, part of the precipitate dissolves to form a deep blue solution. The precipitate that remains is white; upon treatment with excess OH^-, it forms a clear solution. If acid is slowly added to this solution, a white precipitate forms, which dissolves as more acid is added. Write balanced net ionic equations for each reaction that took place.

20 OXIDATION AND REDUCTION: ELECTROCHEMICAL CELLS

Many of the reactions discussed in preceding chapters involve the transfer of electrons from one species to another. A simple example is the formation of Na^+ and F^- from the corresponding atoms. Representing the process in electron dot notation we have:

$$Na \cdot + \cdot \ddot{\underset{\cdot\cdot}{F}} : \rightarrow Na^+ + \left(: \ddot{\underset{\cdot\cdot}{F}} : \right)^- \tag{20.1}$$

It is clear that in this process a sodium atom has transferred an electron to a fluorine atom.

$$Na \cdot \rightarrow Na^+ + e^- \tag{20.1a}$$

$$\cdot \ddot{\underset{\cdot\cdot}{F}} : + e^- \rightarrow \left(: \ddot{\underset{\cdot\cdot}{F}} : \right)^- \tag{20.1b}$$

Any reaction between atoms which leads to the formation of ions may be analyzed similarly. For example, the reaction between magnesium and oxygen atoms

$$\cdot Mg \cdot + : \dot{\underset{\cdot\cdot}{O}} : \rightarrow Mg^{2+} + \left(: \ddot{\underset{\cdot\cdot}{O}} : \right)^{2-} \tag{20.2}$$

may be broken down into two half-reactions:

$$\cdot Mg \cdot \rightarrow Mg^{2+} + 2\ e^- \tag{20.2a}$$

and

$$: \dot{\underset{\cdot\cdot}{O}} : + 2\ e^- \rightarrow \left(: \ddot{\underset{\cdot\cdot}{O}} : \right)^{2-} \tag{20.2b}$$

Processes such as 20.1a and 20.2a, which involve the **loss of electrons,** are referred to as **oxidation** half-reactions; atoms of sodium and magnesium lose electrons to become positively charged ions. The processes represented by 20.1b and 20.2b, which involve the **gain of electrons,** are referred

535

to as **reduction** half-reactions; in gaining electrons atoms of fluorine and oxygen are said to be reduced. The overall Reactions 20.1 and 20.2, in which oxidation and reduction occur simultaneously, are called **oxidation-reduction** reactions or simply "redox" reactions. Since no electrons are created or destroyed in the process, in any oxidation-reduction reaction there can be no net gain or loss of electrons. For example, when lithium reacts with oxygen, two lithium atoms are oxidized to Li^+ for every oxygen atom reduced to O^{2-}.

oxidation:
$$2 \text{ Li·} \rightarrow 2 \text{ Li}^+ + 2 \text{ e}^- \qquad (20.3a)$$

reduction:
$$:\dot{\text{O}}: + 2 \text{ e}^- \rightarrow \left(:\ddot{\text{O}}:\right)^{2-} \qquad (20.3b)$$

overall reaction:
$$2 \text{ Li·} + :\dot{\text{O}}: \rightarrow 2 \text{ Li}^+ + \left(:\ddot{\text{O}}:\right)^{2-} \qquad (20.3)$$

We shall shortly find that it is possible to broaden the meaning of the terms oxidation and reduction so that they can be applied to a wide variety of reactions, including many in which no ionic species are involved. To accomplish this, it is necessary to introduce a new concept—oxidation number.

20.1 OXIDATION NUMBER

The chemical equation written for the reaction between hydrogen and fluorine

$$\tfrac{1}{2} \text{ H}_2(g) + \tfrac{1}{2} \text{ F}_2(g) \rightarrow \text{HF}(g) \qquad (20.4)$$

resembles that for the reaction of sodium with fluorine

$$\text{Na}(s) + \tfrac{1}{2} \text{ F}_2(g) \rightarrow \text{NaF}(s)$$

Indeed, the two reactions themselves have much in common. In both there is an exchange of electrons between atoms; the major difference lies in the extent to which electrons are transferred. In Reaction 20.4 the valence electron of hydrogen is shared with fluorine rather than being transferred to it. This distinction is one of degree rather than kind. The electrons in the H—F covalent bond are displaced strongly toward fluorine. So far as "electron bookkeeping" is concerned, it would seem reasonable to assign these electrons to the fluorine atom

$$\text{H} \ \Big| \ :\ddot{\text{F}}:$$

By assigning electrons in this way we have, in a sense, given a −1 charge to fluorine which now has one more valence electron (8) than an isolated fluorine atom (7). The hydrogen atom, stripped of its valence electron by this assignment, has in effect acquired a +1 charge.

The accounting system that we have just illustrated is widely used in inorganic chemistry. The concept of **oxidation number** is introduced to refer to the charge an atom would have if, as here, the bonding electrons were assigned arbitrarily to the more electronegative element. In the HF molecule, hydrogen is said to have an oxidation number of +1, fluorine an oxidation number of −1. In water the bonding electrons are assigned to the more electronegative oxygen atom:

$$H \left| :\ddot{\underset{\cdot\cdot}{O}}: \right| H$$

This gives oxygen an oxidation number of −2 (8 valence e^- vs 6 e^- in the neutral atom) and hydrogen an oxidation number of +1 (0 e^- vs 1 e^- in the neutral atom). In a nonpolar covalent bond, the bonding electrons are split evenly between the two atoms:

$$:\ddot{\underset{\cdot\cdot}{F}}\cdot \left| \cdot\ddot{\underset{\cdot\cdot}{F}}: \right. \quad \text{oxidation no. F} = 0$$

It should be emphasized that the oxidation number of an atom in a covalently bonded substance is an artificial concept. Unlike the charge of an ion, the oxidation number of an element cannot be determined experimentally. The hydrogen atom in the HF or H_2O molecule does not carry a full positive charge; its oxidation number of +1 in these molecules may be regarded as a "pseudocharge."

In species like HF and H_2O the oxidation numbers show the direction of electron shift, but not the amount

RULES FOR ASSIGNING OXIDATION NUMBERS

In practice, oxidation numbers are seldom determined by assigning bonding electrons in the manner we have just described. Instead, they are obtained by applying certain rules which, while consistent with this scheme, are much simpler to apply.

1. **The oxidation number * of an element in an elementary substance is 0.** For example, the oxidation number of chlorine in Cl_2 or of phosphorus in P_4 is zero.

2. **The oxidation number of an element in a monatomic ion is equal to the charge of that ion.** In the ionic compound NaCl, sodium has an oxidation number of +1, chlorine an oxidation number of −1. The oxidation numbers of aluminum and oxygen in Al_2O_3 (Al^{3+}, O^{2-} ions) are +3 and −2, respectively.

3. **Certain elements have the same oxidation number in all or almost all their compounds.** The 1A metals always exist as +1 ions in their compounds and, hence, are assigned an oxidation number of +1. By the same token, the 2A elements always have oxidation numbers of +2 in their compounds. Fluorine, the most electronegative of all elements, has an oxidation number of −1 in almost all its compounds. Oxygen, second only to fluorine in electronegativity, is ordinarily assigned an oxidation number of −2 (certain exceptions will be pointed out later).

Hydrogen, in its compounds with metals (as in NaH and CaH_2), exists as a −1 ion and, therefore, has an oxidation number of −1. In its compounds with the nonmetals, hydrogen is assigned an oxidation number of +1.

4. **The sum of the oxidation numbers of all the atoms in a neutral species is 0; in an ion, it is equal to the charge of that ion.** The application of this very useful principle is illustrated in Example 20.1.

---

* Many authors prefer to speak of the **oxidation state** of an element in contrast to the **oxidation number** of an atom. We shall follow the common practice of using these terms interchangeably.

Example 20.1 What is the oxidation number of selenium in Na_2Se? of manganese in MnO_4^-?

Solution. For Na_2Se, knowing that the oxidation number of sodium must be $+1$, we have:

$$2(+1) + \text{oxid no. Se} = 0; \text{ oxid no. Se} = -2$$

In the MnO_4^- ion, taking the oxidation number of oxygen to be -2 and realizing that the sum must be -1:

$$\text{oxid no. Mn} + 4(-2) = -1; \text{ oxid no. Mn} = +7$$

OXIDATION STATES OF THE ELEMENTS

The common oxidation states of the elements are tabulated in Figure 20.1. It may be helpful to point out some general principles that are perhaps hidden in this maze of numbers.

1. The metallic elements show only positive oxidation numbers in the compounds they form with nonmetals. This reflects the fact that metals tend to lose rather than to gain electrons when they react with nonmetals. Negative oxidation numbers are found with only a few elements, all of which are strongly electronegative nonmetals. For these elements the minimum oxidation number is equal to the charge of their monatomic anion (N^{3-}, O^{2-}, F^-).

Cite a compound in which N has an oxidation no. of -3; $+5$

2. As pointed out previously, the 1A and 2A metals show only one oxidation state (1A $= +1$, 2A $= +2$). In contrast, transition metals commonly show a variety of oxidation states. As we move down a given group in the transition series, higher oxidation states become more stable. For example, nickel commonly shows an oxidation state of $+2$ and is never oxidized above the $+3$ state; the lower members of this group, palladium and platinum, form a great many stable compounds in the $+4$ state.

3. With some exceptions, the maximum oxidation number of an element is given by its group number in the Periodic Table. In a few families the elements achieve this oxidation state by forming monatomic ions (1A, 2A, 3A, 3B). More frequently, elements show their maximum oxidation numbers only in compounds in which they are covalently bonded to oxygen or another highly electronegative element. For example, the $+6$ oxidation state of sulfur is found in the species SO_3, H_2SO_4, SO_4^{2-}, and SF_6. We find "+7 chlorine" in the perchlorate ion, ClO_4^-, in which the bonding electrons have all been arbitrarily assigned to oxygen.

$$\left[\begin{array}{c} :\ddot{O}: \\ :\ddot{O}: \boxed{Cl} :\ddot{O}: \\ :\ddot{O}: \end{array} \right]$$

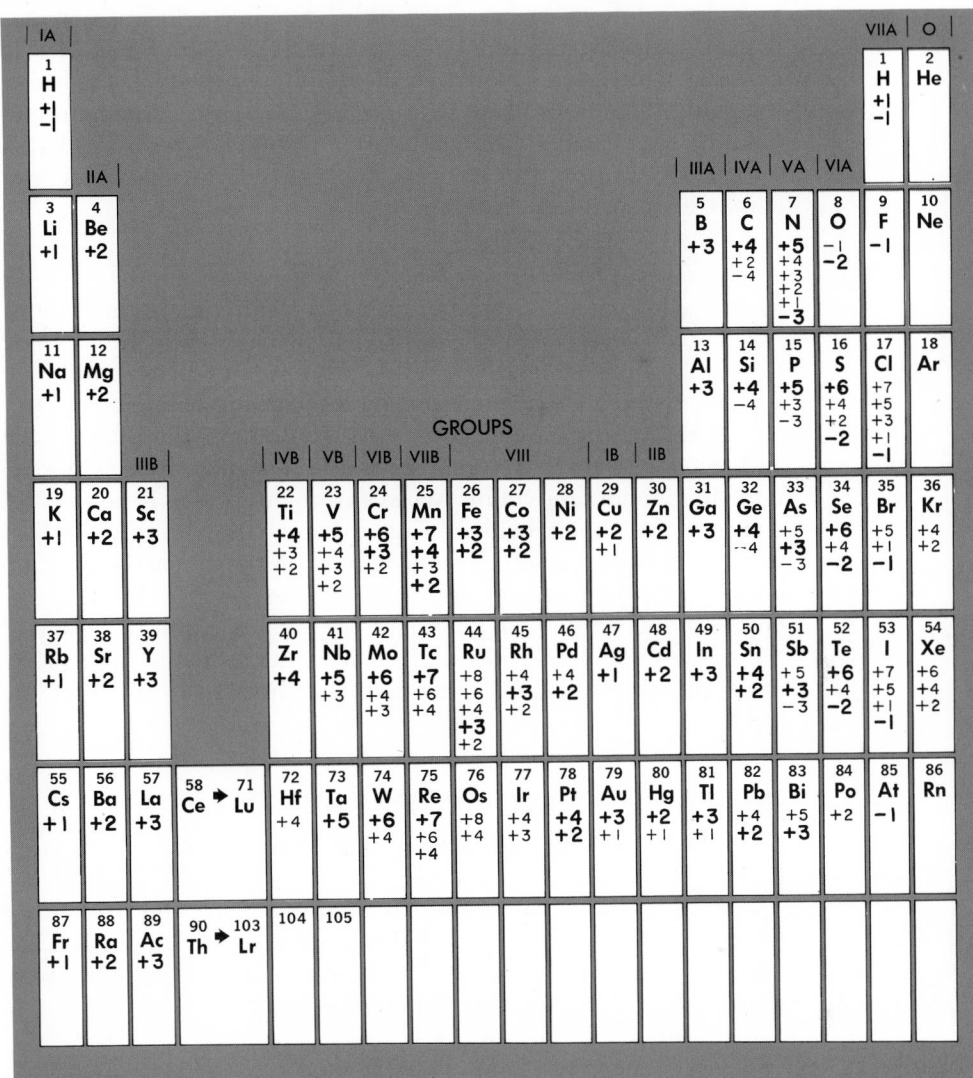

Figure 20.1 Oxidation states of the elements. The most common or stable states are shown in heavy type.

OXIDATION AND REDUCTION. A WORKING DEFINITION

The concept of oxidation number leads directly to a working definition of the terms oxidation and reduction. **Oxidation** is defined as an **increase in oxidation number, reduction** as a **decrease in oxidation number.** Reactions in which one element increases in oxidation number at the expense of another are referred to as oxidation-reduction reactions. Two simple examples follow:

$$2 \text{ Al(s)} + 3 \text{ Cl}_2\text{(g)} \rightarrow 2 \text{ AlCl}_3\text{(s)} \quad \begin{array}{l} \text{Al oxidized (oxidation no. } 0 \rightarrow +3) \\ \text{Cl reduced (oxidation no. } 0 \rightarrow -1) \end{array} \quad (20.5)$$

$$4 \text{ As(s)} + 5 \text{ O}_2\text{(g)} \rightarrow 2 \text{ As}_2\text{O}_5\text{(s)} \quad \begin{array}{l} \text{As oxidized (oxidation no. } 0 \rightarrow +5) \\ \text{O reduced (oxidation no. } 0 \rightarrow -2) \end{array} \quad (20.6)$$

The change in
oxidation no.
tells us whether
an element is oxi-
dized or reduced

These definitions are compatible, of course, with our earlier interpretation of oxidation and reduction in terms of loss and gain of electrons. An element which loses electrons inevitably increases in oxidation number; the gain of electrons always results in a decrease in oxidation number. By defining oxidation and reduction in terms of changes in oxidation number, we greatly simplify the electron bookkeeping in redox reactions. For example, analysis of the reaction

$$HCl(g) + HNO_3(l) \rightarrow NO_2(g) + \tfrac{1}{2} Cl_2(g) + H_2O(l) \qquad (20.7)$$

in terms of oxidation numbers reveals immediately that chlorine is oxidized (oxidation no. $= -1$ in HCl, 0 in Cl_2) while nitrogen is reduced (oxidation no. $= +5$ in HNO_3, $+4$ in NO_2). It is much more difficult to decide precisely which atoms are "losing" or "gaining" electrons.

In discussing oxidation-reduction reactions, the phrases **oxidizing agent** and **reducing agent** are frequently used to designate the species responsible for oxidation and reduction. We speak of chlorine in Reaction 20.5 and oxygen in 20.6 as being oxidizing agents, since they bring about the oxidation of aluminum and arsenic, respectively. In these reactions, aluminum and arsenic act as reducing agents, being responsible for the reduction of chlorine and oxygen. In the more complex reaction represented by 20.7, the species which is oxidized, HCl, acts as a reducing agent; HNO_3, which undergoes reduction, is the oxidizing agent.

20.2 BALANCING OXIDATION-REDUCTION EQUATIONS

Many of the equations that we write to represent oxidation-reduction reactions are simple enough to be balanced by inspection. Consider, for example, the reaction that occurs when powdered iron is heated with dry chlorine gas. The unbalanced equation for this reaction is

$$Fe(s) + Cl_2(g) \rightarrow FeCl_3(s)$$

With a minimum of mathematical intuition and very little effort, we arrive at the balanced equation

$$2\ Fe(s) + 3\ Cl_2(g) \rightarrow 2\ FeCl_3(s) \qquad (20.8)$$

Frequently, however, we are confronted with a rather more complicated redox reaction, where the corresponding equation is not nearly so easy to balance. To illustrate, consider the reaction that occurs when a solution of potassium permanganate is added to hydrochloric acid. We find in the laboratory that a gas which is readily identified as elementary chlorine is generated in this reaction. At the same time the deep purple color of the MnO_4^- ion fades to be replaced by a light pink color which we associate with Mn^{2+}. To represent this reaction, we might tentatively write the (decidedly) unbalanced equation

$$MnO_4^-(aq) + Cl^-(aq) \rightarrow Mn^{2+}(aq) + Cl_2(g)$$

Clearly, we are dealing with a redox reaction; the oxidation number of Mn decreased from $+7$ to $+2$ (reduction) while that of Cl increased from -1 to

0 (oxidation). It is by no means obvious how we are to balance this equation, which must be accomplished if we are to make any sort of calculation dealing with relative amounts of reactants and products. What is needed is a general approach to the problem of balancing redox equations, particularly those that take place in aqueous solution.

HALF-EQUATION METHOD

A straightforward way of balancing a redox equation for a reaction in aqueous solution involves, as a first step, breaking the equation down into an oxidation half-equation and a reduction half-equation. The two half-equations are balanced separately and then combined in such a way as to arrive at an overall equation in which there is no net change in the number of electrons. To illustrate the method, let us go through the various steps, using the reaction between solutions of potassium permanganate and hydrochloric acid as an example.

1. *Separate into two half-equations, an oxidation and a reduction.* In this case, we have:

oxidation: $$Cl^-(aq) \rightarrow Cl_2(g)$$

reduction: $$MnO_4^-(aq) \rightarrow Mn^{2+}(aq)$$

2. *Balance each half-equation separately, first with respect to mass and then with respect to charge.* Let's try the easy one first: the oxidation of Cl^- to Cl_2. To balance with respect to mass, we need only write a coefficient of two in front of Cl^-

$$2\ Cl^-(aq) \rightarrow Cl_2(g)$$

Charge balance is achieved by introducing the proper number of electrons into the equation. At the moment we have a charge of -2 on the left and 0 on the right. Clearly, we can balance this half-equation by adding two electrons to the right:

$$2\ Cl^-(aq) \rightarrow Cl_2(g) + 2\ e^- \tag{20.9a}$$

Looking at the reduction half-equation, we note that we already have the same number of manganese atoms on both sides. However, we seem to have a surplus of four oxygens on the left. *To balance oxygen, we add an appropriate number of H_2O molecules*, in this case four H_2O to the right side of the half-equation

$$MnO_4^-(aq) \rightarrow Mn^{2+}(aq) + 4\ H_2O$$

We now have eight hydrogen atoms on the right and none on the left. *To balance hydrogen, we add H^+ ions*, in this case eight H^+ to the left side of the half-equation

$$MnO_4^-(aq) + 8\ H^+(aq) \rightarrow Mn^{2+}(aq) + 4\ H_2O$$

Finally, the charge must be balanced; at the moment we have a charge of $+2$ on the right and $+7$ on the left ($-1 +8$). To balance, five electrons are added to the left:

$$MnO_4^-(aq) + 8\ H^+(aq) + 5\ e^- \rightarrow Mn^{2+}(aq) + 4\ H_2O \tag{20.9b}$$

Perhaps we should digress at this point to consider a point that may or may not have bothered you. Why do we use H_2O molecules and H^+ ions to balance the elements oxygen and hydrogen? We do this for one simple reason; they are the only species containing these elements that are available in large quantities in the aqueous solution we are working with. We cannot, for example, introduce oxygen atoms or O_2 molecules into our equation because there is no evidence whatsoever that they are among the products of the reaction. Remember, an equation must always correspond to chemical reality. We can readily show that Equation 20.9b does precisely that. If we measure the concentration of H^+ ions (i.e., the acidity), we find that it drops as the reaction proceeds, indicating that H^+ is indeed a reactant. It is somewhat more difficult to show that H_2O is a product, since there are so many water molecules around to begin with, but it can be done. (Can you suggest how?)

3. *Combine the two balanced half-equations in such a way as to make the electron gain equal the electron loss.* In this case we see that 2 electrons are given off in Equation 20.9a while 5 are gained in Equation 20.9b. To arrive at a final equation in which no electrons appear (chemical reality again!), we multiply 20.9a by 5 and 20.9b by 2 and add.

One *cannot* guess the coefficients in an equation like this, so *learn* the method

$5 \times 20.9a$:

$$10 \text{ Cl}^-(aq) \rightarrow 5 \text{ Cl}_2(g) + 10 \text{ e}^-$$

$2 \times 29.9b$:

$$2 \text{ MnO}_4^-(aq) + 16 \text{ H}^+(aq) + 10 \text{ e}^- \rightarrow 2 \text{ Mn}^{2+}(aq) + 8 \text{ H}_2O$$

$$10 \text{ Cl}^-(aq) + 2 \text{ MnO}_4^-(aq) + 16 \text{ H}^+(aq) \rightarrow 5 \text{ Cl}_2(g) + 2 \text{ Mn}^{2+}(aq) + 8 \text{ H}_2O$$

$$(20.9)$$

This is our final balanced equation for the reaction of potassium permanganate with hydrochloric acid. It is always advisable to check to make sure that:

 — the atoms balance (10 Cl, 2 Mn, 8 O, 16 H on both sides)
 — the charges balance ($-10 -2 +16 = +4$)
 — we do, indeed, have the equation with the simplest whole number coefficients. Sometimes the procedure that we use will result in an equation that can be divided by two or some other integer.

We frequently have occasion to write balanced equations for oxidation-reduction reactions taking place in basic solution. For such reactions it would be inappropriate to write equations in which H^+ ions appear, since this ion is present only at very low concentration in basic solution. Instead, the equations should contain OH^- ions or H_2O molecules. A simple way to accomplish this is to "neutralize" any H^+ ions appearing in the half-equations by adding an equal number of OH^- ions to both sides. To illustrate, consider the oxidation, in basic solution, of iodide by permanganate ions:

$$\text{I}^-(aq) + \text{MnO}_4^-(aq) \rightarrow \text{I}_2(aq) + \text{MnO}_2(s) \text{ (basic solution)}$$

One can proceed exactly as in the foregoing example, to obtain the half-equations

Oxidation:

$$2 \text{ I}^-(aq) \rightarrow \text{I}_2(aq) + 2 \text{ e}^- \qquad (20.10a)$$

Reduction: $\text{MnO}_4^-(aq) + 4 \text{ H}^+(aq) + 3 \text{ e}^- \rightarrow \text{MnO}_2(s) + 2 \text{ H}_2O$

The H^+ ions appearing in the reduction half-equation must now be re-moved to obtain an equation valid in basic solution. To do this, four OH^- ions are added to both sides:

We can't have H^+ ions on one side of an equation and OH^- ions on the other

$$\begin{array}{l} MnO_4^-(aq) + 4\ H^+(aq) \quad + 3\ e^- \rightarrow MnO_2(s) + 2\ H_2O \\ \underline{\qquad\quad + 4\ OH^-(aq) \qquad\qquad \rightarrow \qquad\qquad\qquad\quad + 4\ OH^-(aq)} \\ MnO_4^-(aq) + 4\ H_2O \quad\ + 3\ e^- \rightarrow MnO_2(s) + 2\ H_2O + 4\ OH^-(aq) \end{array}$$

Eliminating two water molecules from each side, we arrive at

$$MnO_4^-(aq) + 2\ H_2O + 3\ e^- \rightarrow MnO_2(s) + 4\ OH^-(aq) \quad (20.10b)$$

for the reduction half-reaction in basic solution. To obtain the overall equation, we proceed as before, combining 20.10a and 20.10b in such a way as to make the electron gain equal the electron loss:

$3 \times 20.10a$: $\qquad\qquad\qquad 6\ I^-(aq) \rightarrow 3\ I_2(aq) + 6\ e^-$

$2 \times 20.10b$:

$$\begin{array}{l} \underline{2\ MnO_4^-(aq) + 4\ H_2O + 6\ e^- \rightarrow 2\ MnO_2(s) + 8\ OH^-(aq)} \\ 6\ I^-(aq) + 2\ MnO_4^-(aq) + 4\ H_2O \rightarrow 3\ I_2(aq) + 2\ MnO_2(s) + 8\ OH^-(aq) \end{array}$$

$$(20.10)$$

OXIDATION NUMBER METHOD OF BALANCING EQUATIONS

The **half-equation** method just described is by no means the only method of balancing oxidation-reduction equations. We have stressed this particular method because it is valuable in studying other aspects of oxidation-reduction reactions. Of the various other methods which can be used to balance redox equations, we shall mention only one—the **oxidation number** method.

To illustrate the application of this method, consider the equation referred to earlier:

$$MnO_4^-(aq) + Cl^-(aq) + H^+(aq) \rightarrow Mn^{2+}(aq) + Cl_2(g) + H_2O$$

To balance this equation by the oxidation number method, we proceed as follows:

1. Determine the oxidation number of each element on both sides of the equation, thereby determining which elements have undergone oxidation and reduction:

	Oxid No. Reactants	Oxid No. Products	
Mn	+7	+2	reduced
O	−2	−2	
Cl	−1	0	oxidized
H	+1	+1	

Why must this condition be obeyed?

2. By adjusting the coefficients of the species being oxidized and reduced, make the total increase in oxidation number equal to the total decrease.

In this case, each Mn atom undergoes a decrease in oxidation number of five units; each Cl atom increases in oxidation number by one unit. To make the increase in oxidation number equal to the decrease, there must be five Cl atoms oxidized for every Mn reduced. Thus

$$MnO_4^-(aq) + 5\ Cl^-(aq) \rightarrow Mn^{2+}(aq) + \tfrac{5}{2}\ Cl_2(g)$$

or, multiplying through by two to eliminate fractional coefficients,

$$2\ MnO_4^-(aq) + 10\ Cl^-(aq) \rightarrow 2\ Mn^{2+}(aq) + 5\ Cl_2(g) \qquad (a)$$

Note that for Equation a, the total increase in oxidation number of Cl is $10 \times 1 = 10$, the total decrease in oxidation number of Mn $= 2 \times 5 = 10$.

3. Having determined the coefficients of the species being oxidized and reduced, balance the number of atoms of the remaining elements in the usual manner.

Here, starting with Equation a, the oxygen is balanced first. The presence of 2 MnO_4^- ions on the left, containing a total of eight oxygen atoms, requires that there be eight H_2O molecules, each with one oxygen atom, on the right:

$$2 \; MnO_4^-(aq) + 10 \; Cl^-(aq) \rightarrow 2 \; Mn^{2+}(aq) + 5 \; Cl_2(g) + 8 \; H_2O \qquad \text{(b)}$$

Finally, to balance the hydrogen, 16 H^+ ions must be added to the left:

$$2 \; MnO_4^-(aq) + 10 \; Cl^-(aq) + 16 \; H^+(aq) \rightarrow 2 \; Mn^{2+}(aq) + 5 \; Cl_2(g) + 8 \; H_2O \qquad \text{(c)}$$

The final balanced equation is, of course, identical to that previously derived by the half-equation method.

A balanced oxidation-reduction equation, like any other balanced equation, can be used to calculate the relative amounts of reactants and products involved in a reaction. Just in case the principles involved, which were introduced in Chapter 3, are hazy in your mind, it may be helpful to review them in Example 20.2.

Example 20.2 In the oxidation of I^- by MnO_4^- in basic solution, it is desired to form 16.0 g of I_2. Assuming a 100% yield

    a. how many moles of MnO_4^- should we start with?
    b. how many grams of NaI should be used?

Solution

    a. From the coefficients of the balanced equation (20.10)

$$2 \text{ moles } MnO_4^- \;\simeq\; 3 \text{ moles } I_2$$

But since 1 mole $I_2 = 2(126.9 \text{ g}) = 253.8$ g, we need

$$16.0/253.8 = 0.0630 \text{ mole } I_2$$

$$\text{no. moles } MnO_4^- \text{ required} = 0.0630 \text{ mole } I_2 \times \frac{2 \text{ moles } MnO_4^-}{3 \text{ moles } I_2}$$

$$= 0.0420 \text{ mole } MnO_4^-$$

    b. The equation tells us that we need 6 I^- to produce 3 I_2, or that 2 moles of I^- will form 1 mole of I_2. To get 0.0630 mole I_2, we will need 2×0.0630 mole or 0.126 mole of I^-. Since 1 mole of NaI contains 1 mole of I^-, we need an equal number of moles of NaI. But

$$1 \text{ mole NaI} = (23.0 + 126.9) \text{ g} = 149.9 \text{ g}$$

$$\text{no. grams NaI} = 0.126 \text{ mole NaI} \times \frac{149.9 \text{ g NaI}}{1 \text{ mole NaI}} = 18.9 \text{ g NaI}$$

20.3 ELECTROLYTIC CELLS

By means of a device known as an electrolytic cell, it is possible to use electrical energy to bring about a nonspontaneous oxidation-reduction reaction. To understand how such a cell operates, let us consider the generalized cell diagram shown in Figure 20.2.

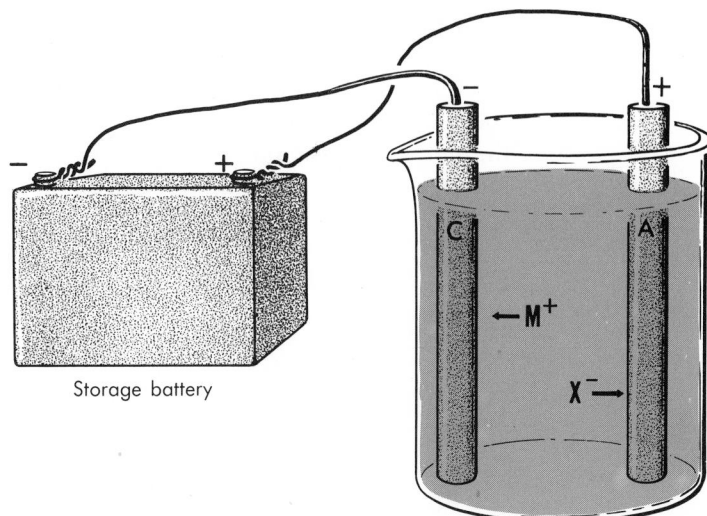

Figure 20.2 General diagram of electrolytic cell. Reduction occurs at cathode (C). Oxidation occurs at anode (A).

In the electrolysis cell the electrode which is attached to the (−) pole of the battery becomes the cathode

Storage battery

Cathode   Anode

The storage battery at the left provides a source of direct electric current; it could be replaced by a simple dry cell or a DC generator. From the terminals of the battery, indicated by + and − signs, two wires lead to the electrolytic cell. This consists of two electrodes, A and C, dipping into a liquid containing ions M^+ and X^-.

By a mechanism which we shall consider in Section 20.4, the battery acts as an electron pump, pushing electrons into C and removing them from A. In order to maintain electrical neutrality, some process must take place within the cell so as to consume electrons at C and liberate them at A. This process is an oxidation-reduction reaction. At electrode C, known as the **cathode,** an ion or molecule undergoes **reduction** by accepting electrons. At the **anode,** A, electrons are produced by the **oxidation** of an ion or molecule. The overall cell reaction is the sum of the two half-reactions occurring at the electrodes. While electrolysis is proceeding, there is a steady flow of ions to the two electrodes. Positive ions (*cations*) move toward the *cathode;* negative ions (*anions*) move toward the *anode.*

Electrolysis may be carried out by passing a direct current through a water solution of an electrolyte or through a molten salt or oxide. The cell reactions are somewhat easier to visualize in the latter case.

COMMERCIAL CELLS

In principle, any oxidation-reduction reaction, no matter how non-spontaneous, may be brought about in an electrolytic cell if the applied voltage is great enough. Many of our most important metals and industrial chemicals are prepared by electrolytic processes of this type. We shall consider three of these: the production of sodium metal by the electrolysis of molten sodium chloride, the preparation of aluminum from bauxite ore (Al_2O_3), and the production of chlorine, hydrogen, and sodium hydroxide by the electrolysis of a water solution of sodium chloride.

Na FROM NaCl

The so-called Downs cell, used commercially to electrolyze molten sodium chloride, is shown in Figure 20.3. The half-reactions occurring in the cell are particularly simple. At the cylindrical iron *cathode,* sodium ions are reduced to sodium metal

$$Na^+ + e^- \rightarrow Na(l) \qquad (20.11a)$$

For every sodium ion reduced at the cathode, a chloride ion is oxidized at the graphite *anode.*

$$Cl^- \rightarrow \tfrac{1}{2} Cl_2(g) + e^- \qquad (20.11b)$$

The total cell reaction, obtained by summing 20.11a and 20.11b, is

$$NaCl(l) \rightarrow Na(l) + \tfrac{1}{2} Cl_2(g) \qquad (20.11)$$

Reaction 20.11 is nonspontaneous; at the operating temperature, 600°C, $\Delta G^{1\,atm}$ is +77.2 kcal. At least this much energy, in the form of electrical work, must be supplied to decompose one mole of sodium chloride. If the products of the cell reaction, elementary sodium and chlorine, are allowed to come into contact with each other, they will combine spontaneously to give the starting material, sodium chloride. To prevent this, the electrodes in the Downs cell are separated by a circular iron screen, which allows the migration of ions but prevents direct contact between the products of electrolysis.

Over 20,000 tons of sodium are made annually in the United States by the electrolysis of molten sodium chloride. Its principal use is in the synthesis of organic compounds, particularly tetraethyl lead, the antiknock ingredient of motor fuels. Molten sodium is used as a coolant to remove heat

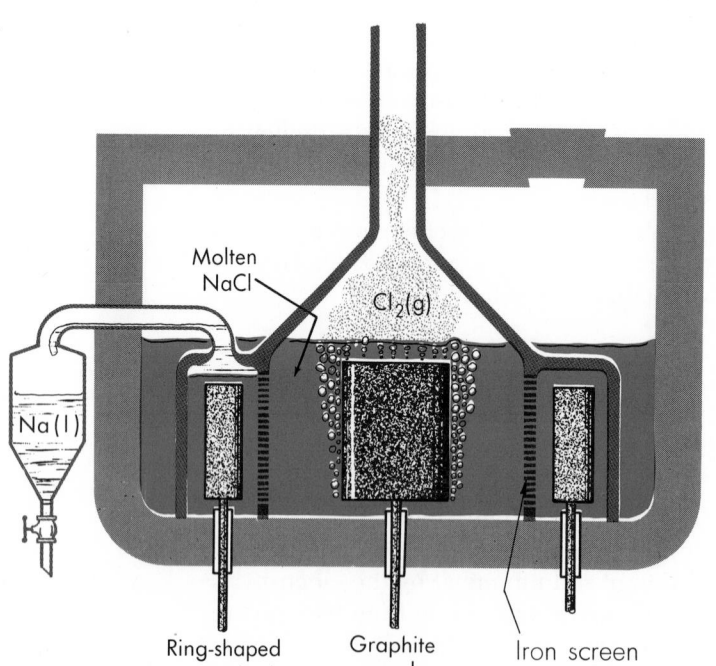

Figure 20.3 Electrolysis of molten sodium chloride. Screen is used to prevent sodium and chlorine from coming in contact with each other.

How do they manage to operate this cell at 600°C when NaCl melts at 800°C?

Molten NaCl

$Cl_2(g)$

Na(l)

Ring-shaped iron cathode

Graphite anode

Iron screen

from nuclear reactors; the excellent heat conductivity of sodium and relatively wide liquid range (m pt = 98°C, b pt = 880°C) makes it ideal for this purpose. The chlorine formed simultaneously in the Downs process is a valuable by-product. There is, however, a cheaper way of making chlorine electrolytically (see p. 548).

Al FROM Al₂O₃

Aluminum is the third most abundant element in the earth's crust. Its importance as a structural material is indicated by the fact that about 2,000,000 tons of aluminum are produced annually in the United States, an amount greater than that of any other metal except copper and iron. Yet, from 1828, when aluminum was first isolated by Wöhler, until 1886, when the Hall electrolytic process for its manufacture was developed, the metal remained little more than a scientific curiosity. In this 58-year period, the price of aluminum never fell below $8 a pound; today it sells for about 30 cents a pound.

The long time lag between the isolation of aluminum and its commercial utilization reflected the difficulties of extracting the metal from its ores. Aluminum occurs in such common minerals as feldspar, granite, and clay, but unfortunately the aluminum in these materials is tightly bound in a network of silicon and oxygen atoms from which its extraction is extremely difficult. The principal source of aluminum has always been bauxite ore, in which the element occurs as the hydrated oxide. Prior to 1886 it was necessary to first convert the oxide to the chloride and then reduce the latter with sodium. The high cost of the sodium in this two-step process made the aluminum prohibitively expensive.

The electrolytic process by which aluminum is produced today from aluminum oxide was worked out by Charles Hall, a graduate student at Oberlin College. After experimenting with a great many materials, he found that the mineral cryolite, Na_3AlF_6, could be used in the molten state as a solvent for Al_2O_3. The use of cryolite makes it possible to reduce the temperature of electrolysis from 2000°C, the melting point of pure Al_2O_3, to about 1000°C. Curiously enough, within a few weeks of the time that Hall produced his first aluminum, a young Frenchman, Heroult, independently worked out an almost identical process for its manufacture.

Some graduate students are more productive than others

The cell used to produce aluminum from aluminum oxide is shown schematically in Figure 20.4. The bauxite ore, before being admitted to the cell, is purified by a two-step process. It is first dissolved away from impurities (principally oxides of iron) by heating under pressure with a concentrated solution of sodium hydroxide

$$Al_2O_3(s) + 6\ OH^-(aq) + 3\ H_2O \rightarrow 2\ Al(OH)_6{}^{3-} \qquad (20.12)$$

and then, after filtering, reprecipitated by adding a weak acid, carbon dioxide, to reverse Reaction 20.12.

$$2\ Al(OH)_6{}^{3-}(aq) + 6\ CO_2(g) \rightarrow Al_2O_3(s) + 6\ HCO_3{}^-(aq) + 3\ H_2O \qquad (20.13)$$

The purified ore is then placed in an electrically heated cell, mixed with cryolite,* and melted. The iron wall of the cell serves as the cathode at which Al^{3+} ions are reduced to form molten aluminum. The anodes, retractable carbon rods, are attacked by the oxygen produced in the cell to form a mixture of carbon dioxide and carbon monoxide. The two half-reactions occurring at the electrodes may be represented most simply as

cathode: $\qquad$ $2\ Al^{3+} + 6\ e^- \rightarrow 2\ Al(l)$ $\qquad$ (20.14a)

anode: $\qquad$ $\dfrac{3\ O^{2-} \rightarrow \frac{3}{2}\ O_2(g) + 6\ e^-}{2\ Al^{3+} + 3\ O^{2-} \rightarrow 2\ Al(s) + \frac{3}{2}\ O_2(g)}$ $\qquad$ (20.14b)
$\qquad$ (20.14)

Production of Al consumes about $\frac{1}{40}$th of all the electrical energy generated in the US

The production of 1 lb of aluminum consumes about 2 lb of aluminum oxide, 0.6 lb of anodic carbon, 0.1 lb of cryolite, and 10 kilowatt-hours (kwhr) of electrical energy.

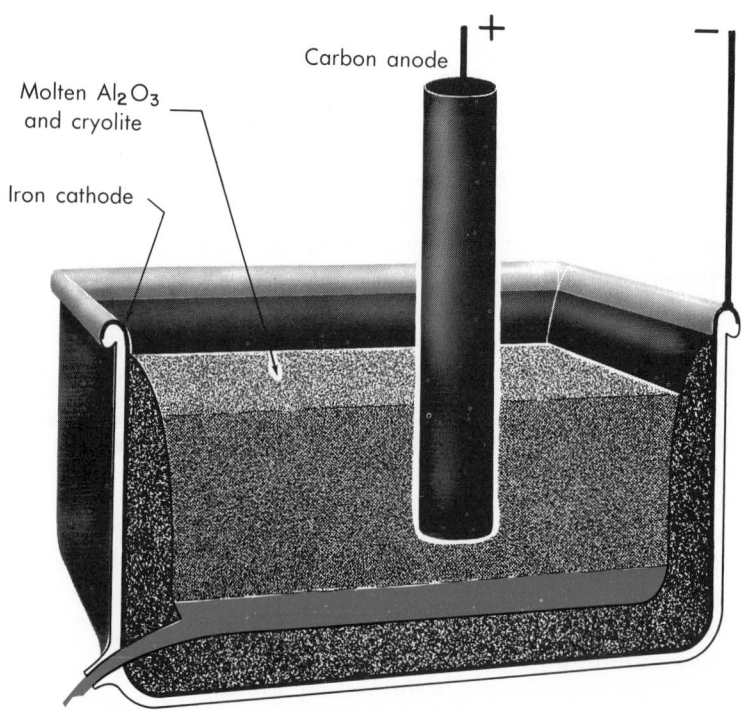

Molten Al_2O_3 and cryolite

Iron cathode

Carbon anode

+

−

Why is it desirable to have the molten Al at the bottom of the cell?

Molten aluminum

Figure 20.4 Electrolytic preparation of aluminum. Aluminum, being more dense than cryolite, collects at the bottom of the cell.

---

* A mixture of the fluorides of aluminum, sodium, and calcium is now used in place of cryolite. This mixture gives a solution with aluminum oxide which has a lower melting point and density than that obtained with cryolite. The lower density facilitates the separation of the molten aluminum, which sinks to the bottom of the cell.

Cl_2, NaOH AND H_2 FROM NaCl(aq)

It is ordinarily less expensive and more convenient to carry out an electrolysis in water solution than in the molten state. However, in the presence of water molecules we may obtain quite different products than we get with a melt. As an example, let us consider the electrolysis of a concentrated aqueous solution of sodium chloride, using platinum or graphite electrodes. The anode reaction is the same as that in the Downs cell, i.e.,

$$2 \ Cl^-(aq) \rightarrow Cl_2(g) + 2 \ e^- \qquad (20.15a)$$

The cathode reaction, however, is quite different. Bubbles of hydrogen are given off at this electrode and the solution surrounding it becomes strongly basic. We deduce that water molecules rather than sodium ions are reduced.

$$2 \ H_2O + 2 \ e^- \rightarrow H_2(g) + 2 \ OH^-(aq) \qquad (20.15b)$$

The overall cell reaction for the electrolysis of an aqueous solution of sodium chloride is obtained bv summing the two half-reactions

$$2 \ H_2O + 2 \ Cl^-(aq) \rightarrow Cl_2(g) + H_2(g) + 2 \ OH^-(aq) \qquad (20.15)$$

Notice that one effect of the cell reaction is to replace Cl^- ions by an equal number of OH^- ions. Evaporation of the solution remaining after electrolysis yields solid sodium hydroxide, mixed with some unreacted sodium chloride. Much of the sodium hydroxide and virtually all the chlorine made in this country are prepared by this process; hydrogen is an important by-product.

By making one modification in this cell it is possible to obtain sodium hydroxide of high purity, uncontaminated by sodium chloride. If a pool of mercury rather than a metal or graphite rod is used as the cathode, Na^+ ions rather than H_2O molecules are reduced. The product obtained directly at the cathode is a dilute solution or *amalgam* of sodium metal in mercury. Portions of this liquid are withdrawn from time to time and reacted with water to form a concentrated solution of sodium hydroxide. The net half-reaction at a mercury cathode is the same as that given above (Equation 20.15b) but, since sodium chloride is insoluble in mercury, it is not present in the recovered sodium hydroxide.

In principle there should be no loss of mercury in this process. It is not involved in the net reaction at the cathode; when the sodium amalgam is reacted with water, the mercury remains at the bottom of the container from whence it can be transferred back to the electrolysis cell. However, it has been obvious for many years that there is a "leak" somewhere in the system. In 1969 plants manufacturing sodium hydroxide and chlorine bought about 1.6 million pounds of mercury, very little of which was used to set up new electrolysis cells. It has now become painfully obvious where much of this mercury is going. It is being discharged into lakes, streams, and the ocean, where part of it is converted by marine organisms into toxic organic compounds such as dimethyl mercury (Chapter 11). This discharge of mercury can be avoided by proper maintenance procedures. Indeed, several of the major companies operating in this field have, under public pressure and/or court order, reduced their mercury losses to negligible quantities.

ELECTROPLATING

One of the most important applications of electrolytic cells is in the process of electroplating, in which a thin layer of metal (seldom as thick as 0.001 inch) is deposited on an electrically conducting surface. Electroplating is used for many different purposes. Sometimes the object is to increase the value or improve the appearance of an object, as is the case in gold and silver plating. Chromium plating is designed to provide an attractive shiny surface with improved wearing properties. Metals such as zinc or tin are plated over steel to protect it against corrosion. In Table 20.1, we list some of the components of electrolytic cells used to plate various metals.

TABLE 20.1   Electroplating Processes

METAL	ANODE	ELECTROLYTE	APPLICATION
Cu	Cu	20% $CuSO_4$, 3% H_2SO_4	electrotype
Ag	Ag	4% AgCN, 4% KCN, 4% K_2CO_3	tableware, jewelry
Au	Au, C, Ni-Cr	3% AuCN, 19% KCN, Na_3PO_4 buffer	jewelry
Cr	Pb	25% CrO_3, 0.25% H_2SO_4	automobile parts
Ni	Ni	30% $NiSO_4$, 2% $NiCl_2$, 1% H_3BO_3	
Zn	Zn	6% $Zn(CN)_2$, 5% NaCN, 4% NaOH, 1% Na_2CO_3, 0.5% $Al_2(SO_4)_3$	galvanized steel
Sn	Sn	8% H_2SO_4, 3% Sn, 10% cresol-sulfuric acid	"tin" cans

One of the simpler electroplating processes is that used with copper (Fig. 20.5). Here, as in most electroplating cells, the metal to be plated is used as the anode (Cu) and the electrolyte contains an ion (Cu^{2+}) derived from that metal. As copper is plated out at the cathode, it goes into solution at the anode

cathode: $\qquad\qquad\qquad$ $Cu^{2+}(aq) + 2\ e^- \rightarrow Cu(s)$
anode: $\qquad\qquad\qquad$ $Cu(s) \rightarrow Cu^{2+}(aq) + 2\ e^-$

thereby maintaining a constant concentration of Cu^{2+} in the electrolyte solution around the electrodes. Sulfuric acid is added to prevent the solution from becoming basic (Reaction 20.15b), which would contaminate the plated copper with such compounds as CuO or $Cu(OH)_2$. Traces of organic materials such as glue or gelatin are also added to the solution because it has been found empirically that their presence leads to a more adherent plate.

You will notice from Table 20.1 that metal cyanides are used in many electroplating processes. The CN^- ion acts as a complexing agent to lower the concentration of free metal ion. This tends to prevent the cation from plating too rapidly, which would give a rough or brittle coating. In silver plating the electrode reactions are

anode: $\qquad$ $Ag(s) + 2\ CN^-(aq) \rightarrow Ag(CN)_2^-(aq) + e^-$
cathode: $\qquad$ $Ag(CN)_2^-(aq) + e^- \rightarrow Ag(s) + 2\ CN^-(aq)$

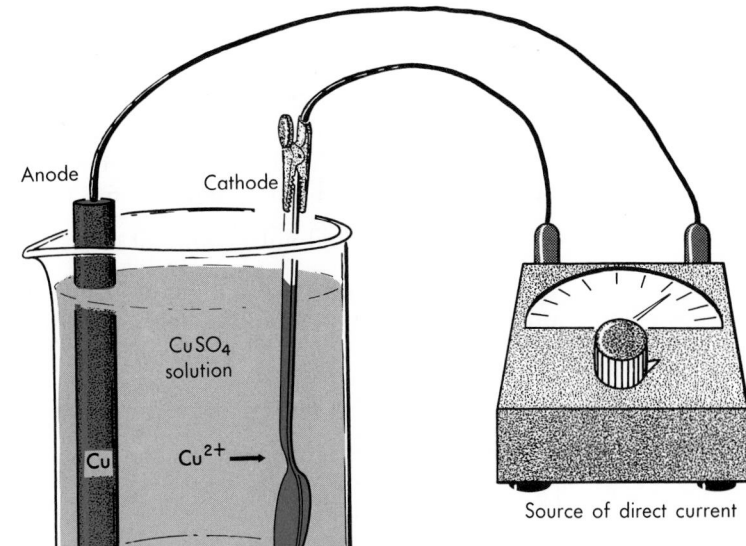

Figure 20.5 Electroplating of copper. Anode is made of copper to keep concentration of Cu^{2+} constant.

Anode

Cathode

$CuSO_4$ solution

Cu $\quad Cu^{2+} \longrightarrow$

Source of direct current

Object to be plated

Solutions containing cyanide ions are extremely toxic: many cases of water pollution have arisen from careless discharge of spent electrolyte by electroplating plants. The problem is magnified by the use of metal cyanides to extract silver and gold from low grade ores. Fish and livestock kills in western streams have occasionally been traced to cyanide solutions used in mining operations.

QUANTITATIVE ASPECTS OF ELECTROLYSIS

From an economic standpoint, one of the most important aspects of an electrochemical process is the relationship between the quantity of electricity passed through the cell and the amounts of substances produced by oxidation and reduction at the electrodes. The nature of this relationship is readily derived from the half-equation for the electrode process.

Half-equation	Quantity of Charge		Amount of Product
$Na^+ + e^- \rightarrow Na$	1 mole e^-	$\simeq$	1 mole Na $= 23.0$ g Na
$Mg^{2+} + 2\ e^- \rightarrow Mg$	2 moles e^-	$\simeq$	1 mole Mg $= 24.3$ g Mg
$Al^{3+} + 3\ e^- \rightarrow Al$	3 moles e^-	$\simeq$	1 mole Al $= 27.0$ g Al
$2\ Cl^- \rightarrow Cl_2 + 2\ e^-$	2 moles e^-	$\simeq$	1 mole $Cl_2 = 71.0$ g Cl

The quantity of electrical charge associated with one mole of electrons is given a special name—the **faraday**, after Michael Faraday, the English scientist who first studied the quantitative aspects of electrochemistry over a century ago. In practical units a faraday, to three significant figures, is equal to 96,500 coulombs.

$$1 \text{ mole of } e^- = 1 \text{ faraday} = 96{,}500 \text{ coulombs} \qquad (20.16)$$

We shall also have occasion to refer to a unit of current flow, the ampere. When a current of one ampere flows through an electrical circuit, one coulomb passes a given point in the circuit in one second. The number of coulombs passing through a cell can be calculated by multiplying the rate of flow in amperes by the elapsed time in seconds

$$\text{no. of coulombs} = (\text{no. of amperes}) \times (\text{no. of seconds}) \qquad (20.17)$$

The relationships which we have just presented can be used in many practical calculations dealing with electrolytic cells, as in Example 20.3.

Example 20.3   Chromium metal can be plated from an acidic solution containing CrO_3 (cf. Table 20.1).
  a. Write a balanced half-equation for the plating process.
  b. How many grams of chromium will be plated by 20,500 coulombs?
  c. How long will it take to plate one gram of chromium using a current of 10.0 amperes?

Solution
  a. The unbalanced half-equation is

$$CrO_3(aq) \rightarrow Cr(s)$$

Proceeding as indicated in Section 20.2, we first add three H_2O molecules to the right to balance oxygen. This requires that we add 6 H^+ to the left side of the equation to balance hydrogen. Finally, to balance the charge, we add 6 e^- to the left.

$$CrO_3(aq) + 6 \text{ } H^+(aq) + 6 \text{ } e^- \rightarrow Cr(s) + 3 \text{ } H_2O$$

  b. From the coefficients of the balanced half-equation, we see that 6 moles of electrons (6 faradays) are equivalent to 1 mole of chromium (52.0 g Cr)

$$6 \text{ faradays} \simeq 52.0 \text{ g Cr}$$

To find the number of grams of chromium plated, we need only convert coulombs to faradays (1 faraday = 96,500 coulombs) and then to grams of chromium (52.0 g Cr $\simeq$ 6 faradays).

$$\text{no. g Cr} = 20{,}500 \text{ coulombs} \times \frac{1 \text{ faraday}}{96{,}500 \text{ coulombs}} \times \frac{52.0 \text{ g Cr}}{6 \text{ faradays}}$$
$$= 1.84 \text{ g Cr}$$

  c. The indicated path here is to first convert grams of chromium to faradays and then faradays to coulombs. From the number of coulombs and the number of amperes (10.0), we can readily calculate the time in seconds, using Equation 20.17.

$$\text{no. of coulombs} = 1.00 \text{ g Cr} \times \frac{6 \text{ faradays}}{52.0 \text{ g Cr}} \times \frac{96{,}500 \text{ coulombs}}{1 \text{ faraday}}$$
$$= 11{,}100 \text{ coulombs}$$

$$\text{no. of seconds} = \frac{\text{no. of coulombs}}{\text{no. of amperes}} = \frac{11{,}100}{10.0}$$
$$= 1110 \text{ sec, or about 18.5 min}$$

We should point out that in (b) and (c) we have, in effect, assumed a 100% yield of chromium in the electrolytic process. In practice we cannot expect to obtain this; some of the electrons will be wasted in such side reactions as

$$2 \; H^+(aq) + 2 \; e^- \rightarrow H_2(g)$$

As a result, if we were to carry out in the laboratory the process indicated in (b), we would obtain considerably less than 1.84 g of Cr. In (c) we would find that the time required to plate a gram of chromium would exceed 18.5 minutes.

The *gram equivalent weight* of a substance involved in an oxidation-reduction reaction is taken to be the weight in grams that is equivalent to one mole of electrons (1 faraday). If the half-equation for the oxidation or reduction is known, the gram equivalent weight can be calculated immediately. For example, for the process referred to in Example 20.3, we see from the half-equation

The GEW of an element was defined by early chemists as that mass which combined with 8 g O_2. Is this consistent with the definition cited here?

$$CrO_3(aq) + 6 \; H^+(aq) + 6 \; e^- \rightarrow Cr(s) + 3 \; H_2O$$

$$6 \text{ faradays} \simeq 52.0 \text{ g Cr}$$

that the gram equivalent weight of chromium must be 52.0 g/6 = 8.67 g. In the laboratory, gram equivalent weight can be determined by measuring the mass of a substance produced by a known quantity of electricity (Example 20.4).

Example 20.4 In the electrolysis of a water solution of indium sulfate, it is found that 5.71 g of In is plated out when a current of 4.00 amperes is applied for one hour. Assuming a 100% yield of indium, what is its gram equivalent weight?

Solution. Let us first find the number of coulombs consumed, convert this to faradays, and then calculate how much indium would be produced by one faraday.

What is the GEW of Zn^{2+} in a reaction where it is reduced to the metal?

no. of coulombs = (no. of amperes) × (no. of seconds) = 4.00 × 3600 = 14,400

$$\text{no. of faradays} = \frac{14,400}{96,500} = 0.149$$

i.e., 0.149 faradays $\simeq$ 5.71 g In

$$\text{GEW In} = 1 \text{ faraday} \times \frac{5.71 \text{ g In}}{0.149 \text{ faraday}} = 38.3 \text{ g In}$$

You will notice that the gram equivalent weight of indium is almost exactly 1/3 of its gram atomic weight (AW In = 114.8). This implies that the indium cation in solution carries a +3 charge, i.e., that the half-equation for its reduction is

$$In^{3+}(aq) + 3 \; e^- \rightarrow In(s)$$

Experiments of this sort offer one of the simplest means of determining the charge of a simple monatomic ion in aqueous solution or, by implication, in a solid ionic compound.

20.4 VOLTAIC CELLS

In an electrolytic cell electrical energy is supplied to bring about a nonspontaneous oxidation-reduction reaction. A voltaic (galvanic) cell is designed to achieve the opposite effect; a spontaneous redox reaction serves as a source of electrical energy. All of us are familiar with certain types of voltaic cells, such as the "dry cell" that is used in flashlights and the lead storage battery that supplies electrical energy to an automobile. To understand how a voltaic cell operates, we shall consider first some very simple cells which are easily constructed in the general chemistry laboratory.

SIMPLE VOLTAIC CELLS, THE
Zn-Cu²⁺ CELL

When a piece of zinc is added to a beaker containing a water solution of copper sulfate, a spontaneous oxidation-reduction reaction takes place:

$$Zn(s) + Cu^{2+}(aq) \rightarrow Zn^{2+}(aq) + Cu(s) \qquad (20.18)$$

Experimentally we observe that a spongy, reddish brown deposit of copper forms on the surface of the zinc; the blue color of the aquated Cu^{2+} ion fades as it is replaced by the colorless Zn^{2+} ion. The temperature of the solution rises as a result of the heat evolved. From the equation we see that the reaction amounts to electron transfer from a zinc atom to a Cu^{2+} ion.

To design a cell that uses Reaction 20.18 as a source of electrical energy, we must make the electron transfer occur indirectly; that is, the electrons given up by zinc atoms must pass through a circuit and do electrical work before they reduce Cu^{2+} ions to copper atoms. Two different ways of accomplishing this are shown in Figure 20.6. Let us first concentrate on the salt-bridge cell shown at the left of the figure and trace the flow of electrical current through it.

1. At the zinc *anode*, electrons are produced by the *oxidation* half-reaction

$$Zn(s) \rightarrow Zn^{2+}(aq) + 2\ e^- \qquad (20.18a)$$

This electrode, which "pumps" electrons into the external circuit, is ordinarily marked as the negative pole of the cell.

2. Electrons generated by Reaction 20.18a move through the external circuit (left to right in Figure 20.6A). This part of the circuit may be a simple resistance wire, a light bulb, an electric motor, an electrolytic cell, or some other device that consumes electrical energy.

3. Electrons pass from the external circuit to the copper *cathode* where they are used in the *reduction* of Cu^{2+} ions in the surrounding solution

$$Cu^{2+}(aq) + 2\ e^- \rightarrow Cu(s) \qquad (20.18b)$$

The copper electrode, which "pulls" electrons from the external circuit, is considered to be the positive pole of the cell.

4. To complete the circuit, ions must move through the aqueous solutions in the cell. As Reactions 20.18a and 20.18b proceed, a surplus of positive ions (Zn^{2+}) tends to build up around the zinc electrode. The region

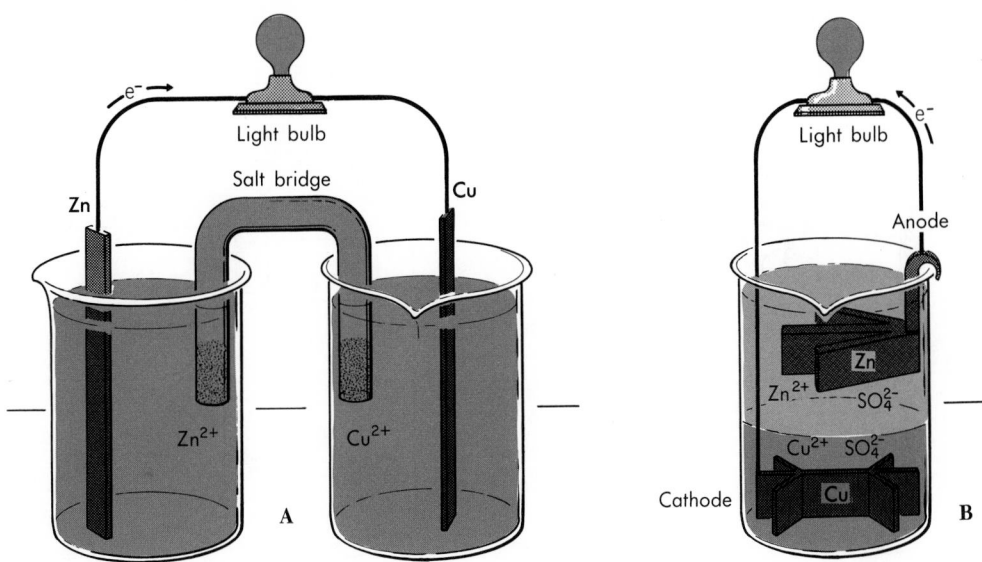

Figure 20.6 Zinc-Cu²⁺ voltaic cell. Salt bridge allows for the transfer of ions but prevents Cu^{2+} ions from coming in contact with zinc.

around the copper electrode tends to become deficient in positive ions as Cu^{2+} ions are discharged. To maintain electrical neutrality, cations must move toward the copper cathode or, alternatively, anions must move toward the zinc anode. In practice, both migrations occur.

In the cell shown in Figure 20.6A, movement of ions occurs through a *salt bridge.* In its simplest form a salt bridge may consist of an inverted U-tube, plugged with glass wool at each end. The tube is filled with a solution of a salt which takes no part in the electrode reactions; potassium nitrate, KNO_3, is frequently used. The salt bridge allows for the passage of ions but prevents direct contact between Cu^{2+} ions and the zinc electrode. If such contact were to occur, Reaction 20.18 would take place directly at the surface of the zinc and the cell would be short-circuited, making it incapable of producing electrical work.

Another way of preventing Cu^{2+} ions from coming in contact with the zinc electrode is illustrated in the so-called gravity cell shown in Figure 20.6B. To form this cell, enough copper sulfate solution is added to the jar to cover the copper electrode. A more dilute, less dense solution of zinc sulfate is then carefully poured over the copper sulfate. So long as the cell is not subjected to vibrations, the boundary between the layers may be maintained over long periods of time. Cells of this design were once used extensively to operate telegraph relays, doorbells, and other stationary electrical apparatus. Since their internal resistance is much lower than that of salt-bridge cells, much larger currents can be drawn from them.

In this cell a small current flow stabilizes the boundary between layers

Cells similar in design to that shown in Figure 20.6A can be set up to take advantage of the spontaneity of many different oxidation-reduction reactions. We can, for example, devise cells (Figure 20.7) in which the following reactions serve as a source of electrical energy.

$$Ni(s) + Cu^{2+}(aq) \rightarrow Ni^{2+}(aq) + Cu(s) \qquad (20.19)$$

$$Zn(s) + 2\,H^+(aq) \rightarrow Zn^{2+}(aq) + H_2(g) \qquad (20.20)$$

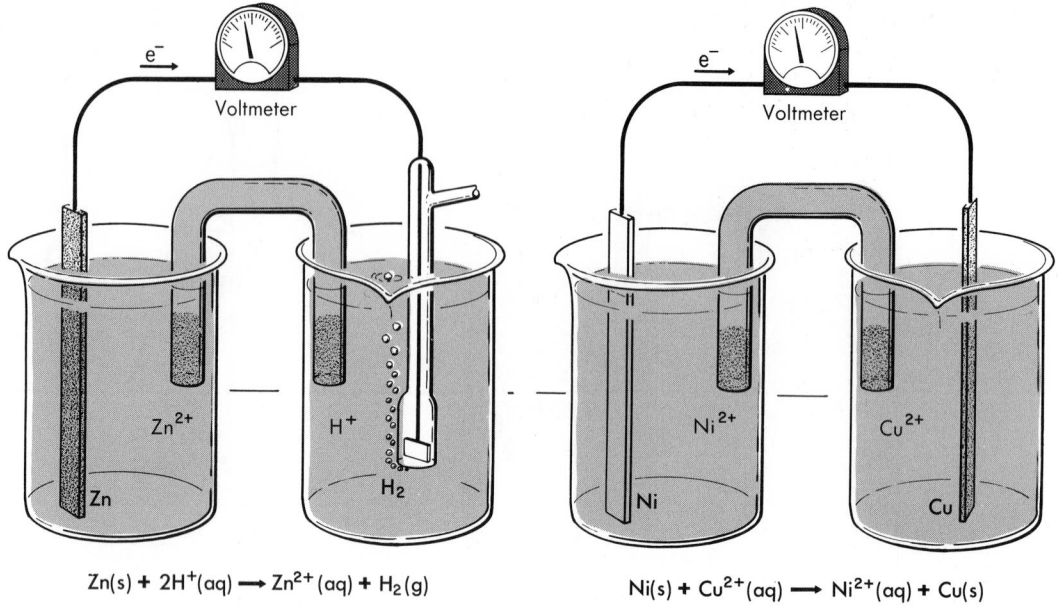

$$Zn(s) + 2H^+(aq) \longrightarrow Zn^{2+}(aq) + H_2(g) \qquad Ni(s) + Cu^{2+}(aq) \longrightarrow Ni^{2+}(aq) + Cu(s)$$

Figure 20.7 Salt bridge cells, Zn-H⁺ and Ni-Cu²⁺, using Reactions 20.19 and 20.20.

Which electrodes are (−) in these cells?

In each case the apparatus consists of two half cells, each containing an electrode dipping into a solution of an appropriate electrolyte (e.g., Ni in NiSO₄). In the H_2-H^+ half cell, hydrogen gas is bubbled over a specially prepared platinum electrode. At the anode of each cell, atoms of the element having the greatest tendency to lose electrons (Ni, Zn) are oxidized. The electrons released travel through the external circuit to the cathode (Cu, Pt) where they reduce the cations (Cu^{2+}, H^+) around that electrode. The circuit is completed by the movement of ions through the salt bridge; cations move to the cathode, anions to the anode.

COMMERCIAL CELLS

As we shall see in Chapter 21, salt-bridge cells can provide us with some very practical information concerning the spontaneity and extent of oxidation-reduction reactants. However, their high internal resistance makes them unsuitable for commercial use. If we attempt to draw an appreciable amount of current from a salt-bridge cell, its voltage drops off sharply. There are on the market today many different types of voltaic cells, all of which are capable of supplying a comparatively large current, at least for a short time.

DRY (LECLANCHÉ) CELL. The construction of the ordinary dry cell used in flashlights and small appliances is shown in Figure 20.8. The zinc wall of the cell serves as the anode; the graphite rod passing through the center of the cell is the cathode. The space between the electrodes is filled with a moist paste containing manganese dioxide, carbon black, and ammonium chloride. When the cell is being used to generate energy, the half-reaction at the anode is

$$Zn(s) \rightarrow Zn^{2+}(aq) + 2\ e^- \qquad (20.21a)$$

At the cathode, manganese dioxide is reduced to a variety of species in which manganese is in the +3 oxidation state. These include Mn_2O_3, $MnO(OH)$, and $Mn_2O_4^{2-}$. For simplicity, we shall write the half-reaction as:

$$MnO_2(s) + 4\ NH_4^+(aq) + e^- \rightarrow Mn^{3+}(aq) + 2\ H_2O + 4\ NH_3(aq) \quad (20.21b)$$

For the overall reaction,

$$Zn(s) + 2\ MnO_2(s) + 8\ NH_4^+(aq) \rightarrow Zn^{2+}(aq) + 2\ Mn^{3+}(aq) + 8\ NH_3(aq) + 4\ H_2O \quad (20.21)$$

If too large a current is drawn from the cell, the ammonia formed by Reaction 20.21b forms a gaseous, insulating layer around the carbon electrode. In normal operation this condition is prevented by the migration of Zn^{2+} ions to the cathode, where they react with ammonia molecules to form complex ions such as $Zn(NH_3)_4^{2+}$, $Zn(NH_3)_2(H_2O)_2^{2+}$, and so on.

LEAD STORAGE BATTERY. The 12-volt storage battery used in automobiles consists of six voltaic cells of the type shown in Figure 20.9 connected in series. A group of lead plates, the grids of which are filled with spongy, gray lead, forms the anode of the cell. The multiple cathode consists of a group of plates of similar design filled with lead dioxide. These two series of plates, alternating with each other throughout the cell, are immersed in a water solution of sulfuric acid, which acts as the electrolyte.

When a lead storage battery is supplying current, the lead in the anode grids is oxidized to Pb^{2+} ions, which immediately precipitate on the plates as lead sulfate, $PbSO_4$. At the cathode the lead dioxide is reduced to Pb^{2+} ions, which also precipitate as $PbSO_4$.

$$\begin{aligned}
Pb(s) + SO_4^{2-}(aq) &\rightarrow PbSO_4(s) + 2\ e^- & (20.22a) \\
\underline{PbO_2(s) + 4\ H^+(aq) + SO_4^{2-}(aq) + 2\ e^- \rightarrow PbSO_4(s) + 2\ H_2O} & & (20.22b) \\
Pb(s) + PbO_2(s) + 4\ H^+(aq) + 2\ SO_4^{2-}(aq) \rightarrow 2\ PbSO_4(s) + 2\ H_2O & & (20.22)
\end{aligned}$$

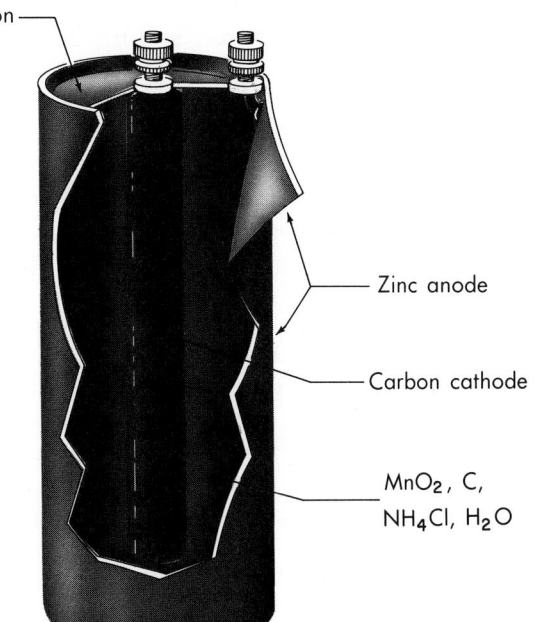

Insulation

Zinc anode

Carbon cathode

MnO_2, C, NH_4Cl, H_2O

Figure 20.8 Section of Zn-MnO₂ dry cell.

A dry cell like this has a potential of 1.5 volts and will deliver a current of 0.5 amperes for about 6 hours

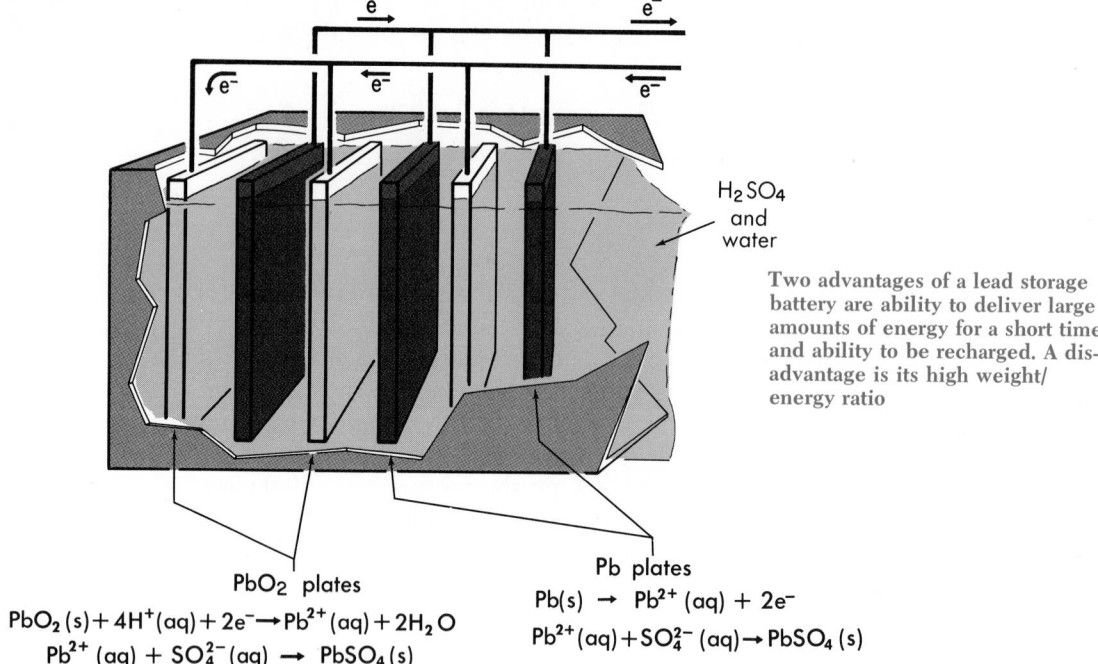

H₂SO₄
and
water

Two advantages of a lead storage
battery are ability to deliver large
amounts of energy for a short time
and ability to be recharged. A disadvantage is its high weight/
energy ratio

PbO₂ plates

$$PbO_2(s) + 4H^+(aq) + 2e^- \rightarrow Pb^{2+}(aq) + 2H_2O$$
$$Pb^{2+}(aq) + SO_4^{2-}(aq) \rightarrow PbSO_4(s)$$

Pb plates

$$Pb(s) \rightarrow Pb^{2+}(aq) + 2e^-$$
$$Pb^{2+}(aq) + SO_4^{2-}(aq) \rightarrow PbSO_4(s)$$

Figure 20.9 Lead storage battery.

Deposits of lead sulfate formed by Reactions 20.22a and 20.22b slowly
build up on the plates, partially covering and replacing the lead and lead
dioxide. As the cell discharges, the concentration of sulfuric acid decreases;
for every mole of lead reacting, two moles of H_2SO_4 (4 H^+, 2 SO_4^{2-}) are replaced by two moles of water. The state of charge of a storage battery can
be checked by measuring the density of the electrolyte. A low density indicates a low sulfuric acid concentration and, hence, a partially discharged
cell.

A lead storage battery, unlike an ordinary dry cell, can be restored to
its original condition by passing a direct current through it in the reverse
direction. While a storage battery is being charged, it acts as an electrolytic
cell; the half-reactions represented by equations 20.22a and 20.22b are reversed:

$$2\ PbSO_4(s) + 2\ H_2O \rightarrow Pb(s) + PbO_2(s) + 4\ H^+(aq) + 2\ SO_4^{2-}(aq) \quad (20.23)$$

The electrical energy required to bring about Reaction 20.23 may be
furnished by a direct-current generator, as in older automobiles, or by an
alternator equipped with a rectifier (in modern cars).

FUEL CELLS. The commercial voltaic cells that we have discussed
are useful for special purposes where the relatively high cost of such
chemicals as Zn, MnO_2, Pb, and PbO_2 is not a crucial factor. From a theoretical point of view, they are interesting in that they convert chemical
energy very efficiently into electrical energy. You may recall from Chapter
12 that the free energy change is a measure of the useful work that can be
obtained from a reaction. In a well-designed voltaic cell, the electrical
work approaches $-\Delta G$ for the cell reaction.

Most of our electrical energy today is produced by generators driven by steam turbines operating on the heat produced by combustion of coal, oil, or natural gas. Here, the conversion of chemical into electrical energy is indirect in that the chemical energy is first converted to heat which is then used to make steam. The indirect process is both theoretically and practically less efficient than the direct conversion that occurs in a voltaic cell. The best power plants convert only about 30 to 40 per cent of the heat of combustion of a fuel into electrical energy. The remainder is dissipated to the air or to bodies of water, where it contributes to thermal pollution.

Scientists and engineers have long speculated on the possibility of converting the chemical energy of fuels directly to electrical energy in a type of voltaic cell known as a fuel cell. In principle there is no reason why this cannot be done. The combustion of a fuel, like the reaction of Zn with Cu^{2+}, is a spontaneous oxidation-reduction reaction and, hence, should serve as a source of electrical energy. To illustrate the intriguing possibilities of a fuel cell, consider the reaction

$$CH_4(g) + 2\ O_2(g) \rightarrow CO_2(g) + 2\ H_2O(l); \quad \Delta H = -213\ kcal; \quad \Delta G^{1\ atm} = -196\ kcal$$

If this reaction is used as a source of heat in a conventional power plant, we are unlikely to obtain more than about 70 kcal of electrical work per mole of methane burned. In contrast, in a fuel cell we could, in principle, extract as much as 196 kcal of electrical work per mole of methane reacting $(-\Delta G^{1\ atm})$. A more realistic estimate might be 176 kcal/mole CH_4, assuming 90 per cent efficiency. Looking at the data in Table 20.2, we see two advantages of the fuel cell. Not only does it produce about 2.5 times as much electrical energy for a given amount of fuel; perhaps more important, the waste heat, which contributes to thermal pollution, is reduced by a factor of about four.

Clearly, oxidizing methane in a fuel cell, at least in principle, is the method of choice for generating electrical power. A cell utilizing this reaction has been designed and is shown in Figure 12.1, p. 324. Molten potassium hydroxide is used as an electrolyte, which means that the temperature must be held at 500°C or higher.

One of the more successful fuel cells is the H_2—O_2 cell shown in Figure 20.10. The two gases enter the cell through porous carbon electrodes surrounded by an aqueous solution of potassium hydroxide. The half-reactions at the electrodes are

anode:	$H_2(g) + 2\ OH^-(aq) \rightarrow 2\ H_2O + 2\ e^-$	(20.24a)
cathode:	$\frac{1}{2}\ O_2(g) + H_2O + 2\ e^- \rightarrow 2\ OH^-(aq)$	(20.24b)
	$H_2(g) + \frac{1}{2}\ O_2(g) \rightarrow 2\ H_2O$	(20.24)

TABLE 20.2   Energy Conversion in a Fuel Cell
vs a Steam Turbine

$$CH_4(g) + 2\ O_2(g) \rightarrow CO_2(g) + 2\ H_2O(l); \quad \Delta H = -213\ kcal$$

	ELECTRICAL ENERGY/ MOLE CH_4	WASTE HEAT/ MOLE CH_4
Steam turbine	70 kcal	143 kcal
Fuel cell	176 kcal	37 kcal

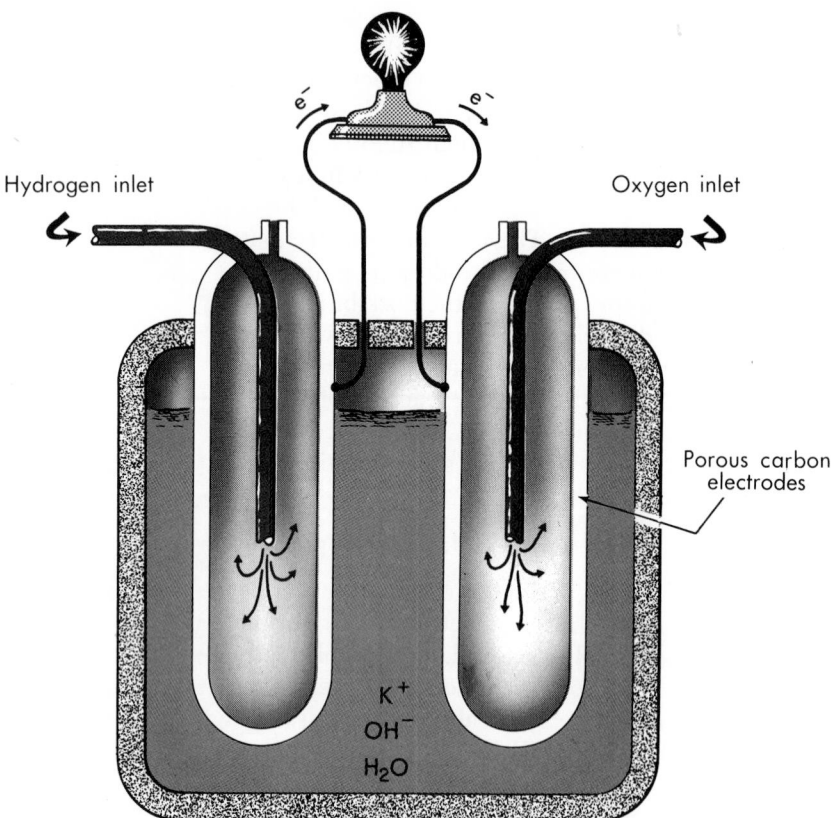

Although several fuel cells of this sort have been developed in recent years, none as yet have proved to be practical for ordinary uses

Figure 20.10 Hydrogen-oxygen fuel cell.

A fuel cell of this type, suitably modified to operate at zero gravity, was used as a source of electrical energy in the Apollo spacecraft.

In spite of extensive industrial research on these devices, we must admit that at present we have no commercially practical high-energy fuel cells. The prototypes that have been made have been subject to all sorts of technical problems, ranging from leaks to excessive corrosion to catalytic failures. No large-capacity cells have been produced, and in the small cells, slow reaction rates have been a limiting factor. At the present time, the development of a practical fuel cell in the moderate or high energy range is one of the most important needs of our society. Fuel cells offer one of the best methods of making more efficient use of fossil fuels and, at the same time, greatly decreasing thermal pollution from power generation.

PROBLEMS

20.1 Give the oxidation number of each atom in

    a. MoO_3
    b. Na_3VO_4
    c. CH_4O
    d. $Mg(ClO_4)_2$
    e. O_2^-
    f. $Cr_2O_7^{2-}$
    g. CoF_6^{3-}

20.21 Give the oxidation number of each atom in

    a. H_2O_2
    b. H_2SO_3
    c. $S_2O_3^{2-}$
    d. $Pt(NH_3)_2Cl_2$
    e. O_2^{2-}
    f. Na_2MnO_4
    g. $ZnWO_4$

20.2   List a neutral species and an oxyanion in which nitrogen shows an oxidation state of +5; +3; +1.

20.3   For each of the following reactions

$$Ni(s) + Cl_2(g) \rightarrow Ni^{2+}(aq) + 2\ Cl^-(aq)$$
$$3\ Cl_2(aq) + 6\ OH^-(aq) \rightarrow$$
$$5\ Cl^-(aq) + ClO_3^-(aq) + 3\ H_2O$$

identify:

    a. the element that is oxidized.
    b. the element that is reduced.
    c. the oxidizing agent.
    d. the reducing agent.

20.4   Using Figure 20.1, predict the formulas of six different compounds which contain two or all three of the elements, Na, Cl, and O (use only an oxidation number of −2 for oxygen).

20.5   Consider the following unbalanced half-equations, all for reactions occurring in acidic solution

$$Zn(s) \rightarrow Zn^{2+}(aq)$$
$$SO_4^{2-}(aq) \rightarrow SO_2(g)$$
$$Sn^{4+}(aq) \rightarrow Sn^{2+}(aq)$$
$$NO(g) \rightarrow NO_3^-(aq)$$

    a. Classify each of these as oxidation or reduction half-reactions.
    b. Balance each half-equation.
    c. Write as many balanced redox equations as possible by combining these half-equations.

20.6   Follow the directions of Problem 20.5 for the following unbalanced half-equations which represent reactions taking place in basic solution

$$S(s) \rightarrow S^{2-}(aq)$$
$$NO(g)(aq) \rightarrow NO_3^-(aq)$$
$$ClO^-(aq) \rightarrow Cl^-(aq)$$
$$Cu(OH)_2(s) \rightarrow Cu(s)$$

20.7   Write balanced equations for each of the following redox reactions

a. $Cu(s) + NO_3^-(aq) \rightarrow Cu^{2+}(aq) + NO(g)$ (acid)
b. $O_2(g) + S^{2-}(aq) \rightarrow S(s)$ (base)
c. $H_2O_2(aq) + Cr_2O_7^{2-}(aq) \rightarrow Cr^{3+}(aq) + O_2(g)$ (acid)
d. $Bi_2S_3(s) + NO_3^-(aq) \rightarrow Bi^{3+}(aq) + S(s) + NO_2(g)$ (acid)
e. $Cl_2(g) \rightarrow ClO_4^-(aq) + Cl^-(aq)$ (base)

20.8   Write a balanced equation to represent the reaction that takes place when gold is treated with aqua regia, a mixture of hydrochloric and nitric acids. The solution takes on a deep brown color due to the formation of $NO_2(g)$; the gold is converted to the $AuCl_4^-$ ion.

20.22   Consider the oxidation states of sulfur listed in Figure 20.1. Give the formula of a molecule and an ion in each state.

20.23   Using Figure 20.1, decide whether the following species can, in redox reactions, act only as oxidizing agents, only as reducing agents, or may be either oxidized or reduced.

    a. SO_3^{2-} d. H_2S
    b. Zn^{2+} e. CrO_4^{2-}
    c. F^- f. Cr^{3+}

20.24   Write the formulas of as many compounds as you can which contain no elements other than potassium, sulfur, and fluorine. (Consider only binary compounds.)

20.25   Consider the following half-equations for reactions in acidic solution

$$Cd^{2+}(aq) \rightarrow Cd(s)$$
$$Pb(s) + SO_4^{2-}(aq) \rightarrow PbSO_4(s)$$
$$Cr_2O_7^{2-}(aq) \rightarrow Cr^{3+}(aq)$$
$$Cl^-(aq) \rightarrow Cl_2(g)$$

    a. Classify each of these as oxidation or reduction processes.
    b. Balance each half-equation.
    c. Write as many balanced redox equations as possible by combining these half-equations.

20.26   Follow the directions of Problem 20.25 for the following unbalanced half-equations in basic solution

$$Cl_2(g) \rightarrow Cl^-(aq)$$
$$Cl_2(g) \rightarrow ClO_3^-(aq)$$
$$O_2(g) \rightarrow OH^-(aq)$$
$$Ag(s) \rightarrow Ag_2O(s)$$

20.27   Write balanced equations for each of the following redox reactions

a. $Pb(s) + PbO_2(s) + SO_4^{2-}(aq) \rightarrow PbSO_4(s)$ (acid)
b. $Se(s) + NO_3^-(aq) \rightarrow SeO_2(s) + NO(g)$ (acid)
c. $Sn^{2+}(aq) + O_2(g) \rightarrow SnO_2(s)$ (acid)
d. $Al(s) + H_2O \rightarrow Al(OH)_6^{3-}(aq) + H_2(g)$ (base)
e. $MnO_4^-(aq) + CN^-(aq) \rightarrow MnO_2(s) + CNO^-(aq)$ (base)

20.28   When air is bubbled through an aqueous solution of $CoCl_2$ in the presence of ammonia, the complex ion $[Co(NH_3)_6]^{3+}$ is formed. Write balanced equations for the oxidation and reduction reactions and for the overall reaction.

20.9 Chlorine is often generated in the laboratory by reacting MnO_2 with hydrochloric acid; the manganese is reduced to the +2 state.

    a. Write a balanced equation for this reaction.

    b. How many grams of MnO_2 are required to form 12.0 g of Cl_2?

    c. What volume of concentrated hydrochloric acid (12.0 M) is required to form one liter of $Cl_2(g)$, measured at 25°C and one atm?

20.10 When a sample of an alloy containing tin is treated with nitric acid, the tin is oxidized to $SnO_2(s)$; the NO_3^- ion is reduced to $NO(g)$.

    a. Write a balanced equation for this reaction.

    b. A sample of alloy weighing 1.000 g gives 0.382 g of SnO_2. What is the percentage of tin in the sample?

    c. How many moles of H^+ and NO_3^- are consumed in (b)?

20.11 Consider the reaction: $Co^{2+}(aq) + 2\ Cl^-(aq) \rightarrow Co(s) + Cl_2(g)$, for which $\Delta G = +75.6$ kcal.

    a. Could this reaction occur in an electrolytic cell? a voltaic cell?

    b. Sketch a cell in which this reaction could take place, label the anode and cathode, and indicate the flow of current throughout the circuit.

20.12 Explain the purpose of

    a. the screen shown around the electrodes in Figure 20.3.

    b. adding cryolite in the electrolysis of aluminum oxide.

    c. using mercury in the cell for the electrolysis of aqueous sodium chloride.

    d. the CN^- ions used in the electroplating of Ag and Au.

20.13 A certain cell is producing aluminum from Al_2O_3 at the rate of 125 g/day. Assuming a yield of 100%,

    a. how many electrons must pass through the cell in one day?

    b. what is the current passing through the cell?

    c. how much oxygen is being produced simultaneously?

20.29 When a solution of sodium iodide is treated with sulfuric acid, elementary iodine and $SO_2(g)$ are formed.

    a. Write a balanced equation for this reaction.

    b. How many ml of 3.0 M H_2SO_4 are required to form one gram of iodine?

    c. In (b), how many ml of SO_2, at 20°C and 720 mm Hg, are formed simultaneously?

20.30 In one type of CH_4—O_2 fuel cell, the electrolyte is molten KOH. At the anode, CH_4 is oxidized to CO_3^{2-}; at the cathode, O_2 is reduced to OH^-.

    a. Write balanced half-equations for the two electrode reactions and combine them to get the overall cell reaction.

    b. How many liters of O_2, at 22°C and 718 mm Hg, are required to react with two liters of CH_4, measured at the same temperature and pressure?

20.31 Consider the equilibrium

$$Ni^{2+}(aq) + Pb(s) \rightleftharpoons Ni(s) + Pb^{2+}(aq)$$

ΔG^{1M} for the forward reaction is +5.5 kcal. If a voltaic cell is set up, using the two metals in contact with 1 M solutions of their ions, which way will the reaction proceed? How could you show experimentally that the reaction in the voltaic cell goes in the direction you have predicted?

20.32 What changes would be observed if

    a. $MgCl_2$ were substituted for NaCl in the Downs cell?

    b. the density of the molten electrolyte in the Hall process were greater than that of aluminum?

    c. a solution of sodium chloride were made acidic before electrolysis?

    d. the copper anode in a copper-plating cell were replaced by zinc?

20.33 In the electrolysis of an aqueous solution of sodium chloride, 1.45×10^{22} electrons pass through a cell.

    a. How many faradays are used?

    b. How many coulombs does this represent?

    c. How many grams of H_2, Cl_2, and OH^- are produced, assuming 100% yield?

20.14 A cell used to electrolyze sodium chloride solution is filled with 1.00 liter of 2.00 M NaCl. A current of five amperes is passed through this solution for eight hours. Assuming that the only reaction that occurs is 20.15,

    a. what volume of $Cl_2(g)$ is produced at 25°C and one atm?

    b. what fraction of the Cl^- ions has been converted to OH^- ions?

20.15 A spoon with an area of four cm² is plated with silver from a $Ag(CN)_2^-$ solution to a depth of 0.0020 cm.

    a. What is the total mass of silver that must be plated (density = 10.5 g/cm³)?

    b. If the current efficiency is 60%, how long will it take to plate the spoon using a current of 0.205 amperes?

20.16 What are the gram equivalent weights of the underlined species in the unbalanced half-equations

    a. $\underline{Fe^{3+}} \rightarrow Fe^{2+}$

    b. $\underline{Au(CN)_4^-} \rightarrow \underline{Au}$

    c. $\underline{Cr_2O_7^{2-}} \rightarrow Cr^{3+}$

20.17 A sample of a certain rhodium salt deposits 5.97 g of Rh when 1.68×10^4 coulombs are passed through an electrolytic cell.

    a. Calculate the gram equivalent weight of rhodium.

    b. What is the charge on the rhodium cation?

20.18 Design salt-bridge cells in which the following reactions occur. In each case, label the anode and cathode and indicate the direction of flow of current through the circuit.

    a. $Ni(s) + Pb^{2+}(aq) \rightarrow Ni^{2+}(aq) + Pb(s)$

    b. $H_2(g) + Cl_2(g) \rightarrow 2\ H^+(aq) + 2\ Cl^-(aq)$

    c. $2\ Fe^{2+}(aq) + Cl_2(g) \rightarrow 2\ Fe^{3+}(aq) + 2\ Cl^-(aq)$

20.19 Explain why

    a. migration of Zn^{2+} ions in a dry cell can prevent a layer of gas building up around the cathode.

    b. sulfuric acid is used in the lead storage battery.

    c. fuel cells offer one approach to the problem of reducing thermal pollution.

20.34 A current of 5.00 amperes is drawn from a dry cell for a period of ten minutes. Assuming that the only reaction taking place is 20.21,

    a. how many grams of zinc are consumed?

    b. what volume of ammonia, at 25°C and one atm, would be produced in the gas state?

20.35 It is desired to plate a coin, which has a diameter of 1.05 in. and a thickness of 0.0500 in., with a layer of gold 0.0010 in. thick.

    a. How many grams of gold are required? (d = 19.3 g/cc.)

    b. How long will it take to plate the coin from AuCN, using a current of 0.100 ampere and assuming 100% yield?

20.36 What is the gram equivalent weight of nitrogen in each of the following processes?

    a. $NO_3^- \rightarrow N_2$

    b. $NO \rightarrow N_2$

    c. $N_2 \rightarrow NH_3$

20.37 A solution of a platinum salt deposits 1.22 g of Pt when 2.40×10^3 coulombs are passed through the cell. Assuming that the only reaction that occurs at the cathode is the reduction of a Pt-containing cation to the metal,

    a. calculate the GEW of platinum.

    b. determine the oxidation number of the platinum in the solution species.

20.38 Follow the directions of Problem 20.18 in sketching voltaic cells in which the following reactions occur.

    a. $Co(s) + 2\ Ag^+(aq)$
        $\rightarrow Co^{2+}(aq) + 2\ Ag(s)$

    b. $Pb(s) + PbO_2(s) + 4\ H^+(aq) + 2\ SO_4^{2-}(aq)$
        $\rightarrow 2\ PbSO_4(s) + 2\ H_2O$

    c. $Sn^{2+}(aq) + Cl_2(g)$
        $\rightarrow Sn^{4+}(aq) + 2\ Cl^-(aq)$

20.39 a. Suppose HCl were used rather than H_2SO_4 in a lead storage battery. Write balanced equations for the half-reactions at each electrode and the overall reaction.

    b. Suppose the reduction product in the dry cell were Mn^{2+} rather than Mn^{3+}. Write balanced equations for the half-reactions and the overall reaction.

20.20 The so-called nickel cadmium battery uses electrodes of Cd and NiO_2 and an electrolyte of KOH. Upon discharge, insoluble hydroxides of the two metals in their most common oxidation state (+2) build up on both electrodes. Write balanced half-equations for the reactions occurring at the two electrodes and combine them to give the equation for the overall cell reaction.

20.40 The reaction $Zn(s) + 2H^+(aq) \rightarrow Zn^{2+}(aq) + H_2(g)$ is spontaneous: the reaction $2\ Ag(s) + Cu^{2+}(aq) \rightarrow 2\ Ag^+(aq) + Cu(s)$ is nonspontaneous. Sketch a two-cell electrical system in which the first reaction could be harnessed to bring about the second. Label anodes and cathodes and indicate the flow of electrons throughout.

*20.41 Using appropriate conversion factors and the definition: 1 volt coulomb = 1 joule, calculate the value of the faraday in kcal/volt.

*20.42 Suggest a design for a fuel cell in which coal, which you may take to be pure carbon, is burned, using a metal oxide as an electrolyte. Using data in Table 12.2, Chapter 12, calculate the maximum amount of electrical energy that could be obtained from the combustion of one ton of coal in such a fuel cell.

*20.43 One kilogram of the silver electroplating solution listed in Table 20.1 is flushed into a water supply containing 2.0×10^5 kg of water. Assuming uniform mixing, what will be the resultant concentration of CN^- in parts per million (i.e., g CN^- per 10^6 g of water)?

21 OXIDATION-REDUCTION REACTIONS: SPONTANEITY AND EXTENT

Voltaic cells such as those described in Chapter 20 are of interest to chemists for reasons that go beyond their practical importance as a source of electrical energy. The property of a voltaic cell which is of particular interest to us is its voltage, or potential, which is a measure of the driving force behind the reaction occurring within the cell. By properly interpreting cell voltages, it is possible to obtain vital information as to the spontaneity and extent of oxidation-reduction reactions taking place in the general chemistry laboratory, in the industrial plant, or in the world around us. We can, for example, obtain answers to questions such as the following:

What reagent could be used to generate hydrogen from a dilute solution of hydrochloric acid? to prepare chlorine gas from sodium chloride?

How can we extract the elements bromine and magnesium from sea water, where they occur as Br^- and Mg^{2+} ions?

How can we inhibit the corrosion of iron or steel structures exposed to the atmosphere?

21.1 STANDARD POTENTIALS

In a properly designed cell, we find that the voltage * measured at a given temperature, let us say 25°C, depends upon two factors: the nature

---

* Strictly speaking, the quantity of interest is the *maximum* voltage, which is obtained when the cell reaction takes place reversibly at an infinitesimal rate. The voltage read on the simple type of voltmeter commonly used in the general chemistry laboratory is ordinarily less than this maximum value because a significant amount of current is drawn from the cell. To avoid this problem, we can use a voltmeter with a very high internal resistance. Alternatively, we can use an instrument called a potentiometer, which measures the voltage which must be applied to just prevent the cell reaction from occurring.

of the cell reaction, and the concentrations of the various species (ions or molecules) participating in the reaction. The *standard voltage* (E°) corresponding to a given cell reaction is that obtained when all such species have an activity of one. This condition is approximately fulfilled when all ions or molecules in solution are at a concentration of 1 M and all gases at a partial pressure of 1 atmosphere. To illustrate, consider the Zn-H⁺ cell shown in Figure 20.7. We find that when the pressure of hydrogen gas is 1 atm and the concentrations of both Zn^{2+} and H⁺ are 1 M, the cell voltage is about +0.76 V. This quantity, +0.76 V, is referred to as the standard cell voltage.

$Zn(s) \rightarrow$
$Zn^{2+}(1 \text{ M}) + 2 \text{ e}^-$
$E_1^\circ = \text{SOP Zn}$
$2 \text{ H}^+(1 \text{ M}) + 2 \text{ e}^- \rightarrow$
$H_2(1 \text{ atm})$
$E_2^\circ = \text{SRP H}^+$
$E_{cell}^\circ = E_1^\circ + E_2^\circ$

$$Zn(s) + 2 \text{ H}^+(aq, 1 \text{ M}) \rightarrow Zn^{2+}(aq, 1 \text{ M}) + H_2(g, 1 \text{ atm}); \quad E^\circ = +0.76 \text{ V} \quad (21.1)$$

Just as one can split a cell reaction such as 21.1 into two half-reactions of reduction and oxidation, so it is possible to divide a standard cell voltage into two parts, one corresponding to the reduction half-reaction, the other to the oxidation half-reaction. The potential corresponding to the reduction half-reaction is referred to as a **standard reduction potential**; associated with the oxidation half-reaction is a quantity which we shall call a **standard oxidation potential.**° The standard cell voltage is the sum of these two potentials.

By methods which we shall now consider, it is possible to establish a table of standard electrode potentials corresponding to a wide variety of oxidation and reduction half-reactions.

ASSIGNMENT OF POTENTIALS

To illustrate how standard potentials can be established, let us consider once again the Zn-H₂ cell. It has been pointed out that at standard concentrations, the voltage of this cell is +0.76 V and, further, that this quantity is the sum of the two half-cell potentials involved. That is,

$$E° = +0.76 \text{ V} = \text{SOP Zn} + \text{SRP H}^+ \quad (21.2)$$

It is experimentally impossible to determine the absolute value of either of the two quantities on the right-hand side of Equation 21.2. To establish a value for the standard oxidation potential of zinc, one must arbitrarily assign a value to the standard reduction potential of the H⁺ ion. It was agreed many years ago to take this value to be 0.00 V. That is,

$$2 \text{ H}^+(aq, 1 \text{ M}) + 2 \text{ e}^- \rightarrow H_2(g, 1 \text{ atm}) \qquad \text{SRP H}^+ = 0.00 \text{ V} \quad (21.3)$$

Substituting in Equation 21.2, it is clear that, on the basis of this convention, the standard oxidation potential of zinc must be +0.76 V

$$Zn(s) \rightarrow Zn^{2+}(aq, 1 \text{ M}) + 2 \text{ e}^- \qquad \text{SOP} = +0.76 \text{ V}$$

As soon as a few potentials have been established, it becomes relatively easy to determine others. Suppose, for example, we wish to deter-

---

° The International Union of Pure and Applied Chemistry has recently suggested that the term "standard electrode potential" be used to refer to what we have called the "standard reduction potential" and that the use of the term "standard oxidation potential" be abandoned.

mine the standard reduction potential of the Cu^{2+} ion. One way to do this is to set up the cell shown in Figure 20.6, p. 555, using 1 M solutions of Zn^{2+} and Cu^{2+}. When it is found that the voltage of this cell is 1.10 volts, it follows that the standard reduction potential of Cu^{2+} must be +0.34 V. That is,

$$SOP\ Zn + SRP\ Cu^{2+} = +1.10\ V$$
$$+0.76\ V + SRP\ Cu^{2+} = +1.10\ V$$
$$SRP\ Cu^{2+} = +0.34\ V$$

In Table 21.1 are listed standard reduction potentials for a series of species (Li^+, K^+, etc.) which are capable of being reduced. The standard oxidation potentials for the corresponding oxidation half-reactions can be obtained by reversing the sign. For example,

Since SRP $H^+ = 0.00$ V, SOP $H_2 = 0.00$ V

$$Li^+(aq) + e^- \rightarrow Li(s);\ \ SRP = -3.05\ V$$
$$Li(s) \rightarrow Li^+(aq) + e^-;\ \ SOP = +3.05\ V$$

In general, for any redox couple, *the standard potentials for the forward and reverse half-reactions (reduction and oxidation) are equal in magnitude but opposite in sign.*

CALCULATION OF CELL VOLTAGES

As the foregoing discussion implies, **the standard voltage of any cell is the algebraic sum of the standard oxidation potential of the species being oxidized in the cell reaction and the standard reduction potential of the species being reduced.** This simple relationship makes it possible, using Table 21.1, to calculate the standard voltages of more than 3000 different voltaic cells. Example 21.1 illustrates how this may be done for two specific cases.

Example 21.1   Calculate the voltages of cells in which the following reactions occur:

    a. $Cl_2(g,\ 1\ atm) + 2\ I^-(aq,\ 1\ M) \rightarrow 2\ Cl^-(aq,\ 1\ M) + I_2(s)$

    b. $MnO_4^-(aq,\ 1\ M) + 8\ H^+(aq,\ 1\ M) + 5\ Cl^-(aq,\ 1\ M) \rightarrow Mn^{2+}(aq,\ 1\ M) +$
       $4\ H_2O + \frac{5}{2}\ Cl_2(g,\ 1\ atm)$

Solution.   In each case we shall split the reaction into two half-reactions, tabulate the proper electrode potentials, and add to obtain the cell voltage.

    a. reduction: $Cl_2(g) + 2\ e^- \rightarrow 2\ Cl^-(aq)$ SRP $= +1.36$ V
       oxidation: $2\ I^-(aq) \rightarrow I_2(s) + 2\ e^-$ $\underline{SOP = -0.53\ V}$
                                        $E° = +0.83\ V$

    b. reduction:
       $MnO_4^-(aq) + 8\ H^+(aq) + 5\ e^-$
         $\rightarrow Mn^{2+}(aq) + 4\ H_2O$ SRP $= +1.52$ V
       oxidation: $2\ Cl^-(aq) \rightarrow Cl_2(g) + 2\ e^-$ $\underline{SOP = -1.36\ V}$
                                          $E° = +0.16\ V$

How could you set up a cell in which to study the reaction in Ex. 21.1a?

Standard potentials can be used in connection with electrolytic as well as voltaic cells. By adding the proper potentials, we can calculate the minimum applied voltage necessary, at standard concentrations, to bring about a nonspontaneous oxidation-reduction reaction in an electrolytic cell.

TABLE 21.1   Standard Reduction Potentials in Water Solution at 25°C

REDUCTION HALF-REACTION		STANDARD REDUCTION POTENTIAL (VOLTS)
$Li^+(aq) + e^-$	$\rightarrow Li(s)$	−3.05
$K^+(aq) + e^-$	$\rightarrow K(s)$	−2.93
$Ba^{2+}(aq) + 2\ e^-$	$\rightarrow Ba(s)$	−2.90
$Ca^{2+}(aq) + 2\ e^-$	$\rightarrow Ca(s)$	−2.87
$Na^+(aq) + e^-$	$\rightarrow Na(s)$	−2.71
$Mg^{2+}(aq) + 2\ e^-$	$\rightarrow Mg(s)$	−2.37
$Al^{3+}(aq) + 3\ e^-$	$\rightarrow Al(s)$	−1.66
$Mn^{2+}(aq) + 2\ e^-$	$\rightarrow Mn(s)$	−1.18
$Zn^{2+}(aq) + 2\ e^-$	$\rightarrow Zn(s)$	−0.76
$Cr^{3+}(aq) + 3\ e^-$	$\rightarrow Cr(s)$	−0.74
$Fe^{2+}(aq) + 2\ e^-$	$\rightarrow Fe(s)$	−0.44
$Cr^{3+}(aq) + e^-$	$\rightarrow Cr^{2+}(aq)$	−0.41
$Cd^{2+}(aq) + 2\ e^-$	$\rightarrow Cd(s)$	−0.40
$PbSO_4(s) + 2\ e^-$	$\rightarrow Pb(s) + SO_4^{2-}(aq)$	−0.36
$Tl^+(aq) + e^-$	$\rightarrow Tl(s)$	−0.34
$Co^{2+}(aq) + 2\ e^-$	$\rightarrow Co(s)$	−0.28
$Ni^{2+}(aq) + 2\ e^-$	$\rightarrow Ni(s)$	−0.25
$AgI(s) + e^-$	$\rightarrow Ag(s) + I^-(aq)$	−0.15
$Sn^{2+}(aq) + 2\ e^-$	$\rightarrow Sn(s)$	−0.14
$Pb^{2+}(aq) + 2\ e^-$	$\rightarrow Pb(s)$	−0.13
$2\ H^+(aq) + 2\ e^-$	$\rightarrow H_2(g)$	0.00
$AgBr(s) + e^-$	$\rightarrow Ag(s) + Br^-(aq)$	0.10
$S(s) + 2\ H^+(aq) + 2\ e^-$	$\rightarrow H_2S(aq)$	0.14
$Sn^{4+}(aq) + 2\ e^-$	$\rightarrow Sn^{2+}(aq)$	0.15
$Cu^{2+}(aq) + e^-$	$\rightarrow Cu^+(aq)$	0.15
$SO_4^{2-}(aq) + 4\ H^+(aq) + 2\ e^-$	$\rightarrow SO_2(g) + 2\ H_2O$	0.20
$Cu^{2+}(aq) + 2\ e^-$	$\rightarrow Cu(s)$	0.34
$Cu^+(aq) + e^-$	$\rightarrow Cu(s)$	0.52
$I_2(s) + 2\ e^-$	$\rightarrow 2\ I^-(aq)$	0.53
$Fe^{3+}(aq) + e^-$	$\rightarrow Fe^{2+}(aq)$	0.77
$Hg_2^{2+}(aq) + 2\ e^-$	$\rightarrow 2\ Hg(l)$	0.79
$Ag^+(aq) + e^-$	$\rightarrow Ag(s)$	0.80
$2\ Hg^{2+}(aq) + 2\ e^-$	$\rightarrow Hg_2^{2+}(aq)$	0.92
$NO_3^-(aq) + 4\ H^+(aq) + 3\ e^-$	$\rightarrow NO(g) + 2\ H_2O$	0.96
$AuCl_4^-(aq) + 3\ e^-$	$\rightarrow Au(s) + 4\ Cl^-(aq)$	1.00
$Br_2(l) + 2\ e^-$	$\rightarrow 2\ Br^-(aq)$	1.07
$O_2(g) + 4\ H^+(aq) + 4\ e^-$	$\rightarrow 2\ H_2O$	1.23
$MnO_2(s) + 4\ H^+(aq) + 2\ e^-$	$\rightarrow Mn^{2+}(aq) + 2\ H_2O$	1.23
$Cr_2O_7^{2-}(aq) + 14\ H^+(aq) + 6\ e^-$	$\rightarrow 2\ Cr^{3+}(aq) + 7\ H_2O$	1.33
$Cl_2(g) + 2\ e^-$	$\rightarrow 2\ Cl^-(aq)$	1.36
$ClO_3^-(aq) + 6\ H^+(aq) + 5\ e^-$	$\rightarrow \frac{1}{2}\ Cl_2(g) + 3\ H_2O$	1.47
$Au^{3+}(aq) + 3\ e^-$	$\rightarrow Au(s)$	1.50
$MnO_4^-(aq) + 8\ H^+(aq) + 5\ e^-$	$\rightarrow Mn^{2+}(aq) + 4\ H_2O$	1.52
$H_2O_2(aq) + 2\ H^+(aq) + 2\ e^-$	$\rightarrow 2\ H_2O$	1.77
$Co^{3+}(aq) + e^-$	$\rightarrow Co^{2+}(aq)$	1.82
$F_2(g) + 2\ e^-$	$\rightarrow 2\ F^-(aq)$	2.87

Table 21.1   *continued on opposite page.*

TABLE 21.1 (Concluded)

REDUCTION HALF-REACTION		STANDARD REDUCTION POTENTIAL (VOLTS)
BASIC SOLUTION		
$Zn(OH)_4{}^{2-}(aq) + 2\ e^-$	$\rightarrow Zn(s) + 4\ OH^-(aq)$	-1.22
$Fe(OH)_2(s) + 2\ e^-$	$\rightarrow Fe(s) + 2\ OH^-(aq)$	-0.88
$2\ H_2O + 2\ e^-$	$\rightarrow H_2(g) + 2\ OH^-(aq)$	-0.83
$Fe(OH)_3(s) + e^-$	$\rightarrow Fe(OH)_2(s) + OH^-(aq)$	-0.56
$S(s) + 2\ e^-$	$\rightarrow S^{2-}(aq)$	-0.48
$Cu(OH)_2(s) + 2\ e^-$	$\rightarrow Cu(s) + 2\ OH^-(aq)$	-0.36
$CrO_4{}^{2-}(aq) + 4\ H_2O + 3\ e^-$	$\rightarrow Cr(OH)_3(s) + 5\ OH^-(aq)$	-0.12
$NO_3{}^-(aq) + H_2O + 2\ e^-$	$\rightarrow NO_2{}^-(aq) + 2\ OH^-(aq)$	0.01
$Ag_2O(s) + H_2O + 2\ e^-$	$\rightarrow 2\ Ag(s) + 2\ OH^-(aq)$	0.34
$ClO_4{}^-(aq) + H_2O + 2\ e^-$	$\rightarrow ClO_3{}^-(aq) + 2\ OH^-(aq)$	0.36
$O_2(g) + 2\ H_2O + 4\ e^-$	$\rightarrow 4\ OH^-(aq)$	0.40
$ClO_3{}^-(aq) + 3\ H_2O + 6\ e^-$	$\rightarrow Cl^-(aq) + 6\ OH^-(aq)$	0.62
$ClO^-(aq) + H_2O + 2\ e^-$	$\rightarrow Cl^-(aq) + 2\ OH^-(aq)$	0.89

Example 21.2 Calculate the minimum voltage which must be applied in the electrolysis of an aqueous solution of sodium chloride (Reaction 20.15, Chapter 20), assuming standard concentrations.

Solution. Writing the electrode reactions with the appropriate standard potentials

$$2\ Cl^-(aq) \rightarrow Cl_2(g) + 2\ e^- \qquad SOP = -1.36\ V$$
$$2\ H_2O + 2\ e^- \rightarrow H_2(g) + 2\ OH^-(aq) \qquad SRP = \underline{-0.83\ V}$$
$$E° = -2.19\ V$$

$E°$ is negative for nonspontaneous reactions

It appears that in order to make this nonspontaneous reaction take place in an electrolytic cell, a voltage of at least 2.19 volts must be applied.

In practice we ordinarily find that the voltage required to operate an electrolytic cell is somewhat higher than that calculated from electrode potentials. The excess voltage, referred to as *overvoltage,* reflects the fact that the cell reaction is not taking place under reversible conditions. In order to make the reaction go at a reasonable rate, we have to supply a greater driving force (voltage). The overvoltage tends to be particularly large when one of the reaction products is a gas. For example, the formation of hydrogen by reduction of water at a mercury cathode is associated with an overvoltage of about 1.5 volts. This explains in part why, in the electrolysis of an aqueous solution of NaCl using a mercury cathode, Na^+ ions rather than H_2O molecules are reduced.

EASE OF REDUCTION AND OXIDATION

The standard potentials listed in Table 21.1 measure the relative tendencies of different species to be reduced or oxidized. The more positive the potential, the more spontaneous is the corresponding half-reaction. A large negative potential signifies a half-reaction that is difficult to bring about. As an example, consider three species chosen from the left column of the table:

$$Mg^{2+}(aq) + 2\ e^- \rightarrow Mg(s) \qquad SRP = -2.37\ V$$
$$2\ H_2O + 2\ e^- \rightarrow H_2(g) + 2\ OH^-(aq) \qquad SRP = -0.83\ V$$
$$Cu^{2+}(aq) + 2\ e^- \rightarrow Cu(s) \qquad SRP = +0.34\ V$$

The signs and magnitudes of the standard reduction potentials for these species tell us that ease of reduction increases in the order $Mg^{2+} < H_2O < Cu^{2+}$. This order is confirmed experimentally. When we pass a direct current through a solution of $CuSO_4$, as in an electroplating cell, Cu^{2+} ions rather than H_2O molecules are reduced. In contrast, magnesium metal cannot be prepared by electrolysis in aqueous solution: if we pass a direct current through a water solution of $MgCl_2$, H_2O molecules rather than Mg^{2+} ions are reduced. As another example, consider the three oxidizable species:

$$2\ I^-(aq) \rightarrow I_2(s) + 2\ e^- \qquad SOP = -0.53\ V$$
$$2\ H_2O \rightarrow O_2(g) + 4\ H^+(aq) + 2\ e^- \qquad SOP = -1.23\ V$$
$$2\ F^-(aq) \rightarrow F_2(g) + 2\ e^- \qquad SOP = -2.87\ V$$

As these potentials imply, iodine but not fluorine can be prepared by oxidation of the corresponding anion in water solution. If we attempt to prepare fluorine gas from aqueous sodium fluoride, either by electrolytic or chemical oxidation, we succeed only in oxidizing water molecules; O_2 rather than F_2 is the product.

What is the strongest oxidizing agent in the Table? the weakest reducing agent?

From a slightly different point of view, Table 21.1 tells us the relative strengths of various oxidizing and reducing agents. The stronger oxidizing agents are located at the lower left ($F_2 > Co^{3+} > H_2O_2 > MnO_4^-$). The halogens listed above fluorine (chlorine, bromine, and iodine) are successively less powerful oxidizing agents in that order. The species listed in the right-hand column are all capable of acting as reducing agents, at least in principle. The stronger reducing agents are those at the upper right, including the 1A and 2A metals, Al, and a few of the transition metals (Mn, Zn, and Fe, among others). Species such as Ag, Au, and Cl^-, which are located well down in the right-hand column, hold on to their electrons too tightly to be very effective as reducing agents.

21.2 SPONTANEITY AND EXTENT OF REDOX REACTIONS

We have pointed out that a voltaic cell is one in which a spontaneous oxidation-reduction reaction occurs. The converse is also true: any reaction that can occur in a voltaic cell to produce a positive voltage must be spontaneous. To decide whether a given reaction is capable of taking place under a particular set of conditions, all we have to do is to calculate the voltage associated with it. If the calculated voltage is positive, the reaction must be spontaneous. If we calculate a negative voltage, the reaction cannot go by itself; the reverse reaction is spontaneous.

Applying these criteria to Reactions 21.4 and 21.5, we decide that, at standard concentrations, cadmium metal but not copper will react with dilute acid to generate hydrogen gas.

$$\text{Cd(s)} \rightarrow \text{Cd}^{2+}\text{(aq, 1 M)} + 2 \text{ e}^- \qquad\qquad \text{SOP} = +0.40 \text{ V}$$
$$\underline{2 \text{ H}^+\text{(aq, 1 M)} + 2 \text{ e}^- \rightarrow \text{H}_2\text{(g, 1 atm)} \qquad\quad \text{SRP} = \ \ 0.00 \text{ V}}$$
$$\text{Cd(s)} + 2 \text{ H}^+\text{(aq, 1 M)} \rightarrow \text{Cd}^{2+}\text{(aq, 1 M)} + \text{H}_2\text{(g, 1 atm)} \qquad E° = +0.40 \text{ V} \quad (21.4)$$

$$\text{Cu(s)} \rightarrow \text{Cu}^{2+}\text{(aq, 1 M)} + 2 \text{ e}^- \qquad\qquad \text{SOP} = -0.34 \text{ V}$$
$$\underline{2 \text{ H}^+\text{(aq, 1 M)} + 2 \text{ e}^- \rightarrow \text{H}_2\text{(g, 1 atm)} \qquad\quad \text{SRP} = \ \ 0.00 \text{ V}}$$
$$\text{Cu(s)} + 2 \text{ H}^+\text{(aq, 1 M)} \rightarrow \text{Cu}^{2+}\text{(aq, 1 M)} + \text{H}_2\text{(g, 1 atm)} \qquad E° = -0.34 \text{ V} \quad (21.5)$$

Indeed, we can go one step further and predict that any metal with a positive standard oxidation potential should react with 1 M acid to produce hydrogen. Experimentally we find that our prediction is confirmed; metals above hydrogen in Table 21.1 are capable of displacing hydrogen from dilute acid, although in several cases the reaction takes place very slowly. Metals with negative standard oxidation potentials such as Cu (-0.34 V) and Ag (-0.80 V) do not reduce H^+ ions to H_2 gas.

Example 21.3 illustrates the application of the simple principle we have been discussing to a somewhat more complex situation.

Example 21.3 Assuming standard concentrations, will a reaction occur between nitric acid and a solution of iron(II) chloride?

Solution. Before we can answer this question, we must decide what the possible reactions are; only then can we calculate whether or not a reaction will occur. To be sure that we do not neglect any possibilities, let us list separately, using Table 21.1, every possible half-reaction. Noting that the species present are H^+, NO_3^-, Fe^{2+}, and Cl^- ions, we have:

possible oxidations:

$$O_1; \text{ Fe}^{2+}\text{(aq, 1 M)} \rightarrow \text{Fe}^{3+}\text{(aq, 1 M)} + \text{e}^- \qquad \text{SOP} = -0.77 \text{ V}$$
$$O_2; 2 \text{ Cl}^-\text{(aq, 1 M)} \rightarrow \text{Cl}_2\text{(g, 1 atm)} + 2 \text{ e}^- \qquad \text{SOP} = -1.36 \text{ V}$$

possible reductions:

$$R_1; 2 \text{ H}^+\text{(aq, 1 M)} + 2 \text{ e}^- \rightarrow \text{H}_2\text{(g, 1 atm)} \qquad \text{SRP} = \ \ 0.00 \text{ V}$$
$$R_2; \text{NO}_3^-\text{(aq, 1 M)} + 4 \text{ H}^+\text{(aq, 1 M)} + 3 \text{ e}^- \rightarrow$$
$$\rightarrow \text{NO(g, 1 atm)} + 2 \text{ H}_2\text{O} \qquad \text{SRP} = +0.96 \text{ V}$$
$$R_3; \text{ Fe}^{2+}\text{(aq, 1 M)} + 2 \text{ e}^- \rightarrow \text{Fe(s)} \qquad \text{SRP} = -0.44 \text{ V}$$

In principle, the two possible oxidations, O_1 and O_2, taken in combination with the three possible reductions, R_1, R_2, and R_3, could lead to six different overall reactions ($O_1 + R_1$, $O_1 + R_2$, $O_1 + R_3$; $O_2 + R_1$, $O_2 + R_2$, $O_2 + R_3$). In practice, it is readily seen that only one of these combinations, $O_1 + R_2$, will give a positive voltage. We deduce that the spontaneous reaction is

oxidation: $\qquad 3 \text{ Fe}^{2+}\text{(aq, 1 M)} \rightarrow 3 \text{ Fe}^{3+}\text{(aq, 1 M)} + 3 \text{ e}^- \qquad -0.77$ V

reduction:

$$\text{NO}_3^-\text{(aq, 1 M)} + 4 \text{ H}^+\text{(aq, 1 M)} + 3 \text{ e}^-$$
$$\underline{\rightarrow \text{NO(g, 1 atm)} + 2 \text{ H}_2\text{O} \qquad\qquad\qquad\qquad +0.96 \text{ V}}$$
$$3 \text{ Fe}^{2+}\text{(aq, 1 M)} + \text{NO}_3^-\text{(aq, 1 M)} + 4 \text{ H}^+\text{(aq, 1 M)}$$
$$\rightarrow 3 \text{ Fe}^{3+}\text{(aq, 1 M)} + \text{NO(g, 1 atm)} + 2 \text{ H}_2\text{O} \qquad +0.19 \text{ V}$$

In other words, nitric acid, at standard concentrations, will oxidize a solution of iron(II) chloride to a solution of iron(III) chloride, forming NO as a reduction product.

While it is, of course, important to know whether a given oxidation-reduction reaction will take place at standard concentrations, this information alone is not sufficient to the chemist interested in carrying out the reaction in the laboratory. For one thing, it is highly unlikely that we shall wish to carry out the reaction under such conditions that all reactants and products are at standard concentrations. Moreover, we would like to know not only whether a particular reaction will take place, but also the extent to which it will occur.

In order to answer these questions, we need to know the magnitude of the equilibrium constant for the reaction. Fortunately it is possible to calculate K for an oxidation-reduction reaction from the corresponding $E°$ value. To obtain a relationship between these two quantities, it is convenient to work through the standard free energy change for the reaction, $\Delta G°$.

RELATION BETWEEN $E°$ AND $\Delta G°$

You will recall from Chapter 12 that a spontaneous reaction is characterized by a *negative* free energy change. As we have seen in this chapter, a spontaneous oxidation-reduction reaction is always associated with a *positive* voltage. These two statements, taken together, imply a direct relationship between the free energy decrease ($-\Delta G$) and the voltage (E). This relationship can be obtained readily from thermodynamic arguments and is:

$$\Delta G = -nFE \qquad (21.6)$$

where ΔG is the free energy change in *calories*, n is the number of moles of electrons transferred in the reaction, F is the value of the faraday in calories per volt (recall Problem 20.41, Chapter 20), *23,060 cal/volt*, and E is the voltage.

The quantity nFE appearing in Equation 21.6 can be given a simple physical interpretation. It is the maximum amount of electrical work that can be obtained from a voltaic cell when n moles of electrons move through the external circuit under a driving force of E volts. In Chapter 12 we pointed out that $\Delta G = -W_{max}$; all we have done here is to write an explicit relation for W_{max}.

For many reactions $\Delta G°$ is found most easily from cell potentials and Eq. 21.7

We are particularly interested in applying this relationship when reactants and products are at their standard concentrations (conc. of 1 M for species in solution, partial pressure of gases 1 atmosphere). Under these conditions we can write

$$\Delta G° = -nFE° = -23,060 \, nE° \qquad (21.7)$$

where $\Delta G°$ is the standard free energy change in calories and $E°$ is the standard voltage.

Example 21.4 Using Equation 21.7, calculate

    a. $\Delta G°$ for the reaction $Zn(s) + 2\ H^+(aq) \rightarrow Zn^{2+}(aq) + H_2(g)$

    b. $E°$ for the reaction $Ag(s) + \frac{1}{2}\ Cl_2(g) \rightarrow AgCl(s)$, using Table 12.2.

Solution

    a. We have previously shown that $E°$ for this reaction is $+0.76$ V. The number of moles of electrons transferred is clearly two (2 moles of e^- required to reduce 2 moles of H^+; 2 moles of electrons released by the oxidation of one mole of Zn). Consequently,

$$\Delta G° = -23,060(2)(+0.76)\ cal = -35,000\ cal$$

    b. We first note that $\Delta G°$ for this reaction is simply the standard free energy of formation of AgCl. This quantity is found from Table 12.2 to be $-26,200$ cal. In this reaction, n is 1: one mole of electrons is given off when one mole of Ag is oxidized to Ag^+.

$$-26,200 = -23,060(1)E°$$
$$E° = +1.14\ V$$

RELATION BETWEEN E° AND K

As we pointed out in Chapter 13, the standard free energy change, $\Delta G°$, is related to the equilibrium constant through the equation

$$\Delta G° = -2.30\ RT\ \log_{10}K \qquad (21.8)$$

where R is the gas constant and T the absolute temperature. Combining Equations 21.7 and 21.8 we obtain

$$nFE° = 2.30\ RT\ \log_{10}K$$

or
$$\log_{10}K = \frac{nFE°}{2.30\ RT}$$

Substituting the values for the constants F and R in the proper units (23,060 cal/volt and 1.99 cal/°K) and taking T to be 298°K (i.e., 25°C, the temperature at which cell voltages are most commonly measured and tabulated) we obtain

$$\log_{10}K = \frac{nE°}{0.059} \qquad (21.9)$$

This relation is valid only at 25°C

Using Equation 21.9 we can calculate, from the standard cell potential, $\log_{10}K$ and hence K for any redox reaction taking place in water solution. In the expression for K, concentrations of species in solution are expressed in moles/liter. For gases taking part in the reaction, partial pressures in atmospheres are used. Terms for pure solids, pure liquids, or solvent water molecules do not appear in the equilibrium expression.

In Table 21.2 we list values of K calculated from Equation 21.9, corresponding to various values of $E°$, for the particular case where $n = 2$. It turns out that if $E°$ is greater than about 0.2 volt, K will be so large, of the order of 10^7 or greater, that the reaction will go virtually to completion

TABLE 21.2 Relation Between E° and K When n = 2

E°	logK	K	E°	logK	K
+1.00	+34	10^{34}	−1.00	−34	10^{-34}
+0.80	+27	10^{27}	−0.80	−27	10^{-27}
+0.60	+20	10^{20}	−0.60	−20	10^{-20}
+0.40	+14	10^{14}	−0.40	−14	10^{-14}
+0.20	+6.8	6×10^6	−0.20	−6.8	2×10^{-7}
+0.10	+3.4	2500	−0.10	−3.4	0.0004
+0.05	+1.7	50	−0.05	−1.7	0.02
0.00	0.0	1	0.00	0.0	1

under ordinary conditions. This behavior is characteristic of such reactions as

$K = 10^{34}$

$$H_2O_2(aq) + 2\ H^+(aq) + 2\ Fe^{2+}(aq) \rightarrow 2\ H_2O + 2\ Fe^{3+}(aq); \quad E° = +1.00\ V$$

Addition of excess hydrogen peroxide to a solution containing Fe^{2+} ions quantitatively oxidizes them to Fe^{3+}. In contrast, we expect that reactions for which the calculated E° value is less than about −0.2 volt will show little tendency to take place. For example, the reaction

$K = 10^{-26}$

$$2\ H^+(aq) + 2\ Fe^{2+}(aq) \rightarrow H_2(g) + 2\ Fe^{3+}(aq); \quad E° = -0.77\ V$$

does not occur to any detectable extent; iron(II) salts such as $FeSO_4$ or $FeCl_2$ are stable in acidic solution, at least in the absence of dissolved oxygen. Conversely, it is possible to reduce Fe^{3+} to Fe^{2+} by bubbling hydrogen through a water solution of an iron(III) salt, thereby reversing the above reaction.

We are particularly interested in working with equilibrium constants for redox reactions where E° is a relatively small number, perhaps falling in the range +0.2 to −0.2 volt. For such reactions we can expect to find appreciable concentrations of both reactants and products in the equilibrium mixture.

The equilibrium constant for an oxidation-reduction reaction, like any equilibrium constant, can be used to determine

1. The *extent* to which the reaction occurs if we start with pure reactants at known concentrations (Example 21.5).

2. The *direction* in which the reaction will proceed spontaneously at specified concentrations of reactants and products (Example 21.6).

Example 21.5 Consider the reaction: $Ag^+(aq) + Fe^{2+}(aq) \rightleftharpoons Ag(s) + Fe^{3+}(aq)$

    a. Calculate K for this reaction.

    b. If enough $AgNO_3$ is added to a solution 0.10 M in Fe^{2+} to make the final $[Ag^+] = 1.0$ M, what will be the equilibrium concentration of Fe^{3+}?

Solution

    a. We first obtain E° from standard potentials

$$E° = SRP\ Ag^+ + SOP\ Fe^{2+} = +0.80\ V - 0.77\ V = 0.03\ V$$

To calculate K, we note that for this reaction, n = 1

$$\log_{10}K = \frac{(1)(0.03)}{0.059} = 0.5; \quad K = 3$$

b. The expression for K is: $K = \frac{[Fe^{3+}]}{[Ag^+] \times [Fe^{2+}]} = 3$

If we let $[Fe^{3+}] = x$, then $[Fe^{2+}] = 0.10 - x$. From the statement of the problem, $[Ag^+] = 1.0$. Making these substitutions

$$\frac{x}{0.10 - x} = 3; \quad x = 0.07$$

We see that about 70% of the Fe^{2+} ions are converted to Fe^{3+}. Certainly Ag^+ would not be the best reagent to use if we wished to quantitatively oxidize Fe^{2+} ions.

Example 21.6 Consider the reaction

$$MnO_2(s) + 4 H^+(aq) + 2 Cl^-(aq) \rightarrow Mn^{2+}(aq) + Cl_2(g) + 2 H_2O$$

for which $E° = -0.13$ V, $K = 4 \times 10^{-5}$

a. Set up an expression for K.

b. Will this reaction take place in 10 M HCl to produce Mn^{2+} at 1 M concentration and $Cl_2(g)$ at 1 atm?

Solution

a. Recalling that partial pressures should be used for gases and that $MnO_2(s)$ and H_2O will not appear in the equation for K, we have

$$K = 4 \times 10^{-5} = \frac{[Mn^{2+}] \times pCl_2}{[H^+]^4 \times [Cl^-]^2}$$

b. To decide whether or not the reaction goes, we need to compare the actual conc. quotient to K (recall Example 13.1, p. 354). From the statement of the problem

$$\frac{(\text{conc. } Mn^{2+}) \times pCl_2}{(\text{conc. } H^+)^4 \times (\text{conc. } Cl^-)^2} = \frac{1 \times 1}{10^4 \times 10^2} = 1 \times 10^{-6}$$

Since the concentration quotient is *less* than K, reaction will have to take place in the forward direction to establish equilibrium. Notice that, at standard concentrations, the reaction would not go, since $E° = -0.13$ V. By raising the concentration of reactants to 10 M, we make the reaction spontaneous. Our prediction can be confirmed by a simple laboratory experiment. Addition of concentrated hydrochloric acid (12 M) to MnO_2 does indeed generate chlorine gas; in 1 M HCl, the reaction does not go.

21.3 EFFECT OF CONCENTRATION ON VOLTAGE

As we pointed out earlier in this chapter, the voltage of a cell depends to some extent upon the concentrations of reactants and products. To illustrate the effect of reactant and product concentrations upon cell voltages, let us consider a specific cell, the $Zn-Cu^{2+}$ cell shown in Figure 20.6, p. 555. You will recall that the standard voltage for the $Zn-Cu^{2+}$ cell is

TABLE 21.3 Effect of Change in Concentration on
Voltage of the Zn-Cu^{2+} Cell

Cell voltages do
not change very
rapidly with
changes in con-
centration of re-
actants or
products

$\dfrac{\text{Conc. }Zn^{2+}}{\text{Conc. }Cu^{2+}}$	100	10	1	0.1	0.01
Voltage	1.04	1.07	1.10	1.13	1.16

1.10 V and that the reaction involves a transfer of electrons from Zn atoms
to Cu^{2+} ions.

$$Zn(s) + Cu^{2+}(aq) \rightarrow Zn^{2+}(aq) + Cu(s); \quad E° = 1.10 \text{ V} \qquad (21.10)$$

In Table 21.3 we show how this voltage changes as we change the ratio of
the concentration of product (Zn^{2+}) to that of reactant (Cu^{2+}).

Looking at the data in Table 21.3 we conclude that there is an inverse
relationship between the ratio (conc. Zn^{2+}/conc. Cu^{2+}) and the cell voltage.
As we decrease the ratio from 100 to 0.01, the cell voltage increases steadily
from 1.04 to 1.16 V. We can readily understand why this should be the case
if we examine the cell reaction (21.10). Equilibrium considerations suggest
that if we decrease the concentration of product, Zn^{2+}, relative to that of
reactant, Cu^{2+}, the reaction will become more spontaneous; that is, it will
have a greater tendency to proceed in the forward direction. But, as we
know, the voltage of a cell is a measure of the spontaneity of the reaction
taking place within the cell. Any change in conditions such as that discussed here, which makes the reaction more spontaneous, would be expected to increase the voltage.

If we were to carry out voltage-concentration measurements on other
types of voltaic cells, the data might look quite different from that in
Table 21.3. Qualitatively, however, the generalization that we drew from
the behavior of the Zn-Cu^{2+} cell is valid for any cell; that is, when the concentration of products is decreased relative to that of reactants, the cell
voltage increases. Conversely, if we increase the concentration of products
at the expense of reactants, we can expect the cell voltage to decrease.

MAGNITUDE OF EFFECT; THE NERNST EQUATION

It is possible to fit the data in Table 21.3 to a simple mathematical equation. Perhaps you have already guessed the form of that equation.

$$E = 1.10 \text{ V} - 0.03 \log_{10} \frac{(\text{conc. }Zn^{2+})}{(\text{conc. }Cu^{2+})}$$

If not, you can easily convince yourself of its validity by substituting numbers from Table 21.3. Notice, for example, that when (conc. Zn^{2+}/conc.
Cu^{2+}) = 100, $\log_{10}$ (conc. Zn^{2+}/conc. Cu^{2+}) = 2, and we have

$$E = 1.10 \text{ V} - 0.03(2) = 1.04 \text{ V}$$

This equation is a special form of a more general relation known as the
Nernst equation. For the general oxidation-reduction reaction at 25°C

$$aA + bB \rightarrow cC + dD$$

in which A, B, C, and D are species whose concentrations can be varied and a, b, c, and d are the corresponding coefficients of the balanced equation, the Nernst equation has the form

$$E = E° - \frac{0.059}{n} \log_{10} \frac{(\text{conc. C})^c (\text{conc. D})^d}{(\text{conc. A})^a (\text{conc. B})^b} \qquad (21.11)$$

(E = cell voltage, E° = standard voltage, n = no. moles e⁻ transferred in reaction.)

Thus, for the cell reactions

$$Zn(s) + 2\ Ag^+(aq) \rightarrow Zn^{2+}(aq) + 2\ Ag(s); \quad E° = 1.56\ V$$

$$E = 1.56 - \frac{0.059}{2} \log_{10} \frac{(\text{conc. Zn}^{2+})}{(\text{conc. Ag}^+)^2}$$

and for

$$Zn(s) + 2\ H^+(aq) \rightarrow Zn^{2+}(aq) + H_2(g) \quad E° = 0.76\ V$$

$$E = 0.76 - \frac{0.059}{2} \log_{10} \frac{(\text{conc. Zn}^{2+})(pH_2)}{(\text{conc. H}^+)^2} \qquad (21.12)$$

Example 21.7 Calculate the voltage of a cell in which the following reaction occurs

$$Zn(s) + 2\ H^+(aq,\ 0.001\ M) \rightarrow Zn^{2+}(aq,\ 1\ M) + H_2(g,\ 1\ atm)$$

Solution. Substituting in Equation 21.12:

$$E = 0.76 - \frac{0.059}{2} \log_{10} \frac{(1)(1)}{(10^{-3})^2}$$

$$= 0.76 - \frac{0.059}{2} \log_{10} 10^6 = 0.76 - \frac{0.059(6)}{2} = +0.58\ V$$

The Nernst equation can also be applied to determine the effect of changes in concentration on the potential of an individual half-cell. For example, for the half-reaction

$$MnO_4^-(aq) + 8\ H^+(aq) + 5\ e^- \rightarrow Mn^{2+}(aq) + 4\ H_2O; \quad SRP = 1.52\ V$$

we can write: $E = 1.52 - \dfrac{0.059}{5} \log \dfrac{(\text{conc. Mn}^{2+})}{(\text{conc. MnO}_4^-)(\text{conc. H}^+)^8}$

We shall have more to say about this application in Section 21.4.

USE OF THE NERNST EQUATION TO DETERMINE CONCENTRATION OF IONS

From the standpoint of chemistry, the most important application of the Nernst equation is in the experimental determination of the concentrations of species in solution. To illustrate what is involved, let us refer back to Equation 21.12. We used this equation, in Example 21.7, to calculate a cell voltage given the concentrations of all the species involved in the reaction. We can turn this procedure around; if we measure the cell voltage and know the concentrations of all but one species, we can use the

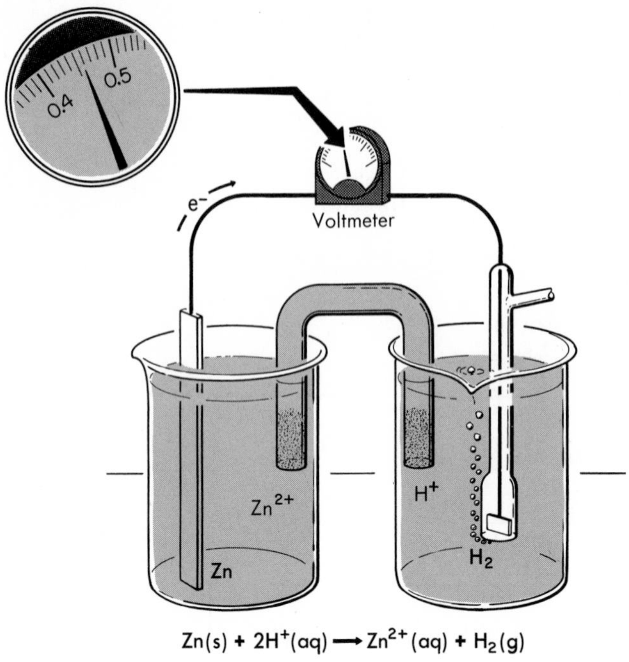
0.4
0.5
e⁻
Voltmeter
Zn^{2+}
H^+
H_2
Zn
$Zn(s) + 2H^+(aq) \longrightarrow Zn^{2+}(aq) + H_2(g)$

Figure 21.1 Use of Zn-H^+ cell to determine the concentration of H^+, using Nernst equation.

equation to obtain the "unknown" concentration. Suppose, for example, that we set up the cell shown in Figure 21.1. Substituting the measured voltage and known concentration in Equation 21.12, we have

$$+0.46 = +0.76 - \frac{0.059}{2} \log_{10} \frac{1}{(\text{conc. } H^+)^2}$$

Simplifying: $\log_{10} \frac{1}{(\text{conc. } H^+)^2} = \frac{2(0.76 - 0.46)}{0.059} = 10; \ \log_{10}(\text{conc. } H^+)^2 = -10$

$$(\text{conc. } H^+)^2 = 10^{-10}; \quad \text{conc. } H^+ = 10^{-5}\ M$$

We see that the cell shown in Figure 21.1 offers a simple and quite precise method of measuring the concentration of H^+ or pH of a solution. Indeed, it is with cells of this type that pH is ordinarily measured in the laboratory. In Figure 21.6 we show a schematic diagram of an instrument known as a pH meter which is specially designed for this purpose. The three essential components of the pH meter are:

—a vacuum-tube voltmeter or potentiometer, capable of measuring voltages to at least ±0.01 volt.

—a reference half cell of known potential.

—a half cell whose potential depends on the concentration of H^+. This consists of a metal electrode dipping into a solution of known pH separated by a *thin, fragile* glass membrane from the solution whose pH is to be determined. The potential across this *glass electrode,* and hence the cell voltage itself, is a linear function of the pH of the solution outside the membrane.

With suitably designed cells it is possible to determine the concentration of virtually any cation or anion in aqueous solution. The sensitivity of this method at very low ion concentrations makes it ideal for measuring

The meter reads pH directly. It is standardized with a buffer of known pH

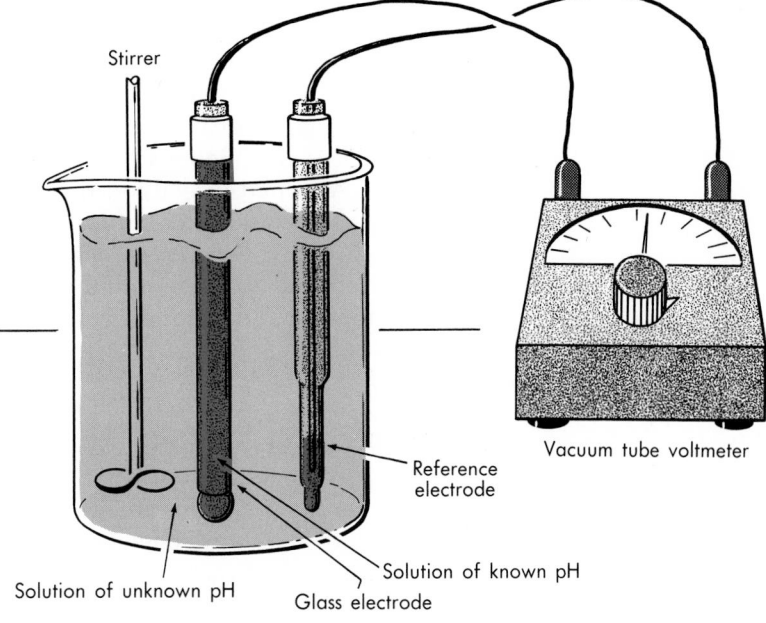

Stirrer

Reference electrode

Vacuum tube voltmeter

Solution of unknown pH

Glass electrode

Solution of known pH

Figure 21.2 pH meter. In many models the reference electrode is incorporated into the probe containing the glass electrode.

ionization constants of weak acids, dissociation constants of complex ions, or solubility products of slightly soluble salts (Example 21.8).

Example 21.8 In order to determine the solubility product of zinc sulfide, a student sets up a voltaic cell consisting of a standard Cu-Cu²⁺ half-cell (conc. Cu²⁺ = 1 M) joined to a Zn-Zn²⁺ half-cell. The concentration of Zn²⁺ is adjusted by adding enough sodium sulfide to precipitate almost all of the Zn²⁺ ions; the equilibrium concentration of S²⁻ is 0.1 M. Under these conditions, the measured cell voltage is +1.75 V. Calculate K_{sp} of ZnS.

Solution. The cell reaction is

$$Zn(s) + Cu^{2+}(aq) \rightarrow Zn^{2+}(aq) + Cu(s)$$

For this reaction, the Nernst equation takes the form

$$E = E° - \frac{0.059}{n} \log_{10} \frac{(\text{conc. } Zn^{2+})}{(\text{conc. } Cu^{2+})}$$

Or, since E° = 1.10 V, n = 2 and conc. Cu²⁺ = 1

$$E = 1.10 - \frac{0.059}{2} \log_{10} (\text{conc. } Zn^{2+})$$

Substituting the measured voltage, 1.75 V, and solving for the concentration of Zn²⁺,

$$1.75 = 1.10 - \frac{0.059}{2} \log_{10} (\text{conc. } Zn^{2+})$$

$$\log_{10} (\text{conc. } Zn^{2+}) = \frac{2(1.10 - 1.75)}{0.059} = -22; \quad \text{conc. } Zn^{2+} = 10^{-22} \text{ M}$$

Knowing that the concentration of S²⁻ is 0.1 M, we have

$$K_{sp} \text{ ZnS} = \text{conc. } Zn^{2+} \times \text{conc. } S^{2-} = 10^{-22} \times 10^{-1} = 1 \times 10^{-23}$$

During the past decade *specific ion electrodes* somewhat similar in design to the glass electrode have been developed to analyze for a variety of cations and anions. One of the first to be used extensively was a fluoride ion electrode which is sensitive to F^- at concentrations as low as 0.1 part per million and, hence, is ideal for monitoring fluoridated water supplies. An electrode which is specific for Cl^- ions is now being used to diagnose for cystic fibrosis. Attached directly to the skin, it detects the abnormally high concentrations of sodium chloride in sweat that are a characteristic symptom of this disorder. Diagnoses that used to require an hour or more can now be carried out in a few minutes; as a result, large numbers of children can be rapidly and routinely screened.

21.4 STRONG OXIDIZING AGENTS

Oxidation-reduction reactions are of considerable importance in industrial, inorganic, and analytical chemistry. They are used to synthesize compounds, to bring otherwise insoluble substances into solution, and to analyze for a variety of ions. Many of these applications make use of one of a relatively small group of strong oxidizing agents.

A "strong" oxidizing agent is a species which tends to pick up electrons readily; in more quantitative terms, it is an ion or molecule that has a large positive standard reduction potential. Looking at Table 21.1, we see that most such species fall into one of two categories:

There are not very many strong oxidizing agents

—molecules of nonmetals (Br_2, O_2, Cl_2, F_2).
—oxyanions (NO_3^-, $Cr_2O_7^{2-}$, ClO_3^-, MnO_4^-).

In this section we shall examine some of the oxidation-reduction reactions in which three of these species, the Cl_2 molecule, the NO_3^- ion, and the $Cr_2O_7^{2-}$ ion, participate. Some of the reactions that we shall be talking about are ones that are frequently carried out in the general chemistry laboratory; others are used in industrial processes. Hopefully, they will serve to illustrate the principles concerning spontaneity and extent of redox reactions that we have emphasized in this chapter.

CHLORINE

As is indicated by the magnitude of its standard reduction potential (+1.36 V), elementary chlorine is a powerful oxidizing agent. It is particularly effective in oxidizing organic compounds, which accounts for its use in water purification (Chapter 11). Chlorination of drinking water oxidizes not only harmful bacteria but many dissolved organic materials which can impart objectionable tastes or odors to the water.

Unfortunately, Cl_2 doesn't taste good either

Perhaps the most familiar reactions in which chlorine acts as an oxidizing agent are those involving bromide and iodide ions.

$$Cl_2(g) + 2 \ Br^-(aq) \rightarrow 2 \ Cl^-(aq) + Br_2(l); \quad E° = +1.36 \ V - 1.07 \ V = +0.29 \ V \tag{21.13}$$

$$Cl_2(g) + 2 \ I^-(aq) \rightarrow 2 \ Cl^-(aq) + I_2(s); \quad E° = +1.36 \ V - 0.53 \ V = +0.83 \ V \tag{21.14}$$

These reactions are frequently used to test for the presence of Br^- or I^- ions. Addition of chlorine to a solution containing either of these ions gives the free halogens Br_2 or I_2. If the water solution is then shaken with a small amount of a nonpolar organic solvent such as carbon disulfide or carbon tetrachloride, the free halogens enter the organic layer, to which they impart their characteristic colors, reddish brown (bromine) or violet (iodine).

Bromine is prepared commercially from sea water, in which it occurs as Br^- ions, by oxidation with chlorine (Reaction 21.13). About 100,000 tons of bromine are produced annually in the United States by this method. The concentration of iodide ions in sea water is so low (conc. $I^- = 4 \times 10^{-7}$ M vs conc. $Br^- = 8 \times 10^{-4}$ M) that it is not feasible to produce iodine in this way. In oil well brines, where the concentration of I^- is considerably higher, Reaction 21.14 can be carried out; most of the iodine used in the United States is prepared by oxidation of the iodides in these brines with chlorine.

Chlorine, unlike oxygen, forms many compounds in which it has a positive oxidation number. These compounds are ordinarily formed by a **disproportionation** reaction in which elementary chlorine is simultaneously oxidized and reduced. An example of one such reaction is that which occurs when chlorine is added to water.

$$Cl_2(g) + H_2O \rightleftharpoons HOCl(aq) + H^+(aq) + Cl^-(aq); \quad K = 3 \times 10^{-5} \quad (21.15)$$

The resulting solution, called *chlorine water,* contains equimolar amounts of the weak acid HOCl (hypochlorous acid) and the strong acid HCl. Half the chlorine (oxid state = 0) is reduced to Cl^- ions (oxid state = -1) while the remainder is oxidized to HOCl (oxid state Cl = $+1$).

The position of the equilibrium in Reaction 21.15 is strongly affected by the concentration of H^+ ions. In basic solution, in which the concentration of H^+ ions is low, chlorine is much more soluble than in pure water. The overall reaction that takes place when chlorine is bubbled through a solution of sodium hydroxide maintained at room temperature is

$$Cl_2(g) + 2\ OH^-(aq) \rightarrow ClO^-(aq) + Cl^-(aq) + H_2O \quad (21.16)$$

We may regard this as the sum of three individual reactions and apply the Multiple Equilibria Rule to find its equilibrium constant:

oxidation-reduction:
$Cl_2(g) + H_2O \rightarrow HOCl(aq) + H^+(aq) + Cl^-(aq);$ $\qquad K_1 = 3 \times 10^{-5}$
dissociation of HOCl:
$HOCl(aq) \rightarrow H^+(aq) + OCl^-(aq);$ $\qquad K_2 = 3 \times 10^{-8}$
neutralization:
$\underline{2\ H^+(aq) + 2\ OH^-(aq) \rightarrow 2\ H_2O;} \qquad\qquad\quad \underline{K_3 = 1 \times 10^{28}}$
$Cl_2(g) + 2\ OH^-(aq) \rightarrow OCl^-(aq) + Cl^-(aq) + H_2O;$ $\qquad K = K_1K_2K_3$
$\qquad\qquad\qquad\qquad\qquad\qquad\qquad\qquad\qquad\qquad\qquad = 9 \times 10^{15}$

The solution formed by the reaction of chlorine with sodium hydroxide via 21.16 is sold under various trade names as a household bleach and disinfectant. It is prepared commercially by electrolyzing a stirred water solution of sodium chloride. Recall that the electrolysis of an NaCl solution gives Cl_2 molecules and OH^- ions (Chapter 20); stirring ensures that these species react with each other. The active ingredient of the resulting solution is the hypochlorite ion, a relatively potent oxidizing agent:

Why would it be dangerous to add acid to household bleach?

$$ClO^-(aq) + H_2O + 2\ e^- \rightarrow Cl^-(aq) + 2\ OH^-(aq) \qquad SRP = +0.89\ V$$

The reaction of chlorine with a hot, concentrated solution of sodium or potassium hydroxide is quite different fro ·1 that observed at room temperature. Any ClO^- ions formed decompose on heating to ClO_3^- and Cl^- ions; the net reaction is

$$3\ Cl_2(g) + 6\ OH^-(aq) \rightarrow ClO_3^-(aq) + 5\ Cl^-(aq) + 3\ H_2O \qquad (21.17)$$

Potassium chlorate is a powerful oxidizing agent (SRP $ClO_3^- = +1.47\ V$) and can react violently with easily oxidized materials, including many organic substances. It is sometimes used as a source of oxygen in the general chemistry laboratory.

$$2\ KClO_3(s) \rightarrow 2\ KCl(s) + 3\ O_2(g) \qquad (21.18)$$

This reaction, in the absence of a catalyst, takes place very slowly below a temperature of 400°C. In the temperature range 350–400°, the principal products are potassium perchlorate, $KClO_4$, and potassium chloride.

$$4\ KClO_3(s) \rightarrow 3\ KClO_4(s) + KCl(s) \qquad (21.19)$$

OXYANIONS, NO_3^- AND $Cr_2O_7^{2-}$

Although oxyanions differ greatly in their oxidizing strength, they have certain characteristics in common. In particular,

1. *Oxyanions are stronger oxidizing agents in acidic than in neutral or basic solution.* We find, for example, that concentrated nitric acid is a powerful oxidizing agent, capable of oxidizing both copper and silver, neither of which reacts with H^+ ions alone:

$$Cu(s) + 2\ NO_3^-(aq) + 4\ H^+(aq) \rightarrow Cu^{2+}(aq) + 2\ NO_2(g) + 2\ H_2O \qquad (21.20)$$

$$Ag(s) + NO_3^-(aq) + 2\ H^+(aq) \rightarrow Ag^+(aq) + NO_2(g) + H_2O \qquad (21.21)$$

In contrast, salts such as KNO_3, containing the NO_3^- ion, are ineffective oxidizing agents in neutral solution. Again, a solution prepared by adding sulfuric acid to potassium dichromate is frequently used as an oxidizing agent to clean laboratory glassware; it removes greases and oils which are impervious to a solution containing only $K_2Cr_2O_7$.

We can readily explain this relationship between oxidizing strength and acidity by examining the half-equations for the reduction of NO_3^- and $Cr_2O_7^{2-}$ ions.

$$NO_3^-(aq) + 2\ H^+(aq) + e^- \rightarrow NO_2(g) + H_2O$$

$$Cr_2O_7^{2-}(aq) + 14\ H^+(aq) + 6\ e^- \rightarrow 2\ Cr^{3+}(aq) + 7\ H_2O$$

We see that, in both cases, the H^+ ion is involved as a reactant. Increasing its concentration should then make the oxidation more spontaneous; i.e., it should make the reduction potential a larger positive number (Example 21.9).

Example 21.9   The standard reduction potential for the half-reaction

$$NO_3^-(aq) + 2\ H^+(aq) + e^- \rightarrow NO_2(g) + H_2O$$

is +0.78 volts. Using the Nernst equation, calculate the reduction potential in 10 M H^+ and in neutral solution, assuming all other species to be at unit concentrations.

Solution.   Applying the Nernst equation to this half-reaction with conc. $NO_3^- = 1$ M and $pNO_2 = 1$ atm, we have

$$RP = +0.78 - \frac{0.059}{1} \log_{10} \frac{1}{(\text{conc. } H^+)^2}$$

simplifying: $RP = +0.78 + 2(0.059) \log_{10} (\text{conc. } H^+)$

Notice that, as predicted, the reduction potential increases as the solution becomes more acidic. In 10 M acid, conc. $H^+ = 10$, $\log_{10} (\text{conc. } H^+) = 1$:

$$RP = +0.78 + 2(0.059)(1) = +0.90 \text{ V}$$

In neutral solution, conc. $H^+ = 1 \times 10^{-7}$, $\log_{10} (\text{conc. } H^+) = -7$

$$RP = +0.78 + 2(0.059)(-7) = -0.05 \text{ V}$$

Clearly, the NO_3^- ion in neutral solution will be a weak oxidizing agent.

In neutral solution the NO_3^- ion is very stable

2. *Oxyanions can be reduced to a variety of species, depending upon the experimental conditions.* Table 21.4 indicates some of the species to which the NO_3^- and $Cr_2O_7^{2-}$ ions can be reduced.

Examining Table 21.4, we can count some 12 different species to which the NO_3^- ion might be reduced and four possible reduction products for $Cr_2O_7^{2-}$. (The actual number of species is somewhat greater; we have left out some of the more exotic ones.) It might seem futile to hope that we could predict in advance which species would be formed in a particular redox reaction. There are, however, some guiding principles that we can use to predict the position of:

a. *Equilibria between species within a given oxidation state.* If we know the equilibrium constant for the appropriate acid-base reaction, this prediction can be made with confidence. Consider, for example, the two

TABLE 21.4   Oxidation States of N and Cr

	NITROGEN			CHROMIUM	
	Acidic Solution	*Basic Solution*		*Acidic Solution*	*Basic Solution*
+5	NO_3^-	NO_3^-	+6	$Cr_2O_7^{2-}$	CrO_4^{2-}
+4	$NO_2(g)$	$NO_2(g)$	+3	Cr^{3+}	$Cr(OH)_3(s)$ *
+3	HNO_2	NO_2^-	+2	Cr^{2+}	$Cr(OH)_2(s)$
+2	$NO(g)$	$NO(g)$			
+1	$N_2O(g)$	$N_2O(g)$			
0	$N_2(g)$	$N_2(g)$			
−1	NH_3OH^+	NH_2OH			
−2	$N_2H_5^+$	N_2H_4			
−3	NH_4^+	$NH_3(g)$			

* In strongly basic solution, complex ions such as $Cr(OH)_6^{3-}$ will form.

species of nitrogen in the +3 oxidation state, HNO_2 and NO_2^-. Their concentrations can be related through the dissociation constant for the weak acid, HNO_2:

$$HNO_2(aq) \rightleftharpoons H^+(aq) + NO_2^-(aq); \quad K_a = 4.5 \times 10^{-4}$$

By rearranging the expression for K_a we obtain:

$$\frac{[HNO_2]}{[NO_2^-]} = \frac{[H^+]}{K_a} = \frac{[H^+]}{4.5 \times 10^{-5}}$$

We see that at H^+ ion concentrations greater than 4.5×10^{-5} M, the ratio $[HNO_2]/[NO_2^-]$ will be greater than one; under these conditions the weak acid HNO_2 will be the predominant species. At H^+ ion concentrations less than 4.5×10^{-5} M, the principal species present will be the weak base, NO_2^-.

A particularly interesting case of acid-base equilibrium is offered by the +6 state of chromium

$$2\ CrO_4^{2-}(aq) + 2\ H^+(aq) \rightleftharpoons Cr_2O_7^{2-}(aq) + H_2O \qquad (21.22)$$
$$\text{yellow} \qquad\qquad\qquad\qquad \text{red}$$

Here, we can determine the position of the equilibrium visually by noting the color of the solution. At a pH less than 7, the $Cr_2O_7^{2-}$ ion predominates; as the solution is made basic, the red color changes to yellow, indicating the formation of CrO_4^{2-}.

b. *Equilibria between different oxidation states.* The reduction of +6 chromium almost always leads to the +3 state. Only by using an excess of a powerful reducing agent such as zinc can we reach the relatively unstable +2 state of chromium.

$$Cr^{3+}(aq) + e^- \to Cr^{2+}(aq); \quad SRP = -0.41\ V$$

In the case of the NO_3^- ion, it is a great deal more difficult to predict which species will be produced on reduction because of the multiplicity of lower oxidation states, ranging from +4 to −3. Indeed, whenever we write an equation showing the reduction of NO_3^- to a single species such as $NO_2(g)$ or $NO(g)$, we are oversimplifying reality; nearly always we get a mixture of reduction products in varying proportions.

When we carry out oxidation-reduction reactions with concentrated nitric acid, nitrogen dioxide, NO_2, can always be detected among the products. For example, treatment of copper metal with concentrated nitric acid yields copious brown fumes of NO_2 in accordance with Reaction 21.20. The formation of NO_2 is readily explained on a kinetic basis. Even though we know very little about the detailed mechanism of the reduction process, it seems certain that the first step is a one-electron transfer to go from the +5 state (NO_3^-) to the +4 state (NO_2).

In principle, reduction of NO_3^- should proceed past NO_2 to lower oxidation states. The standard potentials for the reduction of NO_2 to HNO_2 (+1.08 V), NO (+1.04 V), and N_2 (+1.36 V) are all more positive than that for the reduction of NO_3^- to NO_2 (+0.80 V). However, if NO_2 is produced rapidly, as is the case when concentrated acid is used, it escapes from solution rather than undergoing further reduction. With dilute acid, where reaction occurs more slowly, the situation is quite different; the NO_2 formed in the first step stays around long enough to be further reduced.

Possible overall half equations for the reduction of dilute HNO_3 include:

$$NO_3^-(aq) + 4\ H^+(aq) + 3\ e^- \rightarrow NO(g) + 2\ H_2O \qquad (21.23)$$

$$NO_3^-(aq) + 3\ H^+(aq) + 2\ e^- \rightarrow HNO_2(aq) + H_2O \qquad (21.24)$$

The latter reaction can occur in the stomach (pH 3 to 4) where certain bacteria act as the reducing agent. The HNO_2 produced may react with hemoglobin of the blood, converting it from an iron(II) to an iron(III) complex which cannot act as an oxygen carrier. Infant mortality from this source (blue babies) has led to restrictions on the concentration of nitrate in drinking water.

21.5 OXYGEN, THE CORROSION OF IRON

Of all oxidizing agents, elementary oxygen is the most abundant and, in many ways, the most important. Its presence in air insures that all water supplies including reagent solutions used in the laboratory will ordinarily be saturated with atmospheric oxygen. We often forget this and are puzzled by such phenomena as

— the formation of white or yellow precipitates when hydrogen sulfide is used in qualitative analysis

$$\tfrac{1}{2}\ O_2(g) + H_2S(g) \rightarrow S(s) + H_2O \qquad (21.25)$$

Oxygen is a moderately strong oxidizing agent, particularly in acid solution; SRP = 1.23 V

— the yellow color that solutions of NaI or KI acquire upon standing

$$\tfrac{1}{2}\ O_2(g) + 2\ I^-(aq) + 2\ H^+(aq) \rightarrow I_2(aq) + H_2O \qquad (21.26)$$

— the cloudiness that develops in a solution of tin(II) chloride

$$\tfrac{1}{2}\ O_2(g) + Sn^{2+}(aq) + H_2O \rightarrow SnO_2(s) + 2\ H^+(aq) \qquad (21.27)$$

From an economic standpoint, the most important redox reaction involving dissolved oxygen is the corrosion of iron and steel. It is estimated that the annual cost to this country of corrosion of ferrous metals exceeds five billion dollars. We see the results of corrosion all around us in junk piles and auto graveyards. Perhaps as much as 20 per cent of all the iron produced each year in this country goes to replace products whose usefulness has been destroyed by rust.

In order to understand the mechanism by which iron corrodes, let us consider what happens when a sheet of iron is exposed to a neutral water solution containing an electrolyte such as sodium chloride. The iron tends to oxidize according to the half-reaction

$$Fe(s) \rightarrow Fe^{2+}(aq) + 2\ e^- \qquad (21.28a)$$

For this reaction to take place, some other species must be reduced simultaneously. A reasonable possibility would be the reduction of H^+ ions to elementary hydrogen (SRP H^+ = 0.00 V). This does occur in strongly acidic solution, but cannot take place at an appreciable rate in neutral solution, in which the concentration of H^+ ions is only 10^{-7} M. Instead, oxygen molecules dissolved in the solution are reduced:

$$\tfrac{1}{2}\ O_2(g) + H_2O + 2\ e^- \rightarrow 2\ OH^-(aq) \qquad (21.28b)$$

Adding these two half-equations and noting that iron(II) hydroxide is insoluble in water, we obtain for the primary corrosion reaction,

$$Fe(s) + \tfrac{1}{2} O_2(g) + H_2O \rightarrow Fe(OH)_2(s) \qquad (21.28)$$

There is a great deal of evidence to suggest that Reaction 21.28 represents the first step in the corrosion of iron or steel. However, iron(II) hydroxide is ordinarily further oxidized in the reaction

$$2 Fe(OH)_2(s) + \tfrac{1}{2} O_2(g) + H_2O \rightarrow 2 Fe(OH)_3(s) \qquad (21.29)$$

The final product—the loose, flaky deposit that we call rust—has the reddish brown color of iron(III) hydroxide, $Fe(OH)_3$.

A significant clue to the mechanism of corrosion is the experimental observation that the oxidation half-reaction (21.28a) and the reduction half-reaction (21.28b) do not occur at the same location. If we look at a nail extracted from an old building, we frequently find that the rust is concentrated near the head of the nail, which has been in contact with moist air. The most serious pitting, often amounting to disintegration, is found along the shank of the nail, which is embedded in the wood. These observations lead us to believe that oxidation (21.28a) is occurring along a surface that may be an inch or more away from the point at which oxygen is being reduced (21.28b).

The fact that oxidation and reduction half-reactions take place at different locations suggests that corrosion occurs by an electrochemical mechanism. The surface of a piece of corroding iron may be visualized as consisting of a series of localized voltaic cells. At *anodic areas*, iron is oxidized to Fe^{2+} ions; at *cathodic areas*, elementary oxygen is reduced to OH^- ions. Electrons are transferred through the iron, which acts like the external conductor of an ordinary voltaic cell. The electrical circuit is completed by the flow of ions through the water solution or film covering the iron. The fact that rust ordinarily accumulates at cathodic areas suggests that it is primarily Fe^{2+} ions which move through the solution, from anode to cathode. When these ions arrive at a cathodic area, they are precipitated, first as $Fe(OH)_2$ and ultimately, through further reaction with oxygen, as $Fe(OH)_3$.

Many of the characteristics of corrosion are most readily explained in terms of an electrochemical mechanism. A perfectly dry metal surface is not attacked by oxygen; iron exposed to dry air does not corrode. This seems plausible if corrosion occurs through a voltaic cell, which requires a water solution through which ions can move to complete the circuit. The fact that

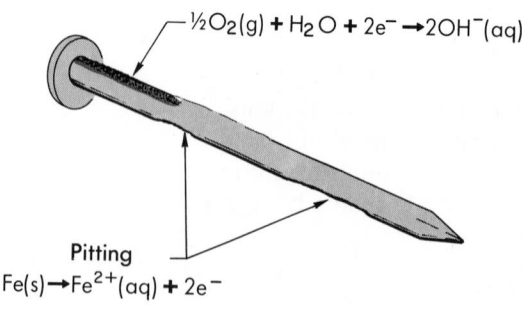

$$\tfrac{1}{2}O_2(g) + H_2O + 2e^- \rightarrow 2OH^-(aq)$$

Pitting

$$Fe(s) \rightarrow Fe^{2+}(aq) + 2e^-$$

Figure 21.3 Corrosion of iron nail driven into wood. Rust collects near head of nail, but pitting takes place near point.

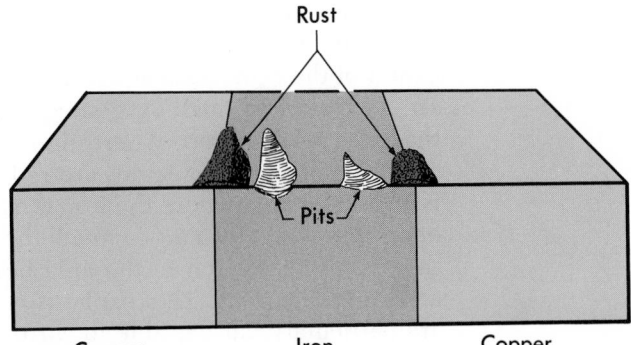

Figure 21.4 Corrosion of iron in contact with copper.

Corrosion often occurs at junctions between dissimilar metals

corrosion occurs more readily in sea water than in fresh water has a similar explanation. The dissolved salts in sea water supply the ions necessary for the conduction of current.

The existence of discrete cathodic and anodic areas on a piece of corroding iron requires that adjacent surface areas differ chemically from each other. There are several ways in which one small area on a piece of iron or steel can become anodic or cathodic with respect to an adjacent area. Two of the most important are:

1. *The presence of impurities at scattered locations along the metal surface.* A tiny crystal of a less active metal such as copper or tin embedded in the surface of the iron acts as a cathode at which oxygen molecules are reduced. The iron atoms in the vicinity of these impurities are anodic and undergo oxidation to Fe^{2+} ions. This effect can be demonstrated on a large scale by immersing in water an iron plate which has been partially copper plated (Fig. 21.4). At the interface between the two metals, a voltaic cell is set up in which the iron is anodic and the copper cathodic. A thick deposit of rust forms at the interface. The formation of rust inside an automobile bumper where the chromium plate stops is another example of this phenomenon.

2. *Differences in oxygen concentration along the metal surface.* To illustrate the effect, consider what happens when a drop of water adheres to the surface of a piece of iron exposed to the air (Fig. 21.5). The metal

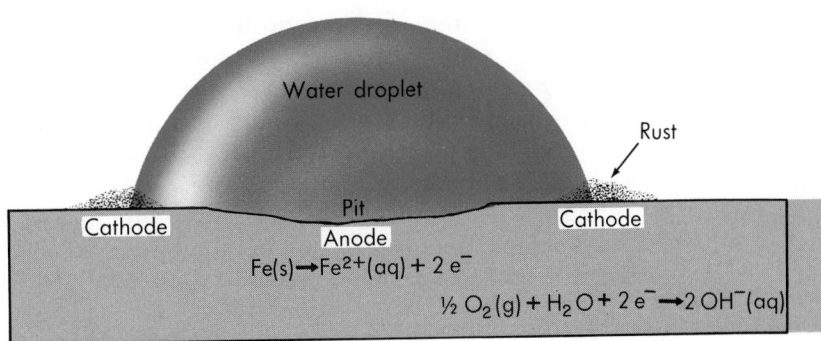

Figure 21.5 Corrosion of iron under drop of water.

around the edges of the drop is in contact with water containing a high concentration of dissolved oxygen. The water touching the metal beneath the center of the drop is depleted in oxygen, since it is cut off from contact with air. As a result, a small oxygen-concentration cell is set up. The area around the edge of the drop, where the oxygen concentration is high, becomes cathodic; oxygen molecules are reduced there. Directly beneath the drop is an anodic area where the iron is oxidized. A particle of dirt on the surface of an iron object can act in much the same way as a drop of water to cut off the supply of oxygen to the area beneath it and thereby establish anodic and cathodic areas. This explains why garden tools left covered with soil are particularly susceptible to corrosion.

Iron or steel objects can be protected from corrosion in several different ways. These include:

1. *Covering the surface with a protective coating.* This may be a layer of paint which cuts off access to moisture and oxygen. Under more severe conditions, it may be desirable to cover the surface of the iron or steel object with a layer of another metal. Metallic plates, applied electrically (Cr, Ni, Cu, Ag, Zn, Sn) or by immersion at high temperatures (Zn, Sn), are ordinarily more resistant to heat and chemical attack than the organic coating left when paint dries. However, if the plating metal is less active than iron, there is a danger that cracks in its surface may enhance the corrosion of the iron or steel. This problem can arise with "tin cans," which are made by applying a layer of tin over a steel base. If the food in the can contains citric acid, some of the tin plate may dissolve, exposing the steel beneath.* When the can is opened, exposing the interior to the air, rust forms spontaneously on the iron surrounding the breaks in the tin surface. A thin coating of lacquer is ordinarily applied over the tin to prevent corrosive effects of this type.

2. *Adding a corrosion inhibitor to the solution in which the object is standing.* These inhibitors may be either inorganic or organic in nature. Salts containing the CrO_4^{2-} ion are particularly effective, probably because they form a thin surface layer of $FeCrO_4$ around anodic areas, thereby preventing further corrosion. An organic inhibitor that is often added to antifreeze to protect against rust formation is tributylamine $(C_4H_9)_3N$, which may be considered a derivative of ammonia, with the three hydrogen atoms replaced by butyl groups (C_4H_9). Organic acids produced by the decomposition of the antifreeze convert tributylamine to salts containing the $(C_4H_9)_3NH^+$ ion (compare the ammonium ion, NH_4^+). These salts are adsorbed on the interior surface of the radiator where they offer protection against corrosion. At the same time, the amine, by neutralizing H^+ ions, prevents the enhanced corrosion that is characteristic of acidic solutions.

3. *Bringing the object into electrical contact with a more active metal* such as magnesium or zinc. Under these conditions, the iron becomes cathodic and, hence, is protected against rusting; the more active metal serves as a sacrificial anode in a large-scale corrosion cell. This method of combating corrosion, known as *cathodic protection,* is particularly useful for steel objects such as cables or pipelines that are buried under soil or water (Fig. 21.6).

---

* Tin forms an extremely stable complex with citrate ion and, hence, is attacked more readily by citric acid than by many stronger inorganic acids.

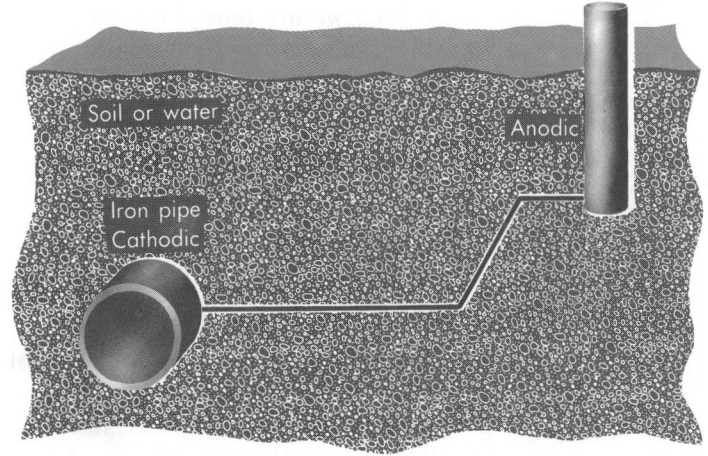

Magnesium bar

Figure 21.6 Cathodic protection of buried iron pipe. Magnesium or zinc bar is oxidized rather than iron.

Similar protection can be provided by using an external electrical source to give the iron a negative potential

Soil or water

Anodic

Iron pipe Cathodic

21.6 REDOX REACTIONS IN ANALYTICAL CHEMISTRY

QUALITATIVE ANALYSIS

At certain stages of the standard schemes of cation and anion analysis, oxidizing or reducing agents are used for one of three different purposes:

1. *To bring a sample into solution.* Solids which are insoluble in both water and dilute nonoxidizing acids can frequently be brought into solution with concentrated nitric acid. It reacts with inactive metals such as copper or silver (Equations 21.20 and 21.21) and with many metal sulfides such as those of copper and bismuth

$$CuS(s) + 2\ NO_3^-(aq) + 4\ H^+(aq)\ \rightarrow Cu^{2+}(aq) + S(s) + 2\ NO_2(g) + 2\ H_2O \quad (21.30)$$

What species are oxidized?

$$Bi_2S_3(s) + 6\ NO_3^-(aq) + 12\ H^+(aq) \rightarrow 2\ Bi^{3+}(aq) + 3\ S(s) + 6\ NO_2(g) + 6\ H_2O \quad (21.31)$$

to form a solution containing the corresponding metal ion.

Mercuric sulfide, which is not attacked by nitric acid, is brought into solution by treatment with *aqua regia*, a mixture of concentrated nitric and hydrochloric acids. The oxidizing agent here is the NO_3^- ion, which converts S^{2-} to free sulfur; the Cl^- ion forms a complex with Hg^{2+}, thereby making the reaction more spontaneous.

$$HgS(s) + 2\ NO_3^-(aq) + 4\ H^+(aq) \rightarrow Hg^{2+}(aq) + S(s) + 2\ NO_2(g) + 2\ H_2O$$
$$Hg^{2+}(aq) + 4\ Cl^-(aq) \rightarrow HgCl_4^{2-}(aq)$$

$$HgS(s) + 2\ NO_3^-(aq) + 4\ H^+(aq) + 4\ Cl^-(aq) \rightarrow HgCl_4^{2-}(aq) + S(s) + 2\ NO_2(g) + 2\ H_2O \quad (21.32)$$

2. *To identify a particular ion.* The mercuric ion, Hg^{2+}, may be tested for by putting its solution on a copper penny or a piece of copper wire; if Hg^{2+} is present, a shiny deposit of metallic mercury forms.

$$Hg^{2+}(aq) + Cu(s) \rightarrow Hg(s) + Cu^{2+}(aq) \quad (21.33)$$

Another test is to add tin(II) chloride, which reduces Hg^{2+} to the insoluble chloride of mercury(I), Hg_2Cl_2

$$2 \; Hg^{2+}(aq) + Sn^{2+}(aq) + 2 \; Cl^-(aq) \rightarrow Hg_2Cl_2(s) + Sn^{4+}(aq) \quad (21.34)$$

3. *To separate ions from one another.* The $\dot{C}r^{3+}$ ion can be separated from Al^{3+} and Zn^{2+} by first treating with hydrogen peroxide in basic solution to oxidize chromium from the +3 to the +6 state

$$2 \; Cr^{3+}(aq) + 3 \; H_2O_2(aq) + 10 \; OH^-(aq) \rightarrow 2 \; CrO_4^{2-}(aq) + 8 \; H_2O \quad (21.35)$$

and then adding Ba^{2+} to precipitate yellow, insoluble barium chromate

$$Ba^{2+}(aq) + CrO_4^{2-}(aq) \rightarrow BaCrO_4(s) \quad (21.36)$$

QUANTITATIVE ANALYSIS

A species that is readily oxidized can be determined quantitatively by titration with an oxidizing agent in much the same way that a base is titrated with an acid. An oxidizing agent that is frequently used in redox titrations is potassium permanganate, $KMnO_4$.

To illustrate the use of MnO_4^- ion, let us consider a specific redox titration, the determination of Fe^{2+} ions with MnO_4^-.

$$MnO_4^-(aq) + 8 \; H^+(aq) + 5 \; Fe^{2+}(aq) \rightarrow Mn^{2+}(aq) + 4 \; H_2O + 5 \; Fe^{3+}(aq) \quad (21.37)$$

$$E^\circ = SRP \; MnO_4^- + SOP \; Fe^{2+} = (+1.52 - 0.77) \; V = 0.75 \; V$$

The large positive E° value for this reaction means that the equilibrium constant is large enough ($K = 10^{64}$) to make the reaction go essentially to completion. What one does in the titration is to start with a known volume of an acidified solution containing Fe^{2+} ions and add from a buret a solution of potassium permanganate of known concentration. At the instant the MnO_4^- ions are added, the solution takes on the pink or purple color characteristic of that ion. As the MnO_4^- ions are used up by Reaction 21.37, the color fades almost immediately. However, when all the Fe^{2+} ions have been titrated, i.e., at the equivalence point, the addition of one or two drops of excess MnO_4^- produces a permanent pink color. The volume of titrant necessary to reach this end point is recorded and the concentration of Fe^{2+} ions calculated as indicated in Example 21.10.

Example 21.10 A 20.0 ml sample containing Fe^{2+} ions requires 18.0 ml of 0.100 M $KMnO_4$ solution for complete reaction. Calculate the concentration of Fe^{2+} ions in the solution.

Solution. Let us first calculate the number of moles of MnO_4^- added. Then, using Equation 21.37, we can calculate the number of moles of Fe^{2+} ion in the sample. Finally, knowing the volume of the sample, we can calculate the concentration of Fe^{2+}.

$$\text{no. moles } MnO_4^- = 0.100 \; \frac{\text{mole}}{\text{lit}} \times 0.0180 \; \text{lit} = 0.00180 \; \text{mole}$$

According to Equation 21.37:

$$1 \; \text{mole } MnO_4^- \backsimeq 5 \; \text{moles } Fe^{2+}$$

Hence,

$$\text{no. moles } Fe^{2+} = 0.00180 \; \text{mole } MnO_4^- \times \frac{5 \; \text{moles } Fe^{2+}}{1 \; \text{mole } MnO_4^-} = 0.00900 \; \text{mole } Fe^{2+}$$

$$\text{conc. } Fe^{2+} = \frac{0.00900 \; \text{mole}}{0.0200 \; \text{lit}} = 0.450 \; M$$

PROBLEMS

21.1 Assuming standard concentrations, calculate the voltage of a cell in which

  a. the reaction: $2 Ag^+(aq) + Cu(s) \rightarrow 2 Ag(s) + Cu^{2+}(aq)$ occurs.
  b. the reaction: $ClO_3^- \rightarrow ClO_4^- + Cl^-$ occurs in basic solution.
  c. two half cells, both containing platinum electrodes, are connected. One half cell contains a solution of $FeSO_4$ and $Fe_2(SO_4)_3$, the other a solution of $KAuCl_4$ and KCl.

21.2 Which of the following reactions can serve as a source of energy in a voltaic cell? Which must be carried out in an electrolytic cell?

  a. $Sn(s) + Sn^{4+}(1 M, aq) \rightarrow 2 Sn^{2+}(1 M, aq)$
  b. $2 I^-(aq, 1 M) + 2 H_2O \rightarrow I_2(s) + 2 OH^-(aq, 1 M) + H_2(g)$
  c. $S(s) + 2 H_2O \rightarrow H_2S(g) + \frac{1}{2} O_2(g)$

21.3 Predict the products that will be obtained upon electrolysis of 1 M solutions of the following (consider the possibility of forming H_2 and/or O_2 from water):

  a. $FeBr_2$ c. $CuSO_4$
  b. NaI d. KNO_3

21.4 Arrange the following species in order of increasing strength as oxidizing agents

$$Ag^+, H^+, Cl_2, Fe^{3+}, Mn^{2+}$$

21.5 Explain the following observations in terms of standard potentials:

  a. Tin added to a solution of a Sn^{2+} salt protects it from oxidation.
  b. The Cu^+ ion is unstable in water solution.
  c. The red color of a solution of $K_2Cr_2O_7$ fades when $FeSO_4$ is added.

21.6 What reactions, if any, will occur when oxygen is bubbled through a strongly acidic solution of

  a. $NaCl$ c. $CrCl_2$
  b. KI d. $FeSO_4$

21.7 Decide what reaction, if any, will occur when the following are mixed (standard concentrations):

  a. SO_4^{2-}, H^+, I^-
  b. $Cr_2O_7^{2-}, H^+, Na^+, Br^-$
  c. Cu^{2+}, Pb^{2+}
  d. $Pb^{2+}, NO_3^-, Fe^{2+}, H^+$

21.21 At standard concentrations, what voltages would you expect for cells in which the following reactions take place?

  a. $2 H_2O + 2 Br^-(aq) \rightarrow H_2(g) + 2 OH^-(aq) + Br_2(l)$
  b. $Fe(s) + Cu^{2+}(aq) \rightarrow Fe^{2+}(aq) + Cu(s)$
  c. $Fe(s) + Cu(OH)_2(s) \rightarrow Fe(OH)_2(s) + Cu(s)$
  d. $H_2O_2(aq) + 2 Ag(s) + 2 H^+(aq) \rightarrow 2 H_2O + 2 Ag^+(aq)$

21.22 A certain half cell consists of a nickel electrode dipping into a 1 M solution of $NiSO_4$. Describe a half cell which when connected to this one will give a voltage of

  a. about 1 volt, with nickel as the anode.
  b. about 1 volt with nickel as the cathode.

21.23 Of the cations listed in Table 21.1, which would you expect to be reduced to the free metal when an aqueous solution is electrolyzed? How do you explain the fact that electrolysis of a sodium chloride solution ordinarily gives Cl_2 rather than O_2 at the anode?

21.24 Arrange the following metals in order of increasing effectiveness as reducing agents

$$Co, Pb, Ag, Zn, Al, Au$$

21.25 Explain, in terms of standard potentials, why

  a. copper is oxidized by nitric acid but not by hydrochloric acid.
  b. Sn^{2+} and Fe^{3+} ions are not found in the same solution.
  c. fluorine cannot be prepared by electrolysis of a water solution.

21.26 Write equations for the reaction, if any, that will occur when hydrogen is bubbled through a strongly acidic solution of

  a. $CuCl_2$ c. $CrCl_3$
  b. $CoCl_3$ d. $Pb(NO_3)_2$

21.27 Write balanced equations for any reaction that will occur when

  a. hydrogen peroxide is added to a solution of sodium iodide.
  b. hydrogen peroxide is added to a solution of iron(II) iodide.
  c. solutions of $Zn(NO_3)_2$ and $Cu(NO_3)_2$ are mixed.

21.8 Calculate $\Delta G°$ for each of the following reactions

  a. $Zn^{2+}(aq) + Cu(s) \rightarrow Zn(s) + Cu^{2+}(aq)$
  b. $4 ClO_3^-(aq) \rightarrow Cl^-(aq) + 3 ClO_4^-(aq)$
  c. $2 Ag(s) + I_2(s) \rightarrow 2 AgI(s)$
  d. $Fe(s) + \frac{1}{2} O_2(g) + H_2O \rightarrow Fe(OH)_2(s)$

21.9 Calculate K for each of the reactions in Problem 21.8.

21.10 Consider the reaction: $Co^{2+}(aq) + Ni(s) \rightarrow Co(s) + Ni^{2+}(aq)$.

  a. What is the equilibrium constant for this reaction?
  b. If excess nickel is added to a solution 0.50 M in Co^{2+}, what will be the equilibrium concentration of Ni^{2+}?

21.11 Consider the reaction: $\frac{1}{2} O_2(g) + 2 Cl^-(aq) + 2 H^+(aq) \rightarrow Cl_2(g) + 2 H_2O(aq)$.

  a. What is K for this reaction?
  b. Calculate the concentration of hydrochloric acid at which this reaction would become spontaneous at 1 atmosphere pressure.
  c. Comment on the possibility of preparing chlorine by this method.

21.12 A voltaic cell consists of a copper electrode immersed in 1 M $CuSO_4$ connected via a salt bridge to a solution of 1 M $AgNO_3$ in which is immersed a silver electrode. By what amount will the voltage change if

  a. the concentration of Cu^{2+} is decreased to 0.01 M?
  b. the silver electrode is replaced by one made of platinum?
  c. the area of the copper electrode is doubled?

21.13 Calculate the voltages of cells in which the following reactions occur.

  a. $Zn(s) + 2 H^+(aq, 0.010 M) \rightarrow Zn^{2+}(aq, 0.1 M) + H_2(g, 1 atm)$
  b. $MnO_2(s) + 4 H^+(aq, 6 M) + 2 Br^-(aq, 6 M) \rightarrow Mn^{2+}(aq, 0.1 M) + Br_2(l) + 2 H_2O$
  c. $Fe^{2+}(aq, 0.10 M) + \frac{1}{2} Cl_2(g, 1 atm) \rightarrow Fe^{3+}(aq, 1.0 \times 10^{-4} M) + Cl^-(aq, 1 M)$.

21.14 When the reaction: $Ni(s) + 2 H^+(aq) \rightarrow H_2(g, 1 atm) + Ni^{2+}(aq)$ is carried out in a certain cell, the voltage is found to be 0.01 V. The concentration of Ni^{2+} is known to be 0.10 M; what is the concentration of H^+?

21.28 The standard free energy of formation of a cation in water solution is defined as ΔG for the reaction $M(s) \rightarrow M^{n+}(aq) + n\ e^-$ Calculate $\Delta G_f°$ for

  a. Fe^{2+} b. Li^+ c. Cu^{2+}

21.29 For the reaction:

$$3 A(s) + 2 B^{3+}(aq) \rightarrow 3 A^{2+}(aq) + 2 B(s)$$

the values of $[B^{3+}]$ and $[A^{2+}]$ are 0.020 M and 0.0050 M, respectively. Calculate K, E°, and $\Delta G°$ for this reaction.

21.30 Consider the reaction

$$Fe(s) + 2 Cr^{3+}(aq) \rightarrow Fe^{2+}(aq) + 2 Cr^{2+}(aq)$$

  a. Calculate K for this reaction.
  b. If excess iron is added to a solution 1 M in Cr^{3+}, what will be the concentration of Cr^{2+} at equilibrium?

21.31 Copper(I) salts tend to disproportionate in water solution to copper(II) salts and copper metal, i.e.,

$$Cu^+(aq) \rightarrow Cu^{2+}(aq) + Cu(s)$$

What fraction of Cu^+ originally present would you expect to find at equilibrium?

21.32 In a cell in which iron is corroding, what would you expect to be the voltage difference between two points which differ only in oxygen concentration if the partial pressure of oxygen at one point is 0.20 atm and, at the other point, 0.001 atm? The reaction is:

$$Fe(s) + \frac{1}{2} O_2(g) + H_2O \rightarrow Fe(OH)_2(s)$$

21.33 Calculate the voltages of cells in which the following reactions occur.

  a. $H_2(g, 1 atm) + 2 H^+(aq, 0.10 M) \rightarrow H_2(g, 1 atm) + 2 H^+(aq, 1 \times 10^{-7} M)$
  b. $Cl_2(g, 1 atm) + 2 Br^-(aq, 8 \times 10^{-4} M) \rightarrow 2 Cl^-(aq, 0.60 M) + Br_2(l)$
     (These are approximately the conditions that apply when bromine is extracted from sea water.)

21.34 A certain voltaic cell contains a standard half cell with a reduction potential of -0.15 V. The other half cell consists of a hydrogen electrode (1 atm) immersed in a solution of unknown pH. The voltage of the cell is found to be 0.12 volts. What is the pH of the solution if:

  a. the hydrogen electrode is cathodic?
  b. the hydrogen electrode is anodic?

21.15 A half cell consisting of a silver electrode dipping into a solution of silver nitrate is connected to one in which a copper electrode is in contact with a 0.10 M solution of $Cu(NO_3)_2$. Excess hydrochloric acid is added to the silver half cell to give a heavy precipitate of AgCl and an equilibrium Cl^- concentration of 2.0 M. Under these conditions it is found that the voltage is 0.10 V and that the silver electrode is the anode. Calculate K_{sp} for AgCl.

21.16 Describe, in some detail, an electrochemical cell in which you could determine

a. the ionization constant of acetic acid.
b. K_d for $Cu(NH_3)_4{}^{2+}$.

21.17 Describe in words how you could accomplish the following conversions

a. $Cl^- \rightarrow Cl_2$ d. $Fe^{2+} \rightarrow Fe^{3+}$
b. $Cu \rightarrow Cu(NO_3)_2$ e. $Bi_2S_3(s) \rightarrow Bi^{3+}(aq)$
c. $NO_3^- \rightarrow NO_2$ f. $Br^- \rightarrow Br_2$

21.18 Write balanced equations to represent

a. a process by which NaOCl can be prepared in solution.
b. the commercial preparation of $KClO_3$.
c. the half-reactions that occur at the Sn-Fe interface which lead eventually to the rusting of a tin can.
d. the reaction that occurs when gold is brought into solution in aqua regia.

21.19 Explain why

a. the MnO_4^- ion is a stronger oxidizing agent in acidic than in basic solution.
b. CuS "dissolves" in nitric but not in hydrochloric acid.
c. a steel bridge support shows a heavy deposit of rust just below the water line.
d. corrosion of automobile bodies is accelerated by salt spread on the roads in the winter.
e. concentrated nitric acid usually gives NO_2 as a reduction product.

21.20 It is possible to determine iodide ions in solution by titration with potassium permanganate.

a. Write a balanced equation for the reaction.
b. If 22.4 ml of 0.100 M $KMnO_4$ is required to titrate 16.0 ml of a solution of sodium iodide, what is the concentration of I^- in that solution?

21.35 A student determines the dissociation constant of $[Ag(NH_3)_2{}^+]$ using a voltaic cell. One half cell consists of a silver cathode dipping into a 0.10 M solution of $AgNO_3$. The other contains a silver anode with a solution prepared by adding excess ammonia to 0.10 M $AgNO_3$. The concentration of NH_3 is 0.7 M; the cell voltage is +0.41 V. Calculate the dissociation constant of the complex ion.

21.36 How could you determine experimentally

a. the solubility product of lead chloride?
b. the dissociation constant of water?

21.37 Describe procedures for preparing

a. NO_2 b. SnO_2 c. $KClO_4$

using redox reactions.

21.38 Write balanced equations to describe

a. the electrolysis of a cold, stirred solution of sodium chloride.
b. the half-reaction that occurs beneath the surface of a dirt particle on an iron tool.
c. the reaction of silver with dilute nitric acid.
d. the reaction of hydriodic acid with a solution of potassium dichromate.

21.39 Explain why

a. silver but not gold is oxidized by nitric acid.
b. plumbers are advised not to replace a section of galvanized iron pipe with copper pipe.
c. iron but not aluminum corrodes readily in moist air.
d. in a window screen, rusting is most severe where the wires meet.
e. transition metal salts are frequently used as corrosion inhibitors.

21.40 Potassium permanganate reacts with $C_2O_4{}^{2-}$ ions to form $CO_2(g)$.

a. Write a balanced equation for the reaction.
b. If 18.4 ml of 0.150 M $KMnO_4$ is required to titrate 26.2 ml of a solution of sodium oxalate, what is the concentration of $C_2O_4{}^{2-}$?

*21.41 In analyzing a sample of hard water for Ca^{2+}, it was first boiled down to exactly 10% of its original volume. Excess sodium oxalate was added to 50.0 ml of this solution to give a precipitate of CaC_2O_4. The precipitate was dissolved in 5 ml of 6 M HCl and titrated to an end point with 6.4 ml of 0.050 M $KMnO_4$. What was the concentration of Ca^{2+} in the hard water (cf. Problem 21.40).

*21.42   The oxidation potential for the reaction: $Fe(s) \rightarrow Fe^{2+}(aq) + 2\ e^-$ is independent of pH in strongly acidic solution. However, as the pH is raised, $Fe(OH)_2$ ($K_{sp} = 1 \times 10^{-15}$) starts to precipitate. Assuming the concentration of Fe^{2+} is originally 0.10 M, calculate the value of the oxidation potential at intervals of 1 pH unit from pH = 5 to pH = 10.

*21.43   Consider the half-reactions

$$
\begin{array}{ll}
M(s) \rightarrow M^{a+}(aq) + a\ e^- & E_1^\circ \\
M^{a+}(aq) \rightarrow M^{b+}(aq) + (b-a)\ e^- & E_2^\circ \\
\hline
M(s) \rightarrow M^{b+}(aq) + b\ e^- & E_3^\circ
\end{array}
$$

a. Write expressions for ΔG° for each of the three half-reactions in terms of E°.
b. Knowing that $\Delta G_3^\circ = \Delta G_1^\circ + \Delta G_2^\circ$, derive a relationship that would enable you to calculate E_3° given E_1° and E_2°.
c. Use the relationship in (b) to obtain E° for the half-reactions below, using Table 21.1.

$$
\begin{array}{l}
Fe(s) \rightarrow Fe^{3+}(aq) + 3\ e^- \\
Sn(s) \rightarrow Sn^{4+}(aq) + 4\ e^-
\end{array}
$$

*21.44   Calculate the minimum concentration of HNO_3 required to dissolve CuS ($K_{sp} = 1 \times 10^{-25}$) which is in equilibrium with 1.0 M Cu^{2+}. Assume the products include NO(g) and S(s).

22   nuclear reactions

The "ordinary chemical reactions" that we have discussed up to this point are ones which involve changes in the outer electronic structure of atoms or molecules. In contrast, there is a large class of processes, called nuclear reactions, which are the result of changes taking place within atomic nuclei. Nuclear reactions differ from ordinary chemical reactions in several important respects. In particular:

1. In ordinary reactions, the different isotopes of an element show virtually identical chemical properties; in nuclear reactions they behave quite differently. Consider, for example, the two isotopes of carbon, $^{12}_{6}C$ and $^{14}_{6}C$. The chemical properties of these isotopes are very similar. Their nuclear properties differ considerably; the $^{12}_{6}C$ nucleus is extremely stable while the $^{14}_{6}C$ nucleus decomposes spontaneously.

2. The nuclear reactivity of an element is essentially independent of its state of chemical combination. In the nuclear chemistry of radium, it makes little difference whether we deal with the element itself or one of its compounds. The radium atom in elementary radium and the Ra^{2+} ion in $RaCl_2$ behave similarly from a nuclear standpoint.

In discussing nuclear reactions or writing equations to represent them, we shall not ordinarily be concerned with what happens to the electrons outside the nucleus. Even though the species taking part in these reactions are atoms, molecules, or ions, the reactions themselves occur within the nucleus.

3. Nuclear reactions frequently involve the conversion of one element to another. Whenever a nuclear process results in a change in the number of protons in the nucleus, a new element of different atomic number is formed. In contrast, elements taking part in ordinary chemical reactions retain their identity.

4. Nuclear reactions are accompanied by energy changes which exceed, by several orders of magnitude, those associated with ordinary chemical reactions. The energy evolved when one gram of radium undergoes radioactive decay (Section 22.1) is about 500,000 times as great as that given off when an equal amount of radium reacts with chlorine to form radium chloride. Still larger amounts of energy are given off in nuclear fission (Section 22.5) and nuclear fusion (Section 22.6).

595

22.1   NATURAL RADIOACTIVITY

The first type of nuclear reaction to be studied was that which is associated with natural radioactivity, in which the nucleus of an unstable, naturally occurring isotope decomposes spontaneously. This phenomenon was discovered, almost accidentally, by a French scientist, Henri Becquerel, in 1896. He found that a certain uranium salt, potassium uranyl sulfate, $K_2UO_2(SO_4)_2$, gave off a powerful type of high energy radiation which was capable of blackening a photographic plate. Curiously enough, this radiation seems never to have been detected previously, even though the element uranium had been isolated more than 100 years before.

Becquerel was able to show that the rate at which radiation was emitted from a uranium salt was directly proportional to the amount of uranium present. There was one apparent exception to this rule: a certain uranium ore known as pitchblende gave off radiation at a rate nearly four times as great as one would calculate on the basis of its uranium content. In July of 1898 Marie and Pierre Curie, colleagues of Becquerel at the Sorbonne, were able to isolate from a ton of pitchblende ore a fraction of a gram of a new element which was much more intensely radioactive than uranium. They named this element polonium, after Marie Curie's native country. Six months later the Curies isolated still another, intensely radioactive, previously unknown element, radium. The Nobel Prize for physics in 1903 was awarded jointly to Henri Becquerel and Marie and Pierre Curie; eight years later, Madame Curie received an unprecedented second Nobel Prize, this time in chemistry.

NATURE OF RADIATION

The radiation emitted by naturally radioactive elements can be separated by an electrical or magnetic field into three distinct parts (Fig. 22.1):

α, β and γ rays were originally so named because their nature was unknown

1. **Alpha rays,** which consist of a stream of positively charged particles (alpha particles) that carry a charge of +2 and have a mass of 4 on the atomic weight scale. These particles are identical with the nuclei of ordinary helium atoms (at. no. = 2, mass no. = 4).

When an alpha particle is ejected from the nucleus, there is a decrease of two units in atomic number and a decrease of four in mass number. For example, the loss of an alpha particle by the nucleus of an ordinary uranium atom (at. no. 92, mass no. 238) gives an isotope of thorium with an atomic number of 90 and a mass number of 234. This nuclear reaction may be represented by the equation

$$^{238}_{92}U \rightarrow {}^{4}_{2}He + {}^{234}_{90}Th \tag{22.1}$$

Note that here, as in all nuclear equations, there is a balance of both atomic number $(90 + 2 = 92)$ and mass number $(4 + 234 = 238)$ on the two sides.

2. **Beta rays,** which are made up of a stream of negatively charged particles (beta particles) that have all the properties of electrons. The ejection of a beta particle (mass $\cong$ 0, charge = −1) results from the transformation of a neutron (mass = 1, charge = 0) at the surface of the nucleus into a proton (mass = 1, charge = +1). Consequently, beta-emission leaves the

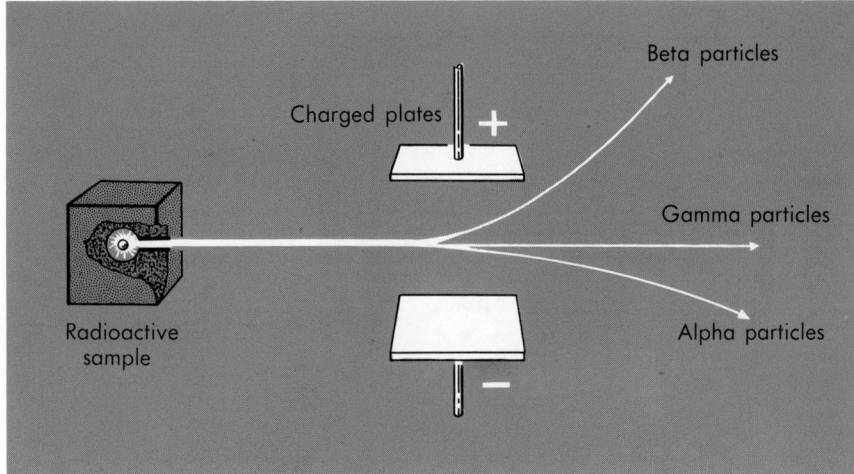

Figure 22.1 Deflection of alpha and beta rays in an electrical field.

mass number unchanged but increases the atomic number by one unit. An example of beta-emission is the spontaneous radioactive decay of thorium-234 (90 protons, 144 neutrons) to protactinium-234 (91 protons, 143 neutrons):

$$\ce{^{234}_{90}Th} \rightarrow \ce{^{0}_{-1}e} + \ce{^{234}_{91}Pa} \tag{22.2}$$

The symbol $_{-1}^{0}e$ is written to represent a beta particle (electron).

3. **Gamma rays,** which consist of electromagnetic radiation of very short wavelength ($\lambda = 0.005$ to 1 Å), i.e., high-energy photons. The emission of gamma rays accompanies virtually all nuclear reactions, as the result of an energy change within the nucleus, whereby an unstable, excited nucleus resulting from alpha- or beta-emission gives off a photon and drops to a lower, more stable energy state. Since gamma-emission changes neither the atomic number nor the mass number, we shall frequently neglect it in writing nuclear equations.

γ rays penetrate matter much more easily than do α or β rays

INTERACTION OF RADIATION WITH MATTER

The alpha, beta, and gamma rays given off during radioactive decay lose their energy when they pass through matter by transferring it to atoms, molecules, or ions with which they collide. These collisions may be elastic, in the sense that only kinetic energy is transferred, thereby raising the temperature of the exposed material. The increased kinetic energy of the target particles is eventually translated into heat, which is given off to the surroundings.

Frequently the interaction of radiation with matter results in inelastic collisions in which the potential energy of the target species is raised. A common type of inelastic collision is one in which an electron is excited to a higher energy level. When the electron drops back to its ground state, energy is given off as electromagnetic radiation which may be visible light ($\lambda = 4000$ to 8000 Å), ultraviolet light ($\lambda = 100$ to 4000 Å), or x-rays ($\lambda =$

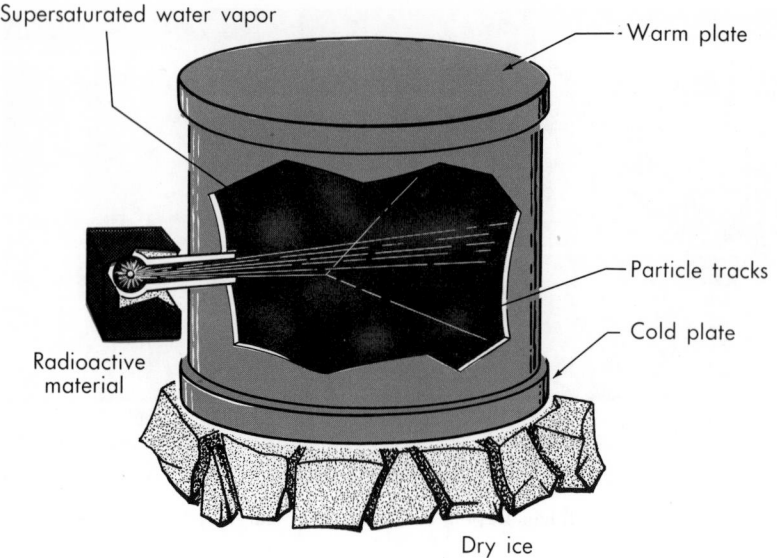

Supersaturated water vapor

Warm plate

Particle tracks

Cold plate

Radioactive material

Dry ice

In recent years bubble chambers containing liquid hydrogen have become popular as particle track detectors. Can you suggest how they work?

Figure 22.2 Cloud chamber. Supersaturated vapor condenses as droplets on ions formed by radiation.

0.05 to 100 Å). Radium salts give off an intense luminescence, which is easily visible in the dark and can even be detected in daylight with a sample containing more than 0.1 g of radium. The dials of luminous watches are painted with a mixture containing a tiny amount of a radium salt and a substance such as zinc sulfide or anthracene which fluoresces upon exposure to radiation. An instrument commonly used to detect and measure radiation, the *scintillation counter*, uses this same principle. Light produced by radiation striking an organic solid or solution activates a photoelectric cell within the counter.

In another type of inelastic collision, an electron is removed from an atom or molecule to form a positively charged ion. The ionizing ability of alpha, beta, and gamma rays is utilized in several instruments used to study radiation. In the Wilson cloud chamber (Fig. 22.2), the ions produced serve as nuclei upon which tiny water droplets condense. The path of an alpha or beta particle through the chamber can be observed by following the condensation trail produced in the vapor. The Geiger-Müller counter (Fig. 22.3) amplifies the electric current produced by a cascading flow of electrons and positive ions to oppositely charged electrodes. The current pulse resulting from each ionization is amplified and detected by means of an electrically activated counting device.

The harmful effect of high-energy radiation on human tissue is caused by its ability to ionize the organic molecules of which body cells are composed. Table 22.1 lists some of the effects to be expected when a human being is exposed to a single dose of radiation of successively higher energies. Exposure to about 100 rads (a rad corresponds to the absorption of 100 ergs of energy per gram of tissue) brings about the typical symptoms of radiation sickness. Doses of 500 rads or more are almost certain to result in death.

Small doses of radiation repeated over long periods of time can also

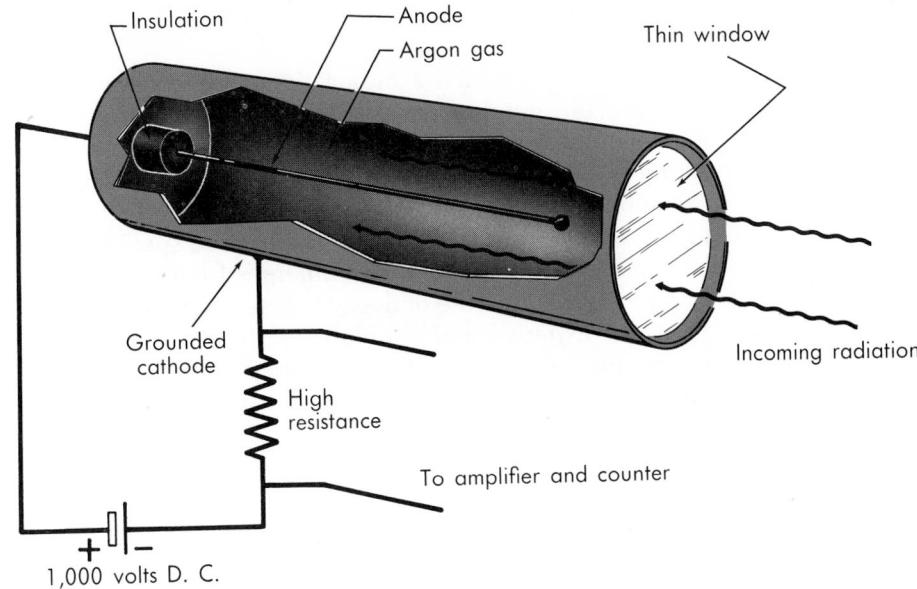

Figure 22.3 Geiger-Müller counter. Ions produced by radiation cause an electrical discharge.

have extremely serious consequences. Many of the early workers in the field of radioactivity developed cancer as a result of chronic overexposure to radiation. Cases are known in which cancers developed as long as 40 years after initial exposure. Recent studies have shown an abnormally large number of cases of leukemia among the survivors of the nuclear bombs detonated at Hiroshima and Nagasaki.

As late as 1940 the dangers from radiation were not commonly recognized

It has long been known that radiation can produce mutations in plants and animals by bringing about changes in chromosomes. There is every reason to suppose that similar genetic effects can arise in human beings as well. Statistical surveys of the children of American radiologists indicate an increased frequency of congenital defects. This is confirmed by studies which have been made of children born to the survivors of Nagasaki and Hiroshima. Perhaps the most disturbing aspect of this problem is that there appears to be no lower limit or "tolerance level" below which the genetic effects of radiation become negligible. Even a small increase in the background level of radiation can be expected to produce a proportional increase in undesirable mutations. Considerations of this type played a major role in achieving the Nuclear Test Ban Treaty of 1963.

TABLE 22.1 Effect of Exposure to a Single Dose of Radiation

DOSE (RADS)	PROBABLE EFFECT
0 to 25	?
25 to 50	Small decrease in white blood cell count
50 to 100	Lesions, marked decrease in white blood cells
100 to 200	Nausea, vomiting, loss of hair
200 to 500	Hemorrhaging, ulcers, possible death
500+	Fatal

22.2 RATE OF RADIOACTIVE DECAY

The rate at which a radioactive sample decays can be measured by counting the number of particles given off in unit time. Modern instruments for measuring radioactivity such as scintillation counters and Geiger counters do this automatically. In interpreting data obtained with such instruments, it is ordinarily necessary to correct for the "background" radiation given off by naturally radioactive species in the environment.

One generalization which emerges from rate studies with radioactive isotopes is that the rate of decay is essentially independent of temperature. From an experimental standpoint, this eliminates the need for precise temperature control in such studies. From a theoretical point of view, it implies that the activation energy for radioactive decay is zero.

FIRST ORDER RATE LAW

In a sample containing a radioactive species, the probability that an atom will decay in a given interval of time is the same for all atoms in the sample. This means that the number of atoms, ΔX, which decay in a given time interval, Δt, is directly proportional to the number of atoms present, X.

$$\frac{\Delta X}{\Delta t} = kX \quad \text{or} \quad \Delta X/X = k\,\Delta t$$

You may recognize this equation as being of the same form as that for a first order reaction (Chapter 14). Upon integration, we arrive at the familiar equation

$$\log_{10} \frac{X_0}{X} = \frac{k\,t}{2.30} \tag{22.3}$$

where X_0 is the amount of radioactive material at zero time (i.e., when the counting process starts), and X is the amount remaining after time t. The first order rate constant, k, is characteristic of the isotope undergoing radioactive decay.

The application of Equation 22.3 to calculations involving the rate of radioactive decay is illustrated in Example 22.1.

Example 22.1   One of the most dangerous radioactive isotopes produced when a nuclear bomb explodes is strontium-90, $^{90}_{38}\text{Sr}$. If one starts with 1.000 g of this isotope, it is found that after five years 0.887 g remains. Determine

    a. the rate constant, k, for the decay of strontium-90.
    b. the amount left after 100 years.

Solution

    a. Substituting into Equation 22.3:

$$\log \frac{1.000}{0.887} = \frac{k\,(5.00 \text{ yr})}{2.30}$$

Solving: $0.0521 = (2.17 \text{ yr}) k; \quad k = 0.0240 \text{ yr}^{-1}$

b. Here, $X_0 = 1.000$ g, $t = 100$ yr, $k = 0.0240$ yr^{-1}

$$\log \frac{1.000}{X} = \frac{(0.0240)(100)}{2.30} = 1.04$$

Taking antilogs: $\frac{1.000}{X} = 11; \quad X = 0.091$ g

HALF-LIFE

You will recall from Chapter 14 that the time required for one half of a sample to decompose via a first order reaction is independent of the amount of sample and, hence, is a characteristic property of that reaction. Decay rates of radioactive isotopes are also frequently expressed in terms of their half-lives. Certain isotopes have extremely long half-lives and consequently produce a very low level of radioactivity. An example is the most common isotope of uranium, which decays according to Equation 22.1, with a half-life of 4.5×10^9 years. At the opposite extreme is the isotope polonium-214, which decays by α-emission with a half-life of 1.6×10^{-4} seconds. Within a second, virtually all the radiation associated with this isotope has been dissipated. Needless to say, isotopes such as these produce a tremendously high level of radiation during their brief existence.

Table 22.2 illustrates how the concept of half-life can be used to estimate the fraction of a radioactive sample remaining after a given number of half-life periods have elapsed.

It is possible to calculate the rate constant for radioactive decay from the half-life, using the equation derived in Chapter 14:

$$k = \frac{0.693}{t_{1/2}} \tag{22.4}$$

Example 22.2 The common radioactive isotope of radium, $^{226}_{88}$Ra, has a half-life of 1620 years. Calculate

    a. the first order rate constant for the decay of radium-226.
    b. the fraction of a sample of this isotope which will remain after 100 years.

Solution

  a. $k = \dfrac{0.693}{1620 \text{ yr}} = 4.28 \times 10^{-4}$ yr^{-1}

  b. $\log \dfrac{X_0}{X} = \dfrac{(4.28 \times 10^{-4} \text{ yr}^{-1})}{2.30} 100 \text{ yr} = 0.0186$

taking antilogs: $X_0/X = 1.044$

The fraction *remaining* is

$$\frac{X}{X_0} = \frac{1}{1.044} = 0.956 \text{ or } 95.6\%$$

This problem illustrates a safety hazard inherent in any attempt to "dispose" of a sample of a relatively long-lived isotope. A sample of a radium salt cannot simply be discharged into the environment in the vain hope that it will decompose rapidly; the level of radiation would be virtually unchanged a century from now. This factor has to be taken into account in any proposed system of radioactive waste disposal.

TABLE 22.2 Rate of Decay of $^{210}_{83}$Bi ($t_{1/2} = 5$ days)

TIME (DAYS)	NO. OF HALF-LIVES	FRACTION LEFT	FRACTION DECAYED
0	0	1	0
5	1	$\frac{1}{2}$	$\frac{1}{2}$
10	2	$\frac{1}{4}$	$\frac{3}{4}$
15	3	$\frac{1}{8}$	$\frac{7}{8}$
20	4	$\frac{1}{16}$	$\frac{15}{16}$
.	.	.	.
.	.	.	.
.	.	.	.
5n	n	$(\frac{1}{2})^n$	$1 - (\frac{1}{2})^n$

AGE OF ROCKS

Certain radioactive isotopes act as "natural clocks" which can help us to determine the time at which various rock deposits solidified or, in other words, to estimate their age. To understand how this can be done, consider a uranium-bearing rock, formed billions of years ago by solidification from a molten matrix. Once the rock became solid, the products of radioactive decomposition of uranium were no longer able to diffuse away and, hence, were incorporated into the rock. Over a long period of time these products, all of which have relatively short half-lives, were ultimately converted to lead-206. The overall equation for the decay process can be written

The decay of $^{238}_{92}$U to $^{206}_{82}$Pb occurs by a long series of steps, of which the slowest has a $t_{1/2}$ of 4.5×10^9 yrs

$$^{238}_{92}\text{U} \rightarrow {}^{206}_{82}\text{Pb} + 8\, {}^4_2\text{He} + 6\, {}_{-1}^{\ 0}\text{e}; \quad t_{1/2} = 4.5 \times 10^9 \text{ yrs} \qquad (22.5)$$

Knowing the half-life for this process, it should then be possible, by measuring the ratio of lead-206 to uranium-238 in the rock today, to calculate the time that has elapsed since the rock solidified. If we should find, for example, that equal numbers of atoms of these two isotopes were present, we would infer that the rock must be about 4.5×10^9 (4.5 billion) years old.

This method of estimating the age of mineral deposits assumes, among other things, that none of the lead-206 has become separated from the parent uranium-238. It is possible to check the validity of this assumption by referring to other "radioactive clocks" which operate in nature. One of these is the β-decay of rubidium-87, which has a half-life of 5.7×10^{10} years.

$$^{87}_{37}\text{Rb} \rightarrow {}^{87}_{38}\text{Sr} + {}_{-1}^{\ 0}\text{e}; \quad t_{1/2} = 5.7 \times 10^{10} \text{ yr} \qquad (22.6)$$

Ages of rocks determined by these and other radioactive methods range from 3 to 4.5×10^9 years; the latter number is often taken as an approximate value for the age of the earth. Interestingly enough, analyses of rock samples taken from the moon's surface indicate ages in this same range. This would seem to eliminate the once prevalent idea that the moon was torn from the earth's surface by a cataclysmic event a long time after the earth solidified.

AGE OF ORGANIC MATERIAL

During the 1950's Professor W. F. Libby and others worked out a method based upon the decay rate of a naturally occurring isotope, carbon-14, for determining the age of organic matter. This method can be applied to objects from a few hundred up to 50,000 years old. It has been used, for example, to check the authenticity of canvases of Renaissance painters and to determine the age of relics left by prehistoric cavemen.

Carbon-14 is produced in the atmosphere by the interaction of neutrons from cosmic radiation with ordinary nitrogen atoms:

$$^{14}_{7}N + ^{1}_{0}n \rightarrow ^{14}_{6}C + ^{1}_{1}H \tag{22.7}$$

The carbon-14 formed by this nuclear reaction is eventually incorporated into the carbon dioxide of the air. A steady-state concentration, amounting to about one atom of carbon-14 for every 10^{12} atoms of carbon-12, is established in atmospheric CO_2. A living plant, taking in carbon dioxide, has this same $^{14}C/^{12}C$ ratio, as do plant-eating animals or human beings.

This method depends on a constant $^{14}C/^{12}C$ ratio over time. Why?

When a plant or animal dies, the intake of radioactive carbon stops. Consequently the radioactive decay of carbon-14

$$^{14}_{6}C \rightarrow ^{14}_{7}N + ^{0}_{-1}e \text{ (half-life = 5720 years)} \tag{22.8}$$

takes over and the ratio of $^{14}C/^{12}C$ drops. By measuring this ratio and comparing it to that in living plants, one can estimate the time at which the plant or animal died (Example 22.3).

Example 22.3 A tiny piece of paper taken from the Dead Sea scrolls, believed to date back to the 1st century, A.D., was found to have a $^{14}C/^{12}C$ ratio 0.795 times that in a plant living today. Estimate the age of the scrolls.

Solution. Knowing the half-life of carbon-14 ($t_{1/2} = 5720$ yrs), we can calculate the first order rate constant from Equation 22.4. Then, using Equation 22.3, we can obtain the elapsed time

$$k = \frac{0.693}{5720 \text{ yrs}} = 1.21 \times 10^{-4} \text{ yr}^{-1}$$

$$\log \frac{X_0}{X} = \frac{1.21 \times 10^{-4} \text{ yr}^{-1} \times t}{2.30}$$

But, $X = 0.795 X_0$, so: $\log \dfrac{X_0}{X} = \log \dfrac{1.000}{0.795} = \log 1.26 = 0.100$

Hence, $0.100 = \dfrac{1.21 \times 10^{-4} \text{ yr}^{-1} \times t}{2.30}$; $t = 1900$ yrs

---

22.3 BOMBARDMENT REACTIONS, ARTIFICIAL RADIOACTIVITY

Prior to 1934 the study of radioactivity was limited to the relatively few radioactive isotopes found in nature. In that year, Irene (daughter of Marie and Pierre) Curie and her husband, Frédéric Joliot, announced the preparation of the first man-made radioactive isotopes. They achieved this by bombarding certain stable isotopes with high-energy α-particles. A typical reaction is:

$$^{27}_{13}Al + ^{4}_{2}He \rightarrow ^{30}_{15}P + ^{1}_{0}n \tag{22.9}$$

The product of this nuclear reaction, phosphorus-30, is radioactive. It decays by emitting a particle called a **positron,** which has the same mass as an electron but a charge of +1 rather than −1.

$$\mathstrut^{30}_{15}\text{P} \rightarrow \mathstrut^{30}_{14}\text{Si} + \mathstrut^{0}_{1}\text{e} \qquad (22.10)$$

In the past 40 years, hundreds of different radioactive isotopes have been prepared in the laboratory by bombardment reactions. At least one such isotope has been formed for every element that occurs in nature; in addition, isotopes of 17 previously unknown elements have been prepared. The bombarding particles used to induce nuclear reactions include the following:

The goal of the alchemist, the transmutation of elements, was achieved in these experiments

POSITIVE IONS (PROTONS, DEUTERONS, α-PARTICLES, . . .). When cations are accelerated to very high velocities, they can acquire sufficient energy to overcome the coulombic repulsion of the target nucleus and bring about a nuclear reaction. Nuclear physicists and engineers have built several different types of instruments to form focused beams of high energy positive ions. One of these, the cyclotron, designed originally by E. O. Lawrence at the University of California, is shown schematically in Figure 22.4.

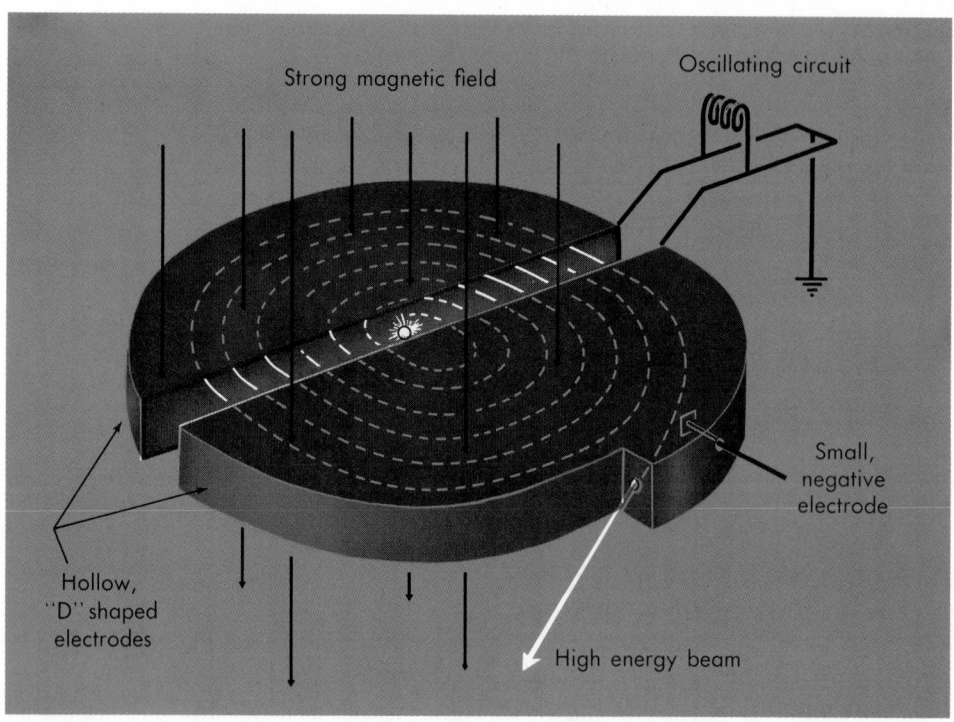

Figure 22.4 The cyclotron consists of two oppositely charged, evacuated "dees" placed between the poles of a powerful electromagnet. Positive ions, originating at the center, enter the upper dee, which is originally at a negative potential. They pass through this dee in a curved path. At the instant they reenter the central corridor, the polarity of the dees is reversed, and the particles enter the lower dee at an increased velocity. This procedure is repeated over and over; the particles move at higher and higher velocities in paths of greater and greater radius. Eventually they are deflected from the periphery of one of the dees to strike the target.

PHOTONS (GAMMA OR X-RAYS). When a beam of high-energy electrons impinges upon a target, photons with very short wavelengths are produced. These are capable of inducing nuclear reactions such as

$$_{35}^{81}Br + _{0}^{0}\gamma \rightarrow _{35}^{80}Br + _{0}^{1}n \tag{22.11}$$

NEUTRONS. Since a neutron experiences no coulombic repulsion when it approaches a nucleus, it need have only a very small kinetic energy to initiate a nuclear reaction. The only important source of neutrons today is the nuclear reactor, described in Section 22.5.

TRANSURANIUM ELEMENTS

Bombardment reactions have proved particularly useful in synthesizing isotopes of elements which do not occur in nature. Thirty some years ago the periodic table ended with uranium (at. no. 92). Before uranium, there were four gaps in the table, corresponding to elements 43, 61, 85, and 87. To be sure, various groups of scientists had reported the discovery of these elements and assigned names to them, but none of these claims had been substantiated. Within a period of about five years, between 1937 and 1942, radioactive isotopes of these elements (technetium, at. no. 43; promethium, at. no. 61; astatine, at. no. 85; francium, at. no. 87) were synthesized in the laboratory.

During the past 30 years, Glenn Seaborg and his colleagues at the University of California at Berkeley have prepared and identified a series of transuranium elements extending now to element 105. The first of these elements to be discovered, neptunium (at. no. 93) and plutonium (at. no. 94), were reported by McMillan and Abelson at Berkeley in 1940. They were produced by bombarding the common isotope of uranium, uranium-238, with low energy neutrons. The unstable isotope uranium-239 formed by this nuclear reaction decays by successive β-emissions to give first neptunium-239 and then plutonium-239.

$$_{92}^{238}U + _{0}^{1}n \rightarrow _{92}^{239}U \rightarrow _{93}^{239}Np + _{-1}^{0}e \tag{22.12}$$
$$\longrightarrow _{94}^{239}Pu + _{-1}^{0}e$$

In principle, neutron bombardment can be used to prepare a wide variety of transuranium elements. The two elements einsteinium (at. no. 99) and fermium (at. no. 100) were first discovered in uranium that had been exposed to the very high neutron density accompanying a thermonuclear explosion (Section 22.5). Unfortunately the yield of transuranium isotopes formed by neutron bombardment decreases exponentially with increasing atomic number. A more practical way to form isotopes with atomic numbers greater than 100 is to bombard appropriate targets with high-energy, positively charged particles. In this way the element mendelevium (at. no. 101) was prepared by bombarding einsteinium with α-particles:

$$_{99}^{253}Es + _{2}^{4}He \rightarrow _{101}^{256}Mv + _{0}^{1}n \tag{22.13}$$

By using heavier nuclei as bombarding particles, it is possible to use target isotopes of lower atomic number, which are available in greater quantities. In 1969 a research group at Berkeley headed by Albert Ghiorso used carbon-12 nuclei to prepare element-104 from californium

$$^{249}_{98}\text{Cf} + ^{12}_{6}\text{C} \rightarrow ^{257}_{104}? + 4^{1}_{0}\text{n} \qquad (22.14)$$

One year later the preparation of element 105 was announced, using the same target isotope but substituting nitrogen-14 for carbon-12 as a bombarding particle. The names of elements 104 and 105 have not been established; the Berkeley group has suggested rutherfordium and hahnium, honoring Ernest Rutherford and Otto Hahn (discoverer of nuclear fission). A Russian group led by G. N. Flerov, which reported prior but unsubstantiated evidence for the formation of these elements, prefers the names bohrium and kurchatovium, after Niels Bohr and the Russian physicist I. V. Kurchatov.

With few exceptions, the isotopes of the transuranium elements have very short half-lives. Moreover, most of them, especially those of very high atomic number, have been formed in extremely minute quantities, amounting in some cases to only a few atoms. One of the greatest achievements of the scientists working in this field has been their ability to work out methods of studying the chemical and physical properties of submicrogram amounts of these elements. Both chemical and physical evidence indicate that the transuranium elements up to atomic number 103 are filling out a second rare-earth series by completing the 5f subshell. Elements 104 and 105 would then be the first members of a new transition series, falling below hafnium and tantalum in the Periodic Table.

The prospects for extending the number of transuranium elements beyond those presently known appear somewhat brighter than they did a few years ago. Progress along this line will depend primarily upon the ability to develop more powerful accelerators so that larger and larger bombarding particles can be used. In this way, Dr. Seaborg has suggested, it should be possible to synthesize elements of considerably higher atomic number than those now known.

DECAY OF ARTIFICIALLY PRODUCED RADIOACTIVE ISOTOPES

The radioactive isotopes produced by bombardment reactions finally decay, often in a sequence of steps, to one or another of the 200-odd stable isotopes of elements ranging from hydrogen to bismuth (at. no. 83). The mode of decay depends upon how the neutron-to-proton ratio in the nucleus of the radioactive isotope compares to that of stable nuclei in the vicinity (Fig. 22.5).

NEUTRON-TO-PROTON RATIO TOO HIGH. An isotope which lies above the "stability curve" shown in Figure 22.5 can become more stable if a neutron in its nucleus is converted to a proton by the process

$$^{1}_{0}\text{n} \rightarrow ^{1}_{1}\text{H} + ^{0}_{-1}\text{e}$$

Beta-emission can be expected to occur whenever the neutron-to-proton ratio is too high; this is almost always the case when the *mass number of the radioactive isotope is greater than the average atomic weight of the element.* An example is the beta-decay of $^{14}_{6}\text{C}$ (at. wt. C = 12):

$$^{14}_{6}\text{C} \rightarrow ^{14}_{7}\text{N} + ^{0}_{-1}\text{e} \qquad (22.15)$$

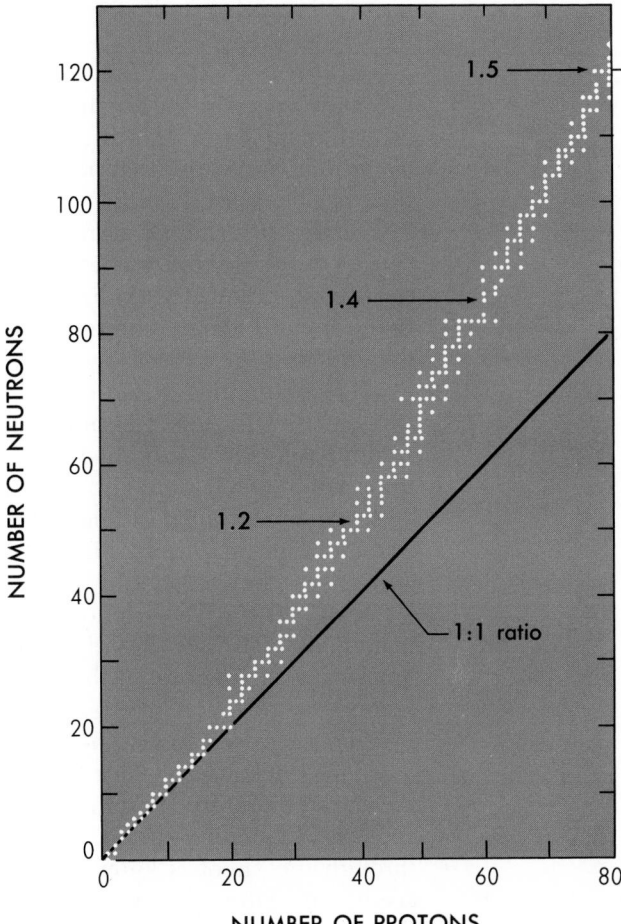

Figure 22.5 Neutron-to-proton ratio in stable isotopes. Ratio increases with increasing atomic number.

NEUTRON-TO-PROTON RATIO TOO LOW. Isotopes lying below the dotted curve in Figure 22.5 can achieve stability by converting a proton into a neutron. One way in which this can happen is by positron emission

$$\mathrm{^1_1H \rightarrow \ ^1_0n + \ ^0_1e}$$

as in Equation 22.10. The same objective can be achieved if the nucleus captures one of the extranuclear electrons

$$\mathrm{^1_1H + \ ^0_{-1}e \rightarrow \ ^1_0n}$$

The electron which enters the nucleus to bring about this conversion is ordinarily one in the K shell closest to the nucleus. The process, known as *K-electron capture*, seems to be preferred to positron emission for isotopes of high atomic number. (Why?) An example is seen in the decay of rubidium-82.

K-electrons are
1s electrons

$$\mathrm{^{82}_{37}Rb + \ ^0_{-1}e \rightarrow \ ^{82}_{36}Kr} \qquad (22.16)$$

USES OF RADIOACTIVE ISOTOPES

The large number of radioactive isotopes now available have been used to study a variety of problems in basic and applied research. Techniques involving radioactive isotopes have proved extremely useful in four major areas.

MEDICINE. The high-energy radiation given off by radium was used for many years in the treatment of cancer to destroy or arrest the growth of malignant tissue. More recently a radioactive isotope of cobalt, cobalt-60, which is cheaper than radium and gives off even more powerful radiation, has been used for this purpose. Certain types of cancer can be treated internally with the aid of radioactive isotopes. If a patient suffering from cancer of the thyroid (malignant goiter) drinks a solution of sodium iodide containing radioactive iodine (^{131}I or ^{128}I), the iodine moves preferentially to the thyroid gland, where the radiation destroys the malignant cells without affecting the rest of the body.

Trace amounts of radioactive samples injected into the blood stream can be used to detect circulatory disorders. For example, by injecting a sodium chloride solution containing a small amount of radioactive sodium into the leg of a patient and measuring the build-up of radiation in the foot, a physician can quickly find out whether the circulation in that area is abnormal. This may help him to decide whether amputation is necessary and, if so, where it should be done.

INDUSTRY. The frictional wear of piston rings can be monitored by making a test ring slightly radioactive by neutron bombardment and measuring the activity of the iron dust in the lubricating oil that circulates around the piston. The rate of corrosion of steel and the locations at which it is most severe can also be measured by a similar technique.

The thickness of very thin sheets of metal, paper, or plastic can be measured by interposing them between a radioactive source and a detector such as a Geiger counter. If the fraction of the radiation which passes through the sheet is known, it is possible to estimate its thickness quite accurately. Thin spots where the sheet might break down in use can also be detected.

ANALYTICAL CHEMISTRY. Techniques involving radioactive isotopes have proved extremely useful for determining trace elements. One of the most important is an approach known as *neutron activation analysis*. This involves bombarding a sample, containing the element to be analyzed for, with neutrons, thereby forming a radioactive isotope within the sample. The resulting level of radioactivity is measured; knowing the efficiency of the bombardment process, it is then possible to calculate how much of the element in question is present.

Neutron activation analysis is now applied routinely for many of the common metals. As little as 10^{-6} g of an element can be readily detected by this technique; in the most favorable cases, an impurity present to the extent of 10^{-10} g shows up. For example, it is possible to analyze for mercury in fish (Chapter 11) at the part per million level or less; the limit of precision appears to be about ±0.01 ppm (10^{-8} g of mercury per gram of sample).

Archaeologists frequently use this technique to investigate the origin of ancient artifacts because it can be applied without destroying the sample. Neutron activation analysis was used to determine the percentages of sodium and manganese in obsidian arrowheads used by the Hopewell Indians who lived in Ohio, Indiana, and Illinois in the period 300 B.C. to 500 A.D. By comparing the percentages of Na and Mn in these arrowheads to those of obsidian deposits in various parts of the United States and

Radiation is still one of the main methods used to treat cancer patients

Mexico, it was shown that they must have come from an area in Yellow- Tourists, maybe?
stone Park, more than 1000 miles away.

MECHANISM DETERMINATION. Chemists studying the rates and mechanisms of reaction frequently use radioactive isotopes to trace the path of a particular element as it passes through various steps from reactant to final product. Organic chemists in particular have learned a great deal about the mechanism of some rather complex reactions by using carbon-14 as a tracer.

Unfortunately several of the more common elements do not have radioactive isotopes sufficiently long-lived to be used as tracers. An important example is oxygen; $^{15}_{8}O$, the longest lived radioactive isotope, has a half-life of only 2 minutes. It is possible to follow reaction mechanisms involving oxygen atoms by enriching with oxygen-18, a stable isotope which is quite rare in nature (abundance $= 0.20\%$). This isotope is readily detected by using a mass spectrometer (Chapter 2) which distinguishes it from the common isotope, oxygen-16.

Inorganic chemists have used oxygen-18 as a tracer to determine the lability or inertness of complex ions. If a sample containing the cation $Fe(H_2O)_6{}^{3+}$ in which the water molecules are enriched in oxygen-18 is added to ordinary water it is found that exchange with the solvent takes place instantaneously. Within a fraction of a second, we reach an equilibrium distribution of $^{18}_{8}O$ between complex and water. In contrast, this same process takes place much more slowly with the inert complex $Cr(H_2O)_6{}^{3+}$. The half-life for the reaction:

$$Cr(H_2O)_5(H_2O^*)^{3+}(aq) + H_2O \rightarrow Cr(H_2O)_6{}^{3+}(aq) + H_2O^* \quad (22.17)$$
$$(^* = {}^{18}_{8}O)$$

is found to be 40 hours. Another reaction which has been studied by tracer techniques using oxygen-18 is the extremely important natural process of photosynthesis:

$$6\ CO_2(g) + 6\ H_2O(l) \rightarrow C_6H_{12}O_6(s) + 6\ O_2(g) \quad (22.18)$$

It was shown many years ago that oxygen-18 introduced into the water consumed in this reaction showed up in the oxygen gas evolved. If, on the other hand, the CO_2 is enriched in oxygen-18, the product is ordinary O_2. Clearly the elementary oxygen given off by green plants must be formed from water rather than from carbon dioxide.

22.4 MASS-ENERGY RELATIONS

We pointed out at the beginning of this chapter that the energy change associated with nuclear reactions is greater by several orders of magnitude than that which accompanies ordinary chemical reactions. The energy change can be calculated from Einstein's equation,

$$\Delta E = \Delta mc^2 \quad (22.19)$$

where Δm is the change in mass,* ΔE is the change in energy, and c is the velocity of light. If one substitutes for c the value 3.00×10^{10} cm/sec, Equa-

---

* Specifically, Δm = mass of products − mass of reactants; ΔE = energy of products − energy of reactants. In most nuclear reactions, the products weigh less than the reactants (Δm negative); in this case, the energy of the products is less than that of the reactants (ΔE negative), and energy is evolved to the surroundings.

tion 22.19 gives directly the relation between the energy change in ergs and the mass change in grams:

$$\Delta E \text{ (in ergs)} = 9.00 \times 10^{20} \times \Delta m \text{ (in grams)}$$

In dealing with nuclear reactions we are frequently interested in obtaining ΔE in units other than ergs. In particular, we may wish to calculate ΔE in kilocalories or in *millions of electron volts*, MeV. (An electron volt is the energy acquired by an electron when it falls through a potential drop of one volt.) Again, it is often convenient to express the mass change in moles or in *atomic mass units*, amu. (An atomic mass unit is 1/12 of the mass of a carbon-12 atom; this means that the mass of a particle in amu is numerically equal to its atomic weight on the carbon-12 scale.) Calculations involving mass-energy conversions are facilitated by conversion factors such as those given in Table 22.3.

TABLE 22.3   Mass-Energy Conversion Factors

TYPE OF CONVERSION	CONVERSION FACTOR
mass-mass	$1 \text{ g} = 6.02 \times 10^{23}$ amu
energy-energy	$1 \text{ erg} = 2.39 \times 10^{-11}$ kcal
	$1 \text{ erg} = 6.24 \times 10^{5}$ MeV
mass-energy	$1 \text{ g} \simeq 9.00 \times 10^{20}$ ergs
	$1 \text{ g} \simeq 2.15 \times 10^{10}$ kcal
	$1 \text{ amu} \simeq 931$ MeV

Using these conversion factors along with the appropriate nuclear masses (Table 22.4), it is possible to calculate the energy change accompanying any nuclear reaction (Example 22.4).

Example 22.4   For the nuclear reaction $^{226}_{88}\text{Ra} \rightarrow {}^{222}_{86}\text{Rn} + {}^{4}_{2}\text{He}$

    a. calculate ΔE, in MeV, when one atom of radium decays.
    b. calculate ΔE, in ergs, when one mole of radium decays.
    c. calculate ΔE, in kcal, when one g of radium decays.

Solution

    a. We shall first calculate Δm, in amu, and then convert this to energy in MeV.

$$\Delta m = \text{mass } {}^{4}_{2}\text{He} + \text{mass } {}^{222}_{86}\text{Rn} - \text{mass } {}^{226}_{88}\text{Ra}$$
$$= 4.0015 \text{ amu} + 221.9703 \text{ amu} - 225.9771 \text{ amu}$$
$$= -0.0053 \text{ amu}$$

(Note that Δm is extremely small; it is necessary to know the masses of the various particles very accurately to obtain an answer accurate to two significant figures.)

$$\Delta E = -0.0053 \text{ amu} \times 931 \frac{\text{MeV}}{\text{amu}} = -4.9 \text{ MeV}$$

i.e., 4.9 MeV of energy are *evolved* when an atom of radium decays.

TABLE 22.4   Nuclear Masses in Atomic Mass Units °

	At. no.	Mass no.	Mass		At. no.	Mass no.	Mass
n	0	1	1.00867	Br	35	79	78.8992
H	1	1	1.00728		35	81	80.8971
	1	2	2.01355		35	87	86.9028
	1	3	3.01550	Rb	37	89	88.8909
He	2	3	3.01493	Sr	38	90	89.8864
	2	4	4.00150	Mo	42	99	98.8849
Li	3	6	6.01348	Ru	44	106	105.8829
	3	7	7.01436	Ag	47	109	108.8789
Be	4	9	9.00999	Cd	48	109	108.8786
	4	10	10.01134		48	115	114.8793
B	5	10	10.01019	Sn	50	120	119.8747
	5	11	11.00656	Ce	58	144	143.8816
C	6	11	11.00814		58	146	145.8865
	6	12	11.99671	Pr	59	144	143.8807
	6	13	13.00006	Sm	62	152	151.8853
	6	14	13.99995	Eu	63	157	156.8914
O	8	16	15.99052	Er	68	168	167.8941
	8	17	16.99474	Hf	72	179	178.9048
	8	18	17.99477	W	74	186	185.9107
F	9	18	17.99601	Os	76	192	191.9187
	9	19	18.99346	Au	79	196	195.9231
Na	11	23	22.98373	Hg	80	196	195.9219
Mg	12	24	23.97845	Pb	82	206	205.9295
	12	25	24.97925		82	207	206.9309
	12	26	25.97600		82	208	207.9316
Al	13	26	25.97977	Po	84	210	209.9368
	13	27	26.97439		84	218	217.9628
	13	28	27.97477	Rn	86	222	221.9703
Si	14	28	27.96924	Ra	88	226	225.9771
S	16	32	31.96329	Th	90	230	229.9837
Cl	17	35	34.95952	Pa	91	234	233.9934
	17	37	36.95657	U	92	233	232.9890
Ar	18	40	39.95250		92	235	234.9934
K	19	39	38.95328		92	238	238.0003
	19	40	39.95358		92	239	239.0038
Ca	20	40	39.95162	Np	93	239	239.0019
Ti	22	48	47.93588	Pu	94	239	239.0006
Cr	24	52	51.92734		94	241	241.0051
Fe	26	56	55.92066	Am	95	241	241.0045
Co	27	59	58.91837	Cm	96	242	242.0061
Ni	28	59	58.91897	Bk	97	245	245.0129
Zn	30	64	63.91268	Cf	98	248	248.0186
	30	72	71.91128	Es	99	251	251.0255
Ge	32	76	75.90380	Fm	100	252	252.0278
As	33	79	78.90288		100	254	254.0331

° Note that these are *nuclear masses*. The masses of the corresponding atoms can be calculated by adding the mass of the extranuclear electrons (mass of electron = 0.000549 amu). For example, the mass of an *atom* of $^4_2$He is

$$4.00150 \text{ amu} + 2(0.000549) \text{ amu} = 4.00260 \text{ amu}$$

Similarly, the mass of an atom of $^{12}_6$C is

$$11.99671 \text{ amu} + 6(0.000549) \text{ amu} = 12.00000 \text{ amu}$$

b. Here, we calculate the mass change in grams and convert to ergs:

$$\Delta m = \text{mass 1 mole } {}_{2}^{4}\text{He} + \text{mass 1 mole } {}_{86}^{222}\text{Rn} - \text{mass 1 mole } {}_{88}^{226}\text{Ra}$$
$$= 4.0015 \text{ g} + 221.9703 \text{ g} - 225.9771 \text{ g} = -0.0053 \text{ g}$$

(Note that Δm in grams for the decay of a mole of radium is numerically equal to Δm in amu for the decay of one atom of radium.)

$$\Delta E = -0.0053 \text{ g} \times 9.0 \times 10^{20} \text{ ergs/g} = -4.8 \times 10^{18} \text{ ergs}$$

i.e., 4.8×10^{18} ergs of energy are *evolved* when a mole of radium decays.

c. Perhaps the simplest way to analyze this problem is to recall from b that the decay of 1 mole ($\sim$226 g) of ${}_{88}^{226}\text{Ra}$ resulted in the loss of 0.0053 g of mass; i.e., $\Delta m = -0.0053$ g when 226 g of Ra decays. Accordingly, when 1 g of Ra decays,

$$\Delta m = \frac{-0.0053 \text{ g}}{226} = -2.3 \times 10^{-5} \text{ g}$$

$$\Delta E = -2.3 \times 10^{-5} \text{ g} \times 2.15 \times 10^{10} \text{ kcal/g} = -4.9 \times 10^{5} \text{ kcal}$$

Energy changes in ordinary chemical reactions are of the order of 1–10 kcal/gram. We see then that ΔE for the radioactive decay of radium is about 100,000 times as great as that for typical non-nuclear processes.

NUCLEAR STABILITY, BINDING ENERGY

With the use of Table 22.4 it is possible to compare the masses of different nuclei to those of the individual neutrons and protons from which they are derived. When we do this, we always find that the nuclear mass is greater than that of the constituent particles. Consider, for example, the ${}_{2}^{4}\text{He}$ nucleus:

One ${}_{6}^{12}\text{C}$ atom weighs exactly 12 amu

$$\text{mass 2 protons} = 2(1.00728) \text{ amu} = 2.01456 \text{ amu}$$
$$\text{mass 2 neutrons} = 2(1.00867) \text{ amu} = \underline{2.01734 \text{ amu}}$$
$$4.03190 \text{ amu}$$
$$\text{mass of } {}_{2}^{4}\text{He} = 4.00150 \text{ amu}$$

TABLE 22.5   Binding Energies of Various Nuclei

	Mass Decrement	Binding Energy (MeV)	Binding Energy per Nucleon (MeV)
${}_{1}^{2}\text{H}$	0.00239	2.22	1.11
${}_{2}^{3}\text{He}$	0.00829	7.72	2.57
${}_{2}^{4}\text{He}$	0.0304	28.3	7.07
${}_{3}^{7}\text{Li}$	0.0421	39.2	5.60
${}_{5}^{10}\text{B}$	0.0695	64.7	6.47
${}_{6}^{12}\text{C}$	0.0989	92.1	7.67
${}_{13}^{27}\text{Al}$	0.2415	224.8	8.33
${}_{27}^{59}\text{Co}$	0.5555	517.2	8.77
${}_{42}^{99}\text{Mo}$	0.9146	851.6	8.60
${}_{63}^{157}\text{Eu}$	1.3815	1286	8.19
${}_{80}^{196}\text{Hg}$	1.6653	1550	7.91
${}_{92}^{238}\text{U}$	1.9342	1801	7.57

In this case there is a decrease in mass of $(4.03190 - 4.00150)$ amu $= 0.03040$ amu when a helium nucleus is formed from two protons and two neutrons. This decrease in mass, called the **mass decrement,** can be calculated for any isotope whose nuclear mass is known. A series of mass decrements, calculated for a few typical isotopes, is given in Table 22.5.

The fact that the formation of a nucleus from protons and neutrons invariably involves a decrease in mass means that any nucleus is stable toward decomposition into these particles. Consider, for example, the helium-4 nucleus; in order to break this up into two protons and two neutrons, one would have to add 0.03040 amu of mass. This would, of course, require the absorption of a large amount of energy. The quantity of energy which would have to be absorbed to decompose a nucleus into protons and neutrons is referred to as the **binding energy** of the nucleus. The binding energy, in MeV, is readily calculated from the mass decrement, making use of the conversion factor 931 MeV $= 1$ amu. For the $_2^4$He nucleus,

$$\text{binding energy (MeV)} = \text{mass decrement (amu)} \times 931 \, \frac{\text{MeV}}{\text{amu}}$$

$$= 0.0304 \text{ amu} \times 931 \, \frac{\text{MeV}}{\text{amu}} = 28.3 \text{ MeV}$$

We can interpret this binding energy to mean that the decomposition of a helium-4 nucleus (an alpha particle) into protons and neutrons would require the absorption of 28.3 MeV of energy. Conversely, 28.3 MeV would be evolved if an alpha particle were formed from two protons and two neutrons.

It will be noted from Table 22.5 that binding energy increases, with rather large fluctuations at first but then more smoothly, as the number of protons and neutrons in the nucleus increases. We might indeed expect this to be the case; the more particles there are in the nucleus, the greater should be the total amount of energy required to break the nucleus apart. To gain a better idea of the relative stabilities of different nuclei, we calculate the binding energy per nuclear particle. For the $_2^4$He nucleus, which contains a total of four nuclear particles (nucleons),

$$\text{binding energy per nucleon} = \frac{\text{binding energy}}{\text{no. nucleons}} = \frac{28.3 \text{ MeV}}{4} = 7.07 \text{ MeV}$$

Looking at the last column of Table 22.5, we note that the binding energy per nucleon is relatively small for very light isotopes such as $_1^2$H and $_2^3$He. It rises to a maximum of about 9 MeV with isotopes of intermediate mass number such as cobalt-59 and copper-63. The binding energy per nucleon then falls off slowly to about 7.5 MeV for the very heavy elements such as uranium (Fig. 22.6).

If we take the binding energy per nucleon to be a measure of the relative stability of a nucleus, it is clear from Figure 22.6 that the most stable nuclei are those of intermediate mass such as $_{27}^{59}$Co, located near the broad maximum of the curve. Nuclei of very heavy elements such as uranium, located at the far right, should be energetically unstable with respect to *fission* into smaller nuclei. Even greater amounts of energy should be obtainable from the *fusion* of very light nuclei such as $_1^2$H at the far left of the curve.

$_1^2$H is often called deuterium and given the symbol D. "Heavy water" is D_2O

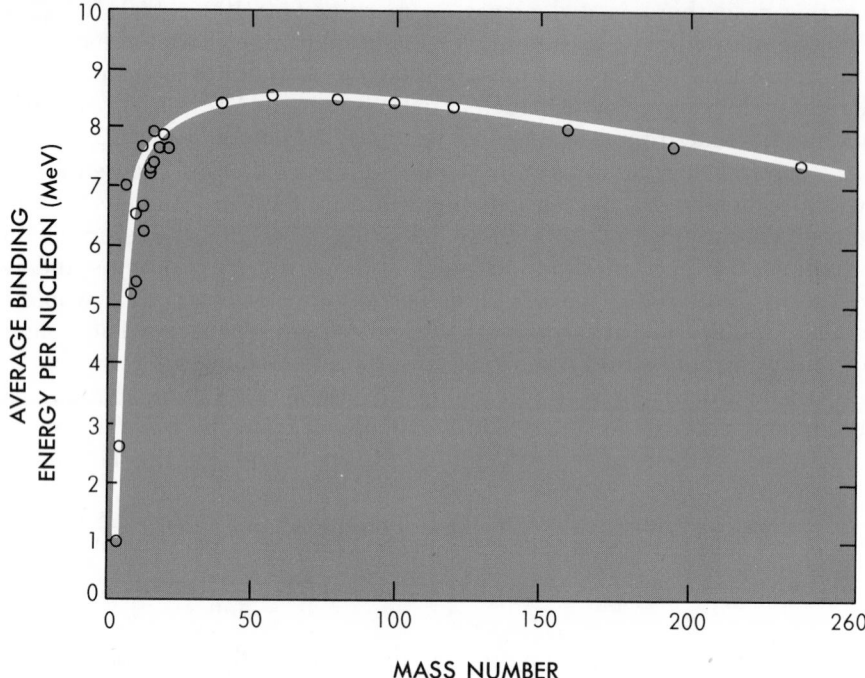

Figure 22.6 Plot of binding energy per nucleon vs mass number. Very light and very heavy nuclei are relatively unstable.

22.5 NUCLEAR FISSION

Shortly before World War II several groups of scientists were studying the products obtained by bombarding uranium with neutrons, hoping to discover new elements with atomic numbers greater than 92. In 1938 Hahn and Strassman in Germany isolated from the products a compound of a group 2A element, which they originally believed to be radium (at. no. 88). Subsequently they were able to show conclusively that this element was barium (at. no. 56), indicating that a uranium atom had been split into fragments. Hahn's first reaction to this discovery was one of disbelief; he later stated, in January of 1939, "As chemists, we should replace the symbol Ra . . . by Ba . . . [but], as nuclear chemists, closely associated with physics, we cannot decide to take this step in contradiction to all previous experience in nuclear physics."

If Hahn was reluctant to admit the possibility of an entirely new type of nuclear reaction, a former colleague of his, Lisa Meitner, was not. In a letter published with O. R. Frisch in January of 1939, she stated: "At first sight, this result seems very hard to understand. . . . On the basis, however, of present ideas about the behavior of heavy nuclei, an entirely different picture of these new disintegration processes suggests itself. . . . It seems possible that the uranium nucleus . . . may, after neutron capture, divide it-self into nuclei of roughly equal size." This revolutionary suggestion was quickly substantiated by experiments carried out in laboratories all over the world.

With the outbreak of World War II, interest in nuclear fission was focused on the enormous amount of energy that could be released in this process. At Los Alamos, in the mountains of New Mexico, a prestigious group of scientists under the direction of J. Robert Oppenheimer worked feverishly to produce the fission or "atomic" bomb. Many of the members of this group were exiles from Nazi Germany; they were spurred on by the fear that Hitler would obtain the bomb first.

This was the so-called Manhattan project

FISSION PROCESS, REACTANTS AND PRODUCTS

Several isotopes of the heavy elements, including platinum, gold, mercury, and lead, are capable of undergoing fission if bombarded by neutrons of sufficiently high energy. In practice, attention has centered upon two particular isotopes, $^{235}_{92}U$ and $^{239}_{94}Pu$, both of which can be split into fragments by low-energy neutrons. During World War II, several different processes were worked out for the separation of uranium-235 from the more abundant isotope, uranium-238, which makes up 99.3 per cent of naturally occurring uranium. The most successful separation technique was that of gaseous diffusion, described in Chapter 5. The element plutonium does not occur in nature; the 239-isotope is made from uranium-238 by the sequence of reactions described by Equation 22.12.

$^{238}_{92}U$ does not undergo the fission reaction

Most of the available data on fission reactions has to do with uranium-235. For this reason, our discussion from this point on will concentrate upon the nuclear reactions that take place when a $^{235}_{92}U$ nucleus interacts with a neutron:

$$^{235}_{92}U + ^{1}_{0}n \rightarrow \text{products} + \text{energy}$$

FISSION PRODUCTS. When a uranium-235 atom undergoes fission, it splits into two unequal fragments and a number of neutrons and beta particles. The fission process is complicated by the fact that different uranium-235 atoms split up in many different ways. For example, while one atom of $^{235}_{92}U$ is splitting to give isotopes of rubidium (at. no. 37) and cesium (at. no. 55), another may break up to give isotopes of bromine (at. no. 35) and lanthanum (at. no. 57), while still another atom yields isotopes of zinc (at. no. 30) and samarium (at. no. 62).

$$^{90}_{37}Rb + ^{144}_{55}Cs + 2 \, ^{1}_{0}n \tag{22.20}$$

$$^{1}_{0}n + ^{235}_{92}U \rightarrow ^{87}_{35}Br + ^{146}_{57}La + 3 \, ^{1}_{0}n \tag{22.21}$$

$$^{72}_{30}Zn + ^{160}_{62}Sm + 4 \, ^{1}_{0}n \tag{22.22}$$

Fission of a macroscopic sample of uranium-235, containing billions of billions of atoms, gives a large number of products; more than 200 isotopes of 35 different elements have been identified among the fission products of uranium-235.

Since the stable neutron-to-proton ratio near the middle of the periodic table, where the fission products are located, is considerably smaller (~1.2) than that of uranium-235 (1.55), the immediate products of the fission process contain too many neutrons for stability. Isotopes such as $^{90}_{37}Rb$ and $^{144}_{55}Cs$ are radioactive, decaying by electron emission. In the case of rubidium-90, three steps are required to reach a stable nucleus.

$$^{90}_{37}Rb \rightarrow {}^{90}_{38}Sr + {}_{-1}^{0}e; \ t_{1/2} = 2.8 \text{ min}$$

$$^{90}_{38}Sr \rightarrow {}^{90}_{39}Y \ + {}_{-1}^{0}e; \ t_{1/2} = 29 \text{ yrs}$$

$$^{90}_{39}Y \rightarrow {}^{90}_{40}Zr + {}_{-1}^{0}e; \ t_{1/2} = 64 \text{ hrs}$$

The radiation hazard associated with nuclear testing arises from the formation of radioactive isotopes such as these. One of the most dangerous of these isotopes is strontium-90, which, in the form of strontium carbonate, is readily incorporated into the bones of animals and human beings.

You will notice from Equations 22.20 to 22.22 that two to four neutrons are produced in the fission process for every one consumed. This creates the possibility of a chain reaction. Once a few atoms of uranium-235 split, the neutrons produced can bring about the fission of many more uranium-235 atoms, which in turn yield more neutrons, capable of splitting more uranium atoms, and so on. This is, of course, precisely what happens in the atomic bomb; the energy evolved in successive fissions escalates to give, within a few seconds, a tremendous explosion.

In order for nuclear fission to result in a chain reaction, the uranium-235 or plutonium-239 sample must be large enough so that most of the neutrons are captured internally. If the sample is too small, most of the neutrons produced will escape from the surface, thereby breaking the chain. The *critical mass* of uranium-235 required to maintain a nuclear chain reaction in a bomb appears to be about 40 kg.

FISSION ENERGY. NUCLEAR REACTORS

Note that in fission only a small fraction of the mass of the reactants is converted into energy

Some idea of the vast amount of energy available from nuclear fission may be obtained by comparing it to the energy evolved in ordinary chemical reactions. The fission of 1 g of uranium-235 evolves about 20,000,000 kcal of energy. The heat of combustion of coal is only about 8 kcal/g; the energy given off when 1 g of TNT explodes is still smaller, about 0.66 kcal. Putting it another way, the fission of 1 g of uranium produces as much energy as the combustion of 5500 lb of coal or the explosion of 33 tons of TNT. The enormous energy change accompanying nuclear fission is directly attributable to the change in mass that takes place in the fission process (Example 22.5, p. 618).

Even before the first atomic bomb was exploded, scientists and politicians began to speculate on the use of nuclear fission as a source of energy for peaceful purposes. The ever-increasing demand for energy coupled with the depletion of our reserves of fossil fuels lends urgency to our search for new sources of energy. Many different kinds of nuclear reactors, in which fission is made to occur at a controlled rate, have been designed to meet this need. The type of reactor which has proved most popular for generating electrical power in the United States is shown in Figure 22.7. Water under a pressure of about 2000 lb/in² is passed through the reactor core in which a nuclear reaction is giving off energy in the form of heat. The water coming out of the reactor at a temperature of 300°C is then

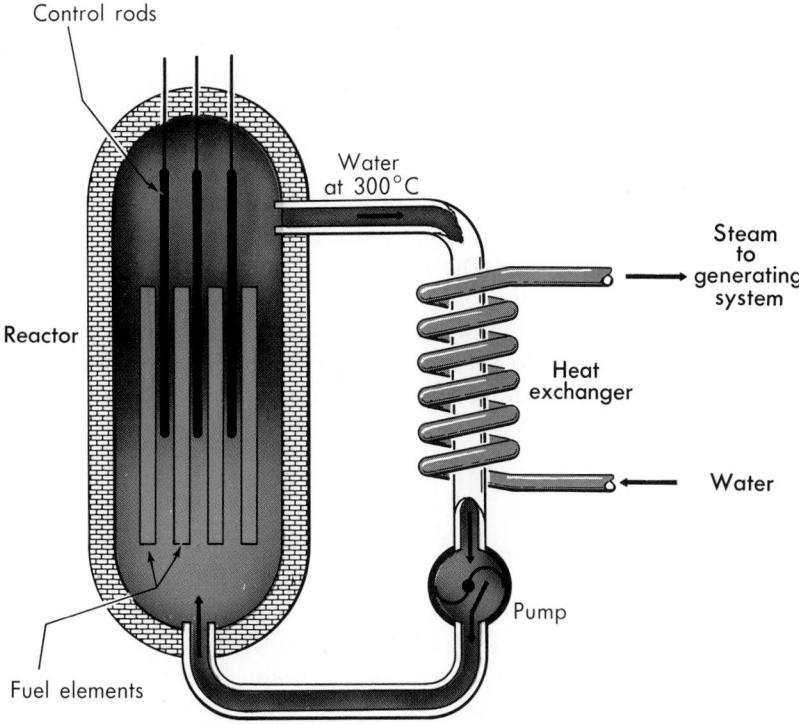

Control rods

Water
at 300°C

The control rods are
made of a material like
cadmium, which strongly
absorbs neutrons. The
rate of fission is very
carefully monitored and
controlled

Reactor

Steam
to
generating
system

Heat
exchanger

Water

Pump

Fuel elements

Recirculated water

Figure 22.7   Nuclear reactor of the pressurized water type.

passed through a closed loop containing a heat exchanger. A second stream
of cold water circulating through the heat exchanger is converted to steam
which is used to drive a turbogenerator. The pressurized water serves both
as a heat transfer medium and a moderator for the nuclear reaction taking
place within the core.

There are several practical problems associated with nuclear reactors
which have caused their development to lag well behind the optimistic
estimates of 20 years ago. One of these is the relatively low efficiency of
heat conversion in a reactor. Another is the hazard posed by the radioactive
isotopes produced by the fission process. It has been necessary to develop
new alloys and other materials of construction which are able to withstand
radiation damage at the high temperatures typical of a nuclear reactor. The
core itself has to be heavily shielded to contain the radiation. This is
readily done in a large nuclear power plant; it can even be accomplished in
a nuclear submarine or ocean liner, but it is hardly practical for a vehicle
as small as an automobile.

Perhaps the most serious problem that must be solved if nuclear fission is to make a significant contribution to our energy needs is that of
disposing of the radioactive waste products. So far, they have simply been
stored, usually as a concentrated aqueous solution, in shielded underground tanks. This poses an eventual danger of corrosion or rupture of the
tanks. Some progress has been made in working out methods for solidify-
ing these wastes, which would make them a great deal less hazardous.

From a long-range standpoint, it seems unlikely that uranium-235 will, by itself, be able to satisfy our future power needs. It has been estimated that the fission of all the uranium-235 in nature could fulfill our energy needs for one century at the most. A much more attractive possibility is to use uranium-235 as a neutron source to convert the more abundant isotope, uranium-238, to plutonium-239, which can then undergo fission. A still more inviting possibility is the conversion of thorium-232 (which is more abundant than uranium) to uranium-233 by a similar process. So-called **breeder reactors** based on these principles have been in existence for many years; the problem is to increase the yield of ^{239}Pu and ^{233}U to the point at which the process becomes feasible on a large scale. Ultimately the hope is to build breeder reactors capable of producing large quantities of fissionable isotopes from more abundant elements such as thorium, lead, and mercury.

22.6 NUCLEAR FUSION

From Figure 22.6 we see that very light isotopes such as those of hydrogen are potentially unstable with respect to fusion into heavier isotopes of elements such as helium. Indeed, from the form of the nuclear stability curve, we would predict that the energy available from nuclear fusion should be considerably greater than that given off in the fission of an equal mass of a heavy element. This is indeed the case; we can calculate that the energy evolved per gram of reactants in a typical fusion process such as

$$^2_1H + ^3_1H \rightarrow ^4_2He + ^1_0n \tag{22.23}$$

is about four times as great as in the fission of uranium-235 (Example 22.5).

Example 22.5 Calculate the amount of energy evolved, in kcal, per gram of reactants in

    a. a fusion reaction: $^2_1H + ^3_1H \rightarrow ^4_2He + ^1_0n$
    b. a fission reaction: $^{235}_{92}U \rightarrow ^{90}_{38}Sr + ^{144}_{58}Ce + ^1_0n + 4\ ^0_{-1}e$

Solution

    a. We first calculate the change in mass, using Table 22.4.

$$\Delta m \text{ (in amu)} = 4.00150 + 1.00867 - 2.01355 - 3.01550$$
$$= -0.01888 \text{ amu}$$

This calculation tells us that, starting with 5 amu of reactants (^{2}H + ^{3}H), there is a mass decrease of 0.01888 amu. It follows that if one were to start with five grams of reactants, Δm would be −0.01888 g. Stating it another way, the mass decrement per gram of reactant would be

$$\frac{-0.01888 \text{ g}}{5} = -0.00378 \text{ g}$$

Using the appropriate conversion factor from Table 22.3,

$$\Delta E = -0.00378 \text{ g} \times 2.15 \times 10^{10} \frac{\text{kcal}}{\text{g}} = -8.13 \times 10^7 \text{ kcal}$$

    b. Proceeding exactly as in a,

$$\Delta m \text{ (in amu)} = 89.8864 + 143.8816 + 1.0087 + 4(0.00055) - 234.9934$$
$$= -0.2145 \text{ amu}$$

$$\text{Mass change per gram of } ^{235}\text{U} = \frac{-0.2145 \text{ g}}{235} = -0.000913 \text{ g}$$

$$\Delta E = -0.000913 \text{ g} \times 2.15 \times 10^{10} \frac{\text{kcal}}{\text{g}} = -1.96 \times 10^7 \text{ kcal}$$

Comparing the answers to a and b, we conclude that nuclear fusion produces about four times as much energy per gram of starting material (8.13×10^7 kcal vs. 1.96×10^7 kcal) as the fission process.

As an energy source, nuclear fusion possesses several additional advantages over nuclear fission. For one thing, fusion is a "clean" process in the sense that the products are stable isotopes such as ^4_2He, rather than the hazardous radioactive isotopes formed by fission. Equally important, light isotopes suitable for fusion are far more abundant than the heavy isotopes required for fission. We can calculate, for example (Problem 22.43), that the fusion of only 2×10^{-13} per cent of the deuterium (^2_1H) in sea water would meet the total annual energy requirements of the world. Finally, the fact that fusion produces small charged particles moving at high speeds raises the possibility of converting nuclear energy directly into electrical energy. If this could be done without first converting to heat, we could avoid the low efficiency typical of all heat engines and thereby reduce thermal pollution of the environment by power plants.

Unfortunately fusion reactions, unlike fission, have very high activation energies. Enormously high temperatures are required to bring about reactions such as 22.23. In the "hydrogen bomb" these temperatures were achieved by carrying out a fission reaction to trigger nuclear fusion. For peacetime applications, it will be necessary to develop equipment in which very high temperatures can be maintained long enough to allow fusion reactions to occur and to give off energy. In any conventional container the reactant nuclei would quickly lose their high kinetic energies by collisions with the walls. Within the past few years promising results have been obtained with "magnetic bottles" in which the charged nuclei are confined within very strong magnetic fields. Another approach to fusion which is being studied is irradiation of solid hydrogen pellets with fantastically energetic pulsed laser beams.

Perhaps the ultimate irony of our time is the fact that we have made so little use of the energy produced in a nuclear fusion process that has been going on since the universe was formed. The energy given off by the sun and other stars results from fusion reactions in which ordinary hydrogen is converted to helium. One mechanism which has been suggested for this process is

$$\begin{aligned}
^1_1\text{H} + ^1_1\text{H} &\rightarrow ^2_1\text{H} + ^0_1\text{e}; & \Delta E &= -0.43 \text{ MeV} \\
^2_1\text{H} + ^1_1\text{H} &\rightarrow ^3_2\text{He}; & \Delta E &= -4.96 \text{ MeV} \\
^3_2\text{He} + ^1_1\text{H} &\rightarrow ^4_2\text{He} + ^0_1\text{e}; & \Delta E &= -19.30 \text{ MeV} \\
\hline
4\,^1_1\text{H} &\rightarrow ^4_2\text{He} + 2\,^0_1\text{e}; & \Delta E &= -24.69 \text{ MeV}
\end{aligned}$$

(22.24)

Each day, processes such as this, occurring within the sun at a temperature of perhaps 10^8°C, produce enormous quantities of energy, of which about 1.5×10^{18} kcal reaches the surface of the earth. This is roughly equivalent to the total amount of energy that man has consumed since the beginning of time.

PROBLEMS

22.1 Explain why

  a. $^{37}_{17}Cl$ and $^{35}_{17}Cl$ have essentially identical chemical properties.
  b. β-rays are deflected to a greater extent than α-rays in an electrical field.
  c. radioactive wastes cannot be disposed of in the same way as chemical wastes.
  d. neutrons do not have to be accelerated to high speeds to bring about nuclear reactions.

22.2 Describe experiments which would enable you to

  a. determine the mass of an alpha particle.
  b. detect the presence of uranium in an ore.
  c. determine the half-life of radium-226.
  d. estimate the age of charcoal found in an archaeological excavation.

22.3 Write balanced nuclear equations for

  a. the loss of an alpha particle by $^{218}_{84}Po$.
  b. the loss of a beta particle by $^{214}_{83}Bi$.
  c. emission of gamma radiation by $^{87}_{38}Sr$.
  d. emission of a positron by $^{82}_{39}Y$.
  e. K-electron capture by $^{44}_{22}Ti$.

22.4 If in Problem 22.3 the reactant species in each case is a neutral atom, what will be the charge of the product isotope?

22.5 The symbolism $^{31}_{15}P$ (p, n) is used to indicate the nuclear reaction in which a proton collides with a phosphorus-31 nucleus to emit a neutron and form another isotope. Write nuclear equations for

  a. the above reaction.
  b. $^{75}_{33}As$ (d, p); d = deuteron
  c. $^{81}_{35}Br$ (γ, n)
  d. $^{82}_{34}Se$ (n, γ)

22.6 Write balanced nuclear equations for

  a. the fission of uranium-235 to give $^{97}_{36}Kr$, another nucleus, and an excess of three neutrons.
  b. the fission of uranium-235 to give $^{141}_{56}Ba$, another nucleus, and an excess of three neutrons.
  c. the production of the 254-isotope of element 102 by bombardment of $^{244}_{96}Cm$ with the common isotope of carbon.
  d. the fusion of two deuterium nuclei.

22.20 Explain why

  a. in the reactor shown in Figure 22.7, the water which passes through the core is not converted directly to steam.
  b. K-electron capture is more common than positron emission among isotopes of high atomic number.
  c. elements of atomic number 102 to 105 were prepared using nuclei heavier than helium-4 as bombarding particles.

22.21 Criticize experiments in which it is proposed to

  a. determine the age of coal deposits by carbon-14 dating.
  b. produce energy by fusing cobalt-60 nuclei together.
  c. dump radioactive wastes into the sea.
  d. bring about nuclear reactions by direct bombardment with electrons.

22.22 Write balanced nuclear equations to represent

  a. loss of an alpha particle by thorium-230.
  b. electron emission by Pa-234.
  c. K-electron capture by gold-187.
  d. loss of a positron by yttrium-84.

22.23 For each of the processes in Problem 22.22, compare reactant and product isotopes as to numbers of protons, neutrons, and electrons. (Assume each reactant isotope is a neutral atom.)

22.24 Write balanced nuclear equations for the following processes

  a. $^{54}_{26}Fe$ (α, n)
  b. $^{9}_{4}Be$ (γ, n)
  c. (α, p)$^{48}_{22}Ti$
  d. (d, 6n)$^{231}_{93}Np$

22.25 Write balanced nuclear equations for

  a. fission of plutonium-239 to give $^{132}_{50}Sn$, another nucleus, and an excess of four neutrons.
  b. formation of an isotope of a transuranium element by bombardment of einsteinium-253 with an α-particle; a neutron is ejected.
  c. fusion of two lithium-6 nuclei.

22.7 A certain natural radioactive series starts with uranium-238 and ends with lead-206. Each step in the series involves the loss of either an alpha- or a beta-particle. In the entire series, how many α-particles are given off? how many β-particles?

22.8 Strontium-90, one of the most hazardous isotopes produced in nuclear fission, has a half-life of 29 years.

    a. Calculate the rate constant for the decay of strontium-90.

    b. What fraction of a strontium-90 sample will remain after 10 years?

22.9 The level of radioactivity associated with a certain isotope is measured with a Geiger counter. Over a one-week period the level drops from 350 counts/minute to 292 counts/minute. Calculate the rate constant for the decay process and the half-life of the isotope.

22.10 Estimate the age of rocks in which

    a. the mole ratio of uranium-238 to lead-206 is 1.00.

    b. the mole ratio of uranium-238 to lead-206 is 1.10.

    c. the weight ratio (in grams) of uranium-238 to lead-206 is 1.10.

Take the half-life of uranium-238 to be 4.5×10^9 years.

22.11 A sample of a funeral shroud taken from an ancient Egyptian tomb is found to have a carbon-14 content 0.590 times that in a living plant. Estimate the age of the linen from which the shroud was made.

22.12 Predict whether the following isotopes will decay by positron or by electron emission.

    a. $^{78}_{32}Ge$ c. $^{12}_{7}N$

    b. $^{65}_{32}Ge$ d. $^{17}_{7}N$

22.13 Explain how isotopes can be used to

    a. analyze a sample of fish for mercury.

    b. treat a patient suffering from cancer.

    c. follow the rate of hydrolysis of the $Cr(H_2O)_6^{3+}$ ion.

    d. measure the rate of self-diffusion of liquid water.

22.26 A certain radioactive series starts with $^{241}_{94}Pu$ and proceeds through 13 steps, eight of which involve alpha-emission and five, beta-emission. What is the end product of the series?

22.27 Cobalt-60, which is widely used in treating cancer patients, has a half-life of 5.26 years.

    a. Calculate the rate constant for the decay of this isotope.

    b. If a hospital purchases 10 mg of this isotope, how much will remain after one year?

22.28 A health officer monitoring the decay of a radioactive sample obtains the following readings on a Geiger counter

9:00 PM July 18	220
5:00 PM July 19	210
7:00 PM July 21	187
4:00 PM July 25	151

Using a graphical method, estimate the time at which the number of counts/min. will be reduced to 50.

22.29 One way of dating rocks is to determine the relative amounts of potassium-40 and argon-40; the decay of ^{40}K has a half-life of 1.27×10^9 years. Analysis of a certain lunar sample gives the following results in mole ratios:

$^{40}Ar/^{40}K = 4.8$ $(t_{1/2}\ ^{40}K = 1.27 \times 10^9\ yrs)$
$^{206}Pb/^{238}U = 0.72$ $(t_{1/2}\ ^{238}U = 4.5 \times 10^9\ yrs)$
$^{87}Sr/^{87}Rb = 0.044$ $(t_{1/2}\ ^{87}Rb = 5.7 \times 10^{10}\ yrs)$

Using these data, obtain the best possible value for the age of the sample. Can you suggest why the K-Ar method gives a low result?

22.30 The radioactive isotope ^3_1H is produced in nature in much the same way as carbon-14. It has a half-life of 12.3 years. Estimate the age of a sample of Scotch whiskey which has a tritium content 0.75 times that of the water in the area in which the whiskey was produced.

22.31 Would you expect the following isotopes to decay by positron or by electron emission?

    a. $^{90}_{42}Mo$ c. $^{3}_{1}H$

    b. $^{102}_{42}Mo$ d. $^{29}_{15}P$

22.32 Describe experiments in which you could use a tracer technique to

    a. determine the solubility product of $AgCl$.

    b. decide whether in the reaction

$$H_3C{-}COOH + CH_3OH \rightarrow$$
$$H_3C{-}COOCH_3 + H_2O$$

the oxygen atom which is eliminated comes from $H_3C{-}COOH$ or CH_3OH.

22.14 Consider the nuclear reaction

$$^{239}_{94}Pu \rightarrow\ ^{235}_{92}U +\ ^{4}_{2}He$$

a. Calculate Δm, in amu, and ΔE, in MeV, for the decay of a plutonium nucleus.

b. How much energy, in kcal, is given off when one gram of plutonium-239 decays?

22.15 Using the data in Table 22.4, calculate the mass of an $^{16}_{8}O$ *atom* in amu.

22.16 Which of the following nuclear reactions would be exothermic? (Show by calculation!)

a. $^{9}_{4}Be +\ ^{10}_{5}B \rightarrow\ ^{19}_{9}F$

b. $^{238}_{92}U \rightarrow\ ^{206}_{82}Pb + 8\ ^{4}_{2}He + 6\ _{-1}^{0}e$

c. $^{109}_{48}Cd \rightarrow\ ^{109}_{47}Ag +\ ^{0}_{1}e$

22.17 For nickel-59, calculate the mass decrement, the total binding energy, and the binding energy per nucleon.

22.18 Consider the fission reaction

$$^{235}_{92}U +\ ^{1}_{0}n \rightarrow\ ^{148}_{58}Ce +\ ^{84}_{34}Se + 4\ ^{1}_{0}n$$

Taking the masses of the product isotopes to be 147.9234 and 83.9211, calculate ΔE in kcal for the fission of one gram of uranium-235. How many tons of TNT would have to be detonated to produce the same amount of energy ($\Delta E = 0.66$ kcal/g)?

22.19 Taking Equation 22.24 to represent the reaction that produces the sun's energy, how many grams of hydrogen would have to be fused to provide the 1.5×10^{18} kcal of energy that reaches the earth each day?

22.33 For the reaction

$$^{210}_{84}Po + 2\ ^{1}_{0}n \rightarrow\ ^{208}_{82}Pb +\ ^{4}_{2}He$$

a. calculate Δm, in amu, and ΔE, in MeV, per atom of polonium-210.

b. what is ΔE, in kcal, when one milligram of polonium-210 reacts?

22.34 Using the data in Table 22.4, calculate the mass of an $^{27}_{13}Al^{3+}$ ion in amu.

22.35 Decide, on the basis of nuclear masses, whether the following reactions should be spontaneous

a. $^{11}_{6}C \rightarrow\ ^{11}_{5}B +\ ^{0}_{1}e$

b. $^{40}_{18}Ar \rightarrow\ ^{40}_{19}K +\ _{-1}^{0}e$

c. $^{230}_{90}Th \rightarrow\ ^{222}_{86}Rn + 2\ ^{4}_{2}He$

22.36 For $^{16}_{8}O$, calculate the mass decrement, the total binding energy, and the binding energy per nucleon.

22.37 Compare the amounts of energy evolved per gram in the two fission processes

$$^{235}_{92}U +\ ^{1}_{0}n \rightarrow\ ^{144}_{58}Ce +\ ^{90}_{38}Sr + 4\ _{-1}^{0}e + 2\ ^{1}_{0}n$$

$$^{239}_{94}Pu +\ ^{1}_{0}n \rightarrow\ ^{144}_{58}Ce +\ ^{90}_{38}Sr + 2\ _{-1}^{0}e + 6\ ^{1}_{0}n$$

22.38 It has been suggested that many of the common isotopes of heavier elements were formed by the fusion of alpha particles.

a. Of the isotopes listed in Table 22.4, which could be formed by direct fusion of α-particles without the emission of any other particle?

b. For the lightest and heaviest isotope listed in (a), calculate ΔE in kcal per gram of α-particles.

*22.39 The half-life of cobalt-60 is 5.26 years. What fraction of a cobalt-60 sample decomposes per second? (Hint: $\ln (1 + x) \approx x$ when $x << 1$.)

*22.40 One curie of a radioactive substance is the amount in which 3.7×10^{10} nuclei undergo radioactive decay in one second. Calculate the mass of one curie of radium (half-life = 1590 yr). Your answer should be one gram.

*22.41 It is possible to obtain an estimate of the activation energy for fusion by calculating the energy required to bring two deuterons close enough together to form an alpha particle. This energy can be estimated by using Coulomb's Law: $E = q_1q_2/r$, where q_1 and q_2 are the charges of the deuterons (4.8×10^{-10} esu), r is the radius of the helium nucleus ($\sim 1 \times 10^{-12}$ cm), and E is the energy in ergs.

a. Estimate E in kcal per mole of α-particles.

b. Using the equation $E = mv^2/2$, estimate the velocity that a deuteron must have if a collision between two of them is to supply the activation energy for fusion.

c. Using the equation $v = (3RT/M)^{1/2}$, estimate the temperature that would have to be reached to achieve fusion ($R = 8.31 \times 10^7$ ergs/mole°K).

*22.42   A certain drug sample weighing 1.000 g is analyzed for penicillin by adding 10.0 mg of penicillin containing carbon-14 with an activity of 120 counts/sec. The radioactive penicillin equilibrates with the inactive penicillin in the drug. A total of 0.140 g of penicillin isolated from the drug shows an activity of 52 counts/sec. What is the percentage of penicillin in the sample?

*22.43   Consider the reaction

$$2\, {}^2_1\text{H} \rightarrow {}^4_2\text{He}$$

a. Calculate ΔE in kcal per gram of deuterium fused.
b. How much energy is potentially available from the fusion of all the deuterium atoms in sea water? The percentage of deuterium in water is about 0.015%. The total mass of water in the oceans is 1.3×10^{27} g.
c. What fraction of the deuterium in the oceans would have to be consumed to supply the annual energy requirement of the world (5.4×10^{16} kcal)?

23 AN INTRODUCTION TO BIOCHEMISTRY

Biochemistry is that science which deals with the chemical substances and reactions that occur in all living organisms, from the smallest of bacteria and viruses to plants and higher animals. It is the field of chemistry which has shown the most development in the past 20 years, and at present is still growing at a tremendous rate. It is through biochemical studies that we are gradually coming to understand, on a molecular level, how the living organism is able to grow, utilize its food, move, reproduce, and carry on the multitude of other activities that are vital to its existence.

Biochemistry is similar to other areas of chemistry in that it deals with chemical substances and their reactions. It differs from them, however, in that the substances which are involved and the reactions they undergo are typically much more complex. Problems of purifying and establishing the identity of reactants and products, and limiting the number of reactions occurring simultaneously, all of which may bother the organic or inorganic chemist, become extremely difficult for the biochemist. He finds he must use virtually all the modern tools available to the chemist, particularly the rather more subtle, nondestructive analytical methods, such as chromatography, electrophoresis, and x-ray diffraction. In this chapter we shall not attempt a broad overview of biochemistry, but shall limit ourselves to just a few areas which illustrate the progress which has been made and, to some extent, the methods by which it was accomplished. Since biochemistry deals with organic compounds, it would be advisable for you to review the section on organic substances in Chapter 8 as preparation for this chapter.

In a single living organism, be it a man, an oak tree, or a bacterial cell, there are a multitude of chemical substances. A few of these are inorganic, like water, carbon dioxide, and sodium ion. Some are rather simple organic species, like lactic acid or acetate ion. For the most part, however, biochemically important substances are among the most complex of organic compounds; they are often polymeric and may contain several different kinds of monomeric units. In spite of their complexity and the fact that they may differ in the details of their structures and reactions, the general kinds of substances found in living organisms are quite similar in species as widely different as beans, grasshoppers, and rabbits. There are several im-

624

portant classes of such substances, but we shall be able to consider only three of them — the carbohydrates, the proteins, and the nucleic acids.

23.1 CARBOHYDRATES

The carbohydrates are compounds of carbon, hydrogen, and oxygen. Their name is derived from the fact that all carbohydrates have molecular formulas which can be written as $C_n(H_2O)_m$. Some common examples of carbohydrates are sugars, including sucrose, $C_{12}H_{22}O_{11}$, and glucose, $C_6H_{12}O_6$. Glucose is present in honey and in many fruits and is the sugar present in our blood. It is an important source of energy in most plants and animals. Starch and cellulose are both long chain polymers of glucose. Because of its wide occurrence in both the free and combined forms, glucose is probably the most abundant of all the organic substances. Since it is a representative carbohydrate, we shall limit our discussion of that class of substances to glucose and its related compounds.

The chemical bonding in glucose can be represented in two dimensions as

$$O{=}\overset{\displaystyle H}{\underset{\displaystyle OH}{C}}{-}\overset{\displaystyle H}{\underset{\displaystyle OH}{C}}{-}\overset{\displaystyle H}{\underset{\displaystyle OH}{C}}{-}\overset{\displaystyle H}{\underset{\displaystyle OH}{C}}{-}\overset{\displaystyle H}{\underset{\displaystyle OH}{C}}{-}\overset{\displaystyle H}{C}{-}H$$

The glucose molecule is typical of the simple sugars. It contains a carbon chain, with one carbonyl, C=O, group and hydroxyl groups on each carbon atom except the one in the carbonyl group. Glucose and most other sugars in solution have the rather remarkable property of being able to rotate the plane of polarized light as it passes through the solution. Such substances are said to be *optically active*. Optical activity is ordinarily found in any organic substance in which there are one or more carbon atoms bonded to four different groups; such carbon atoms are said to be **asymmetric** (Fig. 23.1). In the glucose molecule there are four asymmetric carbon atoms, to each of which are attached nonidentical groups.

Optically active substances exist in left- and right-hand forms, called optical isomers.

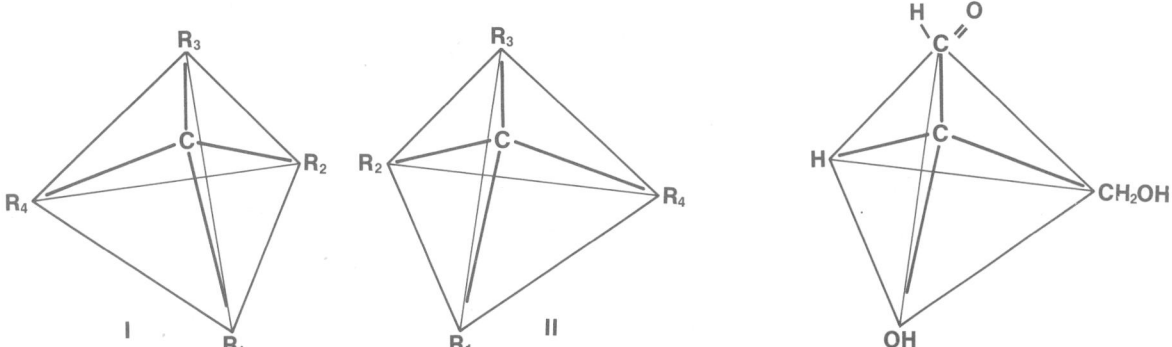

Figure 23.1 The optical isomers (I and II) of $CR_1 R_2 R_3 R_4$ are mirror images of each other. The molecule at the right is the simplest optically active sugar, D-glyceraldehyde.

Because of the asymmetry of the carbon atoms, the two structures below actually represent different substances:

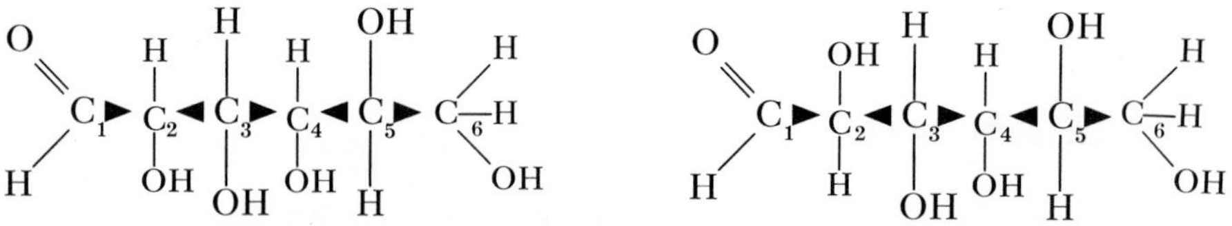

D-Glucose

D-Mannose

(In these drawings, the large and small symbols indicate atoms above and below the plane of the paper.) Both substances are simple sugars but, without breaking bonds, cannot be converted one to the other; glucose and mannose belong to the family of aldohexoses (aldehyde carbonyl, 6-carbon sugar). Since in the aldohexoses there are two different positions for the hydroxyl group on carbon atoms 2, 3, 4, and 5, glucose has 2^4, or 16 possible isomers; these are called *stereoisomers*, to differentiate them from the structural isomers possible in a substance like C_4H_9Cl. Naturally occurring glucose is always found in the form we have indicated, called D-glucose. The mirror image of D-glucose is called L-glucose and is one of the 16 stereoisomers.

D-Glucose and D-mannose are isomeric molecules and differ only in the fact that the positions of the H atom and OH group on the number 2 atom are exchanged. The structure of D-glucose is complicated by the fact that it ordinarily occurs in cyclic forms (these result from the tendency of the number 1 carbon atom at the "head" of the molecule to "eat its tail" and link up with the oxygen atom on the number 5 carbon atom):

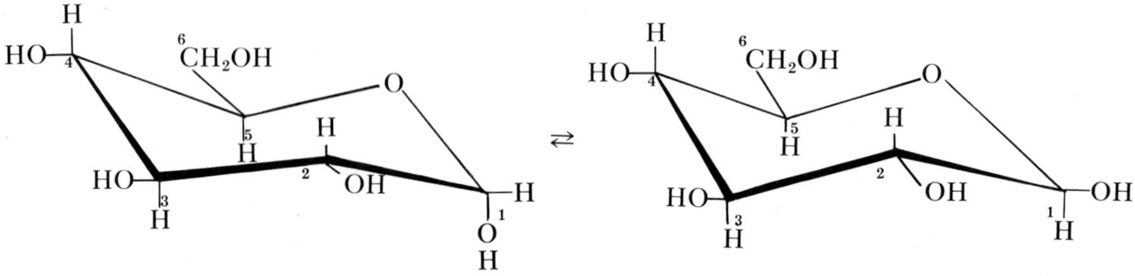

α-D-Glucose

⇌

β-D-Glucose

Each ring is six-membered and contains one oxygen atom. From above each would appear as a hexagon, with the carbon ring atoms (symbols omitted on the ring skeleton) alternately up and down, in a so-called chair conformation. In the rings, the attached H and OH groups will tend to be either perpendicular to the plane of the ring, in which case they are referred to as *axial*, or essentially in the plane of the ring (*equatorial*). (It would be most instructive if at this point you could use a set of models to assemble the glucose molecule, since with a model you can see the spatial relationships much more clearly.) In β-D-glucose, *all* the OH groups on the ring are *equatorial*. This is the most stable isomer of all the aldohexoses, probably because in this form the relatively large OH groups have as much room as possible. The α and β forms of D-glucose differ only in the position of the OH group on the No. 1 carbon atom; it is axial in α and equatorial in β. A solution of D-glucose contains about 36 per cent of the α form and 64 per cent of the β form in equilibrium, and very little of the less stable linear form shown at the top of p. 626.

The problem of determining which of the many stereoisomers of the aldohexoses corresponds to natural glucose was solved by the great German chemist Emil Fischer with the assistance of many hard-working students. By degrading longer chain sugars to shorter ones of known configuration, synthesizing longer chair sugars from shorter, and using optical rotation as a major analytical tool, Fischer, working from about 1885 until 1900, established the configuration of glucose and the other naturally occurring 6-carbon sugars. For this work he received the Nobel Prize in chemistry in 1902. Considering the enormous complexity of the properties of the aldohexoses, particularly the ring equilibria, and the primitive instrumentation available at that time, the determination by Fischer of the configuration of glucose must even now be considered a major achievement of organic chemistry.

Only three of the stereoisomers of glucose occur frequently in natural products

Many sugars contain more than one simple sugar group per molecule. Sucrose, cane sugar, is a **disaccharide**, which cleaves in acid to yield two simple sugars, or **monosaccharides**, of which one is glucose and the other is fructose.

Sucrose

The substances known as starch and cellulose both yield only glucose when subjected to acid hydrolysis. Starch is found in many plants and is stored in roots and seeds. It is present in large amounts in corn, potatoes, and wheat and is one of the main sources of energy in our foods. Cellulose is the substance which makes up the cell walls of plants. Ordinary wood is about 50 per cent cellulose, dry leaves about 10 per cent. Cotton fiber, which contains about 98 per cent cellulose, is the best source of the pure material.

Both starch and cellulose are polymers of glucose, with molecular weights of the order of a million. The glucose units are linked through oxygen bridges, which may be considered to have formed in a polymerization reaction in which one molecule of water was removed for each glucose molecule entering the chain. Cellulose molecules are unbranched; the chains are strongly hydrogen bonded, and this gives the material its high resistance to dissolving in water and alcohols. Starch molecules are occasionally branched and are cross-linked, probably in a more or less random way, once in about 25 units. The structures of starch and cellulose are similar and are indicated in the formulas on the following page.

The essential difference between the cellulose and starch molecules appears to be that in the former the cyclic glucose units are in the β form, whereas in starch the less stable α form is present. Starch is hydrolyzed in the human intestines to glucose, in which form it enters the blood. Cellu-

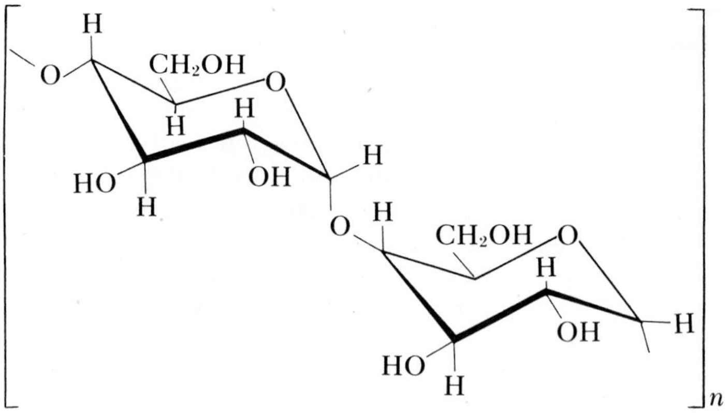

Starch

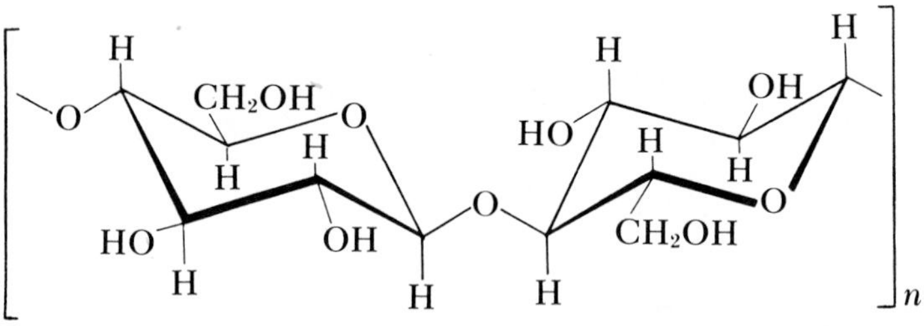

Cellulose

Many animals and some insects can digest cellulose

lose is not affected by human enzymes and, hence, cannot be digested by man. It can, however, be converted to glucose by treatment with 40 per cent HCl at 20°C or by 1 per cent H_2SO_4 at 130°C. Although at present glucose obtained by cellulose hydrolysis is not used as a food, this source may become important as world food requirements increase. (We don't use wood for food as yet, but the day may come when our breakfast cereal will have started out as sawdust.)

On entering the blood stream from the intestine, glucose and other monosaccharides are rapidly absorbed by the liver and muscle tissue and converted into a substance called glycogen. Glycogen is another glucose polymer, similar in structure to starch, but rather more highly branched. Much of the energy from our food is stored in the body in the form of glycogen.

23.2 PROTEINS

Among the most important, and abundant, of the substances which make up a living organism are those compounds called proteins. In our own bodies proteins have many functions. Some are fibrous and are the main components of our muscles, hair, and skin. Others are found in body fluids and serve as carriers of inorganic or organic compounds. A large number of proteins are biological catalysts called enzymes; these facilitate and control the many biochemical reactions which make it possible for us to live (Chapter 14).

Protein is an important component of most foods. It is abundant in lean meats and vegetables like beans and peas, and is present in smaller amounts in nearly everything we eat. In our digestive system proteins are broken down into their components, which are then reassembled in our bodies into other proteins which our system requires. In this section we shall look at the structure of proteins; later we shall examine the manner by which proteins are synthesized in a living organism.

NATURE OF PROTEINS

Proteins, like the polysaccharides starch and cellulose, are polymeric substances. They range in molecular weight from about 6000 for insulin to over a million for some of the more complex enzymes. All proteins, when subjected to treatment with strong acid, will depolymerize into their monomeric components, all of which are **α-amino acids**. α-Amino acids are organic acids in which an NH_2 group is attached to the carbon atom adjacent to a carboxyl group, as indicated in the sketch below:

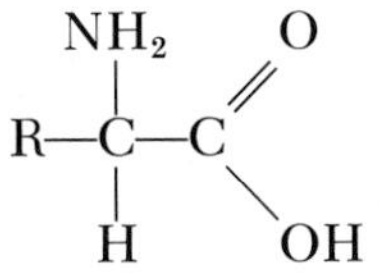

$$\begin{array}{ccc} NH_2 & & O \\ | & & \| \\ R-C-C & & \\ | & & \diagdown \\ H & & OH \end{array}$$

Most α-amino acids are optically active. True or false?

The symbol R represents an organic radical, which may range from an H atom to a large aliphatic or aromatic group. There are about 20 α-amino acids which are obtained by breaking down protein molecules, and in Table 23.1 we have listed their structures.

As you can see from the Table, there are a good many different sorts of R groups. Some, derived from aliphatic or aromatic hydrocarbons, are nonpolar; these are called hydrophobic (water hating) groups, since they decrease the solubility of the amino acid in water. R groups containing substituents like OH, NH_2, SH, and C=O are polar and enhance water solubility (hydrophilic).

Irrespective of the nature of the R group it contains, every α-amino acid contains both a weakly acidic —COOH group and a weakly basic —NH_2 group. These groups behave much like the carboxyl group in acetic acid and the amino group in CH_3NH_2, respectively. In basic solution the carboxyl group becomes —COO^-, whereas in acid the amino group gains a proton, becoming —NH_3^+. Depending on pH, all the α-amino acids can exist as either cations or anions:

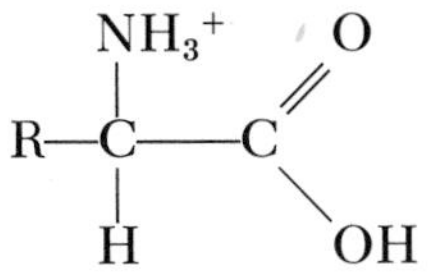

$$\begin{array}{ccc} NH_3^+ & & O \\ | & & \| \\ R-C-C & & \\ | & & \diagdown \\ H & & OH \end{array} \qquad \begin{array}{ccc} NH_2 & & O \\ | & & \| \\ R-C-C & & \\ | & & \diagdown \\ H & & O^- \end{array}$$

Acidic form: a cation 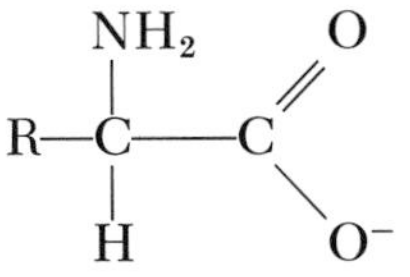   Basic form: an anion

At some pH, called the **isoelectric point**, an amino acid will be uncharged. The particular pH at which this occurs will vary from one acid to another,

TABLE 23.1   The Common α-Amino Acids

(R groups are indicated by shaded areas)

Hydrophobic, Nonpolar R Groups

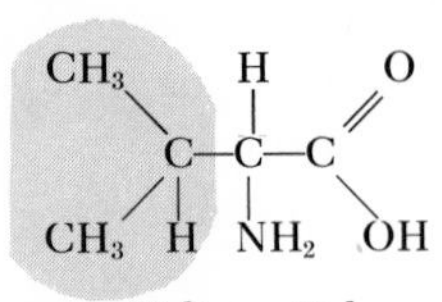

Valine—Val

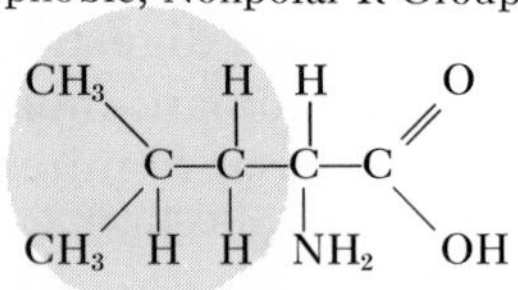

Leucine—Leu

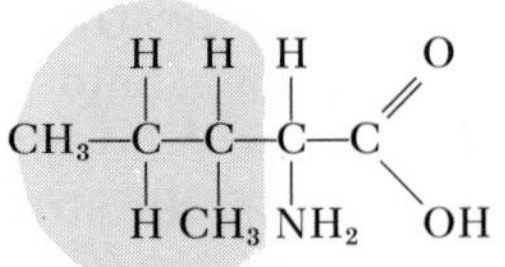

Isoleucine—Ile

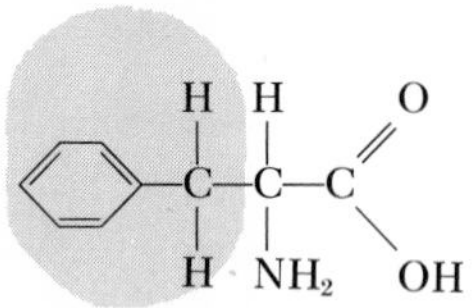

Phenylalanine—Phe

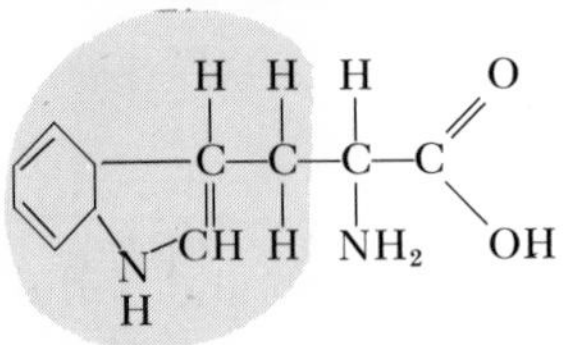

Tryptophan—Trp

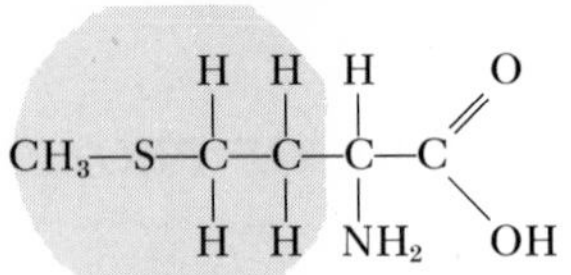

Methionine—Met

Hydrophilic R Groups—Tend To Be Charged

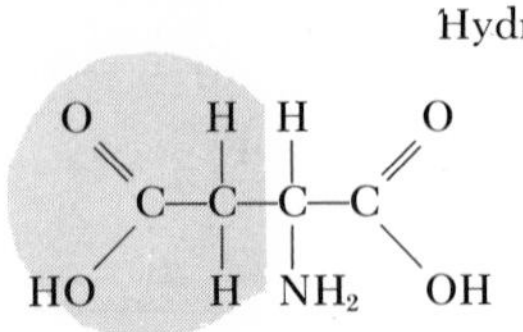

Aspartic Acid—Asp

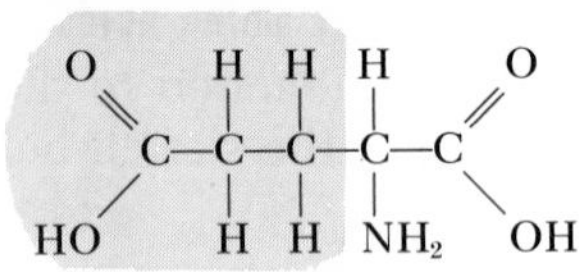

Glutamic Acid—Glu

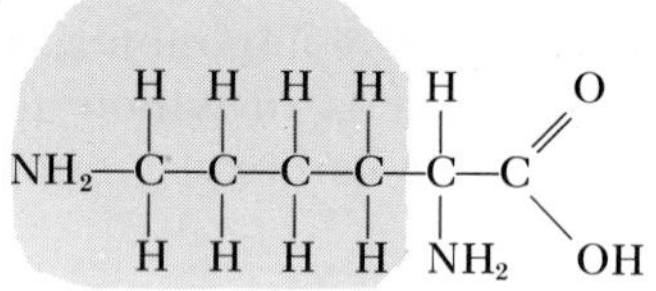

Lysine—Lys

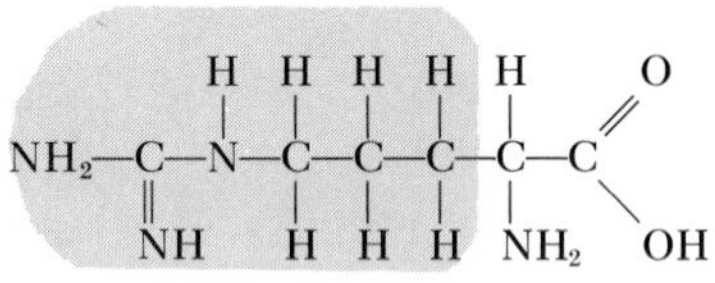

Arginine—Arg

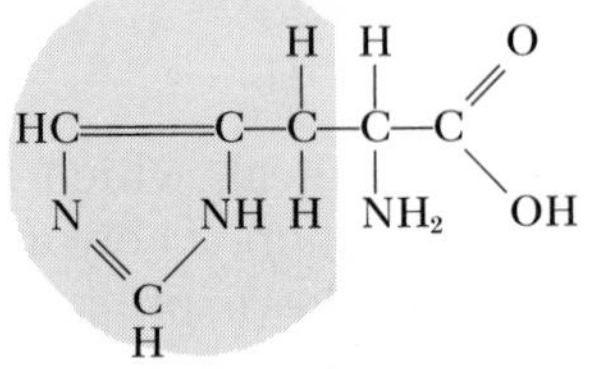

Histidine—His

Slightly Hydrophilic R Groups—Little Tendency To Become Charged

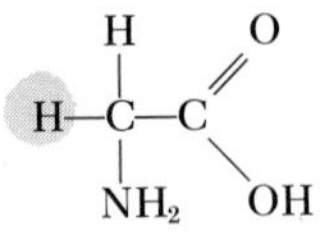

Glycine—Gly

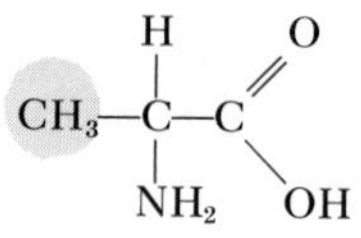

Alanine—Ala

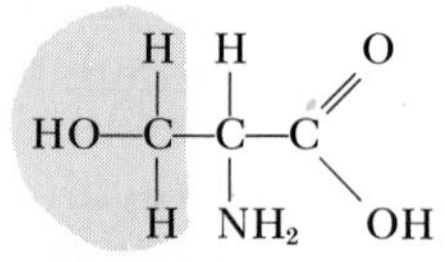

Serine—Ser

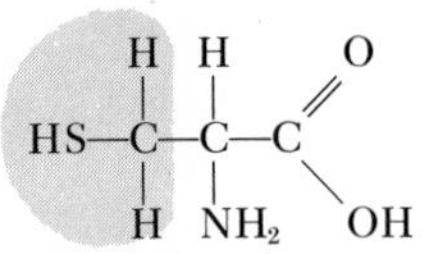

Cysteine—Cys

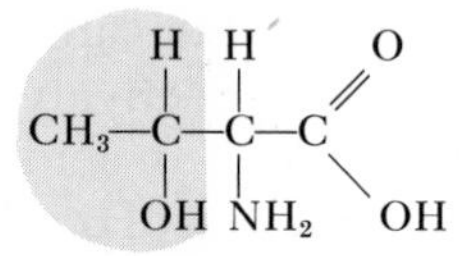

Threonine—Thr

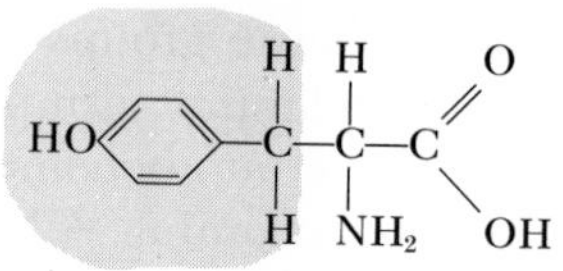

Tyrosine—Tyr

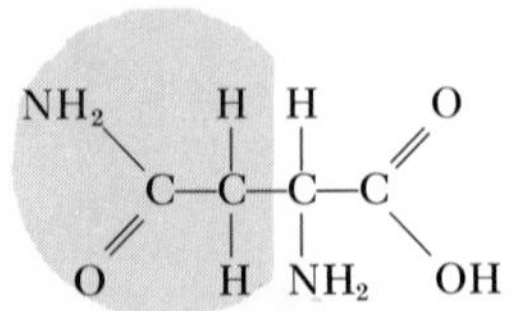

Asparagine—Asn

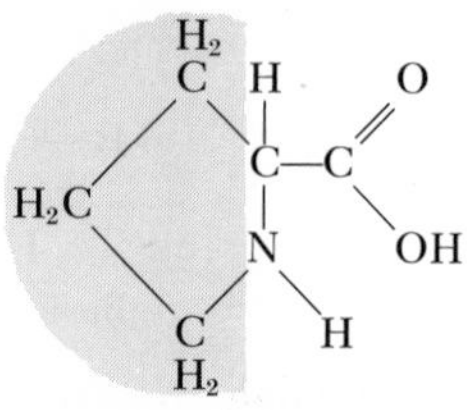

Proline—Pro

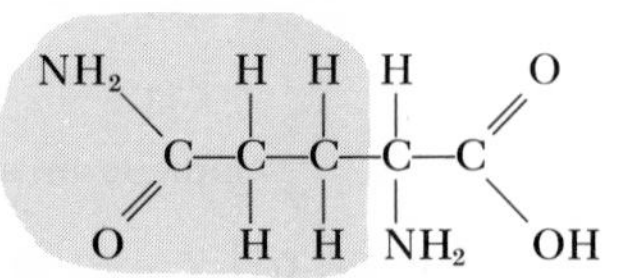

Glutamine—Gln

since many of the polar R groups will tend to take on charges depending on pH. The charges on the α-amino acids, including those on the R groups, affect their properties in important ways, and, indeed, form the basis for many of the procedures used to separate the acids from one another.

Two amino acids may combine by reaction of the acid group on one molecule with the basic group on another:

$$H-N-C-C-OH + H-N-C-C-OH \rightarrow H_2O +$$

Glycine   Alanine

In short form glycylalanine is written as gly-ala

peptide
linkage

$$H-N-C-C-N-C-C-OH$$

Glycylalanine, a dipeptide

In the process a molecule of water is eliminated, and the acid molecules are coupled by a **peptide linkage** into a dipeptide. (You may recall from Chapter 8 that the bonding in nylon polymer involved the same sort of linkage.) It is possible for more amino acids to couple to the dipeptide, forming a tripeptide, or a tetrapeptide, or to continue until a great many amino acids are linked together into a long chain, in which case we have a polypeptide.

Those substances we call **proteins** are **polypeptides** and are polymer chains of amino acid residues linked together by peptide bonds:

$$-N-C-C-N-C-C-N-C-C-N-C-C-N-C-C-$$

Section of a protein molecule

Proteins differ from the synthetic polymers discussed in Chapter 8 in that they are made up from about 20 different monomer units, which are incorporated in the polymer chain in a more or less random order and amount, whereas the synthetic polymers contain in general only one or two possible monomer units. This means that whereas there is one, or at most a few kinds of a polymer like nylon, there are a multitude of possible proteins. Even the little section of a protein molecule we have sketched, containing five amino acid residues, could be put together in about 20^5, or about 3×10^6 ways! Since most protein chains contain at least 100 units, there are many, many, many different proteins.

STRUCTURE OF PROTEINS

The problem of establishing the detailed structures of proteins has been of considerable interest to biochemists for many years. It is only in the last 20 years, however, that the experimental techniques adequate to this task have been available, with many of the most impressive results being obtained only very recently.

The determination of the structure of a typical protein molecule is a gigantic problem, to say the least. One must first prepare a sample of the pure protein. Then one establishes the relative and, hopefully, absolute number of each kind of amino acid unit in the protein chain. Following this, the actual sequence of amino acid residues in the chain is determined. Finally, the conformation of the chain, the manner in which it is actually arranged in space under normal conditions, is established. The fact that we now have detailed structural analyses of several proteins containing over 100 amino acid residues is one of the truly remarkable achievements of modern chemistry. It is possible for you to understand in principle how this was accomplished, and we shall discuss the steps that are involved in some detail.

PURIFICATION OF A PROTEIN

Since the cells of the organism furnishing the protein to be studied may contain hundreds or even thousands of different protein molecules, the isolation of only one of these from the rest is a tremendous task in itself. Proteins tend to be fragile molecules and may be destroyed by extremes of either pH or temperature, so that the methods used must be both gentle and capable of very high resolution.

The sample containing the protein of interest is first homogenized in a suitable solvent; the cell walls and other membranes are ruptured and the cell contents released. After the connective tissue, the blood vessels, and the like have been strained out, the fluid obtained is subjected to a series of procedures, each of which fractionates the sample on the basis of different properties of its components. The procedures used are the following:

Ultracentrifugation. The sample is centrifuged at a series of increasing speeds and experiences gravitational fields of from 1,000 to 100,000 times earth's gravity. Fractionation depends on the fact that the components will settle out at fields that vary inversely with their densities.

Selective Precipitation. The solubilities of proteins depend on the nature of the solvent system and its pH. Like the amino acids, proteins have characteristic isoelectric points, at which they usually have minimum solubility. Resolution depends on selection of that solvent and pH which precipitates selectively a fraction containing the desired protein.

Chromatography. This method was described in Chapter 1. Adsorption and ion exchange columns are both used. The solvent is buffered and the sample resolved into bands on the column on the basis of different tendencies of the components to dissolve in the solvent or remain on the packing.

Electrophoresis. The apparatus is similar to that used in chromatography and contains an adsorbent and a buffered solvent. An electric potential is placed across the ends of the column and the species present migrate toward the ends at speeds which depend on their charges. Components separate into bands as in chromatography.

By applying these procedures one or more times it is often possible to obtain a fraction which contains a single protein. A protein is usually considered to be pure if it can be isolated as a single band when subjected to high resolution electrophoresis.

DETERMINING THE AMINO ACID CONTENT OF A PROTEIN

The first step in the determination of the structure of a protein is to establish the relative numbers of moles of each of the amino acids it contains. A few milligrams of the pure protein are subjected to complete hydrolysis, which breaks all the peptide bonds and frees the individual amino acids present in the chain. The hydrolysis is accomplished by heating with 6 M HCl for as long as 24 hours. The excess HCl is removed and the solution brought to a pH of about 3.0, at which point the amino acids are in the cation form. The sample is adsorbed on a cation exchange column, similar to that used for water purification. An aqueous solvent, buffered at a pH slightly greater than 3.0, is poured through the column and elutes the various amino acids to an extent which is inversely related to the amount of positive charge they carry. Since the stronger acids carry the least + charge, they are eluted from the column first. After these acids, which include aspartic acid and glutamic acid, have washed through, the buffer is made more basic; continued elution removes the weaker acids, still in order of acid strength. The most basic amino acids, like lysine and arginine, are resolved in a separate column with a still more basic solvent. As the buffer comes from the ion exchange column it is treated with a re-

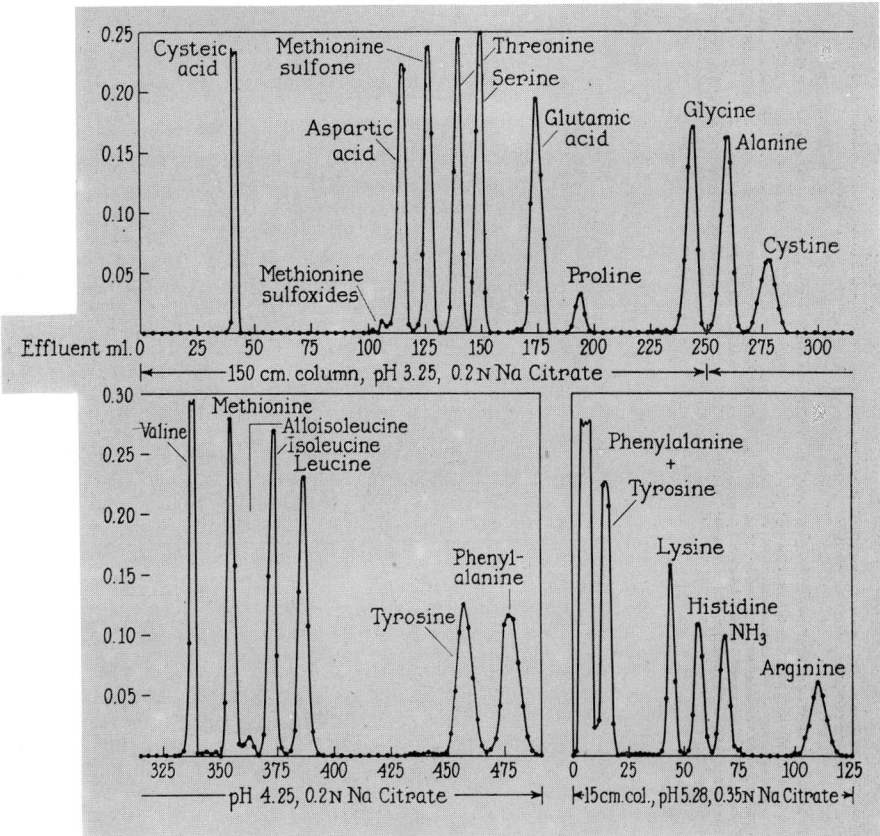

Figure 23.2 Chromatogram showing amino acid analyses of some proteins. (From Moore et al: Ann. Chem. *30:* 1186, 1958.)

TABLE 23.2 Amino Acid Composition of Some Protein Molecules

AMINO ACID	INSULIN (bovine)		CYTOCHROME c (horse)		β-LACTO- GLOBULIN		HUMAN SERUM ALBUMIN	
	A	B	A	B	A	B	A	B
Glycine	0.572	4	0.903	12	0.200	7	0.213	15
Alanine	0.505	3	0.455	6	0.793	30	0.046	3
Valine	0.661	5	0.230	3	0.484	18	0.657	45
Leucine	1.008	6	0.461	6	1.178	44	0.837	58
Isoleucine	0.211	1	0.459	6	0.448	17	0.130	9
Proline	0.252	1	0.302	4	0.457	17	0.443	31
Phenylalanine	0.492	3	0.306	4	0.234	9	0.472	33
Tyrosine	0.690	4	0.298	4	0.204	8	0.259	18
Tryptophan	0.000	0	0.075	1	0.094	3	0.010	1
Serine	0.497	3	0.000	0	0.386	14	0.318	22
Threonine	0.175	1	0.756	10	0.433	16	0.386	27
Cysteine	1.030	6	0.150	2	0.141	6	0.262	20
Methionine	0.000	0	0.153	2	0.215	8	0.087	6
Arginine	0.176	1	0.146	2	0.165	6	0.356	25
Histidine	0.336	2	0.232	3	0.103	4	0.226	16
Lysine	0.176	1	1.428	19	0.795	29	0.865	58
Aspartic acid	0.510	3	0.604	8	0.860	32	0.673	46
Glutamic acid	1.262	7	1.201	12	1.298	48	1.155	80

A = moles amino acid per 1000 grams protein; B = no. amino acid residues per molecule.

So far no regularities have been discovered in the compositions of proteins

agent called ninhydrin, which produces a deep blue color with all amino acids, and passed into a spectrophotometer which measures the color intensity as a function of the total volume of solvent which has come through the column. The whole procedure can be automated to produce a chromatogram such as that shown in Figure 23.2. By calibrating the apparatus with standard amino acid samples it is possible to determine the number of moles of each amino acid present in a given amount of protein.

Some of the results obtained from analysis of some typical proteins are given in Table 23.2. The data in the A columns of the table were obtained from spectrograms such as that in Figure 23.2. To find the number of amino acid residues per protein molecule (column B), one needs only to find a moderately good value of the molecular weight of the protein from, for example, osmotic pressure measurements.

Example 23.1 The molecular weight of insulin was found by various methods to be about 6000. Find the number of serine residues in the insulin molecule, using data in Table 23.2.

Solution. The analysis shows that there are about 0.497 mole of serine produced by hydrolysis of 1000 g of insulin. Since the molecular weight of insulin is about 6000 g, there will be about 6 × 0.497, or three moles serine per mole of insulin and, therefore, three serine residues per insulin molecule.

From the analyses of these and other proteins one can conclude that there is no clear or simple pattern for the relative amounts of the amino acid residues in these substances. It is true that some amino acids, such as glutamic acid, are more often found in a protein than others, such as tryptophan, but there appears to be no way to predict the amount of any given amino acid in any protein, given the amount of any other acid in the same or another protein.

ESTABLISHING THE AMINO ACID SEQUENCE IN A PROTEIN CHAIN

Determining the order in which the various amino acid residues occur on the protein chain is a considerably more difficult problem than finding the numbers of each residue. Many proteins contain more than one chain; when such is the case, the chains may or may not be linked by chemical bonds. If they are not, they may be separated from each other by treatment with acid and resolved by electrophoresis or chromatography into samples containing only one kind of chain. When the chains are linked chemically, it is always through disulfide (—S—S—) bonds between two cysteine molecules; these bonds can be cleaved easily and selectively and the chains then resolved as before and dealt with separately.

The original method of sequence analysis was developed by Sanger, who used it to establish the amino acid order in insulin. For this research, which was completed in 1953, Sanger was awarded a Nobel Prize. The method involved first determining the amino acids at each end of the two chains which make up the insulin molecule. Each chain was then cleaved

TABLE 23.3   Sanger's Method for the Determination of the Amino Acid Sequence in the Short Chain of Insulin

I. End group tests: glycine (Gly) on NH_2 end; aspartic acid (Asp) on COOH end.

II. No. of amino acid residues per molecule:

Gly_1, Ala_1, Val_2, Leu_2, Ile_1, Tyr_2, Ser_2, Cys_4, Asp_2, Glu_4

III. Fragments identified after partial hydrolysis of the chain:

```
No.  1   Gly–Ile–Val–Glu–Glu–
     2                –Glu–Glu–Cys–
     3                     –Glu–Cys–Cys–
     4                          –Cys–Cys–Ala–
     5   –Ser–Val–Cys–Ser–Leu–
     6                –Ser–Leu–Tyr–
     7                     –Leu–Tyr–Glu–
     8                          –Glu–Leu–Glu–
     9                               –Glu–Asp–Tyr–
    10                                    –Tyr–Cys–
    11                                         –Cys–Asp
```

More modern methods for sequence analysis remove and identify amino acids one at a time from the protein chain

IV. Given fragments 1 to 11, it is possible to overlap them *only* as shown if one is to be consistent with overall composition of the chain. Since the two large fragments obtained by overlapping the shorter ones (1–4 and 5–11) include all the residues, the total chain must simply involve the addition of the two large fragments end to end, forming Chain I in Figure 23.3. (The hydrolysis reactions would convert glutamine and asparagine to glutamic acid and aspartic acid, respectively. Further analysis would be required to establish whether the amine or free acid was the actual residue in positions where Asp or Glu was observed.)

randomly into fragments of varying length by treatment with acid. The fragments were separated chromatographically and analyzed chemically. The structure of each chain was finally established by taking advantage of the fact that the known fragments had to overlap in such a way as to be consistent with the sequence and composition in the whole chain. Sanger's method is outlined in Table 23.3. The sequence of amino acid residues in bovine insulin is given in Figure 23.3. There are 21 amino acid residues in one chain and 30 in the other, yielding a molecular weight for the whole molecule of 5733.

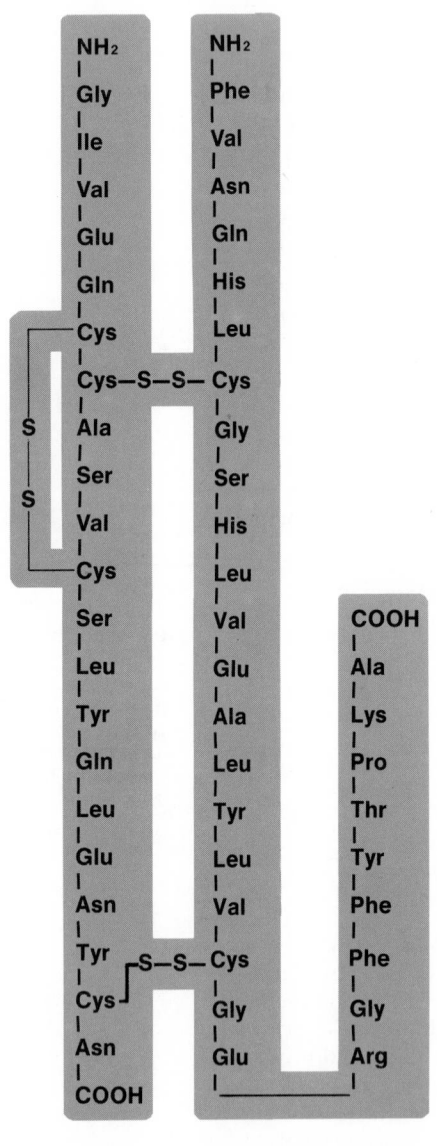

Chain I   Chain II

Figure 23.3 Amino acid sequence in insulin. (See Table 23.1 for the names and structures of the amino acids represented here by 3-letter symbols).

Several of the figures in this chapter are drawn from *Biochemistry*, an excellent text by A. L. Lehninger. We are grateful to Dr. Lehninger and his publisher, the Worth Publishing Company, for permission to adapt these figures for our purposes.

TABLE 23.4   Characteristics of Some Proteins

	MOLECULAR WEIGHT	NO. OF RESIDUES	NO. OF CHAINS
Bovine insulin	5,733	51	2
Human hemoglobin	64,500	574	4
Human serum albumin	68,500	550	1
Bovine chymotrypsinogen	22,600	245	3
Horse γ globulin	149,900	~1250	4
Bovine glutamate dehydrogenase	1,000,000	~8300	40
Horse cytochrome c	13,370	104	1

The amino acid sequences of a large number of protein chains have now been established by methods somewhat more sophisticated than that used by Sanger. Among the larger protein molecules which have been studied are hemoglobin, which has four chains with about 150 residues each, and chymotrypsinogen, an enzyme precursor, with 245 residues.

Perhaps the protein which has been most extensively investigated in different species is cytochrome c (Table 23.4), which serves as an enzyme in one step in a rather complex reaction sequence by which the organism uses oxygen in metabolism. Cytochrome c is found in all organisms which utilize oxygen and has been studied in about 40 species, ranging from vertebrates like monkeys and sheep, to birds, snakes, and fish, to seeds of plants, to insects, and finally to primitive yeasts. Considering the tremendous differences between the species, the molecule is remarkably similar to that in the horse in all the species studied. In 35 positions on the chain, including one section containing 11 residues, the amino acid residue is the same for all the species. In 23 other positions only two different residues occur. In only about 15 positions are there as many as five different residues found. By noting similarities and differences between the amino acids at given positions on the chain it is possible to relate species to one another in much the same way as used by the taxonomist who classifies species on the basis of their morphological features. The family trees of organisms based on cytochrome c differences agree very well with those obtained by the classic methods, indicating that changes in protein structure parallel closely evolutionary changes (Table 23.5).

TABLE 23.5   Cytochrome c Residue Differences Between Species

	NO. OF DIFFERENCES	TIME BRANCHING OCCURRED (millions of years ago)
Man-monkey	1	50
Man-horse	12	70
Man-pigeon	12	250
Man-tuna fish	19	400
Man-baker's yeast	40	1000

CONFORMATION OF PROTEIN MOLECULES

A protein molecule, unlike a molecule of polyethylene, contains many R groups which tend to interact with other groups, either on the same or on adjacent protein molecules. These interactions may involve hydrogen bonding between O atoms on C=O groups and H atoms in nearby N—H groups, forces between charged R groups on the chain, or solvent-R group interactions. All three kinds of interactions can be quite strong and may affect the properties of a protein. In particular, these interactions determine in large measure the conformation of the protein, the manner in which the units in the chain are arranged in three dimensions.

By far the most important experimental method used to establish protein conformation is x-ray diffraction. The x-ray diffraction pattern of a crystal depends on the positions of all atoms in the crystal and can be used to determine those positions. In recent years it has become possible to apply x-ray diffraction techniques to proteins containing upwards of 200 amino acid residues. From a small crystal of such a protein one can obtain about 50,000 spots in an x-ray diffraction pattern; the positions and intensities of these spots are fed into a computer, which can handle the millions upon millions of calculations necessary to relate those spots to the positions of the atoms that caused them. The protein conformations we shall discuss were all determined by x-ray diffraction methods.

There are two ways in which a protein chain can be twisted so as to maximize hydrogen bonding between carbonyl oxygen atoms and amido hydrogen atoms; both of these are observed in natural substances. If the R groups from the amino acid residues are small, as are the H atoms in glycine or the CH_3 groups in alanine, then it is possible for the chains to form a sheet as in Figure 23.5. The sheet lies in the plane of the paper and is made

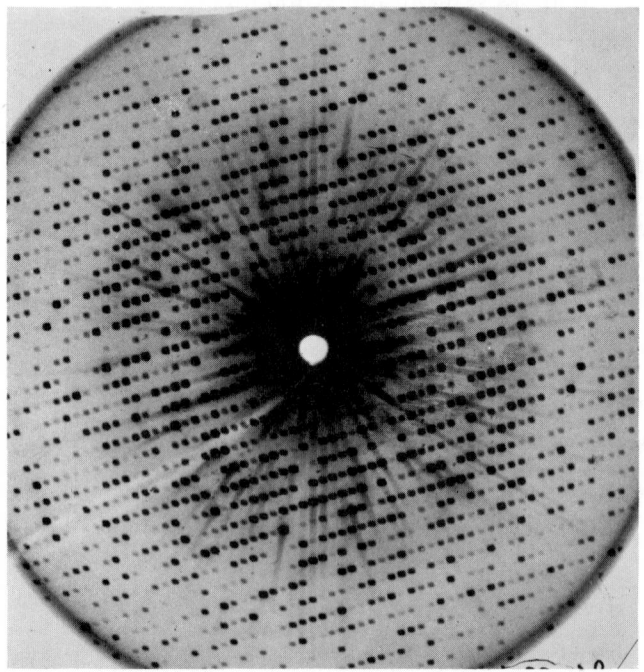

Figure 23.4 X-ray diffraction pattern from whale myoglobin. (From F. C. Kendrew, Scientific American, December, 1961.)

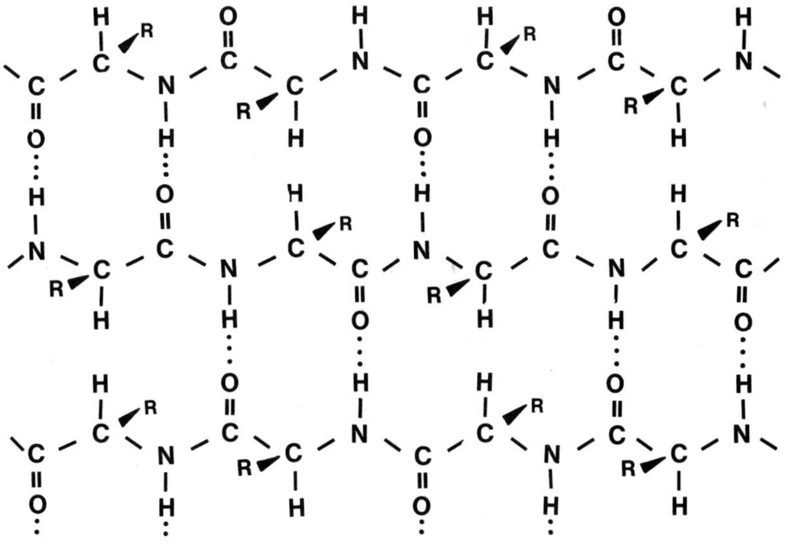

Figure 23.5 β-Keratin structure of silk fiber. The dots represent hydrogen bonds holding protein chains together.

Although the chains lie roughly in a plane, the tetrahedral bonding around N and non-carbonyl C atoms gives the sheet a pleated structure

up from many parallel protein chains held together by hydrogen bonds. As you can see by looking at the middle chain, full hydrogen bonding to adjacent chains is possible at every amino acid residue, making for a very stable fiber. The structure shown is called the pleated sheet, or β-keratin, and is found in silk fiber.

In Figure 23.5 it is clear that there tend to be some R groups in close proximity. In silk the amino acid residues are mainly those of glycine and alanine, where $R = H$ and CH_3, respectively, and it is possible for these small groups to rotate somewhat, all in the same direction, so as to bring alternate R groups in the chain above and below the plane of the sheet, minimizing the steric forces between those groups. (Note that for maximum hydrogen bonding the adjacent chains in the sheet must run in opposite directions.)

In most proteins the R groups are too bulky to allow for formation of sheets such as are found in the silk structure. Pauling and Corey showed that an alternate structure that also maximized H bonding was one in which the protein chain took the form of a helical coil called an α-helix. We have indicated the form of the coil in Figure 23.6, showing first the schematic arrangement of atoms that make up the chain and then a more complete structure to show the positions at which H bonding can occur. Note that the R groups are all on the outside of the helix, where they have the most room, and that the actual structure of the helix is nearly independent of the nature of these groups. The dimensions of the helix correspond quite well to those observed in such natural fibrous proteins as wool, hair, skin, feathers, fingernails, and horn; the structure of these substances is now accepted as being for the most part that of the α-helix.

Although many proteins take the form of a helical coil, the structure can result in significant charge interactions between R groups on the chain. If these groups are highly charged, the chain will tend to spontaneously uncoil into a much less regular structure, called a *random coil*. Polyalanine, $R = CH_3$, exists as an α-helix over wide pH ranges, since the CH_3 group does not readily take on a charge. Polylysine, where $R = (CH_2)_4NH_2$, will readily pick up protons under acidic conditions and be-

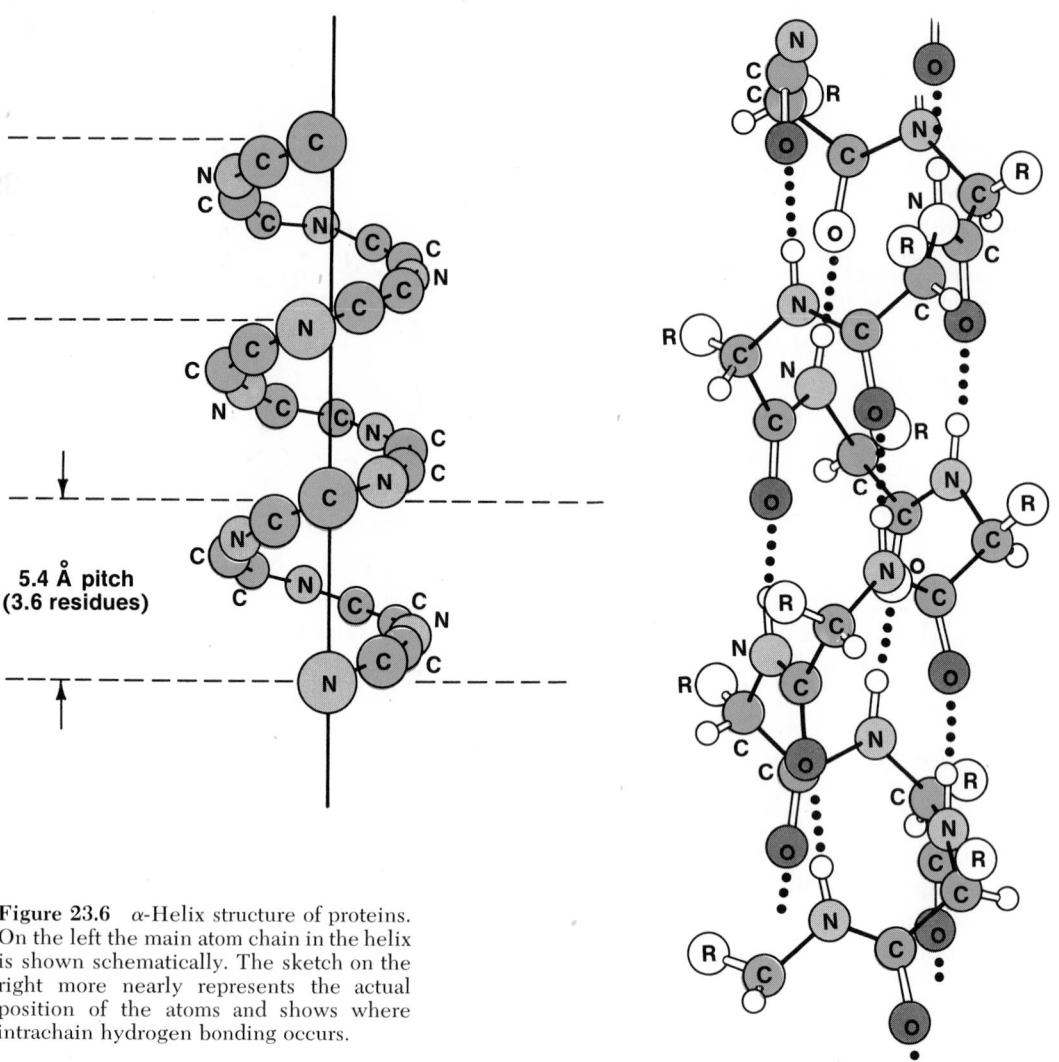

Figure 23.6 α-Helix structure of proteins. On the left the main atom chain in the helix is shown schematically. The sketch on the right more nearly represents the actual position of the atoms and shows where intrachain hydrogen bonding occurs.

5.4 Å pitch
(3.6 residues)

come positively charged; below a pH of 7, polylysine acquires a large enough positive charge to change from an α-helix to a random coil. Similar behavior is observed for other polyamino acids whose R groups can become charged.

Probably the most interesting of the protein structures are those exhibited by the so-called globular proteins, which exist as roughly spherical entities in which the chain weaves its way around the sphere. In Figure 23.7 is shown a typical globular protein structure of cytochrome c isolated from horse heart, as recently determined by Dickerson at the California Institute of Technology. In the drawing each ball represents an α-carbon atom and its associated R group, with the atoms in the peptide linkage denoted by a bond; the actual number of *atoms* in the molecule is about 10 *times* the number of balls shown. The structure begins at the top of the figure with glycine and ends at position 104 with a carboxyl group. Notice that the chain first goes through a few turns of α-helix, moving southeast

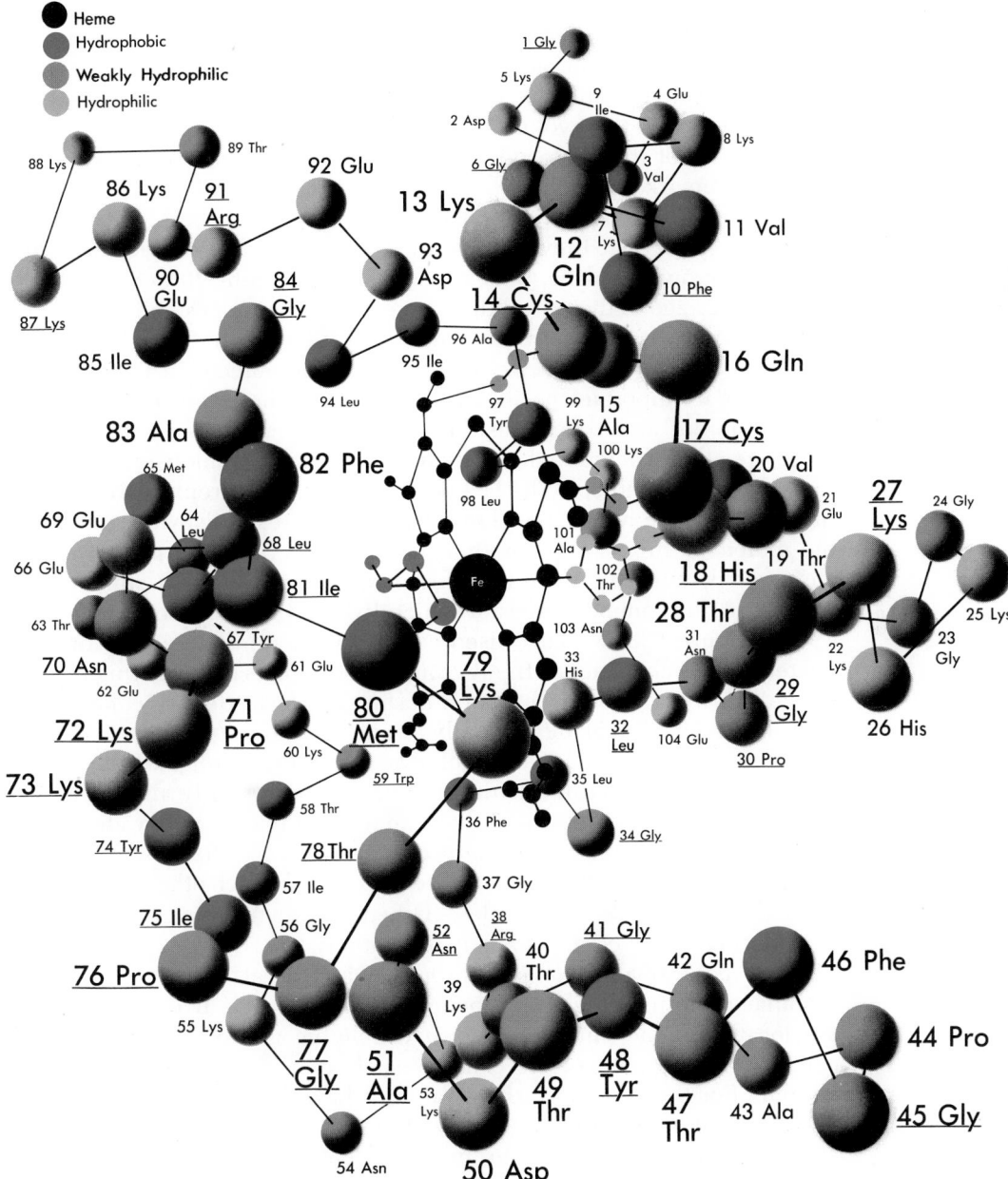

Figure 23.7 Conformation of cytochrome c molecule. Each ball represents an L-carbon atom plus its associated R group. Each line depicts a peptide linkage.

to position 24, west to position 36 behind the molecule, down to a loop forming the lower part of the molecule, then up the left side to position 90, and finally back behind the molecule to the end of the chain. Sheltered within the cytochrome c molecule is the group called *heme*, mentioned earlier in Chapter 19, which contains an Fe atom and which is vital for the electron carrying properties of this protein. (The structure of heme was shown in Figure 19.1.)

In addition to showing the structure of cytochrome c, Figure 23.7 also helps us to rationalize the evolutionary stability observed in this molecule. You will recall that in the many species in which the amino acid sequence in cytochrome c has been determined, there are about 35 positions on the chain in which the amino acid residue does not ever change as one goes from one species to another. In this figure these positions are all underlined. More than a third of the groups in the figure are in this category, indicating that in the evolution of modern organisms from primitive ones, the structure of this molecule, and presumably others of a similar nature, has been remarkably stable.

Noting now the actual positions which have completely resisted change, we see that they include those residues responsible for holding the heme group in place, plus a long chain, positions 70 to 80, across the left front surface of the molecule. These groups, which establish the nature of the surface in the vicinity of the heme group, can easily furnish vital sites of interaction of the enzyme with reactant molecules. There are relatively few unimportant positions in the chain where, for instance, five or more different amino acid residues are observed in the species studied; these are all restricted to the outermost regions of the molecule (e.g., positions 44, 54, 88, and 89). The fact that so many of the irreplaceable residues in cytochrome c are in positions that serve to establish the structure of the molecule supports the notion that the general conformation of cytochrome c is much the same in different species, and is very similar to that present in our common ancestor which lived a billion or so years ago.

> Glycine may be so stable in its position because it is the only residue that was small enough to fit

In Figure 23.7 we have used color to indicate degree of hydrophilic (water loving) character in the R groups. Those balls representing R groups which are most strongly hydrophilic are the most deeply colored, with color intensity decreasing gradually as one goes through the weakly hydrophilic and finally to the hydrophobic (water hating) groups. It is apparent from the drawing that the hydrophilic groups are nearly all on the outer surface of the molecule, where they can interact readily with water. In contrast, the hydrophobic groups are concentrated for the most part in the inner regions, where they are in close proximity to each other but sheltered from the polar water molecules. The molecule is remarkably compact considering that it contains about 1000 atoms.

> The conformation of a protein appears to be fixed mainly by the amino acid sequence in the chain

The conformation of cytochrome c is typical of that of the other globular proteins which have been determined. The hydrophilic groups are located in the outer regions, hydrophobic groups are mostly in the center, and the path of the chain is such as to minimize the molecular volume. For these reasons, these proteins have been said to resemble oily drops. In most proteins the observed atomic arrangement is surprisingly stable. If the chain is made to uncoil, by changing the pH or raising the temperature slightly, it will in many cases revert to its former configuration when normal conditions are restored. If one actually synthesizes a protein, as has been done for a few proteins the size of cytochrome c, the chains spontaneously take on the natural conformation when the pH and temperature have their normal values. Even when more than one chain is present in the protein, the mixture is often able to find the proper conformation. Interchain bonds may stabilize the structures of complex proteins but do not appear to be necessary to bring the chains into the natural arrangement. The observed structure is mainly due to the tendency for hydrophilic and hydrophobic groups to be located, for reasons of thermodynamic stability, in the outer and inner regions, respectively. This arrangement allows for hydrogen bond formation between hydrophilic groups, minimizes repulsive forces between groups of like charge, and minimizes the free energy of interaction of water and the hydrophobic groups.

ENZYMES

In Chapter 14 we discussed briefly the properties of enzymes, the catalysts that are involved in most of the chemical reactions that occur in living systems. Since there are many such reactions, there are many enzymes; well over a thousand have so far been identified and more than a hundred have been isolated as pure crystals. As far as is now known, enzymes are all globular proteins, similar in general structure and conformation to cytochrome c.

As we mentioned in our discussion of catalysis, an enzyme functions by interaction of groups on its surface with reactant molecules. The "lock and key" model, according to which the enzyme and reactant molecules fit together to form a complex in which a particular bond in the reactant is weakened and finally broken, has long been used to qualitatively explain enzyme effect. The availability of detailed information on enzyme structure, such as that we have given for cytochrome c, confirms the general model very well. The fact that enzymes typically are effective for only one reaction, or a few similar reactions, is reasonable, since the physical position of active groups on the enzyme surface limits interaction with reactant species to those which have the proper size and structure. The stability of the structure, and presumably the conformation, of cytochrome c through millions of years, implies the importance of maintaining the detailed properties of the molecular surface. The fact that enzyme action is limited to rather narrow regions of pH and temperature is easily understood in light of the fact that the conformation of globular proteins changes dramatically when they are heated or exposed to highly acidic or basic media.

The most recent studies of enzyme structures indicate that the conformation of an enzyme ordinarily changes when the enzyme-reactant complex forms. This observation may explain the extraordinary catalytic power of some enzymes, since such changes could readily produce a strain in a key chemical bond in the complexed reactant molecule that is sufficient to break the bond.

23.3 NUCLEIC ACIDS

The nucleic acids, the last class of substances we shall consider, differ markedly from the proteins and carbohydrates in their role in living systems. We might, in a general way, say that whereas the proteins make up much of the machinery of an organism, and the carbohydrates furnish a large part of the fuel for constructing the machinery and keeping it going, the nucleic acids are the blueprints that direct its construction and operation. The nucleic acids are the carriers of the genetic information the organism needs to reproduce its kind and to direct the synthesis of all the substances required for its growth and living processes.

The nucleic acids derive their name from the fact that they were first discovered in cell nuclei in 1870. It took many years, however, before their composition was determined, and indeed it was not until about 1950 that the structure and role of these substances were established.

Nucleic acids are complex, long chain, polymeric substances. Their molecular weights are very high, exceeding a million in many cases. Like the proteins, they are made up of several monomeric units. Each unit contains three components: (1) a 5-carbon sugar, (2) phosphoric acid, and (3) a nitrogenous base.

The sugar in a given nucleic acid is always either ribose or deoxyribose, and, depending on which is present, the acid is called either ribonucleic acid, RNA, or deoxyribonucleic acid, DNA.

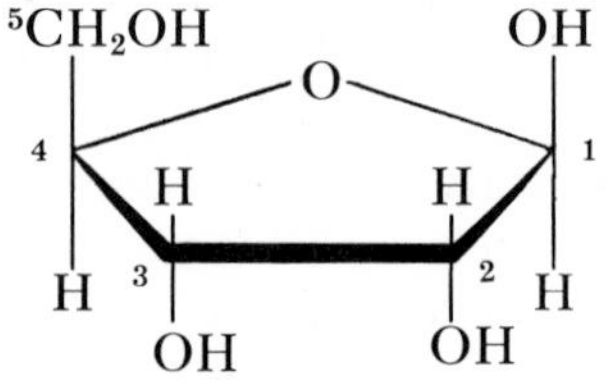

β-D-ribose $\qquad$ β-D-deoxyribose

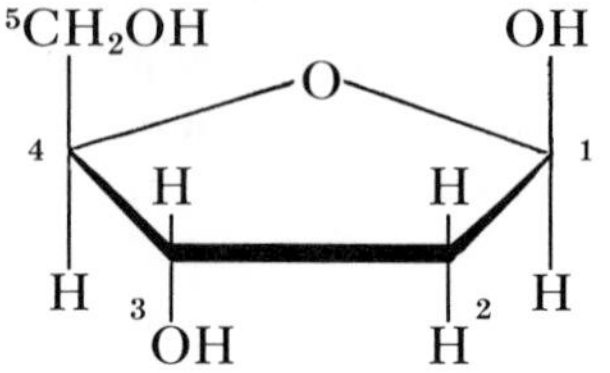

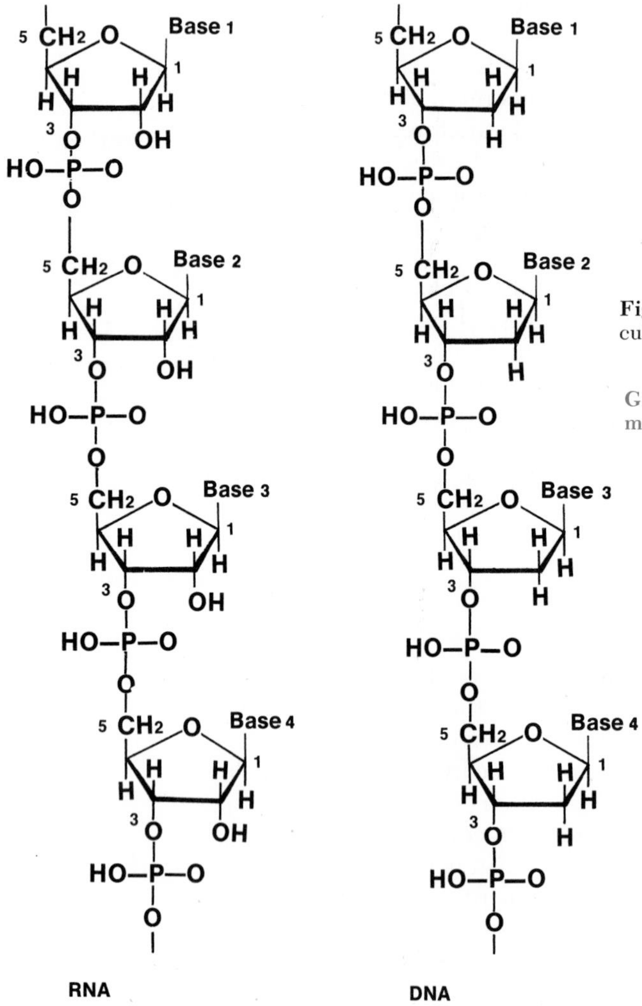

RNA $\qquad$ DNA

Figure 23.8 Fragments of RNA and DNA molecules; four units are shown in both cases.

Given a pure sample of a nucleic acid, how might you determine if it was RNA or DNA?

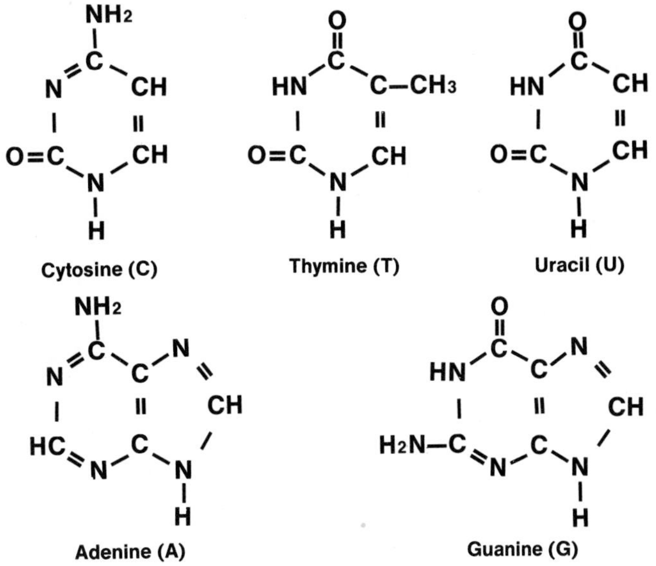

Figure 23.9 The structures of some common nitrogen bases found in nucleic acids.

Cytosine (C) Thymine (T) Uracil (U)

These molecules are all either planar or nearly planar

Adenine (A) Guanine (G)

These two sugars differ only in that the No. 2 carbon atom in ribose has attached OH and H atoms, whereas in deoxyribose it has two attached H atoms. It appears that any given nucleic acid will contain either all ribose or all deoxyribose, with no mixing.

In a nucleic acid the sugar molecules are linked together in a chain by ester bonds formed by phosphoric acid with the hydroxyl groups on the No. 5 and 3 carbon atoms of the sugar. The chains in RNA and DNA are very similar (Fig. 23.8).

Nucleic acids are distinguished from one another by the third component of these substances, the nitrogen bases. There are about five common bases; one or another of these is attached to the No. 1 carbon atom of the sugar molecule, as indicated on p. 644. The structures of the nitrogen bases are shown in Figure 23.9. As you can see, three of these bases are rather similar, with substituted six-membered rings, each containing two nitrogen and four carbon atoms. They fall in the class called pyrimidines and are all planar molecules. The other two bases are also similar to one another; they are both derivatives of a substance called purine. Both adenine and guanine are also nearly planar. Since the names are long, it is convenient to use one letter abbreviations to describe the order of bases in a chain; ATGU would mean that adenine, thymine, guanine, and uracil occurred in the chain, in that sequence. The linkage from the base to the sugar is usually through the nitrogen atom attached to a single H atom (bottom of each drawing) to the No. 1 carbon atom of the sugar. In the linkage reaction, a molecule of water is eliminated and a C—N bond is formed.

The nucleic acids can be considered to be made up of chains of **mononucleotide** units, each unit containing a nitrogen base bonded to a sugar molecule which, in turn, is bonded through an ester linkage to phosphoric acid.

Some mononucleotides are important biochemically

This substance, sometimes called AMP, is involved in many energy-producing processes in an organism

A mononucleotide (adenosine 5-phosphoric acid)

DNA, DEOXYRIBONUCLEIC ACID

DNA is found primarily in the nuclei of cells. It is one of the main components of the chromosomes, where it is usually found in combination with characteristic protein material. The amount of DNA in a cell of a given organism appears to be nearly constant, independent of the tissue from which the cell was obtained. The amount per cell varies from one organism to another, increasing with the complexity of the organism; there is about twice as much DNA in mammal cells as in those from fish or plants, and several hundred times as much per cell in mammals as in bacteria.

It is very difficult, perhaps impossible, to obtain a sample of a pure native DNA from the cell of a vertebrate. DNA molecules characteristically have a tremendous molecular weight; a billion is a good guess. In a higher animal there are probably many different DNA molecules, all with these very high molecular weights. Simple operations designed to isolate intact molecules of DNA are almost doomed to failure, since stirring or pipetting puts enough strain on the long chains to break them up into fragments of molecular weight of a few million. Much of the work on the properties of DNA has been done on bacteria, or viruses, where only one kind of DNA molecule is present, and, at least for the viruses, the molecular weights are relatively low. It has been possible to obtain single DNA molecules from such species and to take their pictures with an electron microscope (Fig. 23.10). It is frequently the case that the chain is closed, as in the micrographs; the chain is long by molecular standards, about 18,000 Å for X174 and 180,000 Å for λ viruses, corresponding to molecular weights of about 3×10^6 and 3×10^7, respectively, which are among the lower values commonly found for DNA molecules.

STRUCTURE OF DNA—THE DOUBLE HELIX

It is possible to hydrolyze DNA completely, obtaining a mixture of the component nitrogen bases as well as the sugar and phosphoric acid. Analysis of the mixture of bases reveals that DNA contains adenine (A),

φX174 DNA

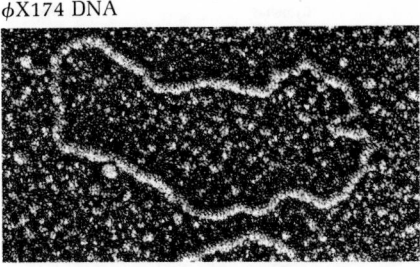

5000 Å

Figure 23.10 Electronmicrographs of DNA molecules isolated from two bacterial viruses.

DNA molecules from mammals are about 1000 times as long as these molecules

λ DNA

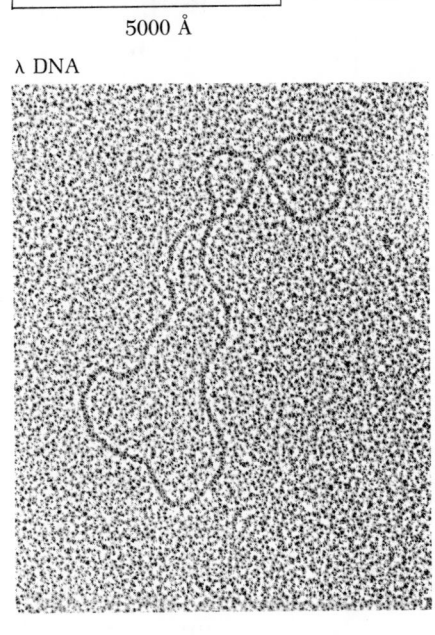

50,000 Å

thymine (T), guanine (G), and cytosine (C), with no uracil (U), and very small amounts of the less common bases. The relative amounts of each base are the same for all the cells of a given species, but vary from species to species (Table 23.6). From the data it is apparent that, whatever the source of DNA, there is very nearly a 1:1 ratio between the number of molecules of adenine and thymine and a 1:1 ratio for the number of molecules

TABLE 23.6   Nitrogen Base Composition of DNA from Different Species

SPECIES	MOLE PER CENT OF BASE			
	A	T	G	C
Man	30.9	29.4	19.9	19.8
Chicken	28.8	29.2	20.5	21.5
Salmon	29.7	29.1	20.8	20.4
Yeast	31.3	32.9	18.7	17.1
E. coli (bacteria)	24.7	23.6	26.0	25.7
T7 virus	26.0	26.0	24.0	24.0

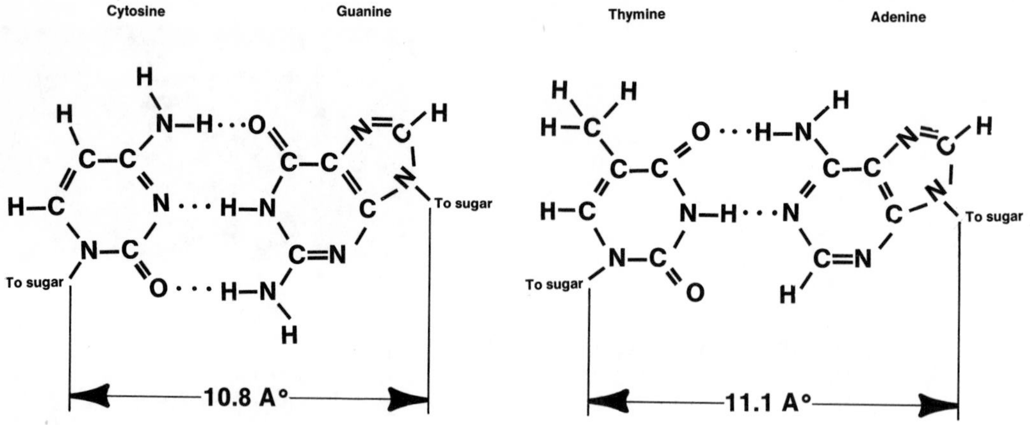

Cytosine   Guanine   Thymine   Adenine

Figure 23.11   Hydrogen bonding between pairs of nitrogen bases.

of guanine and cytosine. This would imply that for some reason these pairs of bases are related in DNA. Determination of the geometry and dimensions of these bases shows that the pairs A-T and G-C can participate readily in hydrogen bonding, forming two planar pairs of about equal overall dimensions (Fig. 23.11). Any other combinations do not fit so well, nor do other pairs have equal dimensions.

Given these facts, plus some relatively crude x-ray data on DNA crystals, Watson and Crick proposed in 1953 the famous *double helix* model for DNA, in which two DNA chains were twisted together to form a long filament. In the double helix the nitrogen bases from separate chains are always paired as A-T and G-C; the helix is held together by the hydrogen bonds between the components of each pair, as in Figure 23.11. The base pairs are in parallel planes and are stacked on top of each other, with each pair perpendicular to the main axis of the helix. The sugar molecules and the phosphate groups are all on the outside of the helix (Fig. 23.12). The double helix model has many attractive features to recommend it. It explains nicely the 1:1 base ratios for A:T and G:C that are observed. It maximizes hydrogen bonding between the bases and so helps to produce a stable structure. The hydrophobic base pairs are on the inside of the helix, while the hydrophilic sugar and phosphate groups are on the outside, in a compact arrangement analogous to that known to be preferred by globular proteins. The dimensions of the helix are essentially those which are observed in x-ray diffraction studies. The actual lengths of DNA molecules in Figure 23.10 can be used with the model to estimate molecular weights which correspond well to those determined by other methods. By these and other experimental tests the model has been found to be very satisfactory and is now accepted. For their work on the structure of DNA Watson and Crick received a Nobel Prize in 1962.

THE CENTRAL DOGMA OF MOLECULAR GENETICS

For many years it has been known that the chromosomes in the nucleus of a cell are the carriers of the genetic information that allows that

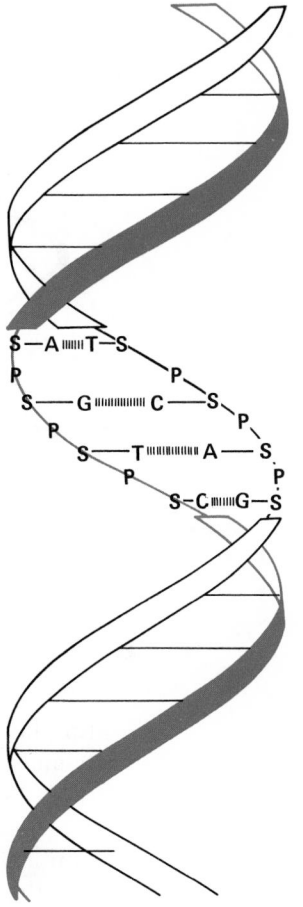

Figure 23.12   Double helix model for DNA.

(A = adenine, T = thymine,
 G = guanine, C = cytosine,
 S = sugar, P = phosphoric acid)

cell to duplicate itself and indeed allows the organism containing that cell to grow, take on the characteristics of its own species, and reproduce its kind. The mechanism by which this is accomplished has long been one of the mysteries of biology. The double helix model for DNA has allowed for the development of a theory, called the central dogma of molecular genetics, which goes a long way toward explaining not only how genetic information in the chromosomes is duplicated, but also how that information is actually used by the cell in creating the chemical species that characterize it and allow it to function as it does.

According to the central dogma, *DNA is the genetically active component of chromosomes.* In the duplication of a cell, the exact duplication of its DNA is vital if the genetic characteristics of the cell are to be maintained.

It is now believed that, in the reproduction process, the two DNA chains in the double helix unwind from each other and separate. As this happens, a new matching chain of DNA is synthesized on each of the original ones, creating two double helices. Since the base pairs in each new double helix must match in the same way as in the original, the two new double helices must both be identical to the original, with each containing one chain from that original. Note that the two chains which make

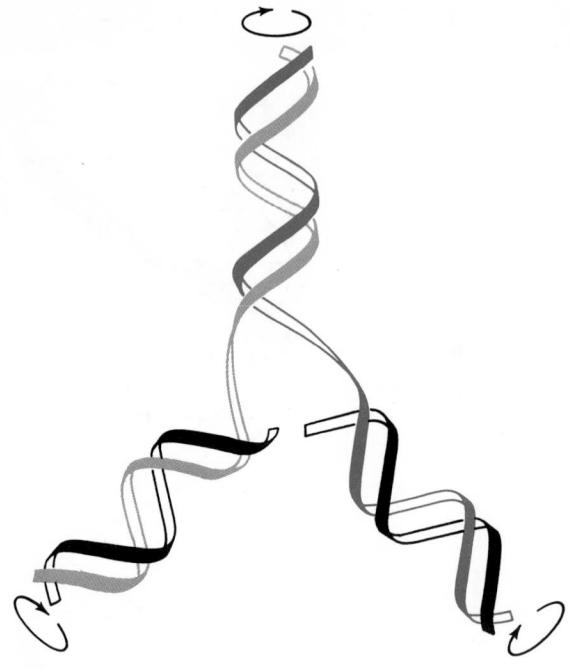

Figure 23.13 Replication of DNA as suggested by Watson and Crick. The complementary strands are separated and each forms the template for synthesis of a complementary daughter strand.

Each new DNA double helix contains one chain from the parent

up the double helix are not themselves identical, but rather complementary, and contain base pairs that match at every point in the chain (Fig. 23.12).

Some very clever experiments strongly support the notion that DNA is duplicated by this method in living cells. Exact replication of genetic data can thereby be accomplished, however complex that data may be. The duplication is controlled by a remarkable enzyme, in whose presence the addition of the proper mononucleotides to the new DNA chain has been observed to occur at a rate of about 10^5 units per minute.

GENES AND ENZYME SYNTHESIS

Each chromosome in a cell has many sections, each section corresponding to a so-called **gene.** The genes govern the detailed characteristics of an organism and differ in major ways from species to species and in small ways from individual to individual within a species. According to the dogma, a gene in a chromosome consists of a section of DNA double helix, in which there is a particular sequence of bases. This sequence of bases contains information which directs the synthesis of an enzyme. The enzyme, in turn, establishes the characteristic associated with the gene; this is called the *one gene–one enzyme* concept.

The problem of storing in a gene the information the organism needs to synthesize an enzyme is analogous to that which exists in a computer which needs to carry out a certain calculation. The information in the DNA in a gene governing the synthesis of an enzyme must specify the order in which the amino acids fit into the protein which comprises the enzyme. In a computer the program must specify the order in which the steps in the calculation are to be carried out. In both cases the problem is handled through a code made up of elements in an information unit. In the computer these elements are magnetic cores, which can be in either one of two possible states: ↑ or ↓. In DNA the elements are the four nitrogen bases: A, T, G, and C. In the computer the operation of addition is coded as a specific sequence of the states of a series of magnetic cores; whenever the control unit encounters the code word for addition, which might for example be ↑ ↑ ↓ ↑ ↑, it will carry out the addition of two numbers. In DNA a particular sequence of nitrogen bases will direct the addition of a specific amino acid to the enzyme chain. The actual sequence of nitrogen bases which is associated with a given amino acid comprises what is known as the *genetic code* for that acid.

A rather surprising amount is known about the nature of the genetic code and the manner in which it is used in enzyme synthesis. We are quite sure, for example, that the code word for a particular amino acid is a certain sequence of three nitrogen bases in the DNA chain, read in a particular direction. ACC and CCA are code words for two different amino acids. The order in which amino acids are added to the protein chain during synthesis is the same as the order in which the code words for the acids appear in the DNA molecule.

The DNA chain which forms a gene consists of a series of base triplets whose order duplicates that of the amino acids in the protein chain coded for by those triplets. If alanine and serine are the third and fourth residues, respectively, which are to be added to the protein chain, the third and fourth triplets in the associated DNA chain would presumably be the code words for alanine and serine, respectively. A protein like cytochrome c, with 104 amino acid residues, is coded for (in the gene governing cytochrome c synthesis) by a particular sequence of 312 bases. A chromosome is a sequence of many genes, laid end to end, with the ends indicated by a special base sequence. (In the DNA in a mammalian cell there are about 5×10^9 base pairs. If an average protein chain in an enzyme is coded for by roughly 500 base pairs in a DNA chain, there are about 1×10^7 genes, and hence enzymes, which can be used to fix the characteristics of that mammal.)

RNA, RIBONUCLEIC ACID

The method by which an organism uses the information contained in the DNA in cell nuclei to synthesize proteins involves the other type of nucleic acid, RNA. RNA serves to carry the genetic information from the cell nucleus to the site of protein synthesis, called the ribosomal region, where it serves as a template in the actual synthesis reactions.

The transfer of information from DNA to RNA molecules is thought to occur by a mechanism similar to that for the duplication of DNA. A section of DNA chain containing a gene coding a particular protein unwinds and, in the presence of the proper enzyme, an RNA chain forms, with base pairs matching in the following way for DNA → RNA: A → U, T → A, C → G, and G → C. The resulting RNA chain complements that for the section of DNA exactly, so that the base sequence in RNA also carries the genetic code, in complementary form, for that protein. Molecules of RNA made in

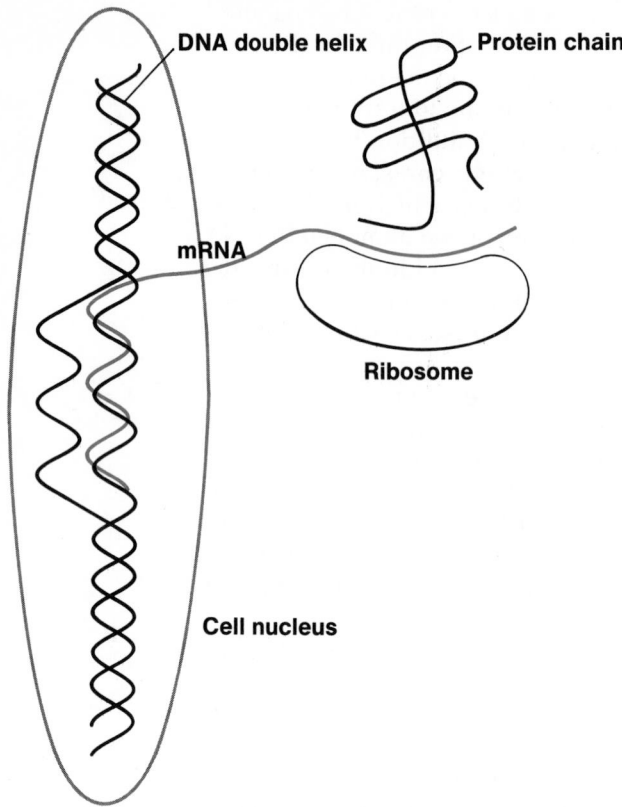

DNA double helix

Protein chain

mRNA

Ribosome

Cell nucleus

Figure 23.14 Role of mRNA in protein synthesis. The mRNA is made on the DNA template and then migrates to another region in the cell where it controls the synthesis of the protein chain for which it carries the code.

this way are called **messenger RNA, mRNA.** Since there is a different mRNA molecule for each enzyme to be synthesized, and each kind of mRNA contains about 500 bases, mRNA molecules are very large and their isolation as single species is not likely by present methods.

At the site of protein synthesis it is necessary that the information on an mRNA molecule be decoded and somehow used to add amino acids in the proper sequence to the protein chain. This is accomplished with the help of a second kind of RNA molecule, called **transfer RNA, tRNA,** which acts both as a decoding unit and as an amino acid carrier.

Transfer RNA molecules are the smallest nucleic acids. Some of these have been isolated in pure form and are known to contain about 75 mononucleotide units. The sequence of bases has been determined in a few cases and, if maximum base pairing within the chain is assumed, the tRNA molecules all take on clover leaf structures shown in Figure 23.15. Each molecule in the natural state is believed to assume a characteristic three-dimensional conformation which is necessary for its proper functioning.

Every amino acid has one or more tRNA molecules which serve to carry it, and only it, through the protein synthesis process. One of the arms of the tRNA molecule contains an **anticodon** group, a series of three bases which indicate, in genetic code, the kind of amino acid to be carried by that tRNA molecule. The tRNA in Figure 23.15 is an alanine carrier, and the anticodon unit, CGG', read from right to left, matches with GCC, which is the genetic code for alanine as it would occur on mRNA. The left and right arms of the molecule are specific for this kind of tRNA and allow it to

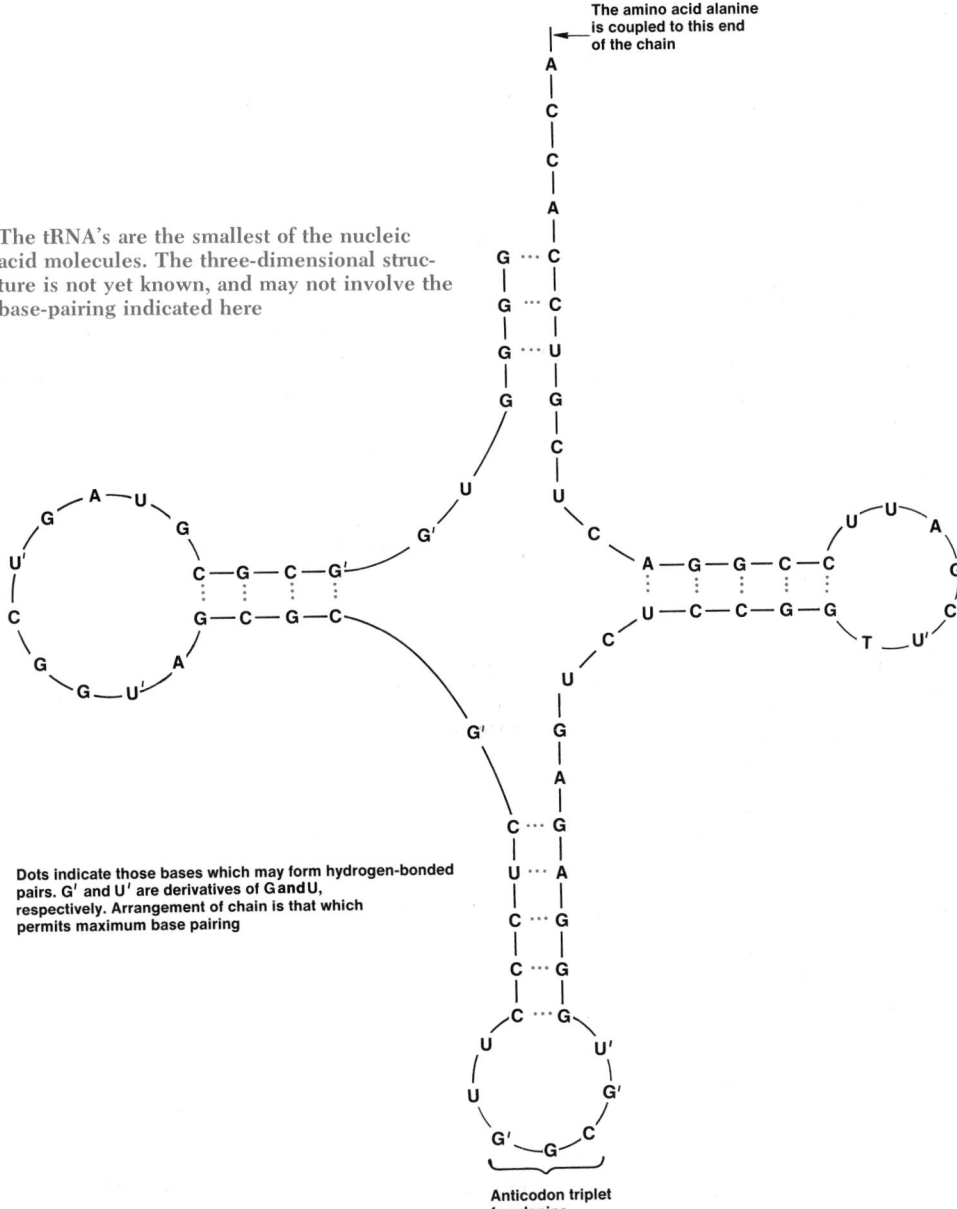

Figure 23.15 Base sequence in yeast alanine tRNA.

recognize a particular enzyme which will couple alanine, and only alanine, to the upper end of the molecule. The specificity of the enzyme in this coupling reaction is nearly perfect, with less than one error in 10,000 reactions.

The actual synthesis of the enzyme, or other protein, occurs on the ribosomes to which the mRNA molecule, carrying the genetic code for the whole enzyme, is attached. Transfer RNA molecules, coded and carrying the proper amino acids, couple to the mRNA through their anticodon sites, add their amino acids to the chain in the order in which they are coded on the mRNA, and depart. The overall synthesis process is indicated in a general way in Figure 23.16. Experiments indicate that the enzyme which

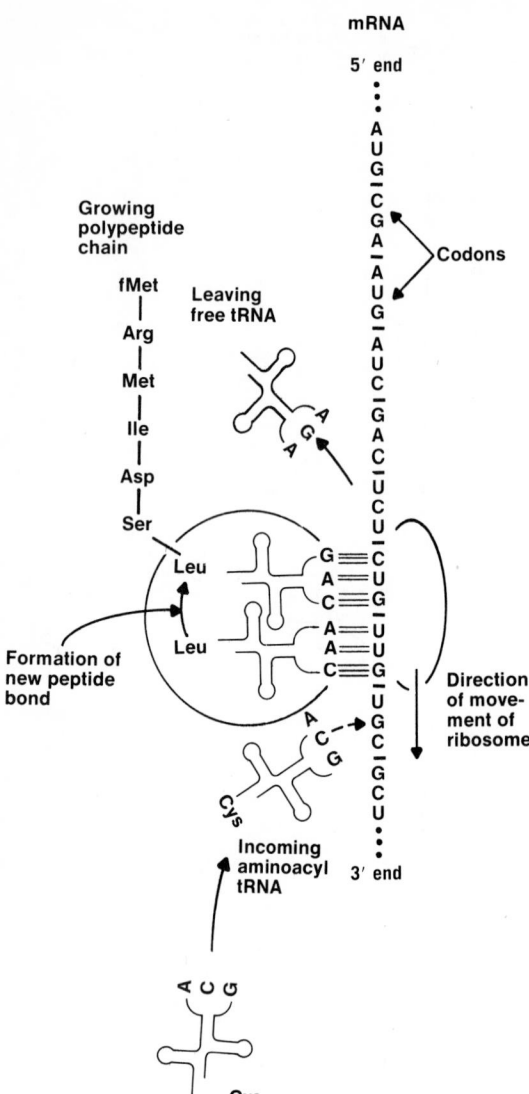

Figure 23.16 General mechanism for protein synthesis. Synthesis of the protein starts when the ribosome is at the 5′ end of the mRNA. Properly coded tRNA molecules couple to mRNA codons and add their amino acids to the protein chain. As synthesis proceeds, the ribosome moves along the mRNA, which picks up the proper tRNA molecules as they are needed and desorbs the tRNA's after they have delivered their amino acids. In the sketch the amino acid residues in the protein chain were added as directed by the top six codons in the mRNA.

Use the genetic code to check whether the amino acid sequence in the protein corresponds to the codon sequence in the mRNA

controls this truly amazing synthesis can add the proper amino acids to the chain at the rate of about one unit per second.

THE GENETIC CODE

Although the concept of a genetic code consisting of certain base sequences in DNA and RNA is supported by many experiments involving enzyme synthesis, it is by no means obvious how one would attack the problem of actually breaking the code and finding which sequence is to be associated with a given amino acid. By some rather simple experiments which are discussed in Problem 23.31 at the end of this chapter, it was shown, as mentioned earlier, that the code word for each amino acid consists of a triplet of bases, taken in a given direction along the nucleic acid

TABLE 23.7   The Genetic Code in mRNA

AMINO ACID	CODING TRIPLETS	AMINO ACID	CODING TRIPLETS
Gly	GGU GGC GGA GGG	Asp	GAU GAC, GAA
Ala	GCU GCC GCA GCG	Glu	GAG
Val	GUU GUC GUA GUG	Asn	AAU AAC
Leu	CUU CUC CUA CUG UUA UUG	Gln	CAA CAG
Ile	AUU AUC AUA	Lys	AAA AAG
Phe	UUU UUC	Arg	CGU CGC CGA
Tyr	UAU UAC		CGG AGA AGG
Trp	UGG	His	CAU CAC
Ser	UCU UCC UCA UCG AGU AGC	Met	AUG
		Cys	UGU UGC
Thr	ACU ACC ACA ACG	Pro	CCU CCC CCA CCG

chain. Further experiments involving synthetic mRNA's of known base sequences, which produced protein-like chains in which amino acid sequences could be determined, allowed Nirenberg and Khorana, in 1964, to determine the code words for all the amino acids. For their deciphering of the genetic code they received the Nobel Prize in 1968. The genetic code as it occurs in mRNA is given in Table 23.7. The code in DNA and tRNA would involve base triplets that are complementary to those in the table.

Figure 23.16 shows some of the complementary triplets

In the genetic code there are, in most cases, several triplets that will serve to indicate a given acid. These triplets frequently differ only in the last letter, so that for about half the acids the first two letters in the code word suffice to define it. Such acids may well have been among the first which received code words, and at that time the third letter may have served as a punctuation mark. Some amino acids with very similar R groups have similar code words; for example, aspartic and glutamic acids, the only two with acidic R groups, have the same code letters in the first two positions.

There are several features of the genetic code that relate directly to evolutionary changes. Errors in translation will, in general, produce mutations rather than nonsense, since a coding error will usually result in the code word for the wrong amino acid rather than a meaningless triplet. Such mutations, which cause changes in the amino acid sequence, may occasionally produce a better organism. These changes are minimized by the fact that third letter errors often do not change the amino acid called for, or, if a change does occur, the substituted acid is similar to the original; these properties tend to give the organism stability against excessive mutations. At present it is believed that the genetic code is essentially universal for all species, from viruses and bacteria to the mammals, including man. The basic elements of the code would appear to date from the earliest bacteria, which lived on earth about three billion years ago.

The development of the central dogma of molecular genetics has indeed been remarkably helpful in opening new avenues of biochemical research. The establishment of the genetic code must be considered to be one of the major achievements of science in this century. As with many important scientific developments, however, it has opened up many new areas for investigation. Molecular genetics of all but the simplest of organisms is still in a primitive state. Such matters as the possible relationships between errors in base sequence in DNA and human diseases, including cancer and the reasons for recessive and dominant characteristics of cell differentiation, offer a continuing challenge to the modern biochemist.

PROBLEMS

23.1 Given the weight percentages of each element in a pure compound isolated from rat liver, how could you decide whether it might be a carbohydrate, a protein, or a nucleic acid?

23.2 Given the structure of D-glyceraldehyde in Figure 23.1, draw in perspective the L-glyceraldehyde molecule.

23.3 Which of the following compounds could have optically active forms? For those which might be active, state the number of asymmetric carbon atoms present in the molecule.

a.

$$H-\underset{\underset{H}{|}}{\overset{\overset{H}{|}}{C}}-\underset{\underset{O}{\|}}{C}-\underset{\underset{H}{|}}{\overset{\overset{H}{|}}{C}}-OH$$

b.

$$H-\underset{\underset{H}{|}}{\overset{\overset{H}{|}}{C}}-\underset{\underset{H}{|}}{\overset{\overset{H}{|}}{C}}-\underset{\underset{H}{|}}{\overset{\overset{OH}{|}}{C}}-C=O$$

c.

d.

$$H-\underset{\underset{H}{|}}{\overset{\overset{H}{|}}{C}}-\underset{\underset{NH_2}{|}}{\overset{\overset{H}{|}}{C}}-\underset{\underset{OH}{|}}{C}=O$$

23.4 What is starch? cellulose? glycogen? How do they differ in structure?

23.16 What is meant by each of the following terms?

a. disaccharide
b. peptide linkage
c. RNA
d. α-helix
e. α-amino acid
f. conformation

23.17 Why are there 16 stereoisomers of a molecule with the following bonding?

$$O=\underset{}{\overset{\overset{H}{|}}{C}}-\underset{\underset{OH}{|}}{\overset{\overset{H}{|}}{C}}-\underset{\underset{OH}{|}}{\overset{\overset{H}{|}}{C}}-\underset{\underset{OH}{|}}{\overset{\overset{H}{|}}{C}}-\underset{\underset{OH}{|}}{\overset{\overset{H}{|}}{C}}-\underset{\underset{OH}{|}}{\overset{\overset{H}{|}}{C}}-H$$

23.18 Which of the following compounds could have optically active forms? How many optically active stereoisomers would each compound have?

a.

$$CH_3-\underset{\underset{OH}{|}}{\overset{\overset{H}{|}}{C}}-\underset{\underset{NH_2}{|}}{\overset{\overset{H}{|}}{C}}-C=O$$

b.

$$O=\underset{\underset{OH}{|}}{C}-\underset{\underset{OH}{|}}{\overset{\overset{H}{|}}{C}}-\underset{\underset{OH}{|}}{\overset{\overset{H}{|}}{C}}-C=O$$

c.

d.

$$HO-\underset{\underset{H}{|}}{\overset{\overset{H}{|}}{C}}-\underset{\underset{H}{|}}{\overset{\overset{OH}{|}}{C}}-\underset{\underset{H}{|}}{\overset{\overset{OH}{|}}{C}}-\underset{\underset{H}{|}}{\overset{\overset{OH}{|}}{C}}-C=O$$

23.19 Sketch in ring form the α and β forms of D-mannose. How many asymmetric carbon atoms would there be in the ring form of mannose? How could you distinguish experimentally between the straight chain and ring forms?

23.5 Predict the sign and magnitude of the charge, if any, on each of the following amino acid molecules in a solution whose pH equals 2 (see Problem 23.20):

a. isoleucine   c. lysine
b. glutamic acid   d. glycine

23.6 A tripeptide contains a glycine, a serine, and a leucine residue. Sketch a possible form of the tripeptide, including all atoms from the residues and the amino and carboxyl end groups. How many different tripeptides could be made using only these three amino acids?

23.7 Establish the amino acid sequence in a polypeptide on the basis of the following information:

Products from complete hydrolysis:
$$Ala_3Pro_2Ser_4Arg_1$$
Terminal residues:
NH$_2$ end: Ser   COOH end: Ala

Fragments obtained on partial hydrolysis:

Ala-Ser-Pro   Ala-Pro-Ser   Ser-Ser-Ala
Ser-Arg   Pro-Ala-Pro   Ser-Pro

23.8 How would you explain the fact that the β-keratin structure, present in silk, is less common in proteins than the α-helix structure?

23.9 In the cytochrome c molecule in man 69 of the amino acid residues are the same as the residues in equivalent positions in the cytochrome c chain from sunflower seeds. At about what time in the past did the evolutionary branches leading to man and to sunflowers diverge? Indicate, by number of the position on the cytochrome c chain, 10 sites which you think would have the same residue in man and sunflowers and 4 positions where the residues would be likely to differ.

23.10 Why is enzymatic catalysis much more sensitive to pH than is catalysis that does not involve enzymes?

23.11 How are RNA and DNA similar? How do they differ?

23.20 The isoelectric points of several amino acids are given below:

Gly 6.06   Lys 9.47   Ala 6.11   His 7.64
Pro 6.30   Asp 2.98   Glu 3.08   Ile 6.04

In an electrophoresis experiment at pH 6.0, which of the above acids would move toward the + pole? the − pole? would not move?

23.21 Predict the order in which the amino acids in Problem 23.20 would be eluted from a cationic exchange column washed with a buffer solution whose pH equals 4.0.

23.22 A short polypeptide chain on complete hydrolysis produced the following numbers of moles of the indicated amino acid per 100 grams of polypeptide:

Ala 0.086   Gly 0.262   Val 0.179
             Ile 0.085   Pro 0.337   Ser 0.588

What is the lowest molecular weight that is reasonable for this polypeptide? How many of each of the amino acid residues would that chain contain? (Hint: What is the minimum number of any of the residues in the chain?)

23.23 Explain why, at pH 7, the following synthetic polypeptides have the conformations that are indicated:

a. polyleucine: α-helix
b. polyglycine: silk structure
c. polyglutamic acid: random coil
d. polylysine: random coil

23.24 Hemoglobin contains four polypeptide chains, each of which is bound to a heme group (Figure 19.1, Chapter 19). The percentage by weight of iron in hemoglobin is 0.33%. Estimate the molecular weight of hemoglobin.

23.25 What simple experiment could you perform on a catalyzed reaction to establish whether the catalyst was or was not an enzyme?

23.26 Sketch a portion of an mRNA chain containing the base triplet GCU. Show all atoms and note that the order of the bases is from top to bottom in Figure 23.8. For what amino acid would this triplet be the code word (see Table 23.7)?

23.12 Given the data in Problem 23.27 and the fact that the DNA molecule in bacteriophage T_2 virus has a length of 56×10^{-4} cm, find the molecular weight of that DNA.

23.13 Cite four assumptions of the central dogma of molecular genetics.

23.14 If nucleotides can be added to the DNA chain during replication at the rate of 10^5 units per minute, how long would it take to duplicate the DNA in a mammalian cell if the DNA were in a single chain containing 5×10^9 base pairs? What does this imply about the number of DNA molecules in a cell if replication occurs at only one point in a given chain and a cell can be completely duplicated, including the DNA, in a matter of a few hours?

23.15 What would be a base sequence in mRNA which would code for the synthesis of the following polypeptide (see Table 23.7)?

Asp-Leu-Ile-Tyr-Ser-Gly-Gly-Glu

23.27 Given the following information, find the molecular weight of λDNA:

In the double helix there are 10 nucleotide pairs per complete turn.
The double helix goes through one complete turn in 34 Å.
The average molecular weight of a nucleotide pair is 700.
The length of the λDNA molecule in Figure 23.10 is 17.2×10^{-4} cm.

23.28 What is the function in the cell of the following nucleic acids?

a. DNA   b. mRNA   c. tRNA

23.29 Given the information in the text and in Problems 23.14 and 23.27, what is the total length of DNA in a mammalian cell? If the diameter of the double helix is about 20 Å and the helix has a conformation which minimizes the volume it occupies, what would be the minimum radius of a sphere containing the DNA in this cell? The radius of a typical mammalian cell is about 1×10^{-3} cm.

23.30 Give the base sequences in the anticodons in tRNA molecules which could serve as carriers for the amino acids listed below. Note that the base code on tRNA is complementary to that on mRNA and will be read in the opposite direction (see Table 23.7 and Fig. 23.16).

a. Leu   c. Phe
b. Trp   d. Met

*23.31 One of the most useful experiments for establishing the general nature of the genetic code was one in which one or more nucleotides were added to or deleted from the DNA chain comprising a gene in viral chromosomes. The following information was obtained:

1. One insertion in the chain caused errors in the code from the point of the insertion. There were no errors before the insertion.
2. One insertion followed by a deletion caused errors in the interval between the two changes but proper reading thereafter.
3. Two insertions caused errors in the code at all points after the first insertion.
4. Three insertions caused errors in the code in the interval between the first and third insertion but proper reading thereafter.

On the basis of the above experiments what can you say about the size of the coding unit, in the DNA molecule?

*23.32 Mutations in species are ordinarily the result of a single base error in the DNA chain. Such mutations usually cause the wrong amino acid to be incorporated into the protein coded for by that part of the DNA chain. Some mutations in human hemoglobin which have been reported are listed below:

	Normal	Abnormal
a.	Gly	Asp
b.	His	Tyr
c.	Lys	Glu
d.	Glu	Gly

Assuming each mutation resulted from a single base error, what base replacements in DNA were responsible for the above mutations? (Note that the genetic code is given for mRNA.)

CONSTANTS, CONVERSION FACTORS AND PROPERTIES OF WATER

CONSTANTS

Acceleration of gravity $= 980.6$ cm/sec^2
Avogadro's number $\quad= 6.023 \times 10^{23}$/mole
Electronic charge $\quad= 4.803 \times 10^{-10}$ esu
$\qquad\qquad\qquad\quad= 1.602 \times 10^{-19}$ coul
Faraday constant $\quad= 96{,}494$ coul/GEW
Gas constant $\qquad\quad= 0.08206$ lit atm/mole°K
$\qquad\qquad\qquad\quad= 1.987$ cal/mole°K
$\qquad\qquad\qquad\quad= 8.315 \times 10^7$ erg/mole°K
Planck's constant $\quad= 6.625 \times 10^{-27}$ erg sec
Velocity of light $\qquad= 2.998 \times 10^{10}$ cm/sec
$\pi \qquad\qquad\qquad\quad= 3.142$
e $\qquad\qquad\qquad\quad= 2.718$
ln x $\qquad\qquad\qquad= 2.303 \log x$
2.303R $\qquad\qquad\quad= 4.576$ cal/mole°K
2.303RT (at 25°C) $\quad= 1364$ cal/mole

CONVERSION FACTORS

Length, volume, and mass: Table 1.1, p. 8.
Mass-energy: Table 22.3, p. 610.
Pressure: 1 atm $= 760$ torr $= 760$ mm Hg $= 14.70$ lb/in^2
$\qquad\qquad\quad= 1.013$ bar $= 1.013 \times 10^6$ dyne/cm$^2 = 1.033$ kg/cm^2
Energy: $\quad$ 1 cal $= 4.184$ joule $= 4.184 \times 10^7$ erg
$\qquad\qquad$ 1 erg/molecule $= 1.439 \times 10^{13}$ kcal/mole
$\qquad\qquad$ 1 ev/atom $= 23.06$ kcal/mole
$\qquad\qquad$ 1 joule/mole $= 2.390 \times 10^{-4}$ kcal/mole

PROPERTIES OF WATER

Density: 0°C $\quad$ 0.99987 g/ml $\qquad\qquad$ 25°C $\quad$ 0.99707 g/ml
$\qquad\quad$ 4°C $\quad$ 1.00000 g/ml $\qquad\qquad$ 100°C $\quad$ 0.95838 g/ml

Heat of fusion: 1.44 kcal/mole at 0°C

Heat of vaporization: $\quad$ 0°C $\quad$ 10.74 kcal/mole
$\qquad\qquad\qquad\qquad$ 25°C $\quad$ 10.52 kcal/mole
$\qquad\qquad\qquad\qquad$ 100°C $\quad$ 9.72 kcal/mole

Vapor pressure (mm Hg)

T(°C)	vp	T(°C)	vp	T(°C)	vp	T(°C)	vp
0	4.58	21	18.65	35	42.2	92	567.0
5	6.54	22	19.83	40	55.3	94	610.9
10	9.21	23	21.07	45	71.9	96	657.6
12	10.52	24	22.38	50	92.5	98	707.3
14	11.99	25	23.76	55	118.0	100	760.0
16	13.63	26	25.21	60	149.4	102	815.9
17	14.53	27	26.74	65	187.5	104	875.1
18	15.48	28	28.35	70	233.7	106	937.9
19	16.48	29	30.04	80	355.1	108	1004.4
20	17.54	30	31.82	90	525.8	110	1074.6

APPENDIX 2

ATOMIC AND IONIC RADII *

Element	Atomic Number	Atomic Radius in Å	Ionic Radius in Å	Element	Atomic Number	Atomic Radius in Å	Ionic Radius in Å
H	1	0.37	(−1) 2.08	Ag	47	1.44	(+1) 1.26
He	2	0.5		Cd	48	1.49	(+2) 0.97
Li	3	1.52	(+1) 0.60	In	49	1.62	(+3) 0.81
Be	4	1.11	(+2) 0.31	Sn	50	1.40	
B	5	0.88		Sb	51	1.41	
C	6	0.77		Te	52	1.37	(−2) 2.21
N	7	0.70		I	53	1.33	(−1) 2.16
O	8	0.66	(−2) 1.40	Xe	54	1.30	
F	9	0.64	(−1) 1.36	Cs	55	2.62	(+1) 1.69
Ne	10	0.70		Ba	56	2.17	(+2) 1.35
Na	11	1.86	(+1) 0.95	La	57	1.87	(+3) 1.15
Mg	12	1.60	(+2) 0.65	Ce	58	1.82	(+3) 1.01
Al	13	1.43	(+3) 0.50	Pr	59	1.82	(+3) 1.00
Si	14	1.17		Nd	60	1.82	(+3) 0.99
P	15	1.10		Pm	61		
S	16	1.04	(−2) 1.84	Sm	62		
Cl	17	0.99	(−1) 1.81	Eu	63	2.04	(+2) 0.97
Ar	18	0.94		Gd	64	1.79	(+3) 0.96
K	19	2.31	(+1) 1.33	Tb	65	1.77	(+3) 0.95
Ca	20	1.97	(+2) 0.99	Dy	66	1.77	(+3) 0.94
Sc	21	1.60	(+3) 0.81	Ho	67	1.76	(+3) 0.93
Ti	22	1.46		Er	68	1.75	(+3) 0.92
V	23	1.31		Tm	69	1.74	(+3) 0.91
Cr	24	1.25	(+3) 0.64	Yb	70	1.93	(+3) 0.89
Mn	25	1.29	(+2) 0.80	Lu	71	1.74	(+3) 0.89
Fe	26	1.26	(+2) 0.75	Hf	72	1.57	
Co	27	1.25	(+2) 0.72	Ta	73	1.43	
Ni	28	1.24	(+2) 0.69	W	74	1.37	
Cu	29	1.28	(+1) 0.96	Re	75	1.37	
Zn	30	1.33	(+2) 0.74	Os	76	1.34	
Ga	31	1.22	(+3) 0.62	Ir	77	1.35	
Ge	32	1.22		Pt	78	1.38	
As	33	1.21		Au	79	1.44	(+1) 1.37
Se	34	1.17	(−2) 1.98	Hg	80	1.55	(+2) 1.10
Br	35	1.14	(−1) 1.95	Tl	81	1.71	(+3) 0.95
Kr	36	1.09		Pb	82	1.75	
Rb	37	2.44	(+1) 1.48	Bi	83	1.46	
Sr	38	2.15	(+2) 1.13	Po	84	1.65	
Y	39	1.80	(+3) 0.93	At	85		
Zr	40	1.57		Rn	86	1.4	
Nb	41	1.43		Fr	87		
Mo	42	1.36		Ra	88	2.20	
Tc	43			Ac	89	2.0	
Ru	44	1.33		Th	90	1.80	
Rh	45	1.34		Pa	91		
Pd	46	1.38		U	92	1.4	

* Radii are taken from Pauling, Linus: "The Nature of the Chemical Bond." 3rd ed., Ithaca, New York, Cornell University Press, 1960. Those of the noble gas atoms are from Journal of Physical Chemistry, 69: 596, 1965.

NOMENCLATURE OF INORGANIC COMPOUNDS

The composition of a compound may be specified by giving either its formula or its name. Throughout this text, we have discussed at some length how one can arrive at the chemical formulas of inorganic compounds. Here, we consider a related problem, that of developing a system of nomenclature for these compounds. In the interest of clarity and simplicity, we shall restrict our discussion to a relatively small number of rules which will suffice to name the great majority of inorganic compounds encountered in an introductory course in chemistry.

IONIC COMPOUNDS

The names of ionic compounds are derived from those of the ions of which they are composed. We shall first consider the nomenclature of individual ions and then the names of the compounds they form.

POSITIVE IONS

Monatomic positive ions take the names of the metal from which they are derived:

Na^+ sodium Ca^{2+} calcium Al^{3+} aluminum

When a metal forms more than one ion, it is necessary to distinguish between these ions. The accepted practice today is to indicate the oxidation number of the ion by a Roman numeral in parentheses immediately following the name of the metal:

Fe^{2+} iron(II) Cu^+ copper(I) Sn^{2+} tin(II)
Fe^{3+} iron(III) Cu^{2+} copper(II) Sn^{4+} tin(IV)

An earlier method, still widely used, adds to the stem of the Latin name of the metal the suffixes -*ous* or -*ic*, representing the lower and higher oxidation states respectively:

Fe^{2+} ferrous Cu^+ cuprous Sn^{2+} stannous
Fe^{3+} ferric Cu^{2+} cupric Sn^{4+} stannic

The only polyatomic cations to be considered here are:

NH_4^+ ammonium Hg_2^{2+} mercury(I) or mercurous

661

NEGATIVE IONS

Monatomic negative ions are named by adding the suffix *-ide* to the stem of the name of the nonmetal from which they are derived:

N^{3-}	nitride	O^{2-}	oxide	F^-	fluoride	H^-	hydride
		S^{2-}	sulfide	Cl^-	chloride		
		Se^{2-}	selenide	Br^-	bromide		
		Te^{2-}	telluride	I^-	iodide		

The nomenclature of polyatomic anions is more complex. The names of some of the more common oxyanions are:

OH^-	hydroxide	NO_2^-	nitrite	ClO_2^-	chlorite
O_2^{2-}	peroxide	SO_4^{2-}	sulfate	ClO^-	hypochlorite
CO_3^{2-}	carbonate	SO_3^{2-}	sulfite	MnO_4^-	permanganate
PO_4^{3-}	phosphate	ClO_4^-	perchlorate	CrO_4^{2-}	chromate
NO_3^-	nitrate	ClO_3^-	chlorate	$Cr_2O_7^{2-}$	dichromate

It will be noted that when a nonmetal such as nitrogen or sulfur forms two oxyanions in different oxidation states, the suffixes *-ate* and *-ite* are used to distinguish between the higher and lower states, respectively. With elements such as chlorine which form more than two oxyanions, the prefixes *per-* (highest oxidation state) and *hypo-* (lowest oxidation state) are used as well.

Oxyanions that contain hydrogen as well as nonmetal and oxygen atoms are properly named as illustrated in the following examples:

HCO_3^-	hydrogen carbonate	HPO_4^{2-}	hydrogen phosphate
HSO_4^-	hydrogen sulfate	$H_2PO_4^-$	dihydrogen phosphate

COMPOUNDS

The name of the positive ion is given first, followed by the name of the negative ion. Examples are

$CaCl_2$	calcium chloride
$FeBr_2$	iron(II) bromide
$(NH_4)_2SO_4$	ammonium sulfate
$Fe(ClO_4)_3$	iron(III) perchlorate
$NaHCO_3$	sodium hydrogen carbonate

In practice, compounds containing metal atoms, regardless of the type of bonding involved, are ordinarily named as if they were ionic. For example, the compounds $AlCl_3$ and $SnCl_4$, in both of which the bonding is primarily covalent, are named as follows:

$AlCl_3$	aluminum chloride	$SnCl_4$	tin(IV) chloride

BINARY COMPOUNDS OF THE NONMETALS

When a pair of nonmetals form only one compound, that compound may be named quite simply. The name of the element whose symbol appears first in the formula is written first. The second portion of the name is formed by adding the suffix -ide to the stem of the name of the second nonmetal. Examples include

$$
\begin{array}{ll}
HCl & \text{hydrogen chloride} \\
H_2S & \text{hydrogen sulfide} \\
NF_3 & \text{nitrogen fluoride}
\end{array}
$$

If more than one binary compound is formed by a pair of nonmetals, as is most often the case, the Greek prefixes, di = two, tri = three, $tetra$ = four, $penta$ = five, $hexa$ = six, and so on, are used to designate the number of atoms of each element. Thus, for the oxides of nitrogen we have

$$
\begin{array}{ll}
{}^*N_2O_5 & \text{dinitrogen pentoxide} \\
{}^*N_2O_4 & \text{dinitrogen tetroxide} \\
NO_2 & \text{nitrogen dioxide} \\
N_2O_3 & \text{dinitrogen trioxide} \\
NO & \text{nitrogen oxide} \\
N_2O & \text{dinitrogen oxide}
\end{array}
$$

A great many of the best-known binary compounds of the nonmetals have acquired common names which are widely and, in some cases, exclusively used. These include

$$
\begin{array}{llll}
H_2O & \text{water} & PH_3 & \text{phosphine} \\
H_2O_2 & \text{hydrogen peroxide} & AsH_3 & \text{arsine} \\
NH_3 & \text{ammonia} & NO & \text{nitric oxide} \\
N_2H_4 & \text{hydrazine} & N_2O & \text{nitrous oxide}
\end{array}
$$

OXYACIDS

The names of some of the more common oxygen acids are listed as follows:

$$
\begin{array}{llll}
H_2CO_3 & \text{carbonic acid} & H_2SO_3 & \text{sulfurous acid} \\
H_3BO_3 & \text{boric acid} & HClO_4 & \text{perchloric acid} \\
HNO_3 & \text{nitric acid} & HClO_3 & \text{chloric acid} \\
HNO_2 & \text{nitrous acid} & HClO_2 & \text{chlorous acid} \\
H_2SO_4 & \text{sulfuric acid} & HClO & \text{hypochlorous acid}
\end{array}
$$

It is of interest to compare the names of these oxyacids to those of the corresponding oxyanions listed previously. Note that oxyanions whose names end in -ate are derived from acids whose names end in -ic. Compare, for example, CO_3^{2-} (carbonate) and H_2CO_3 (carbonic acid); NO_3^- (nitrate) and HNO_3 (nitric acid); ClO_4^- (perchlorate) and $HClO_4$ (perchloric acid). Oxyanions whose names end in -ite are derived from acids whose names end in -ous. Thus we have NO_3^- (nitrite) and HNO_2 (nitrous acid); ClO^- (hypochlorite) and $HClO$ (hypochlorous acid).

COORDINATION COMPOUNDS

The nomenclature of compounds containing complex ions in which a metal atom is held by coordinate covalent bonds to two or more ligands is

---

* Note that in this case the a is dropped from the prefixes penta and tetra in the interests of euphony.

perhaps more involved than that of any other type of inorganic compound. Several rules are required, the more pertinent of which are as follows:

1. As in simple ionic compounds, the cation is named first, followed by the anion.

2. If there is more than one ligand of a particular type attached to the central atom, Greek prefixes are used to indicate the number of these ligands. Where the name of the ligand itself is complex (e.g., ethylenediamine), the number of such ligands is indicated by the prefixes *bis-* or *tris-* instead of *di-* or *tri-* and the name of the ligand is enclosed in parentheses.

3. In naming a complex ion, the names of anionic ligands are written first, followed by those of neutral ligands, and finally that of the central metal atom. This is exactly the reverse of the order in which the groups are listed in the formula of the complex ion: the symbol of the central atom is written first, followed by the formulas of neutral ligands and then those of negatively charged ligands. In writing the formula of a coordination compound, the formula of the complex ion is often set off by brackets.

4. The names of anionic ligands are modified by substituting the suffix *-o* for the usual ending. Thus we have

$$Cl^- \quad \text{chloro} \qquad CO_3^{2-} \quad \text{carbonato}$$
$$OH^- \quad \text{hydroxo} \qquad CN^- \quad \text{cyano}$$

The names of neutral ligands are ordinarily not changed. Two important exceptions are:

$$H_2O \quad \text{aquo} \qquad NH_3 \quad \text{ammine}$$

5. The oxidation number of the central metal atom is indicated by a Roman numeral following the name of the metal. If the complex is an anion, the suffix *-ate* is added, often to the Latin stem of the name of the metal. Examples are

$[Co(NH_3)_6]Cl_3$ — hexamminecobalt(III) chloride
$[Co(en)_3](NO_3)_3$ — tris(ethylenediamine)cobalt(III) nitrate
$[Cr(NH_3)_4Cl_2]Cl$ — dichlorotetramminechromium(III) chloride
$[Pt(H_2O)_3Cl]Br$ — chlorotriaquoplatinum(II) bromide
$K_3[Fe(CN)_6]$ — potassium hexacyanoferrate(III)
$K_4[Fe(CN)_6]$ — potassium hexacyanoferrate(II)

The last two compounds are often referred to as potassium ferricyanide and potassium ferrocyanide, respectively.

appendix 4

EXPONENTS AND LOGARITHMS

(This material is taken from "Mathematical Preparation for General Chemistry," W. L. Masterton and E. J. Slowinski, W. B. Saunders Co., 1970.)

EXPONENTIAL NOTATION

In chemistry we frequently deal with very large or very small numbers. In one gram of the element carbon there are

$$50,150,000,000,000,000,000,000$$

atoms of carbon. At the opposite extreme, the mass of a single carbon atom is

$$0.00000000000000000000001994 \text{ g}$$

Numbers such as these are not only difficult to write; they are very awkward to work with. Imagine how tedious it would be to find the mass of 2150 carbon atoms by carrying out the operation

$$\frac{0.00000000000000000000001994 \text{ g}}{\times 2150}$$

To simplify operations of this type, we use what is known as **exponential notation**. To express a number in exponential notation, we write it in the form

$$C \times 10^n$$

where C is a number between 1 and 10 (e.g., 1, 2.62, 5.8) and n is a positive or negative integer (e.g., 1, -1, -3). To find n, we count the number of places that the decimal point must be moved to give the coefficient, C. If the decimal point must be moved to the *left*, n is a *positive* integer; if it must be moved to the *right*, n is a *negative* integer. Thus we have

$26.23 = 2.623 \times 10^1$ (decimal point moved 1 place to left)
$5609 = 5.609 \times 10^3$ (decimal point moved 3 places to left)
$0.0918 = 9.18 \times 10^{-2}$ (decimal point moved 2 places to right)

Numbers written in exponential notation can be given a very simple interpretation. Recognizing that $10^1 = 10$, $10^3 = 1000$, and $10^{-2} = 1/100 = 0.01$, we could express the three exponentials written above as

$$2.623 \times 10^1 = 2.623 \times 10$$
$$5.609 \times 10^3 = 5.609 \times 1000$$
$$9.18 \times 10^{-2} = 9.18 \times 0.01$$

MULTIPLICATION AND DIVISION. A major advantage of exponential notation is that it simplifies the processes of multiplication and division. To *multiply*, we *add exponents:*

$$10^1 \times 10^2 = 10^{1+2} = 10^3; \quad 10^6 \times 10^{-4} = 10^{6+(-4)} = 10^2$$

To *divide*, we *subtract* exponents:

$$10^3/10^2 = 10^{3-2} = 10^1; \quad 10^{-3}/10^6 = 10^{-3-6} = 10^{-9}$$

To multiply one exponential number by another, we first multiply the coefficients in the usual manner and then add exponents. To divide one exponential number by another, we find the quotient of the coefficients and then subtract exponents. Examples:

$$(5.00 \times 10^4) \times (1.60 \times 10^2) = (5.00 \times 1.60) \times (10^4 \times 10^2) = 8.00 \times 10^6$$

$$(6.01 \times 10^{-3})/(5.23 \times 10^6) = \frac{6.01}{5.23} \times \frac{10^{-3}}{10^6} = 1.15 \times 10^{-9}$$

It often happens that multiplication or division yields an answer which is not in standard exponential notation. Thus we might have

$$(5.0 \times 10^4) \times (6.0 \times 10^3) = (5.0 \times 6.0) \times 10^4 \times 10^3 = 30 \times 10^7$$

The product is not in standard exponential notation since the coefficient, 30, does not lie between 1 and 10. To remedy this situation, we rewrite the coefficient as 3.0×10^1 and then add exponents

$$30 \times 10^7 = (3.0 \times 10^1) \times 10^7 = 3.0 \times 10^8$$

In another case

$$0.526 \times 10^3 = (5.26 \times 10^{-1}) \times 10^3 = 5.26 \times 10^2$$

RAISING TO POWERS AND EXTRACTING ROOTS. To raise an exponential number to a power or extract a root, we make use of the rules

$$(10^n)^a = 10^{na}; \quad (10^n)^{1/a} = 10^{n/a}$$

That is:
$$(10^{-2})^3 = 10^{-6}; \quad (10^{-2})^{1/2} = 10^{-2/2} = 10^{-1}$$

Applying these rules to numbers expressed in exponential notation, we have

$$(2.0 \times 10^{-2})^3 = (2.0)^3 \times (10^{-2})^3 = 8.0 \times 10^{-6}$$

$$(4.0 \times 10^{-2})^{1/2} = (4.0)^{1/2} \times (10^{-2})^{1/2} = 2.0 \times 10^{-1}$$

Here, as in multiplication and division, we operate on the coefficient and exponential terms separately.

Extracting the square root of an exponential number poses a special problem when the exponent is not an even number. Consider, for example,

$$(4.0 \times 10^5)^{1/2}$$

Following the procedure described above, we would obtain

$$(4.0)^{1/2} \times (10^5)^{1/2} = 2.0 \times 10^{5/2}$$

The answer is not in standard exponential notation; indeed $10^{5/2}$ is an awkward expression to work with because it does not readily translate into an ordinary number.

In cases of this type, we rewrite the exponential number to make the exponent an even number. To do this, we may divide the exponential by 10 and multiply the coefficient by 10.

$$(4.0 \times 10^5)^{1/2} = (40 \times 10^4)^{1/2}$$

Now, proceeding in the usual manner, we obtain

$$(40 \times 10^4)^{1/2} = (40)^{1/2} \times (10^4)^{1/2} = 6.3 \times 10^2$$

A similar principle is used in extracting cube roots.

$$(2.0 \times 10^{-5})^{1/3} = (20 \times 10^{-6})^{1/3} = 20^{1/3} \times 10^{-2} = 2.7 \times 10^{-2}$$

(Roots of numbers such as 40 and 20 may be found from tables, from a slide rule, or by using a table of logarithms as described later.)

ADDITION AND SUBTRACTION. Occasionally we may find it necessary to add or subtract two exponential numbers. These processes are extremely simple if both exponents are the same. For example:

$$2.02 \times 10^7 + 3.16 \times 10^7 = (2.02 + 3.16) \times 10^7 = 5.18 \times 10^7$$

$$4.23 \times 10^{-5} - 1.61 \times 10^{-5} = (4.23 - 1.61) \times 10^{-5} = 2.62 \times 10^{-5}$$

If the exponents differ, one of the numbers must be rewritten to make them the same. To add 5.04×10^8 to 3.0×10^7, we might express the latter as a number times 10^8

$$5.04 \times 10^8 + 3.0 \times 10^7 = 5.04 \times 10^8 + 0.30 \times 10^8 = 5.34 \times 10^8$$

LOGARITHMS

We have seen that the processes of multiplication, division, raising to a power, and extracting a root are simplified by expressing the quantities involved in exponential notation. However, even when this is done, a considerable amount of arithmetic may still be required. Suppose, for example, we wish to multiply 6.02×10^{23} by 1.99×10^{-24}. Combining the exponential terms, we obtain

$$(6.02 \times 10^{23}) \times (1.99 \times 10^{-24}) = (6.02 \times 1.99) \times 10^{-1}$$

but we still have to carry out the tedious operation of multiplying 6.02 by 1.99.

We could achieve a further simplification by expressing numbers such as 6.02 and 1.99 as powers of 10; the operation of multiplication could then be replaced by the more rapid process of addition. It is possible to accomplish this by making use of a table of common logarithms, such as that given on pp. 672–3. A common logarithm is simply a power of 10; specifically, *it is the power to which 10 must be raised to give a particular number.* Thus we have

$$\log 100 = \log 10^2 = 2$$

$$\log 0.001 = \log 10^{-3} = -3$$

Again, when we find that the logarithms of 6.02 and 1.99 are, to four significant figures, 0.7796 and 0.2989, respectively, we deduce that

$$6.02 = 10^{0.7796}; \quad 1.99 = 10^{0.2989}$$

FINDING THE LOGARITHM OF A NUMBER. The four-place table of logarithms given on pp. 672–3 allows us to determine directly the logarithm of any three-digit number between 1 and 10. To find the logarithm of 6.02, we follow down the column at the far left of the table until we come to 6.0 and then move across to the column headed "2," reading off the logarithm of 6.02 as 0.7796. Similarly, we find the logarithm of 6.03 to be 0.7803, that of 6.04 to be 0.7810, and so on.

We can readily use the table to estimate accurately the logarithms of four-digit numbers. Suppose, for example, we wish to find the logarithm of 6.023. Since this number is 3/10 of the way between 6.02 and 6.03, its logarithm should be about 3/10 of the way between that of 6.02 (0.7796), and that of 6.03 (0.7803). Expressing this reasoning in the form of an equation

$$\begin{aligned}
\log 6.023 &= \log 6.02 + 0.3 \, (\log 6.03 - \log 6.02) \\
&= 0.7796 + 0.3 \, (0.0007) \\
&= 0.7796 + 0.0002 = 0.7798
\end{aligned}$$

The logarithm of a number less than 1 or greater than 10 can be found by writing it in exponential notation and applying the general rule

$$\log (C \times 10^n) = \log C + n$$

For example, to find the logarithms of 60.2 and 0.0602, we first write these numbers as 6.02×10^1 and 6.02×10^{-2}. Then:

$$\log (6.02 \times 10^1) = \log 6.02 + 1 = 0.7796 + 1 = 1.7796$$

$$\log (6.02 \times 10^{-2}) = \log 6.02 - 2 = 0.7796 - 2 = -1.2204$$

FINDING THE NUMBER CORRESPONDING TO A GIVEN LOGARITHM. The operation of finding an antilogarithm (number corresponding to a given logarithm) is simply the inverse of finding the logarithm of a number. To start with a simple example, let us find the number whose logarithm is 0.4997. We locate 0.4997 in the body of the table, noting that it falls in the horizontal column labeled "3.1" under the vertical column labeled "6." We deduce that

$$0.4997 = \log 3.16$$

By the same procedure, we find the antilogarithm of 0.5011 to be 3.17.

Frequently the logarithm we are working with does not appear directly in the table. Suppose, for example, we wish to obtain the antilogarithm of 0.5000. Even though we cannot locate 0.5000 directly, we can bracket it between 0.4997, the logarithm of 3.16, and 0.5011, the logarithm of 3.17. Since 0.5000 is 3/14 or 0.2 of the way between 0.4997 and 0.5011, its antilogarithm must be about 0.2 of the way between 3.16 and 3.17. That is

$$0.5000 = \log 3.162$$

The process just described can be applied where the logarithm falls between 0 and 1, corresponding to a number between 1 and 10. If we need to find the number associated with a logarithm greater than 1 (e.g., 6.4997) or less than 0 (e.g., −0.6021), we first *rewrite the logarithm in the form of a decimal fraction (mantissa) plus or minus a whole number (characteristic)*. For example, to find the number whose logarithm is 6.4997, we rewrite this quantity as 0.4997 + 6. Now, we look up the mantissa, 0.4997, in the table, finding its antilogarithm to be 3.16. The antilogarithm of the characteristic, 6, is 10^6. Consequently, the number we are looking for must be 3.16×10^6. Summarizing

$$6.4997 = 0.4997 + 6 = \log (3.16 \times 10^6)$$

With a negative logarithm, we proceed in the same manner. To find the number whose logarithm is −0.6021, we first rewrite this quantity as a decimal fraction, between 0 and 1, minus a whole number. A moment's reflection should convince you that the result is 0.3979 − 1. That is

$$-0.6021 = 0.3979 - 1$$

Having cleared this hurdle, we find from the table that the antilogarithm of 0.3979 is 2.50; that of −1 must be 10^{-1}. The number we are looking for is 2.50×10^{-1}. In another case, to find the number whose logarithm is −3.4128, we proceed as follows

$$-3.4128 = 0.5872 - 4 = \log (3.865 \times 10^{-4})$$

OPERATIONS INVOLVING LOGARITHMS. Since logarithms are exponents, the rules governing the use of exponents apply here as well. Specifically:

Multiplication:
$$\log (xy) = \log x + \log y; \qquad \log (6.02 \times 1.99) = \log 6.02 + \log 1.99$$
Division:
$$\log (x/y) = \log x - \log y; \qquad \log (9.17/2.62) = \log 9.17 - \log 2.62$$
Raising to a power:
$$\log (x^n) = n \log x \qquad \log (2.00)^3 = 3 \log 2.00$$

Extracting a root: $\log (x^{1/n}) = \dfrac{1}{n} \log x$ $\log (3.00^{1/2}) = \dfrac{1}{2} \log 3.00$

These relationships enable us to simplify the calculations involved in these four operations. Thus, to multiply 6.02×1.99, we could look up the logarithms of the two numbers, add them together, and look up the antilogarithm of the sum.

$$\log 6.02 + \log 1.99 = 0.7796 + 0.2989 = 1.0785 = 0.0785 + 1$$
$$= \log (1.198 \times 10^1)$$
$$6.02 \times 1.99 = 1.20 \times 10^1$$
$$\log 9.17 - \log 2.62 = 0.9624 - 0.4183 = 0.544 = \log 3.50$$
$$9.17/2.62 = 3.50$$
$$3 \log 2.00 = 3(0.3010) = 0.9030 = \log 8.00; \quad (2.00)^3 = 8.00$$
$$\frac{1}{2} \log 3.00 = \frac{0.4771}{2} = 0.2386 = \log 1.732; \quad (3.00)^{1/2} = 1.732$$

NATURAL LOGARITHMS. For calculation purposes, common logarithms are most convenient since our number system is based upon multiples of 10. However, certain of the equations we use in general chemistry involve a different type of logarithm, taken to the base e, where, to four significant figures,

$$e = 2.718 \ldots$$

Logarithms to the base e are referred to as **natural** logarithms and are often written as "ln"; i.e.,

$$\log_e x \equiv \ln x; \quad \log_{10} x \equiv \log x$$

Tables of natural logarithms, while available, are seldom used. To find the natural logarithm of a number, we first look up the common logarithm and then make use of the relation

$$\ln x = 2.303 \log x$$

To find the natural logarithm of 4.160, we first find the base 10 logarithm from our table to be 0.6191 and then multiply by 2.303:

$$\ln 4.160 = 2.303 \times 0.6191 = 1.426$$

Occasionally we are required to evaluate expressions such as $e^{0.2500}$, where the base of natural logarithms, e, is raised to a power. To do this, we note that if we let $x = e^{0.2500}$:

$$\ln x = 0.2500 = 2.303 \log x$$

$$\log x = 0.2500/2.303 = 0.1086; \quad x = 1.284 = e^{0.2500}$$

PROBLEMS

1   Express the following numbers in standard exponential notation:

a. 2712
b. 0.124
c. 31.6×10^{-3}
d. 0.0045×10^2
e. 0.0000000000000000000001994

2   Carry out the following operations, expressing the answers in standard exponential notation.

a. $(2.61 \times 10^4) \times (3.14 \times 10^{-2})$
b. $(5.18 \times 10^4) \times (4.91 \times 10^{-9})$
c. $\dfrac{6.29 \times 10^{-2}}{3.35 \times 10^6}$
d. $(4.17 \times 10^{-3})/(8.76 \times 10^{-4})$
e. $6.24 \times 10^{-3} + 1.4 \times 10^{-4}$
f. $5.29 \times 10^3 - 1.61 \times 10^2$

3   Find the common logarithms of the following numbers:

a. 6.92
b. 6.92×10^3
c. 4.179×10^2
d. 0.00712
e. 3.022×10^{-5}

4   Find the numbers whose common logarithms are

a. 0.8426
b. 0.4703
c. 0.0053
d. 3.2172
e. 2.9000
f. −6.4178
g. −2.1929
h. −22.0100

5   Evaluate each of the following, giving your answers in standard exponential notation.

a. $(2.16 \times 10^4)^{1/2}$
b. $(6.92 \times 10^{-3})^{1/2}$
c. $(3.40 \times 10^2)^{0.34}$
d. $\ln 12.60$
e. $\ln (9.10 \times 10^{-2})$
f. $\log_{10} e$
g. $e^{0.6204}$
h. $e^{-1.2520}$

6   Making use of the definition: $pH = -\log_{10}$ (conc. H^+), calculate the pH of solutions in which conc. H^+ is:

a. 1.00×10^{-7}
b. 6.24×10^{-6}
c. 5.1×10^{-9}
d. 12

PREFIXES USED WITH SI UNITS

Decimal fractions and multiples of SI units are designated by using the prefixes listed in Table II. Those which are most commonly used in general chemistry are underlined.

TABLE II   SI Prefixes

FACTOR	PREFIX	SYMBOL	FACTOR	PREFIX	SYMBOL
10^{12}	tera	T	10^{-1}	deci	d
10^{9}	giga	G	10^{-2}	centi	c
10^{6}	mega	M	10^{-3}	milli	m
10^{3}	kilo	k	10^{-6}	micro	μ
10^{2}	hecto	h	10^{-9}	nano	n
10^{1}	deka	da	10^{-12}	pico	p
			10^{-15}	femto	f
			10^{-18}	atto	a

To illustrate the use of these prefixes, consider the relation between the familiar units of length in the metric system and the base unit, the meter:

$$1 \text{ kilometer} = 10^3 \text{ meter}; \quad 1 \text{ km} = 10^3 \text{ m}$$
$$1 \text{ centimeter} = 10^{-2} \text{ meter}; \quad 1 \text{ cm} = 10^{-2} \text{ m}$$
$$1 \text{ millimeter} = 10^{-3} \text{ meter}; \quad 1 \text{ mm} = 10^{-3} \text{ m}$$

For a few less familiar units:

$$1 \text{ hectometer} = 10^2 \text{ meter}; \quad 1 \text{ hm} = 10^2 \text{ m}$$
$$1 \text{ micrometer} = 10^{-6} \text{ meter}; \quad 1 \text{ } \mu\text{m} = 10^{-6} \text{ m}$$
$$1 \text{ nanometer} = 10^{-9} \text{ meter}; \quad 1 \text{ nm} = 10^{-9} \text{ m}$$

Notice that one of the base SI units, the kilogram, contains a prefix. Names of decimal multiples and fractions of the unit of mass are derived by attaching the appropriate prefix to the word "gram" and symbol "g". Thus:

$$1 \text{ milligram} = 10^{-3} \text{ gram} = 10^{-6} \text{ kilogram}$$
$$1 \text{ mg} = 10^{-3} \text{ g} \quad = 10^{-6} \text{ kg}$$

DERIVED UNITS

In the International System of Units, all physical quantities are represented by appropriate combinations of the base units listed in Table I. To choose a particularly simple example, the SI unit for volume, the cubic meter, represents the volume of a cube one meter on an edge. Again, in SI, the density of a substance can be expressed by dividing its mass in kilograms by its volume in cubic meters. A list of the derived units most frequently used in general chemistry is given in Table III.

TABLE III SI Derived Units

PHYSICAL QUANTITY	NAME OF UNIT	SYMBOL	DEFINITION
Area	square meter	m^2	
Volume	cubic meter	m^3	
Density	kilogram per cubic meter	kg/m^3	
Force	newton	N	$kg\ m/s^2$
Pressure	pascal	Pa	N/m^2
Energy	joule	J	$kg\ m^2/s^2$
Electric charge	coulomb	C	A s
Electric potential difference	volt	V	J/A s

Perhaps the least familiar of these units to the beginning chemistry student are the ones used to represent force, pressure, and energy.

The *newton* is defined as the force required to impart an acceleration of one meter per second squared to a mass of one kilogram (recall that Newton's second law can be stated as: force = mass × acceleration).

The *pascal* is defined as the pressure exerted by a force of one newton acting on an area of one square meter (recall that pressure = force/area).

The *joule* is defined as the work done when a force of one newton (kg m/s^2) acts through a distance of one meter (recall that work = force × distance).

UNITS OUTSIDE THE INTERNATIONAL SYSTEM

Certain units outside the SI are so widely used, particularly in the United States, that they are likely to be encountered for many years to come, in textbooks and in the scientific literature. Several such units are listed in Table IV, subdivided into three categories:

1. Those units which are used so widely or are so convenient that there is no prospect of abandoning them.

2. Units which are actually decimal fractions or multiples of SI units but have acquired special names. Their use is to be "progressively discouraged," presumably without undue coercion. They seem likely to be around for some time.

3. Units which are not decimally related to SI units, are somewhat awkward to use, and/or appear not to be too strongly entrenched. The use of these units is frowned upon and, sooner or later, they are slated for extinction. Included in this category are the units of the English system and certain common units of pressure and energy. Clearly, the distinction between categories (3) and (1) is an arbitrary one and, as time passes, some units may shift from one to the other.

TABLE IV Units Outside the International System

PHYSICAL QUANTITY	CATEGORY 1	CATEGORY 2	CATEGORY 3
Length		Ångstrom, micron	inch, foot, yard
Mass	amu		ounce, pound
Time	minute, hour, day		
Temperature	°C		°F
Volume	liter		quart, in^3
Force		dyne	kilogram-force
Pressure		bar	atmosphere, torr, mm Hg, lb/in^2
Energy		erg	calorie, BTU

CONVERSIONS BETWEEN SI AND OTHER UNITS

In Table V are listed appropriate conversion factors for translating units from other systems to the International System. The following examples illustrate the use of these conversion factors.

1. The radius of the Al atom is estimated to be 1.43 Å. Express this radius in meters, centimeters, and nanometers.

$$r = 1.43 \text{ Å} \times \frac{10^{-10} \text{ m}}{1 \text{ Å}} = 1.43 \times 10^{-10} \text{ m}$$

$$r = 1.43 \text{ Å} \times \frac{10^{-8} \text{ cm}}{1 \text{ Å}} = 1.43 \times 10^{-8} \text{ cm}$$

$$r = 1.43 \text{ Å} \times \frac{10^{-1} \text{ nm}}{1 \text{ Å}} = 1.43 \times 10^{-1} \text{ nm}$$

2. In the write-up of a certain experiment, the pressure is listed as 1.005 bar. Express this pressure in pascals; in atmospheres; in mm Hg.

$$p = 1.005 \text{ bar} \times \frac{10^5 \text{ Pa}}{1 \text{ bar}} = 1.005 \times 10^5 \text{ Pa}$$

$$p = 1.005 \text{ bar} \times \frac{10^5 \text{ Pa}}{1 \text{ bar}} \times \frac{1 \text{ atm}}{1.013 \times 10^5 \text{ Pa}} = 0.992 \text{ atm}$$

$$p = 1.005 \text{ bar} \times \frac{10^5 \text{ Pa}}{1 \text{ bar}} \times \frac{1 \text{ mm Hg}}{133.3 \text{ Pa}} = 753.9 \text{ mm Hg}$$

3. For the reaction at 25°C and 1 atm: $2NO(g) + O_2(g) \rightarrow 2NO_2(g)$, it is found that ΔH, ΔG and ΔS are respectively −113.2 kJ, −69.8 kJ, and −146 J/K. Calculate these quantities in kcal (ΔH and ΔG) and cal/K (ΔS).

$$\Delta H = -113.2 \text{ kJ} \times \frac{1 \text{ kcal}}{4.184 \text{ kJ}} = -27.1 \text{ kcal}$$

$$\Delta G = -69.8 \text{ kJ} \times \frac{1 \text{ kcal}}{4.184 \text{ kJ}} = -16.7 \text{ kcal}$$

$$\Delta S = -148 \frac{\text{J}}{\text{K}} \times \frac{1 \text{ cal}}{4.184 \text{ J}} = -34.9 \text{ cal/K}$$

TABLE V   Conversion Factors

QUANTITY	SI UNIT	OTHER UNIT	CONVERSION FACTOR
Length	meter	Ångstrom micron inch	$1\ \text{Å} = 10^{-10}\ \text{m} = 10^{-8}\ \text{cm} = 10^{-1}\ \text{nm}$ $1\ \text{micron} = 10^{-6}\ \text{m} = 1\ \mu\text{m}$ $1\ \text{inch} = 2.54 \times 10^{-3}\ \text{m}$
Mass	kilogram	amu pound	$1\ \text{amu} = 1.66041 \times 10^{-27}\ \text{kg}$ $1\ \text{lb} = 0.45359237\ \text{kg}$
Time	second	minute	$1\ \text{min} = 60\ \text{s}$
° Temperature	kelvin	Celsius Fahrenheit	$1\ °\text{C} = 1\ \text{K}$ $1\ °\text{F} = \frac{5}{9}\text{K}$
Volume	cubic meter	liter quart cubic inch	$1\ \ell = 10^{-3}\ \text{m}^3 = 1\ \text{dm}^3 = 10^3\ \text{cm}^3$ $1\ \text{qt} = 9.463 \times 10^{-4}\ \text{m}^3$ $1\ \text{in}^3 = 1.6387 \times 10^{-6}\ \text{m}^3$
Force	newton	dyne kilogram force	$1\ \text{dyn} = 10^{-5}\ \text{N}$ $1\ \text{kgf} = 9.80665\ \text{N}$
Pressure	pascal	bar atmosphere torr mm Hg lb/in²	$1\ \text{bar} = 10^5\ \text{Pa}$ $1\ \text{atm} = 1.01325 \times 10^5\ \text{Pa}$ $1\ \text{torr} = 133.322\ \text{Pa}$ $1\ \text{mm Hg} = 133.322\ \text{Pa}$ $1\ \text{lb/in}^2 = 6893\ \text{Pa}$
Energy	joule	erg calorie BTU	$1\ \text{erg} = 10^{-7}\ \text{J}$ $1\ \text{cal} = 4.184\ \text{J}$ $1\ \text{BTU} = 1055.056\ \text{J}$

° Temperatures in degrees Celsius (t_c) or degrees Fahrenheit (t_f) can be converted to temperatures in degrees Kelvin (T) by using the equations:

$$t_c = T - 273.15; \qquad t_f = \tfrac{9}{5}T - 459.69$$

REFERENCE TABLES

In Tables VI to IX are listed certain constants frequently used in general chemistry, enthalpies of formation, free energies of formation, and enthalpies of bond formation, all in SI units. Table X gives vapor pressures of water in bars as a function of temperature.

TABLE VI   Constants in SI Units

Acceleration of gravity	$= 9.80665\ \text{m/s}^2$
Avogadro's number	$= 6.023 \times 10^{23}/\text{mol}$
Electronic charge	$= 1.602 \times 10^{-19}\ \text{C}$
Faraday constant	$= 96,487\ \text{C/GEW}$
Gas constant	$= 8.314\ \text{J/mol K}$
	$= 8.314\ \text{m}^3\ \text{Pa/mol K}$
	$= 0.08314\ \ell\ \text{bar/mol K}$
Planck's constant	$= 6.626 \times 10^{-34}\ \text{J s}$
Velocity of light	$= 2.998 \times 10^8\ \text{m/s}$
2.303 R	$= 19.15\ \text{J/mol K}$
2.303 RT (at 298 K)	$= 5710\ \text{J/mol}$

TABLE VII Enthalpies of Formation (kJ/mol) at 25°C and 1 atm

AgBr	−99.5	$C_2H_2(g)$	+226.7	$H_2O_2(l)$	−187.6	$NH_3(g)$	−46.2
AgCl(s)	−127.0	$C_2H_4(g)$	+52.3	$H_2S(g)$	−20.1	$NH_4Cl(s)$	−315.4
AgI(s)	−62.4	$C_2H_6(g)$	−84.7	$H_2SO_4(l)$	−811.3	$NH_4NO_3(s)$	−365.1
$Ag_2O(s)$	−30.6	$C_3H_8(g)$	−103.8	HgO(s)	−90.7	NO(g)	+90.4
$Ag_2S(s)$	−31.8	$C_4H_{10}(g)$	−124.7	HgS(s)	−58.2	$NO_2(g)$	+33.8
$Al_2O_3(s)$	−1670.0	$C_5H_{12}(l)$	−173.2	KBr(s)	−392.2	NiO(s)	−244.3
$BaCl_2(s)$	−860.0	$C_2H_5OH(l)$	−277.6	KCl(s)	−435.9	$PbBr_2(s)$	−277.4
$BaCO_3(s)$	−1219.0	CoO(s)	−239.3	$KClO_3(s)$	−391.2	$PbCl_2(s)$	−359.2
BaO(s)	−558.1	$Cr_2O_3(s)$	−1128.0	KF(s)	−562.6	PbO(s)	−217.9
$Ba(OH)_2(s)$	−946.4	CuO(s)	−155.2	KOH(s)	−425.8	$PbO_2(s)$	−276.6
$BaSO_4(s)$	−1465.0	$Cu_2O(s)$	−166.7	$MgCl_2(s)$	−641.8	$Pb_3O_4(s)$	−734.7
$CaCl_2(s)$	−795.0	CuS(s)	−48.5	$MgCO_3(s)$	−1113.0	$PCl_3(g)$	−306.3
$CaCO_3(s)$	−1207.0	$CuSO_4(s)$	−769.9	MgO(s)	−601.8	$PCl_5(g)$	−398.9
CaO(s)	−635.5	$Fe_2O_3(s)$	−822.2	$Mg(OH)_2(s)$	−924.7	$SiO_2(s)$	−859.4
$Ca(OH)_2(s)$	−986.6	$Fe_3O_4(s)$	−1117.0	$MgSO_4(s)$	−1278.0	$SnCl_2(s)$	−349.8
$CaSO_4(s)$	−1433.0	HBr(g)	−36.2	MnO(s)	−384.9	$SnCl_4(l)$	−545.2
$CCl_4(l)$	−139.3	HCl(g)	−92.3	$MnO_2(s)$	−520.9	SnO(s)	−286.2
$CH_4(g)$	−74.8	HF(g)	−268.6	NaBr(s)	−359.9	$SnO_2(s)$	−580.7
$CHCl_3(l)$	−131.8	HI(g)	+25.9	NaCl(s)	−411.0	$SO_2(g)$	−296.9
$CH_3OH(l)$	−238.6	$HNO_3(l)$	−173.2	NaF(s)	−569.0	$SO_3(g)$	−395.2
CO(g)	−110.5	$H_2O(g)$	−241.8	NaI(s)	−288.0	ZnO(s)	−348.0
$CO_2(g)$	−393.5	$H_2O(l)$	−285.9	NaOH(s)	−426.7	ZnS(s)	−202.9

TABLE VIII Free Energies of Formation at 25°C and 1 atm (kJ/mol)

AgBr(s)	−95.9	CO(g)	−137.3	$H_2O(g)$	−228.6	$NH_4Cl(s)$	−203.9
AgCl(s)	−109.7	$CO_2(g)$	−394.4	$H_2O(l)$	−237.2	NO(g)	+86.7
AgI(s)	−66.3	$C_2H_2(g)$	+209.2	$H_2S(g)$	−33.0	$NO_2(g)$	+51.8
$Ag_2O(s)$	−10.8	$C_2H_4(g)$	+68.1	HgO(s)	−58.5	NiO(s)	−216.3
$Ag_2S(s)$	−40.3	$C_2H_6(g)$	−32.9	HgS(s)	−48.8	$PbBr_2(s)$	−259.8
$Al_2O_3(s)$	−1576.0	$C_3H_8(g)$	−23.5	KBr(s)	−379.2	$PbCl_2(s)$	−314.0
$BaCl_2(s)$	−810.9	$C_2H_5OH(l)$	−174.8	KCl(s)	−408.3	PbO(s)	−188.5
$BaCO_3(s)$	−1139.0	CoO(s)	−213.4	$KClO_3(s)$	−289.9	$PbO_2(s)$	−219.0
BaO(s)	−528.4	$Cr_2O_3(s)$	−1047.0	KF(s)	−533.1	$Pb_3O_4(s)$	−617.6
$BaSO_4(s)$	−1353.0	CuO(s)	−127.2	$MgCl_2(s)$	−592.3	$PCl_3(g)$	−286.3
$CaCl_2(s)$	−750.2	$Cu_2O(s)$	−146.4	$MgCO_3(s)$	−1029.0	$PCl_5(g)$	−324.6
$CaCO_3(s)$	−1129.0	CuS(s)	−49.0	MgO(s)	−569.6	$SiO_2(s)$	−805.0
CaO(s)	−604.2	$CuSO_4(s)$	−661.9	$Mg(OH)_2(s)$	−833.7	$SnCl_4(l)$	−474.0
$Ca(OH)_2(s)$	−896.8	$Fe_2O_3(s)$	−741.0	$MgSO_4(s)$	−1174.0	SnO(s)	−257.3
$CaSO_4(s)$	−1320.0	$Fe_3O_4(s)$	−1014.0	MnO(s)	−363.2	$SnO_2(g)$	−519.7
$CCl_4(l)$	−68.7	HBr(g)	−53.2	$MnO_2(s)$	−466.1	$SO_2(g)$	−300.4
$CH_4(g)$	−50.8	HCl(g)	−95.3	NaCl(s)	−384.0	$SO_3(g)$	−370.4
$CHCl_3(l)$	−71.5	HF(g)	−270.7	NaF(s)	−541.0	ZnO(s)	−318.2
$CH_3OH(l)$	−166.2	HI(g)	+1.3	$NH_3(g)$	−16.6	ZnS(s)	−198.3
		$HNO_3(l)$	−79.9				

TABLE IX Enthalpies of Bond Formation (kJ/mol)

Single Bonds

H—H	−436	H—C	−415	C—N	−292	N—O	−175
C—C	−344	H—N	−391	C—O	−350	N—F	−270
Si—Si	−187	H—P	−322	C—S	−259	N—Cl	−200
N—N	−159	H—O	−463	C—F	−441	O—F	−212
P—P	−217	H—S	−368	C—Cl	−328	O—Cl	−210
O—O	−143	H—F	−563	C—Br	−276	O—Br	−217
S—S	−266	H—Cl	−432	C—I	−240	O—I	−241
F—F	−158	H—Br	−366			Cl—F	−251
Cl—Cl	−243	H—I	−299			Br—F	−249
Br—Br	−193					Br—Cl	−218
I—I	−151					I—Cl	−210

Multiple Bonds

C=C	−615	C≡C	−812
N=N	−418	N≡N	−946
O=O	−402	C≡N	−890
C=N	−615		
C=O	−725		
C=S	−477		

TABLE X Vapor Pressure of Water in Bars

t_c	vp	t_c	vp	t_c	vp	t_c	vp
0	0.00611	21	0.02486	35	0.0563	92	0.7559
5	0.00872	22	0.02644	40	0.0737	94	0.8145
10	0.01228	23	0.02809	45	0.0959	96	0.8767
12	0.01403	24	0.02984	50	0.1233	98	0.9430
14	0.01599	25	0.03168	55	0.1573	99.63	1.0000
16	0.01817	26	0.03361	60	0.1992	100	1.0132
17	0.01937	27	0.03565	65	0.2500	102	1.0878
18	0.02064	28	0.03780	70	0.3116	104	1.1667
19	0.02197	29	0.04005	80	0.4734	106	1.2504
20	0.02338	30	0.04242	90	0.7010	108	1.3392
						110	1.4327

appendix 6

ANSWERS TO PROBLEMS

CHAPTER 1

1.14   Use oil, natural gas, nuclear fuel, solar energy.
1.15   **a.** See discussion p. 14.   **b.** p. 12   **c.** p. 19
1.16   **a.** If only part of solution is distilled, distillate < 20% benzene, residue > 20% benzene.
       b. Sample will begin to melt below mpt of pure T.A.; mpt will rise steadily.
       c. Components will come out as separate bands on column.
1.17   **a.** Allow ether to evaporate.   **b.** Electrolyze.
       c. Boil to remove air, condense air, fractionally distil.
1.18   **a.** See discussion p. 10.   **b.** p. 12   **c.** p. 21
1.19   **a.** 8.78×10^{-24} cc   **b.** 12.1 g/cc
       c. Empty space between atoms
1.20   3.78 g/cc
1.21   **a.** 12.20 lit   **b.** 54.8 ml
1.22   **a.** 1.03×10^3 g/cm^2 **b.** 1.01×10^6 dyne/cm^2 **c.** 1.00 atm
1.23 5.2×10^2 tons
1.24 9×10^8 yr
1.25 $°C = 0.75°B + 5$
1.26   58 g, 2 g
1.27   0.022 cm^3
1.28   46.2 mile^2
1.29   **c.**

d(%)	0	10	20	30	40	50	60	70	80	90	100
W(lb)	150	4.1	1.0	0.5	0.3	0.2	0.1	0.1	0.1	0.1	24.7

CHAPTER 2

2.18   **a.** Carry out reaction, compare masses of products to those of reactants.
       b and **c.** Mass spectrometer.
       d. Show that proton + electron combine to form neutral species.
2.19   7, 8, 6; $^{7}_{3}\text{Li}$, 3; $^{19}_{9}\text{F}^-$, 9
2.20   **a.** A.T.   **b.** S.T.   **c.** A.T.
2.21   **a.** 37   **b.** 18   **c.** 1.14×10^{25}
2.22   88.1%

2.23 3rd

2.24 0.0084%

2.25 6.96

2.26 a. 15.9994 b. 0.99972

2.27 a. Mg−24 b. 11% Mg−25, 10% Mg−26

2.28 a. 1.82×10^{24} atoms b. 7.96×10^{-23} g c. 0.376 GAW

2.29 b < d < a < c < e

2.30 2.4×10^{-8} g/cc

2.31 1.5×10^7

2.32 8.1×10^{-10}

2.33 4.6, 7.2

2.34 ½, 0.0618

2.35 14.0

2.36 6.03×10^{23}

2.37 6.07×10^{23}

CHAPTER 3

3.19 a. $C_3H_6NO_2S$ b. 30.0, 5.04, 11.7, 26.7, 26.7 c. 2.18 g

3.20 a. Co_3O_4 b. $Cs_2Cr_2O_7$ c. $C_2H_5NO_2$

3.21 a. Sc_2O_3 b. 0.556 g

3.22 75.0

3.23 a. C_5H_4 b. $C_{10}H_8$

3.24 a. 77.3, 7.60, 15 b. C_6H_7N

3.25 $BiCl_3$

3.26 80.3, 13.6, 6.1; BiOCl

3.27 a. 5.32×10^{-23} g b. 53.2 g

3.28 a. 246.4 g b. 179 g c. 0.0299

3.29 310 g

3.30 a. $2\ B_5H_9(l) + 12\ O_2(g) \rightarrow 5\ B_2O_3(s) + 9\ H_2O(g)$
b. $Al^{3+}(aq) + 3\ OH^-(aq) \rightarrow Al(OH)_3(s)$
c. $2\ C_7H_5N_3O_6(s) \rightarrow 3\ N_2(g) + 5\ H_2O(g) + 7\ CO(g) + 7\ C(s)$

3.31 a. 6.19 b. 0.0441 c. 1.4×10^8

3.32 18

3.33 2.6×10^{-3}

3.34 a. 16.9 b. 13.3

3.35 a. 56.9 g b. 63.3

3.36 29 g, 18 g

3.37 36

3.38 540 days

3.39 a. 1.11 g b. 0.850 g

3.40 1.9

CHAPTER 4

4.13 a. 4.93 kcal b. C_8H_{18}

4.14 a. 0.100 kcal b. 0.136 kcal c. −21.7 kcal

4.15 a. $2\ C_3H_5(NO_3)_3(s) \rightarrow 3\ N_2(g) + \frac{1}{2}\ O_2(g) + 6\ CO_2(g) + 5\ H_2O(l)$
b. −431 kcal c. 90.7 kcal

4.16 $\Delta H = +54.2$ kcal

4.17 **a.** -152.7 kcal   **b.** -11.9 kcal

4.18 **a.** -312 kcal   **b.** $+38$ kcal   **c.** $\Delta H = +93$ kcal

4.19 **a.** 0.067 kcal   **b.** $13.6°C$

4.20 270 cal/°C

4.21 473 kcal

4.22 11 kcal/g

4.23 **a.** 20 kcal   **b.** 0.063

4.24 **a.** $0.17 \times 10^{16}, 0.41 \times 10^{16}$ kcal   **b.** $2.7 \times 10^8, 6.4 \times 10^8$ tons
   c. $3.6, 3.1$ tons

4.25 **a.** 0.090 kcal   **b.** 0.036 cents

4.26 Would get 100% conversion of heat into work at 0°K

4.27 Beef, 2.0 kcal/g; cheese, 4.6 kcal/g; cornflakes, 3.6 kcal/g; potato chips, 5.8 kcal/g; per ounce: beef, 57; cheese, 130; cornflakes, 100; potato chips, 160

4.28 **a.** -202.6 kcal   **b.** $6500°C$ **c.** Yes

4.29 See Table 4.3. $\Delta n_g = -2, -1, 0, +2$

4.30 **a.** n years: $5.4 \times 10^{16}(1.03)^n$ **b.** Total 1971–80 = 61.8×10^{16} kcal
   c. 1981–90 = 84×10^{16} kcal, 1991–2000 = 111×10^{16} kcal, . . .
   d. Appr. 2090

CHAPTER 5

5.20 High compressibility, low density. Basketball, steam engine, balloon.

5.21 p, V, n, T. $V = \text{constant}(t + 273) = a + bt$

5.22 **a.** Raise T or reduce V   **b.** Raise T   **c.** Raise T or reduce p

5.23 1.7×10^{-3} atm

5.24 High velocity of sound, rapid rate of diffusion, rapid rate of leakage through small opening.

5.25 603 mm Hg

5.26 65 gal

5.27 20°K

5.28 114 g

5.29 1.18 g/lit; 28.8

5.30 177

5.31 **a.** 728 mm Hg   **b.** 0.405 g

5.32 16

5.33 150 lb

5.34 0.0169; 40.1

5.35 1.004; 50.1% U-235

5.36 1.92×10^5 cm/sec; 1.45×10^4 cm/sec; same

5.37 Deviations from IGL; less air in flask at end of expt.

5.38 $CHCl_3 < Cl_2 < CO_2 < Ar < N_2$

5.39 H_2: no. O_2: maybe

5.40 MW = 44.010

5.41 **a.** $He < O_2 < C_2H_2 < C_6H_6$ **b.** $O_2 < He < C_2H_2 < C_6H_6$

5.42 $\Delta E = \dfrac{3R}{2} (T_2 - T_1)$; if $T_2 - T_1 = 1$, $\Delta E = \dfrac{3R}{2} = 3$ cal

5.43 Very few gas molecules available to transfer energy to thermometer.

5.44 **a.** $p = 28.7/V$, hyperbola   **b.** $p = 0.00328T$, st. line
c. $pV = 28.7$, st. line, zero slope   **d.** $u = 2980T^{1/2}$
e. $u = 2.95 \times 10^5/M^{1/2}$ **f.** $E = 1040$ cal, st. line, zero slope

5.45 6.8 m, 8.7 m

CHAPTER 6

6.21

	appr diam	mass (AW scale)	charge
a.	10^{-12} cm	1/1837	-1
b.	10^{-12} cm	1 to 260	$+1$ to $+104$
c.	10^{-12} cm	1	$+1$
d.	10^{-12} cm	1	0

6.22 Planets governed by classical mechanics, electrons by quantum mechanics.

6.23 **a.** No more than 2 electrons per orbital.
b. Allows assignment of electrons among equivalent orbitals.

6.24 n dominates in determining energy and distance from nucleus; l has a lesser effect on both; m_l determines orientation of orbitals.

6.25 3.03×10^{-12} erg; 4.36×10^4 cal vs 5.78×10^4 cal

6.26 **a.** 5, 14, 32, 33, 51, 52, 84, 85   **b.** 7   **c.** 1

6.27 Might measure successive ionization potentials. Electronic configuration establishes position in periodic table, hence allows estimation of physical and chemical properties.

6.28 Al $1s^22s^22p^63s^23p^1$ O^- $1s^22s^22p^5$
Ca $1s^22s^22p^63s^23p^64s^2$ F^+ $1s^22s^22p^4$
Na^+ $1s^22s^22p^6$ Mn $1s^22s^22p^63s^23p^64s^23d^5$

6.29 Excited: $1s^22s^22p^53p$, $1s^23d$
Ground: $1s^22s^22p^63s$, $1s^22s^2$
Wrong: $1s^22s^23p^7$, $1s^22s^22p^62d^2$

6.30 **a.** Mg; 0 unp. e^- **b.** None; 4 unp. e^- **c.** N; 3 unp. e^-

6.31 $1s^22s^22p^63s^23p$; Si^+

6.32 1st set: $3,1,0,\frac{1}{2}$ 4th set: $4,0,0,\frac{1}{2}$

6.33 Vertical: 75°C; horizontal: 268°C; observed: 59°C

6.34 **a.** $OsBr_3$ **b.** ICl   **c.** H_2Se **d.** K_3PO_4 **e.** As_2S_3
f. Na_2MoO_4

6.35 $1s^22s^22p^33s^1$; $1s^22s^22p^33p^1$; $1s^22s^22p^34s^1$

6.36 **a.** S   **b.** O   **c.** Mg   **d.** F^- **e.** Nd   **f.** Cl^-

6.37 -1.96×10^{-10} erg; 123 ev

6.38 4860 Å (2nd line Balmer); 1020 Å

6.39 5.49×10^{-63} erg; 3×10^{32}; yes

6.40 2.1×10^6 ev; yes; yes; yes

6.41 9.0×10^7 Å; microwave

6.42 Would behave classically.

6.43 0.726 Å

6.44 3.67×10^{-13} ergs vs 3.37×10^{-12} ergs required.

6.45 **a.** 3, 9   **b.** 12   **c.** $1s^32s^32p^4$; $1s^32s^32p^93s^3$ **d.** 22, 49

CHAPTER 7

7.21 **a.** H—H bond stronger than many ionic bonds.
  b. Cs^+ larger than F^-.
  c. C—Cl bonds longer than C—H.
  d. Recall XeF_4.
7.22 **a.** High charge ions give high melting compound.
  b. Explains paramagnetism, double bond.
  c. Polar only if bent.
  d. Bonds are identical.
7.23 **a.** 3 **b.** 4 **c.** 2 **d.** 2 **e.** 5 **f.** 5
7.24 **a.** $2 K(s) + Te(s) \rightarrow K_2Te(s)$ **b.** $2 Al(s) + 3 S(s) \rightarrow Al_2S_3(s)$
  c. $4 Li(s) + O_2(g) \rightarrow 2 Li_2O(s)$ **d.** $2 Sr(s) + O_2(g) \rightarrow 2 SrO(s)$
  e. $2 Y(s) + 3 Br_2(l) \rightarrow 2 YBr_3(s)$
7.25 **a.** S^{2-}, K^+ **b.** Cr^+, Fe^{3+} **c.** Hg^{2+}, Tl^{3+} **d.** F^-, Mg^{2+}
7.26 F
7.27 HF, 1.7; HCl, 1.0; HBr, 0.7; HI, 0.2

7.28 **a.** :C≡O: **b.** :F̈:   :F̈: **c.** N̈≡Ö:

  d. :N≡O: **e.** :Ö—C̈l—Ö:

7.29 **a.** :C̈l **b.** H   H **c.** :F̈:   F̈:

  d.   :Ö: **e.** :Ö—C—Ö—H

7.30 Check bond angles (60, 109, 120).
7.31 **a.** O—C≡O ↔ O=C=O ↔ O≡C—O
  b. O—C—O ↔ O—C=O ↔ O=C—O

  c. :Ö   :Ö   :Ö·   :Ö

7.32 **a.** HF **b.** SO_3 **c.** SO_2 **d.** CCl_4
7.33 109°: H—C—H, H—C—C, C—O—O, O—O—N; 120°: O—N—O,
  C—C—O, O—C—O

7.34 **a.** Linear   **b.** Trig. pyr.   **c.** Linear   **d.** Linear
　　　 e. Trig. pyr.

7.35 ClF, CO, PCl_3

7.36 **a.** 5   **b.** 6   **c.** 5   **d.** 6

7.37 **a.** sp^3 **b.** sp^2 **c.** sp^2 **d.** sp^2 **e.** sp; one π bond in b,c,d;
　　　 2 in e

7.38 No. of bonds would not change; no. of unpaired electrons would be 0
　　　 for B_2, 2 for C_2.

7.39 Between 2nd and 3rd.

7.40 **a.** Si valence electrons tied down in covalent bond.
　　　 b. Luster; see p. 198.
　　　 c. Ionic crystals brittle; see p. 198.

7.41 −250 kcal

7.42 **a.** 2:1   **b.** Smaller   **c.** $\Delta E = -0.046 n^2 h^2 / m V_a^{2/3}$

7.43 Pentagonal bipyramid.

7.44 All ions have same charge along diagonal plane.

7.45 Could remove two corners of square to give structure with less space
　　　 for unshared electrons.

CHAPTER 8

8.21 **a.** See p. 226, 179.
　　　 b. Monomer is simplest unit of polymer.
　　　 c. H bond is a particularly strong dipole force.

8.22 **a.** CCl_4 **b.** Fe   **c.** NaCl   **d.** CS_2 **e.** NH_3 **f.** C

8.23 **a.** P, D   **b.** H, D   **c.** P, D   **d.** D   **e.** H, D   **f.** P, D

8.24 **a.** Covalent bonds   **b.** Metallic bonds
　　　 c. Dispersion forces   **d.** H bonds

8.25 20 other isomers.

8.26

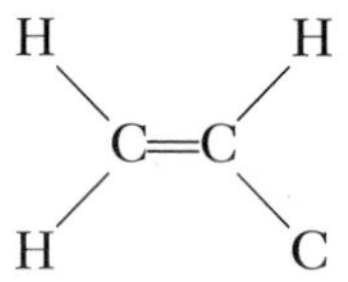

in plane; no

8.27 Cyclopropane under strain (60° bond angles); propene more stable.

8.28 Lowest: —C—C—C—C—C—C—C; highest:

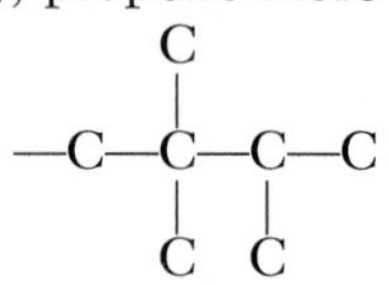

8.29 Double bond formation gives simple molecules in CO_2; SiO_2 macro-
　　　 molecular.

8.30 Cis-trans isomerism gives an extra isomer for $C_2H_2Cl_2$.

8.31 **a.** Ether   **b.** Aldehyde   **c.** Ketone

8.32 **a.**

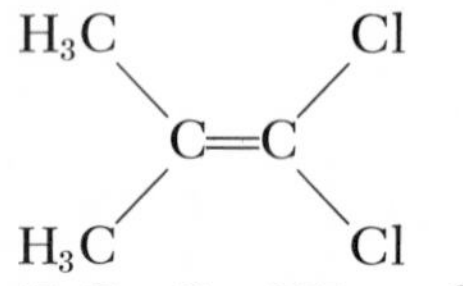

b.

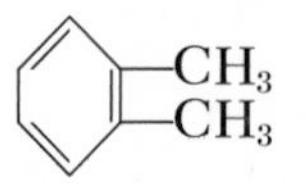

　　　 c. H_3C—C≡CH   **d.** H_3C—CH_2—CH_2—CH_3
　　　 other isomers possible in all cases.

8.33 **a.** Aldehyde or ketone   **b.** Acid
c. Alcohol or ether   **d.** Acid

8.34 **a.** Acetylene   **b.** Paraffin   **c.** Olefin
d. Aromatic   **e.** Aromatic

8.35 H—C—H angle 109°; C—C—C, C—C—O angles 120°; 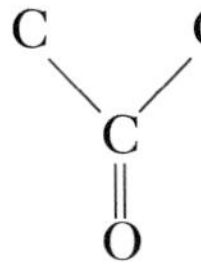
C   C planar

8.36 H_3C—O—CH_3, H_3C—C—C_2H_5, H_3C—C—O—C_2H_5
with C=O groups

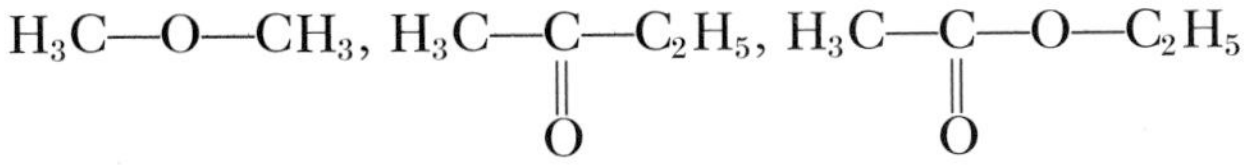

8.37

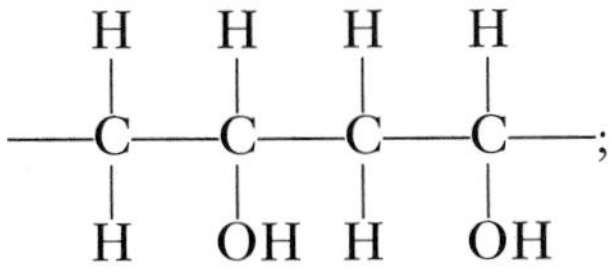

; can H bond with water

8.38

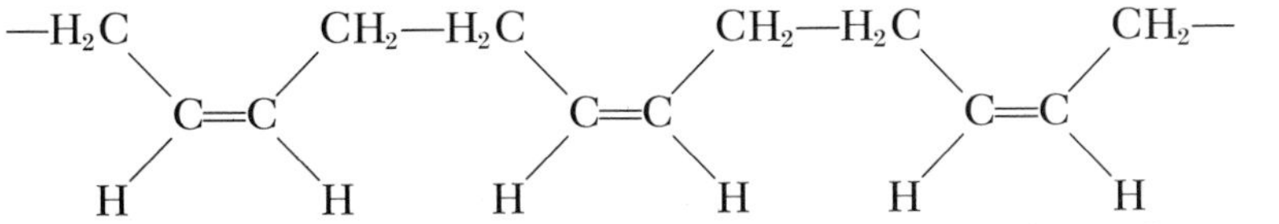

8.39 Two different kinds of atoms in SiO_2, each O bonded to only two Si.

8.40 **a.**

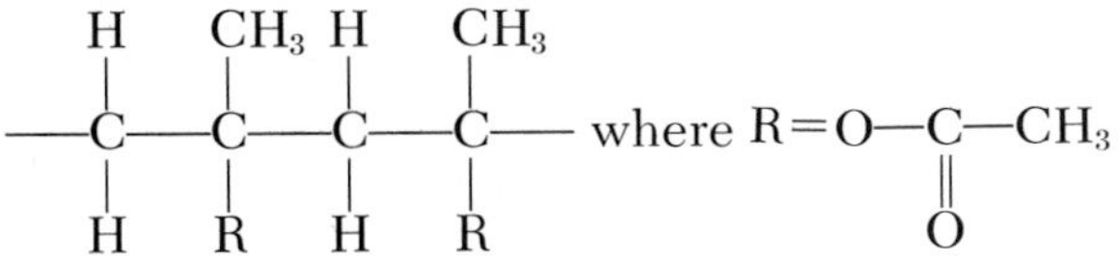

where R=O—C—CH_3 with C=O

b.

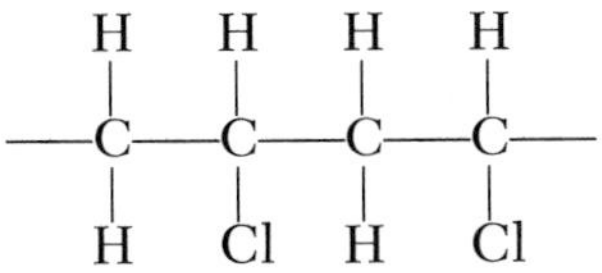

c.

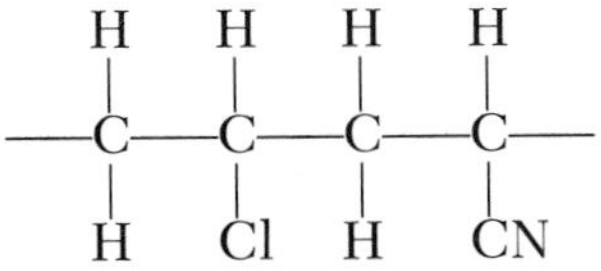

8.41

, H_2C=C—C=CH_2,
with H_3C CH_3

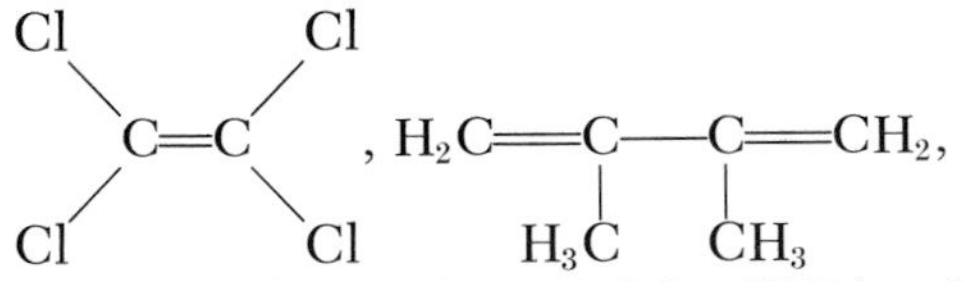

HO—$(CH_2)_3$—OH + HOOC—$(CH_2)_2$—COOH

8.42 Vinyl alcohol polymer in Problem 8.37 could crosslink by eliminating water.

CHAPTER 9

9.20 **a.** Vacuum created when steam condenses.   **b.** Ice sublimes.
c. Closest-packing arrangement sets off centers of spheres in successive layers (see Figure 9.10).

9.21 **a.** Lower p lowers boiling temperature, which slows down cooking process.
b. Ice is sublimed by evacuation.
c. Defects promote formation of Ag image.

9.22 **a.** Compare rate of mixing of two gases with mixing of H_2O, H_2SO_4.
b. X-ray diffraction, use Equation 9.6.
c. Measure vp solid as function of T, plot log vp vs 1/T, take slope.

9.23 **a.** 7.20 cc   **b.** 5.29 cc   **c.** 0.265
9.24 -1.73 kcal
9.25 31 g
9.26 **a.** 470 g   **b.** 24 mm Hg   **c.** 20 mm Hg
9.27 No; yes; 34°C
9.28 8.0 kcal/mole
9.29 **a.** 4.51 atm   **b.** 84°K
9.30 120
9.31 a, c
9.32 **a.** 25 g   **b.** 198 cc
9.33 1.53 Å
9.34 **a.** 6   **b.** 13/3
9.35 1.44 Å
9.36 4.05 Å
9.37 Three curves intersect at triple point (5°C); liquid-solid line inclined to right; liquid-vapor line rises vertically above critical T (289°C).
9.38 Ba^{2+}
9.39 0.15°C; heat transfer
9.40 0.41 radius anion
9.41 **a.** See No. 9.30.
      b. Determine atomic radius, packing pattern by x-ray diffraction. Calculate volume/atom, compare to volume/mole (from measured density).
      c. Same as (b), but assume N, calculate V per mole, then mass per mole using density.
9.42 Simple: 0.476   bcc: 0.320   fcc: 0.260

CHAPTER 10

10.20 **a.** Water would flow into blood cells by osmosis, tending to burst them.
       b. Intermolecular forces differ greatly.
       c. Water passes out of leaves by osmosis.
10.21 **a.** Dilute 17 ml of 6.0 M NaOH to one liter.
       b. Osmotic pressure measurement.
       c. Put under bell jar, observe direction in which water moves.
10.22 **a.** 0.030 g   **b.** 11 ml
10.23 **a.** 25.0   **b.** 0.527   **c.** 7.25   **d.** 6.96
10.24 **a.** 3.4 g   **b.** 25 ml   **c.** 2400 ml
10.25 $V_w = \dfrac{V_c}{M_d} (M_c - M_d)$
10.26 7.5×10^{-4}; 2.6×10^{-5}
10.27 0.14
10.28 **a.** $C_2H_5OC_2H_5 < CH_3OCH_3 < CH_3OH$
       b. $H_2 < CO_2 < H_2O_2$
10.29 **a.** Solubility, boiling point greater   **b.** 100°C
       c. Phenanthrene   **d.** No
10.30 14.6, 3.64
10.31 0.581
10.32 **a.** 3.23   **b.** 9.68   bpt $= 100.28$°C
10.33 0.73 qt

10.34 0.30 M
10.35 a. Sugar in water, methyl alcohol in water   b. None   c. All
10.36 $C_{21}H_{30}O_2$
10.37 $C_{10}H_{14}N_2$
10.38 $2.71 \times 10^4, -0.00032°C$
10.39 7 years
10.40 $300 million
10.41 174
10.42 15
10.43 1.9×10^{-3}

CHAPTER 11

11.17 Ice would be at bottom of lakes; most of water on earth would be frozen.
11.18 Water forms a particularly stable hydrogen-bonded structure in which oxygen is surrounded tetrahedrally by four hydrogens.
11.19 See Figure 11.4.
11.20 a. Ion atmosphere more pronounced for +2 ions.
b. Attractive forces more important when particles are closer together.
11.21 Parathion decomposes more readily but is more toxic.
11.22 Algae growth probably of minor importance; low-phosphate detergents could be harmful if they contain strong alkali.
11.23 1.0×10^{15}
11.24 Contraction when ice melts would be explained in terms of bond bending to give more closely packed structure.
11.25 10%
11.26 3.6 mg/liter
11.27 2.2 mg CN^-/liter
11.28 5.4×10^{-8} M
11.29 1.18×10^6
11.30 a. 0.22, 0.32   b. 0.37 g $Ca(OH)_2$ c. 0.42 g Na_2CO_3
11.31 OH^- reacts with HCO_3^- to form CO_3^{2-}, ppt Ca^{2+}
11.32 a. $-2.1°C$ b. $-4.7°C$
11.33 Measure temperature of maximum density.
11.34 a. 39°C   b. 0.68°C
11.35 One possibility: add 26.7 g NaCl, 3.33 g $MgSO_4$, 2.66 g $MgCl_2$, 1.10 g $CaCl_2$, 0.23 g $KHCO_3$, 0.096 g KBr, 0.037 g K_2CO_3, 0.45 g KCl, 0.016 g $SrCl_2$.

CHAPTER 12

12.15 a. Spont.   b. Nonspont.   c. Spont.   d. Spont.
12.16 Freezing of water below 0°C.
12.17 a. Photosynthesis   b. Reverse osmosis   c. Electrolysis
12.18 41.4 kcal; O_2 would be removed from air.
12.19 +42.5 kcal, +0.0658 kcal/°K; gas formed; gas molecules have more room to move in at lower p.

12.20 **a.** +35.6 kcal   **b.** Heat absorbed, vaporizing water.
       c. Vaporization nonspontaneous at 25°C, 1 atm.
12.21 **b.** −0.038 kcal/°K, −95.5 kcal
12.22 **a.** −   **b.** +   **c.** −   **d.** −
12.23 20.7 cal/°K
12.24 **a.** Usually true   **b.** False   **c.** True   **d.** Usually true
12.25 −23 kcal
12.26 88°C
12.27 25°C: +14.6 kcal, +29.0 kcal; 1000°C: +11.7 kcal, +5.6 kcal
12.28 T = 2390°C
12.29 **a.** Plot of ΔG vs log p is linear.   **b.** 1.4 kcal
       c. +38.3, +47.3, +56.7 cal/°K
       d. $\Delta S^p = \dfrac{\Delta H - \Delta G^p}{T} = \dfrac{\Delta H - \Delta G^{1\,atm}}{T} - \dfrac{a}{T} \log p = \Delta S^{1\,atm} - \dfrac{a}{T} \log p$
12.30 **a.** +690.2 kcal   **b.** No   **c.** Photochemical energy supplied.
12.31 191.0 vs 195.6 kcal
12.32 **a.** $\Delta H = +0.6$ kcal for Ni, +25.4 for Zn, +11.6 for Sn, +7.7 for Fe.
       $\Delta S = +12$ cal/°K for Ni, +13 for Zn, +14 for Sn, +11 for Fe.
       b. Since S of solid is small, $\Delta S \approx S\ H_2O(g) - S\ H_2(g)$.
       c. $T = \Delta H / \Delta S$, T directly proportional to ΔH, large negative value
       of ΔH_f makes ΔH more positive.
       d. $HgO > Cu_2O > Fe_3O_4 > Al_2O_3$
       e. Measure T at which reaction becomes feasible.

CHAPTER 13

13.18 **a.** $\dfrac{[H_2O]^2}{[H_2]^2 \times [O_2]}$ **b.** $\dfrac{[H_2O]}{[H_2] \times [O_2]^{1/2}}$ **c.** $\dfrac{1}{[H_2] \times [O_2]^{1/2}}$
       d. $[H_2]^2 \times [N_2]$
13.19 **a.** 5.0×10^{-6} **b.** 450
13.20 71
13.21 2.6
13.22 **a.** ←   **b.** ←   **c.** ←
13.23 0.27 M
13.24 a and b. $[SO_3] = [NO] = 3.0 \times 10^{-3}$ M; $[SO_2] = [NO_2] = 1.0 \times 10^{-3}$ M
       c. $[SO_3] = [NO] = 6.0 \times 10^{-3}$ M; $[SO_2] = [NO_2] = 2.0 \times 10^{-3}$ M
13.25 **a.** 0.40   **b.** 0.020 M   **c.** 8.0×10^{-3} M
13.26 **a.** 0.70 M   **b.** 0.43
13.27 **a.** 0.012 M   **b.** 1.1×10^{-4} M
13.28 **a.** →   **b.** 0.016
13.29 **a.** No change   **b.** ← (decrease V)   **c.** ←   **d.** →
13.30 **a.** 2.5   **b.** 2.1×10^{-4} moles
13.31 **a.** 0.74   **b.** 1.5
13.32 1900°K
13.33 −2720 cal, 0, +2720 cal
13.34 **a.** −16.6 kcal   **b.** −18.5 kcal   **c.** 4×10^{13}
13.35 $K_1K_2 = \dfrac{[NO]^2}{[N_2] \times [O_2]} \times \dfrac{[NO_2]^2}{[NO]^2 \times [O_2]} = \dfrac{[NO_2]^2}{[N_2] \times [O_2]^2} = K_3$
13.36 0.050 M

13.37 At $M = 1$, $P = 24.4$ atm:
$$\Delta G^{1M} = \Delta G^{1\,atm} + \Delta n_{gas}\ (1.99)(2.30)(298)\ \log_{10} 24.4$$

13.38 $p_x = [x]RT$; $K_p = \dfrac{[C]^c(RT)^c \times [D]^d(RT)^d}{[A]^a(RT)^a \times [B]^b(RT)^b} = K_c(RT)^{c+d-a-b} = K_c^{\Delta n_{gas}}$

13.39 **a.** $\sim 10^{35}$ **b.** 2×10^{-38} M   **c.** $99.99+$

CHAPTER 14

14.16 **a.** Molecules must have sufficient energy and be properly oriented to react.
b. ΔG unchanged.   **c.** Blocks substrate from enzyme site.

14.17 2×10^{-12}; 1.4 hr

14.18 **a.** 1.0×10^{-3} mole/lit sec   **b.** 5.0×10^{-2} sec^{-1}
c. 2.5 lit/mole sec

14.19 1/2, 1/2

14.20 **b.** 0.0158 M

14.21 **a.** 0.040 M; 4.0×10^{-7} M   **b.** 6.0 sec

14.22 0.092 min^{-1}; 0.063 M

14.23 Zero

14.24 10.3 kcal

14.25 1.17

14.26 $E_a' = 41$ kcal

14.27 Enzyme decomposed or deactivated.

14.28 Rate $= k_3(\text{conc. Cl}_2)(\text{conc. COCl})$
$\qquad = k_3(\text{conc. Cl}_2)K_2(\text{conc. Cl})(\text{conc. CO})$
$\qquad\quad k_3K_2K_1^{1/2}(\text{conc. Cl}_2)^{3/2}(\text{conc. CO})$

14.29 c, d

14.30 0.150 mole/lit min; 2.0 lit/mole

14.31 **a.** 2.4×10^{-15} mole/lit; 2.4×10^{-11} mole/lit sec
b. 5.9×10^{-14} **c.** 5.9×10^{-6} pphm; 10 pphm

14.32 Look for I atoms.

14.33 1×10^{-4} oz; 10 oz

14.34 Rate $= k_1(\text{conc. NO}_2)(\text{conc. NO}_3) = k_1K_2(\text{conc. N}_2\text{O}_5{}^*)$
$K_1 = \dfrac{(\text{conc. N}_2\text{O}_5{}^*)(\text{conc. N}_2\text{O}_5)}{(\text{conc. N}_2\text{O}_5)^2} = \dfrac{\text{conc. N}_2\text{O}_5{}^*}{\text{conc. N}_2\text{O}_5}$
rate $= k_1K_1K_2(\text{conc. N}_2\text{O}_5)$

CHAPTER 15

15.21 a, d, e

15.22 **a.** 8000   **b.** 8×10^{-4} **c.** 3×10^{-7}

15.23 **a.** Yes   **b.** No   **c.** N_2, not Ar   **d.** No

15.24 Increase P, use catalyst.

15.25 a, d

15.26 **a.** Burn in excess O_2 at low T.
b. Burn to SO_2, convert to SO_3, add to water.
c. Haber, then Ostwald process.   **d.** UV radiation.

15.27 **a.** $NO(g) \rightarrow NO^+(g) + e^-$
b. $4\ NH_3(g) + 5\ O_2(g) \rightarrow 4\ NO(g) + 6\ H_2O(l)$
c. $Na_2O_2(s) \rightarrow Na_2O(s) + \tfrac{1}{2}\ O_2(g)$

15.28   **a.** $:\overset{..}{\underset{.}{N}}:$ **b.** $:\overset{..}{\underset{..}{O}}—\overset{..}{\underset{..}{O}}:$

        c. 7 pairs e^- around Xe, 6 shared with F atoms

15.29   1240°C

15.30   **a.** 40, 20, 40   **b.** 0.0125

        c. $5.0 \times 10^{-3}, 2.5 \times 10^{-3}, 5.0 \times 10^{-3}$ **d.** 20

15.31   **a.** CO_2 absorbs IR, O_3 UV.

        b. Equil. vp greater at higher T.

        c. Reflect sunlight, act as nuclei for precipitation.

15.32   19°C

15.33   −16°C

15.34   Require increasingly greater amounts of energy to form.

15.35 10^{-46} M

15.36   Almost ($\lambda = 3930$ Å)

15.37   **a.** Cracking   **b.** See Eqns. 15.37, 15.38

        c. Complete combustion removes both.

15.38   Decomposes at lower T.

15.39   See p. 422.

15.40 2×10^{-49} mole/lit sec

15.41   9.1%

15.42 $\Delta T = 1800$°K

15.43   **a.** 4.0 g SO_2 **b.** 1.5 lit   **c.** 1500 ppm

15.44   5; 180°, 120°, 90°

CHAPTER 16

16.18   **a.** $2\ Ag^+(aq) + S^{2-}(aq) \rightarrow Ag_2S(s)$

        c. $Al^{3+}(aq) + 3\ OH^-(aq) \rightarrow Al(OH)_3(s)$

        e. $Ni^{2+}(aq) + 2\ OH^-(aq) \rightarrow Ni(OH)_2(s); Sr^{2+}(aq) + SO_4{}^{2-}(aq) \rightarrow$

          $SrSO_4(s)$

        f. $Pb^{2+}(aq) + 2\ Cl^-(aq) \rightarrow PbCl_2(s)$

16.19   **a.** See p. 434.   **b.** Ni^{2+} has higher charge density.

        c. Conc. Ag^+ in equil. with Ag_2CrO_4 high enough to ppt AgCl.

16.20   **a.** See p. 436.   **b.** pption insoluble chlorides with HCl.

        c. Conc. I^- greater, so that of Ag^+ lower.

16.21   **a.** Always true.   **b.** Sometimes true.

16.22   **a.** $0.060\ Ni^{2+}, 0.12\ Cl^-, 0.060\ K^+, 0.060\ OH^-$

        b. 2.8   **c.** $0.030\ Ni^{2+}, 0.12\ Cl^-, 0.060\ K^+$

16.23   **a.** 25 ml   **b.** 7.05 ml   **c.** 750 ml

16.24   **a.** 1.1×10^{-8} **b.** 2.2×10^{-8}

16.25   **a.** 6.3×10^{-5} **b.** 8×10^{-5}

16.26   **a.** Yes   **b.** Yes   **c.** No

16.27 2×10^{-5} M

16.28   **a.** 0.0005   **b.** 0.0005   **c.** 3×10^{-6}

16.29   AgCl; 1.6×10^{-5}

16.30   3.54

16.31   30.1

16.32   **a.** Add $Cl^-, S^{2-}, SO_4{}^{2-}$ **b.** Add Mg^{2+}, Cu^{2+}, Ag^+

16.33   Group 1 ppted with group 2.

16.34   **a.** Add equivalent amt $BaCl_2$, filter, evaporate.

        b. Add $Mg(NO_3)_2$ **c.** Add $CuCl_2$ **d.** Add $CuSO_4$

16.35 5900 lit

16.36 1400 lit

16.37 a. 1.4×10^{-5} M   b. Al^{3+}, Fe^{3+} c. 99.9 +   d. 5.5 g

16.38 1.64×10^{-10}, 1.77, 1.93, 2.06, 2.27, 2.60

16.39 c. 1.5×10^{-5} M

16.40 13%

CHAPTER 17

17.21 a. See pp. 469, 472.   b. Incomplete diss. to OH^-
    c. Derived from acid by loss H^+ d. Proton acceptor
    e. Electron pair acceptor   f. Complete diss. to H^+
    g. Reacts with water to form acid   h. pH > 7

17.22 a. 2.0   b. -0.3 c. 7.4   d. 2.21

17.23 a. 1×10^{-2}; 1×10^{-12}; A   b. 3×10^{-7}; 3×10^{-8}; A
    c. 10; 1.0×10^{-15}; A   d. 6.3×10^{-14}; 0.16; B

17.24 a. 5×10^{-15}; 2; 14.3   b. 0.3; 3×10^{-14}; 0.5
    c. 0.079; 1.3×10^{-13}; 1.10   d. 2.6×10^{-14}; 0.38; 13.58

17.25 a. H^+, Cu^{2+} b. HCl, HBr   c. H_2SO_4, H_2S
    d. HF, $HC_2H_3O_2$ e. NaCl, KNO_3 f. CaO, Li_2O

17.26 a. HBr   b. H_2SO_4 c. CCl_3COOH
    d. NH_4^+ e. $HBrO_3$

17.27 b. $S^{2-}(aq) + H_2O \rightarrow HS^-(aq) + OH^-(aq)$; basic
    c. $HCO_3^-(aq) + H_2O \rightarrow H_2CO_3(aq) + OH^-(aq)$; basic
    d. $Zn(H_2O)_4^{2+}(aq) \rightarrow Zn(H_2O)_3(OH)^+(aq) + H^+(aq)$; acidic
    f. $C_2H_3O_2^-(aq) + H_2O \rightarrow HC_2H_3O_2(aq) + OH^-(aq)$

17.28 a. Neutral   b. Basic   c. Acidic   d. Basic   e. Basic
    f. Basic

17.29 H–C–O–H; 1.9×10^{-4}
       $\overset{\|}{\underset{O}{}}$

17.30 8×10^{-4}

17.31 a. 1×10^{-7}, 1×10^{-7}, 7.0
    b. 1.3×10^{-4}, 7.7×10^{-11}, 3.89, 0.026
    c. 0.071, 1.4×10^{-13}, 1.15, 14
    d. 1.7×10^{-5}, 5.9×10^{-10}, 4.77, 0.0034

17.32 a. 0.25, 4.0×10^{-14}, 13.40   b. 3.5×10^{-6}, 2.9×10^{-9}, 8.54
    c. 2.1×10^{-3}, 4.8×10^{-12}, 11.32   d. 2.3×10^{-6}, 4.3×10^{-9}, 8.37

17.33 a. 1.9×10^{-11} b. 2.6×10^{-5}

17.34 Conductivity greater, freezing point lower for KOH.

17.35 Ionic M–O bond in KOH, covalent in BrOH.

17.36 a. Acids: H_3O^+, HNO_2; bases NO_2^-, H_2O
    b. Acids: H_2O, H_3PO_4; bases: $H_2PO_4^-$, OH^-
    c. Acids: H_2O, $HC_3H_5O_2$; bases: $C_3H_5O_2^-$, OH^-
    d. Acids: H_2CO_3, HSO_4^-; bases: SO_4^{2-}, HCO_3^-

17.37 a. 2.2×10^3 b. 1.3×10^{-12} c. 7.1×10^{-10} d. 3.5×10^{-5}

17.38 a. BA   b. BB, LB   c. BB, LB   d. LA

17.39 $NH_3 < KF < KCl < NH_4NO_3 < HNO_2 < KHSO_4 < HI < H_2SO_4$

17.40 0.002 mole KOH, 0.12 mole NH_3

17.41 $K_1 \times K_2 = \dfrac{[H^+] \times [HX^-]}{[H_2X]} \times \dfrac{[H^+] \times [X^{2-}]}{[HX^-]} = \dfrac{[H^+]^2 \times [X^{2-}]}{[H_2X]} = K_3$

17.42 pH = 4.0, 3.68

17.43 HZ < HY < HX; 1×10^{-5}, 1×10^{-3}

17.44 0.1, 3×10^{-3}, 1.6×10^{-12}, 3×10^{-3}, 2.5

17.45 Acid: HOOC—CH$_2$—NH$_3{}^+$, base: NH$_2$—CH$_2$—COO$^-$; appr. 7

CHAPTER 18

18.20 All the same.

18.21 **a.** $OH^-(aq) + HCHO_2(aq) \rightarrow H_2O + CHO_2{}^-(aq)$
  b. $OH^-(aq) + NH_4{}^+(aq) \rightarrow NH_3(aq) + H_2O$
  d. $HCN(aq) + OH^-(aq) \rightarrow CN^-(aq) + H_2O$
  e. $HF(aq) + CN^-(aq) \rightarrow HCN(aq) + F^-(aq)$
  f. $H^+(aq) + OH^-(aq) \rightarrow H_2O$

18.22 **a.** 4.5×10^{10} **b.** 1.8×10^{-6}; 8.27

18.23 **a.** 5.6×10^4 **b.** 5.6×10^4 **c.** 2.5×10^8 **d.** 1×10^9

18.24 0.07096

18.25 0.02, 0.06, 5.2; yes

18.26 20 M

18.27 0.5243

18.28 Blue, blue, purple, red, red; pH = 4; NH$_3$ + HBr

18.29 **a.** 0.4 N   **b.** 0.8 N   **c.** 1.2 N   **d.** 0.4 N

18.30 **a.** 2.72   **b.** 4.74 (buffer)   **c.** 7.44   **d.** 8.44   **e.** 9.30
  f. 10.30   **g.** 12.60   **h.** 12.82 (buffer)

18.31 **a.** A   **b.** A   **c.** N   **d.** A; easiest: HNO$_3$—KOH; most difficult: NaC$_2$H$_3$O$_2$—HCl

18.32 **a.** 9.25   **b.** 9.11   **c.** 9.32   **d.** 4.74

18.33 2000; 7.68; 6.8

18.34 **a.** Add NaOH to ppt. Mn(OH)$_2$; filter, treat with H$_2$SO$_4$, evaporate.
  b. Heat with NaOH.
  c. Add HNO$_3$, evaporate.
  d. Add Na$_2$CO$_3$, filter.
  e. Heat with H$_2$SO$_4$.
  f. Neutralize with HI, evaporate.

18.35 Heat with base to convert NH$_4{}^+$ to NH$_3$; add excess acid to convert
  CO$_3{}^{2-}$ to CO$_2$, add Ba(NO$_3$)$_2$ to ppt. BaSO$_4$, leaving Cl$^-$ in solution.

18.36 **a.** $CO_2(g) + 2\,OH^-(aq) \rightarrow CO_3{}^{2-}(aq) + H_2O$, and
     $CO_2(g) + OH^-(aq) \rightarrow HCO_3{}^-(aq)$
  b. $HCO_3{}^-(aq) + H^+(aq) \rightarrow CO_2(g) + H_2O$
  c. $NH_4{}^+(aq) + OH^-(aq) \rightarrow NH_3(g) + H_2O$
  d. $Cu^{2+}(aq) + H_2S(g) \rightarrow CuS(s) + 2\,H^+(aq)$

18.37 2 × 18.28 + 18.29 + 18.30 + 18.31

18.38 ZnS, NiS, CoS, FeS, MnS; virtually all

18.39 Yes

18.40 0.3 mole

18.41 **a.** —COOH; yes   **b.** 6.06   **c.** Yes; 2.3, 9.8

CHAPTER 19

19.21 All +2

19.22 a < d < c < b; replacement of anion by water would increase conductivity.

19.23 $[Co(NH_3)_4ClI]^+$; 1.025 g

19.24 **a.** Stronger base   **b.** Higher charge   **c.** Entropy effect and stability of ring   **f.** Forms stable complex with heme

19.25 **a.** Test chelating ability   **b.** Determine how many other ligands are present by analysis.

19.26 Two positions open above and below square.

19.27 9.06%

19.28 **a.** Octahedral   **b.** Square planar
c. Octahedral with Cl^- at two adjacent corners.
d. Linear   **e.** Octahedral   **f.** Tetrahedral

19.29 **a.** 2   **b.** 2   **c.** 5   **d.** 3   **e.** 2   **f.** 1

19.30

	3d	4s	4p	
Cu^+	(↿⇂)(↿⇂)(↿⇂)(↿⇂)(↿⇂)	(↿⇂)	(↿⇂)( )( )	0
Cu^{2+}	(↿⇂)(↿⇂)(↿⇂)(↿⇂)(↿⇂)	(↿⇂)	(↿⇂)(↿⇂)(↿)	1
Ni^{2+}	(↿⇂)(↿⇂)(↿⇂)(↿⇂)(↿⇂)	(↿⇂)	(↿⇂)(↿⇂)( )	0
Ni^{2+}	(↿⇂)(↿⇂)(↿⇂)(↿)(↿)	(↿⇂)	(↿⇂)(↿⇂)(↿⇂)	2
Cr^{3+}	(↿)(↿)(↿)(↿⇂)(↿⇂)	(↿⇂)	(↿⇂)(↿⇂)(↿⇂)	3

	4d	5s	5p	
Ag^+	(↿⇂)(↿⇂)(↿⇂)(↿⇂)(↿⇂)	(↿⇂)	(↿⇂)( )( )	0

	3s	3p		
Al^{3+}	(↿⇂)	(↿⇂)(↿⇂)(↿⇂)		0

	3s	3p	3d	
Al^{3+}	(↿⇂)	(↿⇂)(↿⇂)(↿⇂)	(↿⇂)(↿⇂)( )( )( )	0

19.31 Valence bond: see above structure. Crystal field predicts only one structure with one electron in each of three lower orbitals.

19.32 Fe^{2+}, Fe^{3+}, Co^{2+}, Co^{3+}, Pt^{4+}

19.33 4

19.34 See Table 19.4.

19.35 Follow conductivity or color; titrate with Ag^+ from time to time.

19.36 4×10^{-3}

19.37 **b.** $Al^{3+}(aq) + 3\ OH^-(aq) \rightarrow Al(OH)_3(s)$;
$Al(OH)_3(s) + 3\ OH^-(aq) \rightarrow Al(OH)_6^{3-}(aq)$
c. $AgBr(s) + 2\ NH_3(aq) \rightarrow Ag(NH_3)_2^+(aq) + Br^-(aq)$
$Ag(NH_3)_2^+(aq) + 2\ H^+(aq) + Br^-(aq) \rightarrow AgBr(s) + 2\ NH_4^+(aq)$

19.38 5×10^{-13}; no

19.39 **a.** Treat with soln $S_2O_3^{2-}$.   **b.** Dissolve in NH_3, add I^-.
c. Add strong acid.   **d.** Add increasing amounts OH^-.

19.40 130 ml

19.41 **a.** Same   **b.** 0.4Δ, 0.8Δ, 1.2Δ, 1.2Δ, 0.6Δ, 0

19.42 **a.** 4E   **b.** 2Δ **c.** Low spin

19.43 $Ni^{2+}(aq) + 2\ NH_3(aq) + 2\ H_2O \rightarrow Ni(OH)_2(s) + 2\ NH_4^+(aq)$
$Ni(OH)_2(s) + 6\ NH_3(aq) \rightarrow Ni(NH_3)_6^{2+}(aq) + 2\ OH^-(aq)$
$Al^{3+}(aq) + 3\ NH_3(aq) + 3\ H_2O \rightarrow Al(OH)_3(s) + 3\ NH_4^+(aq)$
$Al(OH)_3(s) + 3\ OH^-(aq) \rightarrow Al(OH)_6^{3-}(aq)$
$Al(OH)_6^{3-}(aq) + 3\ H^+(aq) \rightarrow Al(OH)_3(s) + 3\ H_2O$
$Al(OH)_3(s) + 3\ H^+(aq) \rightarrow Al^{3+}(aq) + 3\ H_2O$

CHAPTER 20

20.21 a. $+1, -1$ b. $+1, +4, -2$ c. $+2, -2$ d. $+2, -3, +1, -1$
       e. -1 f. $+1, +6, -2$ g. $+2, +6, -2$

20.22 $+6: SO_3, SO_4^{2-}$ $+4: SO_2, SO_3^{2-}$ $+2: SF_2, S_2O_3^{2-}$
       $-2: H_2S, S^{2-}$

20.23 a. O, R   b. O   c. R   d. O, R   e. O   f. O, R

20.24 $KF, K_2S, SF_2, SF_4, SF_6$

20.25 a. R, O, R, O
     b. $Pb(s) + SO_4^{2-}(aq) \rightarrow PbSO_4(s) + 2\ e^-$
        $2\ Cl^-(aq) \rightarrow Cl_2(g) + 2\ e^-$
        $Cd^{2+}(aq) + 2\ e^- \rightarrow Cd(s)$
        $Cr_2O_7^{2-}(aq) + 14\ H^+(aq) + 6\ e^- \rightarrow 2\ Cr^{3+}(aq) + 7\ H_2O$
     c. $Pb(s) + Cd^{2+}(aq) + SO_4^{2-}(aq) \rightarrow PbSO_4(s) + Cd(s)$
        $3\ Pb(s) + 3\ SO_4^{2-}(aq) + Cr_2O_7^{2-}(aq) + 14\ H^+(aq) \rightarrow$
           $3\ PbSO_4(s) + 2\ Cr^{3+}(s) + 7\ H_2O$
        $2\ Cl^-(aq) + Cd^{2+}(aq) \rightarrow Cl_2(g) + Cd(s)$
        $6\ Cl^-(aq) + Cr_2O_7^{2-}(aq) + 14\ H^+(aq) \rightarrow$
           $3\ Cl_2(g) + 2\ Cr^{3+}(aq) + 7\ H_2O$

20.26 a. R, O, R, O
     b. $Cl_2(g) + 12\ OH^-(aq) \rightarrow 2\ ClO_3^-(aq) + 6\ H_2O + 10\ e^-$
        $2\ Ag(s) + 2\ OH^-(aq) \rightarrow Ag_2O(s) + H_2O + 2\ e^-$
        $Cl_2(g) + 2\ e^- \rightarrow 2\ Cl^-(aq)$
        $O_2(g) + 2\ H_2O + 4\ e^- \rightarrow 4\ OH^-(aq)$
     c. $3\ Cl_2(g) + 6\ OH^-(aq) \rightarrow 5\ Cl^-(aq) + ClO_3^-(aq) + 3\ H_2O$
        $2\ Cl_2(g) + 4\ OH^-(aq) + 5\ O_2(g) \rightarrow 4\ ClO_3^-(aq) + 2\ H_2O$
        $2\ Ag(s) + 2\ OH^-(aq) + Cl_2(g) \rightarrow Ag_2O(s) + H_2O + 2\ Cl^-(aq)$
        $4\ Ag(s) + O_2(g) \rightarrow 2\ Ag_2O(s)$

20.27 a. $Pb(s) + PbO_2(s) + 4\ H^+(aq) + 2\ SO_4^{2-}(aq) \rightarrow 2\ PbSO_4(s) + 2\ H_2O$
     b. $3\ Se(s) + 4\ NO_3^-(aq) + 4\ H^+(aq) \rightarrow 3\ SeO_2(s) + 4\ NO(g) + 2\ H_2O$
     c. $2\ Sn^{2+}(aq) + 2\ H_2O + O_2(g) \rightarrow 2\ SnO_2(s) + 4\ H^+(aq)$
     d. $2\ Al(s) + 6\ OH^-(aq) + 6\ H_2O \rightarrow 2\ Al(OH)_6^{3-} + 3\ H_2(g)$
     e. $2\ MnO_4^-(aq) + H_2O + 3\ CN^-(aq) \rightarrow$
        $2\ MnO_2(s) + 3\ CNO^-(aq) + 2\ OH^-(aq)$

20.28 $4\ Co^{2+}(aq) + 24\ NH_3(aq) + O_2(g) + 2\ H_2O \rightarrow$
     $4\ Co(NH_3)_6^{3+}(aq) + 4\ OH^-(aq)$

20.29 a. $2\ I^-(aq) + SO_4^{2-}(aq) + 4\ H^+(aq) \rightarrow I_2(aq) + SO_2(g) + 2\ H_2O$
     b. 2.6 ml   c. 100 ml

20.30 a. $CH_4(g) + 2\ OH^- + 2\ O_2(g) \rightarrow CO_3^{2-} + 3\ H_2O$
     b. Four liters

20.31 Reverse; measure change in weight of Ni electrode.

20.32 a. Get Mg instead of Na.
     b. Al oxidizes at top of cell.
     c. Would not obtain NaOH.
     d. Contaminate with Zn^{2+}, decrease conc. Cu^{2+}.

20.33 a. 0.0241   b. 2.33×10^3 c. 0.0243, 0.856, 0.410

20.34 a. 1.01   b. 3.04 lit

20.35 a. 0.60   b. 2900 sec

20.36 a. 2.80 g   b. 7.00 g   c. 4.67 g

20.37 a. 49.0 g   b. $+4$

20.38 **a.** Co anode surrounded by Co^{2+}; Ag cathode surrounded by Ag^+. Electrons move from Co to Ag.

b. Pb anode, PbO_2 cathode in H_2SO_4 electrolyte. Electrons move from Pb to PbO_2.

c. Pt anode surrounded by Sn^{2+}, Sn^{4+}. Pt cathode surrounded by Cl_2, Cl^-. Electrons move from anode to cathode.

20.39 **a.** $Pb(s) + PbO_2(s) + 4\ H^+(aq) + 4\ Cl^-(aq) \rightarrow 2\ PbCl_2(s) + 2\ H_2O$

b. $Zn(s) + MnO_2(s) + 4\ NH_4^+(aq) \rightarrow$
$Zn^{2+}(aq) + Mn^{2+}(aq) + 4\ NH_3(aq) + 2\ H_2O$

20.40 Voltaic cell: Zn anode, Pt cathode (H_2, H^+). Electrolyte cell: Ag anode, Cu cathode. Connect Zn to Cu, Pt to Ag.

20.41 23.1 kcal/volt

20.42 7.13×10^6

20.43 0.12

CHAPTER 21

21.21 **a.** -1.90 **b.** 0.78   **c.** 0.52   **d.** 0.97

21.22 **a.** Fe^{3+}, Fe^{2+} **b.** $Mn(s)$, Mn^{2+}

21.23 Ions below Mn^{2+}; conc. Cl^- high enough to give Cl_2.

21.24 $Al > Zn > Co > Pb > Ag > Au$

21.25 **a.** NO_3^-, but not H^+ strong enough to oxidize Cu.

b. React to form $Fe^{2+} + Sn^{4+}$.

c. S.O.P. $F^- <$ S.O.P. H_2O

21.26 **a.** $Cu^{2+}(aq) + H_2(g) \rightarrow Cu(s) + 2\ H^+(aq)$

b. $Co^{3+}(aq) + \frac{1}{2}\ H_2(g) \rightarrow Co^{2+}(aq) + H^+(aq)$

21.27 **a.** $H_2O_2(aq) + 2\ H^+(aq) + 2\ I^-(aq) \rightarrow 2\ H_2O + I_2(s)$

b. (a) and $H_2O_2(aq) + 2\ H^+(aq) + 2\ Fe^{2+}(aq) \rightarrow 2\ H_2O + 2\ Fe^{3+}(aq)$

21.28 **a.** -20 kcal   **b.** -70.3 kcal   **c.** $+16$ kcal

21.29 3.1×10^{-4}, -0.034, $+4.7$ kcal

21.30 **a.** $K = 10$ **b.** 0.83

21.31 5×10^{-7}

21.32 0.034 V

21.33 **a.** 0.35 V   **b.** 0.12 V

21.34 **a.** 0.5   **b.** 4.6

21.35 5×10^{-8}

21.36 **a.** Measure E of cell consisting of standard half cell connected to Pb-Pb^{2+} half cell in which conc. $Cl^- = 1$ M.

b. Measure E of cell consisting of standard half cell connected to H_2-H^+ half cell containing 1 M NaOH.

21.37 **a.** React Cu with HNO_3.   **b.** React Sn with HNO_3.

c. Heat $KClO_3$ below mpt.

21.38 **a.** $Cl^-(aq) + H_2O \rightarrow H_2(g) + ClO^-(aq)$

b. $Fe(s) \rightarrow Fe^{2+}(aq) + 2\ e^-$

c. $3\ Ag(s) + NO_3^-(aq) + 4\ H^+(aq) \rightarrow 3\ Ag^+(aq) + NO(g) + 2\ H_2O$

d. $Cr_2O_7^{2-}(aq) + 14\ H^+(aq) + 6\ I^-(aq) \rightarrow 2\ Cr^{3+}(aq) + 7\ H_2O + 3\ I_2(s)$

21.39 **a.** S.O.P. Au too negative.

b. Fe-Cu couple promotes corrosion.

c. Al_2O_3 adheres to surface.

d. O_2 conc. cell set up.

      e. Form insoluble oxide or hydroxide.

21.40   **a.** $2 MnO_4^-(aq) + 5 C_2O_4^{2-}(aq) + 16 H^+(aq) \rightarrow$
        $2 Mn^{2+}(aq) + 10 CO_2(g) + 8 H_2O$

      b. 0.263 M

21.41 1.6×10^{-3} M

21.42   0.47, 0.47, 0.47, 0.53, 0.59, 0.65

21.43   **c.** $+0.04$ V, -0.005 V

21.44 1×10^{-7} M

CHAPTER 22

22.20   **a.** Would be radioactive.   **b.** K electrons closer to nucleus.
      c. See p. 605.

22.21   **a.** Half life C-14 too short.
      b. Would absorb energy (Figure 22.6).
      c. Would contaminate ocean for long period.
      d. Repulsion by extranuclear electrons.

22.22   **a.** $^{230}_{90}Th \rightarrow ^{4}_{2}He + ^{226}_{88}Ra$ **b.** $^{234}_{91}Pa \rightarrow ^{0}_{-1}e + ^{234}_{92}U$
      c. $^{187}_{79}Au + ^{0}_{-1}e \rightarrow ^{187}_{78}Pt$ **d.** $^{84}_{39}Y \rightarrow ^{0}_{1}e + ^{84}_{38}Sr$

22.23   **a.** 90p, 90e, 140n $\rightarrow$ 88p, 90e, 138n
      b. 91p, 91e, 143n $\rightarrow$ 92p, 91e, 142n
      c. 79p, 79e, 108n $\rightarrow$ 78p, 78e, 109n
      d. 39p, 39e, 45n $\rightarrow$ 38p, 39e, 46n

22.24   **a.** $^{54}_{26}Fe + ^{4}_{2}He \rightarrow ^{1}_{0}n + ^{57}_{28}Ni$ **b.** $^{9}_{4}Be + ^{0}_{0}\gamma \rightarrow ^{1}_{0}n + ^{8}_{4}Be$
      c. $^{45}_{21}Sc + ^{4}_{2}He \rightarrow ^{1}_{1}H + ^{48}_{22}Ti$ **d.** $^{235}_{92}U + ^{2}_{1}H \rightarrow ^{231}_{93}Np + 6\,^{1}_{0}n$

22.25   **a.** $^{239}_{94}Pu + ^{1}_{0}n \rightarrow ^{132}_{50}Sn + 5\,^{1}_{0}n + ^{103}_{44}Ru$
      b. $^{253}_{99}Es + ^{4}_{2}He \rightarrow ^{1}_{0}n + ^{256}_{101}Md$
      c. $^{6}_{3}Li + ^{6}_{3}Li \rightarrow ^{12}_{6}C$

22.26 $^{209}_{83}Bi$

22.27   **a.** 0.132 yr^{-1} **b.** 8.8 mg

22.28   Aug. 14

22.29 3.2×10^9, 3.5×10^9, 3.5×10^9; Ar(g) may have escaped.

22.30   5.1 yr

22.31   **a.** p   **b.** e   **c.** e   **d.** p

22.32   **a.** Equilibrate AgCl containing radioactive isotope of Ag or Cl with
      water, measure radioactivity in solution.
      b. Use CH_3OH containing O-18; see whether it shows up in water
      or methyl acetate.

22.33   **a.** -0.0210 amu, -19.6 MeV   **b.** -2150 kcal

22.34   26.97989

22.35   **a.** Spont.   **b.** Nonspont.   **c.** Spont.

22.36   0.1371 amu, 127.6 MeV, 7.98 MeV

22.37 1.96×10^7 kcal, 1.69×10^7 kcal

22.38   **a.** C-12, O-16, Mg-24, Si-28, S-32, Ca-40
      b. -1.40×10^7, -3.41×10^7

22.39 4.18×10^{-9}

22.40   1.00 g

22.41   **a.** 3.3×10^6 kcal   **b.** 2.6×10^8 cm/sec   **c.** 5.5×10^8 °K

22.42   31.3%

22.43   **a.** -1.38×10^8 **b.** 2.8×10^{31} kcal   **c.** 2×10^{-15}

CHAPTER 23

23.16 **a.** See p. 627.   **b.** p. 631   **c.** p. 644   **d.** p. 640
e. p. 629   **f.** p. 638

23.17 4 asymmetric carbons, 2 arrangements around each, so 2^4 isomers.

23.18 **a.** 4   **b.** 4   **c.** 0   **d.** 8

23.19 Interchange H and OH in position 2 in drawing of α and βD-glucose on p. 626. 5 asymmetric carbons. Test for C=O group.

23.20 + pole: Asp, Glu; − pole: Lys, His; others would not move.

23.21 Asp $\approx$ Glu < Ile $\approx$ Gly $\approx$ Ala < His < Lys

23.22 1200; $Ala_1Gly_3Val_2Ile_1Pro_4Ser_7$

23.23 **a.** Relatively large R group which cannot acquire charge.
b. Small R group.
c and **d.** Large R groups which are charged at pH 7.

23.24 68000

23.25 Check effects of pH, temperature (would expect decrease in rate at high T).

23.26 In Figure 23.8, put guanine molecule for Base 1, cytosine for Base 2, uracil for Base 3 (delete lowest unit). In each case, form link from sugar to N at bottom of molecule in Figure 23.9, removing H atom. Would code for alanine.

23.27 3.5×10^7

23.28 See pp. 648–655.

23.29 1.1×10^{-4} cm

23.30 **a.** AAG   **b.** CCA   **c.** AAA   **d.** CAT

23.31 Must be three units long.

23.32 **a.** C $\rightarrow$ T   **b.** G $\rightarrow$ A   **c.** T $\rightarrow$ C   **d.** T $\rightarrow$ C

INDEX

Page numbers in *italics* refer to illustrations; those followed by (t) refer to tables.

A

Absolute zero, 86, 97
Absorption spectrum, 18–19, 19(t)
Acetaldehyde, 220(t)
Acetic acid, 220(t)
  dissociation of, 454
  tritration of, 448
Acetone, 220(t)
Acetylcholine, 386
Acetylene, 60, 215
  bonding orbitals in, 189
  triple bond in, 175
Acid(s), 220(t). See also names of specific
    acids and *Acid-base reactions.*
  amino, 629, 630(t)
  Arrhenius concept of, 469
  Brönsted-Lowry, 469, 471(t), 509
  conjugate, 460
  Lewis, 472, 506
  neutralization of, 477
  organic, 221
  strength of, 452, 457
    determination of, 454
  strong, 451–452, 452(t)
    reaction with solid, 481
    reaction with strong base, 447
    reaction with weak base, 480
    solubility of basic salts in, 482
    titration with strong base, *487*
  volatility of, 494(t)
  weak, 453
    reaction with strong base, 478
    titration with strong base, *488, 489*
Acid-base indicator, 450, 451(t), 485
  choice of, 486
  universal, 451
Acid-base reactions, 477–505
  Brönsted-Lowry concept of, 469
  characteristics of, 481(t)
  end point of, 485
  equivalence point of, 486
  in inorganic synthesis, 493
  in qualitative analysis, 495
  indicators in, 450, 485

Acid-base reactions *(Continued)*
  Lewis concept of, 472
  Solvay Process, 489
  types of, 477–483
Acid-base strength, factors in, 465, *468*
Acid-base titration, 483
Acid dissociation constant, 456, 462(t)
Acrolein, 201, 420
Actinides, 154
Activated carbon, 227, 311
Activated complex, 383
Activation analysis, 608
Activation energy (E_a), 382, 387
Activity coefficient, 439
  equation for, 447
Addition polymers, 223, 223(t)
Air, composition of, 398(t)
  liquified, 399
  pollution of, 339, 416–422
Alcohols, 220(t). See also names of specific
    alcohols.
Aldehyde, 220(t)
Aldohexoses, 626
Alizarine yellow, 450
Alkali metals, 151
Alkenes, 214
Alkyd resins, 217
Alkynes, 214
Allotrope, 226
Alpha (α) particles, 46, 596
Aluminum, 21, 21(t)
  from aluminum oxide, 547
Aluminum nitride, 401
Aluminum oxide, 55–56
Aluminum sulfate, 430
Amines and amides, 224
α-Amino acids, 629, 630(t)
Ammonia, 31, 43(t), 46, 61, 264(t)
  as complexing agent, 506, 526
  as solvent, 471
  basic properties of, 463
  detection of, 496
  formation of, 401, 402–403
  in Solvay Process, 499, 500
  properties of, 292(t)

Ammonia *(Continued)*
  structure of, 182, *183*
  titration of, 490
Ammonium ion, acidic property of, 454
Ampere, 552
Amphoteric compounds, 526
Analysis. See also *Qualitative analysis*
    and *Quantitative analysis.*
  anion, 496
  gravimetric, 440
  neutron activation, 608
  redox reactions in, 589
  volumetric, 440
Angstrom, definition of, 3
Angular momentum quantum number, 138
Anhydrides, 467, 469
Anion, 32
Anion analysis, 496
Anode, 545
Antibonding orbital, 191
Anticodon group, 652, *653*
Antifreeze, 280
Aqua regia, 589
Argon, 407
  separation from air, 399
Aromatic hydrocarbons, 216
Arrhenius, Svandte, 298, 382, 469
Arsenic, in semiconductors, 257
Arsenic(III) sulfide, 436
Asbestos, *50*, 229, 230
Aspirin, 22, 58, 62
Atmosphere, as unit of measurement, 92
  composition of, 398(t)
  mass of, 398
  upper, 411, *412*, *413*
Atmospheric pressure, 89
Atom(s), 25–30
  absolute mass of, 38–39
  components of, 27
  electron cloud in, *136*, 137
  electron structure of, 140–146
  properties of, 26
  quantum state of, 139
  radii of, see *Atomic radii.*
Atomic bomb, 616
Atomic mass unit (amu), 22, 610
Atomic number, 30, 121
Atomic orbital theory, 169
  for complex ions, 517–519
Atomic orbitals, hybrid, 186, 188(t)
Atomic radii, 156–158, 660(t)
  and complexing ability, 508–509
  and electron configuration, 156–158
  and hydrogen bonding, 207
Atomic spectra, 124
Atomic theory, 25–27
Atomic weight, 33–34
  determination of, 34–38
Aufbau principle, 143
Avogadro, Amadeo, 42
Avogadro's Law, 98
Avogadro's number, 38

B

Bakelite, 217, 232
Baking soda, 53
Balance, *4*

Balancing of equations, 54
Balmer, J. J., 126
Balmer series, 127, 127(t)
Band theory, 199
Barbituric acid, 484
Barium, 21, 21(t)
Barium bromide, 430
Barium carbonate, 481
Barium sulfate, 430
  preparation of, 444
Barometer, 90, *91*
Bartlett, Neil, 408
Base(s). See also *Acid-base reactions.*
  Arrhenius concept of, 469
  Brönsted-Lowry, 469, 471(t)
  conjugate, 460
  Lewis, 472, 506
  neutralization of, 477
  strength of, 461, 462(t)
    reaction with weak acid, 478
    reaction with strong acid, 477
  strong, 451–452, 452(t)
  weak, 460
    reaction with strong acid, 480
    titration with strong acid, *490*
Base dissociation constant, 461, 462(t)
  calculations using, 464
Basic salts, 462
Basicity, 449
Battery, solar, 259
  storage, 557, *558*
Bauxite, 547
Becquerel, Henri, 596
Benzene, 216, 244(t)
  as solvent, 271(t)
  bonding in, *217*
  melting point of, 253
  reaction with nitric acid, 57
Benzoic acid, 59, 85, 252
1,2-Benzopyrene, 217
Berthelot, P. M., 322
Berthollet, C. L., 26
Berzelius, J. J., 41
Beta (β) particles, 29, 596
Bicarbonate buffer, 492
Binding energy, 612(t), 613
Biochemical oxygen demand (BOD), 302
Biochemistry, 624
Bjerrum, Niels, 300
Bleach, 581
Body-centered cubic cell, *246*, 247
Bohr, Niels, 127
Bohr model of the atom, *128*
Bohr theory, 127–130
  failure of, 130
Boiling point, 18, 204, 205(t)
  normal, 241, 242(t)
Boiling point elevation, 278, 278(t)
Bombardment, nuclear, 603
Bond(s). See also *Covalent bond, Ionic bond,*
    and *Metallic bond.*
  angle of, 181–183
  electron-sea model, 197
  enthalpy of formation of, 72, 173
  hybrid, 186
  hydrogen, 207
  length of, 173
  multiple, 174
  peptide, 631

Bond(s) *(Continued)*
  pi, 189
  sigma, 189
  types of, 162
Bond energy, 72, 173
Bond formation, heat of, 71, 72(t), 173
Bonding orbital, 191
Boric acid, 60
Boron, 44, 45
  in semiconductors, 257
Boron nitride, 401
Boron trifluoride, as Lewis acid, 472
Borosilicate glass, 229
Boyle, Robert, 94, 95
Boyle's Law, 93
Bragg, W. H., 245
Bragg, W. L., 245
Bragg equation, 245
Breeder reactor, 618
Bromine, 19, 19(t), 23
Bromthymol blue, 485
Brönsted, J. N., 469
Brönsted-Lowry acids and bases, 471(t), 509
Brönsted-Lowry concept, 469
  in non-aqueous solvents, 471
Buffers, 490–493
Butadiene, 232
Butane, 85, 214
Butene, 214
n-Butyl alcohol, solubility in water, 270
Butyric acid, 474

C

Calcium, 20, 21, 21(t)
Calcium carbonate, 62, 65, 65, 312, 347
Calcium chloride, 296
Calcium oxide, 61, 338, 347
Calcium sulfate, 312
Calorie, 63
Calorimeter, 74, 75, 76
Camphor, 60
  freezing point constant of, 278(t), 280, 285
Cannizzaro, Stanislao, 43
Capillary action, 234
Caproic acid, 456
Carbohydrates, 625–628
Carbolic acid, 217
Carbon, 21(t)
  activated, 227
  allotropic forms of, 226
  bond strength of, 210
  electron configuration of, 144(t)
  organic compounds of, 210–225
Carbon black, 227
Carbon dioxide, 50, 61, 347
  as Dry Ice, 252, 409
  in Solvay Process, 499
  preparation of, 409
Carbon-14 isotope, 602–603
Carbon monoxide, 22, 23, 46, 354, 512
  as pollutant, 418
  formation of, 406
  heat of formation of, 69, 69
Carbon tetrachloride, 219
  symmetry of, 185

Carbonates, detection of, 496
Carbonic acid, 479
  two-step dissociation of, 454
Carbonyl group, 221
Carothers, Wallace, 225
Catalyst, 384
  for Haber process, 403
Catalytic muffler, 422
Cathode, 545
Cathode ray tube, 28
Cathodic protection, 588
Cation(s), 32
  as bombarding particles, 604
  electron structures of, 164(t)
  separation of, 443
Cation exchange, 313
Cavendish, Henry, 407
Cell, Downs, 546
  dry, 556, 557
  electrolytic, 544, 545
  voltaic, 21, 554
Cell voltage, calculation of, 567
  effect of concentration on, 575–580
Cellulose, 627
Celsius degree, definition of, 6
Centimeter, definition of, 2
Cesium chloride, crystal structure of, 249, 250
Chain reaction, chemical, 390
  nuclear, 616
Chair conformation, 626
Charles, Jacques, 96, 97
Charles and Gay-Lussac's Law, 95
Chelating agents, 510
Chemical thermodynamics, 371
Chloramines, 311
γ-Chlordane, 16
Chloride ion, 32
Chlorinated hydrocarbons, 305
Chlorine, 21(t), 31, 44
  as disinfectant, 310
  as oxidizing agent, 580
  from brine, 549
Chloroform, 102, 219
  chlorination of, 391
  polarity of, 185
Chlorophyll, 511
Chromatography, 14–17
  column, 15, 15
  in protein separation, 632
  paper, 15
  thin layer, 16, 17
  vapor phase, 17
Chromium, 21(t)
Cinnabar, 21
Cis isomer, 215, 513
Citric acid, 221, 386
Clapeyron, B. P. E., 240
Clausius, Rudolph, 240
Clausius-Clapeyron equation, 240
Cloud chamber, 598, 598
Clouds, electron, 137, 146, 208
  formation of, 410
Coagulation, 309
Coal, 78, 307
Coal tar, 217
Cobalt sulfide, 496
Colligative properties, 276
  of acids, 455

Colligative properties *(Continued)*
  of electrolytes, 296
Collision number (Z), 387
Collision theory, 387
Combining Volumes, Law of, 105
Combustion train, *50*
Common ion effect, 435
Complex, activated, 383
Complex ion(s), 506–531
  bonding in, 517–523
  charge of, 507
  colors of, 522
  coordination number of, 512
  crystal field theory of, 519
  dissociation constants of, 525 (t)
  equilibria of, 524
  high and low spin, 521
  in water solution, 521
  ligands in, 509
  rate of formation, 523
  structure of, 512–516
  valence bond theory of, 517
Compound(s), amphoteric, 526
  coordination, 506–534
  inorganic, nomenclature of, 661
  ionic, 431–434
  macromolecular, 225–230
  organic, 210–225. See also *Organic compounds.*
Concentrated solution, 264, 264(t)
Concentration, bracket symbol for, 349
  change of, by dilution, 268
  effect on cell voltage, 575–580
  measurement using Nernst equation, 578
  units of, 265–268
Conductivity, electrical and thermal, 196, 197(t)
Configuration, electron, 140–146
  and atomic radius, 156
  and chemical properties, 151
  and periodic table, 150–156
  noble gas, 163
  table of, 144–145
Conformation, protein, 638
Congo Red, 503
Conjugate acid and base, 460
Contact process, 407
Conventions, thermochemical, 64
Conversion factors, 8(t), 23, 414, 659
  mass-energy, 610(t)
  use of, 9–11
Cooling tower, 308, *309*
Coordinate covalent bond, 472, 506
Coordination compounds, 506–534
Coordination number, in complex ion, 507, 512, 512(t)
  in crystal, 248
Copper, 21, 45, 47
  atomic radius of, 22
  crystal structure of, 247
  detection of, 525
Copper oxides, 405
Copper sulfate, 19, 506
Copper sulfide, 62, 496
Corrosion, iron, 585–588
  protection from, 588
Coulomb's Law, 156
Covalent bond(s), 162, 167
  atomic orbital theory of, 169

Covalent bond(s) *(Continued)*
  coordinate, 472, 506
  electron densities in, *167*
  electronegativity and, 170, 171(t)
  length of, 173
  molecular orbital theory of, 170
  polarity of, 170
  properties of, 170–175
  stability of, 168
Cracking of hydrocarbons, 214
Cresol red, 503
Cresols, 217
Crick, F., 648
Critical mass, 616
Critical temperature and pressure, 242–243, 243(t)
Cryolite, 547
Crystal(s), *33*
  defects in, 255–259
  ionic, 162, *163, 198*
  macromolecular, 226
  protein, 638
  quartz, 228
Crystal-field model, 519–523
Crystallization, 265
  fractional, 13–14
Cubic closest-packing structure, *248, 249*
Curie, Irene, 603
Curie, Marie and Pierre, 596
Cyanides, use in electroplating, 550
Cyclotron, 604, *604*
Cystine, 59
Cytochrome c, 637, 637(t)
  structure of, *641*

D

Dacron, 224
Dalton, John, 25, 40, 105
Davisson, C. J., 130
Davy, Humphry, 20
DDT, *16,* 22, 31, 305
  accumulation in food chain, 306
  solubility of, 271
de Broglie, Louis, 130
de Broglie wavelength, 130
Debye, Peter, 298
Debye-Hükel model, 298–299, 439
Decane, *209,* 210
Decay, radioactive, rate of, 600–603
Defects, in crystals, 255–259
Deionization, 317
Deliquescence, 296
Democritus, 25
Deoxyribonucleic acid (DNA), 644, 646–648
  composition of, 647(t)
  double helix model of, 648
  reproduction of, 649–650
Deoxyribose, 644
Desalination, 315–318
Detergents, 303
Deuterium, 119
  fusion of, 623
Diamagnetism, 193, 519
Diamond, 226, 227
p-Dichlorobenzene, *219,* 285

Dichlorodiphenyltrichloroethane, see *DDT*
Dichromate ion, 582–585
Dieldrin, *305*
Diethyl ether, 220(t), 221
Diffusion, 111–113
Dilute solution, 264, 264(t)
Dimethyl glyoxime, 526
Dimethyl mercury, 308, 549
Dinitrobenzene, 60
Dinitrogen pentoxide, 377
Dinitrogen tetroxide, 54, 63, 348
Dipeptide, 631
Dipole forces, 206
Dipole moment, 186
Disaccharide, 627
Disinfection, 310
Disorder, and entropy, 331
Dispersion forces, 208
Disproportionation reaction, 581
Dissociation, carbonic acid, 454
    degree of, *460*
    water, 448
Dissociation constant, acid, 455, 457, 462(t)
    base, 461, 462(t)
    complex ions, 525(t)
    water, 449
    weak acid, determination of, 457
Distillation, 11–13
    in desalination, 315
Distillation diagram, *13*
Disulfide bonds, 635
DNA, 644, 646–648
Dolomite, 302
Double bond, 174
Double helix model, 648, *649*
Downs cell, 546
Dry cell, 556, *557*
Dry Ice, 252, 409
Ductility, 197, 198
Dulong, Pierre, 41

E

E°, 566
E$_a$, 382
Earth, age of, 602
EDTA, 528
Efficiency, maximum, 81
Effusion, 112
Einstein, Albert, 125
Einstein's equation, 609
Electrical energy, from fuel cell, 325, 559
    from nuclear reactors, 617
    from solar battery, 259
    from water power, 82
Electrolysis, *324*, 544–553
Electrolyte(s), 295
    solutions of, 295–300
        colligative properties of, 296
        structure of, 298
Electrolytic cell, 544, *545*
Electron(s), 27–29
    charge on, 28, *28*
    emission of, 29, 197
    energy of, 122
    gain and loss of, 535

Electron(s) *(Continued)*
    mass of, 28
    mass-to-charge ratio of, 27
    quantum numbers of, 137–138
    valence, 175
Electron cloud, *136, 137, 146, 147, 148,* 208
Electron configuration, see *Configuration*
Electron levels, see *Levels*
Electron-pair bond, 167
Electron pairs, repulsion of, 181
Electron-sea model, 197
Electron spin, 138
Electron structure, see *Configuration*
Electron volt, 130
Electronegativity, 171, 171(t)
    acid strength and, 467
    partial ionic character and, 172
Electrophoresis, 632
Electroplating, 550, *551*
Electrostatic forces, 156, 300
Electrostatic precipitator, 425
Element(s), abundance of, 21, 21(t)
    chemical properties of, and electron configuration, 151
        periodic nature of, 154
    definition of, 20
    isolation of, *20*
    isotopes of, 30, 611(t)
    oxidation states of, 538, *539*
    transuranium, 605
Empirical formula, 48
End point, 486
Endothermic reactions, 63
Energy, activation, 382
    binding, 612(t), 613
    bond, 72, 173
    Einstein equation for, 609
    electrical, see *Electrical energy*
    electronic, 122
    fission, 616
    free, 326
    ionization, 130
    resonance, 179
    solar, 82
Energy conversion, fuel cell vs. steam turbine, 559(t)
Enthalpy, 64, 321
    of bond formation, 173, 195(t)
    of formation, 71
Enthalpy change (Δ H), and activation energy, 384
    sign of, 322
Entropy, and chemical reactions, 333
    and phase changes, 332
    definition of, 330
    interpretation of, 331
    molar, 331(t)
Entropy change (Δ S), 321, 329–334
Enzymes, 385, 643
Epsom salts, 60
Equations, chemical, 54. See also names of specific equations.
    interpretation of, 55–57
    ideal gas, 93
    net ionic, 430
    nuclear, 596
    oxidation-reduction, 540
    thermochemical, 63

Equilibrium, chemical, 347
    complex ion, 524
    liquid-vapor, 235
    solubility, 434–440
Equilibrium constant (K$_c$), 348, 351
    and free energy change, 364–367
    applications of, 353–357
    for sum of two reactions, 370
    general form of, 351
    relation to E°, 573
Equivalence point, 486
Equivalent, 484
Ester, 220(t)
Ethane, 31, 212
Ethanol, see *Ethyl alcohol*
Ether, 11, 220(t)
Ethyl alcohol, 220, 220(t)
Ethyl bromide, 62
Ethylene, 43(t), 394
    bonding orbitals in, 189
    double bond in, 174
    polymerization of, 222
    structure of, 183
Ethylene glycol, 280, 281(t)
Ethylenediamine, 510
Ethylenediaminetetraacetic acid, 528
Exclusion principle, Pauli, 139
Exothermic reactions, 63
Exponents, use of, 665–667
Extensive properties, 17

Fractional crystallization, 13–14
Fractional distillation, 11–13
Frank, Henry, 295
Free energy, definition of, 326
    of formation (ΔG_f), 328, 328(t)
Free energy change (ΔG), 326
    and equilibrium constant, 364–367
    at one atmosphere pressure ($\Delta G^{1\ atm}$), 327
    calculation at various temperatures, 336
    relation to E°, 572
    variation with T and P, 327(t)
Free path, mean, 415
Free radical, 180
Freeze-drying process, 252
Freezing, in desalination, 316
Freezing point, 253
    lowering of, 278(t), 279
    of electrolytes, 297(t)
Freon, 219
Frisch, O. R., 614
Fuel cell, *324*, 558, *560*
Fuels, consumption of, 79
    fossil, 78
    heating values of, 78(t)
    nuclear, 618
Functional group, 220
Fusion, 253
    heat of, 244, 244(t)
    nuclear, 618–619

F

Face-centered cubic cell, *246*, 247
Fahrenheit degree, definition of, 10
Faraday, 551
Faraday, Michael, 551
Fats, 221
Fatty acid, 222
Filtration, 310
Finreite, Norwald, 308
Fire extinguisher, carbon dioxide, 409
First Law of Thermodynamics, 66
First order radioactive decay, 600
First order reactions, 377
Fischer, Emil, 627
Fission, nuclear, 614–618
    products of, 615
Fixation of nitrogen, 401
Flerov, G. N., 606
"Flickering cluster" theory, 295
Flotation process, 423, *424*
Fluorine, 21(t)
    reactions of, with hydrogen, 536
    with xenon, 408
Fluorocitric acid, 386
Fog dispersal, 411
Foods, caloric values of, 86
Formaldehyde, 177, 221, 394
    as pollutant, 420
Formation, free energy of, 328
    heat of, see *Heat of formation*
Formic acid, 474
Formula, empirical, 48
    determination of, 49–52
    molecular, 48
    determination of, 52
Formula unit, 56

G

ΔG, 326
ΔG_f, 328
Galvanizing, 550(t)
Gamma rays, 597
Gas(es), diffusion of, 111–113
    intermolecular forces in, 108
    kinetic theory of, 109–117
    molecular speed in, 113–116
    noble, 407
        electron structure of, 163
    partial pressure of, 105
    pressure of, 91–95, 110
    real, 107
    solubilities of, 272(t)
Gas constant (R), 93, 99, 114
Gas Law, Ideal, 98
Gasoline, 46, 84
Gassendi, 25
Gauge pressure, 118
Gay-Lussac, J. L., 42, 96
Geiger-Müller counter, 598, *599*
Gene, 650
Genetic code, 651, 654, 655(t)
Geometric isomers, 214, 513, 515
Geometry, molecular, 181–186
    of complex ions, 512–516
Germanium, 155
    as semiconductor, 257
Germer, L. H., 130
Ghiorso, Albert, 606
Gibbs, J. Willard, 323, 341–342
Gibbs-Helmholtz equation, 331, 334
Gillespie, R. J., 181
Glass, manufacture of, 229
Glass electrode, 578

Globin, 511
Globular proteins, 640
Glucose, 625
  ring structure of, 626
Glycerol, 220–221
Glycine, 476, 504
Glycogen, 628
Gold, 21
  as catalyst, 384
  in sea water, 288
Graham, Thomas, 112
Graham's Law, 111
Gram, definition of, 4
Gram atomic weight, 34
Gram equivalent weight, 268, 484, 553
Gram molecular weight, 39
Graphite, 226
Gravimetric analysis, 440
Gravitational force, 24
Gravity cell, 555
Greenhouse effect, 409
Ground state, 125, 128

H

h, 123
ΔH_f, 68
ΔH_{fus}, 244
ΔH_{subl}, 253
ΔH_{vap}, 234
Haber, Fritz, 402
Haber process, 62, 363, 402
Hahn, O., 606, 614
Half cell, 556
Half-equation method, 541
Half-life, of chemical reaction, 379
  of radiactive decay, 601
Half-reactions, 535
Halides, organic, 218–219
Hall, Charles, 547
Hall process, 547, 548
Hard water, 311
Heat, of bond formation, 71, 72(t), 173
  of formation, 68, 70(t)
    definition of, 69
    enthalpy change and, 71
  of fusion, 244, 244(t)
  of sublimation, 253
  of vaporization, 234, 235(t), 242(t)
Heat capacity, 76
  molar, 199
Heat engine, 81
Heat flow (Q), 63
  at constant pressure (Q_p), 64, 76(t), 86
  at constant volume (Q_V), 76(t), 77, 86
Heat pollution, 308
Heating values of fuels, 78(t)
Heitler, W. A., 168
Helium, as eluting agent, 17
  as product of nuclear fusion, 618
  in liquid air, 399
α-Helix, 639, 640
Hematite, 339
Heme, 511, 511, 641
Hemoglobin, 511–512
Henry's Law, 275
Heptachlor, 16, 305
n-Heptane, properties of, 290(t)

Heroult, P. L. T., 547
Hess's Law, 68, 253
Hexagonal closest-packing structure, 248, 249
Hodgkin, Dorothy, 246
Homogeneity, 263
Hückel, Erich, 298
Humor (sick), see marginal comments
Hund, F., 170
Hund's rule, 142
Hybrid atomic orbitals, 186, 188(t), 517
Hydrazine, 54, 61, 85
Hydrocarbons, 210
  aromatic, 216
  chlorinated, 305
  pollutants, 418
  saturated, 212
  unsaturated, 214
Hydrochloric acid, 264(t), 452
Hydroelectric power, 82
Hydrofluoric acid, dissociation of, 453
Hydrogen, 21(t), 29, 31, 33
  atomic spectrum of, 127
  Bohr theory of, 127–130
  fusion of, 619
  liquid, 119
  molecular, bond energy in, 168, 168
  use in fuel cell, 559
Hydrogen bomb, 619
Hydrogen bonds, 207
Hydrogen chloride, 31, 43(t)
  preparation of, 495
Hydrogen cyanide, 43(t)
Hydrogen exchange reaction, 448
Hydrogen fluoride, 43(t), 536
  hydrogen bonding in, 207
  properties of, 292(t)
Hydrogen peroxide, 48, 61, 378
  preparation of, 404
Hydrogen sulfide, 22, 43(t), 61
  detection of, 496
Hydrophilic and hydrophobic groups, 629, 642
Hydroxyl group, 221
Hypochlorite ion, 177
Hypochlorous acid, 581

I

Ice, crystal structure of, 292, 293
  high-pressure forms of, 294
Ideal Gas Law, 93, 348
  use of, 99–105
Ideal solution, 277
Indicator, acid-base, 450, 451(t), 485
  choice of, 486
  universal, 451
Indophenol, 16
Inert complexes, 523
Infrared absorption spectrum, 19
Inhibitor, 386
  corrosion, 588
Initiation, chain, 390
Inner complex, 519
Insect control, 306–307
Insecticides, 305
Insulin, amino acid sequence in, 635, 636
Intensive properties, 17

Intermolecular forces, 204, 206
Iodine, 19, 250, 252, 311
   solubility in carbon tetrachloride, 270
   sublimation of, 252
Iodine chloride, 206
Ion(s), 32
   acid-base properties of, 463(t)
   common ion effect, 435
   complex, 506. See also *Complex ions.*
   electron structure of, 164(t)
   in electrolytic cells, 545
   in solutions, 295
   oxidation number of, 537
   polyatomic, 164(t)
   separation of, 496, 526
   sizes of, 165
Ion exchange, 317, *318*
Ion pairs, 300
Ionic atmosphere, 298, *299*
Ionic bonds, 162
   in complex ions, 519
Ionic character, partial, 172
Ionic compounds, 162
   charges of ions in, 163
   solubility of, 431–434
Ionic crystals, 162
Ionic radii, 660(t)
Ionization constant, 455. See also *Dissociation constant.*
Ionization energy, 130
   related to electron configuration, 149, *150*
Iron, 21(t), 84, 302
   atomic radius of, 22
   complex ions of, 518–521
   corrosion of, 585–588
   detection of, 525
   electron configuration of, 146
   production from ore, 339
Iron(III) hydroxide, 586
Isoelectric point, 629
Isomer, 626
   cis and trans, 215, 513
   complex ion, 515
   geometric, 214, 515
Isomerism, definition of, 212
Isopropyl alcohol, 287
Isotope(s), 30
   masses of, 611(t)
   notation, 30

J

Joliot, Frédéric, 603
Jorgensen, Sophus Mads, 529–531
Joule, 63
Junction, semiconductor, 257

K

Ka, 455
Kb, 461
Kc, 348, 351
Kp, 370
Ksp, 436
Kw, 449
K-electron capture, 607

Kelvin temperature scale, 11
β-Keratin structure, 639
Ketone, 220(t)
Kilocalorie, 63
Kilometer, definition of, 2
Kinetic theory of gases, 109–117
   molecular energy, 115
   molecular speed, 114
   postulates of, 110
Kinetics, chemical, 371
Krypton, 407
Krypton fluoride, 408

L

Labile complexes, 523
Lactic acid, 458
Lanthanides, 154
Law of Combining Volumes, 105
Law of Conservation of Mass, 26
Law of Constant Composition, 26
Law of Dulong and Petit, 41
Law of Mass Action, 358
Law of Multiple Proportions, 26
Law of Partial Pressures, 105
Lawrence, E. O., 604
Lawrencium, 29
Laws of thermochemistry, 66–67
Lead, 21
   in storage battery, 557
Lead chloride, solubility of, 432
Lead poisoning, EDTA treatment for, 529
Lead sulfate, 483
Lead tetraethyl, 307
Leblanc, Nicholas, 501
Le Chatelier's Principle, 363, 435
Leclanché cell, 556
Level(s), electron, 138
   capacities of, 140(t)
   order of, 143
   symbols for, 141
Lewis, G. N., 175, 472
Lewis acid, 506
   restricted meaning of, 473
Lewis base, 506
Lewis concept, 472
Lewis structure, 175–178
Libby, W. F., 602
Ligand, 507, 509(t), 509–512
Ligand field theory, 523
Light, emission by excited atoms, 125
   speed of, 123
Lime, 20
Lime-soda method, 312
Limestone, 66, 229, 302
   in Solvay Process, 500
Limiting density method, 119
Lindane, *16*, 286, 288, *305*
Liquid-vapor equilibrium, 235
Liquids, boiling point of, 241
   freezing point of, 253
   hydrogen bonds in, 207
   properties of, 233
   water, 294
Liter, definition of, 3
Lithium chloride, crystal structure of, 249, *249*

Lithium iodide, 165
  internuclear distances in, *166*
Lithium nitride, 401
Litmus, 450
Logarithms, 672(t)
  use of, 667–670
Lomonosov, Mikhail, 25
London, F., 168
London forces, 208
Lowry, T. M., 469
LSD (D-lysergic acid diethylamide), 62
Lucite, 223(t)
Luster, 197
Lyman series, 127

M

Macromolecular substances, 225–230
Magnesium, 21, 21(t), 45, 64
Magnesium hydroxide, preparation of, 444
Magnesium nitride, 401
Magnetic bottle, 619
Magnetic orbital quantum number, 138
Magnetic spin quantum number, 138
Malic acid, 221
Malleability, 197, 198
Manganese, 21(t)
  oxides of, 405
Manganese dioxide, as catalyst, 384
Manometer, 91, *92, 236*
Mass Action, Law of, 358
Mass decrement, 613
Mass-energy relations, 609–614
Mass number, 30
Mass spectrometer, *35*
Maxwell, James Clerk, 116, 342
Maxwell distribution, 115, 116, 388
Measurement, 2–11. See also *Units of measurement.*
  uncertainty in, see *Significant figures*
Mechanism of reaction, 387
  studies with radioactive tracers, 609
Meitner, Lisa, 614
Melting point, 18
  effect of pressure on, 253
  intermolecular forces and, 204
  molecular weight and, 204, 205(t)
  of macromolecular substances, 226, 228
Mendeleev, Dmitri, 153
Mercury, 21–24, 244(t)
  as water pollutant, 308, 549
  surface tension of, 234
Mercury(II) sulfide, 21
  solubility of, 432
Mesosphere, 413
Messenger RNA, 652
Metal(s), 154
  alkali, 151
    atomic radii of, 157(t)
    electron structure of, 151
  band theory of, 199
  bonding of, 197
  conductivity of, 196
  oxidation number of, 538
  oxides of, 404, 405(t)
Metallic bonds, 162, 196–199
Meter, definition of, 2
Methane, 31, 43(t), 212

Methane *(Continued)*
  heat of vaporization of, 234
  in natural gas, 78
  mass spectrum of, 37, *38*
  reaction with oxygen, 63
  structure of, 181, *183*
  use in fuel cell, 559
Methyl acetate, 220(t), 221
Methyl acetylene, *215*
Methyl alcohol, 22, 85, 220
  as antifreeze, 280
Methyl nitrate, 419
Methyl red, 486
Methyl silicone, 234
Methylcholanthrine, 218
Methylene group, 221
Methylenediamine, 510
Meyer, Lothar, 44
Mica, *229, 230*
Michaelis-Menten equation, 394
Microliter, 17
Millikan, Robert A., 27
Millimeter, definition of, 2
Minimum energy of electron, 133–134
Mohr titration, *441, 442*
Molality, 267
Molarity, 267
Mole, 56
  definition of, 53
Mole fraction, 266
Molecular formula, 48
Molecular genetics, central dogma of, 649
Molecular geometry, 181–186
Molecular orbital theory, 18, 170
Molecular orbitals, 190–196
Molecular substances, general properties of, 203
Molecular weight, 39
  and boiling point, 204, 205(t)
  and Graham's Law, 111
  and volatility, 204, 205(t)
  determination of, 102
  from colligative properties, 284
Molecule(s), 31–32, *32*
  as dipoles, 206
  bond angles in, 181
  dipole moments in, 186
  electron-pair repulsion in, 181
  Lewis structures for, 175
  multiple bonds in, 174, 188
  orbitals in, 190–196
  paramagnetic, 180
  polar, 184
  resonance structures in, 178
  speed of, 114
Monatomic ions, 165
Mond process, 508
Monomers, 222
Mononucleotides, 645
Monosaccharide, 627
Moon, age of, 602
Mulliken, R. S., 170
Multiple bonds, 174
  bond angles in, 183
  hybrid orbitals in, 188, *189*
Multiple Equilibria Rule, 461
Multiple Proportions, Law of, 26
Multistep mechanisms, 388–392

Mutations, 658
Mylar, 224

N

n-p junction, 257
Naphthalene, 217
    density of, 244(t)
    solubility in benzene, 271(t), 273
    sublimation of, 252
Natural gas, 78
Natural logarithm, 670
Neon, 36, *36*, 407
    separation from air, 399
Nernst equation, 576
    use of, 577–580
Net ionic equation, 430
Neutral complex, 508
Neutralization, 477
Neutron(s), 29
    as bombarding particle, 615
Neutron activation analysis, 608
Neutron-to-proton ratio, 606
Newcomen, Thomas, 323
Newton, Isaac, 25
Newton's first law of motion, 4
Nickel, 21, 508
    detection of, 526
Nickel cadmium battery, 564
Nickel chloride, 429
Nickel hydroxide, 429
Nickel oxide, 256
Nicotine, 59, 287
Ninhydrin, 634
Nitrate ion, 179, 507, 582–585
Nitric acid, 57, 264(t), 401
    Ostwald process for, 403
Nitric oxide, 401
    equilibrium concentration of, 355
    failure of octet rule in, 179–180
    heat of formation, *68*, 69
    reaction with ozone, 394
Nitride ion, 401
Nitrilotriacetic acid (NTA), 304
Nitrogen, 21(t), 44, 45
    atmospheric, 398(t)
    bonding in, 175
    chemical properties of, 400
    fixation of, 401
    oxides of, 405
        as pollutant, 418
    separation from air, 399
Nitrogen base, 644, *645*
Nitrogen cycle, *402*
Nitrogen dioxide, 18, 179, 348
    reaction with sulfur dioxide, 359
Nitroglycerine, 84, 402
Noble gas structure, 163
Noble gases, 407
    electron configuration of, 163
Nomenclature, 212, 661–664
Nonelectrolytes, 295
Nonmetals, 154
    oxidation number of, 538
    oxides of, 405, 406(t)
Nonpolar bond, 170
Normality, 268, 484
    dependency on reaction, 485
    relation to molarity, 485

Nuclear fission, 614–618
Nuclear fuel, 80
Nuclear fusion, 618–619
Nuclear reactions, 595–623
    binding energy in, 612(t), 613
    bombardment, 603
    chain, 616
Nuclear reactor, 616, *617*
Nucleic acids, 643–656
Nucleon, 613
Nucleus, atomic, 29–30
    stability of, 612
Nylon, 224

O

Octane, 84
Octet rule, 176
    failure of, 179
Octyl alcohol, solubility in water, 270
Odor, removal from water, 311
Oil drop experiment, 27, *28*
Olefins, 214
Onsager, Lars, 299
Oppenheimer, J. Robert, 615
Optical activity, 625
Orbital, 138
    antibonding, 191
    atomic, 169
        hybrid, 186, 188(t), 517
    bonding, 191
    molecular, 190–196
Orbital diagram, 143, *143*
Order of reaction, 375
    determination of, 379(t)
Organic compounds, 210–225
Osmosis, 281
    entropy change in, 335
    reverse, 316
Osmotic pressure, 281–285
    definition of, 282
    relation to concentration, 283
Ostwald, F. W., 27
Ostwald process, 403
Outer complex, 519
Overlap of electron clouds, *169*
Overvoltage, 569
Oxidation, 535
    definition of, 539
Oxidation number (state), 536
    rules for assignment, 537
Oxidation-reduction, ease of, 569
Oxidation-reduction equations, balancing
        of, 540
    in basic solution, 542
Oxidation-reduction reactions, 536
    equilibrium constant for, 573
    free energy change in, 572
    in analytical chemistry, 589
    in electrolytic cell, 544
    in fuel cell, 558
    in voltaic cell, 554
    Nernst equation in, 576
    spontaneity of, 570–575
    standard potentials in, 568(t)
Oxides, metallic, 404
Oxidizing agent, 540
    strength of, 570
    strong, 580–585

Oxyacids, 467, 469(t)
  strength of, 467(t)
Oxyanions, 177
  as oxidizing agents, 582
Oxygen, abundance of, 21(t)
  as oxidizing agent, 585–588
  in organic compounds, 220
  isotopes of, 45
  molecular, structure of, 180, *195*
  paramagnetism of, 180
  properties of, 403
  separation from air, 399
  solubility in water, 275
  use in fuel cell, 559
Oxyhemoglobin, 511
Ozone, 200, 311, 394, 398
  in upper atmosphere, 413

P

Packing, types of, 248
Paraffins, 212
Paramagnetism, 180, 519
Parathion, 306, 386
Partial ionic character, 172
Partial pressure, 105
  in equilibrium constant, 370
Particles, wave properties of, 131
Parts per million, 399
Pauli, Wolfgang, 139
Pauli exclusion principle, 139
Pauling, Linus, 169, 174
Penicillin, 53
Pentane, 231
Peptide linkage, 631
Perchlorate ion, 538
Periodic table, *inside front cover*
  and electron configuration, 150–156
  structure of, 153–154
Peroxide ion, 404
Peroxyacetyl nitrate, 201, 419
Pesticides, *16*
Petit, Alexis, 41
Petroleum, 78
pH, definition of, 449
  of common materials, 451(t)
pH meter, 578, *579*
pH paper, 451
Phase diagram, 250, *251*
Phase equilibrium, 250
Phenanthrene, 271(t)
Phenol, 217
Phenolphthalein, 450, 486
Phosphate buffer, 492
Phosphorus, 21(t)
Photochemical smog, 418
Photoelectric effect, 29, 197
Photosynthesis, 63
Pi (π) bonds, 189
  in benzene, 216
  in graphite, 227
Pi orbitals, 193
Pitchblende, 596
Planck, Max, 127
Planck's constant, 123
Platinum, as catalyst, 407
  complexes of, 507, 507(t)
Plexiglas, 223(t)
Polar bond, 171, *171*

Polarity, 184–186
  effect on melting and boiling points, 205(t)
  effect on solubility, 269
Pollutants, 38, 416(t)
  sources of, 417, 417(t)
  types of, 303(t)
Pollution, air, 339, 416
  effects of, 419
  reduction of, 422
  water, 300–309
  sources of, 301–303
Polyacrilonitrile (PAN), 223(t)
Polyatomic ions, 164(t), 165
  Lewis structures of, 176
Polyethylene, 45, 222
Polymerization, 215
Polymers, 222
  addition, 223
Polypeptides, 631
Polystyrene, 223(t)
Polyvinyl chloride (PVC), 223(t)
Positron, 604
Potassium, 21, 21(t)
Potassium chlorate, 582
Potassium hexacyanoferrate, 525
Potassium hydrogen phthalate, 502
Potassium thiocyanate, 62, 525
Potassium uranyl sulfate, 596
Potential, standard, 566, 568(t)
  assignment of, 566
Potentiometer, 565
Powell, 181
Precipitation, of proteins, 632
Precipitation reactions, 429–447
  equilibria of, 434–440
  in inorganic preparations, 444
  in qualitative analysis, 442
  in quantitative analysis, 440
  net ionic equations for, 429
  solubility product in, 436
  solubility rules for, 432
Precipitator, electrostatic, 425
Pressure, atmospheric, 89
  effect on boiling point, 241
  effect on solubility, 274–276
  measurement of, 91
  osmotic, 281–285
  partial, 105
  vapor, 235–240
Priestley, Joseph, 82
Principal quantum number, 137
Progesterone, 288
Propagation, chain, 390
Propane, 71, 85, 212
Protein(s), 628–643
  amino acid content of, 633, 634(t)
  conformation of, 638
  molecular weights of, 637(t)
  nature of, 629
  purification of, 632
  structure of, 631
Protein synthesis, mechanism of, *654*
Proton(s), 29
  hydrated, 470
  transfer of, in Brönsted-Lowry acids, 469
Proust, Joseph, 26
Prussian blue, 525

Pump, suction, 90, *90*
Purification, water, 309–318
Pyrex, 229
Pyridine, 260
Pyrites, 423

Q

Q, 63
Q_P, 64
Q_V, 77
Qualitative analysis, acid-base reactions in, 495
  complex ions in, 525
  oxidation-reduction reactions in, 589
  precipitation reactions in, 442
Quantitative analysis, 48
  complex ions in, 528
  oxidation-reduction reactions in, 590
  precipitation reactions in, 440
Quantum number(s), 123, 132, 137
  allowed sets of, 139, 139(t)
  angular momentum, 138
  magnetic orbital, 138
  magnetic spin, 138
  principal, 137
Quantum theory, postulates of, 122
Quartz, 60, 228
Quicklime, 66, 339

R

Rad, 598
Radiation, effect on tissue, 599(t)
  interaction with matter, 597
  mutations caused by, 599
Radical, free, 180
Radioactive isotopes, uses of, 607–609
Radioactivity, artificial, 603–609
  natural, 596–599
Radium, 46
  luminescence of, 598
Radon, 407
Ramsey, William, 408
Random coil, 639
Randomness, and entropy, 331
Raoult's Law, 277
Rare earths, 154
Rate constant, 375
  radioactive decay, 600, 601
Rate-limiting step, 389
Rate of reaction, 372. See also *Reaction rate*.
Rayleigh, Lord (J. W. Strutt), 407
Reaction(s), acid-base, 477–505. See also *Acid-base reactions*.
  catalysts in, 384
  chain, chemical, 390
    nuclear, 616
  collision theory of, 387
  electrochemical, 544–560
  electron transfer in, 535
  endothermic, 63
  enzymes in, 385, 643
  exothermic, 63
  half-life of, 379
  mechanism of, 386–394
  neutralization, 477

Reaction(s) *(Continued)*
  nuclear, 595–623. See also *Nuclear reactions*.
  order of, 375
  oxidation-reduction, 536. See also *Oxidation-reduction reactions*.
  polymerization, 215
  precipitation, 429–447. See also *Precipitation reactions*.
  rate of, see *Reaction rate*
  spontaneity of, 321, 570
  yield of, 58
Reaction rate, 372
  dependence on concentration, 374
  effect of temperature on, 380–384
Reactor, nuclear, 616, *617*
Redox reactions, 536. See also *Oxidation-reduction reactions*.
Reducing agent, 540
  strength of, 570
Reduction, 536
  definition of, 539
Reduction potential, standard, 566, 568(t)
Relative humidity, 409
Resonance, in benzene, 216
Resonance energy, 179
Resonance structure, 178–179
Reverse osmosis, 316
Ribonuclease, 386
Ribonucleic acid (RNA), 644, 651–654
Ribose, 644
Rocks, dating of, 602
Roentgen, W. K., 294
Rubber, synthetic, 232
Rust, 586
Rutherford, Ernest, 27, 127

S

ΔS, 330
Salicylic acid, 58
Salt bridge, 555
Salts, acidity of, 479–481
  preparation of, 493
  water solutions of, 295–300
Sanger, F., 635
Saturated hydrocarbons, 212
  properties of, 213(t)
Savin, 306
Schrödinger, Erwin, 136
Schrödinger wave equation, 136
Scintillation counter, 598
Sea water, ions in, 315(t)
Seaborg, Glenn, 605
Second Law of Thermodynamics, 81
Second order reactions, 379
Sedimentation, 309
Seeding of clouds, 410
Semiconductors, 257
  temperature dependence of, 258
Semipermeable membrane, 282
Separation techniques, 11–17
Sewage, 302
Shell, electron, 138
Shielding, electronic, 156, *157*
Sickle cell anemia, 511
Sidgwick, 181
Sigma (σ) bonds, 189
Sigma orbitals, 191.

Significant figures, in calculations, 7
 definition of, 6
Silica Gel G, *16*
Silicates, 230
Silicon, 21(t), 44, 46
 as semiconductor, 257
Silicon dioxide, 228
Silicone oil, 17
Silicosis, 420
Silt, 302
Silver, 21, 47
Silver acetate, 434
Silver bromide, 46, 255
Silver chloride, 51, *68*
Silver complexes, 509
Silver iodide, cloud seeding with, 410
Silver nitrate, 60
Sizes, of atoms, 156–158
 of ions, 165, *166*
Slater, J. C., 169
Soaps, 221–222, 234, 311
Soda ash, 229
Sodium, 21, 21(t), 44
 atomic spectrum of, *124*
 from sodium chloride, 546
Sodium bicarbonate, 53, 118, 297. See
 also *Sodium hydrogen carbonate.*
Sodium carbonate (washing soda), 463,
 498–501
Sodium chloride, 198, 244(t)
 crystal structure of, 33, 249
 in Solvay Process, 499
Sodium fluoride, 162
 crystal structure of, *163*
Sodium hydrogen carbonate, 498–501
Sodium hydroxide, *50*, 61, 429
 from brine, 549
 preparation of, 444
Sodium stearate, 311
Sodium thiosulfate, 509
Sodium tripolyphosphate, 304
Solar battery, 259
Solar energy, 82
Solar furnace, 82, *83*
Solar radiation, 414
Solar still, *316*
Solids, crystal structure of, 244–250
 macromolecular, 225–230
 metallic, 196
 purity of, 18
 solubility of, 270
 sublimation of, 252
Solubility, 14(t)
 of hydrocarbons, 271
 of ionic compounds, 431–434
 pressure effects on, 274–276
 principles of, 269–272
 rules for, 432(t)
 temperature effects on, 272–274
Solubility equilibrium, 434–440
Solubility product, 436, 438(t)
Solution(s), 263–288, 295–300
 acidic, 264(t)
 basic, 264(t)
 boiling point of, 278
 colligative properties of, 276, 296, 455
 concentration of, 265–268
 freezing point of, 279, 297
 ideal, 277
 ion atmosphere in, 299

Solution(s) *(Continued)*
 gas-liquid, 271
 liquid-liquid, 269
 saturated, 264
 solid-liquid, 270
 structure of, 298
 supersaturated, 265
Solvay Process, 498–501, *501*
Sörensen, S. P. L., 449
Specific heat, 41
Specific ion electrodes, 580
Spectral series, 126
Spectrochemical series, 522
Spectrum, absorption, 18–19, *19*
 atomic, 124
 infrared absorption, *19*
 mass, *36*
 ultraviolet absorption, 514
 x-ray, 161
Spontaneity, criteria of, 321
 of redox reactions, 570
Stability, nuclear, 612
 of covalent bond, 168
 of resonance hybrids, 179
Stability curve, nuclear, 606, *607*
Stack gases, washing of, 424
Standard oxidation potential, 566
Standard reduction potential, 566, 568(t)
Standard voltage ($E°$), 566
 relation to equilibrium constant, 573
 relation to $\Delta G°$, 572
Stannic chloride, 164
Starch, 627
State, changes in, 250–254
 nonequilibrium changes of, 254
 oxidation, 536
 phase diagram of, 250, *251*
 quantum, 122
Stereoisomers, 626
Storage battery, 557, *558*
Strassman, 614
Stratosphere, 412.
Strong acid(s), 451–452, 452(t)
 reaction with solid, 481
 reaction with strong base, 477
 reaction with weak base, 480
 titration with strong base, 487
Strong base(s), 451–452, 452(t)
 reaction with strong acid, 477
 reaction with weak acid, 478
Strong oxidizing agents, 580–585
Strontium-90, 600
 removal with BAETA, 529
Sublimation, 251, 252
 heat of, 253
Substrate, 385
Successive approximations, method of, 459
Sucrose, 625
Sugar, 62, 296, 385, 625
Sulfate ion, 177, 184, 507
Sulfides and sulfides, detection of, 496
Sulfonated polystyrene, 315
Sulfur, 21(t), 34
 oxides of, 406
 as pollutant, 417
Sulfur dioxide, 22, 61
 detection of, 496
 reaction with nitrogen dioxide, 359
 structure of, 178, 184
Sulfur trioxide, 338, 359

Sulfuric acid, 61, 264(t), 359
  commercial form of, 452
  preparation of, 407
Superheating and supercooling, 254
Superoxide ion, 404
Supersaturated solution, 265
Surface reactions, 392
Surface tension, 234, 235(t)
Suspension, 263

T

Taste, removal from water, 311
Teflon, 223(t), 224
Temperature, effect on equilibrium constant, 362
  effect on reaction rate, 380–384
  effect on semiconductors, 258
  effect on solubility, 272–274
  effect on spontaneity, 335(t)
  effect on vapor pressure, 238–240
  in atmosphere, *413*
  measurement of, 10, 258
  of earth, factors determining, 290–291
Termination, chain, 390
Tetrahydrofuran, 287
Thermal pollution, 308
Thermal reactor, automotive, 422
Thermionic effect, 29, 197
Thermistor, 258
Thermite reaction, 86
Thermochemical equations, 63
Thermochemistry, laws of, 66–68
Thermodynamics, chemical, 371
  laws of, 81, 86
Thermoplastic, 222
Thermosphere, 413
Thomsen, Julius, 322
Thomson, J. J., 27
Tin, 21
Titanium, 21(t), 23
Titanium dioxide, 405
Titration, acid-base, 483
  Mohr, *441*, 442
  redox, 590
  with EDTA, 529
TNT, 60, 402
  energy of explosion, 616
Toluene, 217
Torr, 92
Torricelli, Evangelista, 90
Trace elements, as pollutants, 307, 307(t)
Tracer, radioactive, 609
Trans isomer, 215, 513
Transfer RNA, 652
Transistor, 257
Transition metal cations, 164(t)
Transuranium elements, 605
Tributylamine, 588
Trichloroethylene, *219*
Trinitrotoluene, see *TNT*
Triple point, 250
Trona, 501
Troposphere, 411
Trouton's rule, 242

U

Ultracentrifugation, 632
Ultramarine, 230

Ultraviolet absorption spectrum, 514
Ultraviolet radiation, 414
Unit cell, 246
Units of measurement
  concentration, 265–268, 399(t)
  length, 2–3
  mass, 3–5
  pressure, 92
  temperature, 5–6
  volume, 3
  weight, 4
Unsaturated hydrocarbons, 214
Uranium, 30, 113
Uranium hexafluoride, 102, 113
Urea, 287, *393*, 394
Urease, 394
Useful work, 323

V

Valence-bond model, 169
  for complex ions, 517–519
Valence electrons, 175
Vanadium pentoxide, 407
van der Waals equation, 109
van der Waals forces, 208
Vapor phase, see *Gas(es)*
Vapor pressure, 236, 242(t)
  and boiling point, 241
  lowering of, 275–278
  measurement of, *236*
  temperature dependency of, 238–240
Vaporization, heat of, 234, 235(t), 242(t)
Vinyl chloride, *219*
Vitamin B$_{12}$, 246, 511
Vitamin C, *19*, 49, 50, 52
Voltage, cell, calculation of, 567
  effect of concentration on, 575–580
  standard, 566
Voltaic cell, 21, 554
Volume change, effect on equilibrium, 359, 361(t)
Volumetric analysis, 440
Von Laue, Max, 244

W

Washing soda, 463
Water, 31, 40, 44, 46
  "anomalous," 318
  as Brönsted-Lowry acid and base, 470
  as oxidizing agent, 585–587
  bonding in, *183*
  density of, 7, 291
  dissociation of, 448
  distribution of, 289(t)
  heat of fusion of, 244(t)
  heat of vaporization of, 234, 244(t)
  hydrogen bonding in, 207
  in atmosphere, 409
  mass of total supply, 290
  melting point of, 253
  physical properties of, 290, 290(t), 292(t), 659
  polluted, 301
  solid state of, see *Ice*
  structure of, 182, *183*, 294
  triple point of, 250

Water *(Continued)*
  vapor pressure of, 242(t)
Water pollution, 300–309
Water purification, 309–318
Water softening, 311
Watson, J., 648
Wave functions, 131, *132*
Wave properties of particles, 131
Wavelength, de Broglie, 130
Weak acid(s), 453
  reaction with strong base, 478
  titration with strong base, *488, 489*
Weak bases, 460
  reaction with strong acid, 480
  titration with strong acid, *490*
Weather modification, 410–411
Weight percentage, 266
Werner, Alfred, 529–531
Wilson cloud chamber, 598, *598*
Wöhler, F., 547
Wood, 78
Work, maximum useful (Wm), 325
  performed by voltaic cell, 558

X

X-ray diffraction, 47, 165, 244–245, 514, 638
X-ray spectra, 161
Xenon tetrafluoride, 101, 202
  preparation of, 408
  structure of, 184
Xylene, 59, 217

Y

Yield, 58

Z

Zeolite, 313, *314*
Zero order reactions, 379
Zinc-copper cell, 554
Zinc sulfide, 59
  solubility product of, 579
Zirconium, 21(t)
Zwitterion, 504